Chemistry

for the IB Diploma

Second edition

Steve Owen

with
Caroline Ahmed
Chris Martin
Roger Woodward

Cambridge University Press's mission is to advance learning, knowledge and research worldwide.

Our IB Diploma resources aim to:

- encourage learners to explore concepts, ideas and topics that have local and global significance
- help students develop a positive attitude to learning in preparation for higher education
- assist students in approaching complex questions, applying critical-thinking skills and forming reasoned answers.

CAMBRIDGE
UNIVERSITY PRESS

University Printing House, Cambridge CB2 8BS, United Kingdom

One Liberty Plaza, 20th Floor, New York, NY 10006, USA

477 Williamstown Road, Port Melbourne, VIC 3207, Australia

314–321, 3rd Floor, Plot 3, Splendor Forum, Jasola District Centre, New Delhi – 110025, India

103 Penang Road, #05-06/07, Visioncrest Commercial, Singapore 238467

Cambridge University Press is part of the University of Cambridge.

It furthers the University's mission by disseminating knowledge in the pursuit of education, learning and research at the highest international levels of excellence.

Information on this title:
www.cambridge.org/9781107622708 (Paperback)
www.cambridge.org/9781107537637 (Cambridge Elevate enhanced edition, 2 years)
www.cambridge.org/9781316637746 (Paperback + Cambridge Elevate enhanced edition, 2 years)

© Cambridge University Press 2011, 2014

First published 2011
Second edition 2014

20 19 18 17 16 15

Printed in Great Britain by Ashford Colour Press Ltd.

A catalogue record for this publication is available from the British Library

ISBN 978-1-107-62270-8 Paperback
ISBN 978-1-009-33154-8 Paperback + Digital Access, 2 Years
ISBN 978-1-107-53763-7 Digital Access, 2 Years

Additional resources for this publication at education.cambridge.org/ibsciences

Cambridge University Press has no responsibility for the persistence or accuracy of URLs for external or third-party internet websites referred to in this publication, and does not guarantee that any content on such websites is, or will remain, accurate or appropriate. Information regarding prices, travel timetables, and other factual information given in this work is correct at the time of first printing but Cambridge University Press does not guarantee the accuracy of such information thereafter.

All exam-style questions and sample answers have been written by the authors.

Contents

Free online material

The website accompanying this book contains further resources to support your IB Chemistry studies. Visit **education.cambridge.org/ibsciences** and register to access these resources:

Options

Option A Materials

Option B Biochemistry

Option C Energy

Option D Medicinal chemistry

Self-test questions

Assessment guidance

Model exam papers

Nature of Science

Answers to exam-style questions

Answers to Options questions

Introduction

This second edition of *Chemistry for the IB Diploma* is fully updated to cover the content of the IB Chemistry Diploma syllabus that will be examined in the years 2016–2022.

Chemistry may be studied at Standard Level (SL) or Higher Level HL). Both share a common core, and at HL the core is extended with additional HL material. In addition, at both levels, students then choose one Option to complete their studies. Each Option consists of common core and additional HL material. All common core and additional HL material is covered in this print book. The Options are included in the free online material that is accessible via **education.cambridge.org/ ibsciences**

The content is arranged in topics that match the syllabus topics, with core and additional HL material on each topic combined in the book topics. The HL content is identified by 'HL' included in relevant section titles, and by a yellow page border.

Each section in the book begins with learning objectives as starting and reference points. Test yourself questions appear throughout the text so students can check their progress and become familiar with the style and command terms used, and exam-style questions appear at the end of each topic. Many worked examples appear throughout the text to help students understand how to tackle different types of questions.

Theory of Knowledge (TOK) provides a cross-curricular link between different subjects. It stimulates thought about critical thinking and how we can say we know what we claim to know. Throughout this book, TOK features highlight concepts in Chemistry that can be considered from a TOK perspective. These are indicated by the 'TOK' logo, shown here.

Science is a truly international endeavour, being practised across all continents, frequently in international or even global partnerships. Many problems that science aims to solve are international, and will require globally implemented solutions. Throughout this book, International-Mindedness features highlight international concerns in Chemistry. These are indicated by the 'International-Mindedness' logo, shown here.

Nature of Science is an overarching theme of the Chemistry course. The theme examines the processes and concepts that are central to scientific endeavour, and how science serves and connects with the wider community. Throughout the book, there are 'Nature of Science' paragraphs that discuss particular concepts or discoveries from the point of view of one or more aspects of Nature of Science. A chapter giving a general introduction to the Nature of Science theme is available in the free online material.

Free online material

Additional material to support the IB Chemistry Diploma course is available online. Visit education.cambridge.org/ibsciences and register to access these resources.

Besides the Options and Nature of Science chapter, you will find a collection of resources to help with revision and exam preparation. This includes guidance on the assessments, interactive self-test questions and model exam papers. Additionally, answers to the exam-style questions in this book and to all the questions in the Options are available.

Stoichiometric relationships 1

1.1 Introduction to the particulate nature of matter and chemical change

1.1.1 The particulate nature of matter

The three states of **matter** are solid, liquid and gas and these differ in terms of the arrangement and movement of particles. The particles making up a substance may be individual atoms or molecules or ions. Simple diagrams of the three states of matter are shown in Figure **1.1** in which the individual particles are represented by spheres.

Sublimation is the change of state when a substance goes directly from the solid state to the gaseous state, without going through the liquid state. Both iodine and solid carbon dioxide (dry ice) sublime at atmospheric pressure. The reverse process (gas → solid) is often called *deposition* (or sometimes *desublimation*, *reverse sublimation* or occasionally just sublimation).

The properties of the three states of matter are summarised in Table **1.1**.

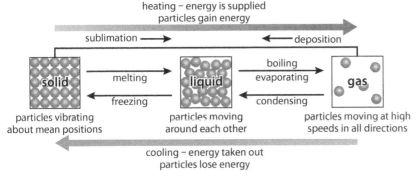

Figure 1.1 The arrangement of particles in solids, liquids and gases and the names of the changes of state. Note that evaporation can occur at any temperature – boiling occurs at a fixed temperature.

	Solids	Liquids	Gases
Distance between particles	close together	close but further apart than in solids	particles far apart
Arrangement	regular	random	random
Shape	fixed shape	no fixed shape – take up the shape of the container	no fixed shape – fill the container
Volume	fixed	fixed	not fixed
Movement	vibrate	move around each other	move around in all directions
Speed of movement	slowest	faster	fastest
Energy	lowest	higher	highest
Forces of attraction	strongest	weaker	weakest

Table 1.1 The properties of the three states of matter.

If a pure substance is heated slowly, from below its melting point to above its **boiling point**, a graph of temperature against time can be obtained (Figure **1.2**).

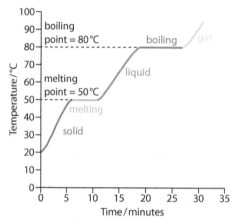

Figure 1.2 A heating curve showing changes of state.

As a solid is heated, its particles vibrate more violently – they gain **kinetic energy** and the temperature of the solid rises. At 50 °C, the solid in Figure **1.2** begins to melt – at this stage there is solid and liquid present together and the temperature remains constant until all the solid has melted. All the **heat energy** being supplied is used to partially overcome the forces of attraction between particles so that they can move around each other. When all the solid has melted, the continued supply of heat energy causes the kinetic energy of the particles to increase so that the particles in the liquid move around each other more quickly. The kinetic energy of the particles increases until the boiling point of the liquid is reached. At this point (80 °C) the continued supply of heat energy is used to overcome the forces of attraction between the particles completely and the temperature of the substance remains constant until all the liquid has been converted to gas. The continued supply of heat energy increases the kinetic energy of the particles of the gas so they move around faster and faster as the temperature of the gas increases.

Both refrigeration and air-conditioning involve changes of state of liquids and gases. In a refrigerator, heat energy is absorbed from the inside of the refrigerator and is used to convert a liquid coolant to a gas – the heat energy is given out to the surrounding as the gas is compressed back to a liquid. Refrigeration is essential in warm countries to preserve food and without it the food would go 'off' much more quickly and be wasted – but how essential is air-conditioning? CFCs (which cause destruction of the ozone layer) have been used as a refrigerant and in making the insulation for refrigerators. In many countries the disposal of old refrigerators is controlled carefully. More environmentally friendly refrigerators are being manufactured using alternatives to CFCs – they also use less electricity.

1.1.2 Chemical change

Elements and compounds

Chemistry is partly a study of how chemical elements combine to make the world and the Universe around us.

Gold is an **element** and all samples of pure gold contain only gold atoms.

> An element is a pure substance that contains only one type of atom (but see *isotopes* in Topic 2).

> An atom is the smallest part of an element that can still be recognised as that element.

> The physical and chemical properties of a compound are very different to those of the elements from which it is formed.

Sodium and chlorine are elements – when they are mixed and heated they combine chemically to form a compound called sodium chloride. Sodium is a grey, reactive metal with a low melting point and chlorine is a yellow-green poisonous gas – but sodium chloride (common salt) is a non-toxic, colourless compound with a high melting point.

Similarly, when iron (a magnetic metal) is heated with sulfur (a non-magnetic yellow solid) a grey, non-metallic solid called iron sulfide is formed (Figure **1.3**).

> **Chemical properties** dictate how something reacts in a chemical reaction.
> **Physical properties** are basically all the other properties of a substance – such as melting point, density, hardness, electrical conductivity etc.

Learning objectives

- Understand that compounds have different properties to the elements they are made from
- Understand how to balance chemical equations
- Understand how to use state symbols in chemical equations
- Describe the differences between elements, compounds and mixtures
- Understand the differences between homogeneous and heterogeneous mixtures

> A compound is a pure substance formed when two or more elements combine chemically.

Figure 1.3 Iron (left) combines with sulfur (centre) to form iron sulfide (right).

The meaning of chemical equations

When elements combine to form compounds, they always combine in **fixed ratios** depending on the numbers of atoms required. When sodium and chlorine combine, they do so in the mass ratio $22.99 : 35.45$ so that $22.99\,g$ of sodium reacts exactly with $35.45\,g$ of chlorine. Similarly, when hydrogen (an explosive gas) combines with oxygen (a highly reactive gas) to form water (liquid at room temperature), $1\,g$ of hydrogen combines with $8\,g$ of oxygen, or $2\,g$ of hydrogen reacts with $16\,g$ of oxygen (using rounded relative atomic masses) – that is, they always combine in a mass ratio of $1:8$.

Figure 1.4 Two carbon atoms react with one oxygen molecule to form two molecules of carbon monoxide.

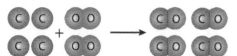

Figure 1.5 Four carbon atoms react with two oxygen molecules to form four molecules of carbon monoxide.

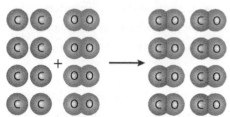

Figure 1.6 Eight carbon atoms react with four oxygen molecules to form eight molecules of carbon monoxide.

Mass is conserved in all chemical reactions.

Elements always combine in the same mass ratios because their atoms always combine in the same ratios, and each type of atom has a fixed mass.

Consider the reaction between carbon and oxygen to form carbon monoxide. This is shown diagrammatically in Figure **1.4**. In this reaction, two carbon atoms combine with one oxygen molecule to form two molecules of carbon monoxide. Now look at Figure **1.5**. If we started with four carbon atoms, they will react with two oxygen molecules to form four molecules of carbon monoxide.

The ratio in which the species combine is fixed in these equations. The number of molecules of oxygen is always half the number of carbon atoms, and the number of carbon monoxide molecules produced is the same as the number of carbon atoms (see Figures **1.4–1.6**). So, we can construct the equation:

$$2C + O_2 \rightarrow 2CO$$

which tells us that two carbon atoms react with one oxygen molecule to form two carbon monoxide molecules, and that this ratio is constant however many carbon atoms react.

Balancing equations

If a reaction involves 5.00 g of one substance reacting with 10.00 g of another substance in a closed container (nothing can be added or can escape), then at the end of the reaction there will still be exactly 15.00 g of substance present. This 15.00 g may be made up of one or more products and some reactants that have not fully reacted, but the key point is that there will no more and no less than 15.00 g present.

A chemical reaction involves atoms joining together in different ways and electrons redistributing themselves between the atoms, but it is not possible for the reaction to involve atoms or electrons being created or destroyed.

When a chemical reaction is represented by a chemical equation, there must be exactly the same number and type of atoms on either side of the equation, representing the same number of atoms before and after this reaction:

	$C_3H_8 + 5O_2$		$\rightarrow$	$3CO_2 + 4H_2O$	
	reactants			**products**	
atoms	C	3		C	3
	H	8		H	8
	O	10		O	10

So this equation is balanced. It is important to realise that only coefficients (large numbers in front of the substances) may be added to balance a chemical equation. The chemical formula for water is H_2O, and this cannot be changed in any way when balancing an equation. If, for instance, the formula is changed to H_2O_2 then it represents a completely different chemical substance – hydrogen peroxide.

State symbols are often used to indicate the physical state of substances involved in a reaction:

(s) = solid
(l) = liquid
(g) = gas
(aq) = aqueous (dissolved in water)

Worked examples

1.1 Balance the following equation

$$\ldots N_2(g) + \ldots H_2(g) \rightarrow \ldots NH_3(g)$$

and work out the sum of the coefficients in the equation.

In the unbalanced equation, there are two N atoms and two H atoms on the left-hand side of the equation but one N atom and three H atoms on the right-hand side. It is not possible for two N atoms to react with two H atoms to produce one N atom and three H atoms; therefore, this equation is not balanced.

It can be balanced in two stages, as follows:

$$\ldots N_2 + \ldots H_2 \rightarrow 2NH_3$$
atoms: 2 N   2 N
          2 H   6 H

$$\ldots N_2 + 3H_2 \rightarrow 2NH_3$$
atoms: 2 N   2 N
          6 H   6 H

This equation is now balanced because there is the same number of each type of atom on both sides of the equation.

The sum of the coefficients in this equation is $1+3+2=6$. The coefficient of N_2 is 1, although we do not usually write this in an equation.

1.2 Balance the following equation:

$$\ldots C_4H_{10}(g) + \ldots O_2(g) \rightarrow \ldots CO_2(g) + \ldots H_2O(l)$$

Compounds are balanced first, then elements:

$$\ldots C_4H_{10}(g) + \ldots O_2(g) \rightarrow 4CO_2(g) + 5H_2O(l)$$

There are two oxygen atoms on the left-hand side of the equation, and O_2 needs to be multiplied by 6.5 to give 13 oxygen atoms, which is the number of oxygen atoms on the other side $[(4 \times 2) + (5 \times 1)]$:

$$\ldots C_4H_{10}(g) + 6.5O_2(g) \rightarrow 4CO_2(g) + 5H_2O(l)$$

The equation is balanced as shown, but it looks much neater when balanced with whole numbers. To achieve this, all the coefficients are multiplied by 2:

$$2C_4H_{10}(g) + 13O_2(g) \rightarrow 8CO_2(g) + 10H_2O(l)$$

Test yourself

1 Balance the following equations:

a $NO + O_2 \rightarrow NO_2$

b $C_3H_8 + O_2 \rightarrow CO_2 + H_2O$

c $CaCO_3 + HCl \rightarrow CaCl_2 + CO_2 + H_2O$

d $C_2H_5OH + O_2 \rightarrow CO_2 + H_2O$

e $WO_3 + H_2 \rightarrow W + H_2O$

f $H_2O_2 \rightarrow O_2 + H_2O$

g $CrO_3 \rightarrow Cr_2O_3 + O_2$

h $Al_4C_3 + H_2O \rightarrow CH_4 + Al_2O_3$

i $HI + H_2SO_4 \rightarrow H_2S + H_2O + I_2$

j $PH_3 + O_2 \rightarrow P_4O_{10} + H_2O$

Mixtures

Elements and compounds are pure substances but most things around us are not pure substances but mixtures. We breathe in air, which is a mixture; all the foods we eat are mixtures; oxygen is carried around our body by blood, another mixture.

The components of a mixture can be elements or compounds – or mixtures! Air is a mixture of mostly elements (nitrogen, oxygen, argon) with smaller amounts of compounds (carbon dioxide, water vapour etc.).

The components of a mixture are not chemically bonded together and so retain their individual properties. In a mixture of iron and sulfur, the iron is shiny and magnetic; the sulfur is yellow and burns in air to form sulfur dioxide. When the mixture is heated and forms the compound iron sulfide, this is not shiny or magnetic or yellow – it is dull and grey and has completely different properties to its elements.

As you saw earlier, when atoms combine to form compounds they do so in fixed ratios, but the components of a mixture can be mixed together in any proportion. For example, ethanol and water can be mixed together in any ratio. Solutions are mixtures and a solution of sodium chloride could be made by dissolving 1 g of sodium chloride in $100\,cm^3$ of water or 2 g of sodium chloride in $100\,cm^3$ water or 10 g of sodium chloride in $100\,cm^3$ of water or many other amounts.

The components of a mixture can be separated from each other by physical means – for example a mixture of sand and salt could be separated by dissolving the salt in water, filtering off the sand and then heating the salt solution to drive off the water.

> A mixture contains two or more substances mixed together.

Homogeneous and heterogeneous mixtures

One example of a **homogeneous mixture** is a solution. No individual particles can be seen in the solution and its concentration is the same throughout. If several $1\,cm^3$ samples of a solution of sodium chloride are taken from a beaker and evaporated separately to dryness, the same mass of sodium chloride will be formed by each sample. Clean air (with no particulates) is also a homogeneous mixture.

One example of a **heterogeneous mixture** is sand in a beaker of water. The sand and water can be distinguished from each other and can also be separated by filtering.

> A homogeneous mixture has the same (uniform) composition throughout the mixture and consists of only one phase.

> A heterogeneous mixture does not have uniform composition and consists of separate phases. Heterogeneous mixtures can be separated by mechanical means.

Mixtures of different solids are also heterogeneous. For example, even though a mixture of iron and sulfur may have been made very carefully so that there are the same masses of iron and sulfur in each cubic centimetre, the composition is not uniform because there are distinct particles of iron and sulfur and each particle of iron and sulfur represents a different phase. The components of the mixture could be separated from each other using a magnet – or even a pair of tweezers to pick out each individual piece of iron and sulfur.

Sea water is a mixture and the process of obtaining fresh water from sea water is called desalination. Desalination is very important in some parts of the world where sufficient fresh water is not available from other sources (for example, in the Middle East). Fresh water obtained by desalination can be used for human consumption, agriculture or in industry.

Nature of science

Data collection is essential in science. The discussion above has used both quantitative (regarding reacting masses) and qualitative data (about the properties of substances). Accurate quantitative data are essential for the advancement of science and scientists analyse such data to make hypotheses and to develop theories. The law of definite proportions governing how elements combine may seem obvious nowadays in the light of the atomic theory but in the seventeenth and eighteenth centuries it was the subject of much debate.

? Test yourself

2 Classify each of the following as an element, a compound or a mixture:
   a water; b oxygen; c potassium iodide; d orange juice; e crude oil; f vanadium; g ammonia; h air; i hydrogen chloride; j magnesium oxide.

3 Classify each of the diagrams below using as many words as appropriate from the list:

element	compound	mixture
solid	liquid	gas

a   b   c   d

Learning objectives

- Define relative atomic mass and relative molecular mass
- Understand what is meant by one mole of a substance
- Calculate the mass of one mole of a substance
- Calculate the number of moles present in a specified mass of a substance
- Work out the number of particles in a specified mass of a substance and also the mass of one molecule

> The relative atomic mass (A_r) of an element is the average of the masses of the isotopes in a naturally occurring sample of the element relative to the mass of $\frac{1}{12}$ of an atom of carbon-12.

> The A_r of carbon is not 12.00, because carbon contains isotopes other than carbon-12 (see page **58**).

> The relative molecular mass (M_r) of a compound is the mass of a molecule of that compound relative to the mass of $\frac{1}{12}$ of an atom of carbon-12.

> The relative formula mass is the mass of one formula unit relative to the mass of $\frac{1}{12}$ of an atom of carbon-12.

1.2.1 Relative masses

Most chemical reactions involve two or more substances reacting with each other. Substances react with each other in certain ratios, and stoichiometry is the study of the ratios in which chemical substances combine. In order to know the exact quantity of each substance that is required to react we need to know the number of atoms, molecules or ions in a specific amount of that substance. However, the mass of an individual ion atom or molecule is so small, and the number of particles that make up even a very small mass so large, that a more convenient method of working out reacting quantities had to be developed.

Relative atomic mass (A_r)

The mass of a hydrogen atom is approximately 1.7×10^{-24} g. Such small numbers are not convenient to use in everyday life, so we use scales of relative mass. These compare the masses of atoms and molecules etc. to the mass of one atom of carbon-12, which is assigned a mass of exactly 12.00. As these quantities are relative, they have no units.

The **relative atomic mass (A_r)** of silver is 107.87. A naturally occurring sample of silver contains the isotopes ^{107}Ag and ^{109}Ag. The 107 isotope is slightly more abundant than the 109 isotope. Taking into account the amount of each isotope present in a sample (the weighted mean) it is found that, on average, the mass of a silver atom is 107.87 times the mass of $\frac{1}{12}$ of a carbon-12 atom. No silver atoms actually exist with the mass of 107.87; this is just the average relative atomic mass of silver.

Relative molecular mass (M_r)

An **relative molecular mass (M_r)** is the sum of the relative atomic masses of the individual atoms making up a molecule. The relative molecular mass of methane (CH_4) is:

$$12.01(A_r \text{ of C}) + 4 \times 1.01(A_r \text{ of H}) = 16.05$$

The relative molecular mass of ethanoic acid (CH_3COOH) is:

$$12.01 + (3 \times 1.01) + 12.01 + (2 \times 16.00) + 1.01 = 60.06$$

If a compound is made up of ions, and therefore does not contain discrete molecules, we should really talk about **relative formula mass**. However, relative molecular mass is usually used to refer to the mass of the formula unit of an ionic compound as well.

? Test yourself

4 Work out the relative molecular masses of the following compounds:
SO_2 NH_3 C_2H_5OH $MgCl_2$ $Ca(NO_3)_2$ $CH_3(CH_2)_5CH_3$
PCl_5 $Mg_3(PO_4)_2$ $Na_2S_2O_3$ $CH_3CH_2CH_2COOCH_2CH_3$

Moles

One mole is the amount of substance that contains the same number of particles (atoms, ions, molecules, etc.) as there are carbon atoms in 12 g of carbon-12. This number is called **Avogadro's constant**, has symbol L (or N_A), and has the value $6.02 \times 10^{23} \, \text{mol}^{-1}$. So, 12.00 g of carbon-12 contains 6.02×10^{23} carbon atoms.

You can have a mole of absolutely anything. We usually consider a mole of atoms (6.02×10^{23} atoms) or a mole of molecules (6.02×10^{23} molecules), but we could also have, for instance, a mole of ping-pong balls (6.02×10^{23} ping-pong balls).

The A_r of oxygen is 16.00, which means that, on average, each oxygen atom is $\frac{16}{12}$ times as heavy as a carbon-12 atom. Therefore 16 g of oxygen atoms must contain the same number of atoms as 12 g of carbon-12, i.e. one mole, or 6.02×10^{23} atoms. Similarly, one magnesium atom is on average $\frac{24.31}{12}$ times as heavy as a carbon-12 atom and, therefore, 24.31 g of magnesium atoms contains 6.02×10^{23} magnesium atoms.

The number of moles present in a certain mass of substance can be worked out using the equation:

$$\text{number of moles } (n) = \frac{\text{mass of substance}}{\text{molar mass}}$$

The triangle in Figure **1.7** is a useful shortcut for working out all the quantities involved in the equation. If any one of the sections of the triangle is covered up, the relationship between the other two quantities to give the covered quantity is revealed. For example, if 'mass of substance' is covered, we are left with number of moles multiplied by molar mass:

$$\text{mass of substance} = \text{number of moles} \times \text{molar mass}$$

If 'molar mass' is covered, we are left with mass of substance divided by number of moles:

$$\text{molar mass} = \frac{\text{mass of substance}}{\text{number of moles}}$$

The molar mass (M) of a substance is its A_r or M_r in grams. The units of molar mass are g mol^{-1}. For example, the A_r of silicon is 28.09, and the molar mass of silicon is $28.09 \, \text{g mol}^{-1}$. This means that 28.09 g of silicon contains 6.02×10^{23} silicon atoms.

When calculating the number of moles present in a certain mass of a substance, the mass must be in grams.

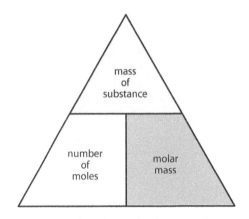

Figure 1.7 The relationship between the mass of a substance, the number of moles and molar mass.

One mole is an enormous number and beyond the scope of our normal experience. How do we understand a number this large? One way is to describe the number in terms of things we are familiar with from everyday life. For instance, one mole of ping-pong balls would cover the surface of the Earth to about 800 times the height of Mount Everest! We know what a ping-pong ball looks like and we may have a rough idea of the height of Mount Everest, so perhaps this description gives us a context in which we can understand 6.02×10^{23}. Another description sometimes used is in terms of a mole of computer paper: one mole of computer printer paper sheets, if stacked one on top of each other, would stretch over 6000 light years (one light year is the distance that light travels in one year) – this is over twice the thickness of our galaxy! Is this description better or worse than the previous one? It certainly sounds more impressive, but does it suffer from the fact that we have no real concept of the size of our galaxy? Can you think of any other ways of describing this number in terms of things you are familiar with from everyday life?

This is an example of a wider idea that we tend to understand things that are beyond our normal experience by reference to things with which we are more familiar.

Worked examples

1.3 Calculate the number of moles of magnesium atoms in 10.0 g of magnesium.

$$\text{number of moles } (n) = \frac{\text{mass of substance}}{\text{molar mass}}$$

$$n = \frac{10.0}{24.31} = 0.411 \text{ mol}$$

10.0 g of magnesium is 0.411 mol.

Note: the unit for moles is mol.

The answer is given to three significant figures, because the mass of substance is given to three significant figures.

1.4 Calculate the mass of 0.3800 mol CH_3COOH.

mass of substance = number of moles × molar mass

mass of substance = $0.3800 \times 60.06 = 22.82$ g

The mass of 0.3800 mol CH_3COOH is 22.82 g.

The answer is given to four significant figures, because the number of moles and the molar mass are given to four significant figures.

? Test yourself

5 Copy and complete the table. The first row has been done for you.

Compound	Molar mass / g mol⁻¹	Mass / g	Number of moles / mol
H_2O	18.02	9.01	0.500
CO_2		5.00	
H_2S			0.100
NH_3			3.50
Q		1.00	0.0350
Z		0.0578	1.12×10^{-3}
$Mg(NO_3)_2$		1.75	
C_3H_7OH		2500	
Fe_2O_3			5.68×10^{-5}

The mass of a molecule

The mass of one mole of water is 18.02 g. This contains 6.02×10^{23} molecules of water. The mass of one molecule of water can be worked out by dividing the mass of one mole (18.02 g) by the number of molecules it contains (6.02×10^{23}):

$$\text{mass of one molecule} = \frac{18.02}{6.02 \times 10^{23}} = 2.99 \times 10^{-23}\,\text{g}$$

$$\text{mass of one molecule} = \frac{\text{molar mass}}{\text{Avogadro's constant}}$$

> **Exam tip**
> Remember – the mass of a molecule is a very small number. Do not confuse the mass of a single molecule with the mass of one mole of a substance, which is a number greater than 1.

The number of particles

When we write '1 mol O_2', it means one mole of O_2 molecules: that is, 6.02×10^{23} O_2 molecules. Each O_2 molecule contains two oxygen atoms; therefore, one mole of O_2 molecules contains $2 \times 6.02 \times 10^{23}$ $= 1.204 \times 10^{24}$ atoms. That is, one mole of O_2 **molecules** is made up of two moles of oxygen **atoms**.

When we talk about '0.1 mol H_2O', we mean 0.1 mol H_2O molecules; i.e. $0.1 \times 6.02 \times 10^{23}$ H_2O molecules; i.e. 6.02×10^{22} H_2O molecules. Each H_2O molecule contains two hydrogen atoms and one oxygen atom. The total number of hydrogen atoms in 0.1 mol H_2O is $2 \times 6.02 \times 10^{22}$; i.e. 1.204×10^{23} hydrogen atoms; i.e. 0.2 mol hydrogen atoms.

Each H_2O molecule contains three atoms. Therefore, the total number of atoms in 0.1 mol H_2O is $3 \times 6.02 \times 10^{22}$; i.e. 1.806×10^{23} atoms; or 0.3 mol atoms.

If you look at Table **1.2** you can see the connection between the number of moles of molecules and the number of moles of a particular atom in that molecule. Figure **1.8** illustrates the relationship between number of particles, number of moles and Avogadro's constant.

oxygen

> **Exam tip**
> You must be clear which type of particle you are considering. Do you have one mole of atoms, molecules or ions?

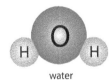
water

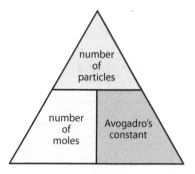

Figure 1.8 The relationship between the number of moles and the number of particles.

Compound	Moles of molecules	Moles of O atoms
H_2O	0.1	0.1
SO_2	0.1	0.2
SO_3	0.1	0.3
H_3PO_4	0.1	0.4
O_3	0.5	1.5
CH_3COOH	0.2	0.4

Table 1.2 The relationship between the number of moles of molecules and the number of moles of particular atoms.

If we multiply the number of moles of molecules by the number of a particular type of atom in a molecule (i.e. by the subscript of the atom), we get the number of moles of that type of atom. Thus, in 0.25 mol H_2SO_4 there are 4×0.25 (i.e. 1.0) mol oxygen atoms.

? Test yourself

6 Work out the mass of one molecule of each of the following:
   a H_2O
   b NH_3
   c CO_2

7 Work out the total number of hydrogen atoms in each of the following:
   a 1.00 mol H_2
   b 0.200 mol CH_4
   c 0.0500 mol NH_3

8 Calculate the total number of atoms in each of the following:
   a 0.0100 mol NH_3
   b 0.200 mol C_2H_6
   c 0.0400 mol C_2H_5OH

9 Calculate the number of moles of oxygen atoms in each of the following:
   a 0.2 mol H_2SO_4
   b 0.1 mol Cl_2O_7
   c 0.03 mol XeO_4

Learning objectives

- Determine the percentage composition by mass of a substance
- Understand what is meant by empirical and molecular formulas
- Calculate empirical and molecular formulas

1.2.2 Empirical and molecular formulas

Percentage composition of a compound

The percentage by mass of each element present in a compound can be worked out using the formula:

$$\text{\% by mass of an element} = \frac{\text{number of atoms of the element} \times \text{relative atomic mass}}{\text{relative molecular mass}}$$

Worked examples

1.5 Find the percentage by mass of each element present in $C_6H_5NO_2$.

The relative molecular mass of $C_6H_5NO_2$ is 123.12.

Percentage of carbon: the relative atomic mass of carbon is 12.01, and there are six carbon atoms present, so the total mass of the carbon atoms is 6×12.01, i.e. 72.06.

$$\% \text{ carbon} = \frac{72.06}{123.12} \times 100 = 58.53\%$$

Percentage of the other elements present:

$$\% \text{ hydrogen} = \frac{5 \times 1.01}{123.12} \times 100 = 4.10\%$$

$$\% \text{ nitrogen} = \frac{14.01}{123.12} \times 100 = 11.38\% \qquad \% \text{ oxygen} = \frac{2 \times 16.00}{123.12} \times 100 = 25.99\%$$

1.6 Calculate the mass of oxygen present in 2.20 g of CO_2.

The relative molecular mass of CO_2 is 44.01. Of this, the amount contributed by the two oxygen atoms is $2 \times 16.00 = 32.00$.

So the fraction of the mass of this compound that is contributed by oxygen is $\dfrac{32.00}{44.01}$

Therefore, in 2.20 g of CO_2, the amount of oxygen is $\dfrac{32.00}{44.01} \times 2.20 = 1.60\,\text{g}$

1.7 What mass of HNO_3 contains 2.00 g of oxygen?

The relative molecular mass of HNO_3 is 63.02. Each molecule contains three oxygen atoms with a total mass of 3×16.00, i.e. 48.00.

The oxygen and the HNO_3 are in the ratio $48.00 : 63.02$.

Therefore the mass of HNO_3 containing 2.00 g of oxygen is:

$$\frac{63.02}{48.00} \times 2.00 = 2.63\,\text{g}$$

Alternative method

The percentage of oxygen in HNO_3 is $\dfrac{3 \times 16.00}{63.02} \times 100 = 76.2\%$

So 76.2% of this sample is oxygen and has a mass of 2.00 g.
We need, therefore, to find the mass of 100%, which is given by

$$\frac{2.00}{76.2} \times 100 = 2.63\,\text{g}$$

Note: in order to obtain this answer, more figures were carried through on the calculator.

10 Calculate the percentage by mass of oxygen in each of the following compounds:

   a C_2H_5OH

   b CH_3CH_2COOH

   c Cl_2O_7

11 Calculate the mass of oxygen in each of the following samples:

   a $6.00\,g$ of C_3H_7OH

   b $5.00\,g$ of SO_2

   c $10.0\,g$ of P_4O_{10}

12 For each of the following compounds work out the mass of substance that will contain $1.00\,g$ of oxygen.

   a CH_3OH

   b SO_3

   c P_4O_6

Empirical and molecular formulas

A molecular formula is a whole number multiple of the empirical formula. Therefore, if the empirical formula of a compound is CH_2, the molecular formula is $(CH_2)_n$ i.e. C_2H_4 or C_3H_6 or C_4H_8, etc.

Empirical formula: the simplest whole number ratio of the elements present in a compound.

Molecular formula: the total number of atoms of each element present in a molecule of the compound. (The molecular formula is a multiple of the empirical formula.)

Worked examples

1.8 If the molecular formulas of two compounds are:

   a $C_4H_{10}O_2$ **b** Re_3Cl_9

what are the empirical formulas?

a We need to find the simplest ratio of the elements present and therefore need to find the highest number that divides exactly into the subscript of each element. In this case, each subscript can be divided by two, and so the empirical formula is C_2H_5O.

b In this case each subscript is divisible by three, and so the empirical formula is $ReCl_3$.

1.9 The empirical formula of benzene is CH. Given that the molar mass is $78.12\,g\,mol^{-1}$, work out its molecular formula.

The mass of the empirical formula unit (CH) is $12.01 + 1.01 = 13.02$. The number of times that the empirical formula unit occurs in the actual molecule (n) is given by:

$$n = \frac{\text{relative molecular mass}}{\text{empirical formula mass}} = \frac{78.12}{13.02} = 6$$

Therefore the molecular formula is $(CH)_6$, which is more commonly written as C_6H_6.

Chemical analysis of a substance can provide the composition by mass of the compound. The empirical formula can then be calculated from these data. In order to work out the molecular formula, the relative molecular mass of the compound is also required.

Worked examples

1.10 A compound has the following composition by mass: C, 0.681 g; H, 0.137 g; O, 0.181 g.
    a Calculate the empirical formula of the compound.
    b If the relative molecular mass of the compound is 88.17, calculate the molecular formula.

a This is most easily done by laying everything out in a table.

	C	H	O
mass/g	0.681	0.137	0.181
divide by relative atomic mass to give number of moles	0.681/12.01	0.137/1.01	0.181/16.00
number of moles/mol	0.0567	0.136	0.0113
divide by smallest to get ratio	0.0567/0.0113	0.136/0.0113	0.0113/0.0113
ratio	5	12	1

Therefore the empirical formula is $C_5H_{12}O$.

b The empirical formula mass of the compound is 88.17. This is the same as the relative molecular mass, and so the molecular formula is the same as the empirical formula ($C_5H_{12}O$).

1.11 If a fluoride of uranium contains 67.62% uranium by mass, what is its empirical formula?

A uranium fluoride contains only uranium and fluorine.

   % fluorine $= 100.00 - 67.62 = 32.38\%$

It makes no difference here that the percentage composition is given instead of the mass of each element present, as the percentage is the same as the mass present in 100 g.

	U	F
percentage	67.62	32.38
mass in 100 g/g	67.62	32.38
divide by relative atomic mass to give number of moles	67.62/238.03	32.38/19.00
number of moles	0.2841	1.704
divide by smallest to get ratio	0.2841/0.2841	1.704/0.2841
ratio	1	6

There are therefore six fluorine atoms for every uranium atom, and the empirical formula is UF_6.

1.12 The experimental set-up shown in the figure can be used to determine the empirical formula of copper oxide. The following experimental results were obtained.

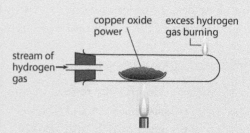

Mass of empty dish / g	24.58
Mass of dish + copper oxide / g	30.12
Mass of dish + copper at end of experiment / g	29.00

Calculate the empirical formula of the copper oxide and write an equation for the reaction.

Hydrogen gas is passed over the heated copper oxide until all the copper oxide is reduced to copper.

mass of copper oxide at start $= 30.12 - 24.58 = 5.54\,g$
mass of copper at end $= 29.00 - 24.58 = 4.42\,g$

The difference in mass is due to the oxygen from the copper oxide combining with the hydrogen.

mass of oxygen in copper oxide $= 5.54 - 4.42 = 1.12\,g$

From now on, the question is a straightforward empirical formula question:

$$\text{number of moles of copper} = \frac{4.42}{63.55} = 0.0696\,mol$$

$$\text{number of moles of oxygen} = \frac{1.12}{16.00} = 0.0700\,mol$$

If each number of moles is divided by the smaller number (0.0696):

Cu **O**

$$\frac{0.0696}{0.0696} = 1 \qquad \frac{0.0700}{0.0696} = 1.01$$

the ratio of copper to oxygen is 1:1, and the empirical formula is CuO.

The equation for the reaction is: $CuO + H_2 \rightarrow Cu + H_2O$

Composition by mass from combustion data

Worked examples

1.13 An organic compound, **A**, contains only carbon and hydrogen. When 2.50 g of **A** burns in excess oxygen, 8.08 g of carbon dioxide and 2.64 g of water are formed. Calculate the empirical formula.

The equation for the reaction is of the form: $C_xH_y + (x + \frac{y}{4})O_2 \rightarrow xCO_2 + \frac{y}{2}H_2O$

All the C in the CO_2 comes from the hydrocarbon **A**.

$$\text{number of moles of } CO_2 = \frac{8.08}{44.01} = 0.184\,mol$$

Each CO_2 molecule contains one carbon atom. Therefore the number of moles of carbon in 2.50 g of the hydrocarbon is 0.184 mol.

All the hydrogen in the water comes from the hydrocarbon **A**.

More significant figures are carried through in subsequent calculations.

$$\text{number of moles of } H_2O = \frac{2.64}{18.02} = 0.147 \, \text{mol}$$

Each H_2O molecule contains two hydrogen atoms, so the number of moles of hydrogen in 2.64 g of H_2O is $2 \times 0.147 = 0.293 \, \text{mol}$. Therefore, the number of moles of hydrogen in 2.50 g of the hydrocarbon is 0.293 mol. The empirical formula and molecular formula can now be calculated.

	C	H
number of moles	0.184	0.293
divide by smaller	0.184/0.184	0.293/0.184
ratio	1.00	1.60

The empirical formula must be a ratio of whole numbers, and this can be obtained by multiplying each number by five. Therefore the empirical formula is C_5H_8.

1.14 An organic compound, **B**, contains only carbon, hydrogen and oxygen. When 1.46 g of **B** burns in excess oxygen, 2.79 g of carbon dioxide and 1.71 g of water are formed.
    a What is the empirical formula of **B**?
    b If the relative molecular mass is 92.16, what is the molecular formula of **B**?

a The difficulty here is that the mass of oxygen in **B** cannot be worked out in the same way as the previous example, as only some of the oxygen in the CO_2 and H_2O comes from the oxygen in **B** (the rest comes from the oxygen in which it is burnt).

$$\text{mass of carbon in 2.79 g of } CO_2 = \frac{12.01}{44.01} \times 2.79 = 0.76 \, \text{g}$$

$$\text{mass of hydrogen in 1.71 g of } H_2O = \frac{2.02}{18.02} \times 1.71 = 0.19 \, \text{g}$$

Note: 2.02 as there are 2 H atoms in a water molecule

mass of oxygen in 1.46 g of **B** is $(1.46 - 0.76 - 0.19) = 0.51 \, \text{g}$

The empirical formula can now be calculated.

	C	H	O
mass / g	0.76	0.19	0.51
moles / mol	0.063	0.19	0.032
ratio	2	6	1

Therefore the empirical formula is C_2H_6O.

b The empirical formula mass is 46.08.

$$\frac{92.16}{46.08} = 2$$

Therefore, the molecular formula is $(C_2H_6O)_2$, i.e. $C_4H_{12}O_2$.

13 Which of the following represent empirical formulas?

C_2H_4	CO_2	CH
HO	C_3H_8	C_4H_{10}
H_2O	H_2O_2	N_2H_4
PCl_5	CH_3COOH	$C_6H_5CH_3$

14 Copy the table below and complete it with the molecular formulas of the compounds, given the empirical formulas and relative molecular masses.

Empirical formula	Relative molecular mass	Molecular formula
HO	34.02	
ClO_3	166.90	
CH_2	84.18	
BNH_2	80.52	

15 Analysis of a sample of an organic compound produced the following composition:

C: 0.399 g   H: 0.101 g

a Calculate the empirical formula.

b Given that the relative molecular mass is 30.08, determine the molecular formula.

16 If an oxide of chlorine contains 81.6% chlorine, calculate its empirical formula.

17 A compound contains 76.0% iodine and 24.0% oxygen. Calculate the empirical formula of the compound.

18 When 4.76 g of an organic compound, **D**, which contains only carbon, hydrogen and oxygen, is burnt in excess oxygen, 10.46 g of carbon dioxide and 5.71 g of water are produced. What is the empirical formula of **D**?

19 When 5.60 g of an iron oxide is heated with carbon, 3.92 g of iron is produced. Calculate the empirical formula of the iron oxide.

Learning objectives

- Solve problems involving masses of substances
- Calculate the theoretical and percentage yield in a reaction
- Understand the terms *limiting reactant* and *reactant in excess* and solve problems involving these

1.3 Reacting masses and volumes

1.3.1 Calculations involving moles and masses

Conservation of mass

The fact that mass is conserved in a chemical reaction can sometimes be used to work out the mass of product formed. For example, if 55.85 g of iron reacts *exactly* and *completely* with 32.06 g of sulfur, 87.91 g of iron sulfide is formed:

$$Fe(s) + S(s) \rightarrow FeS(s)$$

Worked example

1.15 Consider the combustion of butane:

$$2C_4H_{10}(g) + 13O_2(g) \rightarrow 8CO_2(g) + 10H_2O(l)$$

10.00 g of butane reacts exactly with 35.78 g of oxygen to produce 30.28 g of carbon dioxide. What mass of water was produced?

The masses given represent an exact chemical reaction, so we assume that all the reactants are converted to products.

The total mass of the reactants = 10.00 + 35.78 = 45.78 g.

The total mass of the products must also be 45.78 g.

Therefore the mass of water = 45.78 − 30.28 = 15.50 g.

Using moles

We often want to work out the mass of one reactant that reacts exactly with a certain mass of another reactant – or how much product is formed when certain masses of reactants react. This can be done by calculating the numbers of each molecule or atom present in a particular mass or, much more simply, by using the mole concept.

As we have seen, one mole of any substance always contains the same number of particles, so if we know the number of moles present in a certain mass of reactant we also know the number of particles and can therefore work out what mass of another reactant it reacts with and how much product is formed.

There are three main steps in a moles calculation.

1 Work out the number of moles of anything you can.
2 Use the chemical (stoichiometric) equation to work out the number of moles of the quantity you require.
3 Convert moles to the required quantity – volume, mass etc.

Questions involving masses of substances

Worked examples

1.16 Consider the reaction of sodium with oxygen:

$$4Na(s) + O_2(g) \rightarrow 2Na_2O(s)$$

a How much sodium reacts exactly with 3.20 g of oxygen?
b What mass of Na_2O is produced?

a Step 1 – the mass of oxygen is given, so the number of moles of oxygen can be worked out (you could use the triangle shown here).

$$\text{number of moles of oxygen} = \frac{3.20}{32.00} = 0.100\,\text{mol}$$

Note: the mass of oxygen was given to three significant figures, so all subsequent answers are also given to three significant figures.

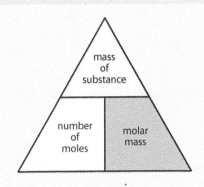

Step 2 – the coefficients in the chemical (stoichiometric) equation tell us that 1 mol O_2 reacts with 4 mol sodium. Therefore 0.100 mol O_2 reacts with 4×0.100 mol sodium, i.e. 0.400 mol sodium.

Step 3 – convert the number of moles to the required quantity, mass in this case:

mass of sodium $= 0.400 \times 22.99 = 9.20\,g$

> Note: the mass of sodium is worked out by multiplying the mass of one mole by the number of moles – the number of moles is *not* multiplied by the mass of four sodium atoms – the four was already taken into account when 0.100 mol was multiplied by 4 to give the number of moles of sodium.

b From the coefficients in the equation we know that 1 mol O_2 reacts with 4 mol sodium to produce 2 mol Na_2O. Therefore 0.100 mol O_2 reacts with 0.400 mol sodium to give 2×0.100 mol Na_2O, i.e. 0.200 mol Na_2O.

The molar mass of $Na_2O = 61.98\,g\,mol^{-1}$
So mass of $Na_2O = 0.200 \times 61.98 = 12.4\,g$

Alternatively, the mass of Na_2O can be worked out using the idea of conservation of mass, i.e.:
mass of Na_2O = mass of O_2 + mass of Na.

Exam tip
Masses may also be given in kilograms or tonnes.

$1\,kg = 1000\,g$

$1\,tonne = 1 \times 10^6\,g$

Before working out the number of moles, you must convert the mass to grams. To convert kilograms to grams, multiply by 1000; to convert tonnes to grams, multiply the mass by 1×10^6.

1.17 Consider the following equation:

$$2NH_3 + 3CuO \rightarrow N_2 + 3H_2O + 3Cu$$

If 2.56 g of ammonia (NH_3) is reacted with excess CuO, calculate the mass of copper produced.

The CuO is in excess so there is more than enough to react with all the NH_3. This means that we do not need to worry about the number of moles of CuO.

Step 1 – the number of moles of NH_3 can be calculated:

$\dfrac{2.56}{17.04} = 0.150$ mol NH_3

Step 2 – two moles of NH_3 produce three moles of copper, so 0.150 mol NH_3 produces $0.150 \times \frac{3}{2}$ mol copper, i.e. 0.225 mol copper.

> The number of moles of copper is therefore 1.5 times the number of moles of NH_3.

Step 3 – the mass of 1 mol copper $= 63.55\,g$, so the mass of copper produced $= 0.225 \times 63.55 = 14.3\,g$.

Formula for solving moles questions involving masses

An alternative way of doing these questions is to use a formula.

$$\frac{m_1}{n_1 M_1} = \frac{m_2}{n_2 M_2}$$

where

m_1 = mass of first substance

n_1 = coefficient of first substance (number in front in the chemical equation)

M_1 = molar mass of first substance

Worked example

1.18 The following equation represents the combustion of butane:

$$2C_4H_{10}(g) + 13O_2(g) \rightarrow 8CO_2(g) + 10H_2O(l)$$

If 10.00 g of butane is used, calculate the mass of oxygen required for an exact reaction.

We will call butane substance 1 and oxygen substance 2 (it doesn't matter which you call what, but you have to be consistent).

$m_1 = 10.00\,g$ $m_2 = ?$

$n_1 = 2$ $n_2 = 13$

$M_1 = 58.14\,g\,mol^{-1}$ $M_2 = 32.00\,g\,mol^{-1}$

$$\frac{m_1}{n_1 M_1} = \frac{m_2}{n_2 M_2}$$

$$\frac{10.00}{2 \times 58.14} = \frac{m_2}{13 \times 32.00}$$

The equation can be rearranged:

$$m_2 = \frac{10.00 \times 13 \times 32.00}{2 \times 58.14} = 35.78$$

Therefore the mass of oxygen required for an exact reaction is 35.78 g.

20 a How many moles of hydrogen gas are produced when 0.4 mol sodium react with excess water?

$$2Na + 2H_2O \rightarrow 2NaOH + H_2$$

b How many moles of O_2 react with 0.01 mol C_3H_8?

$$C_3H_8 + 5O_2 \rightarrow 3CO_2 + 4H_2O$$

c How many moles of H_2S are formed when 0.02 mol HCl react with excess Sb_2S_3?

$$Sb_2S_3 + 6HCl \rightarrow 2SbCl_3 + 3H_2S$$

d How many moles of oxygen are formed when 0.6 mol $KClO_3$ react?

$$2KClO_3(s) \rightarrow 2KCl(s) + 3O_2(g)$$

e How many moles of iron are formed when 0.9 mol CO react with excess iron oxide?

$$Fe_2O_3 + 3CO \rightarrow 2Fe + 3CO_2$$

f How many moles of hydrogen would be required to make 2.4×10^{-3} mol NH_3?

$$N_2 + 3H_2 \rightarrow 2NH_3$$

21 a Calculate the mass of arsenic(III) chloride produced when 0.150 g of arsenic reacts with excess chlorine according to the equation:

$$2As + 3Cl_2 \rightarrow 2AsCl_3$$

b What mass of sulfur is produced when 5.78 g of iron(III) sulfide is reacted with excess oxygen?

$$2Fe_2S_3 + 3O_2 \rightarrow 2Fe_2O_3 + 6S$$

c Calculate the mass of iodine that must be reacted with excess phosphorus to produce 5.00 g of phosphorus(III) iodide according to the equation below.

$$2P + 3I_2 \rightarrow 2PI_3$$

d Consider the reaction shown below. What mass of SCl_2 must be reacted with excess NaF to produce 2.25 g of NaCl?

$$3SCl_2 + 4NaF \rightarrow S_2Cl_2 + SF_4 + 4NaCl$$

The fact that a theory can explain experimental observations does not necessarily make it correct. The explanations presented in this book fit in with experimental observations, but this does not mean that they are 'true' – they just represent our interpretation of the data at this stage in time. Each generation of scientists believes that they are presenting a true description of reality, but is it possible for more than one explanation to fit the facts? You, or indeed I, may not be able to think of a better explanation to fit many of the experimental observations in modern science, but that does not mean that there isn't one. Consider the following trivial example.

Experimentally, when 100 kg of calcium carbonate ($CaCO_3$) is heated, 44 kg of carbon dioxide (CO_2) is obtained. The following calculation can be carried out to explain this.

The equation for the reaction is:

$$CaCO_3 \rightarrow CaO + CO_2$$

$$\text{number of moles of } CaCO_3 = \frac{100}{(20 + 6 + (3 \times 8))}$$

$$= 2 \text{ mol}$$

Two moles of $CaCO_3$ produces two moles of CO_2.

The mass of two moles of CO_2 is $2 \times (6 + (2 \times 8)) = 44$ kg.

Hopefully you can see some mistakes in this calculation, but the result is what we got experimentally. It is also interesting to note that if, in your IB examination, you had just written down the final answer, you would probably have got full marks!

Calculating the yield of a chemical reaction

In any commercial process it is very important to know the **yield** (the amount of desired product) of a chemical reaction. For instance, if a particular process for the preparation of a drug involves four separate steps and the yield of each step is 95%, it is probably quite a promising synthetic route to the drug. If, however, the yield of each step is only 60% then it is likely that the company would look for a more efficient synthetic process.

The yield of a chemical reaction is usually quoted as a percentage – this gives more information than just quoting the yield of the product as a mass. Consider the preparation of 1,2-dibromoethane ($C_2H_4Br_2$):

$$C_2H_4(g) + Br_2(l) \rightarrow C_2H_4Br_2(l)$$

10.00 g of ethene (C_2H_4) will react exactly with 56.95 g of bromine.

The **theoretical yield** for this reaction is 66.95 g – this is the maximum possible yield that can be obtained. The **actual yield** of $C_2H_4Br_2$ may be 50.00 g.

$$\% \text{ yield} = \frac{50.00}{66.95} \times 100 = 74.68\%$$

$$\% \text{ yield} = \frac{\text{actual yield}}{\text{theoretical yield}} \times 100$$

Worked example

1.19 $C_2H_5OH(l) + CH_3COOH(l) \rightarrow CH_3COOC_2H_5(l) + H_2O(l)$
ethanol   ethanoic acid   ethyl ethanoate   water

If the yield of ethyl ethanoate obtained when 20.00 g of ethanol is reacted with excess ethanoic acid is 30.27 g, calculate the percentage yield.

The first step is to calculate the maximum possible yield, i.e. the theoretical yield:

molar mass of $C_2H_5OH = 46.08$ g mol^{-1}

$$\text{number of moles of } C_2H_5OH = \frac{20.00}{46.08} = 0.4340 \text{ mol}$$

> The CH_3COOH is in excess, i.e. more than enough is present to react with all the C_2H_5OH. This means that we do not need to worry about the number of moles of CH_3COOH.

The chemical equation tells us that 1 mol C_2H_5OH produces 1 mol $CH_3COOC_2H_5$. Therefore, 0.4340 mol C_2H_5OH produces 0.4340 mol $CH_3COOC_2H_5$.

The molar mass of $CH_3COOC_2H_5 = 88.12$ g mol^{-1}.

The mass of ethyl ethanoate $CH_3COOC_2H_5 = 0.4340 \times 88.12 = 38.24$ g.

So, the theoretical yield is 38.24 g. The actual yield is 30.27 g (given in the question).

$$\% \text{ yield} = \frac{30.27}{38.24} \times 100 = 79.15\%$$

The percentage yield of $CH_3COOC_2H_5$ is 79.15%.

22 Calculate the percentage yield in each of the following reactions.
  a When 2.50 g of SO_2 is heated with excess oxygen, 2.50 g of SO_3 is obtained.
    $$2SO_2 + O_2 \rightarrow 2SO_3$$
  b When 10.0 g of arsenic is heated in excess oxygen, 12.5 g of As_4O_6 is produced.
    $$4As + 3O_2 \rightarrow As_4O_6$$
  c When 1.20 g of C_2H_4 reacts with excess bromine, 5.23 g of CH_2BrCH_2Br is produced.
    $$C_2H_4 + Br_2 \rightarrow CH_2BrCH_2Br$$

Limiting reactant

Very often we do not use exact quantities in a chemical reaction, but rather we use an excess of one or more reactants. One reactant is therefore used up before the others and is called the **limiting reactant**. When the limiting reactant is completely used up, the reaction stops.

Figure **1.9** illustrates the idea of a limiting reactant and shows how the products of the reaction depend on which reactant is limiting.

Figure 1.9 The reaction between magnesium and hydrochloric acid. In each test tube a small amount of universal indicator has been added. **a** In this test tube, the magnesium is in excess and the reaction finishes when the hydrochloric acid runs out. There is still magnesium left over at the end, and the solution is no longer acidic. **b** In this test tube, the hydrochloric acid is in excess. The magnesium is the limiting reactant, and the reaction stops when the magnesium has been used up. The solution is still acidic at the end.

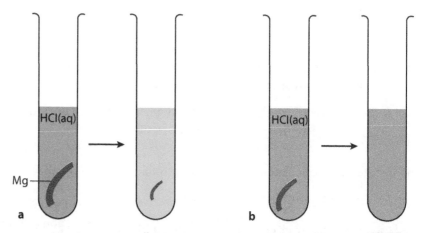

Worked examples

1.20 Consider the reaction between magnesium and nitrogen:

$$3Mg(s) + N_2(g) \rightarrow Mg_3N_2(s)$$

10.00 g of magnesium is reacted with 5.00 g of nitrogen. Which is the limiting reactant?

number of moles of magnesium $= \dfrac{10.00}{24.31} = 0.4114 \, \text{mol}$ number of moles of $N_2 = \dfrac{5.00}{28.02} = 0.178 \, \text{mol}$

The equation tells us that 3 mol magnesium reacts with 1 mol N_2. So 0.4114 mol magnesium reacts with $\dfrac{0.4114}{3}$ mol N_2, i.e. 0.1371 mol N_2.

Therefore, for an exact reaction, $0.1371\,mol\ N_2$ are required to react with $0.4114\,mol$ magnesium. However, $0.178\,mol\ N_2$ are used, which is more than enough to react. This means that N_2 is in excess because there is more than enough to react with all the magnesium present. Magnesium is therefore the limiting reactant.

This can also be seen from working with the number of moles of N_2 – $0.178\,mol\ N_2$ was used in this reaction. This number of moles of N_2 would require $3 \times 0.178\,mol$ magnesium for an exact reaction, i.e. $0.534\,mol$ magnesium. However, only $0.4114\,mol$ magnesium are present; therefore, the magnesium will run out before all the N_2 has reacted.

> **Exam tip**
> If the number of moles of each reactant is divided by its coefficient in the stoichiometric equation, the smallest number indicates the limiting reactant.

1.21 Consider the reaction between sulfur and fluorine: $S(s) + 3F_2(g) \rightarrow SF_6(g)$

$10.00\,g$ of sulfur reacts with $10.00\,g$ of fluorine.

   a Which is the limiting reactant?
   b What mass of sulfur(VI) fluoride is formed?
   c What mass of the reactant in excess is left at the end?

a number of moles of sulfur $= \dfrac{10.00}{32.07} = 0.3118\,mol$ number of moles of $F_2 = \dfrac{10.00}{38.00} = 0.2632\,mol$

The coefficient of sulfur in the equation is 1 and that of F_2 is 3.
$0.3118 / 1 = 0.3118$ and $0.2632 / 3 = 0.08773$, therefore sulfur is in excess (larger number) and F_2 is the limiting reactant (smaller number).

Alternatively we can reason from the chemical equation that $0.2632\,mol\ F_2$ should react with $0.08773\,mol$ sulfur (i.e. $0.2632\,mol$ divided by three). There is more than $0.08773\,mol$ sulfur present, so sulfur is present in excess and F_2 is the limiting reactant.

For the rest of the question we must work with the limiting reactant.

b When the limiting reactant is used up completely, the reaction stops. This means that the amount of product formed is determined by the amount of the limiting reactant we started with.

From the chemical equation, $0.2632\,mol\ F_2$ produces $0.08773\,mol\ SF_6$ (i.e. $0.2632\,mol$ divided by three).

   molar mass of $SF_6 = 146.07\,g\,mol^{-1}$

   mass of SF_6 formed $= 0.08773 \times 146.07 = 12.81\,g$

c From the chemical equation, $0.2632\,mol\ F_2$ reacts with $0.08773\,mol$ sulfur (i.e. $0.2632\,mol$ sulfur divided by three). Originally there were $0.3118\,mol$ sulfur present; therefore the number of moles of sulfur left at the end of the reaction is $0.3118 - 0.08773 = 0.2241$.

The mass of sulfur left at the end of the reaction is $0.2241 \times 32.07 = 7.187\,g$.

> **Exam tip**
> To do a moles question you need to know the mass of just one of the reactants. If you are given the masses of more than one reactant, you must consider that one of these reactants will be the limiting reactant and you must use this one for all calculations.

1.22 For the reaction:

$$4FeCr_2O_4 + 8Na_2CO_3 + 7O_2 \rightarrow 8Na_2CrO_4 + 2Fe_2O_3 + 8CO_2$$

there are 100.0 g of each reactant available. Which is the limiting reactant?

This question could be done by working out the number of moles of each reactant and then comparing them, but there is a shortcut – to work out the masses of each substance if molar quantities reacted:

$$4FeCr_2O_4 \quad + \quad 8Na_2CO_3 \quad + \quad 7O_2 \quad \rightarrow \quad 8Na_2CrO_4 + 2Fe_2O_3 + 8CO_2$$

mass / g =	4×223.85	8×105.99	7×32.00
mass / g =	895.40	847.92	224.00

These are the masses that are required for the exact reaction. Because the highest mass required is that of $FeCr_2O_4$, if the same mass of each substance is taken, the $FeCr_2O_4$ will run out first and must be the limiting reactant.

Nature of science

Science is a constantly changing body of knowledge. Scientists take existing knowledge and try to build on it to improve theories so that they are more widely applicable and have better explanatory power. The concept of the mole developed from the concept of equivalent weight.

? Test yourself

23 What is the limiting reactant in each of the following reactions?
   a 0.1 mol Sb_4O_6 react with 0.5 mol H_2SO_4
      $$Sb_4O_6 + 6H_2SO_4 \rightarrow 2Sb_2(SO_4)_3 + 6H_2O$$
   b 0.20 mol $AsCl_3$ react with 0.25 mol H_2O
      $$4AsCl_3 + 6H_2O \rightarrow As_4O_6 + 12HCl$$
   c 0.25 mol copper react with 0.50 mol dilute HNO_3 according to the equation:
      $$3Cu + 8HNO_3$$
      $$\rightarrow 3Cu(NO_3)_2 + 4H_2O + 2NO$$
   d 0.10 mol NaCl react with 0.15 mol MnO_2 and 0.20 mol H_2SO_4
      $$2NaCl + MnO_2 + 2H_2SO_4$$
      $$\rightarrow Na_2SO_4 + MnSO_4 + 2H_2O + Cl_2$$

24 Boron can be prepared by reacting B_2O_3 with magnesium at high temperatures:
   $$B_2O_3 + 3Mg \rightarrow 2B + 3MgO$$
   What mass of boron is obtained if 0.75 g B_2O_3 are reacted with 0.50 g magnesium?

25 Iron(III) oxide reacts with carbon to produce iron:
   $$Fe_2O_3 + 3C \rightarrow 2Fe + 3CO$$
   What mass of iron is obtained if 10.0 tonnes of Fe_2O_3 are reacted with 1.00 tonne of carbon?

1.3.2 Calculations involving volumes of gases

Real gases and ideal gases

An 'ideal gas' is a concept invented by scientists to approximate (model) the behaviour of real gases. Under normal conditions (around $100\,kPa$ [approximately 1 atmosphere] pressure and $0\,°C$) real gases such as hydrogen behave pretty much like ideal gases and the approximations work very well.

Two assumptions we make when defining the **ideal gas** are that the molecules themselves have no volume (they are point masses) and that no forces exist between them (except when they collide). If we imagine compressing a real gas to a very high pressure then the particles will be much closer together and, under these conditions, the forces between molecules and the volumes occupied by the molecules will be significant. This means that we can no longer ignore these factors and the behaviour of the gas will deviate significantly from our ideal gas model. This will also be the case at very low temperatures when the molecules are moving more slowly. Under conditions of very low temperature and very high pressure a gas is approaching the liquid state and will be least like our predictions for an ideal gas.

The idea that the volumes of the individual gas molecules are zero (so it makes no difference if the gas is H_2 or NH_3) and that there are no forces between the molecules (again no difference between NH_3 and H_2) means that all ideal gases must behave in the same way. This means that the volume occupied by a gas at a certain temperature and pressure depends only on the number of molecules present and not on the nature of the gas. In other words, at a certain temperature and pressure, the volume of a gas is proportional to the number of moles of the gas.

Using volumes of gases

> **Avogadro's law**: equal volumes of ideal gases measured at the same temperature and pressure contain the same number of molecules.

In other words $100\,cm^3$ of H_2 contains the same number of molecules at $25\,°C$ and $100\,kPa$ as $100\,cm^3$ of NH_3, if we assume that they both behave as ideal gases. Under the same conditions, $50\,cm^3$ of CO_2 would contain half as many molecules.

This means that volumes can be used directly (instead of moles) in equations involving gases:

$$H_2(g) + Cl_2(g) \rightarrow 2HCl(g)$$

The above equation tells us that one mole of H_2 reacts with one mole of Cl_2 to give two moles of HCl. Or one volume of H_2 reacts with one volume of Cl_2 to give two volumes of HCl; i.e. $50\,cm^3$ of H_2 reacts with $50\,cm^3$ of Cl_2 to give $100\,cm^3$ of HCl.

Learning objectives

- Understand Avogadro's law and use it to calculate reacting volumes of gases
- Use the molar volume of a gas in calculations at standard temperature and pressure
- Understand the relationships between pressure, volume and temperature for an ideal gas
- Solve problems using the equation
  $$\frac{P_1 V_1}{T_1} = \frac{P_2 V_2}{T_2}$$
- Solve problems using the ideal gas equation

Gases deviate most from ideal behaviour at high pressure and low temperature.

Volume of gas $\propto$ number of moles of the gas

$1\,cm^3$ is the same as $1\,ml$.

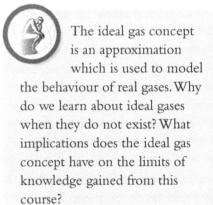
The ideal gas concept is an approximation which is used to model the behaviour of real gases. Why do we learn about ideal gases when they do not exist? What implications does the ideal gas concept have on the limits of knowledge gained from this course?

Worked examples

In both of these worked examples, assume that all gases behave as ideal gases and that all measurements are made under the same conditions of temperature and pressure.

1.23 Consider the following reaction for the synthesis of methanol:

$$CO(g) + 2H_2(g) \rightarrow CH_3OH(g)$$

    a What volume of H_2 reacts exactly with $2.50\,dm^3$ of CO?
    b What volume of CH_3OH is produced?

a From the equation we know that $1\,mol$ CO reacts with $2\,mol$ H_2. Therefore, one volume of CO reacts with two volumes of H_2 – $2.50\,dm^3$ of CO reacts with 2×2.50, i.e. $5.00\,dm^3$, of H_2.
b One volume of CO produces one volume of CH_3OH. Therefore, the volume of CH_3OH produced is $2.50\,dm^3$.

1.24 If $100\,cm^3$ of oxygen reacts with $30\,cm^3$ of methane in the following reaction, how much oxygen will be left at the end of the reaction?

$$CH_4(g) + 2O_2(g) \rightarrow CO_2(g) + 2H_2O(l)$$

From the equation, we know that $1\,mol$ CH_4 reacts with $2\,mol$ O_2. Therefore, one volume of CH_4 reacts with two volumes of O_2 – so $30\,cm^3$ of CH_4 reacts with 2×30, i.e. $60\,cm^3$ of O_2.

The original volume of O_2 was $100\,cm^3$; therefore, if $60\,cm^3$ reacted, the volume of oxygen gas left over at the end of the reaction would be $100 - 60 = 40\,cm^3$.

STP = standard temperature and pressure = $273\,K$, $100\,kPa$ (1 bar)

$100\,kPa = 1.00 \times 10^5\,Pa$

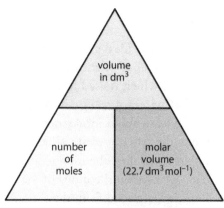

Figure 1.10 The relationship between the number of moles of a gas and its volume.

Converting volumes of gases to number of moles

Because the volume occupied by an ideal gas depends only on the number of particles present (assuming that pressure and temperature are constant) and not on the nature of the particles, the volume occupied by one mole of any ideal gas under a certain set of conditions will always be the same. The volume occupied by one mole of a gas under certain conditions is called the **molar volume**.

molar volume of an ideal gas at STP = $22.7\,dm^3\,mol^{-1}$ or $2.27 \times 10^{-2}\,m^3\,mol^{-1}$

This means that under the same set of conditions, the volume occupied by one mole of NH_3 is the same as the volume occupied by one mole of CO_2 and one mole of H_2, and this volume is $22.7\,dm^3$ at STP.

    The relationship between the number of moles of a gas and its volume is:

$$\text{number of moles} = \frac{\text{volume}}{\text{molar volume}}$$

This is summarised in Figure **1.10**.

The absolute, or Kelvin, scale of temperature starts at absolute zero, which is the lowest temperature possible. It is the temperature at which everything would be in its lowest energy state. **Absolute zero** corresponds to 0 K or −273.15 °C (usually taken as −273 °C) and is also the temperature at which the volume of an ideal gas would be zero. It is not possible to actually reach absolute zero, but scientists have managed to get very close – about 1 nanokelvin!

> A change of 1 °C is the same as a change of 1 K, and 0 °C is equivalent to 273 K

To convert °C to K add 273:
e.g. 25 °C is equivalent to 25 + 273, i.e. 298 K

To convert K to °C subtract 273:
e.g. 350 K is equivalent to 350 − 273, i.e. 77 °C

Volumes of gases are often given in cm^3 and so it is important to know how to convert between cm^3 and dm^3.

Because $1\,dm^3$ (1 litre) is equivalent to $1000\,cm^3$ to convert cm^3 to dm^3 we divide by 1000 (to go from $1000\,(cm^3)$ to $1\,(dm^3)$). The conversion is shown in Figure **1.11**.

 In different countries around the world's different scales of temperature are used – e.g. the Celsius and Fahrenheit scales. The Celsius and Fahrenheit scales are both artificial scales, but the Kelvin scale is an absolute scale. What is the advantage to scientists of using an absolute scale? Why has the **absolute scale of temperature** not been adopted in everyday life?

The Kelvin scale of temperature is named in honour of William Thompson, Lord Kelvin (1824–1907), a Scottish mathematican and physicist, who first suggested the idea of an absolute scale of temperature. Despite making many important contributions to the advancement of science, Kelvin had doubts about the existence of atoms, believed that the Earth could not be older than 100 million years and is often quoted as saying that 'heavier-than-air flying machines are impossible'.

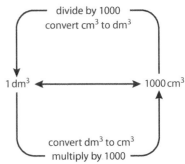

Figure 1.11 Converting between cm^3 and dm^3.

Worked examples

1.25 a Calculate the number of moles in $250\,cm^3$ of O_2 at STP.
 b Calculate the volume of $0.135\,mol$ CO_2 at STP.

a $\text{number of moles} = \dfrac{\text{volume in } dm^3}{22.7}$

$250\,cm^3 = \dfrac{250}{1000}\,dm^3 = 0.250\,dm^3$

$\text{number of moles} = \dfrac{0.250}{22.7} = 0.0110\,mol$

b $\text{volume} = \text{number of moles} \times 22.7 = 0.135 \times 22.7 = 3.06\,dm^3$

1.26 Calculate the volume of carbon dioxide (collected at STP) produced when 10.01 g of calcium carbonate decomposes according to the equation:

$$CaCO_3(s) \rightarrow CaO(s) + CO_2(g)$$

Step 1 – work out the number of moles of $CaCO_3$:

$$\text{number of moles of } CaCO_3 = \frac{10.01}{100.09} = 0.1000 \text{ mol}$$

Step 2 – the chemical equation tells us that 1 mol $CaCO_3$ decomposes to give 1 mol CO_2.

Therefore 0.1000 mol $CaCO_3$ decomposes to give 0.1000 mol CO_2.

Step 3 – convert the number of moles to volume:

1 mol CO_2 occupies 22.7 dm^3 at STP

volume of CO_2 = number of moles × volume of 1 mole (22.7 dm^3)

volume of CO_2 = 0.1000 × 22.7 = 2.27 dm^3

The volume of CO_2 produced is 2.27 dm^3.

1.27 Potassium chlorate(V) decomposes when heated:

$$2KClO_3(s) \rightarrow 2KCl(s) + 3O_2(g)$$

What mass of potassium chlorate(V) decomposes to produce 100.0 cm^3 of oxygen gas measured at STP?

Step 1 – work out the number of moles of O_2. The volume of O_2 must first be converted to dm^3:

$$\text{volume of } O_2 \text{ in } dm^3 = \frac{100.0}{1000} = 0.1000 \text{ dm}^3$$

$$\text{number of moles of } O_2 = \frac{0.1000}{22.7} = 4.405 \times 10^{-3} \text{ mol}$$

Step 2 – the chemical equation tells us that 3 mol O_2 are produced from 2 mol $KClO_3$. Therefore the number of moles of $KClO_3$ is two-thirds of the number of moles of O_2:

$$\frac{2}{3} \times 4.405 \times 10^{-3} = 2.937 \times 10^{-3} \text{ mol}$$

Step 3 – convert the number of moles of $KClO_3$ to mass:

molar mass of $KClO_3$ = 122.55 $g\,mol^{-1}$

mass of $KClO_3$ = 122.55 × 2.937 × 10^{-3} = 0.3599 g

The mass of $KClO_3$ required is 0.3599 g.

Formula for solving moles questions involving volumes of gases

An alternative way of doing these questions is to use a formula.

$$\frac{m_1}{n_1 M_1} = \frac{V_2}{n_2 M_v}$$

Note: this is very similar to the formula that was used earlier with masses.

where:

m_1 = mass of first substance (in g)

n_1 = coefficient of first substance

M_1 = molar mass of first substance

V_2 = volume (in dm^3) of second substance if it is a gas

n_2 = coefficient of second substance

M_v = molar volume of a gas = $22.7\,dm^3$ at STP

This formula can be used if the mass of one substance is given and the volume of another substance is required, or vice versa.

If a volume is given and a volume is required, then an alternative form of this equation is:

$$\frac{V_1}{n_1} = \frac{V_2}{n_2}$$

There is no need to convert units of volume to dm^3 with this equation – but V_2 must have the same units as V_1.

where:

V_1 = volume of first substance if it is a gas

V_2 = volume of second substance

However, with questions involving just gases it is usually easier to work them out using Avogadro's law, as described earlier.

? Test yourself

Assume that all gases behave as ideal gases and that all measurements are made under the same conditions of temperature and pressure.

26 a Calculate the volume of carbon dioxide produced when $100\,cm^3$ of ethene burns in excess oxygen according to the equation:

$$C_2H_4(g) + 3O_2(g) \rightarrow 2CO_2(g) + 2H_2O(l)$$

b Calculate the volume of nitric oxide (NO) produced when $2.0\,dm^3$ of oxygen is reacted with excess ammonia according to the equation:

$$4NH_3(g) + 5O_2(g) \rightarrow 4NO(g) + 6H_2O(g)$$

27 Determine the number of moles present in each of the following at standard temperature and pressure:

a $0.240\,dm^3$ of O_2
b $2.00\,dm^3$ of CH_4
c $0.100\,dm^3$ of SO_2
d $400.0\,cm^3$ of N_2
e $250.0\,cm^3$ of CO_2

28 Work out the volume of each of the following at standard temperature and pressure:

a $0.100\,mol$ C_3H_8
b $100.0\,mol$ SO_3
c $0.270\,mol$ N_2
d $0.8500\,mol$ NH_3
e $0.600\,mol$ O_2

29 Sodium nitrate(V) decomposes according to the equation:

$$2NaNO_3(s) \rightarrow 2NaNO_2(s) + O_2(g)$$

Calculate the volume (in cm^3) of oxygen produced (measured at STP) when 0.850 g of sodium nitrate(V) decompose.

30 Tin reacts with nitric acid according to the equation:

$$Sn(s) + 4HNO_3(aq)$$
$$\rightarrow SnO_2(s) + 4NO_2(g) + 2H_2O(l)$$

If 2.50 g of tin are reacted with excess nitric acid what volume of nitrogen dioxide (in cm^3) is produced at STP?

31 Calculate the mass of sodium carbonate that must be reacted with excess hydrochloric acid to produce 100.0 cm^3 of carbon dioxide at STP.

$$Na_2CO_3(s) + 2HCl(aq)$$
$$\rightarrow 2NaCl(aq) + CO_2(g) + H_2O(l)$$

32 a Oxygen can be converted to ozone (O_3) by passing it through a silent electric discharge:

$$3O_2(g) \rightarrow 2O_3(g)$$

If 300 cm^3 of oxygen are used and 10% of the oxygen is converted to ozone, calculate the total volume of gas present at the end of the experiment.

b Hydrogen reacts with chlorine according to the equation:

$$H_2(g) + Cl_2(g) \rightarrow 2HCl(g)$$

What is the total volume of gas present in the container at the end of the experiment if 100 cm^3 of hydrogen are reacted with 200 cm^3 of chlorine?

Macroscopic properties of ideal gases

So far, all the questions we have dealt with have involved working out volumes of gases at STP. In order to work out volumes of gases under other conditions we must understand a little more about the properties of gases.

The relationship between pressure and volume (Boyle's law)

'Macroscopic' means 'on a large scale'. The opposite is 'microscopic'. The microscopic properties of a gas are the properties of the particles that make up the gas.

> At a constant temperature, the volume of a fixed mass of an ideal gas is inversely proportional to its pressure.

This means that if the pressure of a gas is doubled at constant temperature, then the volume will be halved, and vice versa. This relationship is illustrated in Figure **1.12**.

$$P \propto \frac{1}{V}$$

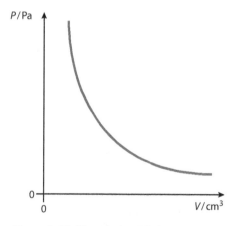

Figure 1.12 The relationship between pressure and volume of a fixed mass of an ideal gas at constant temperature.

The relationship can also be written as:

$$P = \frac{k}{V}$$

where k is a constant.

This can be rearranged to give

$$PV = k$$

This means that the product of the pressure and volume of an ideal gas at a particular temperature is a constant and does not change as the pressure and the volume change.

Other graphs can also be drawn to illustrate this relationship (see Figures **1.13** and **1.14**).

Because pressure is proportional to $\dfrac{1}{\text{volume}}$, a graph of pressure against $\dfrac{1}{\text{volume}}$ would be a straight-line graph that would pass through the origin (although this graph will never actually pass through the origin – the gas would have to have infinite volume at zero pressure). This is shown in Figure **1.13**.

Because $PV = k$, where k is a constant, a graph of PV against pressure (or volume) will be a straight, horizontal line. This is shown in Figure **1.14**.

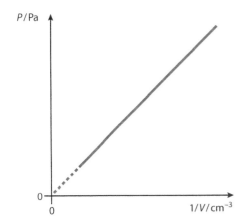

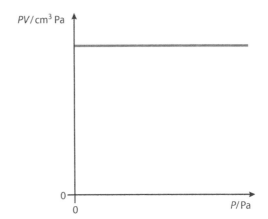

Figure 1.13 The relationship between the pressure and $\dfrac{1}{\text{volume}}$ of a fixed mass of an ideal gas at constant temperature.

Figure 1.14 The relationship between PV and P for a fixed mass of an ideal gas at constant temperature.

The relationship between volume and temperature (Charles' law)

If the temperature is in kelvin, the following relationship exists between the volume and the temperature:

> The volume of a fixed mass of an ideal gas at constant pressure is directly proportional to its kelvin temperature.
>
> $V \propto T$

Therefore, if the kelvin temperature is doubled and the pressure remains constant, the volume of the gas is doubled, and vice versa. This means that if an ideal gas has a volume of $200\,\text{cm}^3$ at $120\,\text{K}$, it will have a volume of $400\,\text{cm}^3$ at $240\,\text{K}$ if the pressure remains constant. This is illustrated in Figure **1.15**.

This relationship does not work for temperatures in °C (Figure **1.16**). For instance, if the volume of an ideal gas at $25\,°\text{C}$ is $500\,\text{cm}^3$, the volume it will occupy at $50\,°\text{C}$ will be about $560\,\text{cm}^3$.

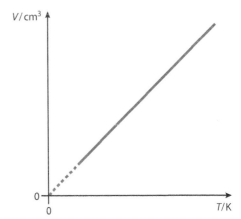

Figure 1.15 The relationship between the volume and temperature (in kelvin) of a fixed mass of an ideal gas at constant pressure.

An ideal gas can never liquefy because there are no forces between the molecules.

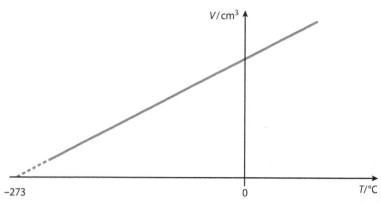

This is a linear relationship but not a proportional one because the graph does not pass through the origin.

Figure 1.16 The relationship between the volume and temperature (in °C) of a fixed mass of an ideal gas at constant pressure. As can be seen, the temperature at which the volume of an ideal gas is zero will be −273 °C. This temperature is **absolute zero**.

The relationship between pressure and temperature

> For a fixed mass of an ideal gas at constant volume, the pressure is directly proportional to its absolute temperature: $P \propto T$

If the temperature (in **kelvin**) of a fixed volume of an ideal gas is doubled, the pressure will also double (Figure **1.17**).

The overall gas law equation

> An ideal gas is one that obeys all of the above laws exactly.

The three relationships above can be combined to produce the following equation:

> $$\frac{P_1 V_1}{T_1} = \frac{P_2 V_2}{T_2}$$

> Note: any units may be used for P and V, so long as they are consistent on both sides of the equation.

> The temperature must be kelvin.

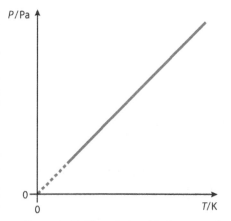

Figure 1.17 The relationship between the pressure and temperature (kelvin) of a fixed mass of an ideal gas at constant volume.

Worked examples

1.28 If the volume of an ideal gas collected at 0 °C and 100 kPa, i.e. at STP, is 50.0 cm³, what would be the volume at 60 °C and 108 kPa?

$P_1 = 100$ kPa $\qquad$ $P_2 = 108$ kPa

$V_1 = 50.0$ cm³ $\qquad$ $V_2 = ?$

The units of P_1 and P_2 are consistent with each other.

$T_1 = 0\,°C = 273\,K$ $\qquad$ $T_2 = 60\,°C = 60 + 273\,K = 333\,K$

Temperature must be in K.

$$\frac{P_1 V_1}{T_1} = \frac{P_2 V_2}{T_2}$$

$$\frac{100 \times 50.0}{273} = \frac{108 \times V_2}{333}$$

Rearranging the equation:

$$V_2 = \frac{100 \times 50.0 \times 333}{273 \times 108} = 56.5 \text{ cm}^3$$

The units of V_2 are the same as those of V_1.

Therefore, the volume occupied by the gas at 60 °C and 108 kPa is 56.5 cm³.

1.29 What temperature (in °C) is required to cause an ideal gas to occupy 1.34 dm³ at a pressure of 200 kPa if it occupies 756 cm³ at STP?

$P_1 = 200$ kPa $\qquad$ $P_2 = 100$ kPa

$V_1 = 1.34$ dm³ $\qquad$ $V_2 = 756$ cm³, i.e. $\dfrac{756}{1000}$ dm³ or 0.756 dm³

The units of P_1 are the same as those of P_2.

$T_1 = ?$ $\qquad$ $T_2 = 273\,K$

$$\frac{200 \times 1.34}{T_1} = \frac{100 \times 0.756}{273}$$

The units of V_1 and V_2 were made consistent with each other. We could have also changed V_1 to cm³.

Rearranging the equation:

$$200 \times 1.34 \times 273 = 100 \times 0.756 \times T_1$$

$$T_1 = \frac{200 \times 1.34 \times 273}{100 \times 0.756} = 968 \text{ K}$$

This must now be converted to °C by subtracting 273.

Temperature = 968 − 273 = 695 °C

The temperature must be 695 °C for the gas to occupy a volume of 1.34 dm³.

The ideal gas equation

If the relationships between P, V and T are combined with Avogadro's law, the ideal gas equation is obtained:

$$PV = nRT$$

Where R is the **gas constant** and n is the number of moles. Although R is a universal constant, it can be quoted with various units and its value depends on these units. The SI units for the gas constant are $J\,K^{-1}\,mol^{-1}$, and this requires the following set of units:

$$R = 8.31\,J\,K^{-1}\,mol^{-1}$$
Pressure: $N\,m^{-2}$ or Pa
Volume: m^3
Temperature: K

A consistent set of units must be used.

Exam tip
A set of units that is equivalent to this uses volume in dm^3 and pressure in kPa – if you use these units you can avoid the problem of converting volumes into m^3.

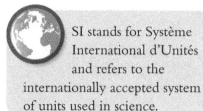 SI stands for Système International d'Unités and refers to the internationally accepted system of units used in science.

$1\,000\,000\,cm^3 \Leftrightarrow 1\,m^3$
$1000\,dm^3 \Leftrightarrow 1\,m^3$
To convert m^3 to cm^3 multiply by $1\,000\,000$.
To convert cm^3 to m^3 divide by $1\,000\,000$.
To convert m^3 to dm^3 multiply by 1000.
To convert dm^3 to m^3 divide by 1000.

Worked examples

1.30 An ideal gas occupies $590\,cm^3$ at $120\,°C$ and 202 kPa. What amount of gas (in moles) is present?

If we use the value of $8.31\,J\,K^{-1}\,mol^{-1}$ for the gas constant, all values must be converted to the appropriate set of units:

$P = 202$ kPa $= 2.02 \times 10^5$ Pa

$V = 590\,cm^3 = \dfrac{590}{1\,000\,000}\,m^3 = 5.90 \times 10^{-4}\,m^3$

$n = ?$

$R = 8.31\,J\,K^{-1}\,mol^{-1}$

$T = 120\,°C = 120 + 273\,K = 393\,K$

$$PV = nRT$$

$$2.02 \times 10^5 \times 5.90 \times 10^{-4} = n \times 8.31 \times 393$$

Rearranging the equation:

$$n = \frac{2.02 \times 10^5 \times 5.90 \times 10^{-4}}{8.31 \times 393} = 0.0365\,mol$$

The number of moles is $0.0365\,mol$.

1.31 A gas has a density of $1.24\,\text{g}\,\text{dm}^{-3}$ at $0\,°\text{C}$ and $1.00 \times 10^5\,\text{Pa}$. Calculate its molar mass.

$$\text{density} = \frac{\text{mass}}{\text{volume}}$$

We know the density, so we know the mass of $1\,\text{dm}^3$ of the gas. If we can find the number of moles in $1\,\text{dm}^3$, we can work out the molar mass.

$P = 1.00 \times 10^5\,\text{Pa}$

$V = 1.00\,\text{dm}^3 = \dfrac{1.00}{1000}\,\text{m}^3 = 1.00 \times 10^{-3}\,\text{m}^3$

$n = ?$

$R = 8.31\,\text{J}\,\text{K}^{-1}\,\text{mol}^{-1}$

$T = 0\,°\text{C} = 273\,\text{K}$

Using $PV = nRT$

$$n = \frac{1.00 \times 10^5 \times 1.00 \times 10^{-3}}{8.31 \times 273} = 0.0441\,\text{mol}$$

This number of moles has a mass of $1.24\,\text{g}$.

$$\text{molar mass} = \frac{\text{mass}}{\text{number of moles}}$$

$$\text{molar mass} = \frac{1.24}{0.0441} = 28.1\,\text{g}\,\text{mol}^{-1}$$

1.32 What is the molar volume of an ideal gas at $18\,°\text{C}$ and $1.10 \times 10^5\,\text{Pa}$? (Give your answer in $\text{m}^3\,\text{mol}^{-1}$ and $\text{dm}^3\,\text{mol}^{-1}$.)

The molar volume of a gas is the volume occupied by one mole of the gas. We are familiar with the value for the molar volume of a gas at STP, which is $22.7\,\text{dm}^3\,\text{mol}^{-1}$.

$P = 1.10 \times 10^5\,\text{Pa}$ $V = ?$

$n = 1.00\,\text{mol}$ $R = 8.31\,\text{J}\,\text{K}^{-1}\,\text{mol}^{-1}$

$T = 18\,°\text{C} = 18 + 273\,\text{K} = 291\,\text{K}$

Using $PV = nRT$:

$$V = \frac{1.00 \times 8.31 \times 291}{1.10 \times 10^5} = 0.0220\,\text{m}^3$$

The molar volume is $0.0220\,\text{m}^3\,\text{mol}^{-1}$ at $18\,°\text{C}$ and $1.10 \times 10^5\,\text{Pa}$. This must be multiplied by 1000 to convert to dm^3 i.e., $22.0\,\text{dm}^3\,\text{mol}^{-1}$.

1.33 When sodium nitrate(V) (often just called sodium nitrate) is heated, it decomposes to give sodium nitrate(III) (also called sodium nitrite) and oxygen gas. When a certain mass of sodium nitrate(V) is heated, 241 cm³ of oxygen is obtained, measured at 97.0 kPa and 22 °C. Calculate the mass of sodium nitrate(III) formed.

$$2NaNO_3(s) \rightarrow 2NaNO_2(s) + O_2(g)$$

We can use $PV = nRT$ to work out the number of moles of O_2:

$P = 97.0 \text{ kPa} = 9.70 \times 10^4 \text{ Pa}$

$V = 241 \text{ cm}^3 = \dfrac{241}{1\,000\,000} \text{ m}^3 = 2.41 \times 10^{-4} \text{ m}^3$

$n = ?$

$R = 8.31 \text{ J K}^{-1} \text{ mol}^{-1}$

$T = 22 \,°C = 295 \text{ K}$

Using $PV = nRT$: $n = \dfrac{9.70 \times 10^4 \times 2.41 \times 10^{-4}}{8.31 \times 295} = 9.54 \times 10^{-3} \text{ mol}$

This gives the number of moles of O_2.

From the chemical equation, the number of moles of O_2 is half the number of moles of $NaNO_2$. Therefore, the number of moles of $NaNO_2$ is $9.54 \times 10^{-3} \times 2 = 1.91 \times 10^{-2} \text{ mol}$.

The molar mass of $NaNO_2$ is 69.00 g mol^{-1}, so the mass of $NaNO_2$ is $69.00 \times 1.91 \times 10^{-2} = 1.32 \text{ g}$.

You would probably say that the room you are sitting in at the moment is full of air. If, however, you do a quick calculation (making a couple of approximations) you should be able to work out that the volume of the molecules of gas in the room is only about 0.01% of the volume of the room – scientific reality is very different from our everyday reality. (There is actually a very small probability that all these molecules could at any one time all end up in the same corner of the room – our survival depends on the fact that this probability is very small!)

Nature of science

A scientific law is a general statement (often in mathematical form) based on observation/experiment of some aspect of the physical world. It will often involve the relationship between various quantities under specified conditions. For example, Boyle's law describes the relationship between the volume and pressure of a fixed mass of an ideal gas at constant temperature. A law does not explain anything – it is just a description of what happens.

A theory is a way of explaining scientific observations or laws. To be accepted, a theory will have been rigorously tested by experiments and observations – for example the particle theories and kinetic theory can be used to explain Boyle's law.

There is no progression from a theory to a law – they are different things.

Avogadro's original hypothesis was that equal volumes of different gases contain the same number of molecules. This was based on deductions from careful measurements and observations made by other scientists such as Gay-Lussac.

In all questions, take the value of the ideal gas constant as $8.31\,J\,K^{-1}\,mol^{-1}$.

33 If a certain mass of an ideal gas occupies $20.0\,cm^3$ at $0\,°C$ and $1.01 \times 10^5\,Pa$, what volume would it occupy at $38\,°C$ and $1.06 \times 10^5\,Pa$?

34 A certain mass of an ideal gas occupies $250.0\,cm^3$ at $20\,°C$ and $9.89 \times 10^4\,Pa$. At what temperature (in $°C$) will it occupy $400.0\,cm^3$ if the pressure remains the same?

35 How many moles of an ideal gas are present in a container if it occupies a volume of $1.50\,dm^3$ at a pressure of $1.10 \times 10^5\,Pa$ and a temperature of $30\,°C$?

36 Calculate the molar mass of an ideal gas if $0.586\,g$ of the gas occupies a volume of $282\,cm^3$ at a pressure of $1.02 \times 10^5\,Pa$ and a temperature of $-18\,°C$.

37 What is the molar volume of an ideal gas at $1.10 \times 10^5\,Pa$ and $100\,°C$?

38 Copper nitrate decomposes when heated according to the equation:
$$2Cu(NO_3)_2(s) \rightarrow 2CuO(s) + 4NO_2(g) + O_2(g)$$
If $1.80\,g$ of copper nitrate is heated and the gases collected at a temperature of $22\,°C$ and $105\,kPa$:
a what volume (in dm^3) of oxygen is collected?
b what is the total volume of gas collected in cm^3?

39 When a certain mass of manganese heptoxide (Mn_2O_7) decomposed, it produced $127.8\,cm^3$ of oxygen measured at $18\,°C$ and $1.00 \times 10^5\,Pa$. What mass of manganese heptoxide decomposed?
$$2Mn_2O_7(aq) \rightarrow 4MnO_2(s) + 3O_2(g)$$

1.3.3 Calculations involving solutions

Solutions

> **Solute**: a substance that is dissolved in another substance.
> **Solvent**: a substance that dissolves another substance (the solute). The solvent should be present in excess of the solute.
> **Solution**: the substance that is formed when a solute dissolves in a solvent.

When a sodium chloride (NaCl) **solution** is prepared, NaCl solid (the **solute**) is dissolved in water (the **solvent**).

Solutions in water are given the symbol (aq) in chemical equations. *aq* stands for *aqueous*.

Note: when a solute is dissolved in a certain volume of water, say $100.0\,cm^3$, the total volume of the solution is not simply $100.0\,cm^3$ or the sum of the volumes occupied by the solute and the volume of the solvent. The total volume of solution produced depends on the forces of attraction between the solute particles and the solvent particles compared with the forces of attraction in the original solvent. This is why concentration is defined in terms of the volume of the solution rather than the volume of the solvent.

Reported values for the concentration of gold in seawater vary greatly. A value of about $2 \times 10^{-11}\,\mathrm{g\,dm^{-3}}$ or $1 \times 10^{-13}\,\mathrm{mol\,dm^{-3}}$ is probably a reasonable estimate. The volume of water in the oceans is estimated to be about $1.3 \times 10^{21}\,\mathrm{dm^3}$, so there is an awful lot of gold in the oceans. Many people (including Nobel Prize-winning scientist Fritz Haber) have tried to come up with ways to extract the gold. The problem is that the concentrations are so low.

The **concentration** of a solution is the amount of solute dissolved in a unit volume of **solution**. The volume that is usually taken is $1\,\mathrm{dm^3}$. The amount of solute may be expressed in g or mol therefore the units of concentration are $\mathrm{g\,dm^{-3}}$ or $\mathrm{mol\,dm^{-3}}$.

Concentrations are sometimes written with the unit M, which means $\mathrm{mol\,dm^{-3}}$ but is described as 'molar'. Thus 2M would refer to a '2 molar solution', i.e. a solution of concentration $2\,\mathrm{mol\,dm^{-3}}$.

The relationship between concentration, number of moles and volume of solution is:

$$\text{concentration } (\mathrm{mol\,dm^{-3}}) = \frac{\text{number of moles (mol)}}{\text{volume } (\mathrm{dm^3})}$$

This is summarised in Figure **1.18**.

If the concentration is expressed in $\mathrm{g\,dm^{-3}}$, the relationship is:

$$\text{concentration } (\mathrm{g\,dm^{-3}}) = \frac{\text{mass (g)}}{\text{volume } (\mathrm{dm^3})}$$

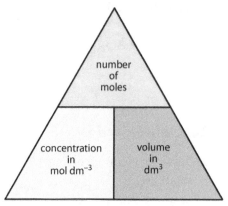

Figure 1.18 The relationship between concentration, number of moles and volume of solution.

Worked examples

1.34 If 10.00 g of sodium hydroxide (NaOH) is dissolved in water and the volume is made up to $200.0\,\mathrm{cm^3}$, calculate the concentration in $\mathrm{mol\,dm^{-3}}$ and $\mathrm{g\,dm^{-3}}$.

Concentration ($\mathrm{g\,dm^{-3}}$)

$$\text{concentration in } \mathrm{g\,dm^{-3}} = \frac{\text{mass}}{\text{volume in } \mathrm{dm^3}}$$

$$\text{volume in } \mathrm{dm^3} = \frac{200.0}{1000} = 0.2000\,\mathrm{dm^3}$$

$$\text{concentration} = \frac{10.00}{0.2000} = 50.00\,\mathrm{g\,dm^{-3}}$$

Concentration (mol dm^{-3})

molar mass of NaOH = 40.00 g mol^{-1}

number of moles = $\dfrac{10.00}{40.00}$ = 0.2500 mol

concentration = $\dfrac{\text{number of moles}}{\text{volume in dm}^3}$

concentration = $\dfrac{0.2500}{0.2000}$ = 1.250 mol dm^{-3}

Alternatively, once we have the concentration in g dm^{-3} we can simply divide by the molar mass to get the concentration in mol dm^{-3}:

concentration = $\dfrac{50.00}{40.00}$ = 1.250 mol dm^{-3}

1.35 Calculate the number of moles of hydrochloric acid (HCl) present in 50.0 cm^3 of 2.00 mol dm^{-3} hydrochloric acid.

number of moles = concentration × volume in dm^3

number of moles = 2.00 × $\dfrac{50}{1000}$ = 0.100 mol

Therefore the number of moles is 0.100 mol.

Square brackets are often used to denote concentrations in mol dm^{-3}. So [HCl] indicates the molar concentration of hydrochloric acid and we could write [HCl] = 2.00 mol dm^{-3} in this worked example.

Concentrations of very dilute solutions

When dealing with very small concentrations, you will occasionally come across the unit *parts per million, ppm*. For instance, if 1 g of a solute is present in 1 million grams of a solution then the concentration is 1 ppm. So, in general, the concentration in ppm is given by:

$$\text{concentration in ppm} = \dfrac{\text{mass of solute} \times 10^6}{\text{mass of solution}}$$

The units of mass of solute and mass of solution must be the same so that they cancel.

The ppm notation is most often used when writing about pollution – e.g. *the concentration of arsenic in drinking water in the US should not exceed 0.010 ppm.*

Worked examples

1.36 If a sample of 252.10 g of water is found to contain 2.03 mg of cyanide, what is the cyanide concentration in ppm?

The mass in mg must first be converted to a mass in g by dividing by 1000:

mass of cyanide = 2.03 × 10^{-3} g

concentration of cyanide in ppm = $\dfrac{2.03 \times 10^{-3} \times 10^6}{252.10}$ = 8.05 ppm

---

Although the concentration in ppm is properly defined as above, it is often used in newspaper articles etc. in a slightly more convenient (but not completely correct) way as the mass of solute, in mg, per dm^3 (litre) of solution. This is a reasonable approximation because the mass of 1 dm^3 of water (which will make up most of the solution) is 1000 g – i.e. 1 000 000 mg. So if the copper concentration in a sample of tap water is 1.2 ppm, this is roughly equivalent to 1.2 mg of copper per dm^3 of water.

The ppm notation is also used when discussing the concentrations of various pollutant gases in air. In this case it is defined as:

$$\text{concentration in ppm} = \frac{\text{volume of gas} \times 10^6}{\text{volume of air}}$$

For instance, the carbon monoxide concentration in a sample of air might be 10 ppm. This indicates that there would be $10\,dm^3$ of CO per $1\,000\,000$ dm^3 of air. At STP this would roughly convert to 12 g CO per million dm^3 of air. So, in a classroom with dimensions of about $5\,m \times 4\,m \times 2\,m$, there would be approximately 0.5 g of carbon monoxide.

Working out the concentration of ions

When ionic substances (see page **119**) dissolve in water, the substance breaks apart into its constituent ions. So, for instance, when copper(II) chloride ($CuCl_2$) dissolves in water, it splits apart into Cu^{2+} and Cl^- ions:

$$CuCl_2(aq) \rightarrow Cu^{2+}(aq) + 2Cl^-(aq)$$

Therefore when 0.100 mol $CuCl_2$ dissolves in water, 2×0.100 mol (i.e. 0.200 mol) Cl^- ions are produced. The concentration of the chloride ions is therefore twice the concentration of the $CuCl_2$.

Worked example

1.37 Calculate the number of moles of chloride ions present in $50.0\,cm^3$ of a $0.0500\,mol\,dm^{-3}$ solution of iron(III) chloride ($FeCl_3$) and the total concentration of all the ions present.

number of moles = concentration × volume in dm^3

$$\text{number of moles of } FeCl_3 = \frac{50.0}{1000} \times 0.0500 = 2.50 \times 10^{-3}\,mol\ FeCl_3$$

$$FeCl_3(aq) \rightarrow Fe^{3+}(aq) + 3Cl^-(aq)$$

So dissolving 2.50×10^{-3} mol $FeCl_3$ produces $3 \times 2.50 \times 10^{-3}$ mol $Cl^-(aq)$, i.e. 7.50×10^{-3} mol $Cl^-(aq)$.

The number of moles of chloride ions present is 7.50×10^{-3} mol.

When one $FeCl_3$ unit dissolves in water, four ions are produced ($Fe^{3+} + 3Cl^-$)

So the total concentration of the ions present is four times the concentration of the $FeCl_3$, i.e. $4 \times 0.0500\,mol\,dm^{-3}$.

The total concentration of ions present is $0.200\,mol\,dm^{-3}$.

Titrations

Titration is a technique for finding the volumes of solutions that react exactly with each other. One solution is added from a burette to the other solution in a conical flask (Figure **1.19**). An indicator is often required to determine the end point of the titration.

In order for the technique to be used to determine the concentration of a particular solution, the concentration of one of the solutions it reacts with must be known accurately – this is a standard solution.

Worked example

1.38 Sulfuric acid (H_2SO_4) is titrated against 25.00 cm^3 of 0.2000 mol dm^{-3} sodium hydroxide solution (NaOH). It is found that 23.20 cm^3 of sulfuric acid is required for neutralisation. Calculate the concentration of the sulfuric acid.

$$2NaOH(aq) + H_2SO_4(aq) \rightarrow Na_2SO_4(aq) + 2H_2O(l)$$

Step 1 – work out the number of moles of NaOH:

$$\text{number of moles} = \text{concentration} \times \text{volume in dm}^3$$

$$\text{number of moles} = 0.2000 \times \frac{25.00}{1000} = 5.000 \times 10^{-3} \text{ mol}$$

Step 2 – the balanced equation tells us that 2 mol NaOH react with 1 mol H_2SO_4. Therefore 5.000 × 10^{-3} mol NaOH react with $\frac{5.000 \times 10^{-3}}{2}$ mol H_2SO_4, i.e. 2.500 × 10^{-3} mol H_2SO_4. This is the number of moles of H_2SO_4 in 23.20 cm^3 of H_2SO_4.

Step 3 – convert number of moles to concentration:

$$\text{concentration} = \frac{\text{number of moles}}{\text{volume in dm}^3}$$

$$23.20 \text{ cm}^3 = \frac{23.20}{1000} \text{ dm}^3 = 0.023\,20 \text{ dm}^3$$

$$[H_2SO_4] = \frac{2.500 \times 10^{-3}}{0.023\,20} = 0.1078 \text{ mol dm}^{-3}$$

The concentration of the H_2SO_4 is 0.1078 mol dm^{-3}.

Figure 1.19 Titration set-up.

A standard solution can be made up from a solid (primary standard). A certain mass of the solute is weighed out accurately and then dissolved in a small amount of distilled water in a beaker. This solution is then transferred to a volumetric flask (washing out the beaker with several lots of distilled water to ensure that all the solute is transferred). Finally, water is added to make the solution up to the mark so that the total volume of the solution is known.

Alternatively, the concentration of a standard solution may be known because it has been titrated against another standard solution.

Equation for solving moles questions involving solutions

The following equation may be used as an alternative method for solving problems:

Exam tip

This equation is useful for solving titration problems in the multiple-choice paper.

$$\frac{c_1 v_1}{n_1} = \frac{c_2 v_2}{n_2}$$

where:

c_1 = concentration of first substance
v_1 = volume of first substance
n_1 = coefficient of first substance
c_2 = concentration of second substance
v_2 = volume of second substance
n_2 = coefficient of second substance

Worked example

1.39 For neutralisation, 25.00 cm^3 of phosphoric(V) acid (H_3PO_4) requires 28.70 cm^3 of sodium hydroxide (NaOH) of concentration 0.1500 mol dm^{-3}. What is the concentration of the phosphoric(V) acid?

$$H_3PO_4(aq) + 3NaOH(aq) \rightarrow Na_3PO_4(aq) + 3H_2O(l)$$

Let H_3PO_4 be substance 1 and NaOH be substance 2.

$c_1 = ?$	$c_2 = 0.1500\ mol\ dm^{-3}$
$v_1 = 25.00\ cm^3$	$v_2 = 28.70\ cm^3$
$n_1 = 1$	$n_2 = 3$

$$\frac{c_1 v_1}{n_1} = \frac{c_2 v_2}{n_2}$$

There is no need to convert the volume to dm^3 when this equation is used so we can use the volume in cm^3 directly.

$$\frac{c_1 \times 25.00}{1} = \frac{0.1500 \times 28.70}{3}$$

Rearranging the equation: $c_1 = \frac{1 \times 0.1500 \times 28.70}{3 \times 25.00} = 0.057\,40\ mol\ dm^{-3}$

The concentration of H_3PO_4 is 0.057 40 mol dm^{-3}.

40 a What mass of sodium sulfate (Na_2SO_4) must be used to make up $250\,cm^3$ of a $0.100\,mol\,dm^{-3}$ solution?

  b What is the concentration of sodium ions in the solution in **a**?

41 Work out the numbers of moles of solute present in the following solutions:

  a $20.0\,cm^3$ of $0.220\,mol\,dm^{-3}$ $NaOH(aq)$

  b $27.8\,cm^3$ of $0.0840\,mol\,dm^{-3}$ $HCl(aq)$

  c $540\,cm^3$ of $0.0200\,mol\,dm^{-3}$ $KMnO_4(aq)$

42 If $29.70\,cm^3$ of sulfuric acid of concentration $0.2000\,mol\,dm^{-3}$ is required for neutralisation of $25.00\,cm^3$ of potassium hydroxide solution, calculate the concentration of the potassium hydroxide solution.

$$2KOH(aq) + H_2SO_4(aq) \rightarrow K_2SO_4(aq) + 2H_2O(l)$$

43 Calcium carbonate is reacted with $50.0\,cm^3$ of $0.500\,mol\,dm^{-3}$ hydrochloric acid.

$$CaCO_3(s) + 2HCl(aq)$$
$$\rightarrow CaCl_2(aq) + CO_2(g) + H_2O(l)$$

  a What mass of calcium carbonate is required for an exact reaction?

  b What volume of carbon dioxide, measured at STP, will be produced?

44 What volume (in cm^3) of $0.0100\,mol\,dm^{-3}$ barium chloride must be reacted with excess sodium sulfate to produce $0.100\,g$ of barium sulfate?

$$BaCl_2(aq) + Na_2SO_4(aq)$$
$$\rightarrow BaSO_4(s) + 2NaCl(aq)$$

45 If $0.100\,g$ of magnesium is reacted with $25.00\,cm^3$ of $0.200\,mol\,dm^{-3}$ hydrochloric acid, calculate the volume of hydrogen gas produced at STP.

$$Mg(s) + 2HCl(aq) \rightarrow MgCl_2(aq) + H_2(g)$$

Water of crystallisation

Some substances crystallise with water as an integral part of the crystal lattice. Examples are hydrated copper sulfate ($CuSO_4 \cdot 5H_2O$) and hydrated magnesium chloride ($MgCl_2 \cdot 6H_2O$). The water is necessary for the formation of the crystals and is called **water of crystallisation**. Substances that contain water of crystallisation are described as hydrated, whereas those that have lost their water of crystallisation are described as anhydrous. So, we talk about 'hydrated copper sulfate' ($CuSO_4 \cdot 5H_2O$) and 'anhydrous copper sulfate' ($CuSO_4$). Hydrated copper sulfate can be obtained as large blue crystals, but anhydrous copper sulfate is white and powdery.

In the case of $CuSO_4 \cdot 5H_2O$, the water can be removed by heating:

$$CuSO_4 \cdot 5H_2O \xrightarrow{\text{heat}} CuSO_4 + 5H_2O$$

However, this is not always the case. When $MgCl_2 \cdot 6H_2O$ is heated, magnesium oxide (MgO) is formed:

$$MgCl_2 \cdot 6H_2O \xrightarrow{\text{heat}} MgO + 2HCl + 5H_2O$$

Worked examples

1.40 When 2.56 g of hydrated magnesium sulfate ($MgSO_4 \cdot xH_2O$) is heated, 1.25 g of anhydrous magnesium sulfate ($MgSO_4$) is formed. Determine the value of x in the formula.

mass of water given off $= 2.56 - 1.25$, i.e. 1.31 g

mass of $MgSO_4 = 1.25$ g

This is now basically just an empirical formula question, and we need to find the ratio between the numbers of moles of $MgSO_4$ and H_2O.

molar mass of $H_2O = 18.02$ g mol^{-1} molar mass of $MgSO_4 = 120.38$ g mol^{-1}

number of moles of $H_2O = \dfrac{1.31}{18.02} = 0.0727$ mol   number of moles of $MgSO_4 = \dfrac{1.25}{120.38} = 0.0104$ mol

Divide by the smaller number to get the ratio:

$$\dfrac{0.0727}{0.0104} = 7$$

The value of x is 7, and the formula of hydrated magnesium sulfate is $MgSO_4 \cdot 7H_2O$.

1.41 a If 10.00 g of hydrated copper sulfate ($CuSO_4 \cdot 5H_2O$) are dissolved in water and made up to a volume of 250.0 cm^3, what is the concentration of the solution?

b What mass of anhydrous copper sulfate would be required to make 250.0 cm^3 of solution with the same concentration as in **a**?

a molar mass of $CuSO_4 \cdot 5H_2O = 249.72$ g mol^{-1}

number of moles $CuSO_4 \cdot 5H_2O = \dfrac{10.00}{249.72} = 0.04004$ mol

concentration $= \dfrac{\text{number of moles}}{\text{volume in dm}^3} = \dfrac{0.04004}{0.2500} = 0.1602$ mol dm^{-3}

> When a hydrated salt is dissolved in water, the water of crystallisation just becomes part of the solvent, and the solution is the same as if the anhydrous salt were dissolved in water.

So dissolving 10.00 g of $CuSO_4 \cdot 5H_2O$ in water and making up the solution to 250.0 cm^3 produces a $CuSO_4$ solution of concentration 0.1602 mol dm^{-3}.

b The number of moles of $CuSO_4$ present in 250.0 cm^3 solution will be exactly the same as above, i.e. 0.04004 mol because the concentration is the same.

> 0.04004 mol $CuSO_4 \cdot 5H_2O$ contains 0.04004 mol $CuSO_4$

molar mass of $CuSO_4 = 159.62$ g mol^{-1}

mass of $CuSO_4 = $ molar mass $\times$ number of moles $= 159.62 \times 0.04004 = 6.391$ g

The mass of $CuSO_4$ required to make 250 cm^3 of a solution of concentration 0.1602 mol dm^{-3} is 6.391 g, as opposed to 10.00 g of $CuSO_4 \cdot 5H_2O$. The two solutions will be identical.

1.42 A 3.92 g sample of hydrated sodium carbonate ($Na_2CO_3 \cdot xH_2O$) was dissolved in water and made up to a total volume of 250.0 cm^3. Of this solution, 25.00 cm^3 was titrated against 0.100 mol dm^{-3} hydrochloric acid, and 27.40 cm^3 of the acid was required for neutralisation. Calculate the value of x in $Na_2CO_3 \cdot xH_2O$.

$$Na_2CO_3(aq) + 2HCl(aq) \rightarrow 2NaCl(aq) + CO_2(g) + H_2O(l)$$

Step 1 – work out the number of moles of HCl:

number of moles = concentration × volume in dm^3

$$\text{number of moles} = 0.100 \times \frac{27.40}{1000} = 2.74 \times 10^{-3}\,\text{mol}$$

Step 2 – the balanced equation tells us that 2 mol HCl react with 1 mol Na_2CO_3. Therefore 2.74×10^{-3} mol HCl react with $\dfrac{2.74 \times 10^{-3}}{2} = 1.37 \times 10^{-3}$ mol Na_2CO_3. This is the number of moles of Na_2CO_3 in 25.00 cm^3.

The original mass of $Na_2CO_3 \cdot xH_2O$ was dissolved in a total volume of 250.0 cm^3. Therefore the number of moles of Na_2CO_3 in 250.0 cm^3 of solution is $1.37 \times 10^{-3} \times 10$, i.e. 1.37×10^{-2} mol.

Step 3 – convert number of moles to mass:

molar mass of $Na_2CO_3 = 105.99$ g mol^{-1}

mass of 1.37×10^{-2} mol Na_2CO_3 = number of moles × molar mass

mass of $Na_2CO_3 = 1.37 \times 10^{-2} \times 105.99 = 1.45$ g

The total mass of $Na_2CO_3 \cdot xH_2O = 3.92$ g.
The mass of this that is due to the water of crystallisation $= 3.92 - 1.45 = 2.47$ g.

$$\text{number of moles of water of crystallisation} = \frac{\text{mass}}{\text{molar mass}} = \frac{2.74}{18.02} = 0.137\,\text{mol}$$

The ratio moles of water of crystallisation : moles of sodium carbonate can be worked out by dividing the number of moles of water by the number of moles of sodium carbonate:

$$\text{ratio} = \frac{0.137}{1.37 \times 10^{-2}} = 10$$

The value of x is 10, and the formula for the hydrated sodium carbonate is $Na_2CO_3 \cdot 10H_2O$.

Back titration

This is a technique by which a known excess of a particular reagent, A, is added to another substance, X, so that they react. Then the excess A is titrated against another reagent to work out how much A reacted with the substance – and therefore how many moles of X were present. This is useful when X is an impure substance.

Worked example

1.43 Limestone is impure calcium carbonate ($CaCO_3$). $2.00\,g$ of limestone is put into a beaker and $60.00\,cm^3$ of $3.000\,mol\,dm^{-3}$ hydrochloric acid (HCl) is added. They are left to react and then the impurities are filtered off and the solution is made up to a total volume of $100.0\,cm^3$. Of this solution, $25.00\,cm^3$ require $35.50\,cm^3$ of $1.000\,mol\,dm^{-3}$ sodium hydroxide (NaOH) for neutralisation. Work out the percentage calcium carbonate in the limestone (assume that none of the impurities reacts with hydrochloric acid).

Let us consider the first part of the question: '$2.00\,g$ of limestone is put into a beaker and $60.00\,cm^3$ of $3.000\,mol\,dm^{-3}$ hydrochloric acid is added':

$$CaCO_3 + 2HCl \rightarrow CaCl_2 + CO_2 + H_2O$$

The limestone is impure, so we cannot work out the number of moles of $CaCO_3$ present, but we do have enough information to work out the number of moles of HCl:

number of moles of HCl = concentration × volume in dm^3

number of moles of HCl = $3.000 \times \dfrac{60.00}{1000} = 0.1800\,mol$

> If the limestone were pure $CaCO_3$, the number of moles present in $2.00\,g$ would be $0.0200\,mol$, which would react with $0.0400\,mol$ HCl.

This is excess HCl, and when the limestone is reacted with it there will be some HCl left over.

The second part of the question is: 'They are left to react and then . . . the solution is made up to a total volume of $100.0\,cm^3$'.

This $100.0\,cm^3$ of solution now contains the HCl left over after the reaction with the $CaCO_3$.

In order to work out the number of moles of HCl that did not react, we must consider the third part of the question: 'Of this solution, $25.00\,cm^3$ require $35.50\,cm^3$ of $1.000\,mol\,dm^{-3}$ sodium hydroxide for neutralisation':

number of moles of NaOH = concentration × volume in dm^3

number of moles of NaOH = $1.000 \times \dfrac{35.50}{1000} = 0.03550\,mol$

This reacts with HCl according to the equation:

$$NaOH + HCl \rightarrow NaCl + H_2O$$

Therefore $0.03550\,mol$ NaOH react with $0.03550\,mol$ HCl. This means that $25.00\,cm^3$ of the HCl solution contained $0.03550\,mol$ HCl. Therefore in $100.0\,cm^3$ of this solution there were 4×0.03550, i.e. $0.1420\,mol$ HCl. This is the number of moles of HCl left over after it has reacted with the $CaCO_3$.

Because $0.1800\,mol$ HCl was originally added to the limestone, the amount that reacted with the $CaCO_3$ was $0.1800 - 0.1420$, i.e. $0.0380\,mol$.

$$CaCO_3 + 2HCl \rightarrow CaCl_2 + CO_2 + H_2O$$

$0.0380\,mol$ HCl reacts with $\dfrac{0.0380}{2}$, i.e. 0.0190, mol $CaCO_3$

molar mass of $CaCO_3 = 100.09\,g\,mol^{-1}$

mass of $CaCO_3$ = number of moles × molar mass = $100.09 \times 0.0190 = 1.90\,g$

% $CaCO_3$ in the limestone = $\dfrac{1.90}{2.00} \times 100 = 95.1\%$

Linked reactions

Sometimes the product of one reaction becomes the reactant in a second reaction. A common example of this is the determination of the concentration of copper ions in solution using sodium thiosulfate.

Worked example

1.44 A $25.0 \, cm^3$ sample of a solution of copper(II) nitrate is added to $10.0 \, cm^3$ of $1 \, mol \, dm^{-3}$ potassium iodide. The iodine produced is titrated against $0.0200 \, mol \, dm^{-3}$ sodium thiosulfate solution using starch indicator near the end point. $22.50 \, cm^3$ of the sodium thiosulfate solution was required for the titration. Calculate the concentration of the copper(II) nitrate solution.

The initial reaction of copper(II) ions with iodide ions is:

$$2Cu^{2+}(aq) + 4I^-(aq) \rightarrow 2CuI(s) + I_2(aq) \quad \textbf{(reaction 1)}$$

> This is a redox titration and these will be considered again in Topic **9**.

A large excess of iodide ions is added to make sure that all the copper ions react. A precipitate of CuI is formed as well as the iodine. If we can determine the number of moles of iodine produced in the solution, we can also find the number of moles of copper ions.

The amount of iodine is determined by titration with sodium thiosulfate solution:

$$2S_2O_3{}^{2-}(aq) + I_2(aq) \rightarrow 2I^-(aq) + S_4O_6{}^{2-}(aq) \quad \textbf{(reaction 2)}$$
thiosulfate ion $\qquad\qquad\qquad$ tetrathionate ion

The number of moles of thiosulfate in $22.50 \, cm^3$ of $0.0200 \, mol \, dm^{-3}$ solution:

$$\text{number of moles} = \text{volume in } dm^3 \times \text{concentration} = \frac{22.50}{1000} \times 0.0200 = 4.50 \times 10^{-4} \, mol \, S_2O_3{}^{2-}$$

From reaction **2** we can see that $2 \, mol \, S_2O_3{}^{2-}$ react with $1 \, mol \, I_2$. Therefore $4.50 \times 10^{-4} \, mol \, S_2O_3{}^{2-}$ react with $\frac{4.50 \times 10^{-4}}{2} \, mol \, I_2$, i.e. $2.25 \times 10^{-4} \, mol \, I_2$. This is the amount of iodine produced in reaction **1**.

From reaction **1**, $2 \, mol \, Cu^{2+}$ produce $1 \, mol \, I_2$, so the number of moles of Cu^{2+} is twice the number of moles of I_2. Therefore the number of moles of Cu^{2+} is $2 \times 2.25 \times 10^{-4}$, i.e. $4.50 \times 10^{-4} \, mol$.

> From reaction **1**, $2 \, mol \, Cu^{2+}$ react to form $1 \, mol \, I_2$. In reaction **2**, $1 \, mol \, I_2$ reacts with $2 \, mol \, S_2O_3{}^{2-}$. Therefore, overall, the number of moles of Cu^{2+} is equivalent to the number of moles of $S_2O_3{}^{2-}$.

The volume of the solution containing Cu^{2+} ions was $25.0 \, cm^3$, and this allows us to work out the concentration:

$$\text{concentration} = \frac{\text{number of moles}}{\text{volume in } dm^3} = \frac{4.50 \times 10^{-4}}{0.0250} = 0.0180 \, mol \, dm^{-3}$$

Therefore the concentration of the copper(II) nitrate solution was $0.0180 \, mol \, dm^{-3}$.

More examples of question types

Some questions can look very difficult at first sight, but a good place to start is to work out the number of moles of whatever you can and see where you can go from there.

Worked examples

1.45 A solution of a chloride of formula MCl_x (concentration $0.0170 \, mol \, dm^{-3}$) reacts with silver nitrate ($AgNO_3$) solution to precipitate silver chloride ($AgCl$). It is found that $25.0 \, cm^3$ of $0.0110 \, mol \, dm^{-3}$ silver nitrate solution reacts with $5.40 \, cm^3$ of the chloride solution.

    a Calculate the number of moles of silver nitrate.
    b Calculate the number of moles of the chloride.
    c Calculate the formula of the chloride.

a number of moles = concentration × volume in dm^3

$$\text{number of moles of } AgNO_3 = \frac{25.0}{1000} \times 0.0110 = 2.75 \times 10^{-4} \, mol$$

b $\text{number of moles of } MCl_x = \dfrac{5.40}{1000} \times 0.0170 = 9.18 \times 10^{-5} \, mol$

c The general equation for the reaction is:

$$MCl_x(aq) + xAgNO_3(aq) \rightarrow xAgCl(s) + M(NO_3)_x(aq)$$

The silver ions in the solution react with the chloride ions to precipitate silver chloride. The ratio of the number of moles of $AgNO_3$ to the number of moles of MCl_x will give us the value of x.

$$\frac{\text{number of moles of } AgNO_3}{\text{number of moles of } MCl_x} = \frac{2.75 \times 10^{-4}}{9.18 \times 10^{-5}} = 3$$

Therefore the value of x is 3, and the formula of the chloride is MCl_3.

1.46 One of the stages in the extraction of arsenic, antimony and bismuth from their ores involves the roasting of the sulfide in oxygen:

$$2M_2S_3 + 9O_2 \rightarrow 2M_2O_3 + 6SO_2$$

A certain mass of the sulfide reacted with $180.0 \, cm^3$ of oxygen gas, measured at $15\,°C$ and 101 kPa pressure to produce $0.335 \, g$ of M_2O_3. Determine the identity of the element M.

We can use $PV = nRT$ to work out the number of moles of oxygen. Substituting the values gives:

$$n = \frac{101000 \times 1.80 \times 10^{-4}}{8.31 \times 288} = 7.60 \times 10^{-3} \, mol$$

We can now use the balanced chemical equation to work out the number of moles of M_2O_3.

From the chemical equation, $9 \, mol \, O_2$ react to form $2 \, mol \, M_2O_3$. Therefore, the number of moles of M_2O_3 is two-ninths of the number of moles of O_2:

$$\text{number of moles of } M_2O_3 = \tfrac{2}{9} \times 7.60 \times 10^{-3} = 1.69 \times 10^{-3} \, mol$$

Now that we have the number of moles and the mass of M_2O_3, we can work out the molar mass:

$$\text{molar mass} = \frac{\text{mass}}{\text{number of moles}} = \frac{0.335}{1.69 \times 10^{-3}} = 198 \, g \, mol^{-1}$$

The formula of the compound is M_2O_3, and its molar mass is $198\,g\,mol^{-1}$. The relative atomic mass of M can be worked out by taking away three times the relative atomic mass of O and then dividing the answer by two:

mass of $M_2 = 198 - (3 \times 16) = 150$

relative atomic mass of $M = \dfrac{150}{2} = 75$

This value is closest to the relative atomic mass of arsenic (74.92) therefore the element M is arsenic.

Exam-style questions

1   What is the total number of atoms in $1.80\,g$ of water?

    A 6.02×10^{22} **B** 6.02×10^{23} **C** 1.80×10^{23} **D** 1.80×10^{24}

2   $88\,kg$ of carbon dioxide contains:

    A $2.0\,mol$ **B** $2000\,mol$ **C** $0.50\,mol$ **D** $3872\,mol$

3   What is the sum of the coefficients when the following equation is balanced with the smallest possible whole numbers?

$$CuFeS_2 + O_2 \rightarrow Cu_2S + SO_2 + FeO$$

    A 7   **B** 8   **C** 11   **D** 12

4   Iron(III) oxide reacts with carbon monoxide according to the equation:

$$Fe_2O_3 + 3CO \rightarrow 2Fe + 3CO_2$$

How many moles of iron are produced when $180\,mol$ carbon monoxide react with excess iron(III) oxide?

    A $120\,mol$ **B** $180\,mol$ **C** $270\,mol$ **D** $360\,mol$

5   Propene undergoes complete combustion to produce carbon dioxide and water.

$$2C_3H_6(g) + 9O_2(g) \rightarrow 6CO_2(g) + 6H_2O(l)$$

What volume of carbon dioxide is produced when $360\,cm^3$ of propene react with $360\,cm^3$ of oxygen at $273\,K$ and $100\,kPa$ pressure?

    A $120\,cm^3$ **B** $240\,cm^3$ **C** $540\,cm^3$ **D** $1080\,cm^3$

6   What mass of $Na_2S_2O_3 \cdot 5H_2O$ must be used to make up $200\,cm^3$ of a $0.100\,mol\,dm^{-3}$ solution?

    A $3.16\,g$ **B** $4.96\,g$ **C** $24.8\,g$ **D** $31.6\,g$

7 $20.00\,cm^3$ of potassium hydroxide (KOH) is exactly neutralised by $26.80\,cm^3$ of $0.100\,mol\,dm^{-3}$ sulfuric acid (H_2SO_4). The concentration of the potassium hydroxide is:

        A $0.0670\,mol\,dm^{-3}$ **C** $0.268\,mol\,dm^{-3}$

        B $0.134\,mol\,dm^{-3}$ **D** $1.34\,mol\,dm^{-3}$

8 Barium chloride solution reacts with sodium sulfate solution according to the equation:

    $BaCl_2(aq) + Na_2SO_4(aq) \rightarrow BaSO_4(s) + 2NaCl(aq)$

When excess barium chloride solution is reacted with $25.00\,cm^3$ of sodium sulfate solution, $0.2334\,g$ of barium sulfate (molar mass $233.40\,g\,mol^{-1}$) is precipitated.

The concentration of sodium ions in the sodium sulfate solution was:

        A $0.08000\,mol\,dm^{-3}$ **C** $0.001000\,mol\,dm^{-3}$

        B $0.04000\,mol\,dm^{-3}$ **D** $0.002000\,mol\,dm^{-3}$

9 When potassium chlorate(V) (molar mass $122.6\,g\,mol^{-1}$) is heated, oxygen gas (molar mass $32.0\,g\,mol^{-1}$) is produced:

    $2KClO_3(s) \rightarrow 2KCl(s) + 3O_2(g)$

When $1.226\,g$ of potassium chlorate(V) are heated, $0.320\,g$ of oxygen gas are obtained. The percentage yield of oxygen is:

        A 100%   **B** 66.7%   **C** 26.1%   **D** 17.4%

10 Elemental analysis of a nitrogen oxide shows that it contains $2.8\,g$ of nitrogen and $8.0\,g$ of oxygen. The empirical formula of this oxide is:

        A NO **B** NO_2 **C** N_2O_3 **D** N_2O_5

11 Nitrogen can be prepared in the laboratory by the following reaction:

    $2NH_3(g) + 3CuO(s) \rightarrow N_2(g) + 3H_2O(l) + 3Cu(s)$

If $227\,cm^3$ of ammonia, when reacted with excess copper oxide, produce $85\,cm^3$ of nitrogen, calculate the percentage yield of nitrogen. All gas volumes are measured at STP.   **[3]**

12 Manganese can be extracted from its ore, hausmannite, by heating with aluminium.

    $3Mn_3O_4 + 8Al \rightarrow 4Al_2O_3 + 9Mn$

    a $100.0\,kg$ of Mn_3O_4 are heated with $100.0\,kg$ of aluminium. Work out the maximum mass of manganese that can be obtained from this reaction.   **[4]**

    b 1.23 tonnes of ore are processed and $200.0\,kg$ of manganese obtained. Calculate the percentage by mass of Mn_3O_4 in the ore.   **[3]**

13 A hydrocarbon contains 88.8% C. 0.201 g of the hydrocarbon occupied a volume of 98.9 cm^3 at 320 K and 1.00×10^5 Pa.

    a Determine the empirical formula of the hydrocarbon.   **[3]**

    b Determine the molecular formula of the hydrocarbon.   **[3]**

14 Limestone is impure calcium carbonate. A 1.20 g sample of limestone is added to excess dilute hydrochloric acid and the gas collected; 258 cm^3 of carbon dioxide were collected at a temperature of 27 °C and a pressure of 1.10×10^5 Pa.

$$CaCO_3(s) + 2HCl(aq) \rightarrow CaCl_2(aq) + CO_2(g) + H_2O(l)$$

    a Calculate the number of moles of gas collected.   **[3]**

    b Calculate the percentage purity of the limestone (assume that none of the impurities in the limestone react with hydrochloric acid to produce gaseous products).   **[3]**

15 25.0 cm^3 of 0.100 mol dm^{-3} copper(II) nitrate solution is added to 15.0 cm^3 of 0.500 mol dm^{-3} potassium iodide. The ionic equation for the reaction that occurs is:

$$2Cu^{2+}(aq) + 4I^-(aq) \rightarrow 2CuI(s) + I_2(aq)$$

    a Determine which reactant is present in excess.   **[3]**

    b Determine the mass of iodine produced.   **[3]**

16 0.0810 g of a group 2 metal iodide, MI$_2$, was dissolved in water and the solution made up to a total volume of 25.00 cm^3. Excess lead(II) nitrate solution (Pb(NO$_3$)$_2$(aq)) was added to the MI$_2$ solution to form a precipitate of lead(II) iodide (PbI$_2$). The precipitate was dried and weighed and it was found that 0.1270 g of precipitate was obtained.

    a Determine the number of moles of lead iodide formed.   **[2]**

    b Write an equation for the reaction that occurs.   **[1]**

    c Determine the number of moles of MI$_2$ that reacted.   **[1]**

    d Determine the identity of the metal, M.   **[3]**

17 0.4000 g of hydrated copper sulfate (CuSO$_4$·xH$_2$O) are dissolved in the solution water and made up to a total volume of 100.0 cm^3 with distilled water. 10.00 cm^3 of this solution are reacted with excess barium chloride (BaCl$_2$) solution. The mass of barium sulfate that forms is 3.739×10^{-2} g.

    a Calculate the number of moles of barium sulfate formed.   **[2]**

    b Write an equation for the reaction between copper sulfate solution and barium chloride solution.   **[1]**

    c Calculate the number of moles of copper sulfate that reacted with the barium chloride.   **[1]**

    d Calculate the number of moles of CuSO$_4$ in 0.4000 g of hydrated copper sulfate.   **[1]**

    e Determine the value of x.   **[3]**

Summary

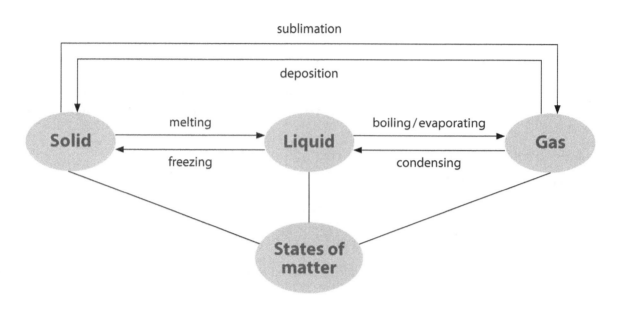

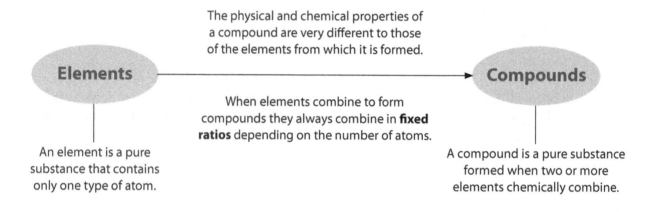

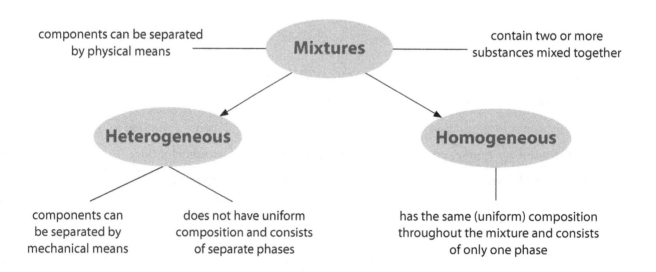

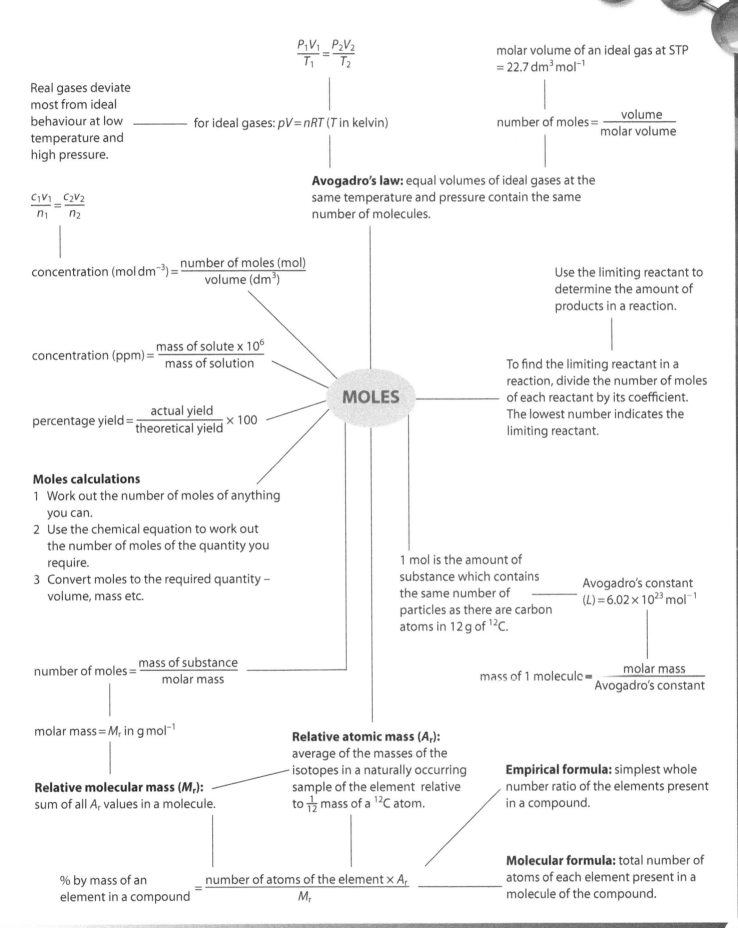

$$\frac{P_1V_1}{T_1} = \frac{P_2V_2}{T_2}$$

molar volume of an ideal gas at STP
$= 22.7\,\text{dm}^3\,\text{mol}^{-1}$

Real gases deviate most from ideal behaviour at low temperature and high pressure. ———— for ideal gases: $pV = nRT$ (T in kelvin)

$$\text{number of moles} = \frac{\text{volume}}{\text{molar volume}}$$

$$\frac{c_1V_1}{n_1} = \frac{c_2V_2}{n_2}$$

Avogadro's law: equal volumes of ideal gases at the same temperature and pressure contain the same number of molecules.

$$\text{concentration (mol dm}^{-3}) = \frac{\text{number of moles (mol)}}{\text{volume (dm}^3)}$$

Use the limiting reactant to determine the amount of products in a reaction.

$$\text{concentration (ppm)} = \frac{\text{mass of solute x } 10^6}{\text{mass of solution}}$$

To find the limiting reactant in a reaction, divide the number of moles of each reactant by its coefficient. The lowest number indicates the limiting reactant.

$$\text{percentage yield} = \frac{\text{actual yield}}{\text{theoretical yield}} \times 100$$

MOLES

Moles calculations
1 Work out the number of moles of anything you can.
2 Use the chemical equation to work out the number of moles of the quantity you require.
3 Convert moles to the required quantity – volume, mass etc.

1 mol is the amount of substance which contains the same number of particles as there are carbon atoms in 12 g of ^{12}C.

Avogadro's constant
$(L) = 6.02 \times 10^{23}\,\text{mol}^{-1}$

$$\text{number of moles} = \frac{\text{mass of substance}}{\text{molar mass}}$$

$$\text{mass of 1 molecule} = \frac{\text{molar mass}}{\text{Avogadro's constant}}$$

molar mass $= M_r$ in g mol^{-1}

Relative atomic mass (A_r): average of the masses of the isotopes in a naturally occurring sample of the element relative to $\frac{1}{12}$ mass of a ^{12}C atom.

Empirical formula: simplest whole number ratio of the elements present in a compound.

Relative molecular mass (M_r): sum of all A_r values in a molecule.

Molecular formula: total number of atoms of each element present in a molecule of the compound.

$$\text{\% by mass of an element in a compound} = \frac{\text{number of atoms of the element} \times A_r}{M_r}$$

2 Atomic structure

Learning objectives

- Understand that an atom is made up of protons, neutrons and electrons
- Define mass number, atomic number and isotope
- Work out the numbers of protons, neutrons and electrons in atoms and ions
- Discuss the properties of isotopes
- Calculate relative atomic masses and abundances of isotopes
- Understand that a mass spectrometer can be used to determine the isotopic composition of a sample

2.1 The nuclear atom

There are approximately 92 naturally occurring elements, plus several more that have been made artificially in nuclear reactions, and probably a few more that have yet to be discovered. As far as we know, there are no more naturally occurring elements – these are the only elements that make up our universe.

Chemistry is the study of how **atoms** of the various elements are joined together to make everything we see around us. It is amazing when one imagines that the entire Universe can be constructed through combinations of these different elements. With just 92 different building blocks (and in most cases many fewer than this), objects as different as a table, a fish and a piece of rock can be made. It is even more amazing when one realises that these atoms are made up of three subatomic ('smaller than an atom') particles, and so the whole Universe is made up of combinations of just three things – protons, neutrons and electrons.

2.1.1 Atoms

> In the simplest picture of the atom the negatively charged electrons orbit around the central, positively charged nucleus (Figure **2.1**). The nucleus is made up of protons and neutrons (except for a hydrogen atom, which has no neutrons).

Protons and neutrons, the particles that make up the nucleus, are sometimes called nucleons.

The actual mass of a proton is 1.67×10^{-27} kg and the charge on a proton is $+1.6 \times 10^{-19}$ C. Relative masses and charges, shown in Table **2.1**, are used to compare the masses of particles more easily. Because the values are relative, there are no units.

From these values it can be seen that virtually all the mass of the atom is concentrated in the nucleus. However, most of the volume of the atom is due to the electrons – the nucleus is very small compared with the total size of the atom.

The diameter of an atom is approximately 1×10^{-10} m and that of a nucleus between about 1×10^{-14} and 1×10^{-15} m, meaning that a nucleus is about 10 000 to 100 000 times smaller than an atom. So, if the nucleus were the size of the full stop at the end of this sentence, the atom would be between 3 and 30 m across.

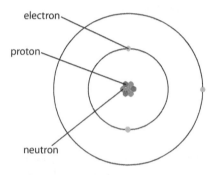

Figure 2.1 A simple representation of a lithium atom (not to scale).

Particle	Relative mass	Relative charge
proton	1	+1
neutron	1	0
electron	5×10^{-4}	−1

Table 2.1 The properties of protons, neutrons and electrons. The mass of an electron is often regarded as negligible.

Protons and neutrons are made up of other particles called quarks.

The **atomic number (Z)** defines an element – it is unique to that particular element. For example, the element with atomic number 12 is carbon and that with atomic number 79 is gold. This means that we could use the atomic number of an element instead of its name. However, the name is usually simpler and more commonly used in everyday speech.

The overall charge on an atom is zero and therefore:

> number of protons in an atom = number of electrons

Atomic number is, however, defined in terms of protons, because electrons are lost or gained when ions are formed in chemical reactions.

> **Mass number (A)** is the number of protons plus neutrons in the nucleus of an atom.

Therefore:

> number of neutrons in an atom = mass number – atomic number

The full symbol of an element includes the atomic number and the mass number (see Figure **2.2**). For example, sodium has an atomic number of 11 and a mass number of 23. The nucleus of a sodium atom contains 11 protons and 12 neutrons (23 − 11). Surrounding the nucleus are 11 electrons. The symbol for sodium is $^{23}_{11}Na$.

Ions

> **Ions** are charged particles that are formed when an atom loses or gains (an) electron(s).

A **positive** ion (cation) is formed when an atom **loses** (an) **electron**(s) so that the ion has more protons(+) than electrons(−) (Figure **2.3**). A **negative** ion (anion) is formed when an atom **gains** (an) **electron**(s) so that the ion has more electrons(−) than protons(+).

> **Atomic number (Z)** is the number of protons in the nucleus of an atom.

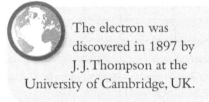

Figure 2.2 Where to place the mass number (A) and atomic number (Z) in the full symbol of an element.

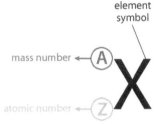

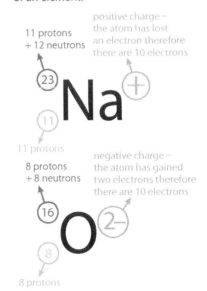

Figure 2.3 The number of subatomic particles in the Na^+ and O^{2-} ions.

2.1.2 Isotopes

The mass number of chlorine is given in many periodic tables as 35.5. It is not possible to have half a neutron – the mass number that is given is an average, taking into account the presence of isotopes.

> **Isotopes** are different atoms of the same element with different mass numbers: i.e. different numbers of neutrons in the nucleus.

The two isotopes of chlorine are ^{35}Cl (chlorine-35) and ^{37}Cl (chlorine-37). Most naturally occurring samples of elements are composed of a mixture of isotopes, but usually one isotope is far more abundant than the others and the mass number of the most common isotope is quoted.

The numbers of protons, neutrons and electrons in some isotopes are shown in Table **2.2**.

> **Isotopes** have the same chemical properties (they react in exactly the same way) but different physical properties (e.g. different melting points and boiling points).

Isotope	Protons	Neutrons	Electrons
$^{1}_{1}H$	1	0	1
$^{2}_{1}H$	1	1	1
$^{3}_{1}H$	1	2	1
$^{12}_{6}C$	6	6	6
$^{13}_{6}C$	6	7	6
$^{14}_{6}C$	6	8	6
$^{35}_{17}Cl$	17	18	17
$^{37}_{17}Cl$	17	20	17

Table 2.2 The numbers of subatomic particles in some common isotopes.

The isotopes of hydrogen are sometimes given different names and symbols: hydrogen-1 is called protium; hydrogen-2 is deuterium (D); and hydrogen-3 is tritium (T).

Isotopes react in the same way because they have the same numbers of electrons, and chemical reactions depend only on the number and arrangement of electrons and not on the composition of the nucleus. For example, both protium and deuterium would react in the same way with nitrogen:

$$N_2 + 3H_2 \rightleftharpoons 2NH_3 \qquad N_2 + 3D_2 \rightleftharpoons 2ND_3$$

Isotopes have different physical properties because, for example, the different masses mean that their atoms move at different speeds. The boiling point of $^{1}H_2$ is $-253\,°C$, whereas that of $^{2}H_2$ is $-250\,°C$. Heavy water (D_2O) has a melting point of $3.8\,°C$ and a boiling point of $101.4\,°C$.

Nature of science

In science, things change! Science is an ever changing and increasing body of knowledge. This is what Richard Feynman meant when he talked about science as '… *the belief in the ignorance of experts*'.

An important example of how knowledge and understanding have changed is the development of atomic theory. Since the days of John Dalton (1766-1844) our view of the world around us has changed dramatically. These developments have happened through careful observation and experiment and have gone hand-in-hand with improvements in equipment and the development of new technology.

Radioactivity was discovered towards the end of the 19th century and this gave scientists a new tool to probe the atom. In a famous experiment Geiger and Marsden subjected a thin film of metal foil to a beam of alpha particles (helium nuclei, $^{4}He^{2+}$) and found that some of the particles were reflected back at large angles. Their experimental data allowed Rutherford to develop the theory of the nuclear atom.

The discovery of subatomic particles (protons, neutrons and electrons) in the late nineteenth and early twentieth centuries necessitated a paradigm shift in science and the development of much more sophisticated theories of the structure of matter.

Test yourself

1 Give the number of protons, neutrons and electrons in the following atoms:

$^{238}_{92}U$ $^{75}_{33}As$ $^{81}_{35}Br$

2 Give the number of protons, neutrons and electrons in the following ions:

$^{40}_{20}Ca^{2+}$ $^{127}_{53}I^{-}$ $^{140}_{58}Ce^{3+}$

3 If you consider the most common isotopes of elements as given in a basic periodic table, how many elements have more protons than neutrons in an atom?

4 The following table shows the number of protons, electrons and neutrons in a series of atoms and ions.

Symbol	Protons	Neutrons	Electrons
D	27	30	25
X	43	54	42
Q	35	44	35
L	27	32	26
M	35	46	36
Z	54	78	54

a Which symbols represent isotopes?
b Which symbols represent positive ions?

Relative atomic masses

Because of the different isotopes present, it is most convenient to quote an average mass for an atom – this is the **relative atomic mass (A_r)**.

> The relative atomic mass (A_r) of an element is the average of the masses of the isotopes in a naturally occurring sample of the element relative to the mass of $\frac{1}{12}$ of an atom of carbon-12.

How to calculate relative atomic mass

Worked examples

2.1 Lithium has two naturally occurring isotopes:

^{6}Li: natural abundance 7% ^{7}Li: natural abundance 93%

Calculate the relative atomic mass of lithium.

Imagine we have 100 Li atoms: 7 will have mass 6 and 93 will have mass 7.

The average mass of these atoms is:

$$\frac{(7 \times 6) + (93 \times 7)}{100} = 6.93$$

Therefore the A_r of Li is 6.93.

2.2 Iridium has a relative atomic mass of 192.22 and consists of Ir–191 and Ir–193 isotopes. Calculate the percentage composition of a naturally occurring sample of iridium.

We will assume that we have 100 atoms and that x of these will have a mass of 191. This means that there will be $(100 - x)$ atoms that have a mass of 193.

The total mass of these 100 atoms will be: $191x + 193(100 - x)$

The average mass of the 100 atoms will be: $\dfrac{191x + 193(100 - x)}{100}$

Therefore we can write the equation: $\dfrac{191x + 193(100 - x)}{100} = 192.22$

$$191x + 193(100 - x) = 19\,222$$

$$191x + 19\,300 - 193x = 19\,222$$

$$-2x = 19\,222 - 19\,300$$

$$-2x = -78$$

Therefore $x = 39$.

This means that the naturally occurring sample of iridium contains 39% Ir–191 and 61% Ir–193.

Alternatively:

$$\frac{A_r - \text{mass number of lighter isotope}}{\text{difference in mass number of two isotopes}} \times 100 = \% \text{ of heavier isotope}$$

In the example here: $\dfrac{(192.22 - 191)}{(193 - 191)} \times 100 = 61\%$

The mass spectrum of an element and relative atomic mass

The proportion of each isotope present in a sample of an element can be measured using an instrument called a **mass spectrometer** (Figure **2.4**).

The readout from a mass spectrometer is called a mass spectrum. In a mass spectrum of an element, we get one peak for each individual isotope. The height of each peak (more properly, the area under each peak) is proportional to the number of atoms of this isotope in the sample tested. The mass spectrum of magnesium is shown in Figure **2.5**.

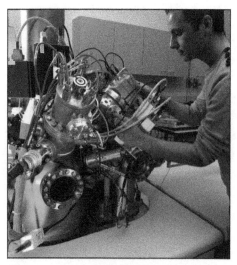

Figure 2.4 Setting up a mass spectrometer.

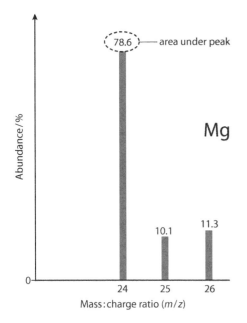

Figure 2.5 The mass spectrum of magnesium showing the amounts of the different isotopes present.

The scale on the x-axis in a mass spectrum is the mass : charge ratio (m/z or m/e). In order to pass through a mass spectrometer, atoms are bombarded with high-energy electrons to produce positive ions. Sometimes more than one electron is knocked out of the atom, which means that the ions behave differently, as if they have smaller masses – hence the use of mass : charge ratio. We can generally ignore this and assume that the horizontal scale refers to the mass of the isotope (the mass of the electron removed is negligible).

The relative atomic mass can be calculated using:

$$A_r = \frac{(78.6 \times 24) + (10.1 \times 25) + (11.3 \times 26)}{100} = 24.3$$

? Test yourself

5 Chromium has four naturally occurring isotopes, and their masses and natural abundances are shown in the table below.

Isotope	Natural abundance (%)
^{50}Cr	4.35
^{52}Cr	83.79
^{53}Cr	9.50
^{54}Cr	2.36

Calculate the relative atomic mass of chromium to two decimal places.

6 Silicon has three naturally occurring isotopes and their details are given in the table below.

Isotope	Natural abundance (%)
^{28}Si	92.2
^{29}Si	4.7
^{30}Si	3.1

Calculate the relative atomic mass of silicon to two decimal places.

7 a Indium has two naturally occurring isotopes: indium-113 and indium-115. The relative atomic mass of indium is 114.82. Calculate the natural abundance of each isotope.

 b Gallium has two naturally occurring isotopes: gallium-69 and gallium-71. The relative atomic mass of gallium is 69.723. Calculate the natural abundance of each isotope.

Learning objectives

- Describe the electromagnetic spectrum
- Describe the emission spectrum of hydrogen
- Explain how emission spectra arise

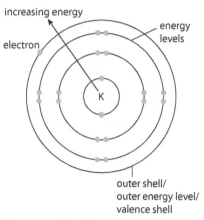

Figure 2.6 The electron arrangement of potassium.

$$\text{frequency} \propto \frac{1}{\text{wavelength}}$$

$$\text{frequency} \propto \text{energy}$$

2.2 Electron configuration

2.2.1 The arrangement of electrons in atoms

At the simplest level of explanation, the electrons in an atom are arranged in **energy levels (shells)** around the nucleus. For example, the electron arrangement of potassium can be represented as shown in Figure **2.6** and is written as 2,8,8,1 (or 2.8.8.1).

The lowest energy level, called the first energy level or first shell (sometimes also called the K shell), is the one closest to the nucleus. The shells increase in energy as they get further from the nucleus. The maximum number of electrons in each main energy level is shown in Table **2.3**.

The main energy level number is sometimes called the **principal quantum number** and is given the symbol n. The maximum number of electrons in each shell is given by $2n^2$.

main energy level number	1	2	3	4	5
maximum number of electrons	2	8	18	32	50

Table 2.3 Distribution of electrons in energy levels.

The general rule for filling these energy levels is that the electrons fill them from the lowest energy to the highest (from the nucleus out). The first two energy levels must be completely filled before an electron goes into the next energy level. The third main energy level is, however, only filled to 8 before electrons are put into the fourth main energy level. This scheme works for elements with atomic number up to 20.

The electromagnetic spectrum

Light is a form of energy. Visible light is just one part of the electromagnetic spectrum (Figure **2.7**).

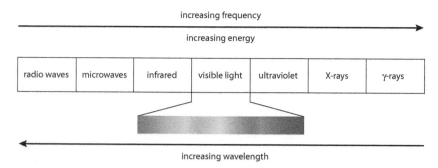

Figure 2.7 The electromagnetic spectrum.

The various forms of electromagnetic radiation are usually regarded as waves that travel at the speed of light in a vacuum $(3.0 \times 10^8 \text{ m s}^{-1})$ but vary in their frequency/energy/wavelength.

Although electromagnetic radiation is usually described as a wave, it can also display the properties of a particle, and we sometimes talk about particles of electromagnetic radiation called **photons**.

White light is visible light made up of all the colours of the spectrum. In order of increasing energy, the colours of the spectrum are: red < orange < yellow < green < blue < indigo < violet.

Evidence for energy levels in atoms

The hydrogen atom spectrum

When hydrogen gas at low pressure is subjected to a very high voltage, the gas glows pink (Figure **2.8**). The glowing gas can be looked at through a **spectroscope**, which contains a diffraction grating and separates the various wavelengths of light emitted from the gas. Because light is emitted by the gas, this is called an **emission spectrum**.

In the visible region, the spectrum consists of a series of sharp, bright lines on a dark background (Figure **2.9**). This is a **line spectrum**, as opposed to a **continuous spectrum**, which consists of all the colours merging into each other (Figure **2.10**).

 Would our interpretation of the world around us be different if our eyes could detect light in other regions of the electromagnetic spectrum?

Are there really seven colours in the visible spectrum?

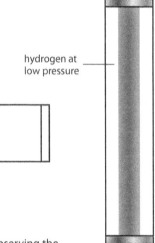
hydrogen at low pressure

spectroscope

discharge tube

Figure 2.8 Observing the emission spectrum of hydrogen.

increasing energy/frequency

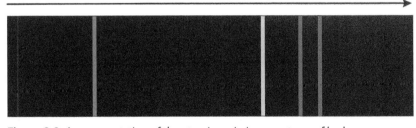

Figure 2.9 A representation of the atomic emission spectrum of hydrogen.

increasing energy/frequency

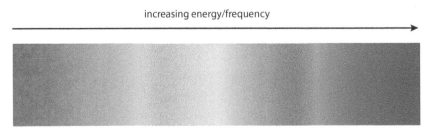

Figure 2.10 A continuous spectrum.

The lines get closer together at higher frequency/energy.

Each element has its own unique emission spectrum, and this can be used to identify the element.

Line spectrum – only certain frequencies/wavelengths of light present.
Continuous spectrum – **all** frequencies/wavelengths of light present.

How an emission spectrum is formed

Passing an electric discharge through a gas causes an electron to be **promoted to a higher energy level** (shell) (Figure **2.11**).

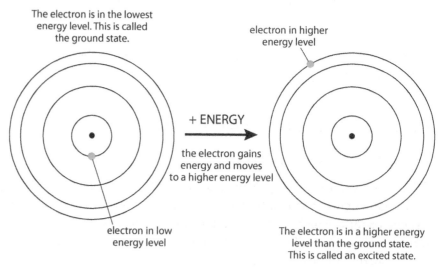

Figure 2.11 An electron can be promoted to a higher energy level in a discharge tube.

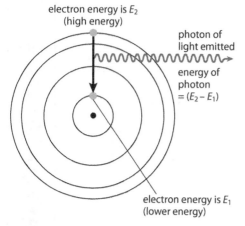

Figure 2.12 When an electron falls from a higher to a lower energy level in an atom, a photon of light is emitted.

Because light is given out, this type of spectrum is an emission spectrum.

Each line in the spectrum comes from the transition of an electron **from a high energy level to a lower one**.

The electron is unstable in this higher level and will **fall to a lower energy level** (Figure **2.12**). As it returns from a level at energy E_2 to E_1, the extra energy $(E_2 - E_1)$ is given out in the form of a photon of light. This contributes to a line in the spectrum.

The energy levels can also be shown as in Figure **2.13**.

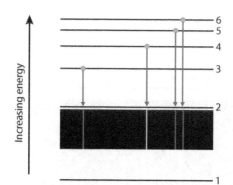

Figure 2.13 How the lines arise in the emission spectrum of hydrogen.

The fact that a line spectrum is produced provides evidence for electrons being in energy levels (shells): i.e. electrons in an atom are allowed to have only certain amounts of energy (Figure **2.14**).

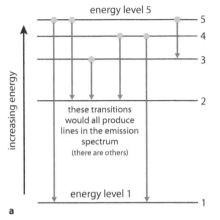

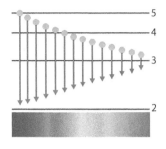

Figure 2.14 a Electrons in energy levels: only transitions between two discrete energy levels are possible, and a line spectrum is produced. **b** If the electrons in an atom could have any energy, all transitions would be possible. This would result in a continuous spectrum.

Different series of lines

Figure **2.15** shows a representation of the emission spectrum of hydrogen across the infrared, visible and ultraviolet regions. The series in each region consists of a set of lines that get closer together at higher frequency. Each series is named after its discoverer.

> The different series of lines occur when electrons fall back down to different energy levels.

Infrared and ultraviolet radiation can be detected only with the aid of technology – we cannot interact with them directly. Does this have implications as to how we view the knowledge gained from atomic spectra in these regions?

Exam tip
The names of the series do not have to be learnt for the examination.

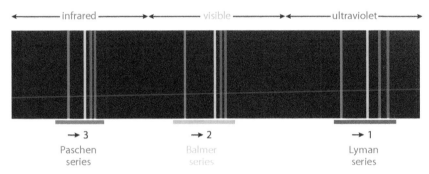

Figure 2.15 A representation of the emission spectrum of hydrogen. The colours and lines in the spectrum in the infrared and ultraviolet regions are just for illustrative purposes.

All the transitions that occur in the visible region of the spectrum (those we can see) involve electrons falling down to level 2 (creating the Balmer series). All transitions down to level 1 occur in the ultraviolet region. Therefore we can deduce that the energy difference between level 1 and any other level is bigger than that between level 2 and any other higher level.

The atomic emission spectrum of hydrogen is relatively simple because hydrogen atoms contain only one electron. Ions such as He^+ and Li^{2+}, which also contain one electron, would have spectra similar to hydrogen – but not exactly the same because the number of protons in the nucleus also influences the electron energy levels.

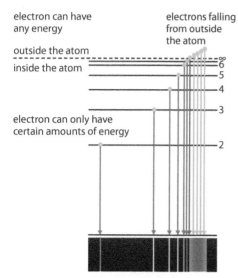

electron can have any energy

outside the atom

inside the atom

electron can only have certain amounts of energy

electrons falling from outside the atom

Figure 2.17 The purple arrow represents the transition giving rise to the convergence limit in the Lyman series for hydrogen.

Convergence

The lines in the emission spectrum get closer together at higher frequency/energy (Figure **2.16**).

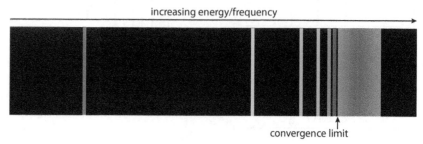

increasing energy/frequency

convergence limit

Figure 2.16 A representation of the Lyman series of hydrogen in the ultraviolet region of the electromagnetic spectrum.

Eventually, at the **convergence limit**, the lines merge to form a **continuum**. Beyond this point the electron can have any energy and so must be free from the influence of the nucleus, i.e. the electron is no longer in the atom (Figure **2.17**).

Extension

The convergence limit is not usually observed, but the frequency can be worked out by plotting a graph of the difference in frequency of successive lines against their frequency and extrapolating the graph to give the frequency when the difference in frequency between successive lines is zero.

Nature of science

Advances in technology are often accompanied by advances in science – Bunsen's development of his burner with a high-temperature flame in the 1850s enabled spectroscopic analysis of substances.

Scientific theories often develop from a need to explain natural phenomena. For instance, Niels Bohr, building on the work of Rutherford, proposed a model for the atom in which electrons orbit the nucleus and only exist in certain allowed energy levels. He then used this model to explain the line spectra of hydrogen and other elements.

? Test yourself

8 Arrange the following in order of:
 a increasing energy
 b decreasing wavelength
 ultraviolet radiation   **infrared radiation**
 microwaves   **orange light**   **green light**

9 Describe how a line in the Lyman series of the hydrogen atom spectrum arises.

10 Draw an energy level diagram showing the first four energy levels in a hydrogen atom and mark with an arrow on this diagram one electron transition that would give rise to:
 a a line in the ultraviolet region of the spectrum
 b a line in the visible region of the spectrum
 c a line in the infrared region of the spectrum.

2.2.2 Full electron configurations

The emission spectra of atoms with more than one electron, along with other evidence such as ionisation energy data (see later), suggest that the simple treatment of considering that electrons in atoms occupy only main energy levels is a useful first approximation but it can be expanded.

Sub-energy levels and orbitals

Each main energy level in an atom is made up of **sub-energy levels** (subshells). The first main energy level consists solely of the 1s sub-level, the second main energy level is split into the 2s sub-level and the 2p sub-level. The sub-levels in each main energy level up to 4 are shown in Table **2.4**.

Main energy level	Sub-levels				Number of electrons in each sub-level			
					s	p	d	f
1	1s				2			
2	2s	2p			2	6		
3	3s	3p	3d		2	6	10	
4	4s	4p	4d	4f	2	6	10	14

Table 2.4 The sub-levels in each main energy level up to level 4.

Within any main energy level (shell) the ordering of the sub-levels (subshells) is always s < p < d < f, but there are sometimes reversals of orders between sub-levels in different energy levels. The relative energies of the subshells are shown in Figure **2.18**.

The Aufbau (building-up) principle (part 1)

The **Aufbau principle** is simply the name given to the process of working out the electron configuration of an atom.

> Electrons fill sub-levels from the lowest energy level upwards – this gives the lowest possible (potential) energy.

So the full electron configuration of sodium (11 electrons) can be built up as follows:
- The first two electrons go into the 1s sub-level → $1s^2$: this sub-level is now full.
- The next two electrons go into the 2s sub-level → $2s^2$: this sub-level is now full.
- The next six electrons go into the 2p sub-level → $2p^6$: this sub-level is now full.
- The last electron goes into the 3s sub level → $3s^1$.
  The full electron configuration of sodium is therefore $1s^2 2s^2 2p^6 3s^1$ (Figure **2.19**). This can also be abbreviated to $[Ne]3s^1$, where the electron configuration of the previous noble gas is assumed and everything after that is given in full.

Learning objectives

- Determine the full electron configuration of atoms with up to 36 electrons
- Understand what is meant by an orbital and a subshell (sub-energy level)

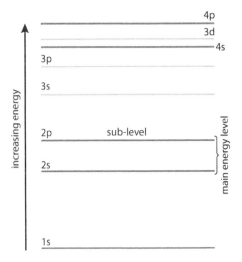

Figure 2.18 The ordering of the energy levels and sub-levels within an atom. The sub-levels within a main energy level are shown in the same colour.

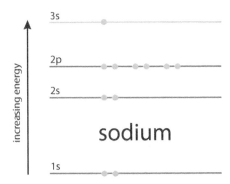

Figure 2.19 The arrangement of electrons in energy levels for a sodium atom.

$[1s^2\ 2s^2\ 2p^6]\ 3s^1$

from neon

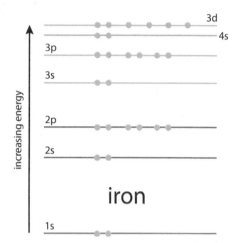

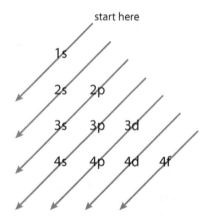

Figure 2.20 The arrangements of electrons in energy levels for an iron atom.

Figure 2.22 Draw out the sub-levels in each main energy level. Starting at 1s, follow the arrows to give the ordering of the sub-levels.

The full electron configuration of iron (26 electrons) is: $1s^2 2s^2 2p^6 3s^2 3p^6 4s^2 3d^6$ (Figure **2.20**). Note that because the 4s sub-level is lower in energy than the 3d sub-level, it is filled first. In other words, two electrons go into the fourth main energy level before the third main energy level is filled. This can be written as $[Ar]4s^2 3d^6$.

This is sometimes written as $[Ar]3d^6 4s^2$ to keep the sub-levels in order of the main energy levels.

The full electronic configuration of germanium (32 electrons) is: $1s^2 2s^2 2p^6 3s^2 3p^6 4s^2 3d^{10} 4p^2$. Or, in abbreviated form: $[Ar]4s^2 3d^{10} 4p^2$.

The order in which the sub-levels are filled can be remembered most easily from the periodic table. For example, Selenium (Se) is in period 4 and 4 along in the p block – therefore the last part of the electron configuration is $4p^4$. The full electron configuration can be worked out by following the arrows in Figure **2.21**:

H→He	$1s^2$					
Li→Be	$2s^2$			B→Ne	$2p^6$	
Na→Mg	$3s^2$			Al→Ar	$3p^6$	
K→Ca	$4s^2$	Sc→Zn	$3d^{10}$	Ga→Se	$4p^4$	

(remember to go down 1 in the d block)

Therefore the electron configuration of selenium is: $1s^2 2s^2 2p^6 3s^2 3p^6 4s^2 3d^{10} 4p^4$.

Figure **2.22** shows an alternative way of remembering the order in which sub-levels are filled.

Note: all the atoms in the same group (vertical column) of the periodic table have the same outer shell electron configuration. For example, all the elements in group 16 (like Se) have the outer shell electron configuration $ns^2 np^4$, where n is the period number.

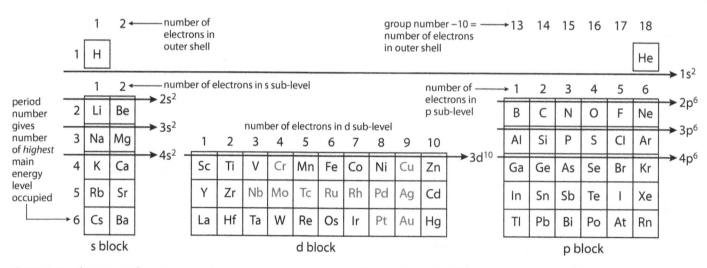

Figure 2.21 Electron configurations can be worked out from the periodic table. The 'p block' is so named because the highest occupied sub-level is a p sub-level. The period number indicates the highest occupied main energy level. Some exceptions to the general rules for filling sub-levels are highlighted in pink. Helium has the configuration $1s^2$ and has no p electrons, despite the fact that it is usually put in the p block.

11 Give the full electron configurations of the following atoms:

    a N   **c** Ar   **e** V

    b Si   **d** As

Orbitals

Electrons occupy atomic orbitals in atoms.

> An **orbital** is a region of space in which there is a high probability of finding an electron. It represents a discrete energy level.

There are four different types of atomic orbital: **s**   **p**   **d**   **f**

The first shell (maximum number of electrons 2) consists of a 1s orbital and this makes up the entire 1s sub-level. This is spherical in shape (Figure **2.23a**).

The 1s orbital is centred on the nucleus (Figure **2.23b**). The electron is moving all the time and the intensity of the colour here represents the probability of finding the electron at a certain distance from the nucleus. The darker the colour the greater the probability of the electron being at that point. This represents the electron density.

The electron can be found anywhere in this region of space (except the nucleus – at the centre of the orbital) but it is most likely to be found at a certain distance from the nucleus.

The second main energy level (maximum number of electrons 8) is made up of the 2s sub-level and the 2p sub-level. The 2s sub-level just consists of a 2s orbital, whereas the 2p sub-level is made up of three 2p orbitals. The 2s orbital (like all other s orbitals) is spherical in shape and bigger than the 1s orbital (Figure **2.24**).

p orbitals have a 'dumb-bell' shape (Figure **2.25**). Three p orbitals make up the 2p sub-level. These lie at 90° to each other and are named appropriately as p_x, p_y, p_z (Figure **2.26**). The p_x orbital points along the x-axis. The three 2p orbitals all have the same energy – they are described as **degenerate**.

> An orbital can contain a maximum of two electrons.

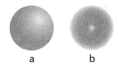

Figure 2.23 a The shape of a 1s orbital; **b** the electron density in a 1s orbital.

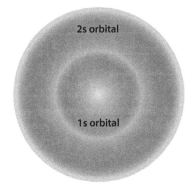

Figure 2.24 A cross section of the electron density of the 1s and 2s orbitals together.

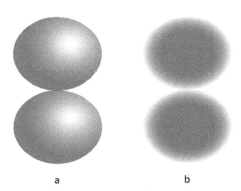

Figure 2.25 a The shape of a 2p orbital; **b** the electron density in a 2p orbital.

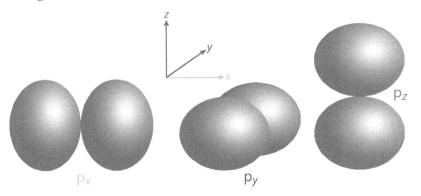

Figure 2.26 The three p orbitals that make up a p sub-level point at 90° to each other.

Figure 2.28 One of the five d orbitals in the 3d sub-level.

Figure 2.29 One of the f orbitals in the 4f sub-level.

Main energy level (shell)	s	p	d	f
1	1			
2	1	3		
3	1	3	5	
4	1	3	5	7

Table 2.5 The number of orbitals in each energy level.

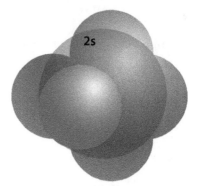

Figure 2.27 The 2s and 2p sub-levels in the second main energy level.

Figure **2.27** shows the orbitals that make up the 2s and 2p sub-levels in the second main energy level.

The third shell (maximum 18 electrons) consists of the 3s, 3p and 3d sub-levels. The 3s sub-level is just the 3s orbital; the 3p sub-level consists of three 3p orbitals; and the 3d sub-level is made up of five 3d orbitals. One of the five 3d orbitals is shown in Figure **2.28**.

The fourth shell (maximum 32 electrons) consists of one 4s, three 4p, five 4d and seven 4f orbitals. The seven 4f orbitals make up the 4f sub-level. One of the f orbitals is shown in Figure **2.29**.

> Within any subshell, all the orbitals have the same energy (they are **degenerate**) – e.g. the three 2p orbitals are degenerate and the five 3d orbitals are degenerate.

The number of orbitals in each energy level is shown in Table **2.5**.

The diagrams of atomic orbitals that we have seen here are derived from mathematical functions that are solutions to the Schrödinger equation. Exact solutions of the Schrödinger equation are only possible for a system involving one electron, i.e. the hydrogen atom. It is not possible to derive exact mathematical solutions for more complex atoms. What implications does this have for the limit of scientific knowledge? When we describe more complex atoms in terms of orbitals, we are actually just extending the results from the hydrogen atom and gaining an approximate view of the properties of electrons in atoms.

Electrons can be regarded as either spinning in one direction (clockwise);

or in the opposite direction (anticlockwise).

Putting electrons into orbitals – the Aufbau principle (part 2)

As well as moving around in space within an orbital, electrons also have another property called **spin**.

There are two rules that must be considered before electrons are put into orbitals.

1 **The Pauli exclusion principle:** the maximum number of electrons in an orbital is two. If there are two electrons in an orbital, they must have opposite spin.

orbital

⊡ ⊡

correct

2 Hund's rule: electrons fill orbitals of the same energy (degenerate orbitals) so as to give the maximum number of electrons with the same spin.

Here we can see how three electrons occupy the orbitals of the 2p sub-level:

2p sub-level

p_x p_y p_z

By contrast, these higher energy situations do not occur:

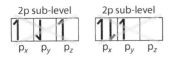

2p sub-level   2p sub-level

p_x p_y p_z p_x p_y p_z

Figures **2.30a** and **b** show the full electron configuration of oxygen and silicon atoms, respectively.

There are a small number of exceptions to the rules for filling sub-levels – i.e. electron configurations that are not quite as expected. Two of these exceptions are **chromium** and **copper**, which, instead of having electron configurations of the form $[Ar]3d^n4s^2$ have only one electron in the 4s sub-level:

$$_{24}Cr: [Ar]3d^54s^1 \qquad _{29}Cu: [Ar]3d^{10}4s^1$$

The reasons for this are complex and beyond the level of the syllabus – but in general, having the maximum number of electron spins the same within a set of degenerate orbitals gives a lower energy (more stable) situation.

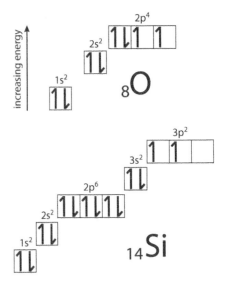

Figure 2.30 Electron configuration: **a** oxygen; **b** silicon.

Nature of science

Scientific theories are constantly being modified, improved or replaced as more data become available and the understanding of natural phenomena improves. The most up-to-date theory of the structure of the atom involves quantum mechanics and this has replaced previous theories.

Developments in apparatus and techniques have been essential in the advancement of science. For instance, J.J. Thomson used electric and magnetic fields to investigate the properties of cathode rays, which led to the discovery of the electron.

? Test yourself

12 Draw out the full electron configurations of the following atoms, showing electrons in boxes:

    a C   **b** P   **c** Cr

Learning objectives

- Solve problems using $E = h\nu$
- Explain successive ionisation data for elements
- Explain the variation in first ionisation energy across a period and down a group

2.3 Electrons in atoms (HL)

2.3.1 Ionisation energy and the convergence limit

As discussed on page **66**, the lines in the emission spectrum of an atom get closer together at higher frequency/energy.

Eventually, at the convergence limit, the lines merge to form a continuum. Beyond this point the electron can have any energy and so must be free from the influence of the nucleus, i.e. the electron is no longer in the atom (Figure **2.17** on page **66**). Knowing the frequency of the light emitted at the convergence limit enables us us to work out the **ionisation energy** of an atom – the energy for the process:

$$M(g) \rightarrow M^+(g) + e^-$$

> The ionisation energy is the minimum amount of energy required to remove an electron from a gaseous atom.

The ionisation energy for hydrogen represents the minimum energy for the removal of an electron (from level 1 to ∞) (Figure **2.31**), and the frequency of the convergence limit in the Lyman series represents the amount of energy given out when an electron falls from outside the atom to level 1 (∞ to 1). These are therefore the same amount of energy.

The relationship between the energy of a photon and the frequency of electromagnetic radiation

Light, and other forms of electromagnetic radiation, exhibit the properties of both waves and particles – this is known as wave–particle duality.

The energy (E) of a photon is related to the frequency of the electromagnetic radiation:

$$E = h\nu$$

where

ν is the frequency of the light (Hz or s^{-1})
h is Planck's constant ($6.63 \times 10^{-34}\,J\,s$)

This equation can be used to work out the differences in energy between various levels in the hydrogen atom.

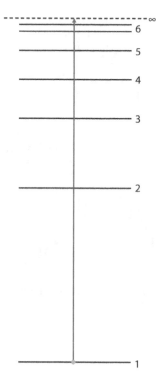

Figure 2.31 The ionisation process in a hydrogen atom.

Worked example

2.3 The frequency of a line in the visible emission spectrum of hydrogen is $4.57 \times 10^{14}\,Hz$. Calculate the energy of the photon emitted.

$E = h\nu$

Therefore $E = 6.63 \times 10^{-34} \times 4.57 \times 10^{14}$

$\qquad = 3.03 \times 10^{-19}\,J$

This line in the spectrum represents an electron falling from level 3 to level 2 and so the energy difference between these two levels is $3.03 \times 10^{-19}\,J$.

The wavelength of the light can be worked out from the frequency using the equation:

$$c = \nu\lambda$$

where

λ is the wavelength of the light (m)
c is the speed of light ($3.0 \times 10^8 \, ms^{-1}$)

The two equations $E = h\nu$ and $c = \nu\lambda$ can be combined:

$$E = \frac{hc}{\lambda}$$

This relates the energy of a photon to its wavelength.

Worked example

2.4 If the frequency of the convergence limit in the Lyman series for hydrogen is 3.28×10^{15} Hz, calculate the ionisation energy of hydrogen in $kJ\,mol^{-1}$.

$E = h\nu$

Therefore $E = 6.63 \times 10^{-34} \times 3.28 \times 10^{15}$

$$= 2.17 \times 10^{-18} \, J$$

This represents the minimum amount of energy required to remove an electron from just one atom of hydrogen, but we are required to calculate the total energy required to remove one electron from each atom in 1 mole of hydrogen atoms – therefore we must multiply by Avogadro's constant.

The energy required is $2.17 \times 10^{-18} \times 6.02 \times 10^{23}$, or $1.31 \times 10^6 \, J\,mol^{-1}$.

Dividing by 1000 gives the answer in $kJ\,mol^{-1}$, so the ionisation energy of hydrogen is $1.31 \times 10^3 \, kJ\,mol^{-1}$.

The ionisation energy of hydrogen can be obtained only from a study of the series of lines when the electron falls back to its ground state (normal) energy level – in other words, only the Lyman series, where the electron falls back down to level 1.

Ionisation energy and evidence for energy levels and sub-levels

The **first** ionisation energy for an element is the energy for the process:

$$M(g) \rightarrow M^+(g) + e^-$$

The full definition is **the energy required to remove one electron from each atom in one mole of gaseous atoms under standard conditions**, but see later.

The **second** ionisation energy is:

$$M^+(g) \rightarrow M^{2+}(g) + e^-$$

The **nth** ionisation energy is:

$$M^{(n-1)+}(g) \rightarrow M^{n+}(g) + e^-$$

The highest energy electrons are removed first. Figure **2.32** shows this for potassium, in which the highest energy electron is the $4s^1$, and this is the first to be removed:

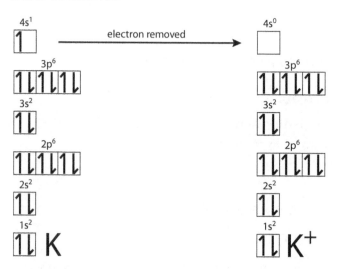

Figure 2.32 The first ionisation of potassium.

The second ionisation energy is always higher than the first, and this can be explained in two ways:

1 Once an electron has been removed from an atom, a positive ion is created. A positive ion attracts a negatively charged electron more strongly than a neutral atom does. More energy is therefore required to remove the second electron from a positive ion.
2 Once an electron has been removed from an atom, there is less repulsion between the remaining electrons. They are therefore pulled in closer to the nucleus (Figure **2.33**). If they are closer to the nucleus, they are more strongly attracted and more difficult to remove.

Successive ionisation energies of potassium

The graph in Figure **2.34** shows the energy required to remove each electron in turn from a gaseous potassium atom.

The simple electron arrangement of potassium is 2,8,8,1 and this can be deduced directly from Figure **2.34**. The large jumps in the graph occur between the main energy levels (shells).

The outermost electron in potassium is furthest from the nucleus and therefore least strongly attracted by the nucleus — so this electron is easiest to remove. It is also **shielded** (screened) from the full attractive force of the nucleus by the other 18 electrons in the atom (Figure **2.35**).

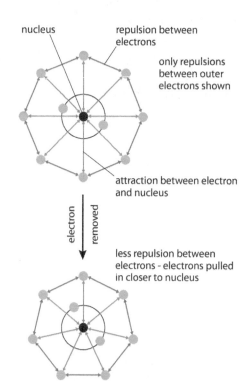

Figure 2.33 When an electron is removed from an atom, the remaining electrons are drawn closer to the nucleus due to reduced repulsion.

74

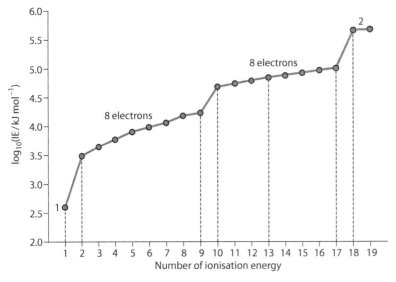

Figure 2.34 Successive ionisation energies (IE) for potassium.

Plotting $\log_{10}$ of these numbers reduces the range. The 1st ionisation energy of potassium is $418\,kJ\,mol^{-1}$, whereas the 19th is $475\,000\,kJ\,mol^{-1}$. It would be very difficult to plot these values on a single graph.

A log scale is used here to allow all the data to be plotted on one graph, but although on one level this has made the data easier to interpret and support the explanations that have been given, it has also distorted the data. The difference between the first and second ionisation energies of potassium is about $2600\,kJ\,mol^{-1}$, but the difference between the 18th and 19th ionisations energies is over $30\,000\,kJ\,mol^{-1}$! How can the way data are presented be used by scientists to support their theories? Can you find examples where the scale on a graph has been chosen to exaggerate a particular trend – is scientific knowledge objective or is it a matter of interpretation and presentation? The arguments for and against human-made climate change are a classic example of where the interpretation and presentation of data are key in influencing public opinion.

Complete shells of electrons between the nucleus and a particular electron reduce the attractive force of the nucleus for that electron. There are three full shells of electrons between the outermost electron and the nucleus, and if this shielding were perfect the **effective nuclear charge** felt by the outer electron would be 1+ (19+ in nucleus −18 shielding electrons). This shielding is not perfect, however, and the effective nuclear charge felt by the outermost electron is higher than +1.

An alternative view of shielding is that the outer electron is attracted by the nucleus but repelled by the inner electrons.

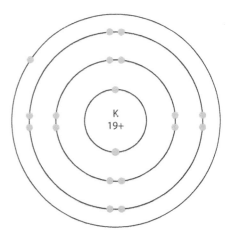

Figure 2.35 The outer electron in a potassium atom is shielded from the full attractive force of the nucleus by the inner shells of electrons (shaded in blue).

In electrostatics, a sphere of charge behaves like a point charge at its centre; therefore relative to the outer electron, spheres of charge inside (the electron shells) behave as if their charge is at the nucleus. The charge felt by the outer electron is $(19+) + (18-) = 1+$ acting at the nucleus.

The electrons do not form perfect spheres of charge, and the movement of the outer electron is not simply in an orbit around the nucleus as shown, and this is why the effective nuclear charge felt by the outer electron in potassium is greater than 1. There are various ways of estimating or calculating the effective nuclear charge for a particular electron in an atom (e.g. Slater's rules). Calculations suggest that the effective nuclear charge felt by the outer electron in potassium is about 3.5+.

Once the first electron has been removed from a potassium atom, the next electron is considerably more difficult to remove (there is a large jump between first and second ionisation energies). This is consistent with the electron being removed from a new main energy level (shell). This electron is closer to the nucleus and therefore more strongly attracted (Figure **2.36a**). It is also shielded by fewer electrons (the ten electrons in the inner main energy levels), because electrons in the same shell do not shield each other very well (they do not get between the electron and the nucleus).

The ionisation energy now rises steadily as the electrons are removed successively from the same main energy level. There is no significant change in shielding, but as the positive charge on the ion increases it becomes more difficult to remove a negatively charged electron (less electron–electron repulsion, so the electrons are pulled in closer to the nucleus).

There is another large jump in ionisation energies between the ninth and the tenth (Figure **2.36b** and **c**) because the ninth electron is the last to be removed from the third main energy level but the tenth is the first to be removed from the second level. The tenth electron is significantly closer to the nucleus and is less shielded than the ninth electron to be removed.

Graphs of successive ionisation energy give us information about how many electrons are in a particular energy level. Consider the graph

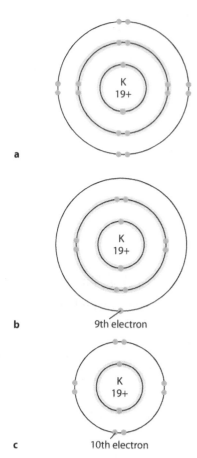

a

b   9th electron

c   10th electron

Figure 2.36 The ionisation energy depends on which main energy level the electron is removed from.

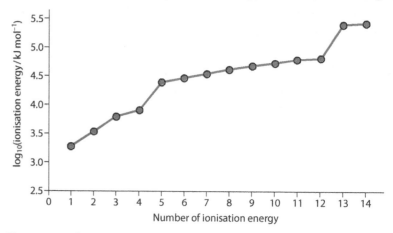

Figure 2.37 The successive ionisation energies of silicon.

for silicon shown in Figure **2.37**. There is a large jump in the ionisation energy between the fourth and the fifth ionisation energies, which suggests that these electrons are removed from different main energy levels. It can therefore be deduced that silicon has four electrons in its outer main energy level (shell) and is in group 14 of the periodic table.

If a graph of ionisation energy (rather than $\log_{10}$ ionisation energy) is plotted for the removal of the first few electrons from a silicon atom, more features can be seen (Figure **2.38**). For example there is a larger jump in the ionisation energy between the second and third ionisation energies.

The full electron configuration of silicon is $1s^2 2s^2 2p^6 3s^2 3p^2$. The first two electrons are removed from the 3p sub-level (subshell), whereas the third electron is removed from the 3s sub-level (Figure **2.39**). The 3p sub-level is higher in energy than the 3s sub-level, and therefore less energy is required to remove the electron. This provides evidence for the existence of sub energy levels (subshells) in an atom.

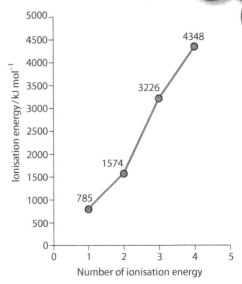

Figure 2.38 The first four ionisation energies of silicon.

? Test yourself

13 **a** The frequency of a line in the emission spectrum of hydrogen is 7.31×10^{14} Hz. Calculate the energy of the photon emitted.

  b The energy of a photon is 1.53×10^{-18} J. Calculate the frequency of the electromagnetic radiation.

14 The table shows the successive ionisation of some elements. Deduce which group in the periodic table each element is in.

Number of ionisation energy	Ionisation energy / kJ mol⁻¹		
	Element X	Element Z	Element Q
1	1 085	736	1 400
2	2 349	1 448	2 851
3	4 612	7 719	4 570
4	6 212	10 522	7 462
5	37 765	13 606	9 429
6	47 195	17 964	53 174

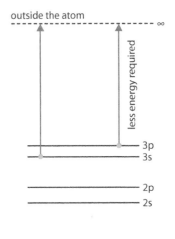

Figure 2.39 More energy is required to remove an electron from the 3s sub-level of silicon than from the 3p sub-level.

We are using reasoning to deduce the existence of energy levels in an atom. Do we **know** that energy levels exist?

The general trend is that ionisation energy increases from left to right across a period.

Variation in ionisation energy across a period

The first ionisation energies for the elements in period 2, from lithium to neon, are plotted in Figure **2.40**.

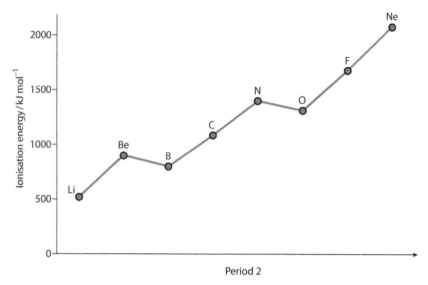

Figure 2.40 The first ionisation energies for the period 2 elements.

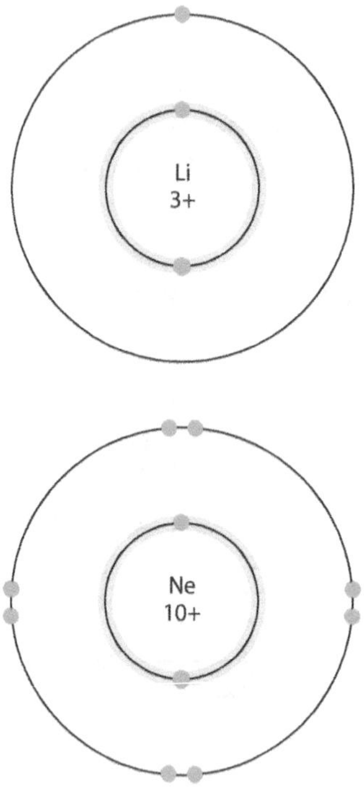

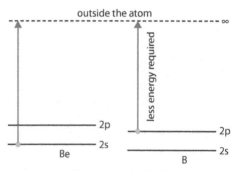

Figure 2.41 Ne has more protons in the nucleus, but the amount of shielding from inner electrons is roughly the same as in lithium.

Figure 2.42 The 2p subshell in boron is higher in energy than the 2s subshell in beryllium.

The nuclear charge increases from lithium (3+) to neon (10+) as protons are added to the nucleus (Figure **2.41**). The electrons are all removed from the same main energy level and, because electrons in the same energy level do not shield each other very well, there is no big change in shielding. Therefore the attractive force on the outer electrons increases from left to right across the period, and the outer electron is more difficult to remove for neon. The neon atom is also smaller than the lithium atom, and so the outer electron is closer to the nucleus and more strongly held.

This can also be explained in terms of the effective nuclear charge felt by the outer electron in neon being higher.

There are two exceptions to the general increase in ionisation energy across a period.

Despite the fact that boron has a higher nuclear charge (more protons in the nucleus) than beryllium the ionisation energy is lower. The electron configurations of beryllium and boron are:

Be $1s^2 2s^2$ B $1s^2 2s^2 2p^1$

The major difference is that the electron to be removed from the boron atom is in a 2p sub-level, whereas it is in a 2s sub-level in beryllium. The 2p sub-level in B is higher in energy than the 2s sub-level in beryllium (Figure **2.42**), and therefore less energy is required to remove an electron from boron.

An alternative, more in-depth, explanation is that the 2p electron in boron is shielded to a certain extent by the 2s electrons, and this increase in shielding from beryllium to boron offsets the effect of the increase in nuclear charge. 2s electrons shield the 2p electrons because there is a significant probability of the 2s electron being closer to the nucleus and therefore getting between the 2p electron and the nucleus.

The second exception is that the first ionisation energy of oxygen is lower than that of nitrogen.

The electron configurations for nitrogen and oxygen are:

N $1s^2 2s^2 2p^3$ O $1s^2 2s^2 2p^4$

The major difference is that oxygen has two electrons paired up in the same p orbital, but nitrogen does not (Figure **2.43**). An electron in the same p orbital as another electron is easier to remove than one in an orbital by itself because of the repulsion from the other electron.

When two electrons are in the same p orbital they are closer together than if there is one in each p orbital. If the electrons are closer together, they repel each other more strongly. If there is greater repulsion, an electron is easier to remove.

Down a group in the periodic table the ionisation energy **decreases**.

The transition metals

The transition metals will be considered in more detail in a later topic, but they are mentioned here for completeness. These elements represent a slight departure from the 'last in, first out' rule for ionisation energy. Although the sub-levels are filled in the order 4s and then 3d, the 4s electrons are always removed before the 3d electrons.

The full electron configuration for an iron atom is $1s^2 2s^2 2p^6 3s^2 3p^6 4s^2 3d^6$. The electron configuration for Fe^{2+} is $1s^2 2s^2 2p^6 3s^2 3p^6 3d^6$. The electron configuration for Fe^{3+} is $1s^2 2s^2 2p^6 3s^2 3p^6 3d^5$.

Nature of science

Theories in science must be supported by evidence. Experimental evidence from emission spectra and ionisation energy is used to support theories about the electron arrangements of atoms.

? Test yourself

15 Work out the full electron configurations of the following ions:
  a Ca^{2+} **b** Cr^{3+} **c** Co^{2+} **d** Rb^+

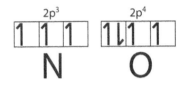

Figure 2.43 Electrons in the 2p sub-level of nitrogen and oxygen.

The variation in first ionisation energy down a group is discussed on page **92**.

When removing electrons, we should not really think about the order they were put into the atom but consider the stability of the final ion. The electron configuration of the final ion will be that which generates the ion of lowest energy. s electrons are generally better at shielding other electrons than are d electrons; by removing the 4s electrons, the shielding of the remaining 3d electrons is reduced, and these are lowered in energy. If the 3d electrons are removed, there is no real energy advantage in terms of reduced shielding – therefore it is less favourable to remove the 3d electrons. Overall, what this all amounts to is that, in the ion, the 3d sub-level is lower in energy than the 4s orbital.

Exam-style questions

1   Which of the following contains 50 neutrons?

   A $^{50}_{23}V$ B $^{89}_{39}Y^+$ C $^{91}_{40}Zr^+$ D $^{86}_{37}Rb^+$

2   Which of the following has more electrons than neutrons?

   A $^{9}_{4}Be^{2+}$ B $^{31}_{15}P^{3-}$ C $^{79}_{35}Br^-$ D $^{40}_{20}Ca^{2+}$

3   Rhenium has two naturally occurring isotopes, ^{185}Re and ^{187}Re. The relative atomic mass of rhenium is 186.2. What are the natural abundances of these isotopes?

   A   40% ^{185}Re and 60% ^{187}Re
   B   60% ^{185}Re and 40% ^{187}Re
   C   12% ^{185}Re and 88% ^{187}Re
   D   88% ^{185}Re and 12% ^{187}Re

4   Which of the following electron transitions in the hydrogen atom will be of highest energy?

   A $n=8 \rightarrow n=4$ C $n=9 \rightarrow n=3$
   B $n=7 \rightarrow n=2$ D $n=6 \rightarrow n=2$

HL 5   Within any main energy level the correct sequence, when the sub-energy levels (subshells) are arranged in order of increasing energy, is:

   A   s   p   f   d   C   s   p   d   f
   B   d   s   f   p   D   s   d   p   f

HL 6   Which of the following electron configurations is **not** correct?

   A   Mg: $1s^22s^22p^63s^2$ C   Ge: $1s^22s^22p^63s^23p^63d^{10}4s^24p^2$
   B   Cu: $1s^22s^22p^63s^23p^64s^23d^9$ D   Br: $1s^22s^22p^63s^23p^64s^23d^{10}4p^5$

HL 7   Planck's constant is 6.63×10^{-34} J s. The energy of a photon of light with frequency 5.00×10^{14} Hz is:

   A 7.54×10^{47} J   C 3.32×10^{-19} J
   B 1.33×10^{-48} J   D 1.33×10^{-20} J

HL 8   Which of the following does **not** have three unpaired electrons?

   A   P   B   V   C Mn^{3+} D Ni^{3+}

HL 9   In which of the following does the second element have a lower first ionisation energy than the first?

   A   Si   C
   B   Na   Mg
   C   Be   B
   D   Ar   Ne

HL 10 The first four ionisation energies of a certain element are shown in the table below.

Number of ionisation energy	Ionisation energy / kJ mol^{-1}
1	418
2	3046
3	4403
4	5866

In which group in the periodic table is this element?

        A group 1   **B** group 2   **C** group 13   **D** group 14

11 a Define the terms **atomic number** and **isotopes**. [3]

   b State the number of protons, neutrons and electrons in an atom of $^{57}_{26}$Fe. [2]

   c A sample of iron from a meteorite is analysed and the following results are obtained.

Isotope	Abundance / %
^{54}Fe	5.80
^{56}Fe	91.16
^{57}Fe	3.04

     i Name an instrument that could be used to obtain this data. [1]

     ii Calculate the relative atomic mass of this sample, giving your answer to two decimal places. [2]

12 a Describe the difference between a **continuous spectrum** and a **line spectrum**. [2]

   b Sketch a diagram of the emission spectrum of hydrogen in the visible region, showing clearly the relative energies of any lines. [2]

   c Explain how a line in the visible emission spectrum of hydrogen arises. [3]

HL d The frequencies of two lines in the emission spectrum of hydrogen are given in the table. Calculate the energy difference between levels 5 and 6 in a hydrogen atom. [2]

Higher level	Lower level	Frequency / Hz
5	3	2.34×10^{14}
6	3	2.74×10^{14}

13 a Write the full electron configuration of an atom of potassium. [1]

HL b Write an equation showing the second ionisation energy of potassium. [2]

HL c Explain why the second ionisation energy of potassium is substantially higher than its first ionisation energy. [3]

HL d State and explain how the first ionisation energy of calcium compares with that of potassium. [3]

14 a Write the full electron configuration of the O^{2-} ion.   [1]

   b Give the formula of an atom and an ion that have the same number of electrons as an O^{2-} ion.   [2]

HL c Explain why the first ionisation energy of oxygen is lower than that of nitrogen.   [2]

HL d Sketch a graph showing the variation of the **second** ionisation energy for the elements in period 2 of
      the periodic table from lithium to neon.   [3]

Summary

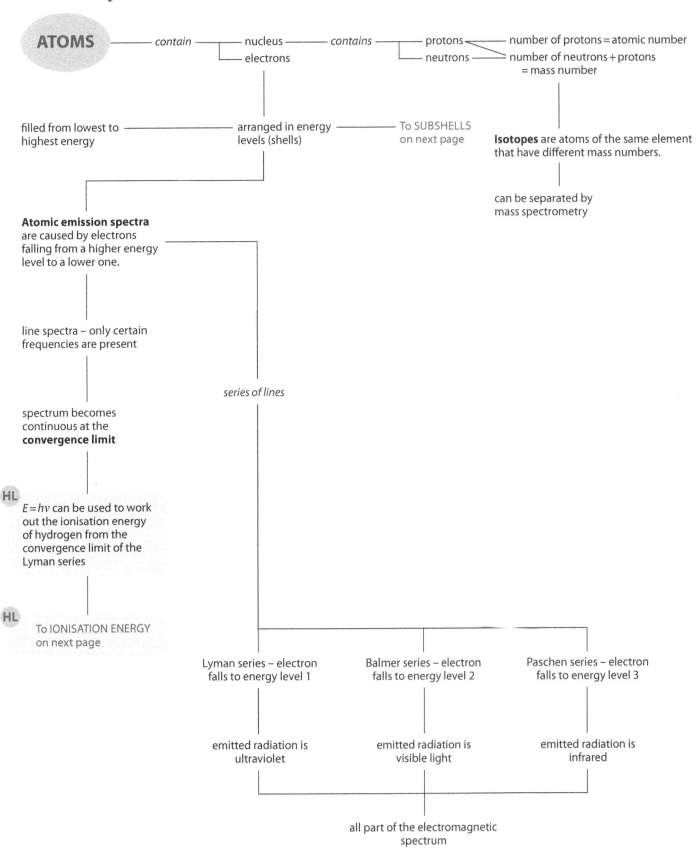

ATOMS —— *contain* —— nucleus —— *contains* —— protons —— number of protons = atomic number

electrons —— neutrons —— number of neutrons + protons = mass number

filled from lowest to highest energy —— arranged in energy levels (shells) —— To SUBSHELLS on next page

Isotopes are atoms of the same element that have different mass numbers.

can be separated by mass spectrometry

Atomic emission spectra are caused by electrons falling from a higher energy level to a lower one.

line spectra – only certain frequencies are present

series of lines

spectrum becomes continuous at the **convergence limit**

HL $E = hv$ can be used to work out the ionisation energy of hydrogen from the convergence limit of the Lyman series

HL To IONISATION ENERGY on next page

Lyman series – electron falls to energy level 1

Balmer series – electron falls to energy level 2

Paschen series – electron falls to energy level 3

emitted radiation is ultraviolet

emitted radiation is visible light

emitted radiation is infrared

all part of the electromagnetic spectrum

Summary – continued

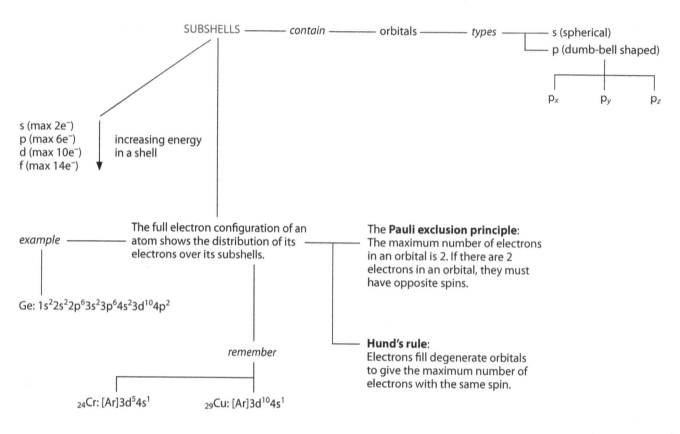

SUBSHELLS ——— *contain* ——— orbitals ——— *types* ——— s (spherical)
p (dumb-bell shaped)

p_x p_y p_z

s (max 2e⁻)
p (max 6e⁻)
d (max 10e⁻)
f (max 14e⁻)

increasing energy in a shell

example ——— The full electron configuration of an atom shows the distribution of its electrons over its subshells.

Ge: $1s^2 2s^2 2p^6 3s^2 3p^6 4s^2 3d^{10} 4p^2$

The **Pauli exclusion principle**: The maximum number of electrons in an orbital is 2. If there are 2 electrons in an orbital, they must have opposite spins.

Hund's rule: Electrons fill degenerate orbitals to give the maximum number of electrons with the same spin.

remember

$_{24}$Cr: $[Ar]3d^5 4s^1$ $_{29}$Cu: $[Ar]3d^{10} 4s^1$

HL

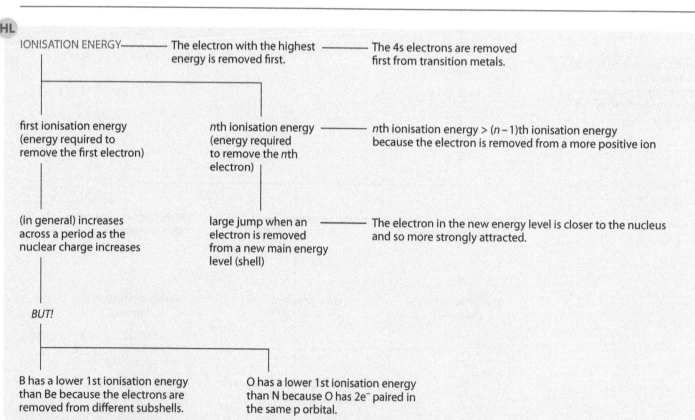

IONISATION ENERGY——— The electron with the highest energy is removed first. ——— The 4s electrons are removed first from transition metals.

first ionisation energy (energy required to remove the first electron)

*n*th ionisation energy (energy required to remove the *n*th electron) ——— *n*th ionisation energy > (*n* – 1)th ionisation energy because the electron is removed from a more positive ion

(in general) increases across a period as the nuclear charge increases

large jump when an electron is removed from a new main energy level (shell) ——— The electron in the new energy level is closer to the nucleus and so more strongly attracted.

BUT!

B has a lower 1st ionisation energy than Be because the electrons are removed from different subshells.

O has a lower 1st ionisation energy than N because O has 2e⁻ paired in the same p orbital.

The periodic table 3

3.1 The periodic table

The elements in the periodic table are arranged in order of atomic number starting with hydrogen, which has atomic number 1. The **groups** are the vertical columns in the periodic table and the **periods** are the horizontal rows (Figure **3.1**).

Most of the elements in the periodic table are metals – these are shown in yellow in Figure **3.1**. The elements shown in pink are non-metals.

Learning objectives

- Understand how the elements in the periodic table are arranged
- Understand the terms **group** and **period**
- Understand how the electron configuration of an element relates to its position in the periodic table

period number \ group number:	1	2	3	4	5	6	7	8	9	10	11	12	13	14	15	16	17	18
1	H 1																	He 2
2	Li 3	Be 4											B 5	C 6	N 7	O 8	F 9	Ne 10
3	Na 11	Mg 12											Al 13	Si 14	P 15	S 16	Cl 17	Ar 18
4	K 19	Ca 20	Sc 21	Ti 22	V 23	Cr 24	Mn 25	Fe 26	Co 27	Ni 28	Cu 29	Zn 30	Ga 31	Ge 32	As 33	Se 34	Br 35	Kr 36
5	Rb 37	Sr 38	Y 39	Zr 40	Nb 41	Mo 42	Tc 43	Ru 44	Rh 45	Pd 46	Ag 47	Cd 48	In 49	Sn 50	Sb 51	Te 52	I 53	Xe 54
6	Cs 55	Ba 56	La 57	Hf 72	Ta 73	W 74	Re 75	Os 76	Ir 77	Pt 78	Au 79	Hg 80	Tl 81	Pb 82	Bi 83	Po 84	At 85	Rn 86
7	Fr 87	Ra 88	Ac 89															

Figure 3.1 The periodic table showing the distribution of metals, metalloids and non-metals.

There are some elements, such as Si, Ge and Sb, that have some of the properties of both metals and non-metals – these are called *metalloids* and are shaded green.

The symbols of the elements that are solid at room temperature and pressure are shown in black in Figure **3.1**, whereas those that are gases are in blue and liquids are in red.

Some of the groups in the periodic table are given names. Commonly used names are shown in Figure **3.2**. The noble gases are sometimes also called the 'inert gases'.

In the periodic table shown in Figure **3.1** it can be seen that the atomic numbers jump from 57 at La (lanthanum) to 72 at Hf (hafnium). This is because some elements have been omitted – these are the lanthanoid elements. The actinoid elements, which begin with Ac (actinium), have also been omitted from Figure **3.1**. Figure **3.2** shows a long form of the periodic table, showing these elements as an integral part.

There is some disagreement among chemists about just which elements should be classified as metalloids – polonium and astatine are sometimes included in the list.

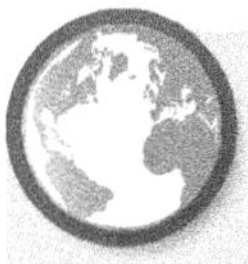

Hydrogen is the most abundant element in the Universe: about 90% of the atoms in the Universe are hydrogen. Major uses of hydrogen include making ammonia and the hydrogenation of unsaturated vegetable oils.

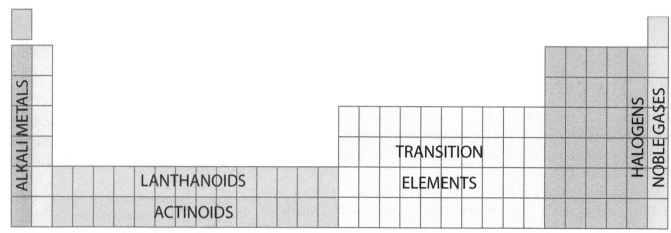

Figure 3.2 The long form of the periodic table, with the names of some of the groups. Hydrogen, though sometimes placed in group 1, does not count as an alkali metal. The lanthanoids and actinoids are also often called 'lanthanides' and 'actinides'.

Mendeleev puzzled over the arrangement of elements in the periodic table until he had a dream in which he claims to have seen the arrangement. Kekulé also came up with the ring structure of benzene after a dream. Does it matter how a scientist comes up with a hypothesis? What is the difference between a scientific and a non-scientific hypothesis? Is it the origins of the hypothesis or the fact that it can be tested experimentally (is falsifiable) that makes it scientific?

Tin has a non-metallic allotrope (grey tin) and prolonged exposure of tin to low temperatures can bring about the transformation to the brittle non-metallic form. The transformation is sometimes called 'tin pest' and there are tales (not necessarily true!) that Napoleon lost the Russian campaign in 1812 because his soldiers had buttons made of tin – in the cold Russian winter the buttons became brittle and their trousers fell down!

Metallic properties

The metallic and non-metallic properties of elements can be related to ionisation energies (see later in this topic).

A metallic structure consists of a regular lattice of positive ions in a sea of delocalised electrons (page **160**). To form a metallic structure, an element must be able to lose electrons fairly readily to form positive ions. Going across a period the ionisation energy increases and so elements lose electrons less easily. So, metallic structures are formed by elements on the left-hand side of the periodic table, which have lower ionisation energies.

Going down a group in the periodic table, ionisation energy decreases, therefore elements are much more likely to exhibit metallic behaviour lower down a group. This can be seen especially well in group 14 going from non-metallic carbon at the top, through the metalloids (Si and Ge) to the metals tin and lead at the bottom.

In general, metallic elements tend to have large atomic radii, low ionisation energies, less exothermic electron affinity values and low electronegativity.

3.1.1 The periodic table and electron configurations

Electrons in the outer shell (the highest main energy level) of an atom are sometimes called **valence electrons**. The group number of an element is related to the number of valence electrons. All the elements in group 1 have one valence electron (one electron in their outer shell); all the elements in group 2 have two valence electrons. For elements in groups 13–18, the number of valence electrons is given by (group number −10); so the elements in group 13 have three valence electrons and so on. The period number indicates the number of shells (main energy levels) in the atom − or which is the outer shell (main energy level).

The periodic table is divided into blocks according to the highest energy subshell (sub-level) occupied by electrons. So in the s block all the elements have atoms in which the outer shell electron configuration is ns^1 or ns^2 (where n is the shell number) and in the p block it is the p subshell that is being filled (Figure **3.3**).

> Four elements are named after the small village of Ytterby in Sweden − yttrium, terbium, erbium and ytterbium.

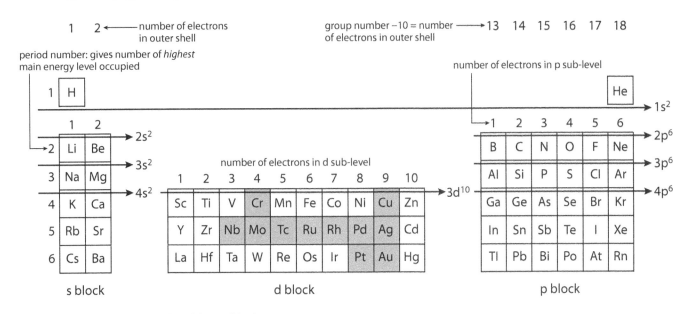

Figure 3.3 Division of the periodic table into blocks.

Consider sulfur − this element is in period 3 and group 16, and so has three shells (the highest occupied shell is the third) and $16 − 10 = 6$ electrons in its outer shell. It is in the p block − therefore its highest energy occupied subshell is a p subshell and the outer shell electron configuration is $3s^2 3p^4$ (six valence electrons).

The noble gases (group 18) have either two (He) or eight electrons (Ne–Rn) in their outer shell. Helium belongs in the s block because its highest energy occupied subshell is 1s, but it is usually put in group 18 with the other noble gases.

Scientists look for patterns in data. They gather evidence, not necessarily just from their own work but also from the published work of other scientists, and analyse the data to discover connections and to try to come up with general laws.

The modern periodic table has developed from one originally conceived by Russian chemist Dmitri Mendeleev in 1869. Mendeleev suggested that the elements were arranged in order of atomic weight (what we would now call relative atomic mass) and produced a table in which elements with similar chemical properties were arranged in vertical groups. Mendeleev took several risks when presenting his data – he suggested that some elements had not been discovered and left spaces for them in his table. Not only did he leave spaces but he also predicted the properties of these unknown elements – he made his hypotheses falsifiable, which added great weight to his theory. The predictions he made were later found to be extremely accurate – the mark of a good theory is that it should be able to be used to predict results that can be experimentally confirmed or refuted. He also suggested that the atomic weights of some elements were incorrect – he realised that tellurium (Te) belonged in the same group as O, S and Se but its atomic weight was higher than iodine and so it should be placed after iodine. Instead of abandoning his theory, he questioned the accuracy of the atomic weight of tellurium and placed it before iodine. This is, of course, the correct place, but Mendeleev's assumption that the atomic weight was lower than that of iodine was not correct.

Henry Moseley, working at the beginning of the 20th century established the connection between atomic number and the periodic table. Like Mendeleev he realised that there were still some elements to be discovered and proposed that three elements between Al and Au were yet to be discovered.

The development of science is not without controversy with regard to who has discovered what. Scientists publish work to make their material available to other scientists and also to establish prior claim on discoveries. For example, the German chemist Julius Lothar Meyer was working on the arrangements of elements at the same time as Mendeleev and came to very similar conclusions – so why is Mendeleev remembered as the father of the modern periodic table rather than Meyer?

Learning objectives

- Understand trends in atomic radius, ionic radius, first ionisation energy, electron affinity and electronegativity across a period
- Understand trends in atomic radius, ionic radius, first ionisation energy, electron affinity and electronegativity down a group

3.2 Physical properties

3.2.1 Variation of properties down a group and across a period

In the next few sections, we will consider how various physical properties vary down a group and across a period in the periodic table.

Atomic radius

The **atomic radius** is basically used to describe the size of an atom. The larger the atomic radius, the larger the atom.

The atomic radius is usually taken to be half the internuclear distance in a molecule of the element. For example, in a diatomic molecule such as

chlorine, where two identical atoms are joined together, the atomic radius would be defined as shown in Figure **3.4**.

Atomic radius increases down a group.

This is because, as we go down a group in the periodic table the atoms have increasingly more electron shells. For example, potassium has four shells of electrons but lithium has only two (Figure **3.5**).

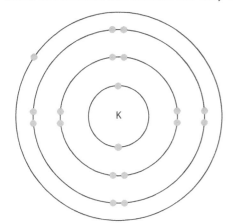

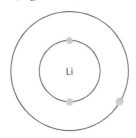

Figure 3.5 Potassium and lithium atoms.

Although the nuclear charge is higher for K, the number of electrons and hence the repulsion between electrons is also greater, and this counteracts any effects due to a greater number of protons in the nucleus.

Atomic radius decreases across a period.

Figure **3.6** shows the variation in atomic radius across period 3 in the periodic table. The reason that atomic radius decreases across a period is that nuclear charge increases across the period with no significant change in shielding. The shielding remains approximately constant because atoms in the same period have the same number of inner shells.

Sodium and chlorine (Figure **3.7**) have the same number of inner shells of electrons (and hence the amount of shielding is similar). However, chlorine has a nuclear charge of 17+ whereas sodium has a nuclear charge of only 11+. This means that the outer electrons are pulled in more strongly in chlorine than in sodium and the atomic radius is smaller.

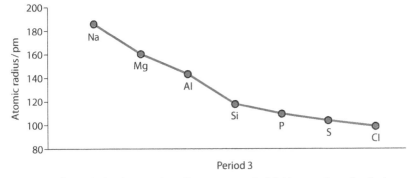

Period 3

Figure 3.6 The variation in atomic radius across period 3. No atomic radius is shown for argon because it does not form covalent bonds and the internuclear distance between atoms bonded together therefore cannot be measured.

atomic radius

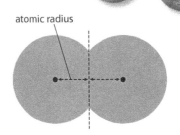

Figure 3.4 The atomic radius of chlorine atoms in a molecule.

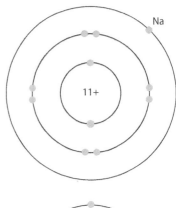

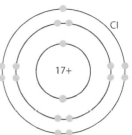

Figure 3.7 Sodium and chlorine atoms. Inner shells, which shield the outer electrons, are highlighted in blue.

Ionic radius

The ionic radius is a measure of the size of an ion.

> In general, the ionic radii of positive ions are smaller than their atomic radii, and the ionic radii of negative ions are greater than their atomic radii.

Figure **3.8** shows a comparison of the atomic and ionic radii (1+ ion) for the alkali metals. Each ion is smaller than the atom from which it is formed (by loss of an electron).

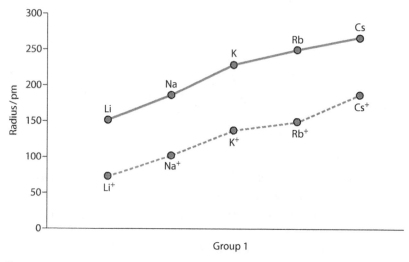

Figure 3.8 Atomic and ionic radii for the alkali metals.

Na is larger than Na^+ because the former has one extra shell of electrons – the electron configuration of Na is 2,8,1, whereas that of Na^+ is 2,8. Also, they both have the same nuclear charge pulling in the electrons (11+), but there is a greater amount of electron–electron repulsion in Na because there are 11 electrons compared with only 10 in Na^+. The electron cloud is therefore larger in Na than in Na^+ because there are more electrons repelling for the same nuclear charge pulling the electrons in.

The fact that negative ions are larger than their parent atoms can be seen by comparing the sizes of halogen atoms with their ions (1−) in Figure **3.9**. Cl^- is larger than Cl because it has more electrons for the same nuclear charge and, therefore, greater repulsion between electrons. Cl has 17 electrons and 17 protons in the nucleus. Cl^- also has 17 protons in the nucleus, but it has 18 electrons. The repulsion between 18 electrons is greater than between 17 electrons, so the electron cloud expands as an extra electron is added to a Cl atom to make Cl^-.

The variation of ionic radius across a period is not a clear-cut trend, because the type of ion changes going from one side to the other – positive ions are formed on the left-hand side of the period and negative ions on the right-hand side.

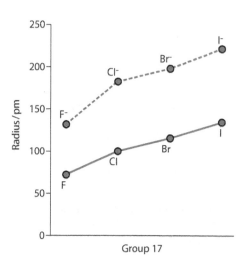

Figure 3.9 A comparison of size between halogen atoms and their ions.

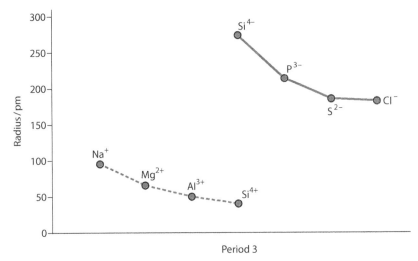

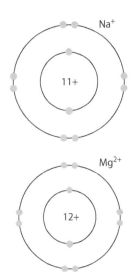

Figure 3.10 Variation of ionic radius of positive and negative ions across period 3.

For positive ions there is a decrease in ionic radius as the charge on the ion increases, but for negative ions the size increases as the charge increases (Figure **3.10**).

Let us consider Na^+ and Mg^{2+} – both ions have the same electron configuration, but Mg^{2+} has one more proton in the nucleus (Figure **3.11**). Because there is the same number of electrons in both ions, the amount of electron–electron repulsion is the same; however, the higher nuclear charge in Mg^{2+} means that the electrons are pulled in more strongly and so the ionic radius is smaller.

Now let us consider P^{3-} and S^{2-}. Both ions have the same number of electrons. S^{2-} has the higher nuclear charge and, therefore, because the amount of electron–electron repulsion is the same in both ions, the electrons are pulled in more strongly in S^{2-} (Figure **3.12**).

Figure 3.11 Mg^{2+} is smaller than Na^+.

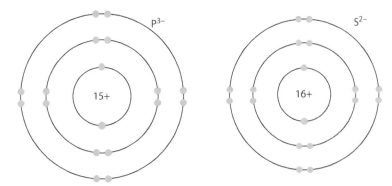

Figure 3.12 S^{2-} is smaller than P^{3-}.

The full definition of first ionisation energy is: the energy required to remove one electron from each atom in one mole of gaseous atoms under standard conditions.

First ionisation energy

The first ionisation energy of an element is the energy required to remove the outermost electron from a gaseous atom – that is, the energy for the process:

$$M(g) \rightarrow M^+(g) + e^-$$

Variation in first ionisation energy down a group

Down any group in the periodic table, the first ionisation energy decreases.

The decrease in first ionisation energy down a group is shown in Figure **3.13**. The size of the atom increases down the group so that the outer electron is further from the nucleus and therefore less strongly attracted by the nucleus (Figure **3.14**).

Although the nuclear charge also increases down a group, this is largely balanced out by an increase in shielding down the group, as there are more electron energy levels (shells). It is the increase in size that governs the change in first ionisation energy.

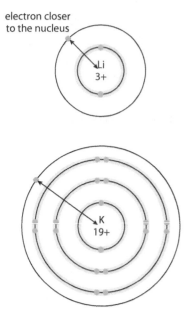

electron closer to the nucleus

Figure 3.14 Potassium has a lower first ionisation energy than lithium. Electrons that shield the outer electron are highlighted in blue.

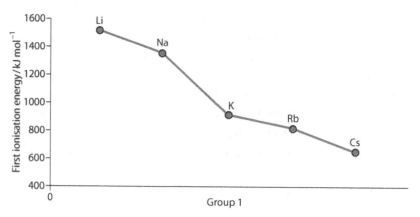

Figure 3.13 First ionisation energy for group 1.

Variation in first ionisation energy across a period

The general trend is that first ionisation energy increases from left to right across a period. This is because of an increase in nuclear charge across the period.

The nuclear charge increases from Na (11+) to Ar (18+) as protons are added to the nucleus. The electrons are all removed from the same main energy level (third shell) and electrons in the same energy level do not shield each other very well. Therefore the attractive force on the outer electrons due to the nucleus increases from left to right across the period

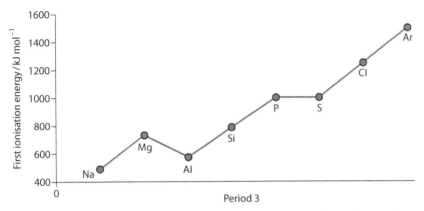

Figure 3.15 The variation in first ionisation energy across period 3 in the periodic table.

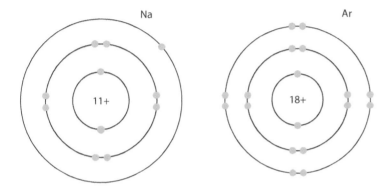

Figure 3.16 Sodium and argon atoms.

The increase in first ionisation energy (Figure **3.15**) can also be explained in terms of the effective nuclear charge felt by the outer electron in argon being higher. The effective nuclear charge felt by the outer electron in a sodium atom would be 11 (nuclear charge) − 10 (number of inner shell electrons), i.e. 1+ if shielding were perfect. The effective nuclear charge felt by the outer electrons in an argon atom would be 18 (nuclear charge) − 10 (number of inner shell electrons), i.e. 8+ if shielding were perfect.

and the outer electron is more difficult to remove from an argon atom (Figure **3.16**). The argon atom is also smaller than the sodium atom and, therefore, the outer electron is closer to the nucleus and more strongly held.

There are two exceptions to the general increase in first ionisation energy across a period, and these are discussed on page **78**.

Exam tip
The exceptions to the trend are required knowledge for all students – refer to page **78**.

Electron affinity

The **first electron affinity** involves the energy change when one electron is added to a gaseous atom:

$$X(g) + e^- \rightarrow X^-(g)$$

It is defined more precisely as the enthalpy change when one electron is added to each atom in one mole of gaseous atoms under standard conditions. Electron affinity is difficult to measure experimentally and data are incomplete.

The first electron affinity is exothermic for virtually all elements – it is a favourable process to bring an electron from far away (infinity) to the outer shell of an atom, where it feels the attractive force of the nucleus.

Variation of electron affinity down group 17

A graph of electron affinity down group 17 is shown in Figure **3.17**. The general trend is that electron affinity decreases down a group, but it can be seen that chlorine has the most exothermic value for electron affinity. A similar trend in electron affinity values is seen going down group 16 and group 14.

The electron affinity becomes less exothermic from Cl to I as the size of the atom increases. The electron is brought into the outer shell of the atom and as the atom gets bigger there is a weaker attraction between the added electron and the nucleus as it is brought to a position which is further from the nucleus.

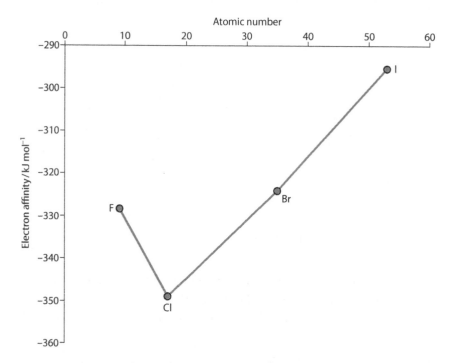

Figure 3.17 Electron affinity values of group 17 elements.

Extension

Electron–electron repulsion also affects the electron affinity and as the atom gets smaller the electrons are, on average, closer together and there is more electron–electron repulsion. This means that the electron affinity should be less exothermic when an electron is added to a smaller atom. Going from F to Cl the electron affinity becomes more exothermic because the decrease in electron–electron repulsion outweighs the fact that there is less attraction between the electron and the nucleus.

Variation in electron affinity across a period

The general trend in electron affinity from group 13 to group 17 is shown in Figure **3.18**.

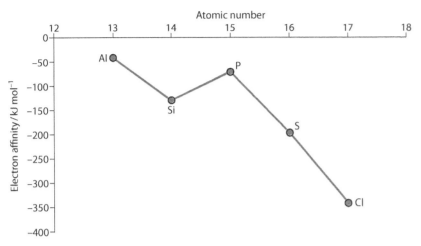

Figure 3.18 Electron affinity values across period 3.

The general trend is that the electron affinity becomes more exothermic. This is because of an increase in nuclear charge and a decrease in atomic radius from left to right across the period. For instance, F has a higher nuclear charge and a smaller radius than O and so the electron will be more strongly attracted when it is brought into the outer shell of the F atom.

Extension

Phosphorus has a less exothermic electron affinity than silicon because of its electron configuration. P has three unpaired electrons in three separate p orbitals and when one electron is added this electron must be paired up in the same p orbital as another electron – this introduces an extra repulsion term that is not present in Si. The arguments being used here are very similar to those for the variation of first ionisation energy across a period discussed in Topic **2**.

Electronegativity

> **Electronegativity** is a measure of the attraction of an atom in a molecule for the electron pair in the covalent bond of which it is a part.

In a covalent bond between two different atoms, the atoms do not attract the electron pair in the bond equally. How strongly the electrons are attracted depends on the size of the individual atoms and their nuclear charge.

Electronegativity decreases down a group – this is because the size of the atoms increases down a group. Consider hydrogen bonded to either F or Cl (Figure **3.19**). The bonding pair of electrons is closer to the F nucleus in HF than it is to the Cl nucleus in HCl. Therefore the electron pair is more strongly attracted to the F nucleus in HF and F has a higher electronegativity than Cl.

Electronegativity is discussed in more detail on page **128**.

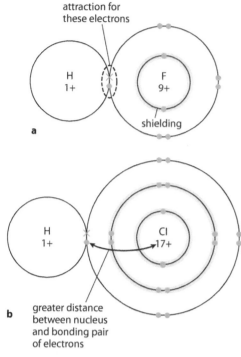

attraction for these electrons

H 1+

F 9+

shielding

a

H 1+

Cl 17+

b

greater distance between nucleus and bonding pair of electrons

Figure 3.19 Hydrogen bonded to **a** fluorine and **b** chlorine.

Chlorine's higher nuclear charge does not make it more electronegative than fluorine because the shielding from inner shells (shown with blue shading in Figure **3.19**) increases from F to Cl such that the effective nuclear charge felt by the bonding electrons is approximately the same in each case (+7 if shielding were perfect).

Electronegativity increases across a period – the reason for this is the increase in nuclear charge across the period with no significant change in shielding. The shielding remains approximately constant because atoms in the same period have the same number of inner shells.

So, if an N–H bond is compared with an F–H bond (Figure **3.20**), the electrons in the N–H bond are attracted by the seven protons in the nucleus, but the electrons in the F–H bond are attracted by the nine protons in the F nucleus. In both cases the shielding is approximately the same (because of two inner shell electrons).

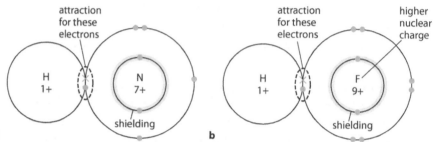

attraction for these electrons

H 1+

N 7+

shielding

a

attraction for these electrons

higher nuclear charge

H 1+

F 9+

shielding

b

Figure 3.20 Hydrogen bonded to **a** nitrogen and **b** fluorine.

? Test yourself

1 Give the names of the following elements:
  a the element in period 3 and group 14
  b the element in period 5 and group 16
  c the element in the same group as sulfur but in period 6
  d a halogen in period 5
  e an element in the same period as potassium that has five outer shell electrons

2 State whether the following properties increase or decrease across a period:
  a electronegativity
  b atomic radius

3 Arrange the following in order of increasing radius (smallest first):
  a Ba   Mg   Sr   Ca
  b O^{2-} Na^+ F^-
  c Na Na^+ K Al^{3+}
  d S   Cl I^- Cl^- S^{2-}

4 Are the following **true** or **false**?
  a A germanium atom is smaller than a silicon atom, but silicon has a higher first ionisation energy.
  b Selenium has a higher first ionisation energy and electronegativity than sulfur.
  c Antimony has a higher first ionisation energy and electronegativity than tin.
  d Cl^- is bigger than Cl, but Se^{2-} is smaller than Se.
  e Iodine has a higher electronegativity than tellurium but a lower electronegativity than bromine.

5 Based on the following data, which element (**X** or **Y**) is more likely to be a metal?

	First ionisation energy / kJ mol^{-1}	Atomic radius / nm	Electronegativity
X	736	0.136	1.3
Y	1000	0.104	2.6

3.2.2 Properties of elements in group 1 and group 17

Group 1 elements

The elements in group 1 are known as the **alkali metals**. They are all highly reactive, soft, low melting point metals (Table **3.1**). They are placed together in group 1 for two reasons – they all have one electron in their outer shell and they react in very similar ways (similar chemical properties).

> The reactions of an element are determined by the number of electrons in the outer shell (highest main energy level) of their atoms. Because elements in the same group in the periodic table have the same number of electrons in their outer shell, they react in basically the same way.

The bonding in all these elements is metallic. The solid is held together by electrostatic attraction between the positive ions in the lattice and the delocalised electrons (see page **160**).

The attraction for the delocalised, negatively-charged, electrons is due to the nucleus of the positive ion. As the ions get larger as we go down the group, the nucleus becomes further from the delocalised electrons and the attraction becomes weaker (Figure **3.22**). This means that less energy is required to break apart the lattice going down group 1.

> Melting point decreases down group 1 (Figure **3.21**).

> Liquid sodium is used as a coolant in some nuclear reactors.

Element	Symbol	Atomic number	Electron configuration	Density / g cm^{-3}	Melting point / °C	Boiling point / °C
lithium	Li	3	[He]2s^1	0.53	180	1330
sodium	Na	11	[Ne]3s^1	0.97	98	890
potassium	K	19	[Ar]4s^1	0.86	64	774
rubidium	Rb	37	[Kr]5s^1	1.53	39	688
caesium	Cs	55	[Xe]6s^1	1.87	29	679

Table 3.1 Similarities in alkali metal properties.

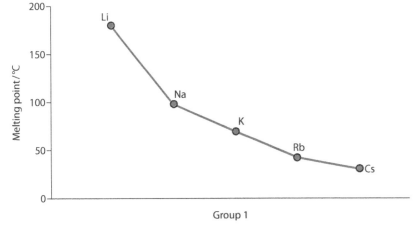

Figure 3.21 Variation in melting point in group 1.

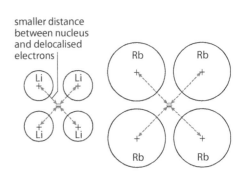

Figure 3.22 The delocalised electrons are attracted more strongly in lithium than in rubidium.

Reactions of the elements in group 1

The elements in group 1 are all reactive metals that react readily with, among other things, oxygen, water and halogens. The atoms all have one electron in their outer shell, and virtually all reactions involve the loss of this outer shell electron to form a positive ion, M^+. The reactions become more vigorous going down the group because the ionisation energy decreases as the size of the atom increases. This means that, for example, caesium loses its outer electron to form a positive ion much more easily than sodium and will react more vigorously.

M_2O is a basic oxide that will dissolve in water to form an alkaline solution, containing M^+ and OH^- ions.

Reaction with oxygen

The alkali metals react vigorously with oxygen and all tarnish rapidly in air. The general equation for the reaction is:

$$4M(s) + O_2(g) \rightarrow 2M_2O(s)$$

Reaction with water

The alkali metals react rapidly with water. The general equation for the reaction is:

$$2M(s) + 2H_2O(l) \rightarrow 2MOH(aq) + H_2(g)$$

An alkaline solution is formed. The alkali metal hydroxides are strong bases and ionise completely in aqueous solution (page **321**).

Lithium, sodium and potassium are all less dense than water.

The reaction with water becomes more vigorous going down the group – sodium melts into a ball, fizzes rapidly and moves around on the surface of the water; potassium bursts into flames (lilac); and caesium explodes as soon as it comes into contact with water.

Group 17 elements

The elements in group 17 are known as the **halogens**. They are all non-metals consisting of diatomic molecules (X_2). Some properties are given in Table **3.2**.

Element	Symbol	Atomic number	Electron configuration	Colour	Melting point/°C	Boiling point/°C	Physical state at room temperature and pressure
fluorine	F	9	$[He]2s^22p^5$	pale yellow	−220	−188	gas
chlorine	Cl	17	$[Ne]3s^23p^5$	yellow–green	−101	−35	gas
bromine	Br	35	$[Ar]3d^{10}4s^24p^5$	deep red liquid, orange vapour	−7	59	liquid
iodine	I	53	$[Kr]4d^{10}5s^25p^5$	grey shiny solid, purple vapour	114	184	solid

Table 3.2 Properties of halogens.

Variation of melting point in group 17

The melting points of the halogens increase going down the group (Figure **3.23**).

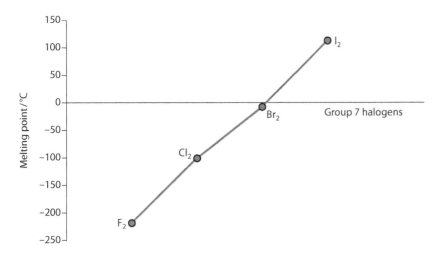

Figure 3.23 Variation in melting point in group 17.

As the relative molecular masses of the X_2 halogen molecules increase, the London forces (page **148**) between molecules get stronger. This means that more energy must be supplied to separate the molecules from each other.

Reactions of the elements in group 17

All the atoms of the elements in group 17 have seven electrons in their outer shell and react either by gaining an electron to form X^- ions or by forming covalent compounds. Reactivity decreases down the group, and fluorine is the most reactive element known, reacting directly with virtually every other element in the periodic table. The variation in reactivity of the halogens cannot be as easily explained as for the alkali metals. The very high reactivity of fluorine can be explained in terms of an exceptionally weak F–F bond and the strength of the bonds it forms with other atoms. The reactivity in terms of the formation of X^- ions can be related to a decrease in electron affinity (energy released when an electron is added to a neutral atom) going down the group as the electron is added to a shell further away from the nucleus, but this is only part of the story and several factors must be considered when explaining the reactivity of the halogens.

The halogens all react with the alkali metals to form salts. The general equation is:

$$2M(s) + X_2(g) \rightarrow 2MX(s)$$

The salts formed are all white/colourless, fairly typical ionic compounds. They contain M^+ and X^- ions. All alkali metal chlorides, bromides and iodides are soluble in water and form colourless, neutral solutions.

Chlorine is produced by the electrolysis of brine. Worldwide annual production is about 60 million tonnes. Chlorine and its compounds are involved in the production of about 90% of the most important pharmaceuticals. Its biggest single use is in the production of PVC.

How vigorous the reaction is depends on the particular halogen and alkali metal used – the most vigorous reaction occurs between fluorine and caesium, and the least vigorous between lithium and iodine.

Displacement reactions of halogens

These are reactions between a solution of a halogen and a solution containing halide ions – they are discussed in more detail on page **404**. A small amount of a solution of a halogen is added to a small amount of a solution containing a halide ion, and any colour changes are observed (see Table **3.3**). Potassium chloride, bromide and iodide solutions are all colourless. The colours of chlorine, bromine and iodine solutions are shown in Figure **3.24**.

	KCl(aq)	KBr(aq)	KI(aq)
Cl_2(aq)	no reaction	orange solution	dark red–brown solution
Br_2(aq)	no reaction	no reaction	dark red–brown solution
I_2(aq)	no reaction	no reaction	no reaction

Table 3.3 Results of reactions between halogen solutions and solutions containing halide ions.

The reactions that occur are:

$$Cl_2(aq) + 2KBr(aq) \rightarrow 2KCl(aq) + Br_2(aq)$$
Ionic equation: $Cl_2(aq) + 2Br^-(aq) \rightarrow 2Cl^-(aq) + Br_2(aq)$

$$Cl_2(aq) + 2KI(aq) \rightarrow 2KCl(aq) + I_2(aq)$$
Ionic equation: $Cl_2(aq) + 2I^-(aq) \rightarrow 2Cl^-(aq) + I_2(aq)$

$$Br_2(aq) + 2KI(aq) \rightarrow 2KBr(aq) + I_2(aq)$$
Ionic equation: $Br_2(aq) + 2I^-(aq) \rightarrow 2Br^-(aq) + I_2(aq)$

The more reactive halogen displaces the halide ion of the less reactive halogen from solution – chlorine displaces bromide ions and iodide ions from solution, and bromine displaces iodide ions from solution.

These reactions are all **redox reactions** (Topic **9**), in which a more reactive halogen oxidises a less reactive halide ion. Chlorine is a stronger oxidising agent than bromine and iodine; it will oxidise bromide ions to bromine, and iodide ions to iodine. Bromine is a stronger oxidising agent than iodine and will oxidise iodide ions to iodine. In terms of electrons, chlorine has the strongest affinity for electrons and will remove electrons from bromide ions and iodide ions.

Orange colour, due to the production of bromine.

Red-brown colour, due to the production of iodine.

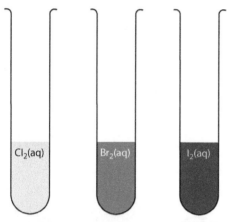

Figure 3.24 Chlorine solution is pale yellow–green (almost colourless if it is dilute), bromine solution is orange, and iodine solution is red–brown.

3.2.3 Oxides of period 2 and period 3 elements

Oxides of elements may be classified as basic, acidic or amphoteric. The nature of the oxides changes across a period and Table **3.4** shows how the oxides change from basic to amphoteric to acidic across period 3.

In general, metallic oxides are **basic** and non-metallic oxides are **acidic**.

A basic oxide is one that will react with an acid to form a salt and, if soluble in water, will produce an alkaline solution. Sodium oxide reacts with water to form sodium hydroxide:

$$Na_2O(s) + H_2O(l) \rightarrow 2NaOH(aq)$$

Sodium oxide reacts with acids such as sulfuric acid to form salts:

$$Na_2O(s) + H_2SO_4(aq) \rightarrow Na_2SO_4(aq) + H_2O(l)$$

Magnesium oxide, because of the relatively high charges on the ions, is not very soluble in water but it does react to a small extent to form a solution of magnesium hydroxide, which is alkaline:

$$MgO(s) + H_2O(l) \rightarrow Mg(OH)_2(aq)$$

Aluminium is on the dividing line between metals and non-metals and forms an amphoteric oxide – these have some of the properties of a basic oxide and some of an acidic oxide. Aluminium is exhibiting properties between those of a metal (basic) oxide and those of a non-metal (acidic) oxide.

Aluminium oxide does not react with water but it does display amphoteric behaviour in that it reacts with both acids and bases to form salts:

reaction with acids: $Al_2O_3 + 6H^+ \rightarrow 2Al^{3+} + 3H_2O$

reaction with alkalis/bases: $Al_2O_3 + 2OH^- + 3H_2O \rightarrow 2Al(OH)_4^-$

Amphoteric oxides react both with acids and with bases.

Learning objectives

- Describe the changes from basic to acidic oxides across a period
- Write equations for the reactions of oxides with water and predict the acidity of the resulting solutions

Exam tip
Reactions highlighted like this must be learnt for examinations.

	Sodium	Magnesium	Aluminium	Silicon	Phosphorus	Sulfur
Formula of oxide	Na_2O	MgO	Al_2O_3	SiO_2	P_4O_{10}	SO_2 SO_3
Nature of element	metal				non-metal	
Nature of oxide	basic		amphoteric	acidic		
Reaction with water	soluble, reacts	sparingly soluble, some reaction	insoluble		soluble, reacts	
Solution formed	alkaline	slightly alkaline	–		acidic	

Table **3.4** The acid–base nature of some period 3 oxides.

The remaining oxides in Table **3.4** are all acidic oxides. An acidic oxide is one that reacts with bases/alkalis to form a salt and, if soluble in water, will produce an acidic solution.

P_4O_6 (phosphorus(III) oxide) and P_4O_{10} (phosphorus(V) oxide) form phosphoric(III) and phosphoric(V) acid, respectively, when they react with water:

$$P_4O_6(s) + 6H_2O(l) \rightarrow 4H_3PO_3(aq)$$

$$P_4O_{10}(s) + 6H_2O(l) \rightarrow 4H_3PO_4(aq)$$

SO_2 (sulfur(IV) oxide) and SO_3 (sulfur(VI) oxide) form sulfuric(IV) and sulfuric(VI) acid, respectively, when they react with water:

$$SO_2(g) + H_2O(l) \rightarrow H_2SO_3(aq)$$

$$SO_3(g) + H_2O(l) \rightarrow H_2SO_4(aq)$$

Nitrogen oxides

There are many oxides of nitrogen, ranging in formula from N_2O to N_2O_5. Two of the most environmentally important are nitrogen(II) oxide (NO) and nitrogen(IV) oxide (NO_2).

Nitrogen reacts with oxygen at very high temperatures to form NO (nitrogen monoxide, nitric oxide or nitrogen(II) oxide):

$$N_2(g) + O_2(g) \rightarrow 2NO(g)$$

This reaction occurs in the internal combustion engine. NO is virtually insoluble in water and is classified as a neutral oxide.

NO can be oxidised in the atmosphere to NO_2, which can react with water to produce nitric(V) acid (HNO_3), which is one of the acids responsible for acid deposition (see Subtopic **8.5**). NO_2 can be classified as an acidic oxide:

$$2NO_2(g) + H_2O(l) \rightarrow HNO_2(aq) + HNO_3(aq)$$

N_2O (nitrogen(I) oxide, nitrous oxide) is another neutral oxide. N_2O is also known as laughing gas and major uses include as an anaesthetic and as the propellant in 'squirty cream'.

Phosphoric(V) acid is an ingredient of Coca-Cola®.

Non-metal oxides such as SO_2 are produced in various industrial processes and when coal is burnt. This can be responsible for **acid rain**, which can, among other things, kill fish in lakes and trees in forests.

Nitrogen oxides (NO_x) may be formed in internal combustion engines, and these are involved in the formation of photochemical smog in cities (Figure **3.25**).

Figure 3.25 A photochemical smog over Hong Kong.

Nature of science

Science and the technology that develops from it have been used to solve many problems – but it can also *cause* them. The development of industrial processes that produce acidic gases led to acid rain being a major environmental problem. Acid rain, and its associated problems, is important to people across the world and it is vital that scientists work to improve the public understanding of the issues involved. Scientists also work as advisors to politicians in developing policies to solve these problems.

Advancements in science have often arisen from finding patterns in data. An understanding of the patterns in physical and chemical properties of elements in the periodic table has allowed chemists to make new substances. For instance, the knowledge that sulfur formed a range of compounds with nitrogen probably led scientists to attempt to make selium-nitrogen and tellurium-nitrogen compounds.

? Test yourself

6 Write balanced equations for the following reactions:
   a rubidium with water
   b potassium with bromine
   c chlorine solution with potassium bromide solution
   d sodium oxide with water
   e sulfur(VI) oxide with water

7 State whether trends down the group in each of the following properties are the same or different when group 1 and group 17 are compared:
   a electronegativity
   b reactivity
   c melting point
   d ionisation energy

8 State whether an acidic or alkaline solution will be formed when each of the following is dissolved in/reacted with water:
   a SO_3
   b MgO
   c Na

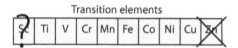

Transition elements

3.3 First-row d-block elements (HL)

3.3.1 The transition elements (d block)

The first-row d-block elements are:

Sc	Ti	V	Cr	Mn	Fe	Co	Ni	Cu	Zn
21	22	23	24	25	26	27	28	29	30

There are also two other rows of d-block elements.

They are called 'd-block' elements because the subshell being filled across this series is the 3d subshell. The electron configurations range from $[Ar]4s^23d^1$ for scandium to $[Ar]4s^23d^{10}$ for zinc:

Sc	Ti	V	Cr	Mn	Fe	Co	Ni	Cu	Zn
$[Ar]4s^23d^1$	$[Ar]4s^23d^2$	$[Ar]4s^23d^3$	$[Ar]4s^13d^5$	$[Ar]4s^23d^5$	$[Ar]4s^23d^6$	$[Ar]4s^23d^7$	$[Ar]4s^23d^8$	$[Ar]4s^13d^{10}$	$[Ar]4s^23d^{10}$

The **transition elements** can be defined as different from 'the d-block elements', and the definition we will use here is:

> a transition element is an element that forms at least one stable ion with a partially filled d subshell

According to this definition, zinc is not counted as a transition element because the only ion it forms is the 2+ ion, with electron configuration $1s^22s^22p^63s^23p^63d^{10}$ (full d subshell). Zinc (Figure **3.26**) does not exhibit some of the typical characteristic properties of transition metals detailed below (e.g. it does not form coloured compounds). The inclusion/exclusion of scandium as a transition element according to this definition is much more controversial. In virtually every compound scandium has oxidation number +3 (no d electrons) however, it also forms a couple of compounds (ScH_2 and $CsScCl_3$) with formal oxidation number +2, but the bonding in these compounds is more complicated and they do not necessarily contain the 2+ ion.

Properties of the transition elements

We have already studied the variation in properties of a set of eight elements across the periodic table when we looked at the properties of period 3 elements. The transition elements also form a set of eight elements across the periodic table, but these are much more similar to each other than the elements across period 3. For instance, they are all metals rather than showing a change from metal to non-metal.

The variation in first ionisation energy and atomic radius of the transition elements and period 3 elements are compared in Figures **3.27** and **3.28**. It can be seen that the variation of ionisation energy and atomic radius across the series of the transition elements is much smaller than across period 3.

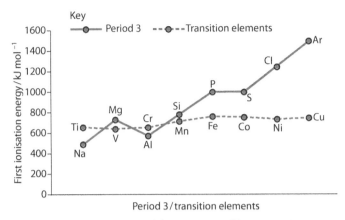

Figure 3.27 A comparison of the variation of first ionisation energy across period 3 with that across the transition metal series.

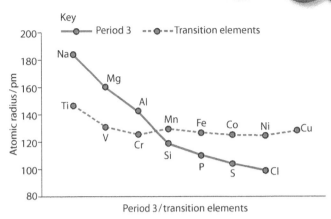

Figure 3.28 A comparison of the variation of atomic radius across period 3 with that across the transition metal series.

Because of their similarity it is possible to draw up a list of characteristic properties of transition elements:

- Transition elements are all typical metals – they have high melting points and densities.
- Transition elements can exhibit more than one oxidation number in compounds/complexes.
- Transition elements form complex ions.
- Transition elements form coloured compounds/complexes.
- Transition elements and their compounds/complexes can act as catalysts in many reactions.
- Compounds of transition elements can exhibit magnetic properties.

Exam tip
The last five properties are the most important for examinations.

Ionisation of transition elements

Transition elements form positive ions. The electron configurations of some transition metal ions are shown in Table **3.5**.

The 4s electrons are always removed before the 3d electrons when an ion is formed.

Element	Electron configuration	Ion	Electron configuration
Cr	$[Ar]4s^1 3d^5$	Cr^{2+}	$[Ar]3d^4$
		Cr^{3+}	$[Ar]3d^3$
Mn	$[Ar]4s^2 3d^5$	Mn^{2+}	$[Ar]3d^5$
Fe	$[Ar]4s^2 3d^6$	Fe^{2+}	$[Ar]3d^6$
		Fe^{3+}	$[Ar]3d^5$
Co	$[Ar]4s^2 3d^7$	Co^{2+}	$[Ar]3d^7$
Cu	$[Ar]4s^1 3d^{10}$	Cu^+	$[Ar]3d^{10}$
		Cu^{2+}	$[Ar]3d^9$

Table 3.5 Electron configurations of transition metals and their ions.

3.3.2 Variable oxidation numbers

The positive oxidation numbers (oxidation states) exhibited by the transition elements are shown in Figure **3.29**. The greatest number of different oxidation numbers and the highest oxidation numbers are found in the middle of the series. From titanium to maganese there is an increase in the total number of electrons in the 4s and 3d subshells, so the maximum oxidation number increases. Maganese has the electron configuration $[Ar]4s^23d^5$ and therefore a maximum oxidation number of +7. Iron has eight electrons in the 4s and 3d subshells and would be expected to have a maximum oxidation number of +8, but the ionisation energy increases from left to right across the transition elements series and it becomes more difficult to reach the highest oxidation numbers towards the right-hand side of the series. The chemistry of copper and nickel is, for the same reason, dominated by the lower oxidation numbers.

> **All transition metals show oxidation number +2.** In most cases this is because they have two electrons in the 4s subshell, and removal of these generates an oxidation number of +2.

Ti	V	Cr	Mn	Fe	Co	Ni	Cu
			7				
		6	6	6			
	5	5	5	5			
4	4	4	4	4	4	4	4
3	3	3	3	3	3	3	3
2	2	2	2	2	2	2	2
1	1	1	1	1	1	1	1
0	0	0	0	0	0	0	0

Figure 3.29 Oxidation numbers of transition metals in compounds. Not all oxidation numbers are common.

Why more than one oxidation number?

The 4s and 3d subshells are close in energy, and there are no big jumps in the successive ionisation energies when the 4s and 3d electrons are removed. Therefore the number of electrons lost will depend on a variety of factors such as lattice enthalpy, ionisation energy and hydration enthalpy. Electrons are not removed in order to generate the nearest noble gas electron configuration. The graph in Figure **3.30** shows a comparison of the first seven ionisation energies of magnesium and manganese. It can be seen that there is a very large jump between the second and third ionisation energies of magnesium, but that there are no such jumps for manganese.

Oxidation numbers are discussed further on page **369**.

Oxidation number and oxidation state are often used interchangeably.

Exam tip
The oxidation numbers highlighted in Figure **3.29** are mentioned specifically in the data booklet.

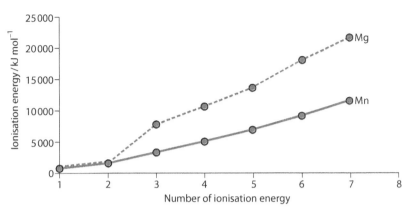

Figure 3.30 Comparison of successive ionisation energies of magnesium and manganese.

Magnetic properties of transition metal compounds

There are two forms of magnetism we need to be concerned with – **paramagnetism** and **diamagnetism**.

> Paramagnetism is caused by unpaired electrons – paramagnetic substances are attracted by a magnetic field.
>
> Diamagnetism is caused by paired electrons – diamagnetic substances are repelled slightly by a magnetic field.

All substances have some paired electrons and so all substances exhibit diamagnetism. However, the diamagnetic effect is much smaller than the paramagnetic effect and so, if there are any unpaired electrons present, the paramagnetic effect will dominate and the substance will be paramagnetic overall and attracted by a magnetic field. The more unpaired electrons, the greater the paramagnetism (magnetic moment).

Consider the electron configurations of two transition metal ions:

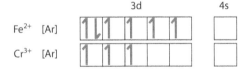

Both contain unpaired electrons and their compounds are paramagnetic – so both $FeCl_2$ and $CrCl_3$ are paramagnetic. Because an Fe^{2+} ion has four unpaired electrons and a Cr^{3+} ion has only three, the iron(II) compound is more paramagnetic (higher magnetic moment).

The Cu^+ ion has the following electron configuration:

All the electrons are paired so compounds of copper(I), such as CuCl, are diamagnetic only.

Extension

The situation is more complicated with complex ions. Depending on the energy difference between the higher and lower set of d orbitals and the amount of energy required to pair up two electrons in the same d orbital (overcoming the repulsions), complexes can be high spin (maximum number of unpaired electrons) or low spin (maximum number of electrons in the lower set of d orbitals). How paramagnetic a substance is then depends on the ligands because they influence the splitting of the d orbitals.

A ligand must possess a lone pair of electrons.

A ligand is a Lewis base.

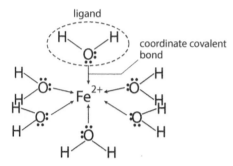

ligand

coordinate covalent bond

Figure 3.31 A complex ion is formed when ligands bond to a transition metal ion. The ligands donate lone pairs into vacant orbitals (3d, 4s or 4p) on the transition metal ion.

Extension

There is a strong case for considering the bonding in a transition metal complex ion as having a significant ionic component. Crystal field theory and ligand field theory consider the bonding from a more ionic point of view.

Complex ions

A complex ion consists of a central metal ion surrounded by **ligands** – transition metal ions form many complexes.

Ligands are negative ions or neutral molecules that have lone pairs of electrons. They use the lone pairs to bond to a metal ion to form a complex ion. **Coordinate covalent bonds** (dative bonds) are formed between the ligand and the transition metal ion.

The structure of $[Fe(H_2O)_6]^{2+}$ is shown in Figure **3.31**. H_2O is the ligand in this complex ion. The shape of this complex ion is octahedral and it is called the hexaaquairon(II) ion. Other ways of drawing this are shown in Figure **3.32**.

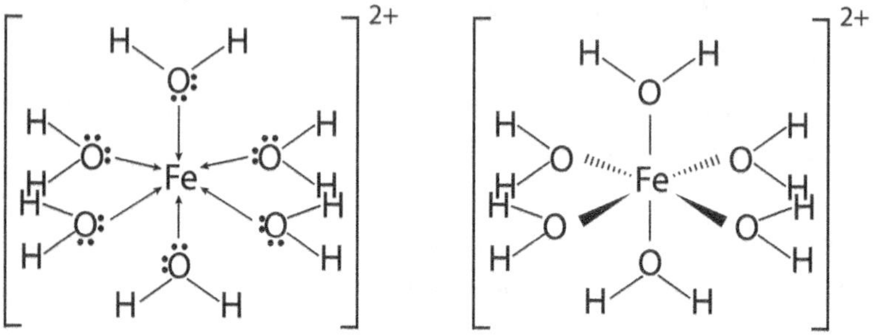

Figure 3.32 Alternative representations of the $[Fe(H_2O)_6]^{2+}$ complex ion.

All transition elements, with the exception of titanium, form an octahedral complex ion with the formula $[M(H_2O)_6]^{2+}$ in aqueous solution.

Complex ions can have various shapes depending on the number of ligands. However, shapes cannot be worked out using the valence shell electron-pair repulsion theory (see page **137**) because more subtle factors also govern the overall shape. If a complex ion contains 6 ligands it will almost certainly be octahedral, but complexes containing 4 ligands may be tetrahedral or square planar (Figure **3.33**).

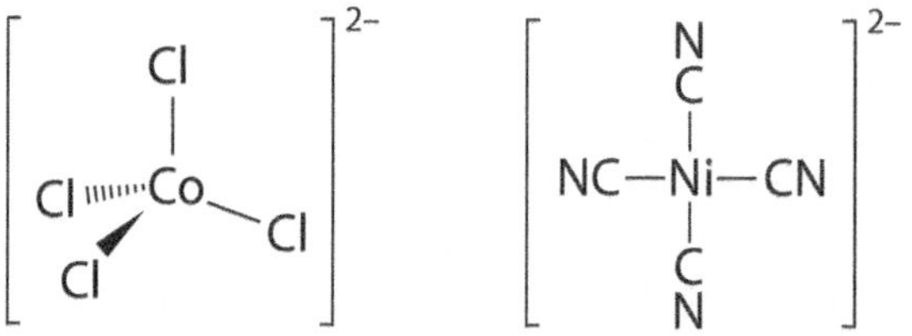

$[CoCl_4]^{2-}$

$[Ni(CN)_4]^{2-}$

Figure 3.33 $[CoCl_4]^{2-}$ is tetrahedral but $[Ni(CN)_4]^{2-}$ is square planar.

Complex ions can undergo substitution reactions in which, for example, H_2O ligands are replaced by other ligands. For example, in the addition of concentrated hydrochloric acid to blue copper(II) sulfate solution:

$$[Cu(H_2O)_6]^{2+}(aq) + 4Cl^-(aq) \rightleftharpoons [CuCl_4]^{2-}(aq) + 6H_2O(l)$$

blue   yellow

As the acid is added, the yellow $[CuCl_4]^{2-}$ complex ion is formed. So, the solution changes colour from blue to green (a mixture of blue and yellow). According to Le Chatelier's principle (see Topic **7**) the position of equilibrium shifts to the right as Cl^- is added.

The oxidation number of a transition metal in a complex ion

The oxidation number of a transition metal in a complex ion can be worked out from the charges on the ligands. Ligands may be either neutral or negatively charged (see Table **3.6**).

In $[Fe(H_2O)_6]^{2+}$ all the ligands are neutral water molecules. The overall charge on the ion is just due to the iron ion, so the oxidation number of iron must be +2.

In $[Ni(CN)_4]^{2-}$ all the ligands have a 1− charge, so the total charge from all four ligands is 4−. The overall charge on the ion is 2−; so, the oxidation number of nickel must be +2 to cancel out 2− from the 4− charge.

Neutral ligands	1− ligands
H_2O	Cl^-
NH_3	CN^-
CO	Br^-

Table 3.6 Charges on ligands.

> **Exam tip**
> Oxidation numbers are discussed in more detail in Topic **9**.

Working out the overall charge on a complex ion

If the oxidation number (charge) of the central transition metal ion and the charges on the ligands are known, the overall charge on the complex ion can be worked out.

Worked example

3.1 Platinum(II) can form a complex ion with 1 ammonia and 3 chloride ligands. What is the overall charge and formula of the complex ion?

Platinum(II) has a charge of 2+
Ammonia is a neutral ligand (NH_3)
Chloride has a 1− charge (Cl^-)

The overall charge is $(2+) + (0) + 3(1-) = 1-$

The formula of the complex ion is: $[Pt(NH_3)Cl_3]^-$

Catalytic ability

Transition elements and their compounds/complexes can act as catalysts. For example, finely divided iron is the catalyst in the **Haber process** in the production of ammonia:

$$N_2(g) + 3H_2(g) \rightleftharpoons 2NH_3(g)$$

Iron in the above reaction is a heterogeneous catalyst (one that is in a different physical state to the reactants) but transition metal compounds often act as homogeneous catalysts (ones that are in the same phase as the reactants). The ability to act as a catalyst relies on a transition metal atom or ion having varying oxidation numbers and also being able to coordinate to other molecules/ions to form complex ions.

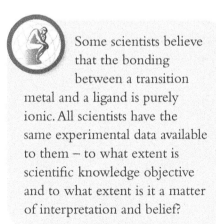 Some scientists believe that the bonding between a transition metal and a ligand is purely ionic. All scientists have the same experimental data available to them – to what extent is scientific knowledge objective and to what extent is it a matter of interpretation and belief?

Science is about finding patterns and these patterns allow us to make predictions. However, it is not always the patterns that are most interesting and careful observation is essential to spot anomalies and exceptions to patterns that could lead to new discoveries and theories. Zinc can be regarded as anomalous in the first row of the d block, for instance, it does not generally form coloured compounds; this has resulted in it not being included as a transition element.

3.4 Coloured complexes (HL)

The colours of some complex ions are shown in Table **3.7**.

In a gaseous transition metal ion, all the 3d orbitals have the same energy – that is, they are degenerate. However, when the ion is surrounded by ligands in a complex ion, these d orbitals are split into two groups. In an octahedral complex ion there are two orbitals in the upper group and three orbitals in the lower groups (Figure **3.34**).

Complex ion	Colour
$[Cu(H_2O)_6]^{2+}$	blue
$[Cu(NH_3)_4(H_2O)_2]^{2+}$	deep blue/violet
$[Fe(SCN)(H_2O)_5]^{2+}$	blood red
$[Ni(H_2O)_6]^{2+}$	green

Table 3.7 The colours of some complex ions.

Extension

All the d orbitals in a complex ion are higher in energy than the d orbitals in an isolated gaseous ion.

Extension

The electrons in d_{z^2} and $d_{x^2-y^2}$ orbitals are repelled more by the ligand electrons because they point directly at the ligands – greater repulsion leads to higher energy.

Extension

The d orbitals are split in different ways in different-shaped complex ions.

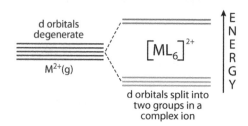

Figure 3.34 The splitting of d orbitals in a complex ion.

The splitting may be regarded as being caused by the repulsion between the electrons in the metal ion d orbitals and the lone pairs on the ligands. Two of the metal ion d orbitals point directly at the ligands and so are raised in energy, whereas the other three d orbitals point between the ligands and are lowered in energy relative to the other two d orbitals.

Energy in the form of a certain frequency of visible light can be absorbed to promote an electron from the lower set of orbitals to the higher set (Figure **3.35**).

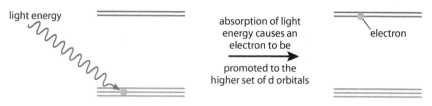

Figure 3.35 Absorption of light by a complex ion.

When white light passes through copper sulfate solution (Figure **3.36**), orange light is absorbed, promoting an electron from the lower set of d orbitals to the higher set. This means that the light coming out contains all the colours of the spectrum except orange and so appears blue, the complementary colour to orange.

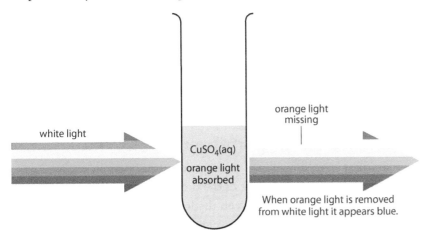

Figure **3.36** Colour and absorption.

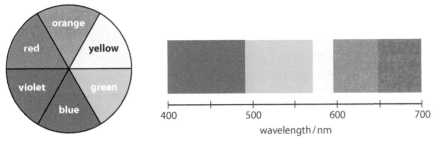

Figure **3.37** A colour wheel – along with the approximate wavelengths of visible light. Complementary colours are opposite each other in the colour wheel, therefore blue is complementary to orange and green is complementary to red.

> For a substance to appear coloured, certain frequencies of light in the visible region of the spectrum must be absorbed.

The colour of a substance will appear to an observer as the **complementary colour** to the light that is absorbed. A colour wheel (Figure **3.37**) shows which pairs of colours are complementary (opposite each other in the colour wheel). If we know the colour of the complex ion, the colour of light that is absorbed can be worked out, and vice versa. For example, because a solution of nickel(II) chloride is green, it must absorb red light – the complementary colour to green.

> The formation of coloured substances requires the presence of a partially filled d subshell.

Let us consider the Sc^{3+} ion or the Ti^{4+} ion. These both have no electrons in the 3d subshell and so are colourless, as it is not possible to absorb energy to promote a 3d electron.

White light is a mixture of all colours (frequencies) of visible light.

Extension

$Cr_2O_7^{2-}$ (orange), CrO_4^{2-} (yellow) and MnO_4^- (purple) are all very highly coloured, but they have no d electrons. They are coloured because of a different mechanism from the one described here.

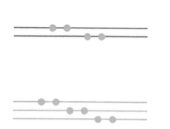

Figure 3.38 A Cu⁺ or Zn²⁺ ion has ten 3d electrons.

The Cu^+ ion and the Zn^{2+} ion both have ten 3d electrons (Figure **3.38**), and as there is no space in the upper set of orbitals it is not possible to promote an electron to the upper set of orbitals. No light in the visible region of the spectrum is absorbed and these ions are colourless.

What do we mean when we say that a solution of copper sulfate is blue? Is blueness a property of copper sulfate solution, or is the blueness in our minds? What colour would copper sulfate solution be in orange light? Or in the dark?

Extension

The actual theory behind the spectra of transition metal complexes is significantly more complex than described here, and the simple idea of an electron being promoted from the lower set of d orbitals to the upper set is only really applicable to transition metal ions with one d electron (d^9 ions also produce relatively simple spectra). This is evident by the fact that transition metal ions will usually absorb more than one frequency of electromagnetic radiation – not just one as predicted by the simple model. For transition metals with more than one d electron, the repulsion between d electrons is important in determining the energies of the various energy states. The absorption of electromagnetic radiation by a transition metal ion could be better described as 'causing a rearrangement of electrons within the d orbitals'.

Most aqueous solutions of Fe^{3+} are actually yellow, but they do not contain $[Fe(H_2O)_6]^{3+}$(aq).

In general, for complex ions containing the same metal and the same ligands, the greater the oxidation number of the transition metal, the greater the splitting of the d orbitals.

Factors that affect the colour of transition metal complexes

At the simplest level, the colours of transition metal complexes can be related to the amount of splitting of the d orbitals. For example, if there is a greater difference in energy between the lower and higher set of d orbitals then a higher frequency (shorter wavelength) of light will be absorbed and the complementary colour will be different. The following factors all have a part to play.

Identity of the metal

Complexes of different metals in the same oxidation state have different colours. For example, Mn^{2+}(aq) ($3d^5$) is very pale pink/colourless but Fe^{2+}(aq) ($3d^6$) is pale green.

Different metal ions have different electron configurations and, because colours are caused by electron transitions, different arrangements of electrons give rise to different colours due to different amounts of repulsion between electrons.

If **isoelectronic** (same number of electrons) transition metal ions complexes are considered, such as $[Mn(H_2O)_6]^{2+}$ and $[Fe(H_2O)_6]^{3+}$ (both metal ions have five 3d electrons) then there will be a greater amount of splitting of d orbitals in $[Fe(H_2O)_6]^{3+}$. A higher nuclear charge on the metal ion (26+ for Fe and 25+ for Mn) for the same number of electrons causes the ligands to be pulled in more closely in an Fe^{3+} complex, so that there is greater repulsion between the ligand electrons and the d electrons of the transition metal ion – and therefore greater splitting of the d orbitals.

Oxidation number

The same metal has different colours in different oxidation states. For example:

$[Fe(H_2O)_6]^{2+}$(aq) is pale green and

$[Fe(H_2O)_6]^{3+}$(aq) is pale violet.

There are two reasons for this:
- the electron configurations of the ions are different
- a higher charge on the metal ion causes the ligands to be pulled in more closely, so that there is greater repulsion between the ligand electrons and the d electrons of the transition metal ion – and therefore greater splitting of the d orbitals.

Nature of the ligand

The same metal ion can exhibit different colours with different ligands. This is mainly because of the different splitting of the d orbitals caused by different ligands.

Ligands can be arranged into a **spectrochemical series** according to how much they cause the d orbitals to split:

$$I^- < Br^- < Cl^- < F^- < OH^- < H_2O < NH_3 < CO \approx CN^-$$

So a chloride ion causes greater splitting of the d orbitals than an iodide ion, and an ammonia molecule causes greater splitting of the d orbitals than a water molecule.

$[Cu(NH_3)_4(H_2O)_2]^{2+}$ has a larger energy gap between the two sets of d orbitals than $[Cu(H_2O)_6]^{2+}$ and absorbs a shorter wavelength (higher frequency) of light (Figure **3.39**). $[Cu(NH_3)_4(H_2O)_2]^{2+}$(aq) is dark blue/violet and absorbs more in the yellow–green (higher frequency) region of the visible spectrum.

A full explanation of the spectrochemical series is difficult at this level. The fact that fluoride ions cause greater splitting of d orbitals than iodide ions can be explained in terms of **charge density** (charge per unit volume) of the ligand – both F^- and I^- have the same charge but the F^- ion is much smaller and therefore causes greater repulsion of the metal ion d electrons and greater splitting of the d orbitals. This explanation cannot, however, be extended to the rest of the spectrochemical series – as can be seen by the fact that CO, a neutral ligand, causes greater splitting of d orbitals than negatively charged ligands that would be expected to have a higher charge density. However, the spectrochemical series can

> Ligands that cause greater splitting of d orbitals are called stronger field ligands.

> A water molecule is more polar than an ammonia molecule and there would be expected to be a higher charge density on the O in H_2O than the N in NH_3.

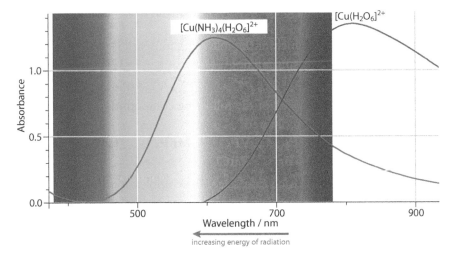

Figure 3.39 $[Cu(H_2O)_6]^{2+}$(aq) is blue and absorbs mostly at the red–orange end of the spectrum. $[Cu(NH_3)_4(H_2O)_2]^{2+}$(aq) is dark blue/violet and absorbs more in the yellow–green (shorter wavelength) region of the visible spectrum.

Why π donation/acceptance should cause changes in the splitting of d orbitals requires a more advanced treatment (using ligand field theory or molecular orbital theory) of the bonding in transition metal complexes.

be explained to a certain extent in terms of π bonding (see Topic **4**) between the ligand and the transition metal ion. Such π overlap of a lone pair on F^- with ligand d orbitals causes the splitting of the d orbitals to be reduced. Ligands that have extra lone pairs of electrons (beyond that required to bond to the transition metal ion), such as water, should cause less splitting of d orbitals than those without extra lone pairs (such as ammonia) due to this π electron donation effect. The CO ligand is a π acceptor and this is why it causes greater splitting of d orbitals.

Nature of science

Models are abstract constructions that allow us to visualise or make sense of theories about particular scientific phenomena. Here we are using a very simple model, in terms of promotion of an electron from a low-energy d orbital to a higher energy d orbital, to explain the colour of transition metal complexes. This model has limited value because it explains *some* aspects of colour but not others. For instance, it does not explain anything about the intensities of the colours absorbed, why transition metal complexes can absorb more than one wavelength of light or why complexes such as MnO_4^- are coloured.

There are, however, much more sophisticated models that have greater explanatory power. Models are a very powerful tool in understanding the real world and making predictions. Models may be qualitative, as here, or quantitative and developed in terms of mathematical equations. When developing a model there is always a balance between simplicity and explanatory/predictive power.

? Test yourself

9 Give the full electron configuration of the following ions:
  a Ni^{2+} c V^{3+}
  b Co^{3+} d Mn^{4+}

10 Give the oxidation number of the transition metal in each of these complex ions:
  a $[Ni(H_2O)_6]^{2+}$ e $[Co(NH_3)_5Br]^{2+}$
  b $[Fe(CN)_6]^{4-}$ f $[Co(NH_3)_4Br_2]^+$
  c $[MnCl_4]^{2-}$ g $[FeO_4]^{2-}$
  d $[Co(NH_3)_6]^{3+}$ h $Ni(CO)_4$

11 Which of the following ions will be paramagnetic?
  Cu^{2+} Fe^{2+} Sc^{3+} Co^{3+} Mg^{2+}

12 Which of the following compounds are likely to be coloured?
  TiF_4 VF_5 MnF_3 CoF_2
  $CuBr$ $ZnCl_2$

13 In each case, select the complex ion that would be expected to absorb the longer wavelength of light:
  a $[Co(H_2O)_6]^{2+}$ or $[Co(H_2O)_6]^{3+}$
  b $[Fe(H_2O)_6]^{2+}$ or $[Fe(NH_3)_6]^{2+}$

Exam-style questions

1   The element in group 13 and period 5 is:

    A   Y   **B**   In   **C**   Tl   **D**   P

2   Which of the following properties decrease in value down group 17?

    A   electronegativity and first ionisation energy
    B   melting point and electronegativity
    C   melting point and atomic radius
    D   ionic radius and first ionisation energy

3   Which of the following is a transition element?

    A   Te   **B**   Np   **C**   Ta   **D**   Sm

4   Which of the following forms an alkaline solution when added to water?

    A SO_3 **B** Na_2O **C** P_4O_{10} **D** SiO_2

5   Which of the following is true for two elements in the same group in the periodic table?

    A   they have the same physical properties
    B   they have similar chemical properties
    C   they have the same electronegativity
    D   they have the same number of shells of electrons

6   A non-metallic element, X, forms a gaseous oxide with the formula X_2O that reacts with water to form an acidic solution. The element X is most likely to be:

    A   Na   **B**   S   **C**   Cl   **D**   P

HL 7   Which of the following is **not** a characteristic property of transition metals?

    A   they form complex ions
    B   they have full d subshells
    C   they exhibit more than one oxidation number in compounds
    D   they form coloured compounds

HL 8   What is the oxidation number of chromium in $K_3[Cr(CN)_6]$?

    A   +6   **B**   −6   **C**   +3   **D**   −3

HL 9   Which of the following complex ions would be expected to be colourless?

    A $[Ni(H_2O)_6]^{2+}$ **C** $[Cu(H_2O)_6]^{2+}$
    B $[Zn(H_2O)_6]^{2+}$ **D** $[Co(H_2O)_6]^{2+}$

HL **10** Which compound will be diamagnetic?

      A $CrCl_3$ **B** $CuCl_2$ **C** $TiCl_4$ **D** VCl_2

11 **a** The atomic and ionic radii of some elements are given in the table:

Element	Atomic radius/pm	Ionic radius/pm
Na	186	98
Al	143	45
Cl	99	181
K	231	133

      i Explain why the atomic radius of aluminium is smaller than that of sodium.   **[2]**

     ii Explain why the ionic radius of aluminium is smaller than its atomic radius but the atomic radius of chlorine is larger than its atomic radius.   **[4]**

   iii Explain why the ionic radius of potassium is smaller than that of chlorine.   **[2]**

  b **i** Write equations for the first electron affinity of chlorine and the first ionisation energy of magnesium. **[2]**

     ii Explain why the first electron affinity of chlorine is more exothermic than the first electron affinity of bromine.   **[2]**

   iii Explain why chlorine has a higher first ionisation energy than magnesium.   **[2]**

HL **12** **a** Write the full electron configuration of:

      i a Cu atom;   **ii** a Cu^{2+} ion   **[2]**

  b One characteristic property of transition metals is that they can form complex ions with ligands. Explain what is meant by the term *ligand*.   **[2]**

  c What is the oxidation number of copper in the complex ion $[CuCl_4]^{2-}$?   **[1]**

  d Explain whether CuBr is diamagnetic or paramagnetic.   **[2]**

  e **i** Explain why solutions containing the complex ion $[Cu(H_2O)_6]^{2+}$ are coloured.   **[3]**

     ii Explain why solutions containing the complex ion $[Cu(NH_3)_2]^+$ are colourless.   **[2]**

   iii When concentrated hydrobromic acid is added to an aqueous solution of copper(II) sulfate the following reaction occurs:

      $[Cu(H_2O)_6]^{2+} + Br^- \rightarrow [Cu(H_2O)_5Br]^+ + H_2O$

      State and explain any differences in the wavelength of light absorbed by $[Cu(H_2O)_6]^{2+}$ and $[Cu(H_2O)_5Br]^+$

        [2]

Summary

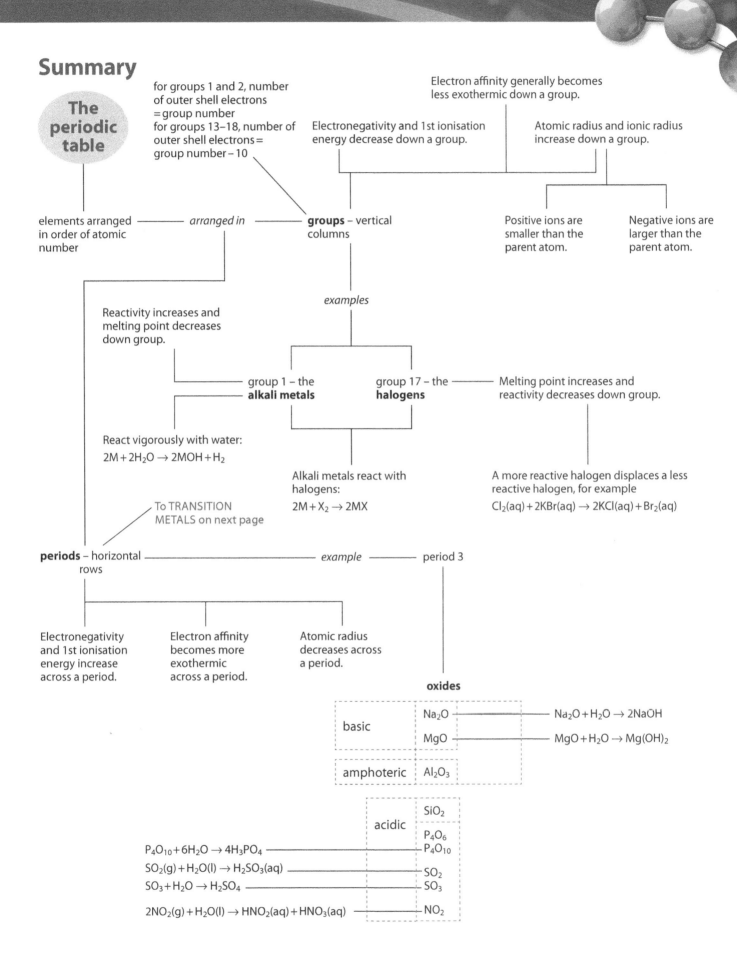

The periodic table

for groups 1 and 2, number of outer shell electrons = group number
for groups 13–18, number of outer shell electrons = group number – 10

Electron affinity generally becomes less exothermic down a group.

Electronegativity and 1st ionisation energy decrease down a group.

Atomic radius and ionic radius increase down a group.

elements arranged in order of atomic number —— *arranged in* —— **groups** – vertical columns

Positive ions are smaller than the parent atom.

Negative ions are larger than the parent atom.

examples

Reactivity increases and melting point decreases down group.

group 1 – the **alkali metals**

group 17 – the **halogens** —— Melting point increases and reactivity decreases down group.

React vigorously with water:
$2M + 2H_2O \rightarrow 2MOH + H_2$

To TRANSITION METALS on next page

Alkali metals react with halogens:
$2M + X_2 \rightarrow 2MX$

A more reactive halogen displaces a less reactive halogen, for example
$Cl_2(aq) + 2KBr(aq) \rightarrow 2KCl(aq) + Br_2(aq)$

periods – horizontal rows —— *example* —— period 3

Electronegativity and 1st ionisation energy increase across a period.

Electron affinity becomes more exothermic across a period.

Atomic radius decreases across a period.

oxides

basic
Na_2O —— $Na_2O + H_2O \rightarrow 2NaOH$
MgO —— $MgO + H_2O \rightarrow Mg(OH)_2$

amphoteric Al_2O_3

acidic
SiO_2
P_4O_6
P_4O_{10}
SO_2
SO_3
NO_2

$P_4O_{10} + 6H_2O \rightarrow 4H_3PO_4$ ——
$SO_2(g) + H_2O(l) \rightarrow H_2SO_3(aq)$ ——
$SO_3 + H_2O \rightarrow H_2SO_4$ ——
$2NO_2(g) + H_2O(l) \rightarrow HNO_2(aq) + HNO_3(aq)$ ——

Summary – continued

HL

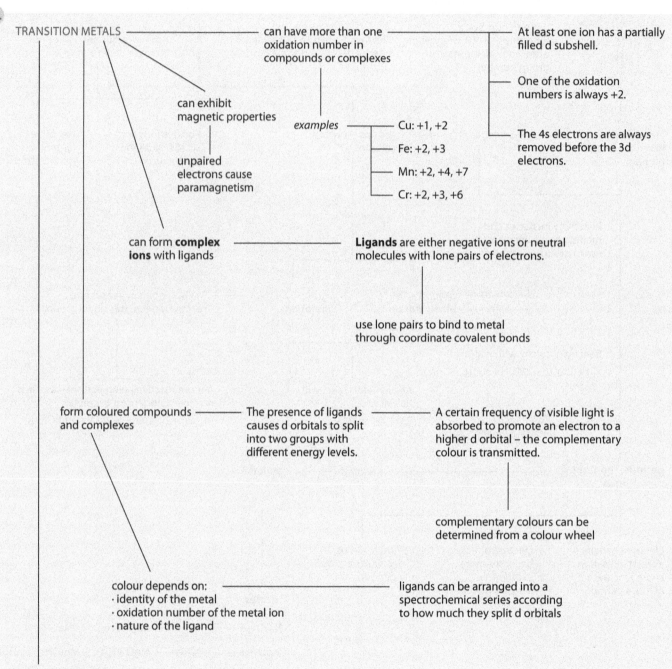

TRANSITION METALS

can have more than one oxidation number in compounds or complexes

- At least one ion has a partially filled d subshell.
- One of the oxidation numbers is always +2.
- The 4s electrons are always removed before the 3d electrons.

examples
- Cu: +1, +2
- Fe: +2, +3
- Mn: +2, +4, +7
- Cr: +2, +3, +6

can exhibit magnetic properties

unpaired electrons cause paramagnetism

can form **complex ions** with ligands

Ligands are either negative ions or neutral molecules with lone pairs of electrons.

use lone pairs to bind to metal through coordinate covalent bonds

form coloured compounds and complexes

The presence of ligands causes d orbitals to split into two groups with different energy levels.

A certain frequency of visible light is absorbed to promote an electron to a higher d orbital – the complementary colour is transmitted.

complementary colours can be determined from a colour wheel

colour depends on:
· identity of the metal
· oxidation number of the metal ion
· nature of the ligand

ligands can be arranged into a spectrochemical series according to how much they split d orbitals

can act as catalysts in form of elements or compounds

Chemical bonding and structure 4

Bonding

Compounds can be divided into two main classes according to the bonding in them – they are either **ionic** or **covalent**. The type of bonding present can usually be deduced by looking at the formula of the compound. Covalent compounds are those formed between two or more non-metallic elements, whereas ionic compounds are usually formed between a metallic element and a non-metallic one. For example, NaCl is an ionic compound but CH_4 is covalent.

There are some compounds for which the distinction is not so clear. For instance, ammonium chloride does not contain any metallic elements but has ionic bonding between the ammonium ions (NH_4^+) and the chloride ions (Cl^-). In addition to this, within the NH_4^+ ion there is covalent bonding.

Generally, as a rough rule of thumb, elements that are close together in the periodic table form covalent compounds but elements that are far apart in the periodic table form ionic compounds. Thus elements from groups 1 and 17 combine to form ionic compounds (CsF being the most ionic) but elements from groups 14, 15, 16 and 17 combine to form covalent compounds. This is discussed in terms of electronegativity on page **129**.

Ionic compound: **metal** and **non-metal**
Covalent compound: two or more **non-metals**

4.1 Ionic bonding and structure

Positive ions are usually formed by metallic elements by the loss of valence (outer shell) electrons. For example, magnesium loses the two electrons in its highest energy level (outer shell) to form a 2+ ion:

$$Mg \rightarrow Mg^{2+} + 2e^-$$

Negative ions are usually formed by non-metallic elements by the gain of electrons. For example, oxygen gains two electrons to fill up its outer shell (highest occupied energy level):

$$O + 2e^- \rightarrow O^{2-}$$

At the simplest level, when elements in the main groups of the periodic table (groups 1, 2, 13 (to a certain extent), 15, 16 and 17) form ions, electrons are gained or lost to achieve the electron configuration of the nearest **noble gas**. That is, electrons are gained or lost to make an ion that is **isoelectronic** (same number of electrons) with the nearest noble gas.

Another way of saying this is that electrons are lost or gained to achieve a full outer shell of electrons. Although this is true for the first 20 elements, it is not generally true after that because of the existence of transition metals (and d orbitals).

Learning objectives

- Recognise the formulas of ionic compounds
- Understand how ions are formed and recall the formulas of some common ions
- Work out the formulas of ionic compounds from the charges on the ions
- Describe the structure of sodium chloride as an example of an ionic lattice
- Explain the physical properties of ionic compounds in terms of structure and bonding

> The transition metals can form more than one ion. For instance, iron can form iron(II), Fe^{2+}, and iron(III), Fe^{3+}.

An Fe^{2+} ion has 24 electrons, an Fe^{3+} ion has 23 electrons, and neither is isoelectronic with a noble gas.

The 4s electrons are lost first when a transition metal atom forms an ion; therefore the electron configurations of Fe^{2+} and Fe^{3+} are:

Fe^{2+} $1s^2 2s^2 2p^6 3s^2 3p^6 3d^6$
Fe^{3+} $1s^2 2s^2 2p^6 3s^2 3p^6 3d^5$

These do not have noble gas electron configurations.

> The number of electrons lost or gained is determined by the electron configuration of an atom.

The number of electrons lost by elements in groups 1 and 2 when they form ions is given by the group number. For instance, magnesium, in group 2 of the periodic table, has two outer shell electrons, and therefore forms a **2+** ion (Figure **4.1**). The number of electrons lost by elements in group 13 is given by the group number −10. For example, aluminium forms a 3+ ion.

The number of electrons gained by elements in groups 15 to 17 when they form ions is given by 18 minus the group number. For instance oxygen, in group 16 of the periodic table, has six outer shell electrons and gains $(18 - 16) = 2$ electrons to form a **2−** ion (Figure **4.2**).

Note: metal atoms do not 'want' to form ions with noble gas electron configurations – for instance, it takes the input of over $2000\,\text{kJ}\,\text{mol}^{-1}$ of energy to remove two electrons from a magnesium atom to form an Mg^{2+} ion.

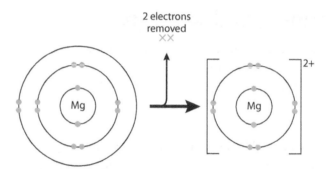

Figure 4.1 The Mg^{2+} ion is isoelectronic with the noble gas atom neon.

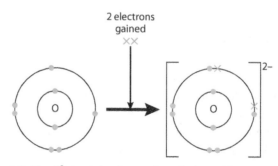

Figure 4.2 The O^{2-} ion is isoelectronic with the noble gas atom neon.

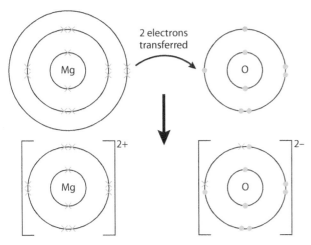

Figure 4.3 Electron transfer in ionic bonding.

When an ionic compound is formed, electrons are transferred from one atom to another to form positive and negative ions. Electrons cannot be created or destroyed; therefore, the total number of electrons lost must always equal the total number gained. You can see this in Figure **4.3**.

In the formation of magnesium fluoride (Figure **4.4**), one magnesium atom must combine with two fluorine atoms, because magnesium loses two electrons when it forms an ion but each fluorine atom can gain only one electron. So the formula of magnesium fluoride is MgF_2 (Figure **4.4**).

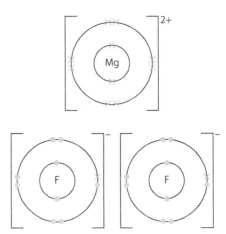

Figure 4.4 The ions in magnesium fluoride.

The formulas of ions

The formulas of commonly encountered positive ions are given in Table **4.1** and of some negative ions in Table **4.2**.

Ion	Symbol	Ion	Symbol	Ion	Symbol
lithium	Li^+	magnesium	Mg^{2+}	iron(III)	Fe^{3+}
sodium	Na^+	calcium	Ca^{2+}	aluminium	Al^{3+}
potassium	K^+	barium	Ba^{2+}		
rubidium	Rb^+	iron(II)	Fe^{2+}		
caesium	Cs^+	copper(II)	Cu^{2+}		
silver	Ag^+	zinc	Zn^{2+}		
ammonium	NH_4^+	nickel(II)	Ni^{2+}		
hydrogen	H^+				

Table 4.1 Positive ions.

Ion	Symbol	Ion	Symbol	Ion	Symbol
fluoride	F^-	oxide	O^{2-}	nitride	N^{3-}
chloride	Cl^-	sulfide	S^{2-}	phosphate(V)	PO_4^{3-}
bromide	Br^-	carbonate	CO_3^{2-}		
iodide	I^-	sulfate(VI)	SO_4^{2-}		
hydroxide	OH^-				
hydrogencarbonate	HCO_3^-				
nitrate(V)	NO_3^-				

Table 4.2 Negative ions. Ions such as NO_3^- are most properly named including the oxidation number (i.e. nitrate(V)) but this is often omitted.

The brackets around the OH are essential because, otherwise the formula would read $MgOH_2$, and would indicate the presence of only one oxygen atom.

Rubies and sapphires are mainly aluminium oxide.

Electrostatic attraction: positive charges attract negative charges and vice versa.

Working out the formulas of ionic compounds

To work out the formula of aluminium fluoride, we need to consider the number of electrons lost by aluminium and gained by fluorine. Aluminium is in group 13 of the periodic table and so forms a 3+ ion by the loss of three electrons, whereas fluorine, in group 17, gains one electron to generate a full outer shell and a 1− ion. The three electrons transferred from the aluminium must be gained by three separate fluorine atoms, therefore the formula of aluminium fluoride is AlF_3.

Another way to look at this is to consider that the overall charge on the compound is zero, so the 3+ charge on the Al^{3+} ion must be cancelled out by the $3 \times 1-$ charge on $3F^-$ ions, i.e. $Al^{3+}(F^-)_3$.

Similarly, the formula of magnesium hydroxide is $Mg(OH)_2$, where the 2+ charge on the Mg^{2+} ion is cancelled out by the $2 \times 1-$ charge on two OH^- ions, i.e. $Mg^{2+}(OH^-)_2$.

A shortcut to working out formulas is to switch over the charges on the ions, for example:

$$Al^{3+} \diagdown\!\!\!\!\diagup\, O^{2-} \longrightarrow Al_2O_3$$

Transition metal ions can form more than one ion, so the oxidation number of the ion is usually given with the name. For example, iron can form iron(II) sulfate or iron(III) sulfate. The Roman numeral in brackets indicates the oxidation number of the ion, which is the same as its charge. Iron(III) sulfate, therefore, contains the Fe^{3+} ion and has the formula $Fe_2(SO_4)_3$. Iron(II) sulfate contains the Fe^{2+} ion and has the formula $FeSO_4$.

The ionic bond and ionic crystals

An ionic bond is an **electrostatic** attraction between oppositely charged ions.

A crystal of sodium chloride consists of a giant lattice of Na^+ and Cl^- ions (Figure **4.5**). All the Na^+ ions in the crystal attract all the Cl^- ions,

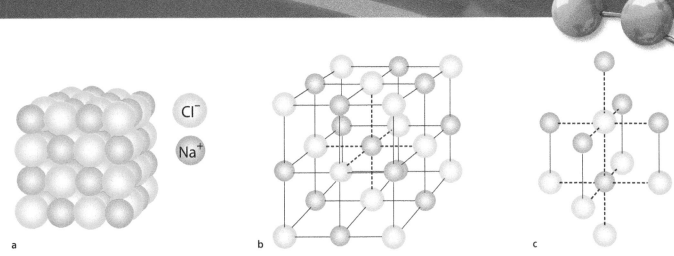

Figure 4.5 **a** A space-filling diagram of the NaCl lattice. The lattice keeps on going in three dimensions – only a tiny part of the structure is shown. **b** An expanded view of the NaCl lattice. **c** Each Na⁺ ion is surrounded by six Cl⁻ ions in an octahedral array and vice versa. The coordination number (number of nearest neighbours) of each ion is 6.

and vice versa, and it is this attraction between oppositely charged ions that holds the crystal together. These electrostatic forces are strong, so it is difficult to break apart the **lattice structure**.

This is a **giant structure** – there are no individual molecules of sodium chloride. Instead, the bonding extends fairly uniformly throughout the whole structure.

It is often better to avoid the term 'ionic bond' and talk about 'ionic bonding' as a whole – the dotted lines in the diagrams of the lattice structures are **not** ionic bonds, they are simply there to give a better idea of the shape. Ionic bonding (electrostatic attractions between oppositely charged ions) extends throughout the whole structure in all directions – there are no individual ionic bonds.

Physical properties of ionic compounds

Melting points and boiling points

Ionic compounds usually have high melting points and boiling points. For instance, sodium chloride has a melting point of 801 °C and a boiling point of over 1400 °C, while magnesium oxide has a melting point of over 2800 °C and a boiling point of about 3600 °C. This means that ionic substances are usually solids at room temperature.

The high melting and boiling points of ionic solids are due to the strong electrostatic forces between the oppositely charged ions. When an ionic solid is melted, the electrostatic forces throughout the giant lattice must be broken and, because these are so strong, a lot of energy is required.

Magnesium oxide has a much higher melting point than sodium chloride because the electrostatic attractions between the 2+ and 2− ions in the magnesium oxide lattice are much stronger than those between the 1+ and 1− ions in sodium chloride. A higher temperature is required to provide sufficient energy to separate the ions in magnesium oxide. The force between ions is proportional to the product of the charges (all other things being equal).

Volatility

Ionic solids have low **volatility** (refers to how readily a substance evaporates). The volatility of ionic substances is low because the electrostatic forces between the ions are strong.

Electrical conductivity of ionic compounds

Ionic substances do not conduct electricity when solid. In the solid state, the ions are held tightly in position in the lattice structure so that they are not free to move around (other than vibrate).

When an ionic substance is melted the ions are able to move freely throughout the liquid. Positive ions can move towards a negative electrode and negative ions towards a positive electrode, so allowing the conduction of electricity.

Solubility in water

Ionic substances are often soluble in water. Water is a polar solvent, and energy is released when the ions are hydrated by being surrounded (ion-dipole attractions) by water molecules (Figure **4.6**). This energy pays back the energy required to break apart the ionic lattice.

Electrical conductivity in ionic solutions

Aqueous solutions (solutions made with water) of ionic substances conduct electricity. This is because the ions are free to move around (Figure **4.7**).

Solubility in non-polar solvents

Ionic solids are not usually soluble in non-polar solvents such as hexane. This is because a great deal of energy is required to break apart the ionic lattice and this is not paid back by the energy released when the non-polar solvent forms interactions with the ions (London forces). This will be considered in more detail on page **157**.

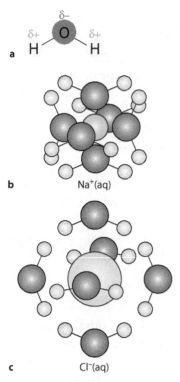

Figure 4.6 a A water molecule is polar.
b A hydrated sodium ion.
c A hydrated chloride ion.

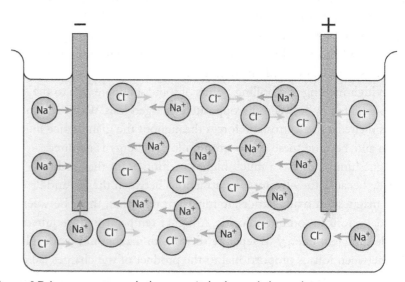

Figure 4.7 Ions move towards the oppositely charged electrode.

Nature of science

Scientists use theories to explain and predict the properties of substances. We can use our theories of the bonding in ionic compounds to explain their properties, for instance, the facts that they have high melting points, conduct electricity when molten etc.

Our theory of bonding in ionic substances predicts that all ionic substances have a common set of properties. Therefore we can predict the properties of any new ionic compound and use experimental observations to test the theory.

? Test yourself

1 State whether the following compounds have ionic or covalent bonding:
    a LiF   **b** CF_4 **c** CaO   **d** NH_3 **e** PCl_3 **f** $CuCl_2$

2 Write the formulas of the following compounds:
    a magnesium oxide
    b barium sulfate
    c calcium hydroxide
    d sodium oxide
    e strontium sulfide
    f aluminium oxide

    g lithium nitride
    h magnesium phosphate
    i magnesium fluoride
    j potassium sulfate
    k ammonium carbonate

    l silver sulfide
    m silver nitrate
    n ammonium chloride
    o copper(II) nitrate
    p rubidium carbonate

4.2 Covalent bonding

Single covalent bonds

> Covalent bonding occurs when atoms share electrons, and a covalent bond is the electrostatic attraction between a shared pair of electrons and the nuclei of the atoms that are bonded.

At the simplest level, electrons are shared to allow the atoms being bonded to achieve a full outer shell of electrons (noble gas electron configuration). One example is the formation of methane (CH_4).

A carbon atom has four electrons in its outer shell (highest occupied energy level) (Figure **4.8**). It will share four electrons so that the number of electrons in its outer shell is eight.

Learning objectives

- Understand that a covalent bond is formed when electrons are shared
- Understand the relationship between bond strength and bond length
- Understand what is meant by electronegativity
- Predict whether a bond will be polar or not

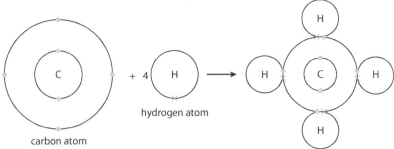

Figure 4.8 The covalent bonding in CH_4.

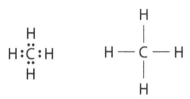

Figure 4.9 Two different types of Lewis structure for methane. In the second one, the line between the atoms represents a shared pair of electrons: that is, a covalent bond.

Figure 4.11 Lewis structures for water.

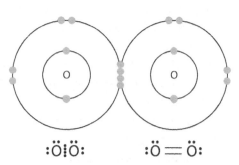

:Ö:Ö: :Ö══Ö:

Figure 4.12 Representations of the covalent bonding in O_2.

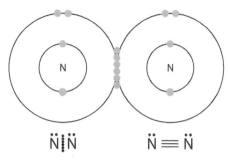

N̈:N̈   N̈══N̈

Figure 4.13 Representations of the covalent bonding in N_2.

A hydrogen atom has one electron in its outer shell and will share one electron with another atom (carbon in this case) to generate an outer shell containing two electrons, i.e. a full outer shell (the same number of electrons as helium).

In methane, the shared electrons are in the outer shell of both atoms making up the bond and so each atom has a full outer shell of electrons (Figure **4.8**).

Two alternative ways of representing the covalent bonding in methane are shown in Figure **4.9**. These are **Lewis structures**.

Consider the bonding in water (H_2O) – an oxygen atom has six electrons in its outer shell and so will share two electrons to generate a full outer shell (eight electrons) or an octet of electrons. The covalent bonding in water is shown in Figure **4.10**.

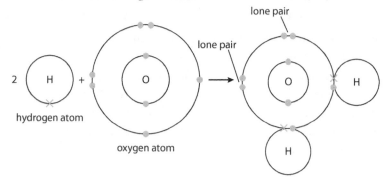

Figure 4.10 Covalent bonding in water. All atoms have full outer shells of electrons.

The pairs of electrons in the outer shell of atoms that are not involved in the covalent bonds are called **lone pairs** of electrons. There are two lone pairs in the outer shell of the oxygen atom in a molecule of water. Alternative representations of the covalent bonding in water are shown in Figure **4.11**.

Multiple covalent bonds

It is possible for more than one pair of electrons to be shared between two atoms. A double bond results from the sharing of two pairs of electrons, and a triple bond arises when three pairs of electrons are shared. For example, the covalent bonding in an oxygen molecule, O_2, is shown in Figure **4.12**. Four electrons – that is, two pairs – are shared and so there is a double bond between oxygen atoms.

The covalent bonding in the nitrogen molecule, N_2, is shown in Figure **4.13**. A nitrogen atom shares three electrons to generate a full outer shell, so six electrons are shared when two nitrogen atoms combine to form N_2. Three shared pairs of electrons between two atoms form a triple bond. More examples of molecules with multiple bonding are shown in Figure **4.14**.

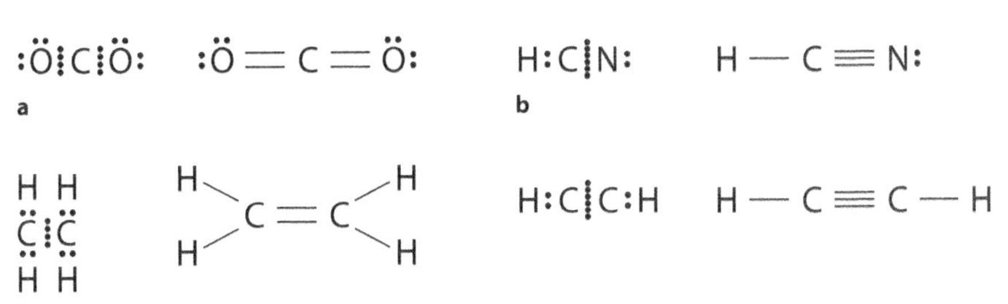

:Ö:C:Ö: :Ö=C=Ö: H:C⁝N: H—C≡N:

a   **b**

H H
C⁝C   ⟍H   H⟍   ⟋H   H:C⁝C:H   H—C≡C—H
H H   C=C
                H⟋   ⟍H

c   **d**

Figure 4.14 Covalent bonding in **a** CO_2, **b** HCN, **c** C_2H_4 (ethene) and **d** C_2H_2 (ethyne).

What holds the atoms together in a covalent bond?

> A covalent bond is the electrostatic interaction between the positively charged nuclei of both atoms and the shared pair of electrons.

The electrons are negatively charged and because the shared electrons are attracted to the nuclei (positively charged) of both atoms simultaneously, this holds the atoms together (Figure **4.15**).

As can be seen from Table **4.3**, triple bonds are stronger than double bonds, which are stronger than a single bond. This is because the attraction of the two nuclei for three electron pairs (six electrons) in a triple bond is greater than the attraction for two electron pairs (four electrons) in a double bond, which is greater than the attraction for one electron pair (two electrons) in a single bond (Figure **4.16**).

It should also be noted that triple bonds are shorter than double bonds, which are shorter than single bonds. This is, again, due to stronger attraction between the bonding electrons and the nuclei when there are more electrons in the bond.

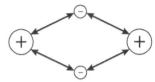

Figure 4.15 The positively charged nuclei attract the negatively charged electrons in the bond.

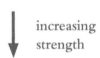

> **Strength**
> single bonds
> double bonds   increasing
> triple bonds   strength

Bond	Length/nm	Bond energy/kJ mol^{-1}
C–C	0.154	348
C=C	0.134	612
C≡C	0.120	837
C–O	0.143	360
C=O	0.122	743

Table 4.3 The relationship between number of bonds and bond length/strength.

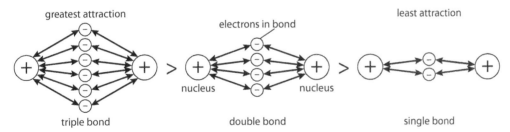

Figure 4.16 The more electrons that make up a covalent bond, the greater the attraction between the electrons and the nuclei, and therefore the stronger the bond.

	Group 14			Group 17	
Bond	Length/nm	Energy/ kJ mol^{-1}	Bond	Length/nm	Energy/ kJ mol^{-1}
C–C	0.154	348	Cl–Cl	0.199	242
Si–Si	0.235	226	Br–Br	0.228	193
Ge–Ge	0.241	188	I–I	0.267	151

Table 4.4 The relationship between length of bonds and bond strength.

In general, when we are comparing just single bonds, the longer the bond the weaker it is. Data for two groups in the periodic table are shown in Table **4.4**.

If we consider the data for group 14, it can be seen that the single bond between the elements gets weaker as the bond gets longer. This is because, as the atoms get bigger, the electron pair in the covalent bond is further away from the nuclei of the atoms making up the bond. If the electron pair is further away from the nuclei, it is less strongly attracted and the covalent bond is weaker (Figure **4.17**). A similar trend can be seen down group 17.

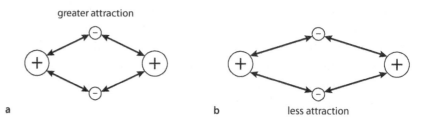

a   b   less attraction

Figure 4.17 The bond in **a** is a shorter bond in which the bonding electrons are closer to the nuclei than in **b**.

> Trends should really only be compared down a group because elements in the same group have the same number of outer shell electrons and therefore any effects due to effective nuclear charge or shielding are most similar. In general, comparisons such as this are most useful and valid when similar molecules, bonds or compounds are considered.

Polarity

Electronegativity

In a covalent bond between two different atoms, the atoms do not attract the electron pair in the bond equally. How strongly the electrons are attracted depends on the size of the individual atoms and their nuclear charge.

In an F_2 molecule, the two fluorine atoms attract the electrons in the bond equally and so the electrons lie symmetrically (Figure **4.18**). This molecule is non-polar.

> **Electronegativity** is a measure of the attraction of an atom in a molecule for the electron pair in the covalent bond of which it is a part.

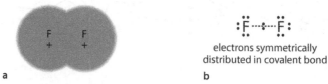

a   b

Figure 4.18 a The electron density in F_2; **b** F_2 is a non-polar molecule.

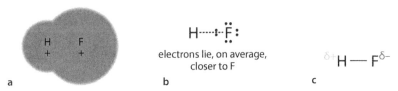

$$H\cdots\vdots\ddot{F}:$$

electrons lie, on average,
closer to F

$$^{\delta+}H-F^{\delta-}$$

a   b   c

Figure 4.19 a The electron density in HF; **b** HF is a polar molecule; **c** usually represented like this

δ− indicates a small negative charge.

However, in HF, fluorine is more electronegative than hydrogen and attracts the electrons in the H–F bond more strongly than the hydrogen atom does. The electrons in the bond lie closer to the fluorine than to the hydrogen (Figure **4.19**) – H–F is a polar molecule. The unsymmetrical distribution of electron density results in small charges on the atoms. Fluorine is δ− because the electrons in the bond lie closer to F, whereas electron density has been pulled away from hydrogen, so it is δ+.

Pauling electronegativities

There are various scales of electronegativity and it is important to realise that, although they are derived from physical quantities (such as bond energies), the numbers themselves are not physical quantities – they have no units and must only be used in a comparative way. The most commonly used scale of electronegativity is that developed by Linus Pauling. The electronegativity values for some elements, worked out using the Pauling method, are shown in Table **4.5**.

Noble gases do not have electronegativity values because they do not form compounds.

Non-metals have higher electronegativities than metals.

H 2.1							
Li 1.0	Be 1.5		B 2.0	C 2.5	N 3.0	O 3.5	F 4.0
Na 0.9	Mg 1.2		Al 1.5	Si 1.8	P 2.1	S 2.5	Cl 3.0
K 0.8	Ca 1.0		Ga 1.6	Ge 1.8	As 2.0	Se 2.4	Br 2.8
Rb 0.8	Sr 1.0		In 1.7	Sn 1.8	Sb 1.9	Te 2.1	I 2.5

Table 4.5 Pauling electronegativities for some elements.

Electronegativity generally decreases down a group and increases across a period. Trends in electronegativity were explained in Topic **3**.

Atoms with similar electronegativities will form covalent bonds. Atoms with widely different electronegativities will form ionic bonds. The difference in electronegativity can be taken as a guide to how ionic or how covalent the bond between two atoms is likely to be.

Exam tip
You do not need to learn these electronegativity values, but you should be aware of trends and recognise highly electronegative atoms such as N, Cl, O and F.

Exam tip
To remember the trends in electronegativity, just remember that fluorine is the atom with the highest electronegativity. Electronegativity must then increase across a period towards fluorine and decrease down a group from fluorine.

Exam tip

If asked about this, assume
that the bonding between
two elements is ionic if the
difference is more than 1.7,
and covalent if the difference is
less than 1.7.

Linus Pauling related the electronegativity difference between two atoms to the ionic character of a bond. He suggested that an electronegativity difference of 1.7 corresponded to 50% ionic character in a bond and reasoned that a higher electronegativity difference than this corresponded to a structure that was more ionic than covalent, whereas if the difference is less than 1.7, the bonding is more covalent than ionic. This is a useful idea, but it must be used with great caution. For instance, KI (electronegativity difference 1.7) would come out of this discussion as having 50% ionic and 50% covalent character, and NaI (difference of 1.6) would appear to be mostly covalent, whereas both behave as predominantly ionic compounds. Pauling, in his original discussion of this, was actually referring to diatomic molecules and not to macroscopic compounds.

? Test yourself

3 Arrange the following atoms in order of electronegativity (lowest first):

Cl   O   H   Br   Na

Learning objectives

- Understand what is meant by a
  coordinate covalent bond
- Work out Lewis structures for
  molecules and ions
- Work out the shapes of
  molecules and ions with up to
  four electron domains
- Predict bond angles in molecules
  and ions
- Predict whether a molecule will
  be polar or non-polar
- Describe the structures and
  bonding of giant covalent
  substances
- Explain the physical properties
  of giant covalent substances in
  terms of structure and bonding

4.3 Covalent structures

The octet rule

In Subtopic **4.2** you met the concept that atoms in covalent bonds have a tendency to have a full valence shell with a total of eight electrons (or two in the case of hydrogen). This is known as the **octet rule**. In most covalent molecules and polyatomic ions (for example, NH_4^+ or CO_3^{2-}), each atom has an octet in its outer shell.

It is not, however, always the case that the formation of covalent bonds results in each atom attaining an octet of electrons. In BF_3 (Figure **4.20**), boron has only six electrons in its outer shell. This is because a boron atom has only three electrons in its outer shell and can therefore share a maximum of three electrons. Similarly, in $BeCl_2$ the beryllium atom only has a total of 4 electrons in its outer shell.

Figure 4.20 Covalent bonding in BF_3.

Figure 4.21 Covalent bonding in SF_6.

It is also possible for atoms to have more than eight electrons in their outer shell. For instance, in SF_6 (Figure **4.21**) the sulfur atom has 12 electrons in its outer shell, and this is described as sulfur having expanded its octet. Only elements in period 3 and beyond (periods 4, 5 …) can expand their octet. Expansion of the octet will be discussed in more detail in the Higher Level section later (see Subtopic **4.6**).

Exam tip
Expansion of the octet is very rarely encountered at Standard Level.

Coordinate covalent bonds (dative covalent bonds)

> A coordinate covalent bond is a type of covalent bond in which both electrons come from the same atom.

Once a coordinate covalent bond has been formed, it is identical to an 'ordinary' covalent bond. For example, NH_4^+ can be formed when H^+ becomes bonded to NH_3:

$$NH_3 + H^+ \rightarrow NH_4^+$$

H^+ does not have any electrons with which to form a covalent bond, but NH_3 has a lone pair of electrons that can be used to form a covalent bond (Figure **4.22**).

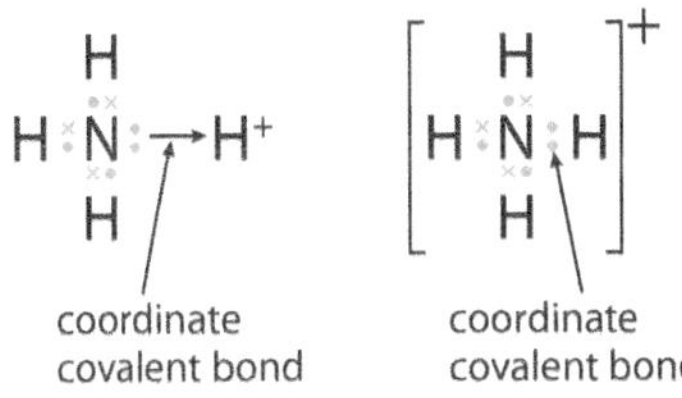

A coordinate covalent bond is sometimes shown as an arrow (Figure **4.23a**). Once it has been formed, a coordinate bond is, however, the same as any other covalent bond. The ammonium ion can be represented as shown in Figure **4.23b**, in which no distinction is made between the individual bonds.

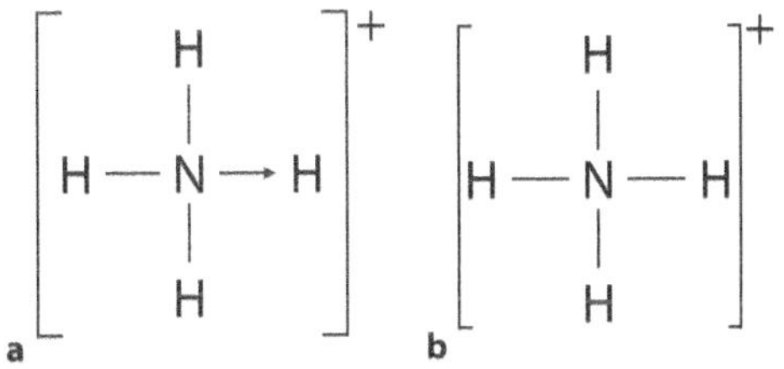

H_3O^+ is formed when a lone pair of electrons is donated from the O in H_2O to the H^+:

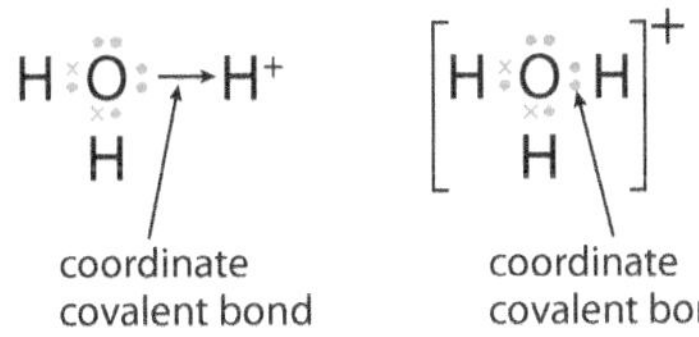

coordinate covalent bond   coordinate covalent bond

NH_3 and BF_3 can combine to form an **adduct** (two molecules bonded together):

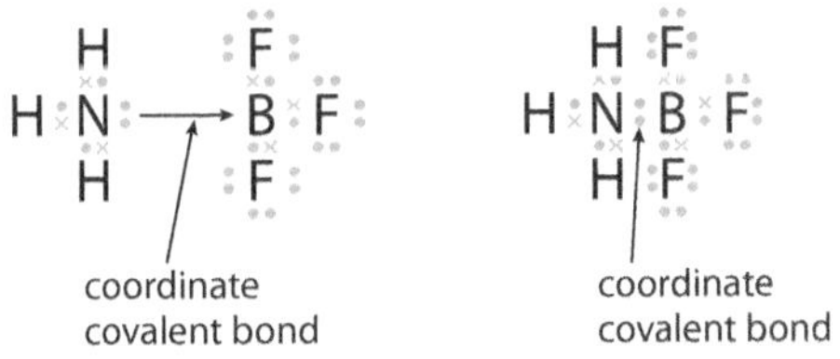

coordinate covalent bond   coordinate covalent bond

In BF_3 there are only six electrons in the outer shell of the boron atom – so there is space for the boron to accept a pair of electrons.

Carbon monoxide
Normally carbon shares four electrons to form four covalent bonds, and oxygen shares two to form two covalent bonds. If a carbon atom combines with an oxygen atom with the formation of two covalent

Figure 4.22 Coordinate covalent bonding in NH_4^+.

Figure 4.23 The ammonium ion, **a** with the coordinate bond shown and **b** with no distinction between the types of bonds.

This is an example of a Lewis acid–base reaction (see Subtopic **8.2**).

Figure 4.24 The structure of carbon monoxide if two 'ordinary' covalent bonds were formed.

Figure 4.25 The structure of carbon monoxide if two 'ordinary' covalent bonds and one coordinate covalent bond were formed.

Figure 4.26 Other ways of showing the bonding in carbon monoxide.

> Dots and crosses can be used in Lewis structures to indicate where the electrons originally came from.

Figure 4.27 Lewis structures for H_2O, CO_2 and CO, showing all outer shell electrons.

> Note: the outer atom(s) is (are) usually the more electronegative atom (except when hydrogen is also present in the molecule).

Figure 4.28 The Lewis structure for NF_3.

bonds, we get the structure shown in Figure **4.24**. However, in this structure, although the oxygen atom has a full outer shell (octet), the carbon atom has only six electrons in its outer shell.

Both atoms can attain an octet if the oxygen atom donates a pair of electrons to carbon in the formation of a coordinate covalent bond. There is now a triple bond between the two atoms, made up of two 'ordinary' covalent bonds and one coordinate covalent bond (Figure **4.25**). Both atoms have a lone pair of electrons.

Other ways of representing the bonding in carbon monoxide are shown in Figure **4.26**.

> Coordinate covalent bonds are important in the formation of transition metal complexes.

Lewis (electron dot) structures

Lewis structures are diagrams showing all the valence (outer shell) electrons in a molecule (or ion). Examples of Lewis structures are shown in Figure **4.27**. Electrons may be shown individually as dots or crosses, or a line may be used to represent a pair of electrons, as in the Lewis structure of CO shown in Figure **4.27**.

Rules for working out Lewis structures:

1 Make sure that the *outer* atoms have eight electrons in their outer shell (except, of course, hydrogen which should have two). This is done by using single bonds, double bonds, triple bonds and + or − charges.

2 If the central atom is from period 2 it should have no more than eight electrons in its outer shell. It should generally (but not always) have a noble gas electron configuration.

3 If the central atom is from period 3 it may have up to 18 electrons in its outer shell.

Let us consider a few examples and go through the steps for drawing the Lewis structures.

NF₃

The central atom is nitrogen and the three outer atoms are fluorine. A fluorine atom has seven electrons in its outer shell and therefore only needs to form one single bond to have a full outer shell.

Therefore, in order for each F atom to have eight electrons in its outer shell, three single bonds between the nitrogen and the fluorine must be formed.

Figure **4.28** shows the Lewis structure for NF_3. Each outer atom has eight electrons in the outer shell and the central atom also has eight electrons in its outer shell. All valence electrons are shown.

CO_3^{2-}

Carbon is the central atom and the oxygens are the outer atoms. An oxygen atom has six electrons in its outer shell and therefore needs two more for a full octet. Oxygen can attain a full outer shell by forming a double bond or by forming a single bond and gaining another electron as a negative charge (Figure **4.29a**).

Because the carbon atom has only four electrons in its outer shell, it does not have enough electrons to form three double bonds with the oxygens, but it does have enough electrons to form one double bond and two single bonds (Figure **4.29b**).

The overall charge on the ion must be included and square brackets drawn around the ion. The Lewis structure for CO_3^{2-} is shown in Figure **4.30**.

As a final check, you should make sure that the central atom has no more than eight electrons in its outer shell, because it is from period 2.

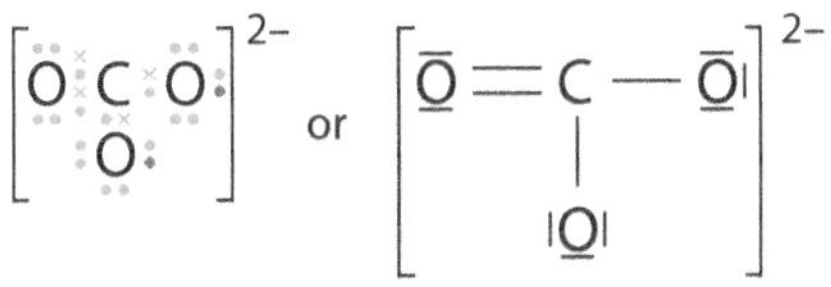

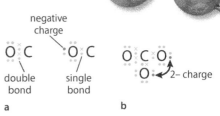

Figure 4.29 a Carbon and oxygen can bond in two different ways. **b** The resultant structure of the CO_3^{2-} ion.

The electron for the negative charge will come from sodium, calcium, etc. – the CO_3^{2-} ion cannot be formed in isolation – a metal atom must lose electrons at the same time to form the 2– charge.

Figure 4.30 The Lewis structure for CO_3^{2-}.

Resonance structures

There is more than one way of drawing the structure of the CO_3^{2-} ion (Figure **4.31**) depending on where we put the C=O (and lone pairs).

There are many other molecules/ions in which it is possible to draw more than one Lewis structure which differ only in the position of double bonds (and lone pairs). Consider ozone O_3 (Figure **4.32**).

Similarly, in the structure of benzene, C_6H_6, the double bonds can be drawn in two ways as shown in Figure **4.33**.

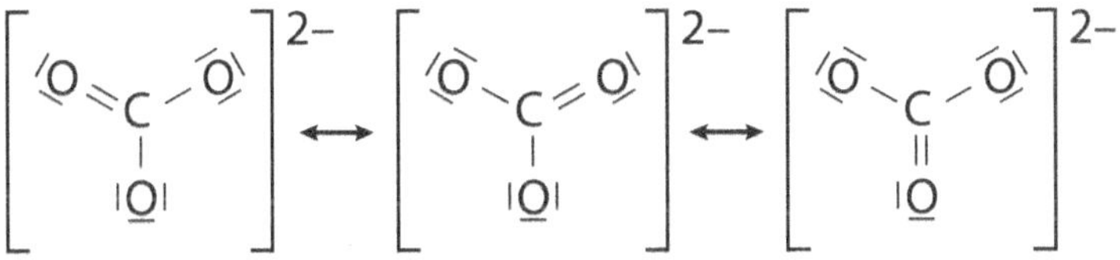

Exam tip
You must remember to show lone pairs. A very common mistake is to forget to show the lone pairs on the outer atoms. The brackets and the charge are also essential.

Figure 4.31 Resonance structures of the carbonate ion.

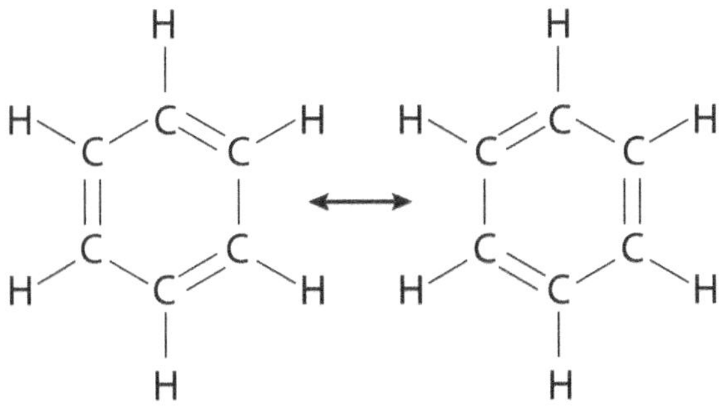

Figure 4.32 Resonance structures of ozone.

Figure 4.33 Resonance structures of benzene.

These individual structures differ only in the position of the double bond and are called **resonance structures**. In CO_3^{2-}, the carbon–oxygen bonds are equal in length, in O_3 the oxygen–oxygen bond lengths are equal and in benzene the carbon–carbon bonds are all identical. This cannot be explained by looking at just one of the structures (in which double bonds would be shorter than single bonds). The actual structure is described as a hybrid of the individual resonance structures. This is shown by the double headed arrow and is considered more fully in the Higher Level section below.

Two possible Lewis structures for the same molecule

SO_2

Two different Lewis structures are possible for SO_2 depending on whether the octet on S is expanded or not.

Approach 1:
Sulfur does not expand its octet

Sulfur has six electrons in its outer shell and can, therefore, form a maximum of two normal covalent bonds. If these are both formed to the same oxygen atom, we get the structure shown in Figure **4.34a**. However, in this structure there is no bond between the sulfur and the right-hand oxygen atom. The structure can be completed by a coordinate bond between the sulfur and the second oxygen, so that all atoms have eight electrons in their outer shell (Figure **4.34b**).

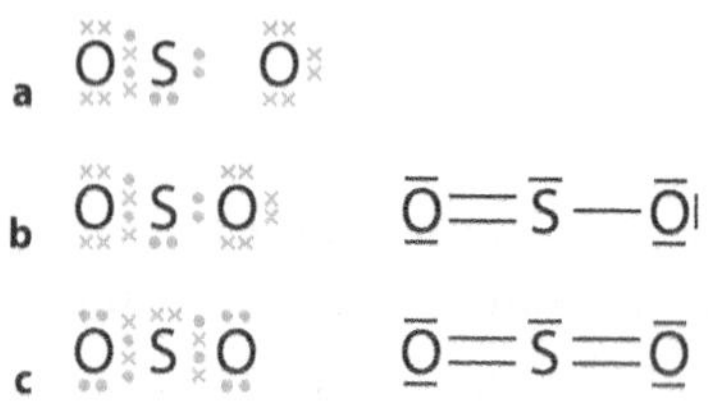

Approach 2:
S expands its octet

Each oxygen atom can achieve an octet by forming a double bond.

A sulfur atom has six electrons in its outer shell and when it forms two double bonds it will have ten electrons in its outer shell and is said to have expanded its octet (Figure **4.34c**).

This is possible for elements in period 3 and beyond because they have d orbitals available for bonding – the maximum number of electrons in the third shell (the outer shell in sulfur) is 18.

Both approaches to working out Lewis structures are valid, although more detailed considerations which involve working out the formal charge on each atom suggest that the second structure is a better representation of the bonding in the molecule (see the Higher Level section later). The bond lengths in SO_2 also suggest the presence of two double bonds in the molecule.

Alternative method for working out Lewis structures

This approach is useful for working out the Lewis structures of molecules/ions just containing period 2 atoms.

1 Add up the total number of valence electrons of all the atoms in the molecule/ion.
2 Divide by two to get the total number of valence electron pairs.
3 Each pair of electrons is represented by a line.

Figure 4.34 Two possible Lewis structures for SO_2. **a** Sulfur forms two covalent bonds with one oxygen atom. **b** In the first approach, the structure is completed with a coordinate bond between sulfur and the second oxygen atom. **c** In the second approach, sulfur expands its octet, forming a double bond with the second oxygen atom.

4 Arrange the lines (electron pairs) so that all the atoms are joined together by at least single bonds and the outer atoms have full outer shells, i.e. are connected to four lines.

5 Rearrange the lines (electron pairs) so that every period 2 atom has four pairs of electrons. The outer atoms already have four pairs, so this should normally involve moving only lone pairs so that they become bonding pairs of electrons.

NO_3^-

total no. electrons = 5   + 3×6 +   1   = $24e^-$
             N $3 \times O$ negative charge

There are therefore $\frac{24}{2} = 12$ pairs of electrons, shown here by 12 lines.

| | | | | | | | | | | | |

Three must be now used to join all the atoms together:

$$
\begin{array}{c}
O \\
| \\
O - N - O
\end{array}
$$

This leaves nine lines (electron pairs) that will be distributed as lone pairs of electrons on the O atoms (as O is the outer atom) to give each an octet.

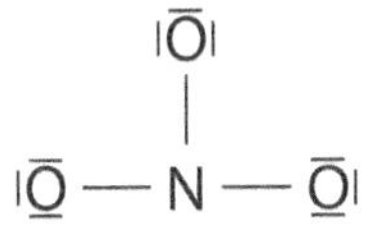

$$
\begin{array}{c}
|\overline{O}| \\
| \\
|\overline{O} - N - \overline{O}|
\end{array}
$$

Now each O is 'attached to four lines' (has four pairs of electrons in its outer shell) and all the electrons have been used. The nitrogen, however, has only three pairs of electrons in its outer shell, and one of the lines needs to be moved from being a lone pair on the O to make a double bond between the N and an O. This does not change the number of electrons in the outer shell of the O but will increase the number of electrons in the outer shell of the N to eight.

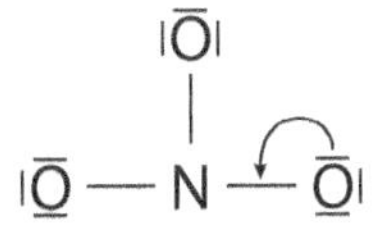

$$
\begin{array}{c}
|\overline{O}| \\
| \\
|\overline{O} - N \curvearrowright \overline{O}|
\end{array}
$$

The final Lewis structure for the NO_3^- ion is shown in Figure **4.35**. As with CO_3^{2-} earlier (with which it is isoelectronic), more than one resonance structure can be drawn with the double bond in different positions.

This is a purely mechanical technique for working out the Lewis structure and does not really involve any understanding of the bonding in the ion. The Lewis structure showing the electrons as dots and crosses is probably clearer (Figure **4.36**).

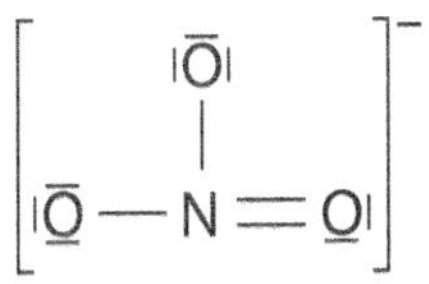

$$
\left[
\begin{array}{c}
|\overline{O}| \\
| \\
|\overline{O} - N = O|
\end{array}
\right]^-
$$

Figure 4.35 The Lewis structure for NO_3^-.

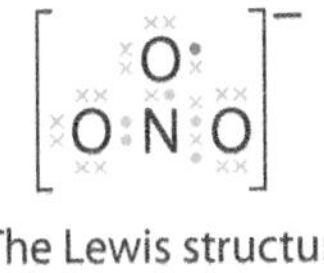

$$
\left[
\begin{array}{c}
O \\
O : N : O
\end{array}
\right]^-
$$

Figure 4.36 The Lewis structure of NO_3^- with the electrons shown individually.

NO_2^-

$$\text{total no. electrons} = \underset{N}{5} + \underset{2 \times O}{2 \times 6} + \underset{\text{negative charge}}{1} = 18e^-$$

There are therefore $\dfrac{18}{2} = 9$ pairs of electrons

$$\mathrm{|\,|\,|\,|\,|\,|\,|\,|\,|}$$

The atoms are joined together using two lines:

$$\mathrm{O - N - O}$$

More lines are added to the outer atoms to give each an octet:

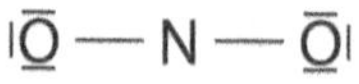

$$\mathrm{I\underline{O} - N - \underline{O}I}$$

So far, only eight lines (electron pairs) have been used and the last line must be put as a lone pair on the N. It cannot be put anywhere else, as the Os already have four electron pairs each.

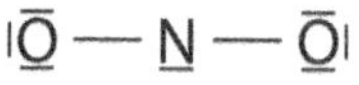

$$\mathrm{I\underline{O} - \underline{N} - \underline{O}I}$$

N only has three electron pairs and so a lone pair is moved from an O to form a double bond:

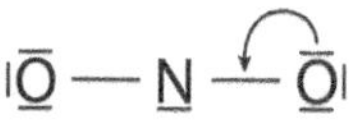

$$\mathrm{I\underline{O} - \underline{N} \overset{\frown}{\,-\,} \underline{O}I}$$

This results in the Lewis structure for NO_2^- shown in Figure **4.37**.

It is important to note that the Lewis structure could be drawn the other way round and that this is entirely equivalent (Figure **4.38**).

O_3

Ozone is isoelectronic with NO_2^- and has the same Lewis structure:

$$\mathrm{I\underline{O} - \underline{O} = \underline{O}I}$$

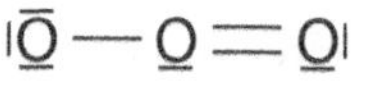

NO_2^+ and CO_2

NO_2^+ and CO_2 are isoelectronic and have the same Lewis structures (Figures **4.39** and **4.40**).

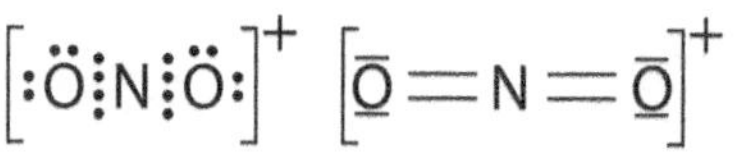

This technique for drawing Lewis structures is useful for molecules/ions involving just period 2 elements, but it also works for compounds involving period 3 elements (and below) if the idea that the atoms do not expand their octet (approach 1, above) is adopted.

Nature of science

Scientists often make careful observations and look for patterns and trends in data, however, noticing discrepancies in these trends can lead to advances in knowledge. For instance, the observation that compounds containing metallic elements had different properties to those containing just non-metallic elements led to the idea of different types of structure and bonding.

Exam tip
NO_2 has been asked about in examinations. It has an odd number of electrons and therefore an unpaired electron. The Lewis structure can be shown with an N=O double bond and an N → O coordinate bond. The unpaired electron would then be on the N.

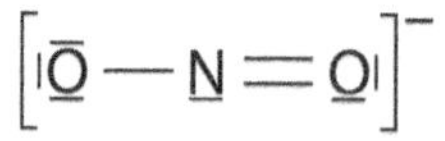

$$\left[\mathrm{I\underline{O} - \underline{N} = \underline{O}I}\right]^-$$

Figure 4.37 The Lewis structure for NO_2^-.

$$\left[\mathrm{I\underline{O} = \underline{N} - \underline{O}I}\right]^-$$

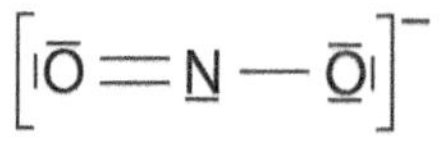

Figure 4.38 An alternative resonance structure for NO_2^-.

$$\left[\mathrm{:\ddot{O}:N:\ddot{O}:}\right]^+ \quad \left[\mathrm{\underline{O} = N = \underline{O}}\right]^+$$

Figure 4.39 The Lewis structures for NO_2^+.

$$\mathrm{:\ddot{O}:C:\ddot{O}:} \qquad \mathrm{\underline{O} = C = \underline{O}}$$

Figure 4.40 The Lewis structures for CO_2.

Lewis introduced the idea of atoms sharing electrons and distinguished between compounds with ionic and covalent bonding (although he did not originally call them that). He also considered the unequal sharing of electrons and polar compounds. Linus Pauling built on this work by developing the concept of electronegativity to quantify, to a certain extent, the unequal sharing of electrons.

? Test yourself

4 Work out Lewis structures for the following:

    a H_2S **d** COF_2 **g** FNO **j** PCl_4^+

    b PCl_3 **e** HCN **h** N_2H_4 **k** NO^+

    c CCl_4 **f** CS_2 **i** H_2O_2 **l** OCN^-

5 Some harder ones – these are likely only to be encountered at Higher Level:

    a XeF_4 **c** BrF_5 **e** N_2O **g** I_3^-

    b PCl_6^- **d** ClF_3 **f** SF_4 **h** N_3^-

Shapes of molecules: valence shell electron pair repulsion theory

We can predict the shapes of molecules using the **valence shell electron pair repulsion (VSEPR) theory**.

> Pairs of electrons (electron domains) in the valence (outer) shell of an atom repel each other and will therefore take up positions in space to minimise these repulsions – to be as far apart in space as possible.

The pairs of electrons may be either non-bonding pairs (lone pairs) or bonding pairs (pairs of electrons involved in covalent bonds).

Basic shapes

The shape of a molecule depends on the number of electron pairs in the outer shell of the central atom. There are five basic shapes (for two to six electron pairs), which are derived from the idea of how a number of things, joined to a central point, can be arranged in space to be as far apart from each other as possible. However, first of all we will just consider molecules with up to four pairs of electrons around the central atom. The basic shapes and bond angles are shown in Table **4.6**.

More precisely, it is a question of how points can be arranged on the surface of a sphere to be as far away from each other as possible.

A double bond is made up of two pairs of electrons, but these electron pairs are constrained to occupy the same region of space. A double bond (or a triple bond) therefore behaves, in terms of repulsion, as if it were just one electron pair and so it is better to talk about the number of **electron domains** – where an electron domain is either a lone pair, the electron pair that makes up a single bond or the electrons pairs that together make up a multiple bond.

No. electron domains	Shape	Diagram	Bond angle	Example
2	linear	Y — X — Y	180°	$BeCl_2(g)$
3	trigonal planar	Y⟍X⟋Y with Y below X	120°	BF_3
4	tetrahedral	Y above X, Y⟍X⟋Y	109.5°	CH_4

Table **4.6** Basic molecule shapes and bond angles.

a   b

Figure 4.41 a This symbol indicates a bond coming out of a plane. **b** This symbol indicates a bond going into a plane.

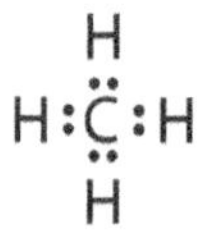

H:C:H (with H above and below)

Figure 4.42 The Lewis structure for CH_4.

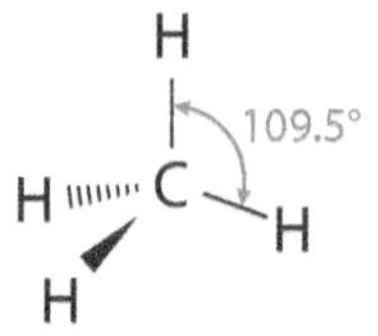

Figure 4.43 CH_4 is tetrahedral. (109.5°)

H:N:H (with H above and below)

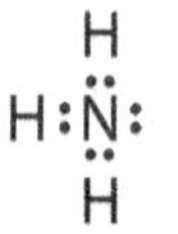

Figure 4.44 The Lewis structure for NH_3.

Figure 4.45 The shape of the NH_3 molecule is based on a tetrahedron, but the actual shape is trigonal pyramidal. (107.3°)

How to predict the shapes of molecules

1 Draw a Lewis structure for the molecule or ion.
2 Count up the number of electron pairs (bonding pairs and lone pairs) in the outer shell of the central atom. A multiple bond counts as a single electron pair, because the electrons are constrained to occupy the same region of space. This gives the total number of electron domains.
3 Look at Table **4.7** to get the basic shape (spatial arrangement of the electron domains).
4 A lone pair is just an electron pair in the outer shell of an atom and, as such, it contributes to the overall shape of the molecule but cannot itself be 'seen'.
5 State the actual shape of the molecule. See Figure **4.41** for how to draw 3D shapes.

CH_4

The Lewis structure for CH_4 is shown in Figure **4.42**. The number of electron pairs in the outer shell of the central atom (C) is four, i.e. there are four electron domains. These four electron domains repel each other and take up positions in space as far away from each other as possible. The shape that allows four things to be as far away from each other as possible is **tetrahedral**. Therefore, the four electron pairs (electron domains) are arranged tetrahedrally around the C atom. The shape of the methane molecule is tetrahedral and the H–C–H bond angle is 109.5° (Figure **4.43**).

NH_3

The Lewis structure for NH_3 is shown in Figure **4.44**.

Bonding pairs of electrons: 3 (in three single bonds)
Non-bonding pairs of electrons: 1 (in one lone pair)
Electron domains: 4

Because these four electron domains repel each other and take up positions in space to be as far apart as possible, the electron pairs are distributed in a tetrahedral arrangement. The basic shape is tetrahedral – but a lone pair is just a pair of electrons in the outer shell of the central atom and, although it repels the other pairs of electrons and influences the shape, it cannot be 'seen'. The shape adopted by the atoms is therefore **trigonal pyramidal** (a pyramid with a triangular base) (Figure **4.45**).

The H–N–H bond angle here is smaller than in a perfect tetrahedron, because the lone pair repels the bonding pairs of electrons more than they repel each other. This will be considered in more detail later.

Here, we talk about it not being possible to 'see' a lone pair of electrons, but of course it is not possible to see **any** of these molecules. There are various techniques for determining the shapes of molecules experimentally. Probably the most important of these is X-ray crystallography, in which a crystal of the substance is placed in an X-ray diffractometer and, from the position and intensity of diffracted X-ray beams, the shape of the molecules, all the angles and bond lengths can be calculated. The lone pairs of electrons cannot be detected by this technique. Do we know or believe the shapes of molecules stated here? Which ways of knowing do we use to interact with this microscopic world? In biology, microscopes are used to 'view' things that are too small to be seen with the naked eye. Is there a difference between the use of a microscope to interact with this invisible world and the use of an X-ray diffractometer? What about the use of electron microscopes?

CO_2

The Lewis structure for CO_2 is shown in Figure **4.46**.

:Ö:C:Ö:

Figure 4.46 The Lewis structure for CO_2.

Bonding pairs of electrons: 4 (in two double bonds)
Non-bonding pairs of electrons: 0
Electron domains: 2 (because a double bond counts as one electron domain – the four electrons that make up the double bond are constrained to occupy the same region of space)

Because these two electron domains repel each other and take up positions in space to be as far apart as possible, the electron domains are distributed in a linear arrangement (Figure **4.47**), and the shape adopted by the atoms is **linear**.

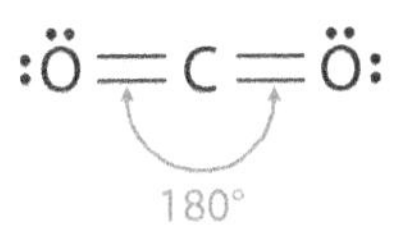

:Ö═C═Ö:

180°

Figure 4.47 The linear CO_2 molecule.

SO_2

The Lewis structure for SO_2 is shown in Figure **4.48**.

Ö:S:Ö

Figure 4.48 The Lewis structure for SO_2.

Bonding pairs of electrons: 4 (in two double bonds)
Non-bonding pairs of electrons: 1 (one lone pair on S)
Electron domains: 3 (because a double bond counts as one electron domain)

Because these three electron domains repel each other and take up positions in space to be as far apart as possible, the electron domains are distributed in a trigonal planar (flat triangle) arrangement (Figure **4.49a**). The lone pair on sulfur is just a pair of electrons in its outer shell and cannot be 'seen'. The shape adopted by the atoms is therefore **bent** (also called 'angular', or 'V-shaped').

Again the bond angle is slightly less than the ideal angle in a trigonal planar structure (120°), and this results from the extra repulsion of a lone pair, so that the bonding pairs are pushed closer together and the bond angle is reduced.

If the Lewis structure had been drawn differently (Figure **4.49b**), with sulfur not expanding its octet, this would have made no difference to the predicted shape.

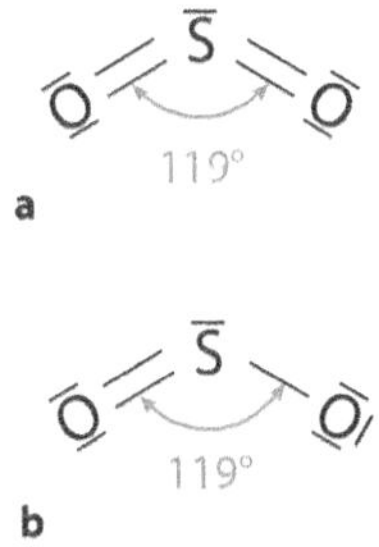

a

119°

b

119°

Figure 4.49 a The bent SO_2 molecule **b** alternative model for SO_2.

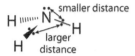

smaller distance

larger distance

Figure 4.50 Lone pairs lie closer to the central nucleus than bonding pairs.

Note: both SO_2 and H_2O have bent structures, but the SO_2 structure is based on trigonal planar whereas the H_2O structure is based on tetrahedral. This means that the bond angle is larger in SO_2.

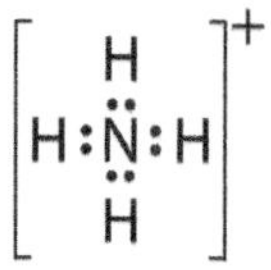

Figure 4.51 The Lewis structure for NH_4^+.

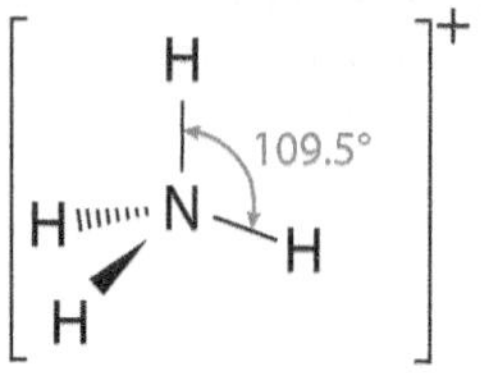

Figure 4.52 Tetrahedral NH_4^+.

Lone pairs and bond angles

The order of repulsion strength for pairs of electrons is:

lone pair–lone pair > lone pair–bonding pair > bonding pair–bonding pair

This is because lone pairs are held closer (Figure **4.50**) to the central nucleus than are bonding pairs (lone pairs are in the outer shell of the central atom, whereas a bonding pair can be imagined as being, on average, halfway between the bonded atoms). The lone pairs are thus closer to the bonding pairs of electrons than the bonding pairs are to each other and repel them more strongly. This means that the repulsion due to lone pairs causes other bond angles to become smaller.

Consider CH_4, NH_3 and H_2O, each of which has four electron pairs in the outer shell of the central atom (Table **4.7**). The basic shape is the arrangement of the electron pairs in the outer shell of the central atom. The more lone pairs present, the smaller the H–X–H bond angle. This is due to greater repulsion from lone pairs than from bonding pairs of electrons. Two lone pairs cause greater repulsion than one, so the bond angle gets smaller as the number of lone pairs increases.

Molecule	Lewis structure	Bonding pairs	Lone pairs	Basic shape	Actual shape	H–X–H bond angle
CH_4	H:C:H (with H above and below)	4	0	tetrahedral	tetrahedral	109.5°
NH_3	H:N: (with H above and below)	3	1		trigonal pyramidal	107.3°
H_2O	:O:H (with H below)	2	2		bent	104.5°

Table 4.7 Structural characteristics of CH_4, NH_3 and H_2O.

Predicting the shapes of ions

The approach to predicting the shapes of ions is exactly the same as for neutral molecules.

NH_4^+

The Lewis structure for NH_4^+ is shown in Figure **4.51**.

Bonding pairs of electrons: 4
Non-bonding pairs of electrons: 0
Electron domains: 4

Because these four electron domains repel each other and take up positions in space to be as far apart as possible (Figure **4.52**), the electron pairs are distributed in a tetrahedral arrangement.

H₃O⁺

The Lewis structure for H_3O^+ is shown in Figure **4.53**.

> Bonding pairs of electrons: 3
> Non-bonding pairs of electrons: 1
> Electron domains: 4

Because these four electron domains repel each other and take up positions in space to be as far apart as possible, the electron pairs are distributed in a tetrahedral arrangement. One of the electron pairs is a lone pair, so the actual shape of the ion is trigonal pyramidal (Figure **4.54**). This structure is based on tetrahedral (bond angle 109.5°) with one lone pair, so a bond angle of about 107° for the H–O–H bond could be predicted (the molecule is isoelectronic with ammonia).

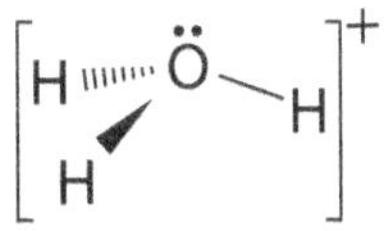

NO₂⁻

The Lewis structure for NO_2^- is shown in Figure **4.55**.

> Bonding pairs of electrons: 3 (one single bond and one double bond)
> Non-bonding pairs of electrons: 1
> Electron domains: 3 (because a double bond counts as one electron domain)

Because these three electron domains repel each other and take up positions in space to be as far apart as possible, the electrons pairs are distributed in a trigonal planar arrangement. One of the electron pairs is a lone pair, so the actual shape of the ion is bent (Figure **4.56**). This structure is based on trigonal planar (bond angle 120°) with one lone pair, so a bond angle of about 117° could be predicted.

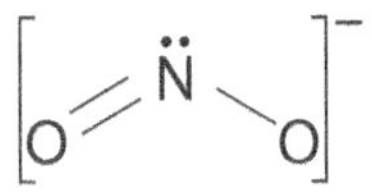

Molecules with more than one central atom

The approach to predicting the shapes of molecules with more than one central atom is the same as for other molecules, except that each 'central atom' must be considered separately.

N₂H₄

The Lewis structure for N_2H_4 is shown in Figure **4.57**. In this molecule, the two N atoms are 'central atoms' and each one must be considered separately (Table **4.8**).

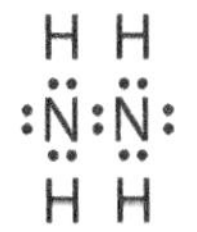

Left-hand nitrogen	Right-hand nitrogen
bonding pairs of electrons: 3	bonding pairs of electrons: 3
non-bonding pairs of electrons: 1	non-bonding pairs of electrons: 1
electron domains: 4	electron domains: 4

Table 4.8 Counting electron domains around the nitrogen atoms in N_2H_4.

It can be seen (Figure **4.58**) that the arrangement of electron pairs around each nitrogen is tetrahedral and, with one lone pair on each nitrogen, the shape about each N atom is trigonal pyramidal.

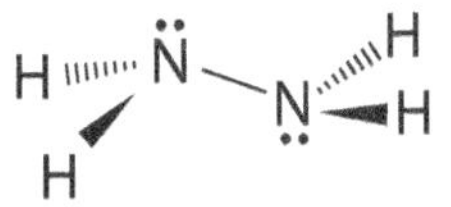

$$\left[H : \overset{\cdot\cdot}{O} : H \atop H \right]^{+}$$

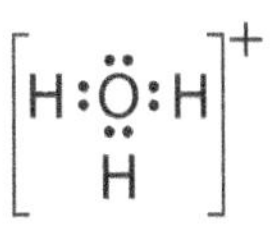

Figure 4.53 The Lewis structure for H_3O^+.

Figure 4.54 Trigonal pyramidal H_3O^+.

> No definitive value for the bond angle in H_3O^+ exists. Values have been measured, but they depend on the actual compound i.e. the negative ion present.

$$\left[:\overset{\cdot\cdot}{\underset{\cdot\cdot}{O}} : N : \overset{\cdot\cdot}{\underset{\cdot\cdot}{O}} \right]^{-}$$

Figure 4.55 The Lewis structure for NO_2^-.

Figure 4.56 NO_2^- is bent.

> No definitive value for the bond angle in NO_2^- exists. Values have been measured, but they depend on the actual compound.

$$\begin{matrix} H & H \\ :N & :N: \\ H & H \end{matrix}$$

Figure 4.57 The Lewis structure for N_2H_4.

Figure 4.58 The structure of N_2H_4.

H:C ⫶ C:H

Figure 4.59 The Lewis structure for C_2H_2.

H—C≡C—H
180° 180°

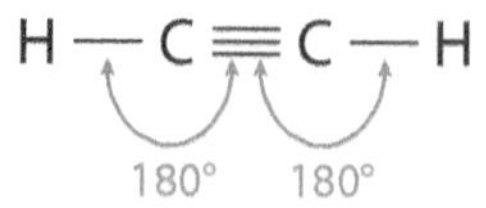

Figure 4.60 Linear C_2H_2.

'Basic shape' is the arrangement of the electron pairs around the central atom.

Extension

VSEPR gives us no information about how the two NH_2 groups in hydrazine are twisted relative to each other, and in the gas phase hydrazine adopts the conformation with the two NH_2 groups staggered relative to each other.

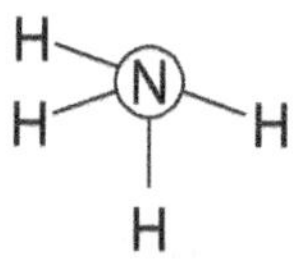

view of a hydrazine molecule
looking down the N—N bond

C_2H_2

The Lewis structure for C_2H_2 is shown in Figure **4.59**. In ethyne, each C is considered separately, but each one is identical. For each C:

Bonding pairs of electrons: 4 (one single bond and one triple bond)
Non-bonding pairs of electrons: 0
Electron domains: 2

The triple bond counts as one electron domain, so, with two electron domains around each C, the shape is linear (Figure **4.60**) about each C and therefore linear overall.

VSEPR summary

The shapes taken up by molecules depend on the number of **pairs** of electrons in the outer shell of the central atom. To take account of molecules containing multiple bonds, this is often expressed in terms of electron domains, where one multiple bond counts as one electron domain.

Pairs of electrons in the outer shell of an atom repel each other. The pairs of electrons may be either bonding pairs or lone pairs. The pairs of electrons take up positions in space to minimise repulsions in a molecule, i.e. to be as far apart from each other as possible. The basic shapes adopted are shown in Table **4.9**.

Lone pairs influence the shapes of molecules but cannot actually be 'seen'. The shapes of some specific molecules are also shown in Table **4.9**.

Lone pairs repel bonding pairs of electrons more than the bonding pairs repel each other. Therefore the presence of lone pairs results in a closing up of bond angles.

Total electron domains	Bonding pairs	Lone pairs	Basic shape	Actual shape	Example
2	2	0	linear	linear	CO_2, NO_2^+, HCN, $BeCl_2$
3	3	0	trigonal planar	trigonal planar	BF_3, SO_3, NO_3^-, CO_3^{2-}
3	2	1	trigonal planar	bent, V-shaped, angular	SO_2, O_3, NO_2^-
4	4	0	tetrahedral	tetrahedral	CCl_4, XeO_4, NH_4^+, BCl_4^-, SO_4^{2-}, PO_4^{3-}
4	3	1	tetrahedral	trigonal pyramidal	NH_3, PCl_3, XeO_3, H_3O^+, ClO_3^-, $SOCl_2$
4	2	2	tetrahedral	bent, V-shaped, angular	H_2O, SCl_2, ClF_2^+, I_3^+

Table 4.9 The basic and actual shapes of some specific molecules – see Subtopic **4.6** for information about atoms with more than four electron domains.

Polar molecules

The electronegativity difference between two atoms covalently bonded together results in the electrons lying more towards one atom than the other. We call such a bond **polar**. However, whether an overall molecule is polar also depends on the shape of the molecule.

For a molecule to be polar it must have a positive end to the molecule and a negative end. For instance HCl, NH_3 and H_2O are all polar (Figure **4.61**). These molecules all have an overall **dipole moment**, and the arrow indicates the direction of the moment.

> Although individual bonds may be polar, a molecule may be non-polar overall if, because of the symmetry of the molecule, the dipole moments of the individual bonds cancel out.

CO_2 is a non-polar molecule (Figure **4.62**). Each C=O bond is polar, because oxygen is more electronegative than carbon, but overall the dipoles cancel so that there is no overall dipole moment and the molecule is non-polar.

BF_3 is also non-polar. Again, each individual bond is polar but the dipoles cancel.

CCl_4 is non-polar, but $CHCl_3$ is polar (Figure **4.63**).

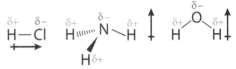

Figure 4.61 These molecules are all polar – one end of the molecule is slightly positive compared with the other end. The arrow indicates the dipole moment direction.

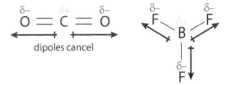

Figure 4.62 Bond polarities can cancel out.

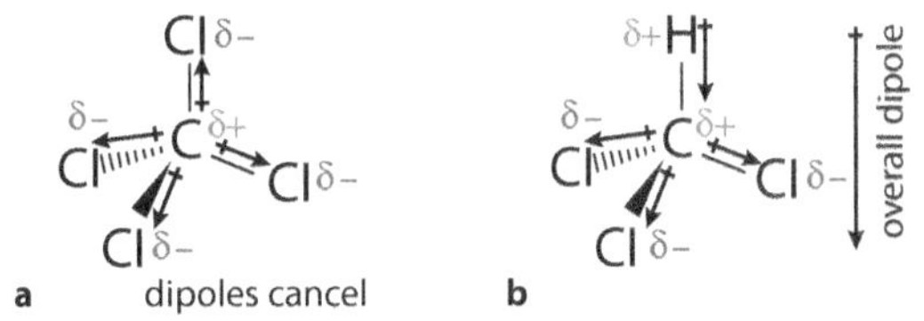

Figure 4.63 a CCl_4 is non-polar because the individual dipoles cancel. **b** $CHCl_3$ is polar because the dipoles do not cancel out; there is a positive end to the molecule and a negative end. Although the C in $CHCl_3$ is shown as δ+, it is not as positive as the H (C is more electronegative than H); therefore, the C is slightly negative compared with the H, although it is positive overall in the molecule.

Some polar and non-polar molecules are shown in Table **4.10**.

Polar	Non-polar
HCl, H_2O, NH_3, SO_2, $CHCl_3$, CH_2Cl_2, CH_3Cl_3, SCl_2, XeO_3, PCl_3, $SOCl_2$, $POCl_3$	CO_2, C_2H_2, C_2Cl_4, BF_3, XeF_2, XeF_4, SF_6, PF_5, XeO_4, SO_3
	These molecules are non-polar, because the symmetry of the molecules causes the individual dipole moments of the bonds to cancel out.

Table 4.10 Some polar and non-polar molecules.

Exam tip

The examination answer as to why CCl_4 is non-polar is 'although each individual bond is polar due to the difference in electronegativity of the atoms, because of the symmetry of the molecule, the dipoles cancel'.

Extension

Dipole moment is the product of the charge and the distance between the charges. The unit is the debye (D).

Many people talk about the scientific method and the idea of testing theories against experimental data – if the experimental results do not agree with the theory then the theory should be discounted. However, scientists do not dismiss theories so easily. The bond angle in an H_2Te molecule is 90°, which is a lot smaller than we would predict, but this does not mean that we dismiss VSEPR completely – rather, theories are modified and improved. It is rare that a theory is completely discounted – this would constitute a paradigm shift and an awful lot of evidence would have to build up to warrant such a radical change.

Nature of science

Scientists often use simplified models as a representation of the real world. A simple picture of balls (atoms) joined together by sticks (covalent bonds) allows us to visualise the molecular world. The VSEPR theory allows us to rationalise and predict the shapes of molecules. Using these fairly simple models we can explain observed properties and make predictions, however, when the models prove inadequate, more sophisticated ones must be developed.

? Test yourself

6 Work out the shapes of the following molecules or ions and predict their bond angles:

a H_2S
b PCl_3
c CF_4
d HCN
e COF_2
f CS_2
g FNO
h PCl_4^+
i OCN^-
j O_3
k C_2F_4
l NO_2^+

7 Select the polar molecules from the following list. For the polar molecules, draw diagrams showing the dipoles.

HBr HCN PH$_3$ SCl$_2$ CF$_4$
N$_2$ OCl$_2$ BCl$_3$ C$_2$Cl$_2$ H$_2$S
CH$_2$Cl$_2$

Giant covalent structures

Allotropes of carbon

Allotropes are different forms of the same element. For instance, diamond, graphite, graphene and fullerenes are all allotropes of carbon. They all contain only carbon atoms, but these atoms are joined together differently in each structure.

Diamond

Diamond has a giant covalent (macromolecular) structure. There are no individual molecules – the whole structure, continuing in three dimensions, represents one giant molecule (Figure **4.64**). Each carbon atom is joined to four others, in a **tetrahedral** array, by covalent bonds.

Diamond has a very high melting point and boiling point (about 4000 °C) because a lot of energy must be supplied to break covalent bonds (very strong) when diamond is melted/boiled. Diamond is very hard for the same reason.

Diamond does not conduct electricity, because all the electrons are held strongly in covalent bonds and are therefore not free to move around in the structure.

Diamond is not soluble in water or organic solvents because the forces between the atoms are too strong. The energy to break these covalent bonds would not be paid back when the C atoms were solvated.

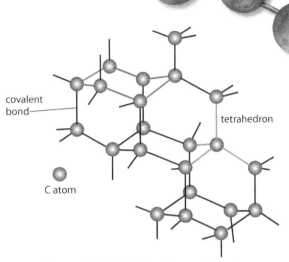

Figure 4.64 Part of the diamond structure, which is based on puckered hexagonal rings. One hexagon is highlighted in red.

Graphite

Like diamond, graphite has a giant covalent structure. Unlike diamond, however, it has a layer structure (Figure **4.65**). Each C is covalently bonded to three others in a trigonal planar array.

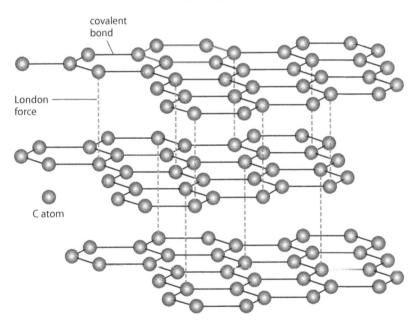

Figure 4.65 Part of the structure of graphite, which is based on planar hexagonal rings. One hexagon is highlighted in red.

There are covalent bonds between the C atoms within a layer but only London forces between the layers (some of these are shown in blue in Figure **4.65**). The presence of weak forces between the layers is usually given as the explanation that graphite is a good lubricant (used in pencils, for example) – not much force is required to separate the layers. However, it has a very high melting/boiling point because covalent bonds within the layers must be broken when it is melted/boiled.

Diamond is the hardest naturally occurring substance. It also has the highest thermal conductivity of any substance – more than five times that of copper. Both these properties make it suitable for use on drill bits. Diamond-encrusted drills can be used to drill through rock.

Substances with giant structures are sometimes called network solids.

Extension

The lubricant properties of graphite are usually explained as being due to the weak forces between layers of carbon atoms. However, graphite is a poor lubricant in a vacuum and it is now believed that the lubricant properties come from adsorbed water molecules.

Because of the strong covalent bonds between atoms, graphite is not soluble in water or non-polar solvents.

Graphite conducts electricity because each C atom forms only three covalent bonds – the extra electrons not used in these bonds (carbon has four outer shell electrons) are able to move within the layers.

Because only three covalent bonds are formed by each carbon atom in the layers, each C atom possesses one p orbital, containing one electron (Figure **4.66**), perpendicular to the plane of the layers. These p orbitals can overlap side-on to give a **π delocalised system** (see Higher Level material on page **170**) extending over the whole layer. Movement of electrons within this system allows the conduction of electricity within layers. Graphite is, however, an electrical insulator perpendicular to the plane of the layers.

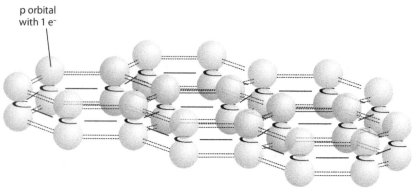

p orbital
with 1 e⁻

Figure 4.66 Delocalisation in graphite.

Graphene

Graphene is a relatively new form of carbon that consists of a single layer of graphite. Only very small pieces have been formed so far but it has some interesting properties that could make it very important in the near future as a new material.

Graphene has a very high tensile strength and would be expected to have a very high melting point because covalent bonds need to be broken to break the sheet. It is also a very good electrical (C forms only three bonds) and thermal conductor.

C_{60} fullerene (buckminsterfullerene)

The fourth allotrope of carbon that will be considered here is a molecular rather than a giant structure. It consists of individual C_{60} molecules, with covalent bonds within the molecule and London forces between the molecules (Figure **4.67**).

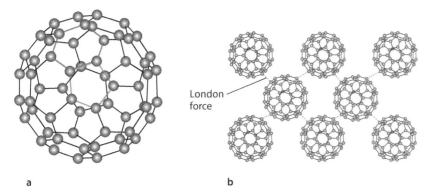

a   b

Figure 4.67 a One molecule of C_{60}. The structure is based on hexagons (one highlighted in red) and pentagons (one highlighted in blue); **b** London forces exist between C_{60} molecules in the solid state.

Diamond and graphite are **giant** structures but fullerene is a **molecular** structure. The melting points of diamond and graphite are therefore substantially higher than that of fullerene, because covalent bonds must be broken when diamond and graphite are melted but only intermolecular forces (London forces) must be overcome when fullerene melts (it actually undergoes sublimation at about $530\,°C$).

C_{60} is insoluble in water but soluble in some organic solvents such as benzene. The energy to overcome the London forces between the C_{60} molecules is paid back by the energy released when London forces are formed between the C_{60} molecules and the solvent.

C_{60} does not conduct electricity. Although each C forms three bonds in the C_{60} molecule, so that there is delocalisation of electrons over the molecule, the molecular structure means that electrons are not able to move from one molecule to the next.

Silicon dioxide

SiO_2 (quartz) has a giant covalent structure (Figure **4.68**). Each silicon atom is bonded to four oxygen atoms in a tetrahedral array. Each oxygen is bonded to two silicon atoms. Due to two lone pairs on each oxygen atom, the basic shape about each Si–O–Si unit is bent (based on a tetrahedron).

The formula is SiO_2 because each Si has a half share of four O atoms, i.e. $Si(O_{\frac{1}{2}})_4 = SiO_2$. The oxide has high melting and boiling points, because covalent bonds between atoms must be broken in order to melt/boil it and this requires a lot of energy.

Nature of science

Science is a highly collaborative subject. The study of graphene lies on the boundary between chemistry and physics. The isolation of graphene was the result of collaboration between scientists of different nationalities and at universities in different countries. Scientists publish work in peer-reviewed journals, which make the knowledge available to other scientists.

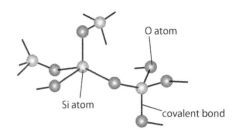

Figure 4.68 Silicon dioxide structure.

Quartz is a piezoelectric material and is used in clocks and watches.

4.4 Intermolecular forces

Learning objectives

- Understand how intermolecular forces arise
- Understand how physical properties of covalent molecular substances depend on the intermolecular forces
- Predict relative boiling points of substances

Van der Waals' forces is the collective name given to the forces between molecules and includes London (dispersion) forces, dipole–dipole interactions and dipole–induced dipole interactions.

Intramolecular forces are forces within a molecule.
Intermolecular forces are forces between molecules.

Intermolecular forces are much weaker than covalent bonds.

London forces are present between all molecules in solid and liquid states.

The nature of intermolecular forces

Br_2 is a liquid at room temperature. It consists of discrete molecules in which the two Br atoms are joined by a covalent bond (an **intra**molecular force). But if Br_2 is to be a liquid, there must be some forces between molecules holding them in the liquid state (otherwise it would be a gas). These forces are **inter**molecular forces (Figure **4.69**).

There are various types of intermolecular forces. The main type between non-polar atoms/molecules is the **London (dispersion) force**. London forces are much weaker than covalent bonds. Therefore, when bromine is heated to form a gas, the Br_2 molecules (held together by covalent bonds) remain intact and it is the London forces that are overcome (Figure **4.70**).

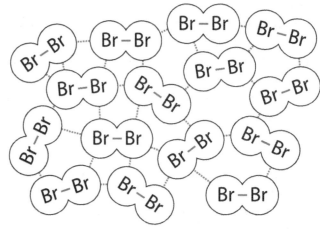

Figure 4.69 Intermolecular and intramolecular forces in liquid bromine. Intramolecular forces are shown in red and intermolecular forces are the dashed lines in blue.

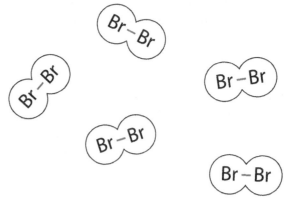

Figure 4.70 Gaseous bromine. Intramolecular forces (covalent bonds) are still present, but the intermolecular forces have been overcome.

How London forces arise

London forces are temporary (instantaneous) dipole–induced dipole interactions.

Consider liquid argon. The electrons in an atom are in constant motion, and at any one time they will not be symmetrically distributed about the nucleus. This results in a temporary (instantaneous) dipole in the atom, which will induce an opposite dipole in a neighbouring atom. These dipoles will attract each other so that there is an attractive force between atoms (Figure **4.71**). Although the dipoles are constantly disappearing and reappearing, the overall force between the argon atoms is always attractive because a dipole always induces an opposite one.

In general, London forces get stronger as the number of electrons in molecules increases. As the number of electrons increases, the relative molecular mass also increases, and we normally talk about 'an increase in the strength of London forces as the relative molecular mass increases'.

A clear correlation between boiling point and relative molecular/ atomic mass can be seen down group 17 (blue line) and down group 18 (red line) in Figure **4.72**.

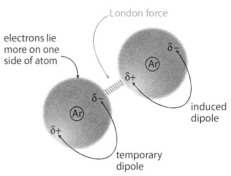

Figure 4.71 The origin of London forces.

In general, London forces get stronger as the relative molecular mass increases.

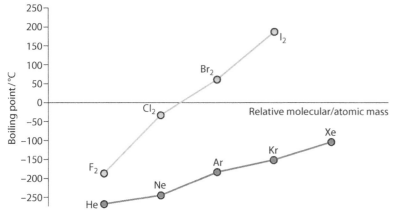

Figure 4.72 Variation in boiling points of elements in groups 17 and 18.

Comparisons are most useful when comparing similar substances – for instance, the elements within a group or a series of similar compounds.

Consider group 17: fluorine (M_r 38.00) is a gas at room temperature, but iodine (M_r 253.80) is a solid. This is because there are more electrons in an iodine molecule, and the atoms making up the molecule are larger. More electrons means that the temporary dipoles will be larger with more electrons moving around. The larger atoms in the molecule means that the outer electrons will be less strongly held, so the molecule is more polarisable, and therefore the induced dipoles will be larger.

Care must be exercised when using this rule of thumb. If the above data for group 17 and group 18 are plotted together against relative molecular mass (relative atomic mass for group 18) we get the graph shown in Figure **4.73**, which does not show a clear correlation.

As the length (and so the relative molecular mass) of a hydrocarbon chain increases, so do the boiling points – butane (C_4H_{10}) has a higher boiling point than ethane (C_2H_6) (Figure **4.74**). A higher boiling point

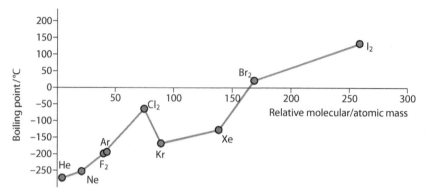

Figure 4.73 Variation in boiling point in groups 17 and 18 with M_r/A_r.

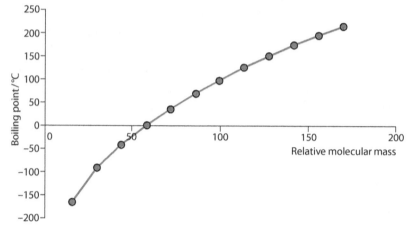

Figure 4.74 Boiling points of alkanes.

means that the London forces between molecules are stronger. The forces between the molecules are stronger because there are more atoms, and therefore more electrons, present in butane than in ethane and also more points of contact between the chains.

The effect of polar molecules

Because of the electronegativity difference between H and Cl, H–Cl molecules are polar.

London forces are present between the molecules in HCl(l) but, because of the polarity of the molecules, there are also other intermolecular forces present (Figure **4.75**). These are called **permanent dipole–permanent dipole** interactions, or usually just **dipole–dipole** attractions.

The intermolecular forces between molecules which are polar (permanent dipole–permanent dipole interactions as well as London forces) are stronger than those between non-polar molecules (only London forces), all other things being equal.

In general, all other things being equal, these permanent dipole interactions would be expected to cause melting points and boiling points to be higher for polar molecules.

If we compare propane and ethanal (Table **4.11**), both have a relative molecular mass of 44 and therefore the strength of London forces should

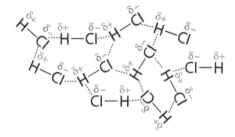

Figure 4.75 Permanent dipole–permanent dipole interactions exist between molecules. These are shown in purple.

All other things being equal – this basically means that we should compare compounds with relative molecular masses as similar as possible.

be similar. However, ethanal is a polar molecule and has dipole–dipole interactions as well as London forces. The intermolecular forces between molecules of ethanal are stronger than those between propane molecules and ethanal has a significantly higher boiling point (more energy must be supplied to overcome the forces between molecules).

Molecule	Lewis structure	M_r	Boiling point/°C
propane		44	−42
ethanal		44	21

Table 4.11 Comparing ethanal and propane.

A similar situation is seen when we compare Br_2 with ICl (Table **4.12**).

Compound	M_r	Boiling point/°C
Br_2	159.80	59
ICl	162.35	100

Table 4.12 Comparing bromine and iodine(I) chloride.

The relative molecular masses of these two compounds are very similar, and so the strengths of the London forces are similar. However, ICl is polar (Cl is more electronegative than I), whereas Br_2 is non-polar. The intermolecular forces are stronger in ICl because of the permanent dipole–permanent dipole interactions. Stronger intermolecular forces means that the boiling point of ICl is higher than that of Br_2.

That we simply cannot say that polar molecules have stronger intermolecular forces and higher boiling points than non-polar molecules can be seen if we compare Br_2 with three other interhalogen compounds (Table **4.13**).

Compound	M_r	Boiling point/°C	Polar/non-polar?
Br_2	159.80	59	non-polar
ClF	54.45	−100	polar
BrF	98.90	20	polar
ClBr	115.35	5	polar

Table 4.13 Comparing bromine with interhalogen compounds.

Br_2 is the only molecule here that is non-polar, but it has the highest boiling point as a result of its highest relative molecular mass and, therefore, strongest London forces.

If molecules with similar relative molecular masses are compared, polar molecules have higher melting and boiling points than non-polar molecules.

Extension

Explain why BrF has a higher boiling point than ClBr, despite ClBr having the higher relative molecular mass.

Comparison of *cis-* and *trans-*1,2-dichloroethene

*Cis-*1,2-dichloroethene and *trans-*1,2-dichloroethene are *cis-trans* isomers – they differ only in the orientation of the Cl atoms about the C=C (Figure **4.76**). (This will be discussed in Topic **10**.)

Both molecules have the same relative molecular mass and are identical, except that one is polar and the other is not. The difference between the boiling points is then solely due to the permanent dipole–dipole interactions in the *cis* form.

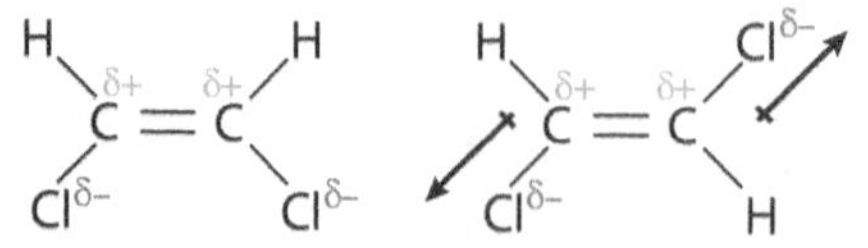

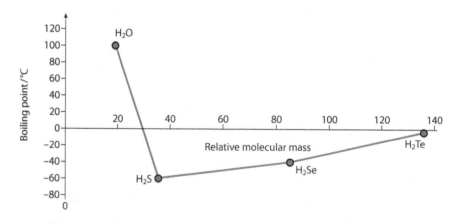

*cis-*1,2-dichloroethene
polar

*trans-*1,2-dichloroethene
non-polar

Figure 4.76 *Cis-* and *trans-*1,2-dichloroethene.

Hydrogen bonding

The third type of intermolecular force that we will examine is hydrogen bonding. **Hydrogen bonding** influences many properties of substances – it is responsible for ice floating on water, it is the force between strands of DNA, it helps maintain the 3D structure of proteins, it is the reason that ethanol is soluble in water. It could be argued that hydrogen bonding is the reason why life on Earth exists as we know it!

The origin of hydrogen bonding

Table **4.14** and Figure **4.77** compare the boiling points of the hydrides of group 16 elements.

Group 16 hydride	Boiling point / °C
H_2O	100
H_2S	−60
H_2Se	−41
H_2Te	−4

Table 4.14 Boiling points of some hydrides.

Figure 4.77 Boiling points of group 16 hydrides.

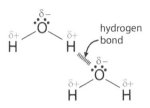

The boiling point increases from H_2S to H_2Te, as would be expected from the increase in the strength of London forces as the relative molecular mass increases.

H_2O has, however, a much higher boiling point than would be expected from its relative molecular mass. This suggests that there must be intermolecular forces other than London forces between H_2O molecules and that these intermolecular forces are stronger than the London forces. These extra intermolecular forces between H_2O molecules are called **hydrogen bonds**. A hydrogen bond between two water molecules is shown in Figure **4.78**.

> It is important to realise that, although hydrogen bonding is the strongest of the intermolecular forces, it is much weaker than covalent bonding.

Hydrogen bonding occurs between molecules when a very electronegative atom (N, O, F) is joined to a hydrogen atom in the molecule. The electronegative atom withdraws electron density from the hydrogen, polarising the bond such that there is a strong interaction between the $\delta+$ hydrogen and the $\delta-$ atom (N, O, F) on the other molecule.

The hydrogen bonding between ammonia molecules and hydrogen fluoride molecules is shown in Figure **4.79**.

We can work out whether there will be hydrogen bonding between molecules simply by looking at whether the molecule contains N/O/F joined directly to a hydrogen atom (Table **4.15**).

Figure 4.78 A hydrogen bond between two water molecules. The hydrogen bond is the dashed purple line between the lone pair of a $\delta-$ oxygen on one molecule and the $\delta+$ hydrogen on the other molecule.

Extension

There is more to hydrogen bonding than just an electrostatic interaction between dipoles. When H is attached to a very electronegative atom, the electron density is pulled away strongly from the H. There is then an interaction between the electron density of the lone pair on N, O or F in one molecule and the H nucleus in a different molecule. There is thus a directional component to hydrogen bonds. If the interaction were purely dipole–dipole then Cl, with the same electronegativity as N, would also be expected to participate in hydrogen bonding – but it does not. This could be explained by the higher energy and more diffuse lone pair on the Cl not interacting effectively with the H.

> The requirements for hydrogen bonding are that the H atom is attached to a very electronegative atom – N, O or F – which possesses at least one lone pair of electrons.

Hydrogen bonding between molecules	No hydrogen bonding between molecules
HF	HCl
H_2O	H_2S
NH_3	PH_3
CH_3CH_2OH	CH_3OCH_3
$CH_3CH_2NH_2$	CH_3CH_2F
Hydrogen bonding between molecules, because there is an H joined directly to an N, O or F atom	No hydrogen bonding between molecules, because H is not joined to N, O or F

Table 4.15 Hydrogen bonding or no hydrogen bonding?

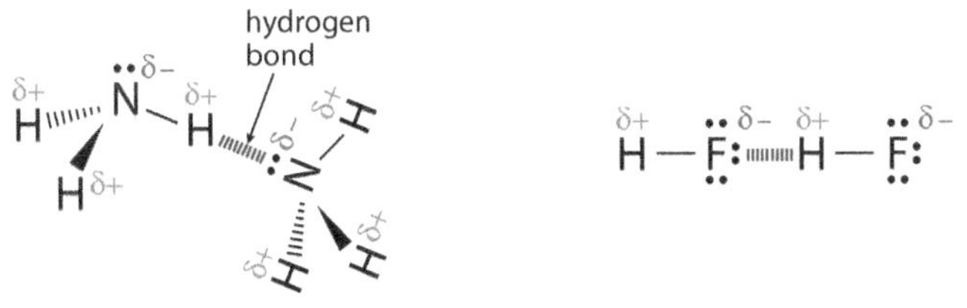

> Hydrogen bonding can influence the solubility of substances in water, and often molecules that are able to hydrogen bond are soluble in water – this will be discussed later.

Figure 4.79 Hydrogen bonding between ammonia and hydrogen fluoride molecules.

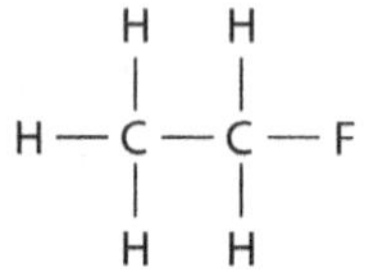

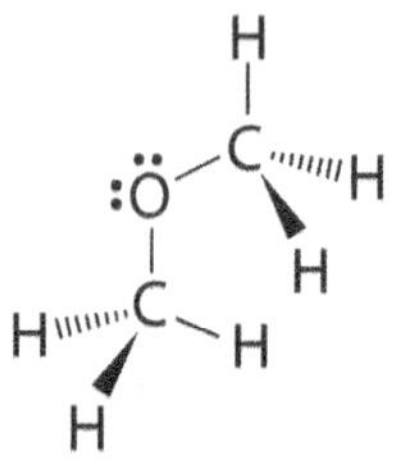

Figure 4.80 The structure of CH_3OCH_3.

Figure 4.81 Hydrogen bonding in ethanol.

cis–trans isomerism is discussed in Topic **10**.

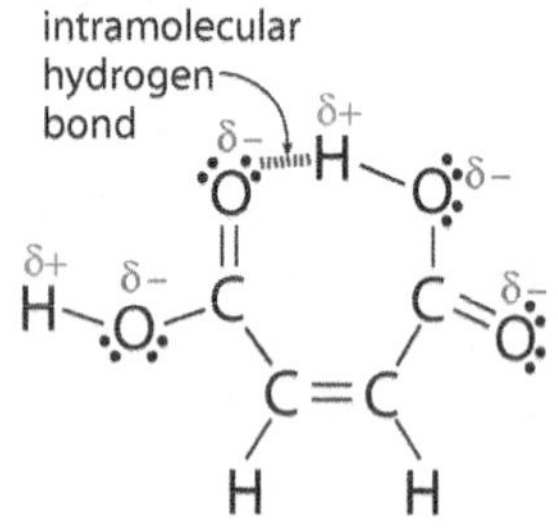

Figure 4.82 Intramolecular hydrogen bonding in *cis*-but-2-ene-1,4-dioic acid.

There is no hydrogen bonding between molecules of CH_3CH_2F because the H is not joined directly to an F atom:

CH_3OCH_3 is an ether with the structure shown in Figure **4.80**. It is a polar molecule, but there is no hydrogen bonding between molecules.

The hydrogen bonding between some molecules of ethanol is shown in Figure **4.81**.

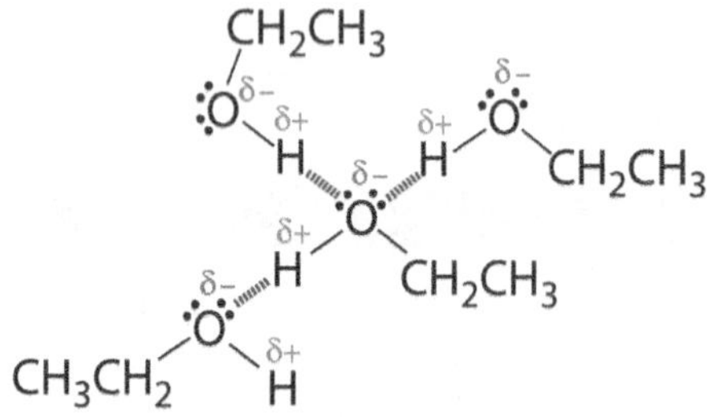

Hydrogen bonding within molecules

Consider the *cis* and *trans* forms of butenedioic acid shown in Table **4.16**.

cis-but-2-ene-1,4-dioic acid (maleic acid)	*trans*-but-2-ene-1,4-dioic acid (fumaric acid)
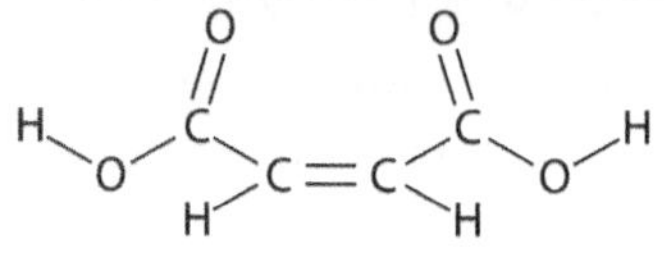	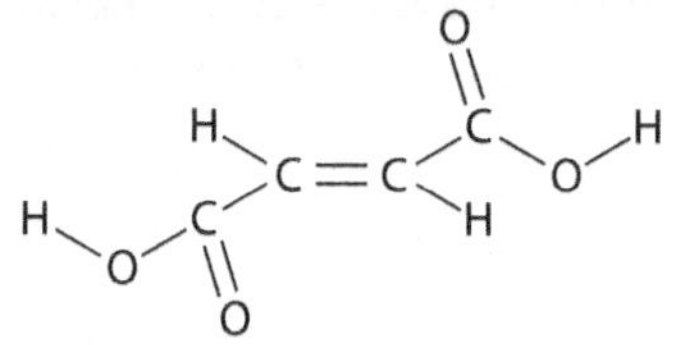
melting point: 130 °C	melting point: 287 °C

Table 4.16 The influence of intramolecular hydrogen bonding within molecules.

The *cis* form has a lower melting point because as well as intermolecular hydrogen bonding it is also able to participate in intramolecular hydrogen bonding (Figure **4.82**). This means that there is less hydrogen bonding between molecules and that the intermolecular forces are weaker than in the *trans* form. The *trans* form participates in only intermolecular hydrogen bonding, because the COOH groups are further away from each other – the intermolecular forces are therefore stronger in the *trans* form.

Melting points and boiling points

> Only intermolecular forces are broken when covalent molecular substances are melted or boiled – covalent bonds are **not** broken.

We have looked at several factors that influence the strength of intermolecular forces, and hence the melting and boiling points of covalent substances. Now let us consider all the above points together with some examples.

The stronger the intermolecular forces, the more energy must be supplied to break them and the higher the boiling point. Differences in melting and boiling points are sometimes explained using this idea about intermolecular forces of attraction:

London < permanent dipole–dipole < hydrogen bonding
WEAKEST ──────────────────────→ **STRONGEST**

However, London forces are present between all molecules and, in some substances, can provide a higher contribution to the intermolecular forces than dipole–dipole interactions – it is therefore important to consider substances with similar relative molecular masses. Table **4.17** summarises the influence of the type of intermolecular force on the boiling point of substances.

Relative molecular mass – generally substances with higher relative molecular masses have higher melting points and boiling points due to stronger London forces.

Type of substance		Intermolecular forces
substances containing H bonded to N, O or F	*decreasing boiling point*	hydrogen bonding permanent dipole–dipole London forces
polar substances		permanent dipole–dipole London forces
non-polar substances		London forces

Table 4.17 If relative molecular masses are approximately equal, then the boiling points usually change in the order shown.

Worked examples

4.1 Arrange sulfur, chlorine and argon in order of increasing boiling point and explain your order.

Sulfur, chlorine and argon are all non-polar substances and therefore the only intermolecular forces are London forces. Any difference between these substances is due to the strength of London forces, which is affected by relative molecular mass (relative atomic mass for argon). Sulfur forms S_8 molecules with a relative molecular mass of 256.56; Cl_2 has a relative molecular mass of 70.90; and Ar has a relative atomic mass of 39.95. Therefore the boiling point of sulfur would be expected to be highest because S_8 has a higher relative molecular mass and therefore stronger London forces than Cl_2 and Ar. More energy is required to overcome the intermolecular forces in sulfur than the intermolecular forces in chlorine and the interatomic forces in Ar. Chlorine would be expected to have a higher boiling point than Ar, again due to the greater mass and the stronger London forces. So, the order of increasing boiling points is argon < chlorine < sulfur.

The actual boiling points for these substances are:
  Sulfur: 445 °C   Chlorine: −34 °C   Argon: −186 °C

4.2 Compare the boiling points of propane ($CH_3CH_2CH_3$), methoxymethane (CH_3OCH_3) and ethanol (CH_3CH_2OH).

Let us compare the relative molecular masses of these substances:

  $CH_3CH_2CH_3$: 44.11 CH_3OCH_3: 46.08 CH_3CH_2OH: 46.08

The relative molecular masses are all very similar and therefore the strengths of the London forces are going to be similar.

The next thing we can look at is whether any of the molecules are polar:

  $CH_3CH_2CH_3$: non-polar CH_3OCH_3: polar CH_3CH_2OH: polar

The presence of the very electronegative O atom in CH_3OCH_3 and CH_3CH_2OH means that these substances are polar. $CH_3CH_2CH_3$ is the only non-polar molecule and will have the lowest boiling point because the only forces between molecules are London forces.

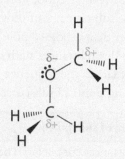

CH_3OCH_3 and CH_3CH_2OH are both polar and have the same relative molecular mass, but CH_3CH_2OH has the higher boiling point because it has an H joined to an O, and therefore there is also hydrogen bonding between molecules.

Hydrogen bonding in CH_3CH_2OH is a stronger intermolecular force than the permanent dipole interactions in CH_3OCH_3; therefore the intermolecular forces are stronger between molecules of CH_3CH_2OH than between molecules of CH_3OCH_3.

The boiling points of these compounds are:

$CH_3CH_2CH_3$: $-42\,°C$ CH_3OCH_3: $-25\,°C$ CH_3CH_2OH: $78\,°C$

? Test yourself

8 Arrange the following sets of molecules in order of increasing boiling point (lowest first):

a CH_4 CCl_4 CF_4
b NH_3 PH_3 AsH_3
c NH_3 N_2H_4 CH_4
d CH_3OH CH_3F C_2H_4
e H_2O H_2S H_2O_2
f $CH_3CH_2CH_2CH_2OH$ $CH_3CH_2OCH_2CH_3$ $CH_3CH_2CH_2CH_2CH_3$
g N_2 F_2 HF   Ne

9 Arrange the following sets of substances in order of increasing boiling point:

a NaCl $SiCl_4$ CCl_4 HCl
b Br_2 HBr $CaBr_2$ PBr_3
c C_4H_{10} C_3H_7OH C_3H_8 CH_3CH_2COOH C_4H_9OH
d SO_2 SiO_2 CO_2

Solubility

It is often said, when referring to solubility, that 'like dissolves like'. What this means is that:

> generally a substance will dissolve in a solvent if the intermolecular forces in the solute and solvent are similar.

Whether or not a substance dissolves depends (in part) on how much energy is needed to overcome intermolecular forces in the solvent and solute and then how much energy is released, to pay back this energy, when intermolecular forces are formed between solvent and solute molecules in the solution.

Solubility is discussed here from the point of view of energy, but entropy changes (discussed in Topic 5) must also will be considered.

Pentane is readily soluble in hexane but insoluble in water

The amount of energy required to overcome the London forces in pure hexane and pure pentane is paid back when London forces are formed between the molecules of hexane and pentane (Figure **4.83**). Pentane does not dissolve in water because there is hydrogen bonding between water molecules. If pentane were to dissolve in water there would be London forces between water molecules and pentane. The energy released if London forces were to form between water molecules and pentane molecules would not pay back the energy required to break the hydrogen bonds between water molecules, because hydrogen bonds are stronger than London forces (Figure **4.84**).

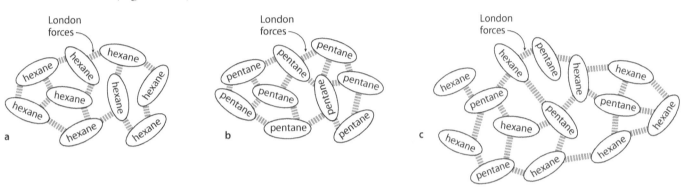

Figure 4.83 a There are London forces between hexane molecules; **b** there are London forces between pentane molecules; **c** there are London forces between hexane and pentane molecules when pentane dissolves in hexane.

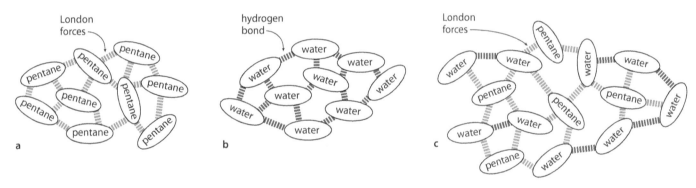

Figure 4.84 a There are London forces between pentane molecules; **b** there are hydrogen bonds between water molecules; **c** pentane does not dissolve in water because only London forces would be formed between pentane and water molecules.

> Substances that are able to participate in hydrogen bonding will generally be soluble in water, because they are able to hydrogen bond to the water.

Ethanol is soluble in water

Ethanol (C_2H_5OH) is very soluble in water, because ethanol is able to hydrogen bond to the water (Figure **4.85**). The hydrogen bonding between water molecules and ethanol molecules in the solution releases energy and pays back the energy to break the hydrogen bonds in pure water and pure ethanol.

Figure 4.85 Hydrogen bonding between water and ethanol.

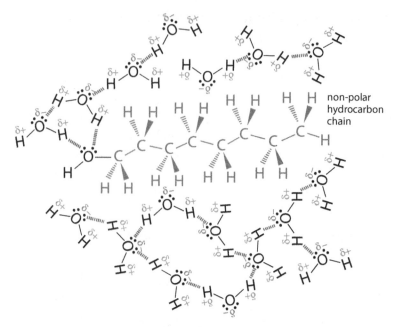

Figure 4.86 The hydrocarbon chain in octan-1-ol prevents hydrogen bonding between water molecules on either side.

Longer chain alcohols become progressively less soluble in water

Octan-1-ol is insoluble in water. Although there is some hydrogen bonding between the O–H group of the alcohol and water molecules, the long hydrocarbon chain prevents water molecules on either side from hydrogen bonding to each other (Figure **4.86**). Energy is needed to break the hydrogen bonds between the water molecules, but this is not paid back as only London forces form between the water molecules and the hydrocarbon part of the molecule.

Dissolving ionic substances

Consider the dissolving of sodium chloride:

$$NaCl(s) \xrightarrow{\text{`+ water'}} NaCl(aq)$$

The aqueous solution contains aqueous ions (Na$^+$(aq) and Cl$^-$(aq)). Ion–dipole interactions form between the water and the ions (Figure **4.87**). Energy is required to overcome the electrostatic forces in the ionic lattice but energy is released when hydrated ions are formed. If the energy released when the hydrated ions are formed is comparable to the energy required to break apart an ionic lattice then the substance is generally soluble.

Ionic substances are not usually soluble in non-polar solvents such as hexane because the interactions between the ions and the hexane molecules would be weak London forces. Sodium chloride is insoluble in hexane because the energy released when a sodium ion or a chloride ion is surrounded by hexane molecules (solvated) is not enough to pay back the energy required to break apart the lattice.

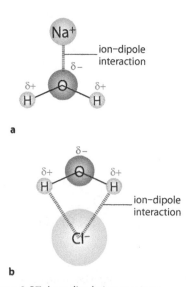

Figure 4.87 Ion–dipole interactions **a** between water and sodium ions; **b** between water and chloride ions.

Comparison of the physical properties of ionic and covalent molecular substances

Table **4.18** shows a comparison of the physical properties of ionic and covalent molecular substances.

Ionic	Covalent molecular
usually solids with high melting points	may be solid, liquid or gas at room temperature
strong electrostatic forces in the giant lattice structure must be overcome	only intermolecular forces are overcome when the substances are melted or boiled; intermolecular forces are weaker than the electrostatic forces in an ionic lattice; no covalent bonds are broken when covalent molecular substances are melted or boiled
non-volatile	many are volatile
strong electrostatic forces in the lattice	weak intermolecular forces
many are soluble in water and usually insoluble in organic solvents	not usually soluble in water but soluble in organic solvents
strong interactions between the ions and the polar water molecules provide the energy to break apart the lattice structure	substances dissolve when intermolecular forces in the solvent and solute are similar; molecules that are able to hydrogen bond to water are usually soluble in water
do not conduct electricity in the solid state but do conduct when molten or in aqueous solution, because of mobile ions	do not conduct electricity in any state
ions not free to move in solid state because they are held tightly in the lattice structure; ions free to move around when molten or dissolved in water	no ions or free electrons present; some substances, such as HCl, dissolve in water with ionisation – HCl(aq) conducts electricity

Table 4.18 The physical properties of ionic and covalent molecular substances compared.

Nature of science

Careful observation and collection of data are important in development of theories. Covalent molecular substances can be liquids or solids which indicates that there must be some force between molecules that holds them in the liquid/solid state (London forces, hydrogen bonds etc.). Calculation indicates that this attraction must be a lot stronger than the gravitational attraction been molecules.

Theories can be used to make predictions. An understanding of the structure and bonding of substances can be used to make predictions about the melting point and solubility of a new substance.

Patterns are important in science but exceptions (such as the boiling point of water) to these often lead to advances in our theoretical understanding. Careful observations from experimental work over many years eventually led to the idea of the hydrogen bond. More and more sophisticated models for the hydrogen bonding in substances are still being developed.

? Test yourself

10 Arrange the following sets in order of solubility in water (least soluble first):

   a NaCl C_6H_{12} $C_5H_{11}OH$
   b CH_3Cl $CaCl_2$ CH_4

Learning objectives

- Describe the structure of, and bonding in, metals
- Explain some of the properties of metals in terms of structure and bonding
- Explain what is meant by an 'alloy' and how the properties of alloys are different from those of pure metallic elements

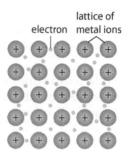

Figure 4.88 A metallic structure. This is a giant structure – there are no individual molecules.

> Metals contain a regular lattice arrangement of positive ions surrounded by a 'sea' of delocalised electrons.

Metallic elements have relatively low ionisation energies and so form positive ions more easily than non–metals.

> **Metallic bonding** is the electrostatic attraction between the positive ions in the lattice and the delocalised electrons.

4.5 Metallic bonding

The structure of a typical metal is shown in Figure **4.88**. The electrons are described as **delocalised** because they do not belong to any one metal atom but, rather, are able to move throughout the structure. The metallic bond is thus electrostatic in nature, resulting from the attraction between the positive metal ions and the negatively charged delocalised electrons:

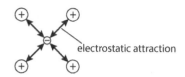

Each electron is attracted by all the positive ions in the structure, so the whole lattice is held together.

The melting points of sodium and magnesium are 98 °C and 649 °C, respectively. There are several reasons why magnesium has a higher melting point than sodium. The first is that magnesium forms a 2+ ion compared with sodium, which forms a 1+ ion (Figure **4.89**). This means that the electrostatic attraction between the ions and the delocalised electrons is stronger in magnesium.

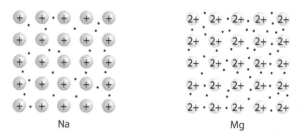

Figure 4.89 Metallic bonding in sodium and magnesium.

The second reason is that the Mg^{2+} ion (65 pm) is smaller than the Na^+ ion (98 pm), and therefore the delocalised electrons are closer to the nucleus of the positive ion in magnesium and are more strongly attracted.

A third reason why magnesium has a higher melting point than sodium is that there are two delocalised electrons per atom in magnesium. Therefore there will be stronger electrostatic attractions between the ions and the delocalised electrons.

Some other physical properties of metals are shown in Table **4.19**.

Physical property	Comment
lustrous	shiny, when freshly scratched/cut
good conductors of electricity	e.g. copper is used in making electrical wires; conduction decreases as temperature increases
good conductors of heat	although the best conductor is diamond
ductile	may be drawn into wires
malleable	may be hammered into shape – many uses arise from the ability to easily shape metals, e.g. making car bodies

Table 4.19 Some physical properties of metals.

Metals conduct electricity because the delocalised electrons are free to move around.

Metals are **malleable/ductile** because of the non-directionality of the bonding. The metal ions in the lattice attract the delocalised electrons in all directions. So, when two layers slide over each other (Figure **4.90**), the bonding in the resulting structure is exactly the same as in the original. This can be compared with ionic solids (Figure **4.91**), which are brittle, as displacement of one layer relative to the other results in like charges repelling each other.

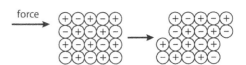

Figure 4.90 Ductility.

Metals such as iron have a wide range of uses, such as in construction (bridges, cars, etc.). Iron is usually used in the form of steel, an alloy of iron and carbon. The fact that mild steel is malleable makes it ideal for the construction of car bodies (and for crushing them afterwards!).

Figure 4.91 Brittleness.

Extension

Actually, metals are more ductile than this simple picture of planes sliding over each other would predict. Ductility is better explained by the movement of dislocations through the lattice.

Alloys

The majority of metals we come across in everyday life are alloys rather than pure metals. Steel, an alloy of iron and carbon, is used much more than pure iron. Other examples of alloys include brass (Cu and Zn), bronze (Cu and Sn) and pewter (Sn, Sb and Cu).

Alloys tend to be stronger and stiffer than pure metals and often combine the desirable properties of the different metals involved. So, aluminium is a light (low density) metal but it is not strong enough to be used in airplane manufacture until it is alloyed with copper (and smaller amounts of magnesium and manganese) to produce duralumin.

The reason why alloys are stronger than pure metals can be explained in terms of structure. At the simplest level, we can imagine that a differently sized atom will prevent planes of metal atoms sliding over each other (Figure **4.92**).

Some properties of alloys are not desirable – e.g. aluminium alloys are more susceptible to corrosion than the pure metal, and the electrical conductivity of copper is reduced by alloying with other metals.

Alloys are homogeneous mixtures of two or more metals, or of a metal with a non-metal.

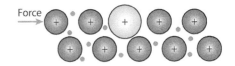

Figure 4.92 The introduction of a larger atom makes it more difficult for the planes of atom/ions to slide over each other – so alloys tend to be stronger and stiffer than pure metals.

Nature of science

Scientific theories can be used to explain natural phenomena. Careful observation established that metals have different properties to ionic and covalent substances – these properties can be explained in terms of a simple model involving a sea of delocalised electrons and non-directional bonding.

? Test yourself

11 Arrange the following sets in order of increasing melting point (lowest first):
  a Na   Li   K
  b Al   Mg   Na

Learning objectives

- Use the concept of formal charge to decide between alternative Lewis structures
- Understand that there are exceptions to the octet rule
- Use VSEPR to predict the shapes of molecules and ions with five or six electron domains
- Explain the formation of sigma and pi bonds in molecules
- Predict the number of sigma and pi bonds in molecules
- Use the concepts of resonance and delocalisation to explain the bonding in molecules and ions
- Explain the absorption of different frequencies of UV light by oxygen and ozone
- Describe the mechanism of CFC and NO_x catalysed ozone depletion

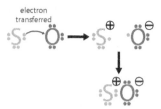

Figure 4.93 Formation of a coordinate covalent bond results in a formal charge on each atom.

Formal charge closest to zero means that there will be the most even distribution of charge in the molecule/ion.

Formal charges closest to zero will generally result in a more stable structure because the separation of charge in a structure requires energy.

4.6 Covalent bonding and electron domains and molecular geometries (HL)

Formal charge and alternative Lewis structures

We have already seen (page **134**) that it is possible to draw more than one Lewis structure for compounds such as SO_2. In this section we will use the concept of **formal charge** to select the best Lewis structure when more than one is possible, that is, the one that provides the best representation of the structure of the actual molecule.

'Formal charge' is the charge that an atom in a molecule would have if we assume that the electrons in a covalent bond are equally shared between the atoms that are bonded – i.e. we assume that all atoms have the same electronegativity. Atoms in a molecule can gain a formal charge either if a coordinate covalent bond is formed or if the molecule has an overall charge (it is an ion).

A coordinate covalent bond is formed when both shared electrons come from the same atom. When the bond forms, it can be imagined that the donor atom gives an electron to the receiver atom, so that each now has one electron, which can be used to form a covalent bond. Because the donor atom has given away an electron, it now has a formal charge of 1+ and the receiver atom has a formal charge of 1− (Figure **4.93**).

The formal charge (FC) on an atom can be worked out from the Lewis structure using the formula:

> FC = (number of valence electrons in the uncombined atom)
> $- \frac{1}{2}$ (number of bonding electrons) − (number of non-bonding electrons)

> In general, the preferred Lewis structure is the one in which the formal charges are closest to zero.

The calculations of the formal charges in the two possible structures for SO_2 are shown in Table **4.20**.

Structure A (no expansion of octet)		Structure B (expansion of octet)	
Ⓞ S Ⓞ		Ⓞ S Ⓞ	
S	$FC = 6 - (\frac{1}{2} \times 6) - 2 = 1+$	S	$FC = 6 - (\frac{1}{2} \times 8) - 2 = 0$
O1	$FC = 6 - (\frac{1}{2} \times 4) - 4 = 0$	O1	$FC = 6 - (\frac{1}{2} \times 4) - 4 = 0$
O2	$FC = 6 - (\frac{1}{2} \times 2) - 6 = 1-$	O2	$FC = 6 - (\frac{1}{2} \times 4) - 4 = 0$

Table 4.20 Calculating formal charges in SO_2.

The formal charge on the sulfur atom in structure A was worked out as follows:

- An unbonded S atom has 6 electrons in its outer shell because it is in group 16; there are 6 electrons involved in covalent bonds (S forms three bonds); and there is one lone pair – two non-bonding electrons
- FC = (number of valence electrons in the uncombined atom) − $\frac{1}{2}$(number of bonding electrons) − (number of non-bonding electrons) = $6 - (\frac{1}{2} \times 6) - 2 = 1+$
- The formal charges on the atoms in SO_2 are shown in Figure **4.94**.

It can be seen that structure B is the preferred one because there are no formal charges on the atoms.

Consider the two alternative Lewis structures for the sulfate(VI) ion shown in Figure **4.95**. The formal charges are worked out in Table **4.21**. There are fewer atoms with a formal charge and the formal charges are closer to zero in structure **D** and this is the preferred structure. In both cases, the sum of the formal charges is 2−, which is the overall charge on the ion.

Structure **A**   Structure **B**

Figure 4.94 Formal charges on the atoms in alternative Lewis structures for SO_2.

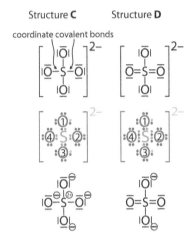

Structure **C**   Structure **D**

coordinate covalent bonds

Figure 4.95 Formal charges for SO_4^{2-}.

Structure C	Structure D
In the first structure, S has not expanded its octet and there are two coordinate covalent bonds. The formal charges are:	In the second structure the S has expanded its octet and there are no coordinate covalent bonds. The formal charges are:
S FC = $6 - \frac{1}{2} \times 8 - 0 = 2+$	**S** FC = $6 - \frac{1}{2} \times 12 - 0 = 0$
O1 FC = $6 - \frac{1}{2} \times 2 - 6 = 1-$	**O1** FC = $6 - \frac{1}{2} \times 2 - 6 = 1-$
O2 FC = $6 - \frac{1}{2} \times 2 - 6 = 1-$	**O2** FC = $6 - \frac{1}{2} \times 4 - 4 = 0$
O3 FC = $6 - \frac{1}{2} \times 2 - 6 = 1-$	**O3** FC = $6 - \frac{1}{2} \times 2 - 6 = 1-$
O4 FC = $6 - \frac{1}{2} \times 2 - 6 = 1-$	**O4** FC = $6 - \frac{1}{2} \times 4 - 4 = 0$

Table 4.21 Calculating formal charges in SO_4^{2-}.

A similar analysis can be carried out on the phosphate(V) ion, PO_4^{3-}. Two possible Lewis structures and formal charges are shown in Figure **4.96**. The formal charges are closer to zero in structure **E** and this is the preferred structure.

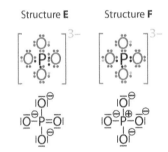

Structure **E**   Structure **F**

Figure 4.96 Formal charges for PO_4^{3-}.

? Test yourself

12 Draw alternative Lewis structures with and without an expanded octet on the central atom for each of the following molecules/ions. Then use the concept of formal charge to decide which structure is probably a better representation of the actual structure.

   a POCl₃   b ClO₄⁻   c XeO₄

The concept of formal charge assumes that the bonding in a molecule/ion is purely covalent. The concept of oxidation number, which will be introduced in Topic **9**, assumes that the bonding in a molecule/ion is purely ionic. Neither provides a complete picture of the bonding; why do we use them?

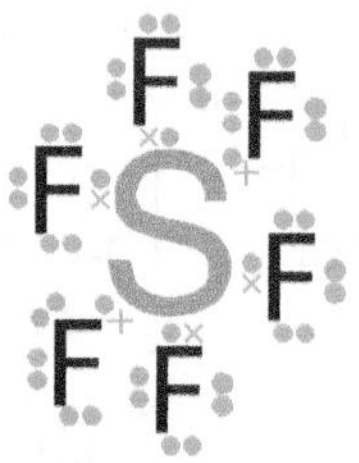

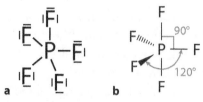

Figure 4.97 The Lewis structure of SF_6.

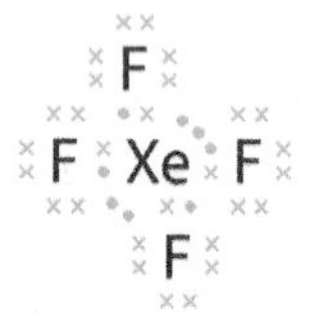

Figure 4.98 The Lewis structure for XeF_4.

Figure 4.99 **a** Lewis structure and **b** molecular shape of PF_5.

Exceptions to the octet rule

As mentioned previously (Subtopic **4.3**), there are some molecules, such as BF_3, BCl_3 and $BeCl_2(g)$, in which the central atom has fewer than 8 electrons in its outer shell – and is said to have an **incomplete octet**. We have also met some molecules and ions, e.g. SO_4^{2-}, PO_4^{3-} and SO_2, for which a structure could be drawn with the central atom having more than eight electrons in its outer shell, – it has **expanded its octet**.

However, for molecules and ions with expanded octets it was always possible to draw structures in which the central atom did not have more than eight electrons in its outer shell if coordinate covalent bonds were used in the Lewis structure instead of double bonds. We will now consider some species in which the central atom is joined to outer atoms with only single bonds – and so the central atom must have an expanded octet. The Lewis structure for sulfur(VI) fluoride, SF_6, is shown in Figure **4.97**. In this structure, the sulfur atom has twelve electrons in its outer shell.

Similarly, in noble gas compounds such as XeF_4 the central atom is usually described as having an expanded octet (Figure **4.98**).

For an atom to be able to expand its octet, it must be in period 3 or beyond (periods 4, 5 …). A period 2 atom cannot expand its octet because the maximum number of electrons in the second shell (main energy level) – the outer shell of a period 2 atom – is 8. For period 3 and beyond, the atoms have d subshells that are available for bonding – so a period 3 atom can have up to 18 electrons in its outer shell (8 + 10 in the 3d subshell). Hence, we can, for example, explain why PF_5 can exist but NF_5 cannot.

Shapes of molecules and ions with five or six electron domains around the central atom

We have already used valence shell electron pair repulsion to explain and predict the shapes of molecules and ions with up to four electron domains (page **138**). Here we will extend the technique to include molecules and ions with five or six electron domains.

The basic shapes for five and six electron domains are shown in Table **4.22**.

Electron domains	Shape	Diagram	Example
5	trigonal bipyramid	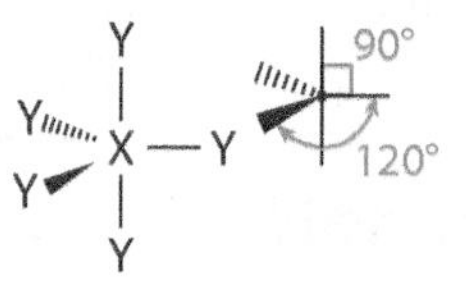	PF_5
6	octahedral	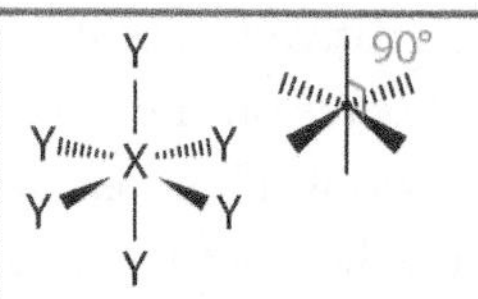	SF_6

Table 4.22 The basic shapes for molecules with five or six electron domains.

PF_5

The Lewis structure and molecular shape of phosphorus(V) fluoride are shown in Figure **4.99**.

Bonding pairs of electrons: 5
Non-bonding pairs of electrons: 0
Electron domains = 5 + 0 = 5

Because these five electron domains repel each other and take up positions in space to be as far apart as possible, the electrons pairs are distributed in a trigonal bipyramidal arrangement. The shape adopted by the atoms is also trigonal bipyramidal (two triangular-based pyramids on top of each other). It is not possible to arrange five things on the surface of a sphere such that they are evenly spaced from each other, so there are two different bond angles in a trigonal bipyramid.

The positions around the middle plane of a trigonal bipyramid are described as equatorial (Figure **4.100**); the positions above and below the central triangle are described as axial.

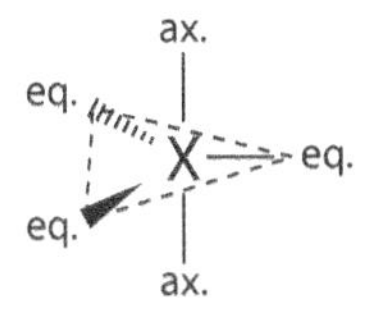

Figure 4.100 Axial and equatorial positions.

> In a trigonal bipyramid, the axial and equatorial positions are not equivalent. Lone pairs always go in an equatorial position. This applies only to trigonal bipyramidal structures.

SF₄

The Lewis structure and molecular shape of sulfur(IV) fluoride are shown in Figure **4.101**.

Bonding pairs of electrons: 4
Non-bonding pairs of electrons: 1
Electron domains: 5

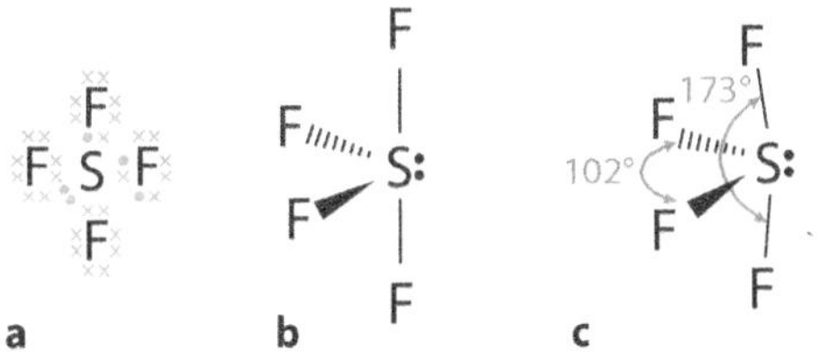

These five electron domains repel each other and take up positions in space to be as far apart as possible – again, the electron pairs are distributed in a trigonal bipyramidal arrangement. One of these electron pairs is a lone pair, and this adopts an equatorial position in a trigonal bipyramid (Figure **4.101b**). The molecule is often described as having a 'see-saw' shape. The lone pair causes the other bonding pairs to bend away from it, and a better representation of the shape is shown in Figure **4.101c**. It can be seen that the lone pair has a very large influence on the bond angles – more than just a couple of degrees.

Figure 4.101 Structure of SF₄: **a** Lewis structure; **b** the see-saw shape; **c** F atoms bend back away from the lone pair.

Exam tip
Care must be exercised here – if you write the angle 102° for the bond angle in your answer it could be marked incorrect, even though it is the correct answer, because it is quite a bit less than 120°.

Extension

The lone pairs always going in an equatorial position is often discussed in terms of the repulsion between the pairs of electrons. The repulsion between electron pairs at 90° to each other is greater than between pairs at 120° to each other. Having the lone pairs (which cause greater repulsion than bonding pairs of electrons) in equatorial positions limits the number of 90° repulsions to two. Having the lone pairs in axial positions would mean that there would be three 90° repulsions – i.e. greater repulsion overall and a less stable molecule.

An alternative explanation is in terms of the hybridisation of the central atom (see Subtopic **4.7**). A trigonal bipyramid involves two different hybridisation schemes – sp² for the equatorial positions and pd for the axial positions. An s orbital has lower energy than a p orbital or a d orbital, and therefore if a lone pair occupies an sp² orbital it is lower in energy than if it were in a pd orbital. Because a lone pair exists solely on the central atom, this should be in the orbital with the lowest available energy – an sp² orbital – in the equatorial position.

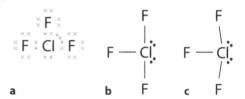

Figure 4.102 Structure of ClF_3: **a** Lewis structure; **b** T-shaped molecule; **c** F atoms bend back away from the lone pairs.

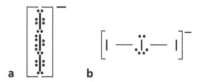

Figure 4.103 Structure of I_3^-: **a** Lewis structure; **b** linear ion.

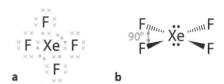

Figure 4.104 Structure of XeF_4: **a** Lewis structure; **b** square planar molecule.

XeF_4 can be prepared by heating xenon with excess fluorine at high temperature and pressure in a nickel container. It is a white, crystalline solid at room temperature.

Figure 4.105 Structure of BrF_5: **a** Lewis structure; **b** slightly distorted square pyramidal molecule.

ClF_3

The Lewis structure and molecular shape of chlorine(III) fluoride are shown in Figure **4.102**.

Bonding pairs of electrons: 3
Non-bonding pairs of electrons: 2
Electron domains: 5

Two of the electron pairs are lone pairs and take up equatorial positions in the trigonal bipyramid (Figure **4.102b**). The actual shape adopted by the molecule is usually described as 'T-shaped' (Figure **4.102c**). The lone pairs cause the other bonding pairs to bend away from them, therefore a better description of the structure might be 'arrow-shaped'

I_3^-

The Lewis structure and molecular shape of the triiodide ion are shown in Figure **4.103**. The ion has three lone pairs and two bonding pairs around the central atom, so the electron pairs are distributed in a trigonal bipyramidal arrangement. The shape is linear, with three lone pairs around the middle of the trigonal bipyramid.

XeF_4

Xenon(IV) fluoride has four bonding pairs and two lone pairs around the central atom. Its Lewis structure and molecular shape are shown in Figure **4.104**. The arrangement of electron pairs is octahedral and the two lone pairs take up positions opposite each other to minimise repulsion. The shape adopted by the molecule is described as square planar.

BrF_5

Bromine(V) fluoride has five bonding pairs and one lone pair around the central atom. Its Lewis structure and molecular shape are shown in Figure **4.105**. The arrangement of the electron pairs is octahedral and the shape of the molecule is square pyramidal. The lone pair causes the four fluorine atoms on the base of the pyramid to bend away from it slightly due to greater repulsion between a lone pair and a bonding pair than between two bonding pairs (Figure **4.105b**).

Table **4.23** summarises the shapes of molecules and ions with five and six electron domains.

Electron domains	Bonding pairs	Lone pairs	Basic shape	Actual shape	Examples
5	5	0	trigonal bipyramidal	trigonal bipyramidal	PF_5, XeO_3F_2
5	4	1	trigonal bipyramidal	see-saw	SF_4, XeO_2F_2
5	3	2	trigonal bipyramidal	T-shape	BrF_3, $XeOF_2$
5	2	3	trigonal bipyramidal	linear	I_3^-, XeF_2
6	6	0	octahedral	octahedral	SF_6, PF_6^-, XeO_6^{4-}, IO_6^{5-}
6	5	1	octahedral	square pyramidal	SF_5^-, BrF_5, XeF_5^+, $XeOF_4$
6	4	2	octahedral	square planar	XeF_4, SF_4^{2-}

Table 4.23 The shapes of molecules and ions with more than four electron domains.

? Test yourself

13 Work out the shapes of the following molecules and ions and predict their bond angles:
- **a** BrF_3
- **b** ClF_5
- **c** SO_2Cl_2 (S central atom)
- **d** SeF_4
- **e** XeF_2
- **f** AsF_6^-
- **g** TeF_5^-
- **h** $F_2ClO_2^-$ (Cl central atom)
- **i** I_3^+
- **j** ICl_2^-

14 Sort the following molecules into polar and non-polar molecules. For the polar molecules, draw diagrams showing the dipoles.

XeF_6 XeF_4 SF_4 PCl_5 SF_2 SF_6 ClF_5 BrF_3 $SOCl_2$

Sigma and pi bonds

A covalent bond is formed when two atomic orbitals, each containing one electron, overlap. When these orbitals overlap **head-on**, the bond formed is a normal single bond and is called a **sigma (σ) bond** (Figure **4.106**). Another way of describing the formation of the sigma bond is that two s atomic orbitals combine to form a σ molecular orbital (Figure **4.107**). Atomic orbitals are found in atoms, but electrons occupy molecular orbitals in molecules. Molecular orbitals, like atomic orbitals, can hold a maximum of two electrons.

The term 'linear combination of atomic orbitals' (LCAO) is used in the syllabus. LCAO is a mathematical approximation that assumes that molecular orbitals can be described as linear combinations (adding or subtracting the wave functions) of the original atomic orbitals. In the context here, the molecular orbitals (sigma and pi bonding orbitals) are formed by adding s and/or p orbitals.

> A covalent bond can also be formed when an atomic orbital containing two electrons overlaps with an empty orbital. This is a coordinate covalent bond.

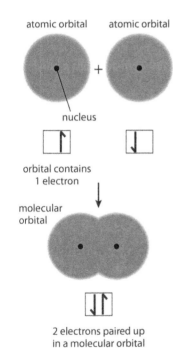

Figure 4.107 The overlap of two s orbitals to form a covalent bond.

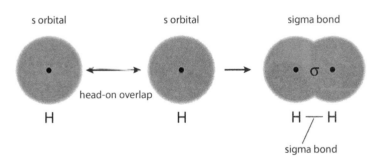

Figure 4.106 The overlap of two s orbitals to form a σ bond.

Extension

When two atomic orbitals combine, two molecular orbitals are formed – one 'bonding' and the other 'antibonding'.

Sigma bonds result from the axial (head-on) overlap of atomic orbitals. The electron distribution in a sigma bond lies mostly along the axis joining the two nuclei.

Sigma bonds can be formed by the axial overlap of any two orbitals – it can be two p orbitals; an s orbital and a p orbital; an sp^3 hybrid orbital (see Subtopic **4.7**) and an s orbital; a d orbital and a p orbital, and so on. Figure **4.108** shows a sigma bond formed from overlap of two p orbitals.

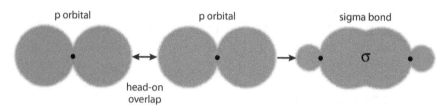

Figure 4.108 Sigma bond formed by head-on overlap of p orbitals.

All covalent bonds (single, double and triple) contain 1 σ bond.

A **pi (π) bond** is formed when two parallel p orbitals overlap side-on (Figure **4.109**).

> A π bond is formed by the sideways overlap of parallel p orbitals. The electron density in the π bond lies above and below the internuclear axis.

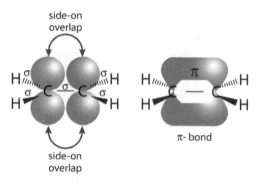

Figure 4.109 Formation of a π bond.

If there are two p orbitals available for bonding on each atom then a triple bond can be formed (Figure **4.110**). A triple bond consists of one σ bond and two π bonds.

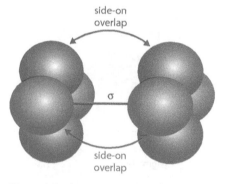

Figure 4.110 A triple bond.

To summarise:
- a single bond consists of a σ bond
- a double bond consists of a σ bond and a π bond
- a triple bond consists of a σ bond and two π bonds.

The bonding in ethene, C_2H_4

When carbon atoms and hydrogen atoms come together to form ethene, sigma bonds are formed between the carbon and hydrogen atoms as shown in Figure **4.111**. There are four orbitals (one s and three p) in the outer shell of each carbon atom. Only three orbitals are used in bonding to the other carbon atom and to two hydrogen atoms, so there is a p orbital (containing 1 electron) remaining on each C atom. These p orbitals can overlap **side-on** to form a **pi (π)** bond (Figure **4.112**).

A double bond has two components:

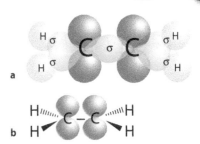

a

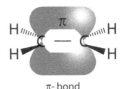

b

Figure 4.111 a The σ bonding in ethene.
b Two p orbitals are not involved in σ bonding.

$$H_{\prime\prime\prime\prime}C\frac{\pi}{\sigma}C^{\prime\prime\prime\prime\prime}H$$

That these components are different can be seen from the bond energies in Table **4.24**. A C=C bond has less than twice the strength of a C–C single bond, meaning that the π bond in ethene is not as strong as the C–C σ bond. Side-on overlap (π bond) is not as effective as head-on overlap – there is a more direct attraction between the electron pair and the nuclei in a σ bond.

	Bond energy / kJ mol^{-1}
C–C	348
C=C	612

Table 4.24 Bond energies in single and double bonds.

Figure 4.112 Bonding in ethene.

Extension

Life is never that simple in chemistry – compare the strength of single and double bonds for O–O and N–N bonds.

? Test yourself

15 Work out the number of sigma and pi bonds in each of the following:

  a O_2
  b N_2
  c BCl_3
  d CO_2
  e H_2CO
  f HCN

  g N_2F_2
  h CO
  i CH_3CHCH_2
  j $SOCl_2$
  k $HCCCH_3$
  l H_2SO_4

Extension

There is one isomer of C_3H_4 that contains two C=C double bonds. Determine the shape of this molecule. Is it planar?

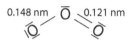

Figure 4.113 The Lewis structure for O_3 with expected bond lengths.

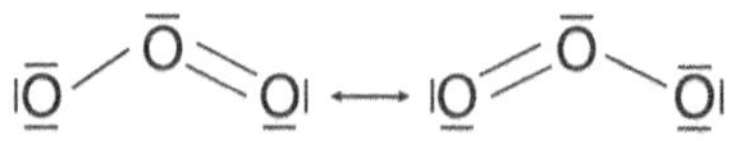

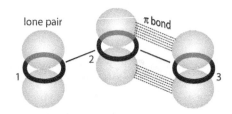

Figure 4.114 Resonance structures for O_3.

Resonance and delocalisation

The Lewis structure of ozone, O_3, is shown in Figure **4.113**. It would be expected from this structure that there would be one short (O=O) and one long (O–O) bond. However, both bond lengths are equal at 0.128 nm. This can be compared with an O=O bond length of 0.121 nm and an O–O bond length of 0.148 nm. The first attempt at explaining the equality of bond lengths in molecules such as O_3 involved the idea of **resonance structures** and **resonance hybrids**. As discussed on page **134**, it is possible to draw an alternative Lewis (resonance) structure for O_3 with the double bond between the other two atoms (Figure **4.114**).

Neither of these structures alone can explain the equal O–O bond lengths, and so the idea of a resonance hybrid is used. This is a hybrid of these two structures – a form that is neither one structure nor the other, but somewhere in between. In this case, with equal contributions from both resonance forms, the bond lengths in the resonance hybrid would be expected to be equal. The double-headed arrow between the resonance structures indicates that the actual structure is a resonance hybrid of the two resonance structures.

> The term 'resonance' is an unfortunate one because it is used in other contexts in conjunction with vibrations. It is important to note that the structure is not constantly flipping between the different resonance forms but is a mixture of them – rather like a mule is a hybrid of a donkey and a horse; it is not a horse one second and a donkey the next.

A more satisfying explanation of the equal bond lengths in O_3 comes from the idea of **delocalisation** of electrons.

> Delocalisation is the sharing of a pair of electrons between three or more atoms.

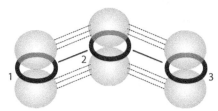

Figure 4.115 The sigma framework and p orbitals of O_3.

If the sigma framework (single bonds) of the O_3 molecule is examined (Figure **4.115**) it can be seen that there is a p orbital on each oxygen atom that is perpendicular to the plane of the molecule. The p orbital on oxygen atom 1 can overlap with the p orbital on oxygen atom 2, which is part of the π bond. In this way, the electron pair of the π bond is spread (delocalised) over all three oxygen atoms (Figure **4.116**).

The delocalised system can be shown using dashed lines between atoms (Figure **4.117**).

Figure 4.116 O_3 has a delocalised π system.

> A single bond is said to have a bond order of 1, a double bond has bond order 2 and a triple bond has bond order 3.

Figure 4.117 Two equivalent representations of the delocalisation in ozone. The dashed lines represent the delocalisation of electrons.

Because the π bond in ozone is shared between two O–O bonds rather than just one, we consider that each bond has a half share of it, and we talk about ozone having a bond order of 1.5. So, the O–O bonds are identical and somewhere between a single bond and a double bond.

We can recognise molecules and ions that are likely to have a delocalised π system by drawing resonance structures. If it is possible to draw more than one resonance structure, where the only difference between them is the position of a double bond (and a lone pair if there is one) then it is likely that this molecule/ion will have a delocalised structure. The delocalisation can be shown using curly arrows, as in Figure **4.118**.

NO_2^- is isoelectronic with O_3 and the bonding can be described in the same way. The two resonance structures for NO_2^- are shown in Figure **4.118**. In each structure, all atoms have a complete octet. The nitrogen–oxygen bonds are equal in length (0.118 nm) in the NO_2^- ions and are between the length of an N–O single bond (0.140 nm) and an N=O double bond (0.114 nm). The two resonance structures are identical except for the position of the double bond (and a lone pair) and, therefore, a delocalised structure is predicted. The arrows show the delocalisation of the π bond between the two N–O bonds. The π bond (shown in orange in Figure **4.118**) is shared between two bonds, therefore the nitrogen–oxygen bond order is 1.5. The delocalised structure of NO_2^- can be represented as shown in Figure **4.119**.

CO_3^{2-} and NO_3^-, which are isoelectronic ions, also have delocalised electrons. Three resonance structures can be drawn for each and these differ only in the position of a double bond (and a lone pair). Figure **4.120** shows the resonance structures of CO_3^{2-}.

All the carbon–oxygen bonds have the same length and are between the length of a single bond and a double bond. This can be explained by the ion having a π delocalised system over the whole ion. The π bond (shown in orange in Figure **4.120**) is shared over three C–O bonds – so the carbon–oxygen bond order is $1 + \frac{1}{3}$ or $1\frac{1}{3}$.

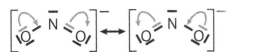

Figure 4.118 The two possible Lewis structures for NO_2^-. The curly arrows (showing movement of a pair of electrons) indicate the delocalisation of electrons.

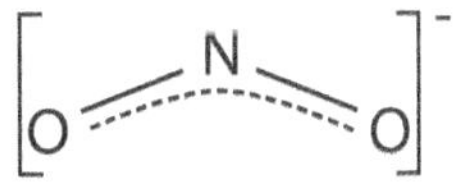

Figure 4.119 NO_2^- is isoelectronic with O_3 and adopts a similar structure.

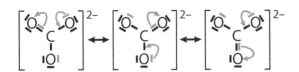

Figure 4.120 Three possible Lewis structures for CO_3^{2-}.

Extension

There are four electrons in the π delocalised system in ozone. To understand why this does not result in two double bonds between the atoms, we have to use a more complex approach to bonding, called molecular orbital theory. Two of the electrons in the π system occupy a bonding molecular orbital and the other two occupy a molecular orbital that is non-bonding.

If asked for a Lewis structure or resonance structure of a species such as NO_2^- in an examination, you must show a structure with all its bonds and lone pairs – one of the structures shown in Figure **4.118.** You must not show the delocalised structure unless specifically asked for this. If you are asked for the delocalised structure you should not include lone pairs of electrons.

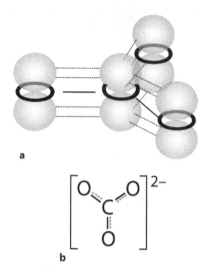

$$\left[\begin{array}{c} O \overset{\cdots}{=} \overset{O}{\underset{C}{\parallel}} \overset{O}{=} O \\ O \end{array} \right]^{2-}$$

b

Figure 4.121 a Overlap of p orbitals in the carbonate ion. **b** Representation of the delocalisation in the carbonate ion.

There is a p orbital on each atom and these overlap side-on to form the π delocalised system. The electrons in this system do not belong to any one atom (Figure **4.121**) but are delocalised over the whole ion (Figure **4.121a**).

The nitrogen–oxygen bond length in the nitrate ion, NO_3^-, is 0.124 nm, which can be compared with the nitrogen–oxygen bond length of 0.118 nm in NO_2^-. The bond order of the nitrogen–oxygen bonds in NO_2^- is 1.5, whereas it is 1.33 in NO_3^-, so the bond lengths would be expected to be longer in NO_3^-.

Benzene

Benzene has the molecular formula C_6H_6. A ring structure for this with alternating double and single bonds was originally proposed by Friedrich Auguste Kekulé (Figure **4.122**). One piece of evidence against this being the best representation of the structure of benzene came when it became possible to measure bond lengths. All the carbon–carbon bond lengths in benzene are equal and, at 0.140 nm, are between the C=C bond length of 0.134 nm and the single bond length of 0.154 nm. This was originally explained by a resonance hybrid with equal contributions from the two structures shown in Figure **4.122**. However, it is explained much better by the idea of delocalisation. The sigma framework (without the double bonds) and p orbitals on each C atom are shown in Figure **4.123**.

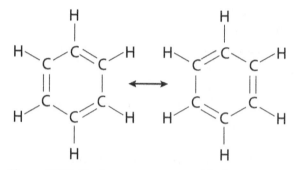

Figure 4.122 The benzene structure with alternating single and double bonds is still known as 'Kekulé benzene' – more systematic names for this would be cyclohexa-1,3,5-triene or 1,3,5-cyclohexatriene.

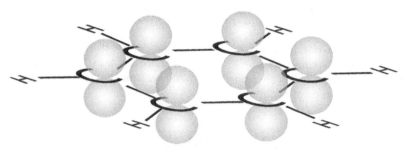

Figure 4.123 The sigma framework for C_6H_6.

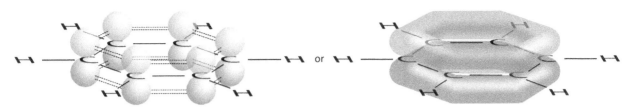

Figure 4.124 Delocalisation over the whole benzene molecule.

Figure **4.124** shows that, instead of the p orbitals just overlapping side-on between adjacent C atoms to give three π bonds, delocalisation can occur over the whole structure.

So benzene has a π delocalised ring of electrons that extends all around the ring of carbon atoms. The structure of benzene is usually drawn (Figure **4.125**) as a hexagon with a ring in the middle indicating the π delocalised system.

Because the π component of each double bond is shared between two carbon–carbon bonds, the bond order in benzene is 1.5.

Figure 4.125 A common way of representing the benzene molecule.

Nature of science

Science is an ever-changing body of knowledge. When new and conflicting evidence comes to light, such as bond length data, theories must be modified. Theories can be superseded by new ones which provide a better explanation of experimental data. The theory of resonance has been largely replaced by the theory of delocalisation.

Test yourself

16 Work out the X–O bond order in each of these delocalised structures:

   a SO_4^{2-} **b** PO_4^{3-} **c** ClO_4^{-}

Absorption of ultraviolet light in the atmosphere

Ozone (O_3) and oxygen (O_2) are important in protecting the surface of the Earth from the damaging effects of ultraviolet radiation. The 'ozone layer' is a region in the stratosphere where there is a higher concentration of ozone. The maximum concentration occurs in the lower regions of the stratosphere between about 15 and 35 km above the Earth's surface. About 90% of the ozone in the atmosphere is found in this region. Even in the ozone layer, the concentration of ozone is very low and there is roughly only one ozone molecule for every 100 000 air molecules.

The atmosphere can be divided into different regions. The troposphere is the region closest to the Earth and the stratosphere is the region between about 12 and 50 km, on average, above the Earth's surface.

λ stands for wavelength

Ultraviolet (UV) light from the Sun reaching the Earth can be divided into three components:
- UV-C $\lambda < 280$ nm (highest energy)
- UV-B $\lambda = 280$–320 nm
- UV-A $\lambda = 320$–400 nm

Free radicals are atoms or groups of atoms with unpaired electrons. An oxygen free radical is simply an oxygen atom and has two unpaired electrons.

UV radiation is absorbed by molecules of O_2 or O_3 as they undergo dissociation. The highest energy UV radiation is absorbed by O_2 molecules in the atmosphere, which causes them to undergo dissociation into oxygen free radicals:

$$O_2 \rightarrow 2O\bullet$$

The UV radiation required to do this is in the UV-C part of the spectrum and must have a wavelength shorter than 242 nm to provide sufficient energy to break the bond. The bond between the atoms in an O_2 molecule is a double bond (bond order 2), but O_3 has a delocalised structure where the bond order is 1.5. This means that the O≐O bond in O_3 is weaker than the O=O in O_2 and so lower energy (longer wavelength) radiation (up to 330 nm) is absorbed by O_3 molecules to break the bond:

$$O_3 \rightarrow O\bullet + O_2$$

So, ozone molecules are particularly effective at absorbing lower energy UV-B radiation (and some UV-C). In the process the molecule is split apart into an oxygen molecule and an oxygen atom, as in the above equation.

All the UV-C and most of the UV-B reaching the Earth from the Sun is absorbed by oxygen (O_2) in the upper parts of the atmosphere and by ozone (O_3) in the stratosphere before it reaches the Earth's surface. Most of the UV radiation reaching the Earth's surface is the least harmful UV-A (Figure **4.126**).

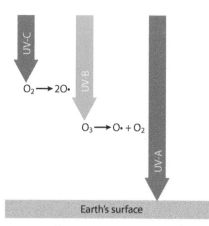

Figure 4.126 Absorption of UV radiation in the atmosphere.

Catalysis of ozone depletion by CFCs and NO$_x$

Ozone molecules are easily destroyed by free radicals. These are present in the stratosphere as nitrogen oxides or are produced when CFCs are broken down by UV light.

CFC stands for chlorofluorocarbon – a molecule containing carbon, chlorine and fluorine.

Chlorofluorocarbons

CFCs such as CCl_2F_2 are very stable compounds at ground level but they are broken down by absorbing UV radiation in the upper atmosphere:

$$CCl_2F_2 \xrightarrow{UV} \underset{\text{free radical}}{\bullet CClF_2} + \underset{\text{free radical}}{Cl\bullet}$$

Homolytic fission – the covalent bond breaks so that one electron goes back to each atom making up the bond.

A C–Cl bond is weaker than a C–F bond and undergoes homolytic fission more readily. The chlorine free radical released in this process

can take part in a chain reaction, which uses up ozone and regenerates chlorine free radicals, Cl• to react with more ozone:

$$\bullet Cl + O_3 \rightarrow ClO\bullet + O_2$$

$$ClO\bullet + O\bullet \rightarrow O_2 \ + \ \underset{\substack{\text{chlorine free} \\ \text{radical regenerated}}}{Cl\bullet}$$

The net effect of these reactions can be seen if the intermediate and the catalyst are cancelled:

$$\begin{array}{l} \bullet\cancel{Cl} + O_3 \rightarrow \cancel{ClO}\bullet + O_2 \\ \underline{\cancel{ClO}\bullet + O\bullet \rightarrow O_2 + \cancel{Cl}\bullet} \\ \quad O_3 + O\bullet \rightarrow 2O_2 \end{array}$$

Nitrogen oxides, NO_x

NO can be formed at high altitudes by aircraft and from the reaction of N_2O or NO_2 with atomic oxygen:

$$N_2O(g) + \bullet O(g) \rightarrow 2\overset{\bullet}{N}O(g)$$

$$\bullet NO_2(g) + \bullet O(g) \rightarrow \bullet NO(g) + O_2(g)$$

Similar types of reactions to those discussed above for CFCs occur with nitrogen oxides. Both NO and NO_2 have unpaired electrons – their total numbers of electrons are odd numbers. The destruction of ozone catalysed by NO can be represented as:

$$\begin{array}{l} \bullet\cancel{NO} + O_3 \rightarrow \bullet\cancel{NO_2} + O_2 \\ \underline{\bullet\cancel{NO_2} + O\bullet \rightarrow O_2 + \bullet\cancel{NO}} \\ \quad O_3 + O\bullet \rightarrow 2O_2 \end{array}$$

NO is a catalyst in this reaction - it is used up in the first step but regenerated in the second step - overall it is not used up.

The cycles involved in the destruction of ozone can stop when the free radicals involved collide:

$$ClO\bullet + \bullet NO_2 \rightarrow ClONO_2$$

Nature of science

Despite the fact that all scientists have the same data available to them, they do not always agree about the interpretation of this data. For example, scientists do not agree about the extent of d orbital involvement in pi bonding (d_π–p_π interactions), for instance in the compound $N(Si(CH_3)_3)_3$ the planar structure can be explained by d_π–p_π or by steric effects.

All reactions occur in the gaseous phase.

Oxygen atoms, O, are available from the dissociation of oxygen or ozone.

The chlorine free radical is a catalyst in these reactions because it is not used up. One Cl• free radical can destroy many thousands of ozone molecules.

Extension

The bond order in NO is 2.5. A simple approach to bonding does not explain this – NO has one electron in a π^* antibonding molecular orbital.

Learning objectives

- Understand what is meant by hybridisation
- Predict the hybridisation of atoms in a molecule

4.7 Hybridisation (HL)

Forming a covalent bond

To form a covalent bond, two orbitals each containing one electron are required. These orbitals overlap to form a covalent bond. Carbon has the outer shell electronic configuration $2s^2 2p^2$:

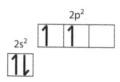

Because it has only two unpaired electrons, carbon should form two covalent bonds. However, it is well known that carbon virtually always forms four covalent bonds. One of the electrons in the 2s orbital must be promoted to the 2p subshell to give four unpaired electrons.

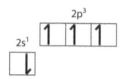

This requires energy. However, bond formation releases energy and the subsequent formation of four bonds instead of two more than pays back the energy needed to promote an electron to a higher subshell.

Carbon now has four unpaired electrons and can form four covalent bonds, but the atomic orbitals do not point in the correct directions for bonding. CH_4 is tetrahedral with bond angles of 109.5°, but the p orbitals are at 90° to each other (Figure **4.127**).

In order to form methane, the four atomic orbitals on carbon (one s and three p) mix to give four **sp³ hybrid orbitals**, which point to the vertices of a tetrahedron. This is the process of **hybridisation** (Figure **4.128**).

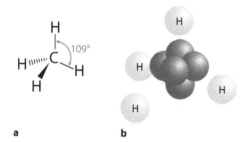

> Hybridisation is the mixing of atomic orbitals in a particular atom to produce a new set of orbitals (the same number as originally) that have characteristics of the original orbitals and are better arranged in space for covalent bonding.

Hybridisation produces orbitals that point towards the atoms to which they are bonding, so it produces more effective bonding. Although promotion requires energy, this is more than paid back by the extra energy released when C forms four bonds as opposed to two.

Figure 4.127 a The bond angles in CH_4; **b** the p orbitals on C are at 90° to each other.

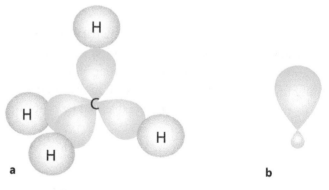

Figure 4.128 a Four sp³ hybrid orbitals point towards the vertices of a tetrahedron and are better set up for bonding to the H atoms. **b** One sp³ orbital.

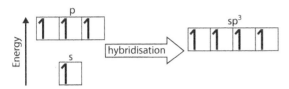

Figure 4.129 The new sp³ orbitals are identical except for direction.

The four sp³ hybrid orbitals have the same energy (Figure **4.129**) – they are degenerate. The overall energy of the s and p orbitals does not change when they undergo hybridisation, so the energy of one sp³ hybrid orbital is the average of the energies of the s and three p orbitals they were formed from. The sp³ orbitals are therefore much closer in energy to the p orbitals than to the s orbitals.

Bonding in methane, CH₄

Methane contains only single bonds between carbon atoms (Figure **4.130**). These are σ bonds formed when sp³ hybrid orbitals on the carbon atom overlap head-on with the s orbitals of each H atom.

Bonding in ethene, C₂H₄

The Lewis structure for ethene is:

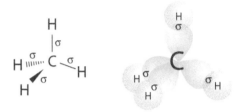

Figure 4.130 Bonding in methane.

$$\begin{array}{cc} H & H \\ \ddot{C} & \vdots \ddot{C} \\ H & H \end{array}$$

The arrangement of atoms around each carbon atom is trigonal planar, and the molecule is planar overall (Figure **4.131a**). Of the three p orbitals on each carbon atom, one is not in the same plane as the hydrogen atoms or the other carbon atom – this p orbital is not involved in hybridisation. Mixing the other two p orbitals and one s orbital produces three sp² orbitals pointing towards the corners of an equilateral triangle (Figure **4.131b**). This leaves one p orbital, containing one electron, perpendicular to the single bond (σ) framework on each C atom (Figure **4.132**). These p orbitals are involved in the formation of the π component of the double bond. The double bond has two different components – a sigma bond, which results from the head-on overlap of two sp² orbitals, and a pi bond which arises from the side-on overlap of two parallel p orbitals.

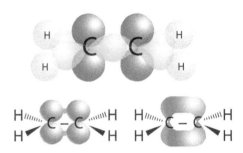

Figure 4.132 The unused p orbitals in ethene overlap side-on to form the π component of the double bond.

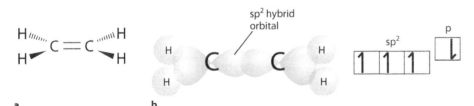

a   b

Figure 4.131 The **a** structure of and **b** hybridisation in ethene.

Figure 4.133 Only the p orbitals that point towards the hydrogen atoms are involved in hybridisation.

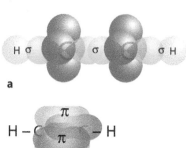

a

b

Figure 4.134 a The formation of σ bonds in ethyne. **b** The formation of π bonds in ethyne; the two π bonds are shown in different shadings for clarity.

The bonding in ethyne, C_2H_2

The Lewis structure for ethyne is:

$$H\!:\!C\!\vdots\!C\!:\!H$$

The linear shape means that one s and one p orbital on each carbon atom are hybridised to produce two sp hybrid orbitals at 180° to each other. The two p orbitals at 90° to the C–H bonds are not involved in the hybridisation – only the p orbitals shown in Figure **4.133** are involved.

Hybridisation of one s and one p orbital produces two sp hybrid orbitals pointing away from each other at 180° on both carbon atoms. These are used to form σ bonds between the carbon atoms and also to the hydrogen atoms (Figure **4.134a**). This leaves two p orbitals on each carbon atom that are not involved in hybridisation. These p orbitals overlap side-on to produce two π bonds (Figure **4.134**).

Table **4.25** shows that the number of hybrid orbitals is always the same as the number of atomic orbitals from which they were produced.

Hybridisation	Orbitals used	Hybrid orbitals produced	Shape
sp	$1 \times s, 1 \times p$	2	linear
sp^2	$1 \times s, 2 \times p$	3	trigonal planar
sp^3	$1 \times s, 3 \times p$	4	tetrahedral

Table 4.25 Hybrid orbitals.

Determining the hybridisation of an atom in a molecule or ion

In order to determine the hybridisation of a particular atom in a molecule or ion, the basic shape – the arrangement of electron domains (electron pairs) on the central atom – must be determined. Table **4.26** shows how the basic shape relates to the hybridisation of the central atom.

Electron domains	Basic shape	Hybridisation
2	linear	sp
3	trigonal planar	sp^2
4	tetrahedral	sp^3

Table 4.26 Relationship between hybridisation and basic shape.

The hybridisation scheme adopted in a molecule depends on the shape of the molecule.

Several examples follow.

With organic molecules this can be simplified:

- when a carbon atom forms just single bonds, the shape is tetrahedral and the hybridisation is sp^3
- when a carbon atom forms a double bond, the shape is trigonal planar and the hybridisation is sp^2
- when a carbon atom forms a triple bond, the shape is linear and the hybridisation is sp.

BF_3

The Lewis structure for boron(III) fluoride is shown in Figure **4.135**. There are three electron domains, which give rise to a trigonal planar shape, and therefore the boron atom is sp^2 hybridised.

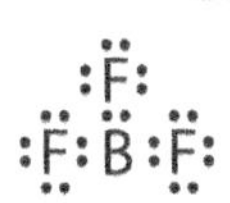

Figure 4.135 The Lewis structure for BF_3.

NH_3

The Lewis structure for ammonia is shown in Figure **4.136**. There are four electron domains – three bonding pairs of electrons and one lone pair. So the basic shape is tetrahedral and the hybridisation of the nitrogen atom is sp^3.

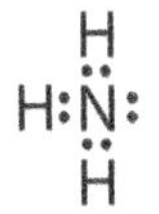

Figure 4.136 The Lewis structure for NH_3.

CO_2

The Lewis structure for carbon dioxide is shown in Figure **4.137**. There are two electron domains around the carbon atom – two double bonds. So the basic shape is linear and the hybridisation of the carbon atom is sp.

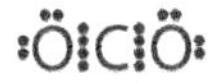

Figure 4.137 The Lewis structure for CO_2.

O_3

The Lewis structure for ozone is shown in Figure **4.138**. There are three electron domains around the central oxygen atom, and so the basic shape is trigonal planar and the hybridisation of the central atom is sp^2.

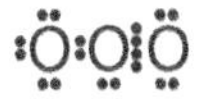

Figure 4.138 The Lewis structure for O_3.

Table **4.27** shows the hybridisation of the central atom in several molecules and ions.

Electron domains	Basic shape	Hybridisation	Examples
2	linear	sp	CO_2, HCN, C_2H_2, NO_2^+
3	trigonal planar	sp^2	BF_3, SO_3, NO_3^-, CO_3^{2-}, SO_2, BF_3, NO_2^-
4	tetrahedral	sp^3	CCl_4, XeO_4, NH_4^+, BCl_4^-, SO_4^{2-}, PO_4^{3-} NH_3, PCl_3, XeO_3, H_3O^+, ClO_3^-, $SOCl_2$ H_2O, SCl_2, ClF_2^+

Table 4.27 Hybridisation of central atoms.

Extension

Other hybridisation schemes, involving d orbitals, can be used to rationalise the bonding in 5- and 6-coordinate molecules – sp^3d^2 for an octahedral shape for example.

Is hybridisation real? There is evidence from ultraviolet photoelectron spectroscopy to suggest that it is not. Why then do we still use this theory? There is a more advanced theory of bonding called molecular orbital theory, but this is much more complicated and does not allow us to draw lines (bonds) between atoms in the same way. Is a more advanced theory always more useful? What are the implications of using something that we know to be nothing more than a mathematical device for making predictions? How can we justify this?

A key feature in many scientific arguments and theories is a clear distinction between cause and effect. There is often some confusion at this level about whether hybridisation is something that causes the shape of a molecule to be how it is or whether hybridisation is just a simplified treatment of bonding that allows us to rationalise the shape of molecules based on a reasonably straightforward and accessible theory.

? Test yourself

17 Predict the hybridisation of the central atom in each of the following

  a BCl$_3$ d CCl$_4$ g PCl$_4^+$ i SeO$_3$

  b NCl$_3$ e BeCl$_2$ h OCN$^-$ j FNO

  c OCl$_2$ f H$_2$S

18 What is the hybridisation of the nitrogen atoms in each of the following?

  a N$_2$H$_4$

  b N$_2$H$_2$

19 What is the hybridisation of the carbon atoms in each of the following?

  a C$_2$F$_4$ b C$_2$F$_2$ c C$_2$F$_6$ d COF$_2$

Exam-style questions

1 What is the formula of the compound formed between lithium and nitrogen?

    A LiN$_2$ **B** LiN$_3$ **C** Li$_3$N   **D** Li$_3$N$_2$

2 Which of the following contains both ionic and covalent bonding?

    A NaCl   **B** NH$_4$Cl   **C** CCl$_4$ **D** PCl$_3$

3 What is the shape of NO$_2^+$?

    A Linear   **C** Trigonal planar

    B Bent   **D** Tetrahedral

4 Which of the following is polar?

    A CO$_2$ **B** CCl$_4$ **C** BF$_3$ **D** PCl$_3$

5 Which of the following molecules exhibit(s) hydrogen bonding?

I NH_3 II CH_3NH_2 III HF   IV CH_3F

A I, II and III only
B IV only

C I and III only
D III and IV only

6 In which of the following are the molecules arranged in order of increasing boiling point (lowest first)?

A NH_3 N_2 Br_2
B H_2O H_2S H_2Se
C CH_3Cl CH_2Cl_2 $CHCl_3$
D C_4H_{10} C_3H_8 C_2H_5OH

7 What is the F–B–F bond angle in BF_4^-?

A 109.5°   B 107°   C 120°   D 90°

8 Which of the following will be the worst conductor of electricity?

A Mg(s)
B $SiCl_4$(l)

C $MgCl_2$(l)
D C(graphite,s)

9 When the compounds C_2H_6, C_2H_4, C_2H_2 and C_6H_6 are arranged in order of increasing C–C bond lengths (shortest first) the correct order is:

A C_2H_6 C_2H_4 C_2H_2 C_6H_6
B C_6H_6 C_2H_4 C_2H_2 C_2H_6
C C_2H_2 C_6H_6 C_2H_4 C_2H_6
D C_2H_2 C_2H_4 C_6H_6 C_2H_6

HL 10 The number of π bonds and hybridisation of C in a molecule of hydrogen cyanide, HCN, are:

	number of π bonds	hybridisation
A	3	sp^2
B	2	sp
C	1	sp^2
D	2	sp^3

HL 11 Which of the following contains delocalised electrons?

A CO_2 B O_3 C SiO_2 D OCl_2

HL 12 In which of the following is the distribution of electron pairs around the central atom octahedral?

  A BCl_4^- **B** SF_4 **C** PCl_4^+ **D** XeF_4

13 **a** Describe the principles of the valence shell electron pair repulsion theory for predicting the shapes of molecules. [4]

  b Predict the shapes and bond angles of the following molecules: [4]
     i PCl_3
     ii CO_2

  c Explain why carbon dioxide is a non-polar molecule but sulfur dioxide is polar. [3]

  d Draw a Lewis structure for carbon monoxide and explain whether it has a shorter or longer C–O bond length than carbon dioxide. [3]

14 Explain the following in terms of structure and bonding.

  a Sodium oxide has a high melting point and does not conduct electricity when solid but conducts electricity when molten. [4]

  b Sodium has a lower melting point than magnesium. [3]

  c Phosphine, PH_3, has a lower boiling point than ammonia, NH_3, and arsine, AsH_3. [3]

  d Silicon dioxide has a much higher melting point than carbon dioxide. [3]

HL 15 **a** Explain the term 'hybridisation'. [2]

  b Predict the hybridisation of the carbon atoms in ethene (C_2H_4) and ethyne (C_2H_2). [2]

  c Explain, by reference to ethene and ethyne, what is meant by the terms 'sigma bond' and 'pi bond'. [4]

HL 16 **a** Use the valence shell electron pair repulsion theory to predict the shapes and bond angles of the following molecules or ions:
     i SO_4^{2-} **ii** XeF_4 **iii** SF_4 [6]

  b Explain whether XeF_4 is polar or non-polar. [2]

  c Explain how you would expect the S–O bond length in SO_4^{2-} to compare with that in SO_2. [2]

Summary

Alloys are homogeneous mixtures of two or more metals or of a metal with a non-metal. —— Alloys have enhanced properties.

BONDING —— metallic bonding —— lattice of positive ions surrounded by delocalised electrons —— electrostatic attraction between positive ions and delocalised electrons

covalent bonding

ionic bonding —— giant lattice structure made up of positive metal ions and negative non-metal ions —— electrostatic attraction between oppositely charged ions

To COVALENT BONDING on next page

can be represented using **Lewis structures**

HL

formal charge can be used to distinguish between different Lewis structures for a molecule/ion

resonance structures differ in the position of double bonds

formal charge (FC) = (number of valence electrons in the uncombined atom) – $\frac{1}{2}$(number of bonding electrons) – (number of non-bonding electrons)

show all outer electrons

the preferred Lewis structure is the one in which the formal charges are closest to zero

characteristics

positive ions:
group 1   1+
group 2   2+
group 13   3+

negative ions:
group 15   3–
group 16   2–
group 17   1–

—— often soluble in water

—— insoluble in non-polar solvents

—— conduct electricity when molten or dissolved in water

—— high melting point

complete transfer of electrons – large difference in electronegativity

pairs of electrons are shared by different atoms

occurs when there is a small difference in electronegativity

Electronegativity
– increases across a period
– decreases down a group
F has the highest electronegativity.

electrostatic attraction between nuclei and shared pair of electrons

In a **coodinate covalent bond** both electrons come from the same atom.

HL

The oxygen–oxygen bond is weaker in O_3 than in O_2, so O_3 absorbs longer wavelength UV radiation.

The destruction of ozone can be catalysed by Cl radicals or NO_x.

In **delocalisation**, a pair of electrons is shared between three or more atoms.

The bond length is intermediate between that of a single and that of a double bond.

achieved by overlap of orbitals

types

σ bond – head-on overlap of 2 orbitals; electron density lies along the internuclear axis

π bond – side-on overlap of 2 p orbitals; electron density lies above and below the internuclear axis

Hybridisation is the mixing of atomic orbitals.

—— 2 electron domains around central atom – basic shape is linear – sp hybridisation.

—— 3 electron domains around central atom – basic shape is trigonal planar – sp^2 hybridisation.

—— 4 electron domains around central atom – basic shape is tetrahedral – sp^3 hybridisation.

A single bond is always 1 σ bond.

A double bond consists of 1 σ and 1 π bond.

A triple bond consists of 1 σ and 2 π bonds.

Summary – continued

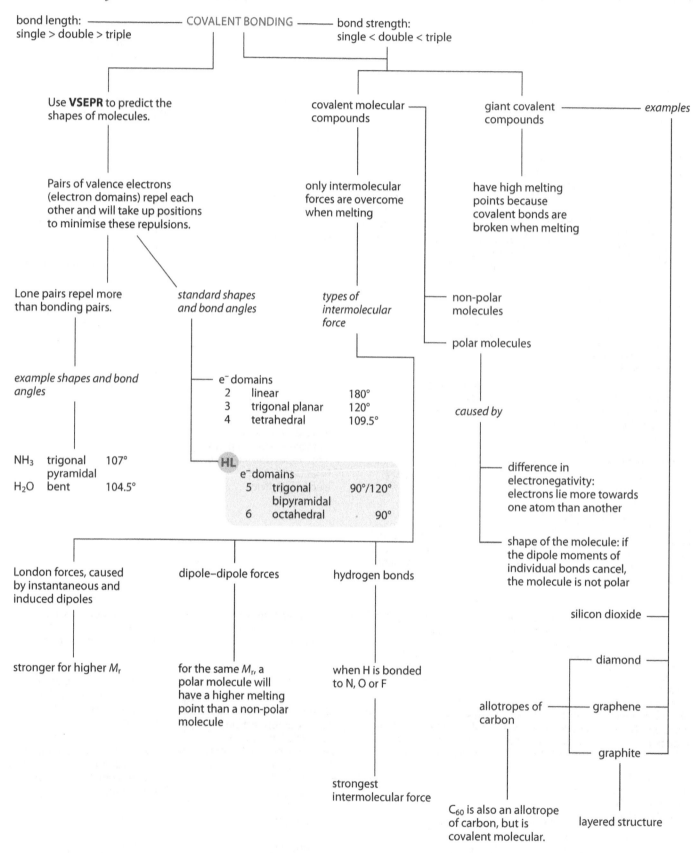

bond length: ————————————— COVALENT BONDING ———— bond strength:
single > double > triple   single < double < triple

Use **VSEPR** to predict the shapes of molecules.

covalent molecular compounds

giant covalent compounds ———— *examples*

Pairs of valence electrons (electron domains) repel each other and will take up positions to minimise these repulsions.

only intermolecular forces are overcome when melting

have high melting points because covalent bonds are broken when melting

Lone pairs repel more than bonding pairs.

standard shapes and bond angles

types of intermolecular force

non-polar molecules

polar molecules

example shapes and bond angles

e⁻ domains
2   linear   180°
3   trigonal planar   120°
4   tetrahedral   109.5°

caused by

NH_3 trigonal pyramidal   107°
H_2O bent   104.5°

HL
e⁻ domains
5   trigonal bipyramidal   90°/120°
6   octahedral   90°

difference in electronegativity: electrons lie more towards one atom than another

shape of the molecule: if the dipole moments of individual bonds cancel, the molecule is not polar

London forces, caused by instantaneous and induced dipoles

dipole–dipole forces

hydrogen bonds

silicon dioxide ————

diamond ————

stronger for higher M_r

for the same M_r, a polar molecule will have a higher melting point than a non-polar molecule

when H is bonded to N, O or F

allotropes of carbon ———— graphene ————

graphite ————

strongest intermolecular force

C_{60} is also an allotrope of carbon, but is covalent molecular.

layered structure

Energetics/thermochemistry 5

5.1 Measuring energy changes

Heat and temperature

Heat and **temperature** are very different things that often get confused.

> Heat is a form of energy that flows from something at a higher temperature to something at a lower temperature.

> Temperature is a measure of the average kinetic energy of particles.

Internal energy is the name given to the total amount of energy (kinetic and potential) in a sample of a substance. If a 100 g block of iron at 100 °C is placed in contact with a 50 g block of iron at 50 °C, heat will flow from the hotter block to the colder one until they are at the same temperature. When they are at the same temperature, the 100 g block of iron will have higher internal energy simply because there is more of it, but the average kinetic energy of the particles in the two blocks will be the same because they are at the same temperature.

We often use the words *system* and *surroundings* to describe what is happening in terms of energy flow in chemical reactions.

> The 'system' is our chemical reaction and the 'surroundings' is everything else in the Universe (Figure **5.1**)!

Exothermic and endothermic reactions and enthalpy changes

Chemical reactions may be classified as either **exothermic** or **endothermic**.

> In an exothermic reaction, heat energy is transferred from a system (chemical reaction) to the surroundings (Figure **5.2a**) – the surroundings get hotter.

> In an endothermic reaction, a system (chemical reaction) takes in heat energy from the surroundings (Figure **5.2b**) – the surroundings get cooler.

Learning objectives

- Understand the difference between heat and temperature
- Explain what is meant by exothermic and endothermic reactions
- Draw enthalpy level diagrams for exothermic and endothermic reactions
- Understand what is meant by stability
- Understand the principle of experimental methods for determining enthalpy changes
- Work out enthalpy changes from experimental data

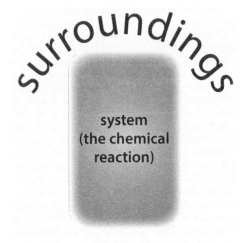

Figure 5.1 The Universe!

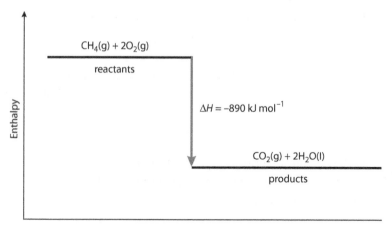

Figure 5.2 **a** The heat energy released in an exothermic reaction comes from the decrease in internal energy (the total energy of all the particles) of the system, for example through the creation of chemical bonds (conversion of chemical energy to heat energy). **b** The heat energy consumed in an endothermic reaction is converted to internal energy, for example through the breaking of chemical bonds (conversion of heat energy to chemical energy).

> To be more precise, the enthalpy change is the heat energy exchanged with the surroundings at constant pressure.

> Δ means a change in a quantity.

> ΔH values are usually quoted under standard conditions. This is discussed in more detail on page **187**.

We talk about the **enthalpy change** of a system. The enthalpy change is basically the amount of heat energy taken in/given out in a chemical reaction. Enthalpy is given the symbol H and enthalpy change is ΔH.

It is not possible to measure the enthalpy (H) of a system (related to the total energy of all the particles making up a substance) only the enthalpy change (ΔH) when the system moves from an initial state to some final state.

> ΔH for an **exothermic** reaction is **negative**.
> ΔH for an **endothermic** reaction is **positive**.

Consider a reaction such as methane burning in oxygen:

$$CH_4(g) + 2O_2(g) \rightarrow CO_2(g) + 2H_2O(l) \qquad \Delta H = -890 \, kJ \, mol^{-1}$$

This reaction is exothermic, because heat is given out to the surroundings.

Figure 5.3 An enthalpy level diagram for the combustion of methane. No scale is shown on the vertical axis because we cannot measure the initial enthalpy or final enthalpy of the system.

Figure **5.3** shows an enthalpy level diagram for the combustion of methane. The reaction is exothermic, and the enthalpy of the products is less than that of the reactants. The enthalpy change of a reaction is the amount of heat given out – this is shown by the red arrow. The negative sign for the enthalpy change indicates a decrease in enthalpy.

Because the energy of the products is less than the energy of the reactants, the difference between the two must equal the amount of heat given out. Another way of stating this for an exothermic reaction is:

- total energy of the reactants = total energy of the products + heat given out
- or ΔH = enthalpy of products − enthalpy of reactants

In an exothermic reaction the products are at a lower energy (enthalpy) level than the reactants, and we say that the products are more stable than the reactants. This will be discussed further on page **213**.

> In this section the terms *enthalpy* and *energy* are being used fairly interchangeably. At this level, the enthalpy of a system can be regarded as essentially the total energy stored in a substance – i.e. basically the same as the internal energy. This is not strictly true but it is sufficient for a good understanding of the concepts.

The reaction between nitrogen and oxygen to form nitrogen(II) oxide (nitric oxide) is endothermic:

$$N_2(g) + O_2(g) \rightarrow 2NO(g) \qquad\qquad \Delta H = +181\,\text{kJ}\,\text{mol}^{-1}$$

Energy (heat) flows from the surroundings into the system because the products have a higher energy than the reactants. The sign of ΔH is positive, indicating the increase in enthalpy in an endothermic reaction. In this case:

- total energy of products = total energy of reactants + heat taken in
- or, as above, ΔH = enthalpy of products − enthalpy of reactants

The enthalpy level diagram for this reaction is shown in Figure **5.4**. The products have higher energy (enthalpy) and are less stable than the reactants.

Some definitions

Enthalpy changes have different values, depending on the conditions under which they are measured. To make them transferable, they are all quoted for the same set of conditions, which is called **standard conditions**. If an enthalpy change is not measured under standard conditions, its value is corrected to that at standard conditions. An enthalpy change under standard conditions is called a **standard enthalpy change** and has the symbol $\Delta H^{\ominus}$, where the symbol $^{\ominus}$ means 'under standard conditions'.

> Total energy is conserved in a chemical reaction.

> The term **stability** is usually used to describe the relative energies of reactants and products in a chemical reaction. If the products have less energy than the reactants then they are more stable.

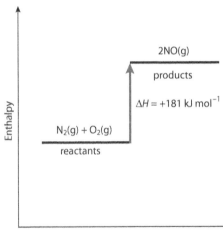

Figure 5.4 An enthalpy level diagram for an endothermic reaction.

Standard conditions:
pressure = 100 kPa (1.00×10^5 Pa).

There is no mention of temperature in the definition of standard conditions and a temperature should always be specified for a standard enthalpy change. This can then be written as $\Delta H^{\ominus}_{298}$ or $\Delta H^{\ominus}(298\,K)$. Where no temperature is stated we will assume that the temperature is 298K.

Standard enthalpy change of reaction ($\Delta H^{\ominus}_r$) is the enthalpy change (heat given out or taken in) when molar amounts of reactants, as shown in the stoichiometric equation, react together under standard conditions to give products.

For example, for the reaction $N_2(g) + 3H_2(g) \rightarrow 2NH_3(g)$, the enthalpy change of reaction is $-92\,kJ\,mol^{-1}$. This means that 92 kJ of heat energy are given out when 1 mol N_2 reacts with 3 mol H_2 to form 2 mol NH_3.

If the equation is written as:

$$\tfrac{1}{2}N_2(g) + \tfrac{3}{2}H_2(g) \rightarrow NH_3(g) \qquad \Delta H^{\ominus}_r = -46\,kJ\,mol^{-1}$$

then the enthalpy change of reaction is for 0.5 mol N_2 reacting, and the enthalpy change is half as much.

Standard enthalpy change of combustion ($\Delta H^{\ominus}_c$) is the enthalpy change when one mole of a substance is completely burnt in oxygen under standard conditions.

A standard enthalpy change of combustion is always negative because combustion is always an exothermic process. For example:

$$CH_4(g) + 2O_2(g) \rightarrow CO_2(g) + 2H_2O(l) \qquad \Delta H^{\ominus}_c = -890\,kJ\,mol^{-1}$$

Nature of science

There are certain unifying principles that are fundamental to the study of science. The idea of conservation of energy is one such principle.

Measuring enthalpy changes

In this section we will consider some experimental methods for measuring enthalpy changes of chemical reactions.

Specific heat capacity

We usually take the definition of **specific heat capacity** (*c*) to be:

The energy required to raise the temperature of 1 g of substance by 1 K (1 °C).

Specific heat capacity has units of $J\,g^{-1}\,K^{-1}$ or $J\,g^{-1}\,°C^{-1}$. Units that are also encountered are $kJ\,kg^{-1}\,K^{-1}$ or $J\,kg^{-1}\,K^{-1}$.

1 K is the same as 1 °C when temperature changes are being considered.

It can also be defined as the energy to raise the temperature of 1 kg of substance by 1 K.

The specific heat capacity of aluminium is $0.90\,J\,g^{-1}\,°C^{-1}$. Therefore, if 0.90 J of heat energy are put into 1 g of aluminium, the temperature is raised by 1 °C (Figure **5.5**).

If 1.80 J of heat energy were put into this block of aluminium, the temperature would go up by 2 °C. If the 1 g block of aluminium were

replaced by a 2 g block of aluminium, then 1.80 J would be required to raise the temperature by 1 °C, because 0.90 J are required to raise the temperature of each 1 g by 1 °C. The amount of heat energy required is therefore proportional to the mass and the temperature change. An equation can be derived for how much heat energy (q) must be supplied to raise the temperature of mass m by ΔT °C:

$q = mc\Delta T$

Specific heat capacity indicates how much energy is required to heat up a substance, therefore substances with higher specific heat capacities are more difficult to heat up than substances with lower specific heat capacities. For example, the specific heat capacity of iron is roughly half that of aluminium – therefore if the same amount of heat energy is supplied to 10 g of each metal, the temperature of the iron will go up by twice the amount.

The specific heat capacity also applies when a substance cools. For instance, when 1 g of aluminium cools from 21 °C to 20 °C, 0.90 J of energy are given out.

Measuring an enthalpy change of combustion

The basic technique is called **calorimetry**. The idea is that the heat given out in a combustion reaction is used to heat another substance of known specific heat capacity, such as water. The equation $q = mc\Delta T$ can be used to calculate the amount of heat given out.

The experimental set-up shown in Figure **5.6** could be used to determine the enthalpy change when one mole of a liquid substance is burnt (the enthalpy change of combustion). The mass and temperature change of the water must be measured, as well as the mass change of the alcohol.

> ## Worked example
>
> **5.1** Use the following experimental data to determine the enthalpy change of combustion of ethanol (C_2H_5OH) given that the specific heat capacity of water is 4.18 J g^{-1} °C^{-1}.
>
> Mass of water = 150.00 g
> Initial temperature of water = 19.5 °C
> Maximum temperature of water = 45.7 °C
> Initial mass of spirit burner = 121.67 g
> Final mass of spirit burner = 120.62 g

The temperature change of the water = 45.7 − 19.5 = 26.2 °C

The amount of heat energy supplied to the water is given by:

$q = mc\Delta T$

$q = 150.00 \times 4.18 \times 26.2 = 16\,400$ J

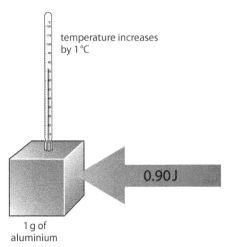

ΔT is the change in temperature.

temperature increases by 1 °C

0.90 J

1 g of aluminium

Figure 5.5 The specific heat capacity of aluminium is 0.90 J g^{-1} °C^{-1}.

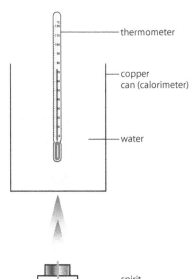

thermometer

copper can (calorimeter)

water

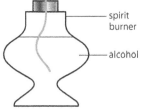

spirit burner

alcohol

Figure 5.6 An experiment to work out the enthalpy change of combustion of an alcohol.

> **Exam tip**
> Note – the mass of water and **not** the mass of ethanol is used here – it is the water that is being heated.

This amount of heat energy is supplied by the burning of the ethanol.

mass of ethanol burnt $= 121.67 - 120.62 = 1.05\,g$

The number of moles of ethanol burnt is given by:

$$\text{number of moles} = \frac{\text{mass}}{\text{molar mass}}$$

The molar mass of ethanol is $46.08\,g\,mol^{-1}$. Therefore:

$$\text{number of moles ethanol burnt} = \frac{1.05}{46.08} = 0.0228\,mol$$

When 0.0228 mol ethanol are burnt, 16 400 J of heat energy are produced. Therefore the amount of energy released when one mole is burnt is given by:

$$\text{energy} = \frac{16\,400}{0.0228} = 721\,000\,J\,mol^{-1}$$

The enthalpy change of combustion of ethanol is therefore:

$$\Delta H = -721\,kJ\,mol^{-1}$$

More than three significant figures were carried through on the calculator to give this final answer.

The value of the enthalpy change is negative because the combustion reaction is exothermic.

When we carry out experiments to measure enthalpy changes we often get unexpected values. What criteria do we use when carrying out these experiments to decide whether there are flaws in the experiment or flaws in the theory? This is an example of a more general problem, and it is sometimes too easy to dismiss an 'anomalous' result as being due to systematic errors in the experiment. The opposite also occasionally happens, and a new theory is put forward from an experiment that is possibly flawed – cold fusion is an example of this (originally announced by Pons and Fleischmann in 1989).

The accepted literature value for the enthalpy change of combustion of ethanol is $-1371\,kJ\,mol^{-1}$, so it can be seen that this experiment does not give a very accurate answer. There are several major flaws (systematic errors) in the experimental set-up. The most major problem is that of heat loss to the surroundings. Of the heat energy released when the ethanol burns, only some of it goes into heating the water – the rest goes into heating the copper can and the surrounding air. The experiment could have been improved by determining the specific heat capacity of the can and taking this into account when doing the calculation, insulating the can so that less heat is lost through the can to the surroundings, using some sort of draught shield to reduce convection currents around the experiment, etc.

Another major problem with the experiment is incomplete combustion of the ethanol. When ethanol undergoes complete combustion the equation for the reaction is:

$$C_2H_5OH + 3O_2 \rightarrow 2CO_2 + 3H_2O$$

However, if there is not a sufficient supply of oxygen to the flame, some of the ethanol can burn to produce carbon monoxide and soot (carbon) as well as water. This is called **incomplete combustion** and gives out less heat than complete combustion. Incomplete combustion causes the flame to be yellow/orange rather than blue because of the presence of soot particles, and the soot can also be seen on the bottom of the copper can.

Other, more minor, problems with the experiment include evaporation of the water and alcohol.

More accurate values for the enthalpy change of a combustion reaction require the use of a **bomb calorimeter**. This is a heavily insulated piece of apparatus in which the substance is ignited electronically in a plentiful supply of oxygen.

Test yourself

The specific heat capacity of water is 4.18 J g^{-1} °C^{-1}.

1 Work out the specific heat capacities of each metal from the data given:
  a Gold – the temperature of 2.00 g of gold is raised by 11.7 °C when 3.00 J of energy is supplied.
  b Silver – the temperature of 100.0 g of silver is raised by 2.12 °C when 50.0 J of energy is supplied.

2 a When 1.20 g of hexane (C_6H_{14}) are burnt, the temperature of 250.0 g of water is raised by 56.0 °C. Calculate the enthalpy change when one mole of hexane is burnt.
  b When 2.00 kg of octane (C_8H_{18}) are burnt, the temperature of 500 kg of water is raised by 46.0 °C. Calculate the enthalpy change when one mole of octane is burnt.

3 Use the following experimental data to determine the enthalpy change of combustion of propan-1-ol (C_3H_7OH):
  Mass of water = 200.00 g
  Initial temperature of water = 18.2 °C
  Maximum temperature of water = 38.6 °C
  Initial mass of spirit burner = 185.51 g
  Final mass of spirit burner = 184.56 g

4 The actual value for the enthalpy change of combustion of propan-1-ol is −2010 kJ mol^{-1}. Account for any differences between this value and the one calculated from the experimental data in question **3**.

Enthalpy changes in solution

A general method for measuring enthalpy changes involving solutions in the laboratory is to measure out known amounts of reagents, record their initial temperatures, mix together the reagents in a polystyrene cup and record the maximum/minimum temperature observed. The specific heat capacity of the final solution is assumed to be the same as for water. Before we look at some examples of how to do this, we must consider a couple of definitions.

Enthalpy change of neutralisation (ΔH_n) is the enthalpy change when one mole of water molecules are formed when an acid (H^+) reacts with an alkali (OH^-) under standard conditions:

$$H^+(aq) + OH^-(aq) \rightarrow H_2O(l)$$

Enthalpy change of solution (ΔH_{sol}) is the enthalpy change when one mole of solute is dissolved in excess solvent to form a solution of 'infinite dilution' under standard conditions, e.g.:

$$NH_4NO_3(s) \xrightarrow{\text{excess } H_2O} NH_4^+(aq) + NO_3^-(aq) \quad \Delta H_{sol} = +25.7 \text{ kJ mol}^{-1}$$

The enthalpy change of neutralisation is always exothermic.

'Infinite dilution' means that any further dilution of the solution produces no further enthalpy change – i.e. the solute particles are assumed not to interact with each other in the solution.

The enthalpy change of solution may be exothermic or endothermic.

Worked examples

5.2 Consider the following experiment: $100.0\,cm^3$ of $1.00\,mol\,dm^{-3}$ potassium hydroxide solution are measured out and poured into a polystyrene cup and the temperature of the potassium hydroxide solution was measured. Then $120.0\,cm^3$ of $1.00\,mol\,dm^{-3}$ hydrochloric acid are measured out and the initial temperature was measured. The hydrochloric acid was in excess to make sure that all the potassium hydroxide reacted. The hydrochloric acid was then poured into the polystyrene cup and the mixture stirred rapidly. The maximum temperature was recorded.

The results of this experiment:

Initial temperature of potassium hydroxide solution = 19.7 °C

Initial temperature of hydrochloric acid = 19.7 °C

Maximum temperature reached = 25.9 °C

Use these data to determine the enthalpy change of neutralisation.

Temperature change of the mixture = 6.2 °C

Total volume of the reaction mixture = $220.0\,cm^3$

We will assume that the density of the mixture is the same as that of water, and so $220.0\,cm^3$ of solution has a mass of 220.0 g.

Assumption: the density of the potassium hydroxide and hydrochloric acid solutions are the same as water, so $1\,cm^3$ of solution has a mass of 1 g.

We can work out how much heat (*q*) has been released in this reaction by using the temperature change of the mixture:

$$q = mc\Delta T = 220 \times 4.18 \times 6.2 = 5700\,J$$

Assumption: the specific heat capacity of the mixture is the same as that of water. This is a fairly reasonable assumption because the reaction mixture is mostly water.

To work out the enthalpy change of neutralisation, we need to know how many moles of water have been formed. The equation for the reaction is:

$$KOH + HCl \rightarrow KCl + H_2O$$

number of moles KOH = concentration × volume in dm^3

$$\text{number of moles KOH} = 1.00 \times \frac{100}{1000} = 0.100\,mol$$

$$\text{number of moles HCl} = 1.00 \times \frac{120}{1000} = 0.120\,mol$$

The HCl is in excess so the number of moles of water produced is 0.100 mol. Therefore 5700 J of energy is released when 0.100 mol water are formed.

The definition of enthalpy change of neutralisation is the enthalpy change when one mole of water is produced. Therefore, for one mole of water formed:

$$\text{heat energy released} = \frac{1}{0.100} \times 5700 = 57\,000\,\text{J}\,\text{mol}^{-1}$$

Therefore, the enthalpy change of neutralisation, $\Delta H_n = -57.0\,\text{kJ}\,\text{mol}^{-1}$.

> This is negative because the reaction is exothermic.

The accepted literature value for the heat of neutralisation of KOH with HCl is $-57.2\,\text{kJ}\,\text{mol}^{-1}$.

Possible errors in this experiment are heat loss to the surroundings and the assumptions that have been made about the specific heat capacities and the density of the solutions. The heat capacity of the calorimeter (polystyrene cup) was also not taken into account – some of the heat energy given out from the reaction was used to heat up the cup.

5.3 a $100.0\,\text{cm}^3$ of $1.00\,\text{mol}\,\text{dm}^{-3}$ potassium hydroxide solution were reacted with $100.0\,\text{cm}^3$ of $1.00\,\text{mol}\,\text{dm}^{-3}$ hydrochloric acid. The temperature rise was $6.82\,°\text{C}$. Calculate the enthalpy change of neutralisation.

b The experiment in part **a** was repeated with $50.0\,\text{cm}^3$ of $1.00\,\text{mol}\,\text{dm}^{-3}$ potassium hydroxide solution and $50.0\,\text{cm}^3$ of $1.00\,\text{mol}\,\text{dm}^{-3}$ hydrochloric acid. Calculate the temperature change of the reaction mixture.

a We can use the method from the previous example to calculate the enthalpy change of neutralisation:

$$q = mc\Delta T = 200 \times 4.18 \times 6.82 = 5700\,\text{J}$$

The number of moles of KOH and HCl in this case are the same, both $0.100\,\text{mol}$. Therefore $0.100\,\text{mol}$ water are formed.

$$\text{heat energy released} = \frac{1}{0.100} \times 5700 = 57\,000\,\text{J}\,\text{mol}^{-1}$$

Therefore the enthalpy change of neutralisation, $\Delta H_n = -57.0\,\text{kJ}\,\text{mol}^{-1}$.

b A shortcut can be used to answer this part. The volume of each solution is half that in part **a** but the concentrations are the same. Therefore we can deduce that the number of moles of water formed will be half as much as in part **a**. This means that half as much heat energy will be given out in the neutralisation reaction. However, the total volume of reaction mixture that is being heated is half the original volume – therefore only half as much heat energy will be required to heat it to the same temperature. So the temperature change in this experiment is the same as in part **a**, i.e. $6.82\,°\text{C}$.

Exam tip
Some people prefer to use the equation $q = -mc\Delta T$. This avoids the problem of forgetting to add the negative sign for the final enthalpy change for an exothermic reaction. The temperature change for an endothermic reaction must then be taken as negative.

5.4 Consider the following experiment: 100.0 cm³ of water are measured out and poured into a polystyrene cup and the temperature of the water was measured. Then 5.20 g of ammonium chloride are measured out. The ammonium chloride was added to the water and the solution stirred vigorously until all the ammonium chloride had dissolved. The minimum temperature was recorded.

The results of this experiment:

Initial temperature of water = 18.3 °C
Minimum temperature = 15.1 °C

Use the experimental data to determine the enthalpy change of solution of ammonium chloride.

Temperature change of the mixture = 3.2 °C

We will assume that the density of the solution is the same as that of water, and so 100.0 cm³ of solution has a mass of 100.0 g. We can work out how much heat has been absorbed in this reaction by using the temperature change of the mixture:

$$q = mc\Delta T = 100.0 \times 4.18 \times 3.2 = 1340 \, \text{J}$$

Assumption: the specific heat capacity of the solution is the same as that of water.

To work out the enthalpy change of solution, we need to know how many moles of ammonium chloride dissolved:

$$\text{number of moles NH}_4\text{Cl} = \frac{\text{mass}}{\text{molar mass}}$$

Therefore, the number of moles of NH_4Cl that dissolve $= \dfrac{5.20}{53.50} = 0.0972 \, \text{mol}$.

Therefore 1340 J of energy is absorbed when 0.0972 mol NH_4Cl dissolve.

The definition of enthalpy change of solution is the enthalpy change when one mole of substance dissolves. Therefore, for one mole of NH_4Cl dissolving:

$$\text{heat energy absorbed} = \frac{1}{0.0972} \times 1340 = 13\,800 \, \text{J mol}^{-1}$$

This is positive because the reaction is endothermic.

Therefore the enthalpy change of solution, $\Delta H_{\text{sol}} = +13.8 \, \text{kJ mol}^{-1}$.

The accepted value for the enthalpy change of solution of ammonium chloride is 15.2 kJ mol⁻¹. Errors in this experiment include the absorption of heat from the surroundings and the assumption about the specific heat capacity of the solution being the same as that of water. The mass of the ammonium chloride was also not taken into account when working out the heat energy released in the experiment (i.e. it was not included in the mass of the solution). The results from this experiment are greatly improved by vigorous stirring of the solution because the ammonium chloride does not dissolve instantaneously. If it is allowed to dissolve slowly, there is more time for heat to be absorbed from the surroundings and the temperature drop is not as large as expected. This effect can be reduced by using the technique of the next experiment.

5.5 The following experiment may be used to determine the enthalpy change of reaction for:

$$Zn(s) + CuSO_4(aq) \rightarrow ZnSO_4(aq) + Cu(s)$$

$50.0\,cm^3$ of $0.200\,mol\,dm^{-3}$ copper(II) sulfate solution are placed in a polystyrene cup. The temperature was recorded every 30 s for 2 min. At 2 min 30 s, 1.20 g of powdered zinc are added. The mixture was stirred vigorously and the temperature recorded every half minute for several minutes. The results obtained were then plotted to give the graph shown in Figure **5.7**.

Use these data to determine the enthalpy change for this reaction.

This type of reaction is called a displacement reaction or a single replacement reaction.

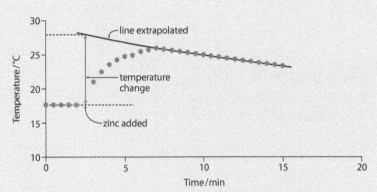

Figure 5.7 Temperature against time for the reaction of zinc with copper sulfate solution.

The problem with this reaction is that it does not occur instantaneously, and so although heat is being given out in the reaction, at the same time the reaction mixture is also cooling down by losing heat to the surroundings. From the graph it can be seen that there is an initial rise in temperature, where heat being given out by the reaction is the major factor, but after that the reaction mixture cools. By extrapolating the line back from this cooling part of the curve, we can estimate the temperature from which the mixture appears to be cooling. If we look at the value of the temperature on this curve at 2 min 30 s, the point at which the zinc was added, we should get an estimate of the temperature rise if the reaction had occurred instantaneously. It can be seen that this temperature rise is greater than the temperature rise actually measured in the experiment.

From the graph we estimate the temperature change as $10.3\,°C$.

The heat given out in the reaction is given by:

$$q = mc\Delta T = 50.0 \times 4.18 \times 10.3 = 2150\,J$$

Assumption: the density of copper sulfate solution is the same as that of water.

To work out the enthalpy change of reaction, we need to know how many moles of copper sulfate reacted (the zinc was in excess).

number of moles of $CuSO_4$ = concentration × volume in dm^3

number of moles of $CuSO_4 = 0.200 \times \dfrac{50.0}{1000} = 0.0100\,mol$

Therefore 2150 J of energy are released when 0.0100 mol copper sulfate reacts.

For the reaction of one mole of copper sulfate:

heat energy released $= \dfrac{1}{0.0100} \times 2150 = 215\,000\,J\,mol^{-1}$

This is negative because the reaction is exothermic.

Therefore the enthalpy change of reaction $= -215\,kJ\,mol^{-1}$.

The extrapolation of the line is very much a matter of judgement and could introduce errors into the calculation.

Experimental work and the collection of data is a major part of science. The best data for making accurate predictions are quantitative data. These can be analysed mathematically to allow theories to be developed. Scientists must, however, be aware of the errors and uncertainties in their data and report the results of their experiments appropriately.

? Test yourself

5 a $200.0\,cm^3$ of $0.150\,mol\,dm^{-3}$ hydrochloric acid is mixed with $100.0\,cm^3$ of $0.300\,mol\,dm^{-3}$ sodium hydroxide solution. The temperature rose by $1.36\,°C$. If both solutions were originally at the same temperature, calculate the enthalpy change of neutralisation.

b Predict the temperature rise if the experiment in part **a** is repeated using:

  i $400.0\,cm^3$ of $0.150\,mol\,dm^{-3}$ hydrochloric acid and $200.0\,cm^3$ of $0.300\,mol\,dm^{-3}$ sodium hydroxide solution

  ii $200.0\,cm^3$ of $0.300\,mol\,dm^{-3}$ hydrochloric acid and $100.0\,cm^3$ of $0.600\,mol\,dm^{-3}$ sodium hydroxide solution

  iii $50.0\,cm^3$ of $0.300\,mol\,dm^{-3}$ hydrochloric acid and $25.0\,cm^3$ of $0.600\,mol\,dm^{-3}$ sodium hydroxide solution.

6 a When $1.00\,g$ of magnesium chloride is dissolved in $50.0\,cm^3$ of water the temperature goes up from $21.5\,°C$ to $29.1\,°C$. Calculate the enthalpy change of solution of magnesium chloride.

b Predict the temperature change when $2.00\,g$ of magnesium chloride is dissolved in $100\,cm^3$ of water.

c Predict the temperature change when $2.00\,g$ of magnesium chloride is dissolved in $50.0\,cm^3$ of water.

Learning objectives

- Use Hess's law to calculate enthalpy changes
- Define enthalpy change of formation
- Calculate enthalpy changes from enthalpy change of formation data

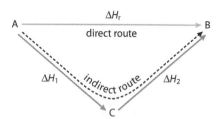

Figure 5.8 An enthalpy cycle.

5.2 Hess's law

It is not always possible to design experiments to measure certain enthalpy changes and so we often have to use data from reactions in which the enthalpy change *can* be measured to work out the enthalpy change for a particular reaction in which it *cannot* be measured directly. In order to do this, we use **Hess's law**:

> The enthalpy change accompanying a chemical reaction is independent of the pathway between the initial and final states.

What this basically means is that if we consider the conversion of A into B, the enthalpy change for the reaction is the same if we go directly from A to B or indirectly via other reactions and intermediates (Figure **5.8**).

If we know the value for ΔH_1 (for the conversion of A to C) and ΔH_2 (for the conversion of C to B) we can work out the value of the enthalpy change ΔH_r for the conversion of A into B using this cycle. Hess's law states that the enthalpy change for the direct conversion of A to B is exactly the same as the enthalpy change for the indirect route between A and B. Therefore, in this case:

$$\Delta H_r = \Delta H_1 + \Delta H_2$$

Let us consider a slightly different situation in which the enthalpy changes we know are:

$$A \rightarrow C = \Delta H_1 \qquad B \rightarrow C = \Delta H_2$$

This produces a slightly different cycle, in which the arrow for ΔH_2 is the other way around, because the reaction we know is from B to C (Figure **5.9**). Now the enthalpy change from A to B is given by $\Delta H_r = \Delta H_1 - \Delta H_2$. We have $-\Delta H_2$ in this case, because the conversion $C \rightarrow B$ on the indirect route ($A \rightarrow C \rightarrow B$) goes in the opposite direction to the arrow we have drawn for ΔH_2.

Another possible situation is one in which we know:

$$C \rightarrow A = \Delta H_1 \qquad C \rightarrow B = \Delta H_2$$

This is shown in Figure **5.10**. Here the enthalpy change for the direct route is given by $\Delta H_r = -\Delta H_1 + \Delta H_2$. In going from $A \rightarrow C \rightarrow B$, we go the wrong way along the ΔH_1 arrow (therefore the sign is negative) but the right way along the ΔH_2 arrow (therefore the sign is positive).

Why Hess's law works

The reason Hess's law works can be understood from Figure **5.11**.

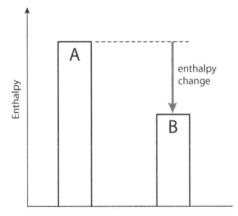

Figure 5.11 Enthalpy is basically a property of a substance under a certain set of conditions. This means that the enthalpy depends on the substance and the conditions. So A has a certain amount of enthalpy, B has a certain amount of enthalpy and these are fixed for a particular set of conditions. The route that is taken between A and B cannot affect the amount of enthalpy A or B has, and therefore the difference in enthalpy between A and B is constant.

Working out enthalpy changes

We are not always able to determine enthalpy changes directly from experiments and Hess's law can then be used to work out unknown enthalpy changes from ones that are known.

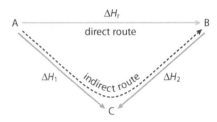
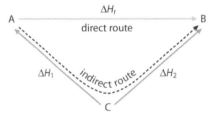

Figure 5.9 If we know the enthalpy change for the conversion $B \rightarrow C$ rather than $C \rightarrow B$, the arrow between B and C is the other way around.

Figure 5.10 If we know the enthalpy change for the conversion $C \rightarrow A$ rather than $A \rightarrow C$, the arrow between A and C is the other way around.

Worked examples

5.6 Given these enthalpy changes:

$$2C(s) + O_2(g) \rightarrow 2CO(g) \qquad \Delta H^{\ominus} = -222\,kJ\,mol^{-1} \qquad \Delta H_1 \qquad \textbf{Reaction 1}$$
$$C(s) + O_2(g) \rightarrow CO_2(g) \qquad \Delta H^{\ominus} = -394\,kJ\,mol^{-1} \qquad \Delta H_2 \qquad \textbf{Reaction 2}$$

calculate the enthalpy change for the reaction:

$$2CO(g) + O_2(g) \rightarrow 2CO_2(g)$$

Two methods will be considered for working out the enthalpy change.

Method A (using a cycle)

The enthalpy changes given are used to construct a cycle. The enthalpy change that we have to find is put at the top of the cycle.

Only the first reaction has been added here. **Reaction 1** is highlighted in yellow and shows the reaction of 2 C and 1 O_2 to form 2 CO. The other O_2, highlighted in orange, remains unchanged, and there is no enthalpy change for this. This means that the total enthalpy change along the red arrow is ΔH_1.

If **Reaction 2** is now added to complete the cycle, it must be multiplied by two, because we need to form $2\,mol\,CO_2$. The enthalpy change must also be multiplied by two, hence $2\Delta H_2$:

$$2C(s) + 2O_2(g) \rightarrow 2CO_2(g) \qquad 2\Delta H_2$$

Reaction 2 is highlighted in green (Figure **5.12**).

The values can now be put into the cycle:

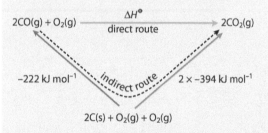

The overall enthalpy change is given by

$$\Delta H^{\ominus} = -\Delta H_1 + 2\Delta H_2$$

$$\Delta H^{\ominus} = -(-222) + (2 \times -394) = -566\,kJ\,mol^{-1}$$

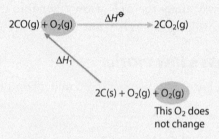

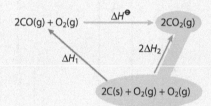

Figure 5.12 Reaction cycles.

This is worked out from Hess's law: the enthalpy change for the direct route is the same as that for the indirect route. Because the direction of the indirect route arrow is opposite to that of the red arrow, the sign of the quantity on the red arrow is reversed.

Method B (manipulating the equations)

In this method we rearrange the equations for the data given to produce the equation for the reaction corresponding to the enthalpy change we have to find.

First, **Reaction 1** is reversed to give 2CO on the left-hand side, as in the overall equation:

$$2CO(g) \rightarrow 2C(s) + O_2(g) \qquad \Delta H^{\ominus} = +222\,kJ\,mol^{-1} \qquad \text{sign changed}$$

Now **Reaction 2** is multiplied by two to give $2CO_2$ on the right-hand side, as in the overall equation:

$$2C(s) + 2O_2(g) \rightarrow 2CO_2(g) \qquad \Delta H^{\ominus} = -788\,kJ\,mol^{-1} \qquad \text{enthalpy change multiplied by two}$$

We now have 2CO on the left-hand side and $2CO_2$ on the right-hand side, as in the overall equations. The two equations and their enthalpy changes are now added together and common terms cancelled to produce the overall equation and its enthalpy change:

$$2CO(g) \rightarrow 2\cancel{C}(s) + \cancel{O_2}(g) \qquad \Delta H^{\ominus} = +222\,kJ\,mol^{-1}$$
$$\underline{2\cancel{C}(s) + 2O_2(g) \rightarrow 2CO_2(g) \qquad \Delta H^{\ominus} = -788\,kJ\,mol^{-1}}$$
$$2CO(g) + O_2(g) \rightarrow 2CO_2(g) \qquad \Delta H^{\ominus} = -566\,kJ\,mol^{-1}$$

> This is very similar to methods for solving simultaneous linear equations in mathematics.

The relationship between the various enthalpy changes in the above example can be seen in an enthalpy level diagram (Figure **5.13**). Downward arrows represent exothermic processes.

> **Exam tip**
> Only one of these methods needs to be understood. Find a method that you are happy with and ignore the other!

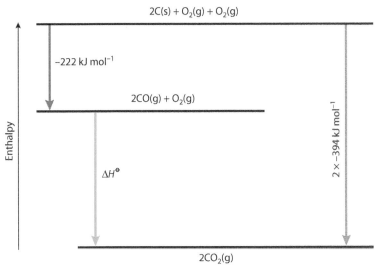

Figure 5.13 An enthalpy level diagram. Exothermic processes are shown as downward arrows – endothermic processes would be shown as upward arrows.

Worked example

5.7 Use the following enthalpy change data:

$$2H_2(g) + O_2(g) \rightarrow 2H_2O(l) \qquad \Delta H_1 = -572 \text{ kJ mol}^{-1} \qquad \textbf{Reaction 1}$$
$$2H_2(g) + O_2(g) \rightarrow 2H_2O(g) \qquad \Delta H_2 = -484 \text{ kJ mol}^{-1} \qquad \textbf{Reaction 2}$$

to work out the enthalpy change for the process:

$$H_2O(l) \rightarrow H_2O(g)$$

Method A (enthalpy cycle)

Each enthalpy change given should be divided by two, as the enthalpy change for the reaction we have to find just involves one H_2O:

$$H_2(g) + \tfrac{1}{2}O_2(g) \rightarrow H_2O(l) \qquad \Delta H_3 = -286 \text{ kJ mol}^{-1}$$

$$H_2(g) + \tfrac{1}{2}O_2(g) \rightarrow H_2O(g) \qquad \Delta H_4 = -242 \text{ kJ mol}^{-1}$$

A cycle can be constructed.

The enthalpy change for the direct route is the same as that for the indirect route:

$$\Delta H = -\Delta H_3 + \Delta H_4$$

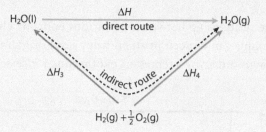

The indirect route goes in the opposite direction to the ΔH_3 arrow, so the sign is negative.

Substituting values we get:

$$\Delta H = -(-286) + (-242) = +44 \text{ kJ mol}^{-1}$$

Method B (manipulating equations)

Divide each equation by two to get one H_2O in each:

$$H_2(g) + \tfrac{1}{2}O_2(g) \rightarrow H_2O(l) \qquad \Delta H = -286 \text{ kJ mol}^{-1}$$

$$H_2(g) + \tfrac{1}{2}O_2(g) \rightarrow H_2O(g) \qquad \Delta H = -242 \text{ kJ mol}^{-1}$$

The first equation must be reversed so that $H_2O(l)$ is on the left-hand side, as in the overall equation:

$$H_2O(l) \rightarrow H_2(g) + \tfrac{1}{2}O_2(g) \qquad \Delta H = +286 \text{ kJ mol}^{-1}$$

We now have the correct numbers of H_2O on the correct sides and can add the two equations (and the two enthalpy changes) together:

$$H_2O(l) \rightarrow \cancel{H_2(g)} + \cancel{\tfrac{1}{2}O_2(g)} \qquad \Delta H = +286 \text{ kJ mol}^{-1}$$
$$\underline{\cancel{H_2(g)} + \cancel{\tfrac{1}{2}O_2(g)} \rightarrow H_2O(g) \qquad \Delta H = -242 \text{ kJ mol}^{-1}}$$
$$H_2O(l) \rightarrow H_2O(g) \qquad \Delta H = +44 \text{ kJ mol}^{-1}$$

An enthalpy level diagram for the reactions in the above example is shown in Figure **5.14**.

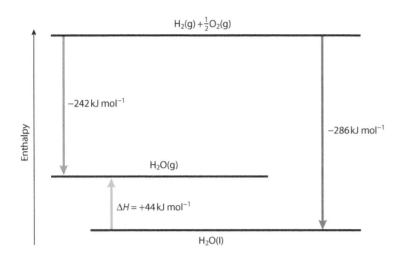

Enthalpy

$H_2(g) + \frac{1}{2}O_2(g)$

$-242\,kJ\,mol^{-1}$

$-286\,kJ\,mol^{-1}$

$H_2O(g)$

$\Delta H = +44\,kJ\,mol^{-1}$

$H_2O(l)$

Figure 5.14 An enthalpy level diagram. Exothermic processes are shown as downwards arrows and endothermic reactions are upwards arrows.

Worked example

5.8 Use the information given below to calculate the enthalpy change for the reaction:

$$C_2H_4(g) + H_2O(g) \rightarrow C_2H_5OH(l)$$

$H_2O(l) \rightarrow H_2O(g)\ \Delta H = +44\,kJ\,mol^{-1}$	ΔH_1	**Reaction 1**
$\Delta H_c^{\ominus}\,[C_2H_5OH(l)] = -1371\,kJ\,mol^{-1}$	ΔH_2	**Reaction 2**
$\Delta H_c^{\ominus}\,[C_2H_4(g)] = -1409\,kJ\,mol^{-1}$	ΔH_3	**Reaction 3**

We are given standard enthalpy change of combustion data and so, knowing the definition, we can write out the full chemical equations for the enthalpy changes:

Reaction 2: $C_2H_5OH(l) + 3O_2(g) \rightarrow 2CO_2(g) + 3H_2O(l)$ $\qquad \Delta H_2 = -1371\,kJ\,mol^{-1}$

Reaction 3: $C_2H_4(g) + 3O_2(g) \rightarrow 2CO_2(g) + 2H_2O(l)$ $\qquad \Delta H_3 = -1409\,kJ\,mol^{-1}$

Method A (enthalpy cycle)

The following cycle can be constructed:

$C_2H_4(g) + H_2O(g)\ (+3O_2(g)) \xrightarrow{\ \Delta H^{\ominus}\ } C_2H_5OH(l)\ (+3O_2(g))$

ΔH_3

$2CO_2(g) + 2H_2O(l) + H_2O(g)$

ΔH_2

ΔH_1

$2CO_2(g) + 2H_2O(l) + H_2O(l)$

$3O_2(g)$ has been added to each side of the original equation, but because the same thing has been added to both sides this does not affect the overall enthalpy change.

ΔH_3 is the enthalpy change for the combustion of ethene – $H_2O(g)$ does not change in this process.

This cycle shows the individual enthalpy changes involved. All arrows are drawn in the same direction as the equations given. The values can be substituted into the cycle:

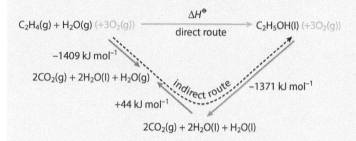

The box on the right states:
The arrow for the indirect route goes in the opposite direction to the arrow for the enthalpy changes ΔH_1 and ΔH_2.

The enthalpy change for the direct route is the same as that for the indirect route.

$$\Delta H^{\ominus} = -1409 - (+44) - (-1371) = -82\,\text{kJ}\,\text{mol}^{-1}$$

The cycle could have been simplified by reversing the enthalpy change ΔH_1 and incorporating ΔH_1 and ΔH_3 into one step, but care must be then taken with the sign of ΔH_1.

Method B (manipulating equations)

Reactions 1 and **2** are reversed to give $H_2O(g)$ and C_2H_5OH on the correct sides:

$$H_2O(g) \rightarrow H_2O(l) \qquad\qquad\qquad \Delta H_1 = -44\,\text{kJ}\,\text{mol}^{-1}$$
$$2CO_2(g) + 3H_2O(l) \rightarrow C_2H_5OH(l) + 3O_2(g) \qquad \Delta H_2 = +1371\,\text{kJ}\,\text{mol}^{-1}$$

All the species are now on the correct sides, as in the overall equation. These equations are now added and common species cancelled:

$$H_2O(g) \rightarrow \cancel{H_2O}(l) \qquad\qquad\qquad\qquad \Delta H_1 = -44\,\text{kJ}\,\text{mol}^{-1}$$
$$2\cancel{CO_2}(g) + 3\cancel{H_2O}(l) \rightarrow C_2H_5OH(l) + 3\cancel{O_2}(g) \qquad \Delta H_2 = +1371\,\text{kJ}\,\text{mol}^{-1}$$
$$\underline{C_2H_4(g) + 3\cancel{O_2}(g) \rightarrow 2\cancel{CO_2}(g) + 2\cancel{H_2O}(l)} \qquad \Delta H_3 = -1409\,\text{kJ}\,\text{mol}^{-1}$$
$$C_2H_4(g) + H_2O(g) \rightarrow C_2H_5OH(l) \qquad\qquad \Delta H = -82\,\text{kJ}\,\text{mol}^{-1}$$

? Test yourself

7 **a** Use the data below to calculate the enthalpy change for the process:
$$C_2H_5OH(l) \rightarrow C_2H_5OH(g)$$

$$C_2H_5OH(l) + 3O_2(g) \rightarrow 2CO_2(g) + 3H_2O(l) \qquad \Delta H = -1371\,\text{kJ}\,\text{mol}^{-1}$$
$$C_2H_5OH(g) + 3O_2(g) \rightarrow 2CO_2(g) + 3H_2O(l) \qquad \Delta H = -1415\,\text{kJ}\,\text{mol}^{-1}$$

   b Use the data below to calculate the enthalpy change for the process:
$$C_6H_6(l) \rightarrow C_6H_6(g)$$

$$2C_6H_6(l) + 15O_2(g) \rightarrow 12CO_2(g) + 6H_2O(l) \qquad \Delta H = -6544\,\text{kJ}\,\text{mol}^{-1}$$
$$2C_6H_6(g) + 15O_2(g) \rightarrow 12CO_2(g) + 6H_2O(l) \qquad \Delta H = -6606\,\text{kJ}\,\text{mol}^{-1}$$

8 **a** Use the data below to calculate the enthalpy change for the reaction:
$$BrF(g) + 2F_2(g) \rightarrow BrF_5(l)$$

$$BrF(g) + F_2(g) \rightarrow BrF_3(l) \qquad\qquad\qquad \Delta H = -242\,\text{kJ}\,\text{mol}^{-1}$$
$$BrF_3(l) + F_2(g) \rightarrow BrF_5(l) \qquad\qquad\qquad \Delta H = -158\,\text{kJ}\,\text{mol}^{-1}$$

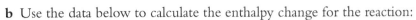

b Use the data below to calculate the enthalpy change for the reaction:

$$ClF_3(g) + F_2(g) \rightarrow ClF_5(g)$$

$$Cl_2(g) + 3F_2(g) \rightarrow 2ClF_3(g) \qquad \Delta H = -328 \, kJ \, mol^{-1}$$
$$Cl_2(g) + 5F_2(g) \rightarrow 2ClF_5(g) \qquad \Delta H = -510 \, kJ \, mol^{-1}$$

9 Use the data below to work out an enthalpy change for the reaction:

$$C_2H_2(g) + 2H_2(g) \rightarrow C_2H_6(g)$$

$$C_2H_2(g) + 2\tfrac{1}{2}O_2(g) \rightarrow 2CO_2(g) + H_2O(l) \qquad \Delta H = -1300 \, kJ \, mol^{-1}$$
$$H_2(g) + \tfrac{1}{2}O_2(g) \rightarrow H_2O(l) \qquad \Delta H = -286 \, kJ \, mol^{-1}$$
$$C_2H_6(g) + 3\tfrac{1}{2}O_2(g) \rightarrow 2CO_2(g) + 3H_2O(l) \qquad \Delta H = -1560 \, kJ \, mol^{-1}$$

10 Calculate the enthalpy change for the following reaction, given the data in the table:

$$C_4H_8(g) + H_2(g) \rightarrow C_4H_{10}(g)$$

	$\Delta H_c^{\ominus}/kJ \, mol^{-1}$
$C_4H_8(g)$	−2717
$C_4H_{10}(g)$	−2877
$H_2(g)$	−286

Standard enthalpy change of formation

> **Standard enthalpy change of formation** ($\Delta H_f^{\ominus}$) is the enthalpy change when one mole of a substance is formed from its elements in their standard states under standard conditions.

The equation representing the enthalpy change of formation of methane is:

$$C(s) + 2H_2(g) \rightarrow CH_4(g)$$

The carbon here is graphite, which is the most stable form of carbon.

For ammonia, the enthalpy change of formation is for the reaction:

$$\tfrac{1}{2}N_2(g) + \tfrac{3}{2}H_2(g) \rightarrow NH_3(g)$$

which shows the formation of **one mole** of ammonia.

> **Standard state** refers to the pure substance at 100 kPa and a specified temperature (assume 298 K unless another temperature is specified).

The standard state of nitrogen is $N_2(g)$, whereas that of iodine is $I_2(s)$. These are the states in which these substances exist at 25 °C and 100 kPa pressure.

The standard enthalpy change of formation of any element in its standard state is zero because, by definition, no heat energy is taken in or given out when one mole of an element in its standard state is formed from one mole of the element in its standard state:

$$I_2(s) \rightarrow I_2(s) \quad \Delta H_f^{\ominus} = 0$$

State symbols must always be included in equations for enthalpy changes because the enthalpy change will be different depending on the state of a substance. For example, the standard enthalpy change of formation of $I_2(g)$ is not zero.

$\Delta H_f^\ominus$ for any element in its standard state is zero.

A standard enthalpy change of formation may be exothermic or endothermic, depending on the substance.

Using standard enthalpy changes of formation to work out enthalpy changes

Worked example

5.9 Given the enthalpy changes ($kJ\,mol^{-1}$):

$\Delta H_f^\ominus$ [Fe$_2$O$_3$(s)] = -822; $\Delta H_f^\ominus$[CO(g)] = -111; $\Delta H_f^\ominus$[CO$_2$(g)] = -394

calculate the standard enthalpy change for the reaction:

$Fe_2O_3(s) + 3CO(g) \rightarrow 2Fe(s) + 3CO_2(g)$

This reaction is important in the extraction of iron from its ore.

First we set up an enthalpy change cycle. The equations of the reactions for which we know the enthalpy changes are:

$2Fe(s) + \frac{3}{2}O_2(g) \rightarrow Fe_2O_3(s)$ $\Delta H_f^\ominus = -822\,kJ\,mol^{-1}$
$C(s) + \frac{1}{2}O_2(g) \rightarrow CO(g)$ $\Delta H_f^\ominus = -111\,kJ\,mol^{-1}$
$C(s) + O_2(g) \rightarrow CO_2(g)$ $\Delta H_f^\ominus = -394\,kJ\,mol^{-1}$

These can be used to construct a cycle (Figure **5.15a**) in which the equation for the reaction with the enthalpy change we want to find is along the top and arrows go **up** from the elements to the compounds. The total enthalpy change for the formation of the reactants, Fe$_2$O$_3$(s) and 3CO(g), from their elements is:

$\Delta H_f^\ominus$[Fe$_2$O$_3$(s)] + $3\Delta H_f^\ominus$[CO(g)] = $-822 + (3 \times -111) = -1155\,kJ\,mol^{-1}$

The total enthalpy change for the formation of the products, 2Fe(s) and 3CO$_2$(g), from their elements is:

$2\Delta H_f^\ominus$[Fe(s)] + $3\Delta H_f^\ominus$[CO$_2$(g)] = $0 + (3 \times -394) = -1182\,kJ\,mol^{-1}$

Note that the standard enthalpy change of formation of Fe(s) is 0 because it is an element in its standard state. These values can be added to the cycle (Figure **5.15b**).

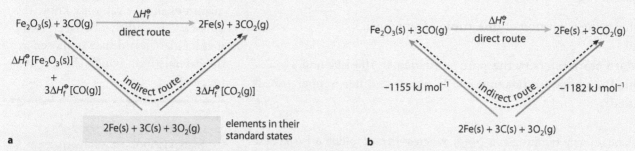

Figure 5.15 Working out an enthalpy change from $\Delta H_f^\ominus$ values. **a** Enthalpy cycle for the reaction Fe$_2$O$_3$(s) + 3CO(g) $\rightarrow$ 2Fe(s) + 3CO$_2$(g). **b** The same enthalpy cycle with $\Delta H_f^\ominus$ values included.

The enthalpy change for the direct route is the same as the enthalpy change for the indirect route. The indirect route goes in the opposite direction to the blue arrow and therefore the sign of this enthalpy change is reversed:

$\Delta H_r^\ominus = -(-1155) + (-1182) = -27\,kJ\,mol^{-1}$

Therefore the enthalpy change of reaction is $-27\,kJ\,mol^{-1}$.

What we have actually done in this cycle is to subtract the total enthalpy change of the reactants from the total enthalpy change of the products to give the overall enthalpy change. This can be shown as:

$$\Delta H_r = \Sigma \Delta H_f(\text{products}) - \Sigma \Delta H_f(\text{reactants})$$

Σ means 'sum of'.

Using this equation is by far the easiest way of solving problems involving enthalpy changes of formation.

Worked example

5.10 Given these standard enthalpy change values ($kJ\,mol^{-1}$):

$$\Delta H_f^{\ominus}[NH_3(g)] = -46; \Delta H_f^{\ominus}[NO(g)] = 90; \Delta H_f^{\ominus}[H_2O(g)] = -242$$

calculate the enthalpy change for the reaction:

$$4NH_3(g) + 5O_2(g) \rightarrow 4NO(g) + 6H_2O(g)$$

Using $\Delta H_r = \Sigma \Delta H_f^{\ominus}(\text{products}) - \Sigma \Delta H_f^{\ominus}(\text{reactants})$

$\Sigma \Delta H_f (\text{products}) = (4 \times 90) + (6 \times -242) = -1092\,kJ\,mol^{-1}$
$\Sigma \Delta H_f (\text{reactants}) = 4 \times -46 = -184\,kJ\,mol^{-1}$

Note: $\Delta H_f^{\ominus}$ for $O_2(g)$ is zero because it is an element in its standard state.

$\Delta H_r^{\ominus} = -1092 - (-184) = -908\,kJ\,mol^{-1}$

So the enthalpy change for this reaction is $-908\,kJ\,mol^{-1}$.

Determining an enthalpy change of formation from an enthalpy change of reaction

If you are given an enthalpy change of reaction and some enthalpy changes of formation, it may be possible to work out a missing enthalpy change of formation.

Worked example

5.11 Given that the standard enthalpy change of formation of ethane ($C_2H_6(g)$) is $-85\,kJ\,mol^{-1}$ and this equation:

$$C_2H_2(g) + 2H_2(g) \rightarrow C_2H_6(g) \qquad \Delta H^{\ominus} = -311\,kJ\,mol^{-1}$$

work out the enthalpy change of formation of ethyne, $C_2H_2(g)$.

$\Delta H^{\ominus} = \Sigma \Delta H_f^{\ominus}(\text{products}) - \Sigma \Delta H_f^{\ominus}(\text{reactants})$

$-311 = -85 - (\Delta H_f^{\ominus}[C_2H_2(g)] + [2 \times 0])$

Rearranging the equation gives:

$\Delta H_f^{\ominus}[C_2H_2(g)] = -85 + 311 = +226\,kJ\,mol^{-1}$.

Nature of science

Hess carried out a series of experiments looking at how the amount of heat energy given out in chemical reactions depended on whether the reaction occurred directly or indirectly in a series of steps. His experiments required careful measurements and an awareness of the limitations of his data. He published his work in 1840 and proposed a general law that for a chemical reaction, the heat energy evolved is constant and does not depend on whether substances combine directly or indirectly. His hypothesis could be tested by further experimental work and a theory based on the principle of conservation of energy developed.

? Test yourself

11 Write equations for the enthalpy change of formation of the following:
   a $HF(g)$
   b $CH_3Cl(g)$
   c $H_2O(l)$
   d $C_5H_{11}OH(l)$

12 Calculate the enthalpy change for the following reaction, given the enthalpy changes of formation in the table:
$$S_2Cl_2(l) + Cl_2(g) \rightarrow 2SCl_2(g)$$

	$\Delta H_f^\ominus / kJ\,mol^{-1}$
$S_2Cl_2(l)$	−59.4
$SCl_2(g)$	−19.7

13 Calculate the enthalpy change for the following reaction, given the enthalpy changes of formation in the table:
$$4BCl_3(l) + 3SF_4(g)$$
$$\rightarrow 4BF_3(g) + 3SCl_2(g) + 3Cl_2(g)$$

	$\Delta H_f^\ominus / kJ\,mol^{-1}$
$BCl_3(l)$	−427
$SCl_2(g)$	−19.7
$SF_4(g)$	−775
$BF_3(g)$	−1137

14 Calculate the enthalpy change of formation of $NO_2(g)$ from the following data:
$$2Pb(NO_3)_2(s) \rightarrow 4NO_2(g) + 2PbO(s) + O_2(g)$$
$$\Delta H = +602\,kJ\,mol^{-1}$$

	$\Delta H_f^\ominus / kJ\,mol^{-1}$
$Pb(NO_3)_2(s)$	−452
$PbO(s)$	−217

600

500

400

300

200

206

5.3 Bond enthalpies

Enthalpy changes for reactions in the gas phase can be worked out if we know the amount of energy required/released when bonds are broken/made. To be able to do this we must first consider the definition of bond enthalpy (also just called **bond energy**):

> **Bond enthalpy** is the enthalpy change when one mole of covalent bonds, in a gaseous molecule, is broken under standard conditions.

For example, the H–H bond enthalpy is $436 \, \text{kJ} \, \text{mol}^{-1}$, which can be represented by the equation:

$$H_2(g) \rightarrow 2H(g) \qquad\qquad \Delta H^\ominus = +436 \, \text{kJ} \, \text{mol}^{-1}$$

One mole of H_2 molecules is broken apart to give two moles of gaseous hydrogen atoms.

The H–Cl bond enthalpy is $431 \, \text{kJ} \, \text{mol}^{-1}$, which is represented by:

$$HCl(g) \rightarrow H(g) + Cl(g) \qquad\qquad \Delta H^\ominus = +431 \, \text{kJ} \, \text{mol}^{-1}$$

> Bond enthalpies are used for reactions occurring in the gaseous state.

If we consider the process:

$$Br_2(l) \rightarrow 2Br(g) \qquad\qquad \Delta H^\ominus = +224 \, \text{kJ} \, \text{mol}^{-1}$$

this does not represent the bond enthalpy of the Br–Br bond, because this reaction can be broken down into two processes:

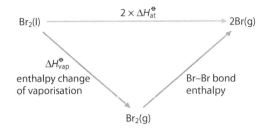

The process $Br_2(l) \rightarrow 2Br(g)$ is called atomisation, and energy must be supplied to overcome the London forces between the bromine molecules as well as to break the Br–Br bonds.

Energy must be supplied to break apart two atoms in a molecule against the attractive force holding them together. Therefore, the process of breaking bonds must be endothermic – the enthalpy change is positive.

Energy is released when two atoms come together to form a bond. Therefore, the process of making bonds is exothermic – the enthalpy change is negative.

Some average bond enthalpies are shown in Table **5.1**.

Learning objectives

- Define average bond enthalpy
- Use bond energies to calculate enthalpy changes

Exam tip
$HCl(g) \rightarrow \frac{1}{2}H_2(g) + \frac{1}{2}Cl_2(g)$ does **not** represent the bond enthalpy of HCl. $\frac{1}{2}H_2(g)$ represents half a mole of H_2 molecules and is not the same as H(g), which is one mole of gaseous H atoms.

The Br–Br bond enthalpy is $193 \, \text{kJ} \, \text{mol}^{-1}$, and the **standard enthalpy change of vaporisation** (for the formation of one mole of gaseous molecules) is $31 \, \text{kJ} \, \text{mol}^{-1}$.

The enthalpy change for the process $Br_2(l) \rightarrow 2Br(g)$ is actually **twice** the enthalpy change of atomisation (ΔH_{at}) as the enthalpy change of atomisation is defined as the enthalpy change for formation of one mole of gaseous atoms.

Bond breaking requires energy (endothermic): ΔH positive.
Bond making releases energy (exothermic): ΔH negative.

Bond	Bond enthalpy /kJ mol^{-1}	Bond	Bond enthalpy /kJ mol^{-1}	Bond	Bond enthalpy /kJ mol^{-1}	Bond	Bond enthalpy /kJ mol^{-1}
H–H	436	C–H	412	O–O	146	C–O	360
C–C	348	Si–H	318	O=O	496	C=O	743
C=C	612	N–H	388	Si–O	374	C≡O	1070
C≡C	837	P–H	322	F–F	158	C–N	305
Si–Si	226	O–H	463	Cl–Cl	242	C=N	613
N–N	163	S–H	338	Br–Br	193	C≡N	890
N=N	409	F–H	562	I–I	151	C–F	484
N≡N	944	Cl–H	431	N–Cl	200	C–Cl	338
P–P	172	Br–H	366	Si–F	590	C–Br	276
S–S	264	I–H	299	N–F	278	C–I	238

Table 5.1 Some average bond enthalpies.

The bond energies given in tables are **average** values. For example, the C–H bond energy in ethane (C_2H_6) is slightly different from that in benzene (C_6H_6) so the value quoted in the table is an average value, averaged over the values for C–H bonds in many compounds. This can introduce some inaccuracies into calculations involving bond energies. Bond energy calculations are therefore most accurate when they involve breaking/making only a few bonds.

> **Average bond enthalpy** is the average amount of energy required to break one mole of covalent bonds, in gaseous molecules under standard conditions. 'Average' refers to the fact that the bond enthalpy is different in different molecules and therefore the value quoted is the average amount of energy to break a particular bond in a range of molecules.

Extension

Why is bond making exothermic?

When two atoms are a long way apart, the electrical potential energy of the two together is a maximum (consider separating the two atoms from where they are bonded together to where they are an infinite distance apart – work must be done against the force holding them together, so the potential energy increases). As they approach closer together, because of the attractive force between them, the potential energy decreases and this energy is released as heat. This can be likened to a ball falling to Earth – as the ball and the Earth come together, the potential energy of the system is first of all converted to kinetic energy and then to heat as the two collide.

Using bond enthalpies to work out enthalpy changes for reactions

Example

Consider the reaction between ethene and bromine to produce 1,2-dibromoethane:

$$C_2H_4(g) + Br_2(g) \rightarrow C_2H_4Br_2(g)$$

What is the enthalpy change for this reaction?

If the species are drawn as structural formulas, then all the bonds can be seen clearly:

We imagine the reaction happening with all the bonds in the reactants being broken:

Then new bonds form to make the products:

The bond energies can be added up (Tables **5.2** and **5.3**) to work out the enthalpy change.

Bond broken	Bond energy / kJ mol^{-1}	Number of bonds	Total energy / kJ mol^{-1}
C–H	412	4	1648
C=C	612	1	612
Br–Br	193	1	193
Total energy to break all bonds			2453

Table 5.2 Calculation of the enthalpy change when all bonds are broken in ethene and bromine.

The total enthalpy change when all the bonds are broken is $+2453\,\text{kJ mol}^{-1}$. This is positive because breaking bonds is an endothermic process.

Bond made	Bond energy / kJ mol^{-1}	Number of bonds	Total energy / kJ mol^{-1}
C–H	412	4	1648
C–C	348	1	348
C–Br	276	2	552
Total energy released when bonds made			2548

Table 5.3 Calculation of the enthalpy change when bonds are formed in $C_2H_4Br_2$.

Bromine and 1,2-dibromoethane are both liquids under standard conditions. The equation here is for the reaction in the gas phase, and the enthalpy change will not be the same as for the reaction involving liquid bromine to form $C_2H_4Br_2(l)$.

The total enthalpy change when all the bonds are made is $-2548\,\text{kJ}\,\text{mol}^{-1}$. This is negative because making bonds is an exothermic process.

If we add up these enthalpy changes, we get the overall enthalpy change for the reaction:

$$\Delta H^{\ominus} = 2453 - 2548 = -95\,\text{kJ}\,\text{mol}^{-1}$$

The overall process is sometimes summarised as:

$$\Delta H_{\text{r}} = \Sigma(\text{bonds broken}) - \Sigma(\text{bonds made})$$

where the bond energies are taken directly from Table **5.1** and no signs changed, i.e. all values are put in as positive.

This approach would give:

$$\Delta H^{\ominus} = [(4 \times 412) + (1 \times 612) + (1 \times 193)] - [(4 \times 412) + (1 \times 348) + (2 \times 276)]$$

A shortcut for doing this question would be to realise that there are four C–H bonds in ethene and four C–H bonds in 1,2-dibromoethane, and therefore there is no need to make or break these bonds:

If the enthalpy change for:

$$C_2H_4(g) + Br_2(g) \rightarrow C_2H_4Br_2(g)$$

is calculated using enthalpy changes of formation the value obtained is $-121\,\text{kJ}\,\text{mol}^{-1}$, which is different from the value of $-95\,\text{kJ}\,\text{mol}^{-1}$ obtained using bond enthalpy calculations. The value obtained using enthalpy changes of formation is expected to be more reliable, because enthalpy changes of formation are specific to the particular substances, whereas bond enthalpies are average values. The C–H bond energies in C_2H_4 and $C_2H_4Br_2$ would not be expected to be the same, and the values used for the C=C and C–C bonds are not necessarily the actual bond energies in these compounds.

Using a cycle in calculations involving bond enthalpies

Example

Using bond enthalpies given in Table **5.4**, we can calculate the enthalpy change for the reaction:

$$H_2(g) + \tfrac{1}{2}O_2(g) \rightarrow H_2O(g)$$

Values for enthalpy changes of reaction obtained using bond enthalpies are less reliable than those calculated using experimental data such as the enthalpy change of combustion or formation.

Extension

The hybridisation of the C atom can significantly affect the C–H bond enthalpy.

Bond	Bond enthalpy / kJ mol^{-1}
H–H	436
O=O	496
O–H	463

Table 5.4 Bond enthalpies.

A cycle can be drawn:

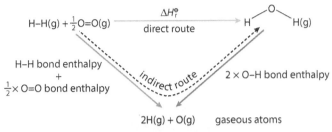

The values can be substituted into the cycle:

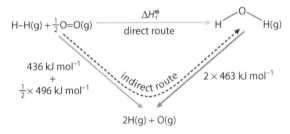

The enthalpy change for the direct route is the same as that for the indirect route. The direction of the arrow for the indirect route is opposite to that of the red arrow, and therefore the sign of this enthalpy change must be reversed:

$$\Delta H_r^{\ominus} = (436 + \tfrac{1}{2} \times 496) - (2 \times 463) = -242\,\text{kJ}\,\text{mol}^{-1}$$

In order to use bond enthalpies, the substances must be in the gas phase. This is because intermolecular forces must also be taken into account in the liquid and solid states. So, in order to use bond enthalpies with solids and liquids we must first generate gaseous species. If we wanted to find the enthalpy change for the formation of *liquid* water from its elements – i.e. for the process:

$$H_2(g) + \tfrac{1}{2}O_2(g) \rightarrow H_2O(l)$$

we would also have to know the enthalpy change of vaporisation of water – i.e. the enthalpy change for the process:

$$H_2O(l) \rightarrow H_2O(g) \qquad \Delta H_{vap} = +41\,\text{kJ}\,\text{mol}^{-1}$$

Figure **5.16** shows how this can be incorporated into the cycle. The enthalpy change for the reaction is: $\Delta H_r^{\ominus} = 684 - 926 - 41 = -283\,\text{kJ}\,\text{mol}^{-1}$

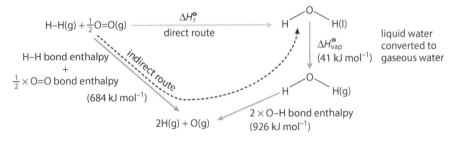

Figure 5.16 Including the enthalpy of vaporisation in an enthalpy cycle.

A reaction in the gas phase will be endothermic if less energy is released when bonds are formed (exothermic) than is required to break bonds (endothermic). This could be the case if stronger/more bonds are broken than are formed.

A reaction in the gas phase will be exothermic if more energy is released when bonds are formed (exothermic) than is required to break bonds (endothermic). This could be the case if stronger/more bonds are made than are broken.

It is important to note that bond energies can only be used to decide whether a reaction in the **gas** phase is exothermic or endothermic. For reactions involving solids and liquids, other interactions, such as intermolecular forces, must also be taken into account.

Worked example

5.12 Calculate the enthalpy change for the process:

$$3C(s) + 4H_2(g) \rightarrow C_3H_8(g)$$

using the following bond energies ($kJ\,mol^{-1}$):

C–H = 412; H–H = 436; C–C = 348

and the enthalpy change $C(s) \rightarrow C(g)$ $\Delta H^{\ominus} = +715\,kJ\,mol^{-1}$

A cycle can be used to work out the overall enthalpy change:

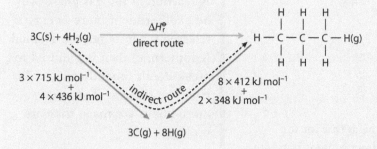

In the left-hand process, represented by the blue arrow, three moles of solid carbon are converted into gaseous atoms and four moles of H–H bonds are broken. In the right-hand process (red arrow), eight moles of C–H bonds and two moles of C–C bonds are broken to generate gaseous atoms.

The value of the unknown enthalpy change is given by:

$$\Delta H^{\ominus} = [(3 \times 715) + (4 \times 436)] - [(8 \times 412) + (2 \times 348)]$$
$$= -103\,kJ\,mol^{-1}$$

The published value for this enthalpy change is $-104\,kJ\,mol^{-1}$, so in this case there is fairly good agreement between data worked out using bond enthalpies and in other ways.

The enthalpy change:
$$C(s) \rightarrow C(g) \qquad \Delta H_{at}^{\ominus} = +715\,kJ\,mol^{-1}$$
is the enthalpy change of atomisation for carbon – the enthalpy change for the formation of one mole of gaseous atoms.

Working out a bond enthalpy from an enthalpy change of reaction

Example

Using the bond enthalpies for the F–F and Br–Br bonds and the enthalpy change of reaction given, we can calculate the mean Br–F bond energy in BrF_3:

$$Br_2(g) + 3F_2(g) \rightarrow 2BrF_3(g) \qquad\qquad \Delta H^\ominus = -545\,kJ\,mol^{-1}$$

This is most easily done using the equation:

$$\Delta H_r = \Sigma(\text{bonds broken}) - \Sigma(\text{bonds made})$$

$$-545 = [193 + (3 \times 158)] - [6 \times \text{Br–F}]$$

Rearranging this we get:

$$6 \times \text{Br–F} = 193 + (3 \times 158) + 545$$

$$6 \times \text{Br–F} = 1212$$

Therefore the Br–F bond enthalpy is $\dfrac{1212}{6}$, i.e. $202\,kJ\,mol^{-1}$.

Potential energy profiles and stability

A potential energy diagram (profile) can be used to show the energy changes occuring during a chemical reaction. The potential energy diagrams for exothermic and endothermic reactions are shown in Figure **5.17**. A reaction coordinate represents the progress of a reaction from reactants to products. As reactant particles approach each other, repulsion between the particles and the partial breaking of bonds causes the potential energy of the system to increase. The activation energy is the minimum energy that colliding particles must have before a collision results in a chemical reaction – this will be considered in more detail on page **246**.

In an exothermic reaction, the products are at lower energy than the reactants and are therefore said to be **more stable** than the reactants. For an endothermic reaction, the products are at higher energy and are **less stable** than the reactants.

Potential energy diagrams are often used in discussions about rate of reaction and will be discussed later, on page **246**.

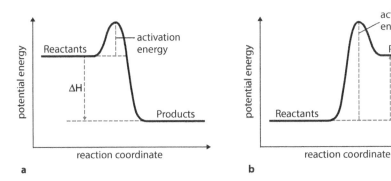

Figure 5.17 a Potential energy diagram for an exothermic reaction; **b** potential energy diagram for an endothermic reaction.

Absorption of UV light in the atmosphere

Ozone (O_3) and oxygen (O_2) in the atmosphere are important in protecting the surface of the Earth from the damaging effects of ultraviolet radiation. UV radiation is absorbed by molecules of O_2 and/or O_3 as they undergo dissociation – breaking the bond between oxygen atoms:

$$O_2 \xrightarrow{\lambda < 242\,nm} 2O$$
molecule $\qquad$ atoms

$$O_3 \xrightarrow{\lambda < 330\,nm} O + O_2$$
molecule $\qquad$ atom $\quad$ molecule

O_2 molecules absorb higher energy UV radiation than O_3 because the bond between the oxygen atoms in O_2 is stronger than that between the oxygen atoms in O_3.

The UV radiation required to break the bond in O_2 (double bond) is in the UV-C part of the spectrum and must have a wavelength shorter than 242 nm to provide sufficient energy to break the bond. The bond in ozone is between a single bond and double bond (see page **170**) and the absorption of lower energy, UV-B, radiation will break the bond. A more detailed account is given in the Higher Level section on page **174**.

Nature of science

Scientists use models to explain experimental data and to make predictions. Here we are using a very simple model in terms of breaking bonds and making bonds to make predictions about enthalpy changes for reactions. Agreement between a theoretical model and the experimental data is used to evaluate models.

? Test yourself

15 Use the bond enthalpies in Table **5.1** on page **208** to work out the enthalpy changes for the following reactions:
a $CH_4(g) + Cl_2(g) \rightarrow CH_3Cl(g) + HCl(g)$
b $C_2H_2(g) + 2H_2(g) \rightarrow C_2H_6(g)$
c $CO(g) + Cl_2(g) \rightarrow COCl_2(g)$

16 Calculate the mean Cl–F bond enthalpy, given the following data:
$Cl_2(g) + 3F_2(g) \rightarrow 2ClF_3(g) \quad \Delta H^\ominus = -164\,kJ\,mol^{-1}$
and the Cl–Cl and F–F bond enthalpies in Table **5.1** (page **208**).

Both ClF_3 and ClF_5 just contain single bonds between Cl and F.

17 Calculate the mean Cl–F bond enthalpy, given the following data:
$ClF_3(g) + F_2(g) \rightarrow ClF_5(g) \qquad \Delta H^\ominus = -91\,kJ\,mol^{-1}$
and the F–F bond enthalpy in Table **5.1** (page **208**).

18 Work out the enthalpy change for the reaction:
$S(s) + F_2(g) \rightarrow SF_2(g)$
given the bond energies:
$S–F = 327\,kJ\,mol^{-1}$ and $F–F = 158\,kJ\,mol^{-1}$
and this enthalpy change:
$S(s) \rightarrow S(g) \qquad \Delta H = +223\,kJ\,mol^{-1}$

5.4 Energy cycles (HL)

So far, the energy cycles we have drawn have been for covalent substances. In this section we will consider an enthalpy level diagram for ionic substances. Before we look at this we must consider some definitions.

> **Standard enthalpy change of atomisation ($\Delta H_{at}^{\ominus}$)** – this is the enthalpy change when **one mole of gaseous atoms** is formed from an element under standard conditions.

ΔH_{at} is always endothermic – for example:

$$\tfrac{1}{2}H_2(g) \rightarrow H(g) \ \Delta H_{at}^{\ominus} = +218 \, kJ \, mol^{-1}$$

$$\tfrac{1}{2}Cl_2(g) \rightarrow Cl(g) \ \Delta H_{at}^{\ominus} = +121 \, kJ \, mol^{-1}$$

$$Na(s) \rightarrow Na(g) \ \Delta H_{at}^{\ominus} = +109 \, kJ \, mol^{-1}$$

The values for the enthalpy changes of atomisation of hydrogen and chlorine are half the bond enthalpy values given in Table **5.1** on page **208**. The bond enthalpy for Cl_2 refers to breaking of one mole of covalent bonds to produce two moles of gaseous atoms:

$$Cl_2(g) \rightarrow 2Cl(g) \qquad \text{bond enthalpy} = 242 \, kJ \, mol^{-1}$$

whereas the enthalpy change of atomisation for Cl_2 refers to breaking half a mole of covalent bonds to produce one mole of gaseous atoms.

> **First ionisation energy** is the enthalpy change when one electron is removed from each atom in one mole of gaseous atoms under standard conditions:
>
> $$M(g) \rightarrow M^+(g) + e^-$$

> **Second ionisation energy** is the enthalpy change for the process:
>
> $$M^+(g) \rightarrow M^{2+}(g) + e^-$$

> **First electron affinity** is the enthalpy change when one electron is added to each atom in one mole of gaseous atoms under standard conditions:
>
> $$X(g) + e^- \rightarrow X^-(g)$$

> **Second electron affinity** is enthalpy change for the process:
>
> $$X^-(g) + e^- \rightarrow X^{2-}(g)$$

Learning objectives

- Define the enthalpy changes involved in the formation and dissolution of ionic substances
- Understand how to draw a Born–Haber cycle
- Understand how to construct energy cycles for ionic substances dissolving in water

All ionisation energies are endothermic.

The (g) symbol is essential in these equations.

The first electron affinity is exothermic for virtually all elements; it is a favourable process to bring an electron from infinity to where it feels the attractive force of the nucleus in an atom.

The second electron affinity is always endothermic. It is an unfavourable process to add an electron to an ion which is already negatively charged, due to repulsion between the negative charges.

Look carefully to determine whether lattice enthalpy is exothermic or endothermic – as it can be defined in either direction, i.e. as the making or breaking of the lattice. Here it is defined as the breaking of the lattice and is an endothermic process because energy must be supplied to separate the ions against the attractive forces holding them in the lattice.

The Born–Haber cycle was developed by two Nobel-prize-winning German scientists: Fritz Haber (1868–1934) and Max Born (1882–1970).

Lattice enthalpy ($\Delta H_{latt}^{\ominus}$) is the enthalpy change when one mole of an ionic compound is broken apart into its constituent gaseous ions under standard conditions.

For example, for NaCl:

$$NaCl(s) \rightarrow Na^+(g) + Cl^-(g) \qquad \Delta H_{latt}^{\ominus} = +771\,kJ\,mol^{-1}$$

Note, for ammonium nitrate the equation is:

$$NH_4NO_3(s) \rightarrow NH_4^+(g) + NO_3^-(g)$$

Born–Haber cycles

A **Born–Haber cycle** is an enthalpy level diagram breaking down the formation of an ionic compound into a series of simpler steps. For example, for sodium chloride, we will show the stages in construction of the cycle. The first step we put in is the equation for the enthalpy change of formation:

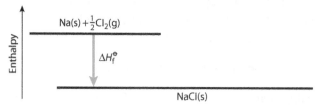

The enthalpy change of formation for NaCl(s) is exothermic and the (green) arrow is shown in the downward, negative, direction.

The aim now is to construct a cycle that gives us an alternative route between the elements, $Na(s) + \frac{1}{2}Cl_2(g)$, and the ionic compound NaCl(s).

NaCl(s) can be broken apart into its constituent gaseous ions – this is the lattice enthalpy:

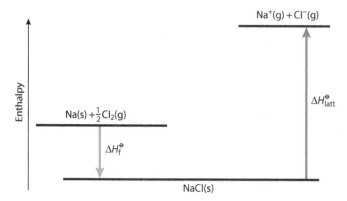

The arrow for ΔH_{latt} goes upwards, indicating an endothermic process.

Putting the lattice enthalpy into the cycle gives us an idea of what we are aiming at so, to complete the cycle, Na(s) and $\frac{1}{2}Cl_2(g)$ must be converted into gaseous ions. This and the other stages are detailed in the full Born–Haber cycle shown in Figure **5.18**. Na(s) is first converted into $Na^+(g)$ in two steps – these are shown as steps 1 and 2 in Figure **5.18**.

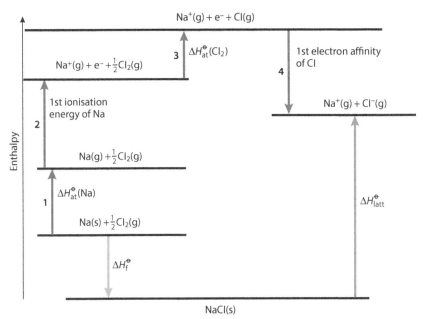

Figure 5.18 The Born–Haber cycle for NaCl. Upward arrows represent endothermic processes; downward arrows represent exothermic processes.

Step 1 involves the enthalpy change of atomisation for sodium:

$$Na(s) \rightarrow Na(g)$$

Step 2 involves the first ionisation energy of sodium:

$$Na(g) \rightarrow Na^+(g) + e^-$$

The sodium must be converted to gaseous atoms before it can be ionised because the definition of ionisation energy involves removal of an electron from a **gaseous** atom.

The sodium has been converted into the required species, $Na^+(g)$, and now we must do the same with the chlorine. Step 3 is the atomisation of chlorine:

$$\tfrac{1}{2}Cl_2(g) \rightarrow Cl(g)$$

The cycle is then completed in step 4 by adding the electron removed from the sodium atom to the chlorine atom. This process is the first electron affinity of chlorine and is exothermic, so the arrow goes downwards:

$$Cl(g) + e^- \rightarrow Cl^-(g)$$

If all the quantities in the cycle except one are known, then application of Hess's law allows the missing quantity to be worked out. For example, let us calculate the lattice enthalpy of sodium chloride using the data in Table **5.5**.

The values can be put into the diagram (Figure **5.19**) and Hess's law used to calculate the missing lattice energy.

ΔH_{at} [Na(s)]	109 kJ mol^{-1}
ΔH_{at} [Cl$_2$(g)]	121 kJ mol^{-1}
first ionisation energy (Na)	494 kJ mol^{-1}
first electron affinity (Cl)	−364 kJ mol^{-1}
ΔH_f [NaCl(s)]	−411 kJ mol^{-1}

Table 5.5 Data for the calculation of the lattice enthalpy of sodium chloride.

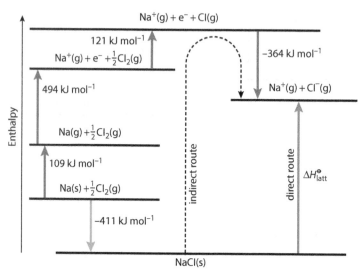

Figure 5.19 Working out a missing value – the lattice energy.

The enthalpy change for the direct route is the same as that for the indirect route. Therefore:

$$\Delta H^{\ominus}_{latt} = 411 + 109 + 494 + 121 - 364$$

$$\Delta H^{\ominus}_{latt} = 771 \, kJ \, mol^{-1}$$

The sign of −411 was reversed because the direction of the indirect route arrow goes against the direction of this arrow.

Worked example

5.13 Draw a Born–Haber cycle for magnesium oxide and use it to work out the second electron affinity of oxygen.

Enthalpy term	Enthalpy change / kJ mol^{-1}
$\Delta H^{\ominus}_{at}$ [Mg(s)]	150
$\Delta H^{\ominus}_{at}$ [O$_2$(g)]	248
first ionisation energy (Mg)	736
second ionisation energy (Mg)	1450
first electron affinity (O)	−142
$\Delta H^{\ominus}_{f}$(MgO)	−602
$\Delta H^{\ominus}_{latt}$ (MgO)	3889

The Born–Haber cycle is shown in Figure **5.20**.

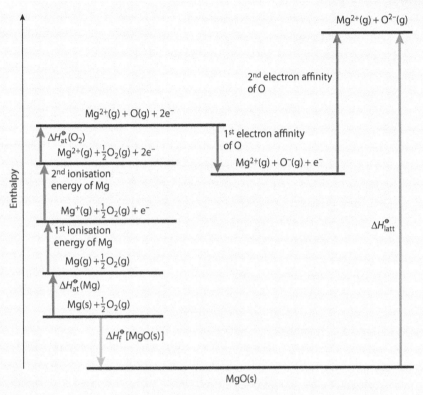

Figure 5.20 A Born–Haber cycle for MgO. The first electron affinity for O(g) is negative, but the second electron affinity is positive.

The second electron affinity of O is worked out most easily by applying Hess's law and looking at the enthalpy change for the direct route and the indirect route (Figure **5.21**).

The signs of the enthalpy changes highlighted in Figure **5.21** are reversed, as the direction of the indirect route is in the opposite direction to the arrows on the diagram. The enthalpy change for the direct route is the same as that for the indirect route:

second electron affinity of O = 142 − 248 − 1450 − 736 − 150 − 602 + 3889 = 845 kJ mol^{-1}.

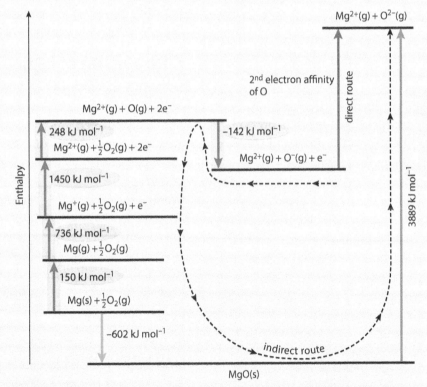

Figure 5.21 The Born–Haber cycle for MgO, showing the enthalpy changes for which the sign must be changed to work out the second electron affinity of oxygen.

Magnesium chloride

The Born–Haber cycle for $MgCl_2$ is shown in Figure **5.22**.

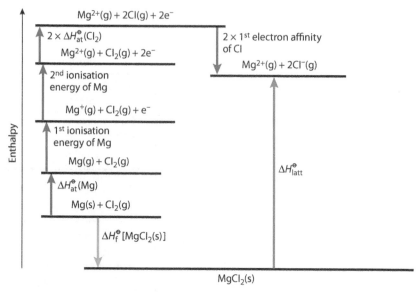

Figure 5.22 A Born–Haber cycle for $MgCl_2$.

The only difference here from the other Born–Haber cycles considered above is that the enthalpy change of atomisation and the electron affinity of chlorine are both multiplied by two. Remember that the definition of the enthalpy change of atomisation refers to the formation of one mole of gaseous atoms and not the breaking of one mole of covalent bonds.

Comparison of lattice enthalpies

Some lattice enthalpies are shown in Table **5.6**.

Lattice enthalpy is the result of electrostatic attractions between oppositely charged ions in the giant lattice. The greater the electrostatic attraction between the ions, the more energy has to be supplied to break the lattice apart.

> The electrostatic attraction between ions depends on the **charge on the ions** and the **size of the ions** (ionic radii).

Substance	NaCl		MgCl₂		MgO	
lattice enthalpy / kJ mol⁻¹	+771		+2526		+3791	
charges on ions	1+	1–	2+	1–	2+	2–
radius of metal ion / pm	95		65		65	
Substance	**CsCl**		**BaCl₂**		**BaO**	
lattice enthalpy / kJ mol⁻¹	+645		+2056		+3054	
charges on ions	1+	1–	2+	1–	2+	2–
radius of metal ion / pm	169		135		135	

Table 5.6 Some lattice enthalpies.

The effect of charge

$MgCl_2$ has a higher lattice enthalpy than NaCl (Figure **5.23**).

> The higher the charges on the ions, the more strongly they will attract each other – and therefore the greater the lattice enthalpy.

The force between ions is proportional to the product of the charges, so the force of attraction increases along the series:

$$1+/1- < 1+/2- < 2+/2-$$

The effect of size

CsCl has a smaller lattice enthalpy than NaCl.

Ions act like point charges – a positive ion such as Na^+ behaves as if its ionic charge all acts at its centre and a negative ion such as Cl^- behaves as if a $1-$ charge exists at its centre. The bigger the ionic radii, the greater the distance between the centres of the ions and, therefore, the smaller the attraction between the ions and the smaller the lattice enthalpy (Figure **5.24**).

> Lattice enthalpy is larger for smaller ions.

We can see that $MgCl_2$ has a larger lattice enthalpy than NaCl, not just because the Mg^{2+} ion has a higher charge than the Na^+ ion, but also because the Mg^{2+} ion is smaller.

Worked example

5.14 Arrange the following compounds in order of increasing lattice enthalpy:

$BaCl_2$ LiF   ZnS

The most important factor in determining lattice enthalpy is the charge on the ions. ZnS contains 2+ and 2− ions, $BaCl_2$ contains 2+ and 1− ions and LiF contains 1+ and 1− ions. The order of increasing lattice enthalpy is:

$$LiF < BaCl_2 < ZnS$$

The attraction between 2+/2− ions in ZnS is greater than that between 2+/1− ions in $BaCl_2$, which is greater than between 1+/1− ions in LiF.

Because the melting point of an ionic substance depends on the force of attraction between ions, a substance with a high lattice enthalpy would also be expected to have a high melting point (Table **5.7**).

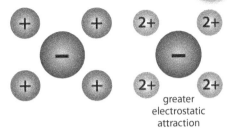

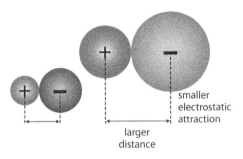

Figure 5.23 NaCl contains 1+ and 1− ions, whereas $MgCl_2$ contains 2+ and 1− ions.

Figure 5.24 Smaller ions have a greater attraction between them.

Exam tip

The largest lattice enthalpy is obtained for small, highly charged ions. When comparing different compounds, the effect of charge causes a larger change in lattice enthalpy than do size variations.

Substance	Lattice enthalpy / kJ mol^{-1}	Melting point / °C
NaCl	+771	801
MgO	+3791	2852
KCl	+711	770
CaO	+3401	2614
CsCl	+645	645
BaO	+3054	1918

Table 5.7 The lattice enthalpies and melting points of some substances.

? Test yourself

19 Use the data in the table to calculate the first electron affinity of Br.

	/ kJ mol^{-1}
ΔH_{at} [K(s)]	90
ΔH_{at} [Br$_2$(l)]	112
first ionisation energy (K)	418
ΔH_{latt} (KBr)	670
ΔH_f (KBr(s))	−394

20 Use the data in the table to calculate the lattice enthalpy of BaF$_2$.

	/ kJ mol^{-1}
ΔH_{at} [Ba(s)]	176
ΔH_{at} [F$_2$(g)]	79
first ionisation energy (Ba)	502
second ionisation energy (Ba)	966
first electron affinity (F)	−348
ΔH_f (BaF$_2$(s))	−1201

21 Arrange the following in order of increasing lattice enthalpy (smallest first):

LiF   KCl   CaO
CaS   CaCl$_2$

22 State whether the following enthalpy changes are *always exothermic, always endothermic* or *sometimes exothermic and sometimes endothermic*:
  a first ionisation energy
  b second ionisation energy
  c enthalpy change of atomisation
  d enthalpy change of formation
  e first electron affinity
  f second electron affinity

Dissolving ionic substances

In this section, we will consider the enthalpy changes that occur when ionic substances dissolve in water. The equation for sodium chloride dissolving in water is:

$$NaCl(s) \xrightarrow[\text{excess H}_2\text{O}]{\Delta H_{sol}} Na^+(aq) + Cl^-(aq)$$

The enthalpy change represented here is called the **enthalpy change of solution ($\Delta H_{sol}^{\ominus}$)** and is defined as:

> The enthalpy change when one mole of solute is dissolved in excess solvent to form a solution of 'infinite dilution' under standard conditions.

The dissolving of sodium chloride in water can be broken down in to two separate processes and an energy cycle drawn (Figure **5.25**).

Stage 1: lattice enthalpy (ΔH_{latt}) – breaking apart the lattice into gaseous ions – an endothermic process.

Stage 2: hydration of the ions (ΔH_{hyd}) – surrounding the gaseous ions by water molecules – an exothermic process.

Enthalpy change of hydration (ΔH_{hyd}) is defined as:

> The enthalpy change when one mole of gaseous ions is surrounded by water molecules to form an 'infinitely dilute solution' under standard conditions.

For example:

$$Na^+(g) \rightarrow Na^+(aq) \quad \Delta H_{hyd} = -390 \, kJ \, mol^{-1}$$

Ion–dipole forces (Figure **5.26**) are formed between the ions and water molecules and hydration enthalpies are always exothermic. So the dissolving process (Figure **5.27**) involves an endothermic process (ΔH_{latt}) and an exothermic process (ΔH_{hyd}).

Overall, an enthalpy change of solution will be exothermic if the total enthalpy change of hydration is more negative than the lattice enthalpy is positive.

Working out the value of the enthalpy change of solution for sodium chloride:

$$\Delta H_{sol} = \Delta H_{latt} + \Delta H_{hyd}(Na^+) + \Delta H_{hyd}(Cl^-)$$
$$\Delta H_{sol} = 771 + (-406) + (-364)$$
$$\Delta H_{sol} = +1 \, kJ \, mol^{-1}$$

'Infinite dilution' means that any further dilution of the solution produces no further enthalpy change – i.e. the solute particles are assumed not to interact with each other in the solution.

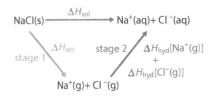

Figure 5.25 Dissolving and energy.

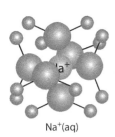

Na$^+$(aq)

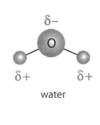

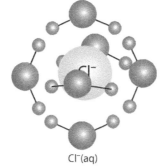

Cl$^-$(aq)

Figure 5.26 Hydrated Na$^+$ and Cl$^-$ ions.

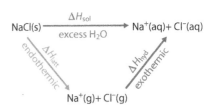

Figure 5.27 The energy changes involved in dissolving.

So, the dissolving of sodium chloride in water is an endothermic process (just). For dissolving magnesium chloride ($MgCl_2$) the enthalpy change of solution would be given by the equation:

$$\Delta H_{sol} = \Delta H_{latt} + \Delta H_{hyd}(Mg^{2+}) + 2\Delta H_{hyd}(Cl^-)$$

In general:

$$\Delta H_{sol}(X_nY_m) = \Delta H_{latt} + n\Delta H_{hyd}(X^{m+}) + m\Delta H_{hyd}(Y^{n-})$$

Hydration enthalpies

Some hydration enthalpies are shown in Table **5.8**. It can be seen from the values in the table that the enthalpy change of hydration depends on the size of an ion and its charge.

Positive ion	ΔH_{hyd} / kJ mol^{-1}	Negative ion	ΔH_{hyd} / kJ mol^{-1}
Li$^+$	−519	Cl$^-$	−364
Na$^+$	−406	Br$^-$	−335
K$^+$	−322	I$^-$	−293
Mg^{2+}	−1920		
Ca^{2+}	−1650		

Table 5.8 Enthalpy of hydration values for some ions.

More highly charged ions have a more exothermic enthalpy change of hydration because of a stronger electrostatic attraction between the ion and water molecules. Ca^{2+} and Na^+ ions are roughly the same size, but Ca^{2+} has a much more exothermic enthalpy change of hydration because of its higher charge.

Smaller ions have more exothermic enthalpy changes of hydration because of a stronger electrostatic attraction between the ion and water molecules. Ca^{2+} and Mg^{2+} ions have the same charge but Mg^{2+} has a more exothermic enthalpy change of hydration because of its smaller size.

As for lattice enthalpy, the magnitude of the charge has a greater effect on the enthalpy change of hydration than does the size of the ions. So, the hydration enthalpies of all 2+ ions will be more exothermic than those of 1+ ions.

Nature of science

Data is very important in science. The use of energy cycles allows the calculations of energy changes for reactions which cannot be measured directly.

? Test yourself

23 Use the values given in the table to work out the enthalpy change of solution of potassium iodide.

Lattice enthalpy of KI(s) / kJ mol^{-1}	629
Enthalpy change of hydration of K$^+$(g) / kJ mol^{-1}	−322
Enthalpy change of hydration of I$^-$(g) / kJ mol^{-1}	−293

24 Use the values given in the table to work out the enthalpy change of solution of magnesium chloride.

Lattice enthalpy of MgCl$_2$(s) / kJ mol^{-1}	2493
Enthalpy change of hydration of Mg^{2+}(g) / kJ mol^{-1}	−1920
Enthalpy change of hydration of Cl$^-$(g) / kJ mol^{-1}	−364

5.5 Entropy and spontaneity (HL)

The dissolving of sodium chloride (common salt) in water is an example of an endothermic process that occurs spontaneously at room temperature. The idea of an endothermic reaction occurring spontaneously goes against our experience from everyday life that things do not seem to move spontaneously from a lower to a higher energy state (a book does not jump from a lower shelf (lower potential energy) to a higher one (higher potential energy) but will fall to a lower shelf if the shelf it is on breaks. Endothermic processes such as the melting of ice at room temperature and water evaporating are relatively common, however, and this suggests that it is not just a consideration of the energy changes involved that must be used to predict whether a reaction occurs spontaneously. The examples of endothermic processes mentioned here all have one thing in common: they all involve an increase in **disorder**. In a sodium chloride solution, the Na^+ and Cl^- ions are dispersed throughout the solution and are moving around, which is a much more disordered, or random, arrangement than a separate crystal of sodium chloride and a beaker of pure water. Similarly, liquid water is much more disordered than a solid lump of ice, as in the liquid form all the molecules are moving around each other randomly. To understand why these processes occur we need to introduce a new term called **entropy** (S).

> **Entropy** is a measure of how the available energy is distributed among the particles.

When a system becomes more disordered, the energy in it can be distributed in more ways among the available particles and the entropy of the system is higher.

Imagine 100 people sitting on chairs in a hall – there is only a limited number of ways in which they can move but still remain seated. However, if you have the same 100 people running around in the hall there is an absolutely enormous number of different ways in which they can move.

The first situation can be likened to a solid and the second to a gas. In a gas, all the particles are moving at different speeds in different directions, and also rotating and vibrating, and there is a huge number of ways (high entropy) in which the energy could be distributed among all the different particles still giving the same total energy. In a solid, the particles are more constrained and there are fewer ways in which the energy could be distributed among the particles (lower entropy).

Entropy is given the symbol **S**. The units of entropy are $J K^{-1} mol^{-1}$. $S^{\ominus}$ is called **standard entropy**. It is possible to work out values for standard entropies for substances – for example, the standard entropy of $H_2(g)$ is $131 J K^{-1} mol^{-1}$, and the standard entropy of NaCl(s) is $72.4 J K^{-1} mol^{-1}$.

Learning objectives

- Understand what is meant by entropy
- Predict the sign of the entropy change for a given reaction
- Work out entropy changes from standard entropy values
- Calculate values of ΔG for a reaction
- Work out whether or not a reaction is spontaneous from the sign of ΔG
- Predict how the spontaneity of a reaction varies with temperature

Extension

A better (more correct) way of saying this is that entropy is a measure of the ways in which the available energy could be distributed among the available energy states. All molecular motion (translational, vibrational, rotational) is quantised – entropy relates to the probability of distributing the energy across the energy microstates that are available.

Joules and not kilojoules.

Actual values of entropy can be calculated from experimental data – as opposed to enthalpy, for which only enthalpy *changes* can be measured.

An entropy change is represented by the symbol $\Delta S^{\ominus}$.

A positive value for $\Delta S^{\ominus}$ indicates an increase in entropy – the energy is more spread out (can be distributed over a larger number of possible energy states). This corresponds to an increase in disorder.

For example, $\Delta S^{\ominus}$ for the process $H_2O(l) \rightarrow H_2O(g)$ is $+119\,J\,K^{-1}\,mol^{-1}$. The disorder of the system has increased. The particles in a gas have complete freedom of movement in three dimensions and, therefore, the energy can be distributed in more ways between the particles (higher entropy). The movement of the particles in a liquid is more constrained and so there are not as many ways to distribute the energy among the particles (lower entropy).

A negative value for $\Delta S^{\ominus}$ indicates a decrease in entropy – a decrease in the number of ways the available energy can be distributed among the particles. This corresponds to a decrease in disorder.

For example, $\Delta S^{\ominus}$ for the process $NH_3(g) + HCl(g) \rightarrow NH_4Cl(s)$ is $-285\,J\,K^{-1}\,mol^{-1}$. Two moles of gas being converted into a solid results in a decrease in disorder. The particles in a solid have less freedom of movement, and so there are fewer ways of distributing the energy among the particles and, therefore, the solid has lower entropy than the gas.

How to predict the sign of an entropy change

Gases have higher entropy than liquids, which have higher entropy than solids (Figure **5.28**).

Table **5.9** shows the values of standard entropies for elements across period 2 in the periodic table. Lithium to carbon are all solid elements and have low entropy values at 298 K, but nitrogen to neon are all gases and have much higher entropy values.

increasing entropy

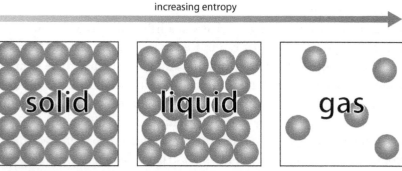

Figure 5.28 In a solid the particles vibrate about mean positions; in a liquid the particles move around each other; in a gas the particles move at high speeds in all directions.

Element	Li	Be	B	C	N_2	O_2	F_2	Ne
State	solid	solid	solid	solid	gas	gas	gas	gas
$S^{\ominus}/J\,K^{-1}\,mol^{-1}$	29	10	6	6	192	205	203	146

Table 5.9 Standard entropies for elements across period 2.

Reaction	Entropy	$\Delta S^\ominus$	Explanation
$N_2(g) + 3H_2(g) \rightarrow 2NH_3(g)$	decrease	–	Four moles of gas on the left-hand side are converted to two moles of gas on the right-hand side; a decrease in the number of moles of gas is a decrease in disorder; therefore the energy can be distributed in fewer ways in the products (energy is less spread out).
$CaCO_3(s) \rightarrow CaO(s) + CO_2(g)$	increase	+	One mole of solid becomes one mole of solid and one mole of gas; the number of moles of gas increases.
$CH_4(g) + 2O_2(g) \rightarrow CO_2(g) + 2H_2O(l)$	decrease	–	Three moles of gas are converted to one mole of gas.
$C_2H_4(g) + H_2(g) \rightarrow C_2H_6(g)$	decrease	–	Two moles of gas are converted to one mole of gas.

Table 5.10 Entropy changes for some reactions.

To predict the sign of the entropy change in a reaction, the easiest way to think about it is in terms of **disorder**. If there is an increase in disorder the energy is more spread out at the end than at the beginning, and so the entropy is higher at the end. Because gases have significantly higher entropy than solids and liquids, the most important factor in determining whether a chemical reaction involves an increase or decrease in entropy is whether there is an increase or decrease in the number of moles of gas (Table **5.10**).

- An **increase** in number of moles of **gas**: $\Delta S^\ominus$ +ve (entropy **increases**).
- A **decrease** in number of moles of **gas**: $\Delta S^\ominus$ −ve (entropy **decreases**).

If the number of moles of gas is the same on both sides of an equation, for example:

$$F_2(g) + Cl_2(g) \rightarrow 2ClF(g)$$

the prediction could be made that the entropy change would be approximately zero.

The entropy of a system is also changed by other factors as follows:
- entropy increases as the temperature increases
- entropy increases as the pressure decreases
- the entropy of a mixture is higher than the entropy of pure substances.

Calculating an entropy change for a reaction

Values of $\Delta S^\ominus$ can be worked out from standard entropies, $S^\ominus$:

entropy change = total entropy of products − total entropy of reactants

$$\Delta S^\ominus = \Sigma S^\ominus_{products} - \Sigma S^\ominus_{reactants}$$

Example

Calculate the standard entropy change for this reaction:

$$N_2(g) + 3H_2(g) \rightarrow 2NH_3(g)$$

$$\Delta S^\ominus = (2 \times 193) - [192 + (3 \times 131)]$$

$$\Delta S^\ominus = -199\,J\,K^{-1}\,mol^{-1}$$

	$S^\ominus / J\,K^{-1}\,mol^{-1}$
N_2	192
H_2	131
NH_3	193

The decrease in entropy corresponds to the prediction made above based on the number of moles of gas (four moles of gas go to two moles of gas).

The standard entropy values are multiplied by the appropriate coefficients in the equation.

25 Work out whether each of the following processes involves an increase or a decrease in entropy:

a $C_2H_2(g) + 2H_2(g) \rightarrow C_2H_6(g)$

b $2C_2H_6(g) + 7O_2(g) \rightarrow 4CO_2(g) + 6H_2O(l)$

c $COCl_2(g) \rightarrow CO(g) + Cl_2(g)$

d $2C(s) + O_2(g) \rightarrow 2CO(g)$

26 Use the entropy values in the table to calculate the standard entropy change in each of the following reactions:

a $CH_4(g) + 2O_2(g) \rightarrow CO_2(g) + 2H_2O(l)$

b $2Cu(NO_3)_2(s) \rightarrow 2CuO(s) + 4NO_2(g) + O_2(g)$

c $4BCl_3(l) + 3SF_4(g) \rightarrow 4BF_3(g) + 3SCl_2(g) + 3Cl_2(g)$

Substance	$S^{\ominus}/J K^{-1} mol^{-1}$
$CH_4(g)$	186
$O_2(g)$	103
$CO_2(g)$	214
$H_2O(l)$	70
$Cu(NO_3)_2(s)$	193
$CuO(s)$	43
$NO_2(g)$	240
$BCl_3(l)$	206
$SF_4(g)$	292
$BF_3(g)$	254
$SCl_2(g)$	282
$Cl_2(g)$	83

Predicting whether a reaction will be spontaneous

> **Spontaneous reaction:** one that occurs without any outside influence.

A spontaneous reaction does not have to happen quickly.

If sodium and oxygen are put together in an isolated container at 25 °C, they will react spontaneously to produce sodium oxide:

$4Na(s) + O_2(g) \rightarrow 2Na_2O(s)$

This reaction will occur by itself – nothing has to be done to make the reaction occur. This is an example of a spontaneous reaction.

If methane and oxygen are put into an isolated container at 25 °C, they will react together spontaneously to form carbon dioxide and water. This reaction, although it is spontaneous, is not a very fast reaction at room temperature (unless a spark is supplied) and would have to be left for a very long time before a significant amount of carbon dioxide and water could be detected. Similarly, the conversion of diamond to graphite at room temperature is a spontaneous process, but luckily occurs immeasurably slowly!

The first law of thermodynamics is basically a statement of the principle of conservation of energy: the energy of the Universe remains constant.

Whether a reaction will be spontaneous or not under a certain set of conditions can be deduced by looking at how the **entropy of the Universe** changes as the reaction occurs. The second law of thermodynamics states that for a process to occur spontaneously it must result in an increase in the entropy of the Universe.

This can be understood in terms of the spreading out of energy. A system will tend to move spontaneously from where the energy is concentrated (low entropy) to where it is more spread out (higher entropy). Where the energy is concentrated (low entropy) it is useful and can bring about change, but where it is spread out (higher entropy) it is less useful and cannot bring about change.

The Universe may be regarded as being composed of the system (the chemical reaction) and the surroundings. The entropy change of the Universe is given by:

$$\Delta S_{\text{Universe}} = \Delta S_{\text{surroundings}} + \Delta S_{\text{system}}$$

If, during a reaction, the value of $\Delta S_{\text{Universe}}$ is positive, the entropy of the Universe increases and the reaction occurs spontaneously.

When heat is given out in a chemical reaction, the surroundings get hotter and the particles move around more – therefore the entropy of the surroundings increases. So the entropy change of the surroundings can be related to the enthalpy change of the system. A new equation can be derived from the equation of the entropy change of the Universe given above. The new equation is:

$$\Delta G = \Delta H - T\Delta S$$

ΔG is called the change in **Gibbs free energy**, or just the **free energy change**. Under standard conditions, we have $\Delta G^{\ominus}$, which is the **standard free energy change**.

ΔG is related to the entropy change of the Universe and, from the condition that for a reaction to occur spontaneously the entropy of the Universe must increase, we can derive the condition:

for a reaction to be spontaneous, ΔG for the reaction must be **negative**.

Calculating $\Delta G^{\ominus}$

The standard free energy change for a reaction can be calculated using the equation:

$$\Delta G^{\ominus} = \Delta H^{\ominus} - T\Delta S^{\ominus}$$

The units of ΔG are usually kJ mol^{-1}.

T must be in K.

The subscripts are now omitted, because both ΔH and ΔS refer to the system, i.e. the chemical reaction.

Extension

The entropy change of the surroundings depends on the temperature and is given by:

$$\Delta S_{\text{surroundings}} = \frac{-\Delta H_{\text{system}}}{T}$$

[$-\Delta H$ because an exothermic reaction (ΔH negative) causes the entropy of the surroundings to increase ($\Delta S_{\text{surroundings}}$ positive)]

$$\Delta S_{\text{Universe}} = \Delta S_{\text{surroundings}} + \Delta S_{\text{system}}$$

$$\Delta S_{\text{Universe}} = \frac{-\Delta H_{\text{system}}}{T} + \Delta S_{\text{system}}$$

$$-T\Delta S_{\text{Universe}} = \Delta H_{\text{system}} - T\Delta S_{\text{system}}$$

$-T\Delta S_{\text{Universe}}$ is given the symbol ΔG.

Example

We can calculate $\Delta G^{\ominus}$ at 298 K for:

$$C_2H_2(g) + 2H_2(g) \rightarrow C_2H_6(g)$$

given the following information:

$$\Delta H^{\ominus} = -313 \, \text{kJ K}^{-1}\text{mol}^{-1} \qquad \Delta S^{\ominus} = -233 \, \text{J K}^{-1}\text{mol}^{-1}$$

$\Delta H^{\ominus}$ is in kJ but $\Delta S^{\ominus}$ is in J, and in order to combine them they must be converted so they are both in kJ or J. As $\Delta G^{\ominus}$ is usually quoted in kJ, $\Delta S^{\ominus}$ will be converted to kJ. Therefore:

$$\Delta S^{\ominus} = \frac{-233}{1000} = -0.233 \, \text{kJ K}^{-1}\text{mol}^{-1}$$

$$\Delta G^{\ominus} = -313 - 298 \times (-0.233) = -244 \, \text{kJ mol}^{-1}$$

Because the value of ΔG is negative, the reaction is spontaneous.

Worked example

5.15 Consider the decomposition of $Mg(NO_3)_2(s)$:

$$2Mg(NO_3)_2(s) \rightarrow 2MgO(s) + 4NO_2(g) + O_2(g)$$

	$Mg(NO_3)_2(s)$	$MgO(s)$	$NO_2(g)$	$O_2(g)$
$\Delta H_f^{\ominus}$/kJ mol^{-1}	−790	−602	34	0
$S^{\ominus}$/J K^{-1}mol^{-1}	164	27	240	205

a Use the following data to work out $\Delta G^{\ominus}$ and hence whether or not the reaction will be spontaneous at 25 °C.

b As the temperature is increased from absolute zero, work out the temperature (in °C) at which the reaction first becomes spontaneous.

a To calculate the enthalpy change, we have been given ΔH_f and so can use:

$$\Delta H = \Sigma \Delta H_f(\text{products}) - \Sigma \Delta H_f(\text{reactants})$$

$$\Delta H^{\ominus} = [(2 \times -602) + (4 \times 34) + 0] - [(2 \times -790)] = 512 \, \text{kJ mol}^{-1}$$

To calculate the entropy change we use:

$$\Delta S = \Sigma S(\text{products}) - \Sigma S(\text{reactants})$$

$$\Delta S^{\ominus} = [(2 \times 27) + (4 \times 240) + 205] - [(2 \times 164)] = 891 \, \text{J K}^{-1}\text{mol}^{-1}$$

At 298 K:

$$\Delta G^{\ominus} = \Delta H^{\ominus} - T\Delta S^{\ominus}$$

$$\Delta G^{\ominus} = 512 - 298 \times \frac{891}{1000} = 246 \, \text{kJ mol}^{-1}$$

$\Delta S^{\ominus}$ is divided by 1000 to convert to kJ.

At 298 K the reaction is **not** spontaneous because ΔG is positive.

b Because ΔS is positive, as the temperature is increased $T\Delta S$ will become larger until it is eventually bigger than ΔH and the reaction will become spontaneous.

We can estimate the temperature at which this reaction is going to become spontaneous. As the temperature is increased, ΔG will become less and less positive, until it becomes zero, and then it will be negative. So if we work out the temperature at which ΔG becomes zero, the reaction will be spontaneous at any temperature above that.

Using

$$\Delta G = \Delta H - T\Delta S$$

$$0 = 512 - T \times \frac{891}{1000}$$

This calculation is only approximate because the values of ΔH and ΔS change with temperature, and we have used $\Delta H^{\ominus}$ and $\Delta S^{\ominus}$, i.e. the values at 298 K.

Rearranging the equation gives:

$$T = 575 \, \text{K}$$

This means that this reaction is likely to become spontaneous above 575 K, that is 302 °C.

Non-spontaneous reactions

If a reaction is non-spontaneous it does not mean that it can never happen – it just means that it will not happen without external influence. For instance, in the above worked example we can increase the temperature to make the reaction spontaneous.

Consider the following reaction:

$$2H_2O(l) \rightarrow 2H_2(g) + O_2(g) \qquad \Delta G^{\ominus} = +474 \, \text{kJ} \, \text{mol}^{-1}$$

This reaction is not spontaneous at 25 °C, but it can be made to happen at this temperature by the continuous passage of an electric current (electrolysis).

The effect of changing temperature on the spontaneity of a reaction

We will use the equation $\Delta G = \Delta H - T\Delta S$ to explain the effect of temperature on the value of ΔG and hence on the spontaneity of a reaction.

Consider a reaction for which ΔH is positive and ΔS is positive. If $T\Delta S$ is smaller than ΔH, ΔG will be positive and the reaction will **not** be spontaneous. So a reaction like this is non-spontaneous at low temperatures (when $T\Delta S$ is smaller than ΔH). As the temperature is raised, $T\Delta S$ becomes larger and, because this is being subtracted from ΔH, ΔG becomes less positive/more negative – the reaction becomes more spontaneous as the temperature is raised. Eventually, when $T\Delta S$ becomes greater than ΔH, ΔG will be negative and the reaction will be spontaneous.

Now consider a reaction for which ΔH is positive and ΔS is negative. Because ΔS is negative $-T\Delta S$ in the equation is positive. ΔH is also positive, so overall ΔG is positive and the reaction is not spontaneous. Because both ΔH and $-T\Delta S$ are positive, this reaction is non-spontaneous at all temperatures: ΔG can never be negative.

Both reactions considered so far have been endothermic and it can been seen that an endothermic reaction can only occur spontaneously if it involves an increase in entropy (and the temperature is sufficiently high).

ΔH	ΔS	−TΔS	ΔG	Spontaneous?
negative	positive	negative	negative	at all temperatures
positive	positive	negative	becomes more negative as temperature increases	becomes more spontaneous as temperature increases
negative	negative	positive	becomes less negative as temperature increases	becomes less spontaneous as temperature increases
positive	negative	positive	positive	never

Table 5.11 Determining if a reaction will be spontaneous.

Reactions in which ΔS is positive become more spontaneous as temperature increases, but reactions in which ΔS is negative become less spontaneous as temperature increases.

An exothermic reaction (ΔH negative) will always be spontaneous at some temperature or other. If the reaction involves an increase in entropy (ΔS positive) then −TΔS will be negative. Because ΔH is also negative, ΔG will always be negative and the reaction will always be spontaneous.

If the reaction involves a decrease in entropy (ΔS negative), the reaction will be spontaneous at low temperatures, when ΔH is more negative than −TΔS is positive. It will, however, become less spontaneous as the temperature increases (because −TΔS is positive and becomes more positive as the temperature increases). At higher temperatures, −TΔS will be more positive than ΔH is negative, and therefore ΔG will be positive and the reaction will be non-spontaneous.

Table **5.11** gives an overview of this discussion.

Gibbs free energy and equilibrium

Consider the Haber process for the production of ammonia:

$$N_2(g) + 3H_2(g) \rightleftharpoons 2NH_3(g) \qquad \Delta G = -33\,\text{kJ}\,\text{mol}^{-1}$$

According to the discussion above, this reaction will proceed from left to right – nitrogen and hydrogen will spontaneously become ammonia – but the reverse reaction ($\Delta G = +33\,\text{kJ}\,\text{mol}^{-1}$) will not occur spontaneously. However, equilibrium can be reached in either direction – if we start with nitrogen and hydrogen, the system will form an equilibrium mixture in which nitrogen, hydrogen and ammonia are present; if we start with pure ammonia, some will spontaneously react to form nitrogen and hydrogen so that all three are present in the equilibrium mixture. This does not, however, violate the second law of thermodynamics because the value of ΔG that was calculated was for complete conversion of one mole of nitrogen and three moles of hydrogen to two moles of ammonia. The equilibrium mixture always has a lower Gibbs free energy (higher entropy) than either the pure reactants or the pure products (a mixture has higher entropy than pure substances) therefore the conversion of either reactants **or** products into the equilibrium mixture results in a process in which ΔG is negative (Figure **5.29**).

The overall Gibbs free energy of a system (note that we are looking at the Gibbs free energy (G) here and not the **change** in Gibbs free energy (ΔG)) depends on how much of each substance is present, and the equilibrium mixture represents the composition that gives the minimum value of the Gibbs free energy (maximum value of entropy). When the system is at equilibrium, the Gibbs free energy of the amount of reactants

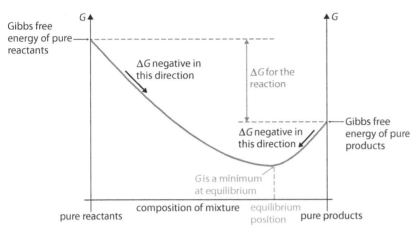

Figure 5.29 Variation in the Gibbs free energy for a reaction for which ΔG is negative overall.

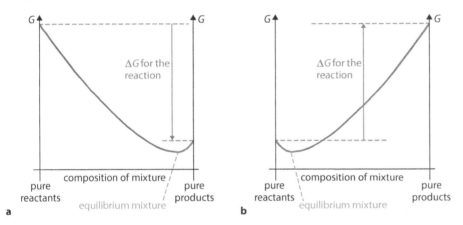

Figure 5.30 a ΔG is negative and the position of equilibrium lies closer to the products; **b** ΔG is positive and the position of equilibrium lies closer to the reactants.

present is the same as that of however much of the products is present, and so ΔG is zero and there is no tendency to spontaneously move in either direction away from equilibrium. Any shift away from the equilibrium position results in an increase in G and therefore a process for which ΔG is positive – i.e. non-spontaneous.

The value and sign of ΔG give us information about the position of equilibrium. If ΔG is negative then the position of equilibrium will lie closer to the products than the reactants (Figure **5.30a**). The more negative the value of ΔG, the closer the position of equilibrium lies towards the products. If ΔG is numerically very large and negative then the position of equilibrium lies very close to pure products, which corresponds to the idea discussed above – a reaction for which ΔG is negative proceeds spontaneously from reactants to products. If ΔG is positive then the position of equilibrium lies more towards the reactants (Figure **5.30b**) – the more positive the value, the closer the position of equilibrium lies towards pure reactants.

A value of ΔG of -30 kJ mol^{-1} (or more negative) indicates that a reaction essentially goes to completion at around room temperature; one of $+30$ kJ mol^{-1} (or more positive) indicates a reaction that does not really proceed towards products at all.

Nature of science

The word **spontaneous** has a specific meaning in science which is not the same as how it is used in everyday life. Words sometimes have a different meaning in science from the same word in everyday life. Scientists must agree on the meaning of these words and many professional bodies, such as IUPAC, provide the final word on the use of scientific terms. The meaning of terms can also change – for instance, 'standard conditions' used to be defined in terms of atmospheric pressure but has now changed to 100 kPa. The whole scientific community must agree to such changes because scientists must understand what other scientists mean – everybody cannot use their own definition!

Science involves a constantly changing body of knowledge. Our understanding of entropy has evolved since the term was first used by Rudolf Clausius in a paper in 1865 and now we discuss entropy from a probability/statistical point of view.

? Test yourself

27 Given the data below, calculate $\Delta G^{\ominus}$ for the following reaction at 298 K and state whether it is spontaneous or not:

$C_2H_4(g) + H_2(g) \rightarrow C_2H_6(g)$
$\Delta H^{\ominus} = -137 \, kJ \, mol^{-1}$ and
$\Delta S^{\ominus} = -55.3 \, J \, K^{-1} \, mol^{-1}$

28 Given the data below, calculate $\Delta G^{\ominus}$ for the following reaction and state whether it is spontaneous or not at 298 K:

$C_3H_8(g) + 5O_2(g) \rightarrow 3CO_2(g) + 4H_2O(l)$
$\Delta H^{\ominus} = -2219 \, kJ \, mol^{-1}$ and
$\Delta S^{\ominus} = -373.3 \, J \, K^{-1} \, mol^{-1}$

29 Consider the decomposition of $Pb(NO_3)_2(s)$:

$2Pb(NO_3)_2(s) \rightarrow 2PbO(s) + 4NO_2(g) + O_2(g)$
$\Delta H^{\ominus} = +598 \, kJ \, mol^{-1}$ $\Delta G^{\ominus} = +333 \, kJ \, mol^{-1}$

a Work out the value for $\Delta S^{\ominus}$ at 298 K.

b Assuming that ΔH and ΔS do not change with temperature, calculate the temperature, in °C, above which this reaction will become spontaneous.

30 For each of the following reactions, predict whether it becomes more or less spontaneous as temperature increases:

a $N_2O_4(g) \rightarrow 2NO_2(g)$
b $N_2(g) + 3H_2(g) \rightarrow 2NH_3(g)$
c $2KNO_3(s) \rightarrow 2KNO_2(s) + O_2(g)$
d $2AgNO_3(s) \rightarrow 2Ag(s) + 2NO_2(s) + O_2(g)$

31 Consider the equilibrium
$N_2O_4(g) \rightleftharpoons 2NO_2(g) \, \Delta G^{\ominus} = 5 \, kJ \, mol^{-1}$
Does the position of equilibrium lie more towards the reactants (N_2O_4) or the products (NO_2)?

Exam-style questions

1   The specific heat capacity of a liquid is $4.00 \, \text{J} \, \text{g}^{-1} \, \text{K}^{-1}$; 2000 J of heat energy are supplied to 100.0 g of the liquid. By how much would the temperature of the liquid increase?

           A 278 K   **B** 5 K   **C** 80 K   **D** 20 K

2   Which of the following is correct about endothermic reactions?

           A   Heat energy is taken in and the temperature increases.
           B   Heat energy is given out and the temperature increases.
           C   Heat energy is taken in and the temperature decreases.
           D   Heat energy is given out and the temperature decreases.

3   Use the following information:

   $$2H_2(g) + O_2(g) \rightarrow 2H_2O(l) \qquad \Delta H = -572 \, \text{kJ} \, \text{mol}^{-1}$$
   $$2H_2(g) + O_2(g) \rightarrow 2H_2O(g) \qquad \Delta H = -484 \, \text{kJ} \, \text{mol}^{-1}$$

   to calculate the enthalpy change for the process:

   $$H_2O(g) \rightarrow H_2O(l)$$

           A $-88 \, \text{kJ} \, \text{mol}^{-1}$ **C** $+88 \, \text{kJ} \, \text{mol}^{-1}$
           B $-44 \, \text{kJ} \, \text{mol}^{-1}$ **D** $+44 \, \text{kJ} \, \text{mol}^{-1}$

4   Use the bond enthalpies in the table to calculate the enthalpy change (in $\text{kJ} \, \text{mol}^{-1}$) for the reaction:

   $$CH_4(g) + 2Cl_2(g) \rightarrow CH_2Cl_2(g) + 2HCl(g)$$

Bond	Bond enthalpy / kJ mol^{-1}
C–H	410
Cl–Cl	240
C–Cl	340
H–Cl	430

           A −720   **B** +240   **C** +620   **D** −240

5   Use the enthalpy change of formation values in the table to calculate the enthalpy change for the following reaction:

   $$4NH_3(g) + 3O_2(g) \rightarrow 2N_2(g) + 6H_2O(l)$$

Substance	$\Delta H_f^{\ominus}$ / kJ mol^{-1}
$NH_3(g)$	−46
$H_2O(l)$	−286

           A $-240 \, \text{kJ} \, \text{mol}^{-1}$ **C** $-1532 \, \text{kJ} \, \text{mol}^{-1}$
           B $-332 \, \text{kJ} \, \text{mol}^{-1}$ **D** $-1900 \, \text{kJ} \, \text{mol}^{-1}$

6 Use the enthalpy changes given in the table to calculate the enthalpy change (in $kJ\,mol^{-1}$) for the reaction:

$$C_4H_6(g) + 2H_2(g) \rightarrow C_4H_{10}(g)$$

Substance	$\Delta H_c^\ominus / kJ\,mol^{-1}$
$C_4H_6(g) + 5\frac{1}{2}O_2(g) \rightarrow 4CO_2(g) + 3H_2O(l)$	−2595
$C_4H_{10}(g) + 6\frac{1}{2}O_2(g) \rightarrow 4CO_2(g) + 5H_2O(l)$	−2875
$H_2(g) + \frac{1}{2}O_2(g) \rightarrow H_2O(l)$	−285

    A 290   **B** −290   **C** −5   **D** 5

HL **7** Which of the following processes is exothermic?

    A $Br_2(l) \rightarrow 2Br(g)$ **C** $CaF_2(s) \rightarrow Ca^{2+}(g) + 2F^-(g)$

    B $Na(g) \rightarrow Na^+(g) + e^-$ **D** $Cl(g) + e^- \rightarrow Cl^-(g)$

HL **8** Which of the following reactions has a $\Delta S^\ominus$ value that is negative?

    A $2H_2O_2(aq) \rightarrow 2H_2O(l) + O_2(g)$

    B $CaCO_3(s) \rightarrow CaO(s) + CO_2(g)$

    C $CaCO_3(s) + 2HCl(aq) \rightarrow CaCl_2(aq) + CO_2(g) + H_2O(l)$

    D $2C_4H_{10}(g) + 13O_2(g) \rightarrow 8CO_2(g) + 10H_2O(l)$

HL **9** Which of the following will have the largest value of lattice enthalpy?

    A NaCl   **B** MgO   **C** CaO   **D** $MgBr_2$

HL **10** Using the data given below, calculate the value of $\Delta G^\ominus$ for the following reaction at 298 K:

$$SO_2(g) + Cl_2(g) \rightarrow SO_2Cl_2(l)$$
$$\Delta H^\ominus = -97.3\,kJ\,mol^{-1} \text{ and } \Delta S^\ominus = -254.4\,J\,K^{-1}\,mol^{-1}$$

    A $-21.5\,kJ\,mol^{-1}$ **C** $-173.1\,kJ\,mol^{-1}$

    B $75\,714\,kJ\,mol^{-1}$ **D** $-97.0\,kJ\,mol^{-1}$

11 **a** Explain what you understand by the term 'average bond enthalpy'. [2]

b Use the average bond enthalpies given in the table to calculate the enthalpy change for the combustion of ethanol vapour, according to the equation:

$$C_2H_5OH(g) + 3O_2(g) \rightarrow 2CO_2(g) + 3H_2O(g)$$ [3]

Bond	C–H	C–C	C–O	O–H	O=O	C=O
Bond enthalpy / kJ mol^{-1}	412	348	360	463	496	743

c Explain why bond enthalpies could not be used to work out the enthalpy change for the reaction:

$$C_2H_5OH(l) + 3O_2(g) \rightarrow 2CO_2(g) + 3H_2O(l)$$ [2]

d Consider the reaction:

$$N_2O_4(g) \rightarrow 2NO_2(g) \qquad \Delta H^{\ominus} = +57\,kJ\,mol^{-1}$$

Draw a potential energy profile for this reaction and explain whether NO_2 or N_2O_4 is more stable. [3]

12 **a** Define 'standard enthalpy change of formation'. [2]

b Write a chemical equation for the standard enthalpy change of formation of propan-1-ol. [2]

c The equation for the combustion of propan-1-ol is:

$$CH_3CH_2CH_2OH(l) + 4.5O_2(g) \rightarrow 3CO_2(g) + 4H_2O(l)$$

Use the enthalpy change of formation values in the table to calculate the enthalpy change for this reaction.

Substance	$\Delta H_f^{\ominus}$ / kJ mol^{-1}
$CH_3CH_2CH_2OH(l)$	−316
$CO_2(g)$	−394
$H_2O(l)$	−286

[3]

HL **d** Use the standard entropy values in the table below to calculate the entropy change for the complete combustion of propan-1-ol and justify the sign of the entropy change. [3]

Substance	$S^{\ominus}$ / J K^{-1} mol^{-1}
$CH_3CH_2CH_2OH(l)$	196.6
$CO_2(g)$	214
$H_2O(l)$	69.9
$O_2(g)$	205

HL **e** Calculate the standard free energy change, $\Delta G^{\ominus}$, for the complete combustion of propan-1-ol and explain whether the reaction will be spontaneous at 25 °C. [3]

13 Hydrazine, N_2H_4, has been used as a rocket fuel.

   a Draw a Lewis structure for hydrazine.   **[1]**

   b Write a chemical equation to represent the enthalpy change of formation of gaseous hydrazine.   **[2]**

   c Use bond enthalpies from the table to calculate the enthalpy change of formation of gaseous hydrazine.   **[3]**

Bond	N≡N	N=N	N–N	H–H	N–H
Bond enthalpy / kJ mol^{-1}	944	409	163	436	388

   d The equation for the combustion of liquid hydrazine is:

$$N_2H_4(l) + O_2(g) \rightarrow N_2(g) + 2H_2O(l) \quad \Delta H = -622 \, kJ \, mol^{-1}$$

      The enthalpy change of formation of $H_2O(l)$ is $-286 \, kJ \, mol^{-1}$.

      Use these data to calculate the enthalpy change of formation of liquid hydrazine.   **[3]**

   e Calculate the enthalpy change for the process:

$$N_2H_4(l) \rightarrow N_2H_4(g)$$

      [2]

HL 14 a Define 'lattice enthalpy'.   **[2]**

   b Write chemical equations to represent the following enthalpy changes: **[3]**
      i the lattice enthalpy of potassium chloride
      ii the first electron affinity of chlorine
      iii the first ionisation energy of potassium

   c Construct a Born–Haber cycle for the formation of potassium chloride and use it and the values in the table to calculate the lattice enthalpy of potassium chloride.   **[5]**

Process	Enthalpy change / kJ mol^{-1}
ΔH_{at} [K(s)]	90
ΔH_{at} [Cl_2(g)]	121
first ionisation energy (K)	418
first electron affinity (Cl)	−364
ΔH_f [KCl(s)]	−436

   d Explain why the value of the lattice enthalpy for calcium chloride is substantially greater than that for potassium chloride.   **[2]**

   e The enthalpy changes of hydration of potassium ions and chloride ions are given in the table. Use these values and other value(s) from part **c** to calculate the enthalpy change of solution of potassium chloride.   **[2]**

Ion	Enthalpy change of hydration / kJ mol^{-1}
K$^+$(g)	−340
Cl$^-$(g)	−359

Summary

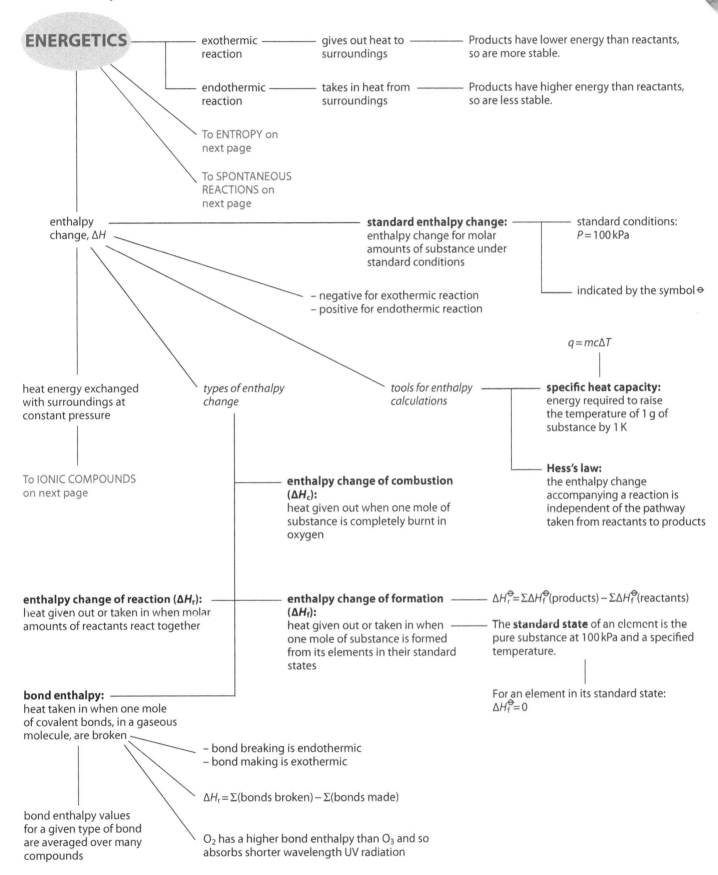

ENERGETICS ── exothermic reaction ── gives out heat to surroundings ── Products have lower energy than reactants, so are more stable.

endothermic reaction ── takes in heat from surroundings ── Products have higher energy than reactants, so are less stable.

To ENTROPY on next page

To SPONTANEOUS REACTIONS on next page

enthalpy change, ΔH ── **standard enthalpy change:** enthalpy change for molar amounts of substance under standard conditions

standard conditions: $P = 100\,kPa$

indicated by the symbol ⊖

– negative for exothermic reaction
– positive for endothermic reaction

$q = mc\Delta T$

heat energy exchanged with surroundings at constant pressure

types of enthalpy change

tools for enthalpy calculations ── **specific heat capacity:** energy required to raise the temperature of 1 g of substance by 1 K

To IONIC COMPOUNDS on next page

enthalpy change of combustion (ΔH_c): heat given out when one mole of substance is completely burnt in oxygen

Hess's law: the enthalpy change accompanying a reaction is independent of the pathway taken from reactants to products

enthalpy change of reaction (ΔH_r): heat given out or taken in when molar amounts of reactants react together

enthalpy change of formation (ΔH_f): heat given out or taken in when one mole of substance is formed from its elements in their standard states

$\Delta H_r^{\ominus} = \Sigma \Delta H_f^{\ominus}(\text{products}) - \Sigma \Delta H_f^{\ominus}(\text{reactants})$

The **standard state** of an element is the pure substance at 100 kPa and a specified temperature.

For an element in its standard state: $\Delta H_f^{\ominus} = 0$

bond enthalpy: heat taken in when one mole of covalent bonds, in a gaseous molecule, are broken

– bond breaking is endothermic
– bond making is exothermic

$\Delta H_r = \Sigma(\text{bonds broken}) - \Sigma(\text{bonds made})$

bond enthalpy values for a given type of bond are averaged over many compounds

O_2 has a higher bond enthalpy than O_3 and so absorbs shorter wavelength UV radiation

Summary – continued

HL

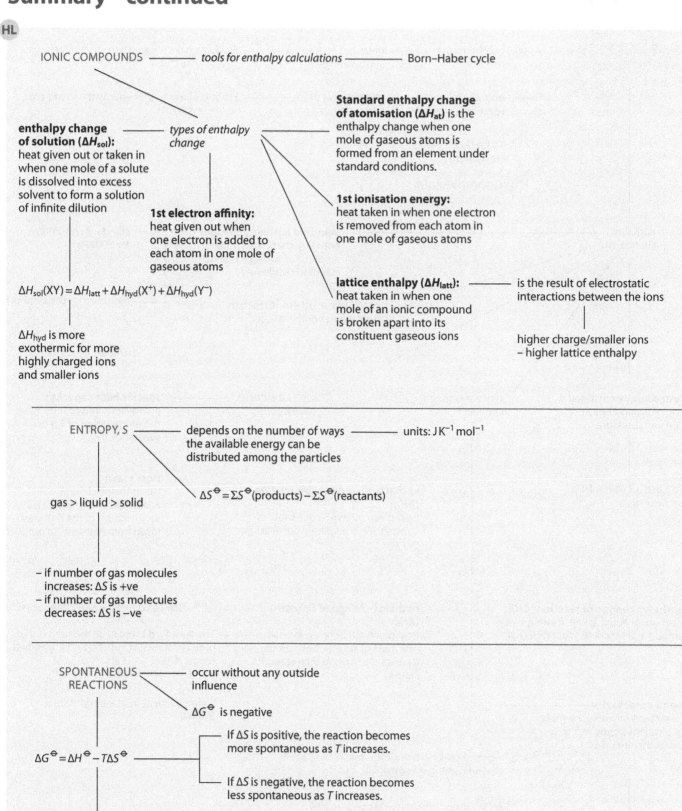

IONIC COMPOUNDS ———— *tools for enthalpy calculations* ———— Born–Haber cycle

types of enthalpy change

enthalpy change of solution (ΔH_{sol}): heat given out or taken in when one mole of a solute is dissolved into excess solvent to form a solution of infinite dilution

Standard enthalpy change of atomisation (ΔH_{at}) is the enthalpy change when one mole of gaseous atoms is formed from an element under standard conditions.

1st electron affinity: heat given out when one electron is added to each atom in one mole of gaseous atoms

1st ionisation energy: heat taken in when one electron is removed from each atom in one mole of gaseous atoms

lattice enthalpy (ΔH_{latt}): heat taken in when one mole of an ionic compound is broken apart into its constituent gaseous ions

is the result of electrostatic interactions between the ions

higher charge/smaller ions – higher lattice enthalpy

$$\Delta H_{sol}(XY) = \Delta H_{latt} + \Delta H_{hyd}(X^+) + \Delta H_{hyd}(Y^-)$$

ΔH_{hyd} is more exothermic for more highly charged ions and smaller ions

---

ENTROPY, S ———— depends on the number of ways the available energy can be distributed among the particles ———— units: $J\,K^{-1}\,mol^{-1}$

gas > liquid > solid

$$\Delta S^{\ominus} = \Sigma S^{\ominus}(\text{products}) - \Sigma S^{\ominus}(\text{reactants})$$

– if number of gas molecules increases: ΔS is +ve
– if number of gas molecules decreases: ΔS is −ve

SPONTANEOUS REACTIONS ———— occur without any outside influence

$\Delta G^{\ominus}$ is negative

$$\Delta G^{\ominus} = \Delta H^{\ominus} - T\Delta S^{\ominus}$$

If ΔS is positive, the reaction becomes more spontaneous as T increases.

If ΔS is negative, the reaction becomes less spontaneous as T increases.

ΔG = Gibbs free energy change ———— The value of ΔG is related to the position of equilibrium.

Chemical kinetics 6

6.1 Collision theory and rate of reaction

6.1.1 What is 'rate of reaction'?

When we consider the rate of a chemical reaction what we are looking at is how fast or slow the reaction is (Figures **6.1** and **6.2**). This can be thought of in terms of how quickly the reactants are used up or how quickly the products are formed.

> **Rate of reaction** is the speed at which reactants are used up or products are formed.

Experiments to measure the rate of reaction

Consider the reaction between calcium carbonate and hydrochloric acid:

$$CaCO_3(s) + 2HCl(aq) \rightarrow CaCl_2(aq) + CO_2(g) + H_2O(l)$$

The rate of this reaction can be measured in various ways, two of which will be considered for this experiment:

1 measurement of the rate at which CO_2 is produced
2 measurement of the rate at which the mass decreases.

Measurement of the rate at which CO_2 is produced

The apparatus is set up as shown in Figure **6.3** without the calcium carbonate. The bung in the conical flask (Erlenmeyer flask) is removed, the calcium carbonate added, the bung quickly replaced and the timer started. This experiment can be used to generate a graph of volume of carbon dioxide produced against time by noting the volume on the measuring cylinder every ten seconds and then plotting the data. Sample data for this experiment are shown in Table **6.1**.

Figure 6.1 Rusting is a very slow chemical reaction but one that costs economies billions of dollars each year.

Figure 6.2 An explosion is a very fast reaction – gases and a great deal of heat are generated very quickly.

> Instead of a measuring cylinder, a gas burette or a gas syringe could be used.

Learning objectives

- Understand what is meant by and define the rate of a chemical reaction
- Describe experimental methods for measuring the rates of various types of chemical reactions
- Analyse numerical and graphical data from rate experiments

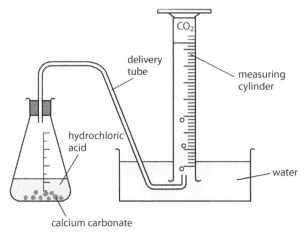

Figure 6.3 An experiment to measure rate of CO_2 production.

Time/s	Volume of CO_2/cm^3
0	0.0
10	19.0
20	33.0
30	44.0
40	50.0
50	54.0
60	56.5
70	58.5
80	59.5
90	60.0
100	60.0
110	60.0

Table 6.1 Sample data for the experiment shown in Figure 6.3.

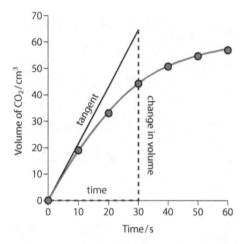

Figure 6.5 The tangent is drawn at the initial point to determine the initial rate.

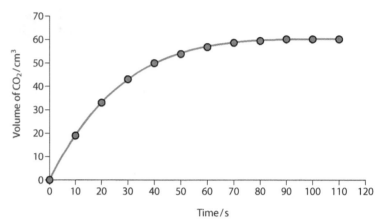

Figure 6.4 Change in volume with time.

The reaction appears to finish at 90 s (Figure 6.4) because no more gas is produced after that. The average rate of reaction during the first 90 s can then be worked out as:

$$\text{average rate} = \frac{\text{change in volume}}{\text{time}} = \frac{60.0}{90} = 0.67 \, cm^3 \, s^{-1}$$

The rate at any particular time is given by the slope (gradient) of the graph at that time. This can be worked out by drawing a tangent to the curve at that point (Figure 6.5).

The gradient of the tangent is given by:

$$\text{gradient} = \frac{\text{change in volume}}{\text{time}} = \frac{64}{30} = 2.1 \, cm^3 \, s^{-1}$$

Therefore the initial rate of reaction is $2.1 \, cm^3 \, s^{-1}$, which means that, initially, the gas is being produced at a rate of $2.1 \, cm^3$ per second.

It can be seen from the graph in Figure 6.4 that the gradient decreases as time goes on – in other words, the rate is fastest at the beginning and gets slower. At 90 s, the reaction has stopped and the gradient of the graph is zero.

The volume of carbon dioxide produced can be used to calculate the concentration of hydrochloric acid in the flask at any time. The more gas that has been produced, the lower the concentration of the remaining hydrochloric acid. The actual concentration of acid at any time could be worked out using a moles calculation, assuming that the initial volume and concentration of the acid are known. These data could, then be used to plot a graph of concentration of hydrochloric acid against time.

Possible problems with experiments like this include the fact that some gas is likely to escape before the bung is put on the flask (resulting in all values for the volume of carbon dioxide being lower than expected) and variations in the sizes of the calcium carbonate pieces.

The same experimental set-up can be used for investigating the rate of reaction between magnesium and hydrochloric acid. This reaction is strongly exothermic and the reaction mixture becomes hotter during the experiment. This will cause the rate to be higher than expected.

Measurement of the rate at which the mass decreases

The rate of this reaction can also be determined by measuring the speed at which the mass decreases. The experimental set-up for this is shown in Figure **6.6**. The mass decreases as carbon dioxide is given off. The data for this experiment are shown in Table **6.2**, along with the resulting graph in Figure **6.7**.

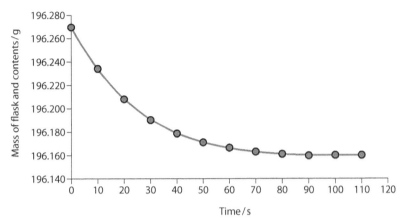

Figure 6.7 Change in mass with time.

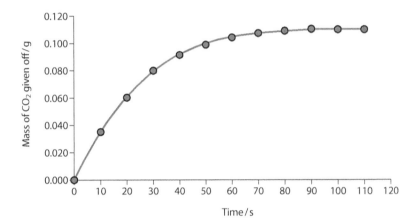

Figure 6.8 Loss in mass with time.

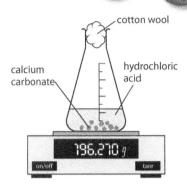

Figure 6.6 The cotton wool allows the gas to escape but stops mass being lost as a result of splashes.

Time / s	Mass of flask / g
0	196.270
10	196.235
20	196.210
30	196.189
40	196.178
50	196.171
60	196.166
70	196.163
80	196.161
90	196.160
100	196.160
110	196.160

Table 6.2 Sample data for the experiment shown in Figure **6.6**.

Time / s	Mass lost / g
0	0.000
10	0.035
20	0.060
30	0.081
40	0.092
50	0.099
60	0.104
70	0.107
80	0.109
90	0.110
100	0.110
110	0.110

Table 6.3 Calculating the carbon dioxide lost.

Alternatively, the mass of carbon dioxide lost can be worked out (196.270 − mass at any time, see Table **6.3**), and this can be plotted as shown in Figure **6.8**.

The average rate of reaction and initial rate of reaction can be worked out from either graph, using the same techniques as above. In this case we get:

$$\text{average rate} = \frac{\text{change in mass}}{\text{time}} = \frac{0.110}{90} = 1.22 \times 10^{-3}\,\text{g s}^{-1}$$

The initial rate is given by drawing a tangent at the initial point and in this case the initial rate is $4.0 \times 10^{-3}\,\text{g s}^{-1}$.

As above, the mass of carbon dioxide produced can be related to the concentrations of the hydrochloric acid or calcium chloride in the flask at any time and a graph of concentration against time could be plotted.

Rate of reaction defined

Although the above reactions were followed by looking at changes in volume and mass, rates are most often considered in terms of changing concentrations. We can define the rate of reaction:

> **Rate of reaction** is the change in concentration of reactants or products per unit time.

Unit time could be 1 s, 1 min, etc.

Units for rate of reaction are therefore $mol\,dm^{-3}\,s^{-1}$, $mol\,dm^{-3}\,min^{-1}$, etc. The average rate over a period of time can be worked out as:

> $$average\ rate = \frac{change\ in\ concentration}{time}$$

For the reaction $A + B \rightarrow C$, the rate at which the reactants are used up is equal to the rate at which the products are produced, i.e. if the rate of reaction with respect to A is $0.12\,mol\,dm^{-3}\,s^{-1}$, the rate of reaction with respect to C will also be $0.12\,mol\,dm^{-3}\,s^{-1}$.

However, for the reaction $A \rightarrow 2D$, the rate at which D is produced will be twice the rate at which A is used up, because one mole of A will produce two moles of D. Therefore, if the rate of reaction with respect to A is $0.16\,mol\,dm^{-3}\,s^{-1}$, the rate of reaction with respect to D will be $0.32\,mol\,dm^{-3}\,s^{-1}$.

The rate of reaction at any time can be found from a graph of concentration against time by drawing a tangent at the particular time and finding the gradient (slope) of the tangent, in a similar way to that shown for a volume against time graph in Figure **6.5**.

Figure **6.9a** shows the how the concentration of A varies in the reaction $A \rightarrow 2D$. The initial rate of reaction can be determined from the gradient of the tangent shown ($\frac{2}{24} = 0.083\,mol\,dm^{-3}\,s^{-1}$). The graph in Figure **6.9b** shows how the concentration of the product, D, changes in the same reaction. The rate at 40 s can be determined from the graph using the tangent shown ($\frac{2}{76} = 0.026\,mol\,dm^{-3}\,s^{-1}$).

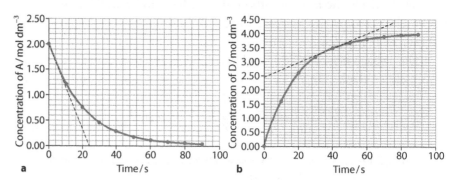

Figure 6.9 Graphs showing how the concentrations of **a** the reactant and **b** the product vary for the reaction $A \rightarrow 2D$.

Extension

The rate of reaction can also be defined in terms of calculus notation. For the reaction:

$$A + B \rightarrow C + D$$

The rate of reaction can be given as $\dfrac{d[C]}{dt}$, where [C] is the concentration of C.

Because the concentration of A is decreasing, $\dfrac{-d[A]}{dt}$ is equal to $\dfrac{d[C]}{dt}$ and for this reaction:

$$\text{rate} = \frac{-d[A]}{dt} = \frac{-d[B]}{dt} = \frac{d[C]}{dt} = \frac{d[D]}{dt}$$

Following the rate of a chemical reaction using changes in colour

The iodination of propanone in the presence of an acid catalyst can be followed conveniently using a colorimeter (Figure **6.10**) to monitor the changes in colour that occur as iodine is used up.

$$\underset{\text{propanone}}{CH_3COCH_3(aq)} + I_2(aq) \xrightarrow{H^+} \underset{\text{iodopropanone}}{CH_3COCH_2I(aq)} + H^+(aq) + I^-(aq)$$

The iodine is brown and all the other species are colourless. The reaction mixture fades from brown to colourless as iodine is used up in the reaction and the decrease in the absorption of light can be measured using the colorimeter (Figure **6.11**).

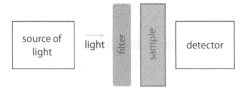

Figure 6.10 A colorimeter can be used to measure the amount of light absorbed at a particular wavelength – the darker the colour of the sample, the more light is absorbed.

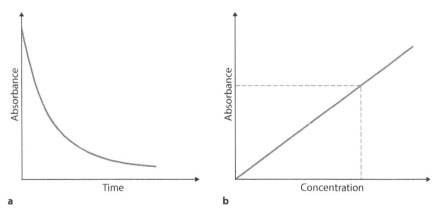

a   b

Figure 6.11 a The absorbance falls with time as iodine is used up. **b** A calibration curve is used to convert the absorbance of iodine to a concentration. It can be constructed by measuring the absorbance of several solutions of known iodine concentration.

Learning objectives

- Describe and explain the collision theory
- Define the activation energy of a reaction
- Understand the effects of surface area of solid reactants, temperature, catalysts, concentration and pressure on rate of reaction
- Understand that the average energy of particles in a gas is proportional to its temperature in kelvin
- Sketch the Maxwell–Boltzmann distribution and use it to explain the effect of a change in temperature on the rate of a reaction and how a catalyst speeds up a reaction

Particles must collide to react

Reaction rates are generally discussed in terms of **collision theory**. This states that for a reaction to occur particles must collide; however, for these collisions to result in a reaction, two conditions must be fulfilled:

- a collision must involve more than a certain **minimum amount of energy**
- molecules must collide with the **correct orientations**.

Note: not all collisions result in a reaction.

The collision must involve more than a certain minimum amount of energy

Particles must collide with sufficient energy in order to react. The minimum amount of energy that colliding particles must possess to result in a reaction is called the **activation energy (E_a)**. If two particles with less than the activation energy collide, they will just bounce off each other and no reaction will result; however, if the particles have energy greater than or equal to the activation energy then, assuming the orientation of the collision is also correct, the particles will react. A collision that results in a reaction is called a **successful** or **effective** collision.

Activation energy is the energy needed to overcome repulsions, to start breaking bonds, to deform molecules and to allow rearrangement of atoms, electrons etc.

The activation energy for an exothermic reaction is shown on the potential energy profile in Figure **6.12**.

Imagine the reaction between two particles, A and BC:

$$A + B-C \rightarrow A-B + C$$

As the two particles approach, repulsion between the atoms (internuclear and between electrons) causes an increase in the potential energy. The B–C bond begins to break and the A–B bond begins to form. The highest point along the curve is called the **transition state** (activated complex) and all three atoms are joined together by partial bonds (A⋯B⋯C). As the A–B bond continues to form, the potential energy falls (remember, bond-making releases energy).

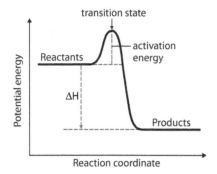

Figure 6.12 A potential energy profile, showing the activation energy for an exothermic reaction.

Molecules must collide with the correct orientation

If molecules do not collide with the correct orientation they will not react (Figure **6.13**).

> Not every collision with energy greater than the activation energy results in a reaction.

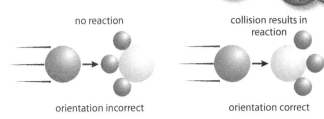

Figure 6.13 Orientation is important.

Factors affecting reaction rate

The main factors that affect the rate of a chemical reaction are:

- concentration of reactants
- pressure for (reactions involving gases)
- surface area of solid reactants
- temperature
- catalysis.

These will be considered in turn.

The effect of concentration on the reaction rate

With more particles in a certain volume, the particles collide more often (the collision frequency is higher) and therefore there is greater chance of a successful collision (i.e. one that results in a reaction) occurring in a certain time (Figure **6.14**).

The effect of pressure on the reaction rate

The effect of increasing the pressure is essentially the same as that of increasing the concentration of gaseous reactants. As the pressure is increased, the collision frequency increases (Figure **6.15**).

> Only reactions involving gases are significantly affected by changing the pressure.

Exam tip
We are dealing here with how quickly a reaction occurs and you must therefore have the idea of **time** in your explanation – it is not correct here to say that 'the particles collide more' you must write something like 'the particles collide more often/more frequently' or 'there are more collisions in a certain time'.

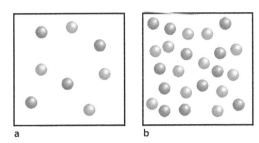

Figure 6.14 a Lower concentration – the particles are further apart and collide less frequently; **b** higher concentration – the particles are closer together and collide more frequently.

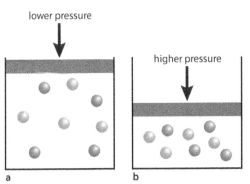

Figure 6.15 a Lower pressure – the particles are further apart and collide less frequently; **b** higher pressure – the particles are closer together and collide more frequently.

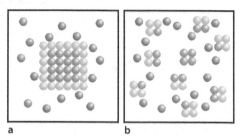

a   b

Figure 6.16 a Low surface area – only the particles coloured green are exposed on the surface and able to collide with the red particles; **b** high surface area – particles coloured both green and blue are exposed and are able to collide with the red particles.

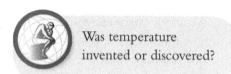

Was temperature invented or discovered?

At the same temperature lighter particles travel faster than heavier ones.

kinetic energy $= \frac{1}{2}mv^2$

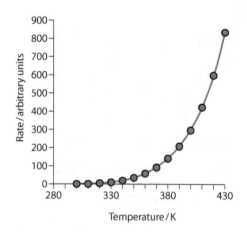

Figure 6.17 The rate of a reaction increases exponentially as the temperature rises.

The effect of surface area of solid reactants

Reactions generally only occur at the surface of a solid. Making a solid more finely divided increases the surface area and therefore the number of particles exposed at the surface. The effective concentration of the particles of the solid has thus been increased and there is a greater chance of a particle of the other reactant colliding with a particle on the surface and reaction occurring (Figure **6.16**).

The relationship between temperature and the energy of particles in a gas

Before we can understand the effect of temperature on the rate of a chemical reaction, we must look at how changing the temperature affects how the particles in a gas move.

For an ideal gas:

> The average kinetic energy of the particles in a gas is proportional to its temperature in kelvin.

Therefore if a sample of oxygen is heated from 300 K to 600 K, the **average** energy of the particles is doubled.

This relationship does not depend on the identity of the gas. So the average kinetic energy of the particles in a sample of helium at 300 K is the same as the average kinetic energy of the particles in oxygen at 300 K. However, because the mass of an O_2 molecule is eight times the mass of a helium atom, the helium atoms will be travelling substantially faster at the same temperature.

A consequence of this relationship between temperature and kinetic energy is that a large increase in temperature is required for a significant increase in the average speed of the particles and hence in the collision frequency – you will see the importance of this in the next section.

The effect of temperature on rate of reaction

Increasing the temperature has a major effect on the rate of the reaction. As the temperature increases, the rate of reaction increases exponentially (Figure **6.17**).

It is often said that, as a rough rule of thumb, a rise in temperature of 10 K causes the reaction rate to be approximately doubled.

As the temperature increases, the molecules have more energy and therefore move faster. This means that the collision frequency increases, i.e. the particles collide more often. This is, however, only a minor effect and can explain only a small increase in rate (approximately 2% for a 10 K rise in temperature) as the temperature increases. The major cause of the increase in rate as the temperature increases is that, not only do the particles collide more often, but they also collide harder, that is, with more energy, so that there is greater chance that a collision will result in reaction.

Let us consider a sample of gas – the molecules are constantly colliding with each other and, therefore, do not all have the same speed and hence energy. This is shown in Figure **6.18**, which represents the **Maxwell–Boltzmann distribution** of molecular kinetic energies at a particular temperature. It can be seen that there are only a few particles with high energy and only a few with very low energy. Most particles have energy around the average.

Features to note on Figure **6.18**:
- it is **not** symmetrical
- no molecules have zero kinetic energy
- at higher energy the line does not reach the energy axis
- the area under the curve represents the total number of particles and will not change as the temperature changes.

> The vertical axis could be labelled as 'number of particles with a certain amount of energy' or 'proportion of particles with a certain amount of energy'.

> The main reason that the rate of reaction increases with temperature is an increase in the number of particles with energy greater than or equal to the activation energy.

As the temperature is increased, this distribution of energies changes (Figure **6.19**). At higher temperatures the curve is flatter and the maximum has moved to the right. So there are fewer particles with lower energy and more particles with higher energy. With more particles having energy greater than or equal to the activation energy (E_a) at the higher temperature, a greater proportion of collisions will be successful, and therefore the rate of reaction will increase. The areas underneath the curves are the same because the number of particles (amount of substance) does not change if the temperature is increased.

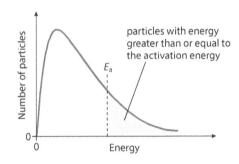

Figure 6.18 The Maxwell–Boltzmann distribution of the energy of the particles in a sample of gas. The shaded area represents the particles that have sufficient energy to react when they collide.

Extension

The graph in Figure **6.19** is actually a histogram, where each bar in the histogram represents the number of molecules in a certain narrow range of kinetic energies.

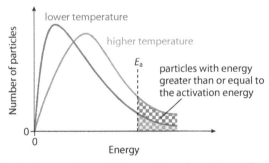

Figure 6.19 The shaded area represents the number of particles with energy greater than or equal to the activation energy at the lower temperature. The checked area represents the number of particles with energy greater than or equal to the activation energy (E_a) at the higher temperature.

How do we know when we encounter a new reaction that the rate will increase when we increase the temperature? This comes from the idea of inductive reasoning – inductive reasoning is a fundamental tool of scientists, and to a certain extent chemistry would not exist in the form it does without inductive reasoning. Inductive reasoning allows us to generalise from the specific – for instance, if we carry out a series of experiments and each one gives us the result that increasing the temperature increases the rate of reaction, then we postulate a general law that the rate of reaction increases with temperature. How is it possible to do this when we have not studied every single chemical reaction? Indeed, a philosopher would say that this is not rational and there is no logical basis for being able to do this – this is the problem with induction. Common sense, however, would tell us that the rate of reaction for every simple reaction should increase as temperature increases – we rely on the uniformity of nature and we cannot imagine a situation in which it would not be true. But can we know this, or only believe it to be true? Is there a difference between a scientist having faith in induction and religious faith?

Catalysis

A **catalyst** is a substance that increases the rate of a chemical reaction without itself being used up in the reaction.

An example of a catalyst is manganese(IV) oxide in the decomposition of hydrogen peroxide solution:

$$2H_2O_2(aq) \xrightarrow{\text{MnO}_2} 2H_2O(l) + O_2(g)$$

Without the catalyst the reaction occurs very slowly, but it is very rapid once the catalyst has been added.

A catalyst acts by allowing the reaction to proceed by an alternative pathway of lower activation energy.

This is shown on the potential energy profile in Figure **6.20**.

If we look at the Maxwell–Boltzmann distribution we can understand why a lower activation energy results in a faster reaction (Figure **6.21**).

The catalyst is often written above the reaction arrow and does not appear in the chemical equation because it does not change in the reaction.

Catalysts are important in many industrial processes.

Higher Level only: the mechanism is different for a catalysed reaction (see page **264**).

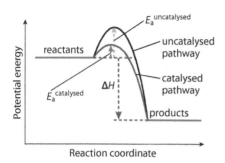

Figure 6.20 Effect of a catalyst on the activation energy of a reaction.

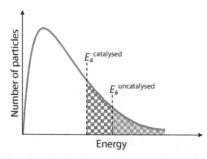

Figure 6.21 The shaded area represents the number of particles with energy greater than or equal to the activation energy for the uncatalysed reaction. The checked area represents the number of particles with energy greater than or equal to the activation energy for the catalysed reaction. A larger number of particles have energy greater than the activation energy; therefore a greater proportion of collisions results in reaction and the reaction rate increases.

Nature of science

Although collision theory is consistent with experimental results, this does not prove that the theory is correct. The mark of a scientific theory, and what distinguishes it from a non-scientific one, is the idea of falsifiability – so far, collision theory has been supported by experimental evidence, but if new experimental data are produced that cannot be explained using the current collision theory, then the theory will have to be modified or dismissed in favour of a new theory that does explain all the available experimental data. Collision theory is the best explanation (at this level) of the experimental data produced so far. Other explanations may be possible, but this interpretation of the results is widely accepted at the moment, and the theory is used to make predictions and explain phenomena.

Test yourself

1 A series of experiments was carried out to measure the volume of gas produced when magnesium reacts with dilute hydrochloric acid. The equation for the reaction is:

$$Mg + 2HCl \rightarrow MgCl_2 + H_2$$

In the first experiment, 0.10 g of Mg ribbon was reacted with 30 cm³ of 0.50 mol dm⁻³ HCl. The data for this experiment are recorded in the table. The reaction was carried out at 20 °C.

a Draw a graph of these data and state and explain, in terms of the collision theory, how the rate of reaction changes with time.

b Use your graph to calculate the initial rate of the reaction with units.

c Calculate the average rate for the first 120 s.

d The experiment was repeated under the same conditions, except that 0.10 g of powdered Mg were used. On the same set of axes you used in part **a**, sketch the graph that would be obtained. Label this graph **X**.

e The original experiment was repeated, except that 0.05 g of Mg ribbon was used. On the same axes sketch the graph that would be obtained. Label this graph **Y**.

f The original experiment was repeated at 10 °C. On the same axes sketch the graph that would be obtained. Label this graph **Z**.

g Sketch the Maxwell–Boltzmann distribution for the original experiment and the experiment at 10 °C and use this to explain the effect of a change in temperature on the rate of this reaction.

Time / s	Volume of gas / cm³
0	0.0
15	18.6
30	32.3
45	44.3
60	54.8
75	62.7
90	68.4
105	72.6
120	74.9
135	75.4
150	75.6
165	75.6
180	75.6

Learning objectives

- Explain the terms rate constant and order of reaction
- Work out the rate expression (rate equation) from numerical data and solve problems related to it
- Sketch and explain graphs of concentration against time and rate against time for zero-, first- and second-order reactions

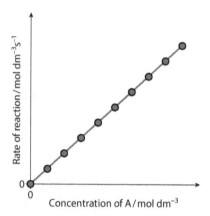

Figure 6.22 Possible rate-concentration graph for a reaction A → B.

A rate equation is also called a rate law.

Note: small k – not capital.

6.2 Rate expression and reaction mechanism (HL)

6.2.1 The rate equation/rate expression

In this section, we will consider the quantitative effect of changes in conditions on the rate of reaction.

A rate of reaction is usually affected by a change in concentration of the reactants. Consider the reaction A → B. If a series of experiments is carried out in which the concentration of A is varied and the rate of reaction measured, a graph like the one in Figure **6.22** might be plotted.

From this graph it can be seen that the rate of reaction is directly proportional to the concentration of A. For this reaction we can write:

$$\text{rate} \propto [A]$$

where [A] is the concentration of A. By adding a constant of proportionality, we can write:

$$\text{rate} = k[A]$$

This equation is called a **rate equation** or **rate expression** – k is called the **rate constant**.

> The rate equation is an **experimentally determined** equation that relates rate of reaction to the concentrations of substances in the reaction mixture.

We can write a rate equation for any reaction. For example, for the reaction:

$$W + X \rightarrow Y + Z$$

we can write:

> $$\text{rate} = k[W]^m[X]^n$$

> The **rate constant** is a constant of proportionality relating the concentrations in the experimentally determined rate equation to the rate of a chemical reaction.

The rate constant is only a constant for a particular reaction at a particular temperature.

> The **order of a reaction** with respect to a particular reactant is the power of the reactant's concentration in the experimentally determined rate equation.

In the rate equation above, the order with respect to W is m and the order with respect to X is n. The overall order is $m + n$.

If we consider our original rate equation, rate $= k[A]$, again: this reaction is **first order with respect to A** (the power of [A] is 1) and first order overall.

A rate equation can be determined only from experimental data – that is, from a series of experiments in which the effect of changing the concentration of the reactants on the rate of reaction is investigated. The important thing to realise is that there is no connection between the chemical (stoichiometric) equation for a reaction and the rate equation – i.e. we can not simply look at an equation such as $2A + B \rightarrow C + D$ and deduce that the order with respect to A is two and that with respect to B is one. The reason for this is that the reaction may not occur in one single step – this will be considered in more detail below.

The order of a reaction with respect to a particular reactant or overall order may have an integral or fractional value. Fractional orders are only found in complex reactions, for instance the pyrolysis of ethane to ethene and hydrogen:

$$C_2H_6(g) \rightarrow C_2H_4(g) + H_2(g)$$

involves a complex chain reaction and has a rate equation of the form:
rate $= k[C_2H_6]^{\frac{1}{2}}$

Experimental determination of a rate equation

Consider the reaction $A + B \rightarrow C$. The dependence of the reaction rate on the concentrations of A and B can be determined by conducting the following set of experiments.

First, a series of experiments is carried out using a fixed amount of B and changing the concentration of A each time. Each experiment should yield a graph of concentration of A against time. The initial rates can be worked out from these graphs by drawing tangents at the initial points. From these data the order of the reaction with respect to A can be determined by plotting a graph of initial rate against concentration of A.

Next, the experiments are repeated using a fixed amount of A and varying the concentration of B. This allows the order of reaction with respect to B to be calculated.

When the orders with respect to A and B are known, the rate equation, and hence a value for the rate constant, can be worked out.

6.2.2 Determining the order of reaction and the rate expression from experimental data

Example

Consider the data for the reaction $2A \rightarrow B$, given in Table **6.4**.
We want to determine:

An **overall order of reaction** is the sum of the powers of the concentration terms in the experimentally determined rate equation.

> **Exam tip**
> Remember that the rate equation can be determined only from experimental data and not from the chemical (stoichiometric) equation.

The initial rate is taken because this is the only point at which we know the concentration of A and concentration of B – none has been used up.

Experiment	[A]/ mol dm^{-3}	Rate/ mol dm^{-3} s^{-1}
1	1.0	0.60
2	2.0	1.2
3	5.0	3.0

Table 6.4 Experimental data for the reaction 2A → B.

1 the order with respect to A
2 the rate equation
3 the value of the rate constant (with units)
4 the rate of reaction when [A] = 1.3 mol dm^{-3}.

1 If we consider experiments 1 and 2, we can see that as the concentration of A is doubled from 1.0 mol dm^{-3} to 2.0 mol dm^{-3}, the rate of reaction also doubles from 0.60 mol dm^{-3} s^{-1} to 1.2 mol dm^{-3} s^{-1}. So concentration is multiplied by a factor of two and the rate goes up by a factor of 2^1. This means that the order with respect to A is one – in other words, the reaction is first order with respect to A. This can be summarised:

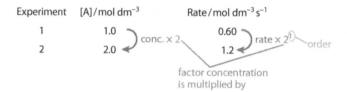

The fact that the order of reaction with respect to A is 1 can be further confirmed by looking at experiments 1 and 3. As the concentration of A is multiplied by a factor of five, the rate of reaction is multiplied by a factor of 5^1.

2 The reaction is first order with respect to A, so the rate equation is:
rate = k[A]

3 To find the value of k, we substitute the values from any of the experiments into the rate equation. If this is done using the values from experiment 1, we get:

$$0.60 = k \times 1.0$$

This can be rearranged to give $k = 0.60$.
k has units, and these can be worked out by substituting units into the rate equation:

$$\text{rate} = k[A] \quad \rightarrow \quad \text{mol dm}^{-3}\text{s}^{-1} = k \times \text{mol dm}^{-3}$$

'mol dm^{-3}' can be cancelled from each side:

$$\cancel{\text{mol dm}^{-3}}\text{s}^{-1} = k \times \cancel{\text{mol dm}^{-3}}, \text{ so } \text{s}^{-1} = k$$

Therefore the units of k are s^{-1} in this case, and the rate constant is 0.60 s^{-1}.

4 The rate of reaction when [A] = 1.3 mol dm^{-3} can be worked out by substituting this value into the rate equation along with the value of k:

$$\text{rate} = k[A]$$

$$\text{rate} = 0.60 \times 1.3$$

The rate of reaction is 0.78 mol dm^{-3} s^{-1}.

Because the order with respect to A is 1, the order is omitted from the rate equation.

This could also have been worked out by realising that the reaction is first order with respect to A, and that 1.3 mol dm^{-3} is 1.3 times the concentration of A in experiment 1, and therefore the rate of reaction is 1.3^1 times the rate of reaction in experiment 1.

Worked examples

Experiment	[A]/mol dm^{-3}	[B]/mol dm^{-3}	Rate/mol dm^{-3} h^{-1}
1	0.10	0.10	0.50
2	0.30	0.10	4.50
3	0.30	0.20	4.50

6.1 Given these data for the reaction $3A + B \rightarrow C + D$ determine:

   a the order with respect to A

   b the order with respect to B

   c the overall order of the reaction

   d the rate equation

   e the value of the rate constant (with units)

   f the rate of reaction when $[A] = 1.60\,\text{mol dm}^{-3}$ and $[B] = 0.30\,\text{mol dm}^{-3}$

a To find the order with respect to A, we must consider experiments 1 and 2, because the only thing that changes in these two experiments is the concentration of A (the concentration of B remains constant). From experiment 1 to experiment 2, the concentration of A is multiplied by a factor of three and the rate goes up by a factor of nine, which is 3^2. This means that the order with respect to A is two – the reaction is second order with respect to A.

Experiment	[A]/mol dm^{-3}	[B]/mol dm^{-3}	Rate/mol dm^{-3} h^{-1}
1	0.10	0.10	0.50
2	0.30	0.10	4.50

conc. × 3   rate × 3^2 order

factor concentration is multiplied by

b To find the order with respect to B, we must consider experiments 2 and 3, because the only thing that changes in these two experiments is the concentration of B (the concentration of A remains constant). From experiment 2 to experiment 3, the concentration of B is multiplied by a factor of two and the rate does not change, i.e. it is multiplied by a factor of 2^0. This means that the order with respect to B is zero – the reaction is zero order with respect to B.

> Any number to the power zero is one.

c The overall order of reaction is the sum of the orders with respect to A and B – in this case $2 + 0$. Therefore the overall order is 2.

d The rate equation is: $\text{rate} = k[A]^2[B]^0$, which is usually just written as: $\text{rate} = k[A]^2$.

e The value of the rate constant can be calculated by substituting values from any experiment into the rate equation. It doesn't matter which experiment is taken, but values must not be taken from different experiments. If we use the data from experiment 2:

$$\text{rate} = k[A]^2$$

$$4.50 = k \times 0.30^2$$

$$k = 50$$

To work out the units for k, the units of concentration and rate are substituted into the rate equation:

$$\mathrm{mol\,dm^{-3}\,h^{-1}} = k(\mathrm{mol\,dm^{-3}})^2$$

$\mathrm{mol\,dm^{-3}}$ can be cancelled from each side:

$$\cancel{\mathrm{mol\,dm^{-3}}}\,\mathrm{h^{-1}} = k(\mathrm{mol\,dm^{-3}})^{\cancel{2}}$$

$$\mathrm{h^{-1}} = k\,\mathrm{mol\,dm^{-3}}$$

This can be rearranged to give:

$$\frac{\mathrm{h^{-1}}}{\mathrm{mol\,dm^{-3}}} = k$$

i.e. $k = \mathrm{mol^{-1}\,dm^3\,h^{-1}}$

Alternatively, at this stage it can be seen that the units of k must include $\mathrm{mol^{-1}\,dm^3}$ for the powers of mol and dm to be zero on both sides.

When a quantity with a power is brought from the bottom to the top of an expression, the sign of the power changes, i.e. $\dfrac{1}{x^2}$ is equivalent to x^{-2}.

Therefore the value of the rate constant, k, is $50\,\mathrm{mol^{-1}\,dm^3\,h^{-1}}$.

It is good practice to write any positive powers first, so this is better written as $50\,\mathrm{dm^3\,mol^{-1}\,h^{-1}}$.

f   The rate of reaction when $[A] = 1.60\,\mathrm{mol\,dm^{-3}}$ and $[B] = 0.30\,\mathrm{mol\,dm^{-3}}$ can be worked out by substituting these values together with the value of k into the rate equation:

$$\text{rate} = k[A]^2 = 50 \times 1.60^2 = 128\,\mathrm{mol\,dm^{-3}\,h^{-1}}.$$

6.2   Given these data for the reaction $2P + Q \rightarrow R + S$ determine:

a   the order with respect to P
b   the order with respect to Q
c   the overall order of the reaction
d   the rate equation
e   the value of the rate constant (with units)

Experiment	[P]/mol dm^{-3}	[Q]/mol dm^{-3}	Rate/mol dm^{-3} s^{-1}
1	1.20	2.00	5.00×10^{-3}
2	2.40	2.00	1.00×10^{-2}
3	6.00	8.00	0.100

a   To find the order with respect to P, we must consider experiments 1 and 2, because the only thing that changes in these two experiments is the concentration of P (the concentration of Q remains constant). From experiment 1 to experiment 2, the concentration of P is multiplied by a factor of two and the rate goes up by a factor of two, i.e. 2^1. This means that the order with respect to P is one – the reaction is first order with respect to P.

b   It is a more difficult problem to find the order with respect to Q because there are no two experiments in which the concentration of P remains constant, and so we cannot easily see how just changing [Q] affects the rate. One way of getting around this is to add another row to the table:

Experiment	[P]/mol dm^{-3}	[Q]/mol dm^{-3}	Rate/mol dm^{-3} s^{-1}
1	1.20	2.00	5.00×10^{-3}
2	2.40	2.00	1.00×10^{-2}
2A			
3	6.00	8.00	0.100

We can fill in the values in this new row by realising that the order with respect to P is one. If the concentration of P in experiment 2A is five times that in experiment 1, and because [Q] is the same in both experiments, the rate in experiment 2A will be $5 \times 5.00 \times 10^{-3}$ i.e. 2.50×10^{-2} mol dm^{-3} s^{-1}.

Experiment	[P]/mol dm^{-3}	[Q]/mol dm^{-3}	Rate/mol dm^{-3} s^{-1}
1	1.20	2.00	5.00×10^{-3}
2	2.40	2.00	1.00×10^{-2}
2A	**6.00**	**2.00**	**2.50×10^{-2}**
3	6.00	8.00	0.100

The concentration of P has been chosen to be the same as that in experiment 3.

We can now consider experiments 2A and 3 and see the effect of just changing the concentration of Q on the rate of reaction. From experiment 2A to experiment 3, the concentration of Q is multiplied by a factor of four and the rate changes by a factor of 4^1. This means that the order with respect to Q is one.

Another way to approach this, without adding another row to the table, is to just consider experiments 1 and 3.

Experiment	[P]/mol dm^{-3}	[Q]/mol dm^{-3}	Rate/mol dm^{-3} s^{-1}
1	1.20	2.00	5.00×10^{-3}
2	2.40	2.00	1.00×10^{-2}
3	6.00	8.00	0.100

We know that going from experiment 1 to experiment 3 the concentration of P has increased by a factor of five. Because the reaction is first order with respect to P, the result of this will be to multiply the rate of reaction by a factor of 5^1. If this were done without any change in the concentration of Q, the rate of reaction would be $5 \times 5.00 \times 10^{-3}$, i.e. 2.50×10^{-2} mol dm^{-3} s^{-1}. However, the rate of reaction in experiment 3 is 0.100, which is four times 2.50×10^{-2}. Thus the effect of multiplying the concentration of Q by four is that the rate of reaction is multiplied by 4^1; therefore the order with respect to Q is one. This approach is, of course, equivalent to adding an extra row to the table.

c   The order with respect to P is 1 and the order with respect to Q is 1, so the overall order is $1 + 1$, i.e. 2.

d   The rate equation is: rate $= k[P]^1[Q]^1$, which is usually just written: rate $= k[P][Q]$.

e   The value of the rate constant can be calculated by substituting values from any one experiment into the rate equation. If we use the data from experiment 3:

$0.100 = k \times 6.00 \times 8.00$

$0.100 = k \times 48.0$

$k = 2.08 \times 10^{-3}$

To work out the units for k, the units of concentration and rate must be substituted into the rate equation:

mol dm^{-3} s^{-1} $= k \times$ mol dm^{-3} $\times$ mol dm^{-3}

mol dm^{-3} can be cancelled from each side:

$\cancel{\text{mol dm}^{-3}}$ s^{-1} $= k \times$ mol dm^{-3} $\times$ $\cancel{\text{mol dm}^{-3}}$

So s^{-1} $= k \times$ mol dm^{-3}.

This can be rearranged to give:

$$\frac{s^{-1}}{mol\,dm^{-3}} = k$$

$$k = mol^{-1}\,dm^3\,s^{-1}$$

Therefore the rate constant, k, is $2.08 \times 10^{-3}\,mol^{-1}\,dm^3\,s^{-1}$.

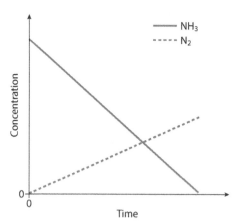

Figure 6.23 Concentration of reactant and product against time for a zero-order reaction.

> The rate is independent of the concentration.

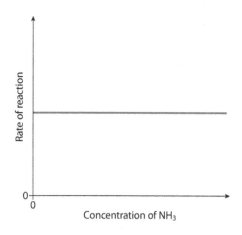

Figure 6.24 Rate against concentration for a zero-order reaction.

Zero-order reactions

Consider a zero-order reaction – the decomposition of ammonia on a tungsten surface:

$$2NH_3(g) \rightarrow N_2(g) + 3H_2(g)$$

The concentration of ammonia decreases at a constant rate in this reaction. The constant rate is indicated by the graph of concentration against time being a straight line (Figure **6.23**). A straight line has a constant gradient (slope) and indicates that the rate does not change as the concentration of NH_3 changes.

The concentration of the nitrogen increases at a constant rate, indicated by the straight dashed line in Figure **6.23**.

The two graphs of concentration against time were drawn to the same scale, and it can be seen that the magnitude of the gradient of the line for N_2 is half that for NH_3. This is because in the chemical equation for this reaction the coefficient of NH_3 is 2 but that of N_2 is 1 – so N_2 is produced at half the rate at which NH_3 is used up.

Changing the concentration of ammonia has no effect on the rate of the reaction (Figure **6.24**).

The rate equation is rate $= k$, which shows that the rate is constant. The units of the rate constant are the same as the rate – i.e. concentration $\times$ time^{-1}. A set of units for the rate constant could therefore be $mol\,dm^{-3}\,s^{-1}$.

> The rate equation for this reaction is rate $= k$.
> The units of k are concentration $\times$ time^{-1} (i.e. units could be $mol\,dm^{-3}\,s^{-1}$ or $mol\,dm^{-3}\,h^{-1}$ etc.)

First-order reactions

Let us consider a first–order reaction – the decomposition of hydrogen iodide on a platinum surface:

$$2HI(g) \rightarrow H_2(g) + I_2(g)$$

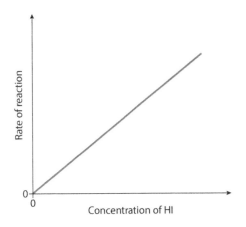

Figure 6.25 Rate against concentration for a first-order reaction.

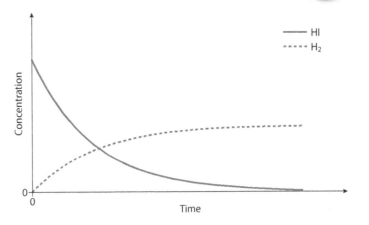

Figure 6.26 Concentration of reactant and product against time for a first-order reaction.

The rate of this reaction is directly proportional to the concentration of HI, as shown by the straight line through the origin in Figure **6.25**.

The graph in Figure **6.26** shows how the concentration of HI decreases with time. This shows an exponential decay and a constant **half-life**. The time taken for the concentration to drop by half does not depend on concentration in a first-order reaction – the time taken for the concentration to fall from $0.1\,mol\,dm^{-3}$ to $0.05\,mol\,dm^{-3}$ is the same as the time taken for the concentration to fall from $0.08\,mol\,dm^{-3}$ to $0.04\,mol\,dm^{-3}$.

The dashed line in Figure **6.26** shows the increase in concentration of one of the products (H_2) with time. The rate of production of H_2 is half the rate at which HI is used up, which can be seen from the coefficients in the chemical equation:

$$2HI \rightarrow H_2 + I_2$$

Second-order reactions

Consider a second-order reaction – the decomposition of hydrogen iodide without a catalyst:

$$2HI(g) \rightarrow H_2(g) + I_2(g)$$

Figure **6.27** shows how the rate of the reaction varies with the concentration of hydrogen iodide.

It can be proved that a reaction is second order (rather than third order, etc.) by plotting a graph of rate against concentration of HI squared (Figure **6.28**). As the rate is proportional to $[HI]^2$, this graph is a straight line through the origin.

> The rate of reaction is proportional to concentration squared.

> The rate is directly proportional to the concentration.

> The rate equation for this reaction is: rate = $k[HI]$. The units of k are $time^{-1}$.

> The half-life is related to the rate constant by the equation
> $$rate\ constant = \frac{\ln 2}{half\text{-}life}$$

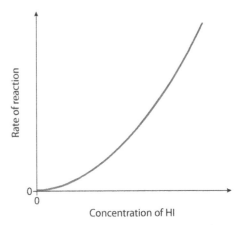

Figure 6.27 Rate against concentration for a second-order reaction.

The rate equation for this reaction is rate = $k[\text{HI}]^2$.
The units of the rate constant are concentration^{-1}time^{-1} (i.e. units could be $mol^{-1}dm^3s^{-1}$).

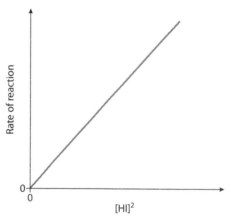

Figure 6.28 Rate against concentration^2 for a second-order reaction.

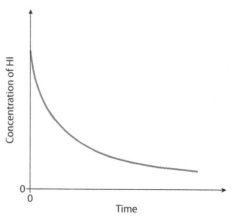

Figure 6.29 Concentration of reactant against time for a second-order reaction.

Figure **6.29** shows how the concentration of HI changes with time. This is not an exponential relationship and does not have a constant half-life.

The units of the rate constant

It can be seen from the treatment above that the units of the rate constant are related to the overall order of the reaction. This is summarised in Table **6.5**.

Overall order	Units of k	Example of units
0	concentration time^{-1}	$mol\,dm^{-3}s^{-1}$
1	time^{-1}	s^{-1}
2	concentration^{-1} time^{-1}	$mol^{-1}dm^3s^{-1}$
3	concentration^{-2} time^{-1}	$mol^{-2}dm^6s^{-1}$

Table 6.5 The relationship between overall order of the reaction and the units of the rate constant.

For a zero-order reaction, the units of k are the same as that of the rate. Each time the overall order increases by 1 the units of k are divided by concentration.

Generally the units of k are concentration$^{(1 - \text{overall order})}$ time^{-1}.

? Test yourself

2 Consider the following data for the reaction:

$A + 2B \rightarrow C + D$

a Deduce the order of reaction with respect to A and with respect to B.
b What is the overall order of reaction?
c Work out a value for the rate constant of this reaction with units.
d What will be the rate of reaction when the concentration of A is $0.100\,mol\,dm^{-3}$ and that of B is $0.0500\,mol\,dm^{-3}$?

Experiment	[A] / $mol\,dm^{-3}$	[B] / $mol\,dm^{-3}$	Rate / $mol\,dm^{-3}s^{-1}$
1	0.200	0.100	0.0200
2	0.400	0.100	0.0400
3	0.400	0.200	0.160

3 Consider the following data for the reaction:

$$2X + 4Y \rightarrow Q + 2R$$

a Write the rate equation for this reaction.
b Calculate a value for the rate constant from these data.
c What are the units of the rate constant?

Experiment	[X]/mol dm^{-3}	[Y]/mol dm^{-3}	Rate/mol dm^{-3} s^{-1}
1	1.50×10^{-2}	3.00×10^{-2}	1.78×10^{-3}
2	4.50×10^{-2}	3.00×10^{-2}	5.34×10^{-3}
3	4.50×10^{-2}	1.20×10^{-1}	2.14×10^{-2}

4 A reaction is zero order with respect to P and second order with respect to Q. What would be the effect of doubling the concentration of P and the concentration of Q on the overall rate of reaction?

5 Consider the following data for the reaction:

$$2D + 3E \rightarrow 2C + B$$

a Work out the order of reaction with respect to D.
b Work out the order of reaction with respect to E.

Experiment	[D]/mol dm^{-3}	[E]/mol dm^{-3}	Rate/mol dm^{-3} s^{-1}
1	2.50×10^{-3}	5.00×10^{-3}	4.28×10^{-4}
2	1.00×10^{-2}	5.00×10^{-3}	1.71×10^{-3}
3	4.00×10^{-2}	2.00×10^{-2}	6.84×10^{-3}

c What is the rate equation for the reaction?
d Work out a value, with units, for k, the rate constant for this reaction.
e What is the rate of reaction when the concentration of D is 0.0600 mol dm^{-3} and that of E is 0.0300 mol dm^{-3}?

6.2.3 Mechanisms of reactions

In this section we consider why a rate equation cannot be derived directly from the chemical equation for a reaction. Consider the reaction:

$$2NO_2(g) + F_2(g) \rightarrow 2NO_2F(g)$$

If this reaction were to occur in one single step, all three molecules must collide together at exactly the same time, and we would expect that doubling the concentration of any one of the three reactant molecules would double the chance of a collision and therefore the rate of the reaction. The rate of reaction in this case would therefore depend on $[NO_2]^2$ and $[F_2]$, and the rate equation would be:

$$\text{rate} = k[NO_2]^2[F_2]$$

The actual rate equation obtained from experiment is, however,

$$\text{rate} = k[NO_2][F_2]$$

The fact that these two equations are different suggests that the reaction does not occur in just one step in which all three molecules collide and break apart to form the products. This was always going to be unlikely because the chance of three gas molecules all colliding at exactly the same time is extremely small. This reaction must occur in a series of steps, and it is most likely that each step involves just two molecules colliding. A mechanism that has been proposed for this reaction is:

Learning objectives

- Understand what is meant by the mechanism of a reaction and the rate-determining step
- Work out reaction mechanisms from experimental data and relate a given mechanism to the experimental data

$[NO_2]^2$ because there are two NO_2 molecules and doubling the concentration of each will cause the rate to double.

A **reaction mechanism** consists of a series of steps that make up a more complex reaction. Each simple step involves a maximum of two molecules colliding.

$$NO_2 + F_2 \rightarrow NO_2F + F \qquad \textbf{step 1}$$
$$NO_2 + F \rightarrow NO_2F \qquad \textbf{step 2}$$

The first thing that must be checked with the mechanism is that it agrees with the overall chemical equation. In order to do this, species that are the same on both sides of the equations are cancelled and then the two equations are added together:

$$NO_2 + F_2 \rightarrow NO_2F + \cancel{F} \qquad \textbf{step 1}$$
$$\underline{NO_2 + \cancel{F} \rightarrow NO_2F} \qquad \textbf{step 2}$$
$$2NO_2 + F_2 \rightarrow 2NO_2F \qquad \textbf{overall equation}$$

The mechanism must be consistent with the overall chemical equation.

F is produced in step 1 and used up again in step 2. F is an intermediate.

Now we need to see whether or not this mechanism agrees with the experimental rate equation. Each step involves just two species colliding, and therefore we can derive the rate equation for each step directly from its chemical equation:

Step 1: $\quad$ rate $= k_1[NO_2][F_2]$

Step 2: $\quad$ rate $= k_2[NO_2][F]$

k_1 is the rate constant for step 1.

It can be seen that the rate equation for step 1 is the same as the experimental rate equation, and so it would seem that this step governs the overall rate of reaction and that the second step has no apparent effect on the rate. Step 1 is called the **rate-determining step** of the mechanism and occurs significantly more slowly than step 2.

The slowest step in a reaction mechanism is called the rate-determining step.

$$NO_2 + F_2 \xrightarrow{\text{slow}} NO_2F + F \qquad \textbf{step 1} \qquad \textbf{rate-determining step}$$

$$NO_2 + F \xrightarrow{\text{fast}} NO_2F \qquad \textbf{step 2}$$

Step 2 is fast compared with the rate-determining step and has, effectively, no influence on the overall rate of reaction. This means that changing the concentrations of the species present in this step does not affect the rate of the reaction to any great extent, so the concentrations of these species do not occur in the rate equation.

The potential energy profile for this reaction is shown in Figure **6.30** and it can be seen that the rate-determining step (step 1) has a much higher activation energy than the other step.

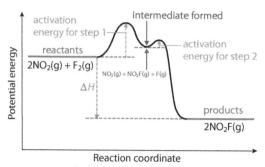

Figure 6.30 Potential energy profile for a two-step reaction. The formation of an intermediate is shown by the potential energy well in the profile.

The idea of a rate-determining step can be seen by analogy with a football stadium. The owners of the football club Rapid Kinetics have considered various options for speeding up the process of getting the spectators to their seats. They have broken down the process of getting to the seats into three separate steps:

1 getting to the stadium by road
2 getting into the stadium
3 getting to your seat once you are in the stadium.

It was suggested that the owners of the club could apply to the local council to improve the roads and the traffic flow leading to the stadium, and someone else suggested that they could install lifts and escalators to improve the process of getting around the stadium, but then some bright spark noticed the main problem – there was only one gate to get into the stadium! Improving the roads around the stadium and installing lots of lifts and escalators would have very little effect on the rate of the overall process of people getting to their seats because the rate-determining step is getting people through the one turnstile into the stadium. They need to work on improving the rate of the rate-determining step, and it could be expected that doubling the number of gates to get into the stadium would double the speed of the overall process.

Let us consider another mechanism, this time for the reaction $A + 2B \rightarrow C$:

$B + B \rightarrow Q$	**step 1**	**rate-determining step**
$Q + A \rightarrow C$	**step 2**	**fast**

Q is an intermediate.

Let us first check that the mechanism agrees with the chemical equation:

$B + B \rightarrow \cancel{Q}$	**step 1**
$\cancel{Q} + A \rightarrow C$	**step 2**
$\overline{2B + A \rightarrow C}$	**overall equation**

Step 1 is the rate-determining step, so the concentrations of the reactants involved in this affect the rate of the overall reaction and occur in the rate equation. Step 2 is a fast step that occurs after the rate-determining step, and therefore the species involved do not affect the rate of reaction or occur in the rate equation. This means that B occurs twice in the rate equation and A not at all. The rate equation consistent with this mechanism is:

$$\text{rate} = k[B]^2$$

In both examples we have considered, the rate-determining step is the first step. Now consider a mechanism in which the rate-determining step is the second step.

For the same overall equation, $A + 2B \rightarrow C$, another possible mechanism could be:

$B + B \rightleftharpoons Q$	**step 1**	**fast**
$Q + A \rightarrow C$	**step 2**	**rate-determining step**

Step 1 is an equilibrium reaction. At equilibrium, the rate of the forward reaction is the same as the rate of the reverse reaction, i.e.

rate of forward reaction $= k_f[B]^2$

rate of reverse reaction $= k_r[Q]$

$k_f[B]^2 = k_r[Q]$

which can be rearranged to give:

$$[Q] = \frac{k_f[B]^2}{k_r}$$

So concentration of Q is proportional to the concentration of B squared.

S is an intermediate.

Intermediates do not appear in a rate equation.

This is basically the same as the previous mechanism, except that the second step is the rate-determining step. The species in step 2 influence the rate of the reaction and we can write the rate equation as:

rate $= k[Q][A]$

However, Q is produced by the reaction between two molecules of B, and we can replace [Q] with $[B]^2$ in the rate equation. Therefore the rate equation consistent with this mechanism would be:

rate $= k[B]^2[A]$

Which of the two above mechanisms is more likely to be the actual mechanism can be worked out by experimentally determining the rate equation for the reaction.

From the treatment above we can see that:

> The rate equation contains concentrations of reactants involved up to and including the rate-determining step.

This can be further seen with another possible mechanism for this reaction:

$A + B \rightleftharpoons S$ **step 1**   **fast**
$S + B \rightarrow C$ **step 2**   **rate-determining step**

The reactants involved up to and including the rate-determining step are A once and B twice, so the rate equation would also be:

rate $= k[B]^2[A]$

There is no simple way of distinguishing between the two above mechanisms experimentally.

Reactions involving a catalyst

$$CH_3COCH_3(aq) + I_2(aq) \rightarrow CH_3COCH_2I(aq) + HI(aq)$$

This reaction is acid (H^+) catalysed.

The experimental rate equation is:

rate $= k[CH_3COCH_3][H^+]$

The rate equation does not include I_2, so this must be involved only after the rate-determining step.

At a simple level, the mechanism could be proposed as:

X is an intermediate.

$CH_3COCH_3 + H^+ \rightarrow X$ **rate-determining step**
$X + I_2 \rightarrow CH_3COCH_2I + HI + H^+$ **fast**

H^+ will cancel out when the equations are added together.

The catalyst is involved in the rate-determining step but is regenerated in the second step and therefore does not appear in the overall chemical equation.

Catalysts change a reaction mechanism, allowing the reaction to occur via an alternative pathway that has a lower activation energy. A homogeneous catalyst (in the same phase as the reactants) usually works by forming an intermediate with one or other of the reactant molecules – this requires a lower activation energy than the original reaction.

Consider the reaction $A \rightarrow X + Y$, which occurs in a single step. If a catalyst (C) is introduced, the reaction will happen by a different mechanism involving two steps:

$A + C \rightarrow A–C$ **rate-determining step** (activation energy E_2)
$A–C \rightarrow X + Y + C$ **fast** (activation energy E_3)

The rate equation for the original reaction would be rate $= k[A]$ and that for the catalysed reaction would be rate $= k[A][C]$. It can be seen that the catalyst appears in the rate equation. The potential energy profile for this reaction is shown in Figure **6.31**.

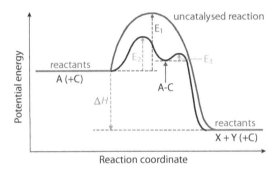

Figure 6.31 The effect of a catalyst on activation energy. E_1 is the activation energy for the uncatalysed reaction, E_2 is the activation energy for the first step of the catalysed reaction (the rate-determining step) and E_3 is the activation energy for the second step (fast) of the catalysed reaction.

S_N1 versus S_N2 mechanisms

In this section, we will consider the kinetics of the reactions that will be studied further in Subtopic **10.3.1**.

Consider the reaction:

$$(CH_3)_3CBr + OH^- \rightarrow (CH_3)_3COH + Br^-$$

This is a nucleophilic substitution reaction of 2-bromo-2-methylpropane. The experimentally determined rate equation for this reaction is:

$$\text{rate} = k[(CH_3)_3CBr]$$

OH^- does not occur in the rate equation and therefore can be involved only in a fast step after the rate-determining step. The mechanism for this reaction has been suggested as:

$(CH_3)_3CBr \rightarrow (CH_3)_3C^+ + Br^-$ **rate-determining step**
$(CH_3)_3C^+ + OH^- \rightarrow (CH_3)_3COH$ **fast**

> **Molecularity** is the number of 'molecules' that react in a particular step (usually the rate-determining step).

The reaction is described as an S_N1 mechanism, where S stands for substitution, N for nucleophilic (the attacking species is a nucleophile in this case) and 1 is the molecularity of the rate-determining step – the molecularity is 1 because **one** molecule reacts in the rate-determining step.

$CH_3CH_2CH_2Br$ (1-bromopropane) also undergoes a nucleophilic substitution reaction:

$$CH_3CH_2CH_2Br + OH^- \rightarrow CH_3CH_2CH_2OH + Br^-$$

The rate equation is different:

$$rate = k[CH_3CH_2CH_2Br][OH^-]$$

With both reactants from the original equation appearing once only in the rate equation, this suggests that this reaction occurs in just one step and the mechanism is simply:

$$CH_3CH_2CH_2Br + OH^- \rightarrow CH_3CH_2CH_2OH + Br^-$$

Because there is only one step it is, of course, also the rate-determining step. This reaction mechanism is described as S_N2, where the 2 refers to the molecularity of the single step.

Summary of 'rules' for writing mechanisms

1 The mechanism must agree with the overall stoichiometric equation.
2 A maximum of two particles can react in any one step.
3 All species in the rate equation must appear in the mechanism in or before the rate-determining step.
4 The power of a particular reactant's concentration in the rate equation indicates the number of times it appears in the mechanism up to and including the rate-determining step.

> A mechanism being consistent with the experimental rate equation does not mean that the mechanism is correct. It can never be proved that a mechanism is correct, only that it is incorrect. A mechanism is accepted so long as it agrees with the experimental data, but if new experimental data are produced that are not consistent with the mechanism, the mechanism is disproved and a new mechanism must be developed that agrees with these and other experimental data.

Nature of science

The principle of Occam's (Ockham's) razor is often used in science to decide between different theories. It involves the idea that if there are two theories that give equally acceptable explanations of experimental data, the simpler one should be adopted. This is often used to decide between different reaction mechanisms. Consider the reaction $A + 2B \rightarrow X + Y$, which has the rate equation: $rate = k[A][B]$.

A mechanism that is consistent with the rate equation and the stoichiometric equation is:

A+B → Q   rate-determining step
B+Q → X+Y   fast

However, it is possible to come up with lots of other possible mechanisms (an infinite number if we are just dealing with random letters and not actual chemicals!), such as:

A+B → Q   rate-determining step
Q → R+S   fast
S+B → Z   fast
Z+R → X+Y   fast

This mechanism also fits the experimental data, but unless there is experimental evidence for the transient existence of, for instance S, in the reaction mixture, this mechanism will be rejected in favour of the first one – the extra steps are not needed to fit the experimental data. It is important when using Occam's razor to realise that it works only when both theories provide a full explanation of the data – if there were extra data that suggested the formation of S in the reaction mixture then the first mechanism must be rejected because it does not account for this. In that case, we should seek the simplest possible mechanism that includes the formation of S as an intermediate.

? Test yourself

6 Consider the reaction:
   $$2A + 3B \rightarrow 4C + D$$
   The rate equation is: rate $= k[B]^2$.
   A proposed mechanism for this reaction is:

   A+B → 2C+D   **step 1**   slow
   A+2B → C+D   **step 2**   fast

   Suggest **three** reasons why this is not a suitable mechanism for this reaction.

7 Consider the reaction:
   $$P + 2Q \rightarrow R + S$$
   A student has suggested some possible two-step mechanisms for this reaction:

   Mechanism 1

   $Q+Q \xrightarrow{\text{slow}} X$

   $P+X \xrightarrow{\text{fast}} R+S$

   Mechanism 3

   $Q+P \xrightarrow{\text{slow}} Y+S$

   $Q+Y \xrightarrow{\text{fast}} R$

   Mechanism 2

   $Q+P \underset{}{\overset{\text{fast}}{\rightleftharpoons}} Z+R$

   $Q+Z \xrightarrow{\text{slow}} S$

   Mechanism 4

   $Q+P \xrightarrow{\text{slow}} Y+S$

   $Q+Z \xrightarrow{\text{fast}} R$

a Write the rate equation that would be consistent with Mechanism 1.
b Explain why Mechanism 4 cannot be the mechanism for this reaction.
c The experimentally determined rate equation for this reaction is: rate $= k[P][Q]$. Which mechanism is consistent with the experimental data?

8 Consider the reaction:
   $$2NO + Br_2 \rightarrow 2NOBr$$
   The rate equation is rate $= k[NO]^2[Br_2]$. Suggest two different mechanisms that are consistent with this rate equation.

9 Consider the reaction:
   $$2X + Y \rightarrow 2Z$$
   The rate equation for this reaction is:
   rate $= k[X][Y]$. Suggest a mechanism for this reaction.

Learning objectives

- Understand that increasing the temperature causes the rate constant to increase
- Work out values of activation energy and the frequency factor (pre-exponential factor) using the Arrhenius equation

6.3 Activation energy (HL)

The Arrhenius equation

In a rate equation such as rate $= k[A][B]$, the effect of temperature variation is accounted for by a change in the value of the rate constant.

> As the temperature increases, the rate constant increases exponentially.

The **Arrhenius equation** can be used to model the variation of the rate constant with temperature:

$$k = Ae^{\frac{-E_a}{RT}}$$

$e^{\frac{-E_a}{RT}}$ represents the fraction of collisions that have $E \geq E_a$. However, not all collisions with $E \geq E_a$ result in reaction. The molecules must collide in the correct orientation, and A contains a factor that allows for this.

A is called the frequency factor (also called the pre-exponential factor or A-factor) and takes account of the **frequency** of collisions and the **orientation** of the collisions. A can be regarded as the product of two terms — one representing the frequency of molecular collisions and the other taking account of the fact that not all collisions with $E \geq E_a$ result in reaction because the molecules must collide with the correct orientation. A can be regarded as a **constant** (it actually varies very slightly with temperature).

The Arrhenius equation may also be written in the form:

$$\ln k = \frac{-E_a}{R} \times \frac{1}{T} + \ln A$$

The Arrhenius equation in this form can be used to work out a value for the activation energy for a reaction. In order to do this, the following procedure must be followed:

1 Conduct a series of experiments at a range of temperatures.
2 Calculate a rate constant (k) for each temperature.
3 Plot a graph of $\ln k$ (y-axis) against $\frac{1}{T}$ (x-axis), where T is the absolute temperature (in kelvin). This graph should be a straight line. The gradient of the graph is $\frac{-E_a}{R}$, where R is the gas constant.
4 The intercept of the graph on the $\ln k$ axis (y-axis) is ln A.

After we have carried out the series of experiments, we could have data such as those listed in the first two columns of Table **6.6**. $\frac{1}{T}$ and $\ln k$ are then calculated.

R is the gas constant, i.e. $8.31\,J\,K^{-1}\,mol^{-1}$.

T is the temperature in kelvin.

T/K	k/s⁻¹	$\frac{1}{T}$/K⁻¹	ln k
300	0.00088	0.003 33	−7.03
350	0.0037	0.002 86	−5.60
400	0.0108	0.002 50	−4.53
450	0.0250	0.002 22	−3.69
500	0.0487	0.002 00	−3.02
550	0.0842	0.001 82	−2.47
600	0.133	0.001 67	−2.02
650	0.195	0.001 54	−1.63
700	0.272	0.001 43	−1.30

Table 6.6 Sample experimental data and derived values.

A graph of ln k (y-axis) against $\frac{1}{T}$ (x-axis) produces a straight-line graph (Figure **6.32**).

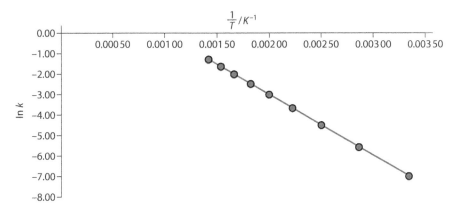

Figure 6.32 ln k (y-axis) against $\frac{1}{T}$ (x-axis).

Strictly speaking, what we have worked out is $\ln\left(\frac{k}{s^{-1}}\right)$. The natural log of the rate constant divided by its units is worked out to produce a pure number with no units.

We can understand why a straight-line graph is produced by comparing the Arrhenius equation with the equation of a straight line ($y = mx + c$, where m is the gradient and b is the intercept on the y-axis):

$$\ln k = \frac{-E_a}{R} \times \frac{1}{T} + \ln A$$
$$y = m \quad x + \quad c$$

Extension

The Arrhenius equation is a mathematical model that works well to explain variation of the rate of most reactions over a limited temperature range. There are actually a few reactions that get slower as temperature increases, which would correspond to a negative activation energy using this equation.

So, the Arrhenius equation is the equation of a straight line, where $\ln k$ is y and $\frac{1}{T}$ is x. From this equation we can see that the gradient of this straight line is $\frac{-E_a}{R}$. From the graph in Figure **6.32** we can work out the gradient of the straight line:

$$\text{gradient} = \frac{(-1.3) - (-7.0)}{0.001\,43 - 0.003\,33}$$

i.e. the gradient $= -3000\,K$.

The gradient of the line is equal to $\frac{-E_a}{R}$, so:

$$\frac{-E_a}{R} = -3000$$

$$E_a = 3000 \times R$$

$$E_a = 3000 \times 8.31 = 24\,900\,J\,mol^{-1}$$

i.e. the activation energy $= 24.9\,kJ\,mol^{-1}$.

From the equation of a straight line above it can be seen that the intercept on the $\ln k$ axis (y-axis) is $\ln A$. The intercept occurs at 3.00 and therefore $\ln A = 3.00$. The value of A can be obtained by raising each side as a power of e (e^x is the inverse function of $\ln x$).

Therefore $A = e^{3.00}$ – i.e. $A = 20.1\,s^{-1}$.

The units of A will be the same as the unit of k.

The values of E_a and A can also be worked out without using a graph if the values of the rate constant are known at two different temperatures. For example, if the values are:

$$T_1 = 300\,K \qquad k_1 = 8.80 \times 10^{-4}\,s^{-1}$$

$$T_2 = 550\,K \qquad k_2 = 8.42 \times 10^{-2}\,s^{-1}$$

and:

$$\ln k_1 = -\frac{E_a}{RT_1} - \ln A$$

$$\ln k_2 = -\frac{E_a}{RT_2} - \ln A$$

Subtracting the first equation from the second we get:

$$\ln k_2 - \ln k_1 = -\frac{E_a}{R}\left(\frac{1}{T_2} - \frac{1}{T_1}\right)$$

In our example:

$$\ln k_1 = -7.03 \text{ and } \ln k_2 = -2.47$$

$$\frac{1}{T_1} = 3.33 \times 10^{-3}\,K^{-1} \text{ and } \frac{1}{T_2} = 1.82 \times 10^{-3}\,K^{-1}$$

Substituting the numbers in to the above equation gives:

$$-2.47 + 7.03 = -\frac{E_a}{8.31}(1.82 \times 10^{-3} - 3.33 \times 10^{-3})$$

Rearranging this gives $E_a = 25\,100\,\text{J}\,\text{mol}^{-1}$ or $25.1\,\text{kJ}\,\text{mol}^{-1}$.

The value of A is obtained by substituting the values of k_1, T_1 and E_a (in $\text{J}\,\text{mol}^{-1}$) into $\ln k_1 = -\dfrac{E_a}{RT_1} - \ln A$. This gives $A = 20.8\,\text{s}^{-1}$.

The values calculated here are different from those worked out using a graphical method because only two points have been used from the graph and there is no guarantee that they lie exactly on the line of best fit. Also, working out gradients and intercepts by hand from a graph introduces more uncertainties.

Interpretation of activation energy and frequency factor values

In general, all other things being equal, the higher the activation energy for a reaction the slower it will be. $e^{-E_a/RT}$ can be rewritten as $\dfrac{1}{e^{E_a/RT}}$ – so if E_a is larger, $e^{E_a/RT}$ will be larger and $\dfrac{1}{e^{E_a/RT}}$ will be smaller and k will be smaller.

All other things being equal, the value of the frequency factor A will be smaller for reactions between more complicated molecules because these can collide in many different ways and only a small proportion of collisions will be in the correct orientation for reaction. However, if two atoms collide virtually all the collisions will be in the correct orientation.

The effect of a catalyst on the value of the rate constant

The effect of a catalyst on the rate equation, rate $= k[A][B]$, is to **increase the value of the rate constant**.

A catalyst provides an alternative pathway of lower activation energy. E_a is smaller and, because e is raised to a negative power, this makes the value of $e^{-E_a/RT}$ bigger and hence k is bigger.

Nature of science

Scientific theories must explain experimental data. The first process in developing a theory of the effect of temperature on reaction rate would involve collecting data. Many different systems would have to be studied to make sure that a general law could be established. At the simplest level these data would suggest the law: increasing the temperature increases reaction rate. A simple theory based on the idea of particles colliding more often could explain this law. However, as more quantitative data are collected a more mathematical description of the effect of temperature could be developed. Mathematical analysis of the effect of temperature on collision rate would indicate problems with the theory and require development of a more sophisticated model based on the Arrhenius equation.

? Test yourself

The value of the gas constant is 8.31 J K^{-1} mol^{-1}.

10 Use the following data to calculate values for the activation energy and frequency factor for the second-order reaction $P + Q \rightarrow Z$

$\frac{1}{T}$/K^{-1}	ln k
0.002 50	2.27
0.002 22	4.14
0.002 00	5.64
0.001 82	6.86
0.001 67	7.88
0.001 54	8.75
0.001 43	9.49
0.001 33	10.1
0.001 25	10.7

11 Use the following data to calculate values for the activation energy and frequency factor for the reaction $A + B \rightarrow C$

Temperature / K	k / mol^{-1} dm^3 s^{-1}
400	1.74×10^{-2}
450	3.53×10^{-1}
500	3.92
550	28.1
600	145
650	581
700	1.91×10^3
750	5.35×10^3
800	1.32×10^4

Exam-style questions

1 Which of the following best explains why an increase in temperature causes the rate of a reaction to increase?

    A the particles collide more
    B the particles collide more frequently
    C more particles have energy greater than the activation energy
    D the activation energy is lower at higher temperature

2 An experiment was carried out to measure the rate of decomposition of hydrogen peroxide according to the equation:

$$2H_2O_2(aq) \rightarrow 2H_2O(l) + O_2(g)$$

$56.0 \, cm^3$ of gas was produced in 30.0 s. The average rate of reaction during this time was:

    A $1.87 \, cm^3 s^{-1}$ **C** $0.536 \, s \, cm^{-3}$
    B $28.0 \, cm^3 min^{-1}$ **D** $112 \, min \, cm^{-3}$

3 Which of the following will **not** increase the rate of the reaction between magnesium and hydrochloric acid?

$$Mg(s) + 2HCl(aq) \rightarrow MgCl_2(aq) + H_2(g)$$

      A increasing the surface area of the magnesium
      B increasing the volume of hydrochloric acid used
      C increasing the concentration of the hydrochloric acid
      D increasing the temperature

4 In the decomposition of hydrogen peroxide, manganese(IV) oxide is a catalyst. Which of the following best describes the function of a catalyst and its mode of action?

      A it speeds up the reaction by increasing the activation energy
      B it slows down the reaction by decreasing the collision frequency of particles
      C it speeds up the reaction by allowing the reaction to occur by an alternative pathway of lower activation energy
      D it speeds up the reaction by increasing the average energy of the particles

5 In the reaction between $1.00\,g$ marble chips (calcium carbonate) and $25.0\,cm^3$ hydrochloric acid, which of the following sets of conditions should give the fastest rate of reaction?

      A $0.50\,mol\,dm^{-3}$ HCl(aq) and small marble chips at $20\,°C$
      B $0.10\,mol\,dm^{-3}$ HCl(aq) and small marble chips at $30\,°C$
      C $0.30\,mol\,dm^{-3}$ HCl(aq) and small marble chips at $70\,°C$
      D $0.50\,mol\,dm^{-3}$ HCl(aq) and large marble chips at $30\,°C$

HL 6 The rate equation for the reaction $CO + NO_2 \rightarrow CO_2 + NO$ is: rate $= k[NO_2]^2$. When the concentration of CO is increased by a factor of 2 and the concentration of NO_2 is increased by a factor of 3, the rate of reaction is increased by a factor of:

      A 3   **B** 6   **C** 9   **D** 18

HL 7 Consider the following experimental data for this reaction:

$$2NO + Br_2 \rightarrow 2NOBr$$

$[NO]/mol\,dm^{-3}$	$[Br_2]/mol\,dm^{-3}$	Rate $/mol\,dm^{-3}\,s^{-1}$
0.10	0.10	0.010
0.20	0.10	0.040
0.20	0.30	0.12

The rate equation for this reaction is:

      A rate $= k[NO]^2[Br_2]$ **C** rate $= k[NO_2]^2[Br_2]^3$
      B rate $= k[NO_2] + [Br_2]$ **D** rate $= k[NO_2][Br_2]$

HL 8 The activation energy for a reaction can be calculated from a graph of:

      A $\ln k$ vs T **C** $\ln k$ vs $\frac{1}{T}$
      B $\frac{1}{k}$ vs $\ln T$ **D** $\ln k$ vs $\ln T$

9 Consider the reaction:

$$2NO(g) + O_2(g) \rightarrow 2NO_2(g)$$

Some possible mechanisms for this reaction are:

I	$NO(g) + NO(g) \rightleftharpoons N_2O_2(g)$	fast
	$N_2O_2(g) + O_2(g) \rightarrow 2NO_2(g)$	slow
II	$NO(g) + NO(g) \rightarrow N_2O_2(g)$	slow
	$N_2O_2(g) + O_2(g) \rightarrow 2NO_2(g)$	fast
III	$NO(g) + O_2(g) \rightleftharpoons NO_3(g)$	fast
	$NO_3(g) + NO(g) \rightarrow 2NO_2(g)$	slow
IV	$NO(g) + O_2(g) \rightarrow NO_3(g)$	slow
	$NO_3(g) + NO(g) \rightarrow 2NO_2(g)$	fast

The rate equation for this reaction is: rate $= k[NO]^2[O_2]$. Which of these mechanisms is/are consistent with this rate equation?

A **I** only
B **I** and **IV** only

C **II** and **III** only
D **I** and **III** only

10 The units of k for a reaction with rate equation rate $= k[A]^2$ could be:

A $mol\,dm^{-3}\,s^{-1}$
B $mol^{-1}\,dm^3\,h^{-1}$

C $mol^2\,dm^{-6}\,s^{-1}$
D $mol^{-2}\,dm^6\,h^{-1}$

11 Explain by reference to the Maxwell–Boltzmann distribution why the rate of a reaction in the gas phase increases as temperature increases. **[4]**

12 The data in the table refer to the reaction:

$$CaCO_3(s) + 2HCl(aq) \rightarrow CaCl_2(aq) + CO_2(g) + H_2O(l)$$

Time/s	Volume of CO_2 produced/cm^3
0	0.0
10	16.0
20	30.0
30	41.0
40	47.0
50	51.0
60	53.5
70	55.5
80	56.5
90	57.0
100	57.0
110	57.0

a Explain, with the aid of a diagram, how these data could be obtained experimentally. **[3]**

b Plot these data on graph paper and label the line **A**. **[2]**

c Use the graph that you have plotted in part **b** to state and explain where the rate of reaction is fastest. **[2]**

d The original data were collected for an experiment using 1.00 g of calcium carbonate and 20.0 cm^3 of 0.300 mol dm^{-3} hydrochloric acid at 20 °C. The experiment was then repeated using exactly the same conditions, except that the temperature of the HCl(aq) was 30 °C.

Sketch, on the same axes as your graph in part **b**, the curve that would be obtained. Label this graph **B**. **[2]**

e i Calculate the maximum volume of carbon dioxide (in cm^3) that should have been collected in the original experiment if 1.00 mol CO$_2$ occupies 24.0 dm^3 under these conditions. **[3]**

ii Explain why the volume of gas collected is less than you predicted in part **i**. **[1]**

HL **13** The following data refer to the reaction:

$$X + 2Y \rightarrow Z$$

Experiment	Concentration of X / mol dm^{-3}	Concentration of Y / mol dm^{-3}	Rate of reaction / mol dm^{-3} s^{-1}
1	0.500	0.500	3.20×10^{-3}
2	0.250	0.500	1.60×10^{-3}
3	0.250	0.250	8.00×10^{-4}

a Explain what is meant by the term 'order of reaction'. **[2]**

b Deduce the rate equation for this reaction. **[4]**

c Calculate the rate constant with units for this reaction. **[2]**

d What is the rate of reaction when the concentrations of X and Y are both 0.100 mol dm^{-3}? **[2]**

e State and explain how the value of the rate constant for this reaction will change as the temperature increases. **[2]**

HL **14** Under certain conditions the decomposition of phosphine, PH$_3$, is zero order. The equation for the reaction is:

$$4PH_3(g) \rightarrow P_4(g) + 6H_2(g)$$

a Sketch a graph showing how the concentration of phosphine varies with time. **[2]**

b Sketch a graph showing how the rate of decomposition of phosphine varies as the concentration of phosphine changes. **[2]**

HL **15** Consider the gas-phase reaction between nitrogen(II) oxide and oxygen:

$$2NO(g) + O_2(g) \rightarrow 2NO_2(g) \qquad \Delta H = -113 \, kJ \, mol^{-1}$$

The rate equation for the reaction is: rate $= k[NO]^2$.

a Explain, by reference to this reaction, why the rate equation cannot be derived from the stoichiometric equation. **[2]**

b Explain what is meant by the 'rate-determining step' in a chemical reaction. **[1]**

c Suggest a two-step mechanism for this reaction and state the molecularity of the rate-determining step. **[4]**

d If the total volume of the reaction container is doubled at constant temperature, state and explain the effect on the rate of this reaction. **[2]**

e Sketch, on the same axes, a potential energy profile for this reaction with and without a catalyst. Clearly label the curves and the activation energies of the catalysed and uncatalysed reactions. **[4]**

Summary

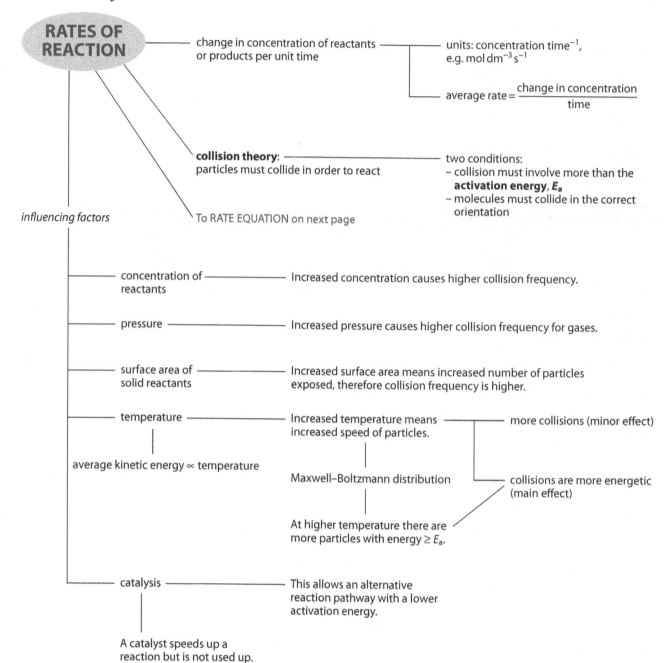

RATES OF REACTION

change in concentration of reactants or products per unit time

units: concentration time^{-1}, e.g. mol dm^{-3} s^{-1}

average rate = $\dfrac{\text{change in concentration}}{\text{time}}$

collision theory: particles must collide in order to react

two conditions:
- collision must involve more than the **activation energy, E_a**
- molecules must collide in the correct orientation

influencing factors

To RATE EQUATION on next page

concentration of reactants — Increased concentration causes higher collision frequency.

pressure — Increased pressure causes higher collision frequency for gases.

surface area of solid reactants — Increased surface area means increased number of particles exposed, therefore collision frequency is higher.

temperature — Increased temperature means increased speed of particles.

average kinetic energy $\propto$ temperature

more collisions (minor effect)

Maxwell–Boltzmann distribution

collisions are more energetic (main effect)

At higher temperature there are more particles with energy $\geq E_a$.

catalysis — This allows an alternative reaction pathway with a lower activation energy.

A catalyst speeds up a reaction but is not used up.

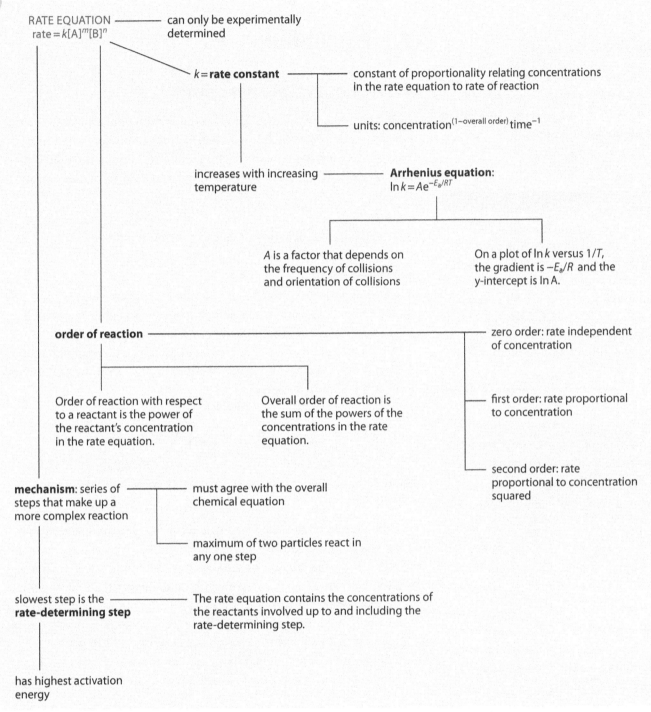

RATE EQUATION ——————— can only be experimentally
rate $= k[A]^m[B]^n$ determined

$k =$ **rate constant** ——————— constant of proportionality relating concentrations
in the rate equation to rate of reaction

units: concentration$^{(1-\text{overall order})}$ time^{-1}

increases with increasing ——————— **Arrhenius equation**:
temperature $\ln k = Ae^{-E_a/RT}$

A is a factor that depends on
the frequency of collisions
and orientation of collisions

On a plot of $\ln k$ versus $1/T$,
the gradient is $-E_a/R$ and the
y-intercept is $\ln A$.

order of reaction ——————— zero order: rate independent
of concentration

Order of reaction with respect
to a reactant is the power of
the reactant's concentration
in the rate equation.

Overall order of reaction is
the sum of the powers of the
concentrations in the rate
equation.

first order: rate proportional
to concentration

second order: rate
proportional to concentration
squared

mechanism: series of ——————— must agree with the overall
steps that make up a   chemical equation
more complex reaction

maximum of two particles react in
any one step

slowest step is the ——————— The rate equation contains the concentrations of
rate-determining step   the reactants involved up to and including the
rate-determining step.

has highest activation
energy

7 Equilibrium

Learning objectives

- Understand that a reversible reaction can come to a state of equilibrium
- Explain what is meant by dynamic equilibrium

Extension

At the microscopic level all reactions are reversible – if two particles come together with the formation of a bond, that bond could also break as the two components move apart. This is often likened to the idea of running a film backwards, so that, at the molecular level, the 'film of a reaction' can always be run backwards.

Calcium oxide is also known as quicklime or lime. When heated strongly it glows bright white. This was used as theatre lighting, which gave rise to the phrase *in the limelight*.

A system has reached equilibrium when no further change appears to occur – all concentrations remain constant.

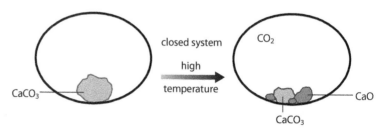

$$\text{CaCO}_3 \xrightarrow{\text{forward}} \text{CaO} + \text{CO}_2$$
$$\text{CaCO}_3 \xleftarrow{\text{reverse}} \text{CaO} + \text{CO}_2$$

$$\text{Rate}_{\text{forward}} = \text{Rate}_{\text{reverse}}$$

7.1 Equilibrium

7.1.1 Reversible reactions

As the name suggests, reversible reactions are reactions that can go either way. In a common reversible reaction, calcium carbonate, when heated strongly, decomposes to form calcium oxide and carbon dioxide. But calcium oxide also reacts with carbon dioxide to form calcium carbonate:

$$\text{CaCO}_3(s) \rightleftharpoons \text{CaO}(s) + \text{CO}_2(g)$$

The double arrow ($\rightleftharpoons$) shows that the reaction is reversible.

Equilibrium

The decomposition of CaCO_3 is reversible, but what happens if we put solid calcium carbonate in an open container and keep it at a constant high temperature for an extended period of time (Figure **7.1**)?

All the calcium carbonate is converted to calcium oxide because the carbon dioxide escapes and is not available to react with the calcium oxide to re-form calcium carbonate.

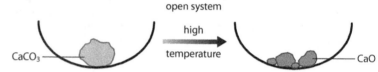

Figure 7.1 In an open container, all the calcium carbonate is converted to calcium oxide.

If we do exactly the same experiment, at the same temperature, but with the calcium carbonate in a sealed container (Figure **7.2**), we find that after the same amount of time we still have some calcium carbonate present. No matter how long we continue the experiment (keeping it at a constant temperature), the amounts of calcium carbonate, calcium oxide and carbon dioxide stay the same. The reaction appears to have stopped, and we say that the system has reached a state of **equilibrium**.

The reaction has not actually stopped but is proceeding in both directions at equal rates. In other words, the calcium carbonate is

Figure 7.2 In a closed system, a state of equilibrium is attained.

decomposing to give calcium oxide and carbon dioxide at exactly the same rate as the calcium oxide and carbon dioxide are recombining to form calcium carbonate. This type of equilibrium is called a **dynamic equilibrium**. All equilibria in chemistry are dynamic.

> In **dynamic equilibrium**, macroscopic properties are constant (concentrations of all reactants and products remains constant) and the rate of the forward reaction is equal to the rate of the reverse reaction.

Exam tip
When asked to define dynamic equilibrium you must address both the 'dynamic' part and the 'equilibrium' part.

Nature of science

Radioisotopes can be used to provide evidence for the dynamic nature of equilibrium. If, for instance a small amount of calcium carbonate labelled with radioactive ^{14}C is introduced into the equilibrium system shown in Figure **7.2**, there will be no overall change in the amount of CO_2 present but the radioactive ^{14}C will be distributed between the $CaCO_3$ and the CO_2, indicating that reactions are still going on.

Equilibrium and rate of reaction

Consider the reaction $H_2(g) + I_2(g) \rightleftharpoons 2HI(g)$. If we start with just hydrogen and iodine vapour in a closed container at a certain temperature and follow how the concentration of hydrogen and hydrogen iodide change with time, we obtain a graph of the form shown in Figure **7.3**.

The concentration of H_2 decreases at first, until it levels off when equilibrium is reached. The concentration of HI is initially zero, but it increases until it flattens off and does not change any more after equilibrium has been reached. If we plot a graph of rate against time for the forward and reverse reactions we get a graph of the form shown in Figure **7.4**.

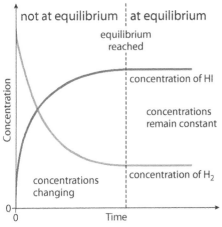

Figure 7.3 Graph showing how the concentrations of hydrogen and hydrogen iodide change with time.

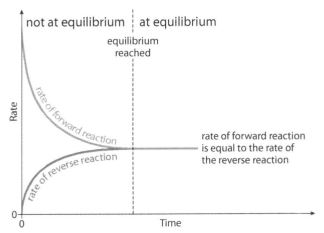

Figure 7.4 Graph showing how the rates of the forward and reverse reactions change as a reversible reaction comes to equilibrium.

The characteristics of the equilibrium state

Some of these have been discussed above.

1 **Macroscopic properties are constant at equilibrium** – at equilibrium the concentrations of all reactants and products remain constant.

2 **At equilibrium the rate of the forward reaction is equal to the rate of the reverse reaction**.

3 **Equilibrium can be attained only in a closed system** – as we saw above, if calcium carbonate is heated in an open container, equilibrium is never reached because the carbon dioxide escapes and has no opportunity to recombine with the calcium oxide. In effect, a reaction going on in solution, but not involving the production of a gas, represents a closed system.

4 **All species in the chemical equation are present in the equilibrium reaction mixture** – for example, if nitrogen and hydrogen are allowed to come to equilibrium according to the equation $N_2(g) + 3H_2(g) \rightleftharpoons 2NH_3(g)$, the reaction mixture at equilibrium contains nitrogen, hydrogen and ammonia.

5 **Equilibrium can be attained from either direction** – consider the equilibrium:

$$\underset{\text{ethanoic acid}}{CH_3COOH(l)} + \underset{\text{ethanol}}{C_2H_5OH(l)} \rightleftharpoons \underset{\text{ethyl ethanoate}}{CH_3COOCH_2CH_3(l)} + \underset{\text{water}}{H_2O(l)}$$

If we mix together ethanoic acid and ethanol in the presence of an acid catalyst and incubate them at 60 °C, they come to equilibrium. If we do the same starting with ethyl ethanoate and water, they also come to equilibrium. A state of equilibrium, in which all four species are present, can be reached by mixing:

- ethanoic acid and ethanol
- ethyl ethanoate and water
- all four substances
- any three substances.

In each case, equilibrium is reached but the actual concentrations present at equilibrium depend on how much of each substance we started with.

Physical equilibria

All the above reactions involved **chemical equilibria** – equilibria established as a result of chemical reactions. There are also **physical equilibria** (equilibria involving a change in state), for example, the equilibrium between a liquid and its vapour.

Evaporation

When a liquid is in an open container it evaporates. At the molecular level, the particles must overcome the forces holding them in the liquid (intermolecular forces) in order to escape into the gas phase – evaporation is an endothermic process.

The spread of molecular kinetic energies in a liquid is shown in Figure **7.5**. Molecules in the shaded region have sufficient energy to escape and evaporate. The faster moving molecules (higher kinetic energy) are able to overcome the intermolecular forces, so they escape first and the average kinetic energy of the particles in the liquid phase drops. Average kinetic energy is an indication of the temperature and, therefore, the temperature of the liquid falls. Heat is drawn from the surroundings to allow further molecules to evaporate. If the container is open then all the liquid will eventually evaporate.

Liquid–vapour equilibrium

Consider a volatile liquid, such as bromine, in a closed container (Figure **7.6**). At the beginning there are no molecules of vapour above the liquid – molecules escape from the liquid (evaporation) but there is no condensation (Figure **7.6a**). As molecules of vapour appear, these strike the surface of the liquid and some re-enter it – this process is condensation. At first the rate of condensation is low, but as the number of molecules in the vapour phase increases the rate of condensation increases (Figure **7.6b**). Eventually the rate of condensation becomes equal to the rate of evaporation and nothing more appears to change (Figure **7.6c**).

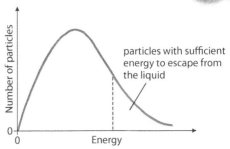

Figure 7.5 The spread of molecular energies in a liquid.

Figure 7.6 Equilibrium is attained when the rate of evaporation is equal to the rate of condensation.

Because the rate of evaporation and the rate of condensation are equal, this is a dynamic equilibrium.

When the rate of evaporation (vaporisation) is equal to the rate of condensation, the colour of the vapour remains constant (it does not get any darker or lighter) and a state of equilibrium has been reached. This is known as a **phase equilibrium** because it involves a change of phase (state).

7.1.2 Position of equilibrium

The 'position of equilibrium' refers to the relative amounts of reactants and products present at equilibrium. Some reactions go almost to completion, for example:

$$2NO(g) \rightleftharpoons N_2(g) + O_2(g)$$

Learning objectives

- Understand what is meant by the position of equilibrium
- Apply Le Chatelier's principle to predict the effect of changes in conditions on the position of equilibrium

At 700 K, the position of equilibrium lies a long way to the **right**. There is a large amount of N_2 and O_2 and not very much NO at equilibrium – roughly one and a half million times as many N_2 and O_2 molecules as NO molecules. However, for:

$$H_2(g) + CO_2(g) \rightleftharpoons H_2O(g) + CO(g)$$

at 1100 K, the total number of H_2 and CO_2 molecules at equilibrium is roughly equal to the total number of H_2O and CO molecules – the equilibrium is evenly balanced.

Water dissociates according to the equation:

$$H_2O(l) \rightleftharpoons H^+(aq) + OH^-(aq)$$

At 298 K the number of water molecules present at equilibrium is over 250 million times greater than the total number of H^+ and OH^- ions. The position of this equilibrium lies a long way to the **left** – not many H^+ and OH^- ions are present at equilibrium.

The effect of changing conditions on the position of equilibrium – Le Chatelier's principle

Le Chatelier's principle

If a system at equilibrium is subjected to a change, the position of equilibrium will shift in order to minimise the effect of the change.

This means that, if we take a particular system at equilibrium under a certain set of conditions and change one of those conditions, such as the temperature or the pressure, the system will move to a new position of equilibrium. **Le Chatelier's principle** allows us to predict in which direction the position of equilibrium will shift.

The effect of temperature

Consider the equilibrium:

$$2HI(g) \rightleftharpoons H_2(g) + I_2(g) \qquad\qquad \Delta H = +10\,\text{kJ}\,\text{mol}^{-1}$$

At room temperature (298 K), the number of molecules of HI is roughly 28 times the total number of molecules of H_2 and I_2:

$$2HI(g) \rightleftharpoons H_2(g) + I_2(g)$$
$$96.6\% \qquad\quad 3.4\%$$

However, when the system reaches equilibrium at 700 K, the number of HI molecules is only approximately seven times the total number of H_2 and I_2 molecules:

$$2HI(g) \rightleftharpoons H_2(g) + I_2(g)$$
$$88\% \qquad\quad 12\%$$

Henri Louis Le Chatelier (1850–1936) was a French chemist. His original statement on his principle is:
Every system in stable equilibrium submitted to the influence of an exterior force which tends to cause variation, either in its temperature or its condensation (pressure, concentration, number of molecules in the unit of volume) in its totality or only in some one of its parts can undergo only those interior modifications which, if they occur alone, could produce a change of temperature, or of condensation, of a sign contrary to that resulting from the exterior force.

This means that there are relatively more H_2 and I_2 molecules present at equilibrium at 700 K than at 298 K. In this case, as the temperature increases, the position of equilibrium shifts to the right (there is more H_2 and I_2 present).

The effect of a temperature change on the position of an equilibrium can be considered in terms of Le Chatelier's principle. The value of ΔH given above refers to the forward direction – so in this case the forward reaction is endothermic and the reverse reaction is exothermic.

As the temperature is increased, the position of equilibrium shifts in the direction that will minimise the effect of the change. So to minimise the effect of the increase in temperature in this reaction, the position of equilibrium shifts in the endothermic direction to take in the heat that is added (heat energy is converted to chemical energy). The endothermic direction is to the right and, therefore, as the temperature is increased the position of equilibrium shifts to the right to produce relatively more H_2 and I_2.

Consider another reaction:

$$N_2(g) + 3H_2(g) \rightleftharpoons 2NH_3(g) \qquad\qquad \Delta H = -92\,\text{kJ}\,\text{mol}^{-1}$$

This time, at 300 K and 10 atmosphere pressure, we have:

$$N_2(g) + 3H_2(g) \rightleftharpoons 2NH_3(g)$$
$$85\%15\%$$

At 700 K and the same pressure:

$$N_2(g) + 3H_2(g) \rightleftharpoons 2NH_3(g)$$
$$99.8\%0.2\%$$

In this case, increasing the temperature causes the position of equilibrium to be shifted to the left, i.e. there is less ammonia present at equilibrium at the higher temperature.

As the temperature is increased, the position of equilibrium shifts in the endothermic direction to take in heat and minimise the effect of the change. This time, the endothermic direction is to the left and, therefore, as the temperature is increased the position of equilibrium shifts to the left:

> **HEAT reaction mixture:** position of equilibrium is shifted in the **endo**thermic direction.
> **COOL reaction mixture:** position of equilibrium is shifted in the **exo**thermic direction.

endothermic
$$2HI(g) \rightleftharpoons H_2(g) + I_2(g)$$
exothermic

> This is analogous to the idea that, if the heating is turned up you might take off your jumper in order to minimise the change and keep your temperature roughly the same as it was before.

exothermic
$$N_2(g) + 3H_2(g) \rightleftharpoons 2NH_3(g)$$
endothermic

The effect of pressure

Consider this equilibrium:

$$2NO_2(g) \rightleftharpoons N_2O_4(g)$$
$$\text{brown}\quad\text{colourless}$$

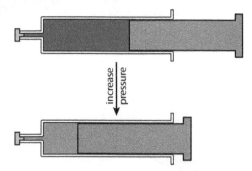

Figure 7.7 The effect of increasing pressure on the $2NO_2(g) \rightleftharpoons N_2O_4(g)$ equilibrium.

$PV = nRT$, so the number of moles of gas is proportional to the volume and pressure of the gas. If we imagine this reaction going to completion and two moles of $NO_2(g)$ being completely converted to one mole of $N_2O_4(g)$, the volume of gas at the end would be half what we started with. If this reaction were carried out in a vessel of constant volume, the conversion of two moles of $NO_2(g)$ to one mole of $N_2O_4(g)$ would involve the pressure decreasing to half its original value.

Exam tip
You must remember the word 'gas' – solids and liquids are affected very little by changes in pressure.

Some NO_2 (brown) is put into a sealed gas syringe (Figure **7.7**). As the pressure is increased, the colour initially gets slightly darker because the same number of molecules are squeezed into a smaller space. The mixture then becomes less brown as a new position of equilibrium is established. At higher pressure there is less NO_2 (brown) and more N_2O_4 (colourless) present in the equilibrium mixture, and therefore it is a paler brown than at lower pressure. So it can be seen that, in this case, as the pressure increases, the position of equilibrium shifts to the right (more N_2O_4 is present at equilibrium).

This reaction involves a decrease in the number of gaseous molecules, from two on the left-hand side to one on the right-hand side. Two moles of gas take up more space than one mole of gas, so as the pressure is increased the position of equilibrium shifts in order to minimise the effect of this pressure change. This is achieved by the position of equilibrium shifting to the side with fewer gaseous molecules and therefore lower volume – that is, the right-hand side.

Consider the following reaction:

$$2SO_3(g) \rightleftharpoons 2SO_2(g) + O_2(g)$$

This reaction involves the conversion of two molecules of **gas** (on the left-hand side) to three molecules of **gas** (on the right-hand side). As the pressure is increased, the position of equilibrium shifts to the left-hand side – i.e. the side with **fewer gas molecules** – to minimise the effect of the change.

Now consider this equilibrium:

$$2HI(g) \rightleftharpoons H_2(g) + I_2(g)$$

Because there is the same number of molecules of **gas** on both sides, changing the pressure has **no effect** on the position of equilibrium.

In these three examples, it has been stressed that we have to consider the number of molecules of **gas** when predicting the effect of a change in pressure on the position of equilibrium.

Consider once again the decomposition of calcium carbonate:

$$CaCO_3(s) \rightleftharpoons CaO(s) + CO_2(g)$$

There is one molecule of gas on the right-hand side but none on the left-hand side – therefore increasing the pressure causes the position of equilibrium to shift to the left.

> If a reaction involves a change in the number of gas molecules, an increase in pressure results in the position of equilibrium shifting in the direction that gives a decrease in the number of gas molecules.

The effect of concentration

Consider this system at equilibrium:

$$2[CrO_4]^{2-}(aq) \; + \; 2H^+(aq) \; \rightleftharpoons \; [Cr_2O_7]^{2-}(aq) \; + \; H_2O(l)$$

chromate(VI) ion   from acid   dichromate(VI) ion
yellow   orange

The fact that the colour of the solution on the left of Figure **7.8** is yellow indicates that the position of equilibrium lies to **left**, so there is significantly more of the yellow ion present than the orange ion.

When some acid (H^+) is added to the mixture in the flask, the colour of the solution changes to orange. There is now much more of the orange dichromate(VI) ion present, which means that the position of equilibrium has shifted to the right. This can be explained in terms of Le Chatelier's principle – as more acid is added, the position of equilibrium shifts to the right to use up the excess acid and so minimise the effect of the change.

If we now add an alkali (OH^-) to the solution, the colour changes back to yellow. The OH^- ions react with the H^+ ions to form water. So adding alkali reduces the concentration of H^+ ions in the solution and the position of equilibrium must shift to the left in order to minimise the effect of the change by replacing the H^+ ions.

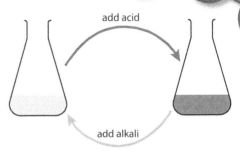

add acid

add alkali

Figure 7.8 The chromate(VI)/dichromate(VI) equilibrium.

$$H^+(aq) + OH^-(aq) \rightleftharpoons H_2O(l)$$

In general, if the concentration of one of the species in an equilibrium mixture is increased, the position of equilibrium shifts to the opposite side to reduce the concentration of this species.

? Test yourself

1 Predict the effect of increasing pressure on the position of equilibrium in the following systems:
   a $CH_4(g) + 2H_2O(g) \rightleftharpoons CO_2(g) + 4H_2(g)$
   b $N_2O_5(g) + NO(g) \rightleftharpoons 3NO_2(g)$
   c $NO(g) + NO_2(g) \rightleftharpoons N_2O_3(g)$

2 Predict the effect of increasing temperature on the position of equilibrium in the following systems:
   a $NO(g) + NO_2(g) \rightleftharpoons N_2O_3(g)$
   $$\Delta H = -40\,kJ\,mol^{-1}$$
   b $CH_4(g) + H_2O(g) \rightleftharpoons CO(g) + 3H_2(g)$
   $$\Delta H = +206\,kJ\,mol^{-1}$$
   c $CO(g) + 2H_2(g) \rightleftharpoons CH_3OH(g)$
   $$\Delta H = -90\,kJ\,mol^{-1}$$

3 Predict the effect of the following changes on the position of equilibrium:
   a Removing the CO_2 from the equilibrium:
   $$CaCO_3(s) \rightleftharpoons CaO(s) + CO_2(g)$$
   b Adding acid (H^+) to the system:
   $$NH_4^+ \rightleftharpoons H^+ + NH_3$$
   c Adding sodium hydroxide to the system:
   $$CH_3COOH(aq) \rightleftharpoons CH_3COO^-(aq) + H^+(aq)$$
   d Adding sodium hydroxide to the system:
   $$3I_2(aq) + 3H_2O(l) \rightleftharpoons 5HI(aq) + HIO_3(aq)$$

- Write the expression for the equilibrium constant for a given reversible reaction
- Understand what is meant by reaction quotient
- Understand the connection between the size of the equilibrium constant and the extent of reaction
- Understand that it is possible to write more than one equilibrium constant for a particular reaction
- Understand how changing conditions affect the value of the equilibrium constant and the position of equilibrium

A series of experiments is carried out on the reaction:

$$CH_3COOH(l) + C_2H_5OH(l) \rightleftharpoons CH_3COOCH_2CH_3(l) + H_2O(l)$$

ethanoic acid   ethanol   ethyl ethanoate   water

Various known amounts of ethanol and ethanoic acid are reacted together and allowed to come to equilibrium at the same temperature. The **equilibrium concentrations** of each component of the reaction mixture are determined and it is found that the ratio:

$$\frac{[CH_3COOCH_2CH_3(l)][H_2O(l)]}{[CH_3COOH(l)][C_2H_5OH(l)]}$$

is constant for all the experiments

where $[CH_3COOH(l)]$ represents the concentration of CH_3COOH at equilibrium.

If the same procedure is repeated for the reaction:

$$N_2(g) + 3H_2(g) \rightleftharpoons 2NH_3(g)$$

it is found that the following ratio is constant at a particular temperature:

$$\frac{[NH_3(g)]^2}{[N_2(g)][H_2(g)]^3}$$

All concentrations are measured at equilibrium.

This leads us to the general **equilibrium law**. For the reaction $aA + bB \rightleftharpoons cC + dD$ (where all reactants are in the same phase), the value of the following ratio is constant at a particular temperature:

$$\frac{[C]^c[D]^d}{[A]^a[B]^b}$$

[A] represents the concentration of A at equilibrium.

Because this ratio is constant at a particular temperature we can write:

$$2SO_2(g) + O_2(g) \rightleftharpoons 2SO_3(g)$$

$$K_c = \frac{[SO_3(g)]^2}{[SO_2(g)]^2\,[O_2(g)]} \begin{array}{l}\leftarrow \text{products}\\ \leftarrow \text{reactants}\end{array}$$

$$K_c = \frac{[C]^c[D]^d}{[A]^a[B]^b}$$

where K_c is the **equilibrium constant**. The 'c' indicates that this equilibrium constant is expressed in terms of concentrations.

K_c **is constant for a particular reaction at a particular temperature.**

Extension

The equilibrium constant K_p is often calculated. This is the equilibrium constant expressed in terms of partial pressures. The partial pressure of a gas in a mixture of gases is the pressure that the gas would exert if it were present in the container by itself.

The expressions for the equilibrium constant for some reactions are given below:

$N_2O_4(g) \rightleftharpoons 2NO_2(g)$ $\quad K_c = \frac{[NO_2(g)]^2}{[N_2O_4(g)]}$

$N_2O_5(g) + NO(g) \rightleftharpoons 3NO_2(g)$ $\quad K_c = \frac{[NO_2(g)]^3}{[N_2O_5(g)][NO(g)]}$

$CO(g) + 3H_2(g) \rightleftharpoons CH_4(g) + H_2O(g)$ $\quad K_c = \frac{[CH_4(g)][H_2O(g)]}{[CO(g)][H_2(g)]^3}$

Extension

You only need to write equilibrium constants for homogeneous equilibria (all reactants and products in the same physical state). Reactions involving solids and gases, for example, are more complicated and the concentrations of pure solids do not appear in the equilibrium constant expression. The equilibrium constant for the decomposition of $CaCO_3$ (page **278**) is $K_c = [CO_2]$.

What use is an equilibrium constant?

An equilibrium constant provides information about how far a reaction proceeds at a particular temperature.

$K_c >> 1$ – the reaction proceeds almost totally towards the products.
$K_c << 1$ – the reaction hardly proceeds at all towards the products.

The values of the equilibrium constants for a series of reactions at 298 K are given in Table **7.1**. These equilibrium constants are all very much greater than one. These reactions proceed almost totally towards the products, so that there is virtually no hydrogen and halogen in the equilibrium mixture. However, for the reaction $N_2(g) + O_2(g) \rightleftharpoons 2NO(g)$ the value of the equilibrium constant is 10^{-31} at 298 K. This value is very much less than one, indicating that the reaction hardly proceeds at all towards the products – that is, the position of equilibrium lies a long way to the left.

Reaction	K_c
$H_2(g) + F_2(g) \rightleftharpoons 2HF(g)$	10^{95}
$H_2(g) + Cl_2(g) \rightleftharpoons 2HCl(g)$	10^{33}
$H_2(g) + Br_2(g) \rightleftharpoons 2HBr(g)$	10^{19}

Table 7.1 Equilibrium constants for reactions at 298 K.

The reaction quotient

The reaction quotient, Q, is the ratio of the concentrations of the reactants and products (raised to the appropriate powers) at any point in time. An expression for Q is exactly the same as that for the equilibrium constant – except that the concentrations are **not** equilibrium concentrations. So, for the reaction:

$H_2(g) + I_2(g) \rightleftharpoons 2HI(g)$

$$Q = \frac{[HI(g)]^2}{[H_2(g)][I_2(g)]}$$

When the system is at equilibrium, the concentrations are all equilibrium concentrations and the value of Q is equal to the value of K_c at that temperature.

The value K_c for the above reaction is 54 at 700 K. If we mix together 2 mol HI, 1 mol H_2 and 1 mol I_2 in a 1 dm^3 vessel then the concentrations of the species present in the reaction mixture are $[HI] = 2\,mol\,dm^{-3}$, $[H_2] = 1\,mol\,dm^{-3}$ and $[I_2] = 1\,mol\,dm^{-3}$.

So the value of Q is $\frac{2^2}{1 \times 1} = 4$

$$\text{concentration} = \frac{\text{number of moles}}{\text{volume}}$$

Because the value of Q is not equal to the value of the equilibrium constant, we can see immediately that the system is not at equilibrium. We can also see that, because the value of Q is less than that of K_c, the reaction must proceed to the right (to produce more HI) to reach equilibrium. If the reaction proceeds to the right, there will be more HI and less H_2 and I_2 in the reaction mixture and the value of Q will increase.

If 0.5 mol of the H_2 reacts with 0.5 mol I_2 to produce an extra 1 mol HI then the new concentrations are:

$$[HI] = 3 \, mol \, dm^{-3}, \, [H_2] = 0.5 \, mol \, dm^{-3} \text{ and } [I_2] = 0.5 \, mol \, dm^{-3}$$

So the new value of Q is $\dfrac{3^2}{0.5 \times 0.5} = 36$

The system is still not at equilibrium because Q is not equal to K_c, but it is closer to equilibrium. Because the value of Q is still smaller than K_c, more H_2 and I_2 must react together to form HI.

In general:

- If $Q < K_c$ then the reaction must proceed towards the products (to the right) to reach equilibrium.
- If $Q > K_c$ then the reaction must proceed towards the reactants (to the left) to reach equilibrium.

Two different values of the equilibrium constant for the same reaction under the same conditions.

There can be different values of K_c for the same reaction under the same conditions.

Consider the equilibrium $2NO_2(g) \rightleftharpoons N_2O_4(g)$. The expression for its equilibrium constant is:

$$K_c = \frac{[N_2O_4(g)]}{[NO_2(g)]^2}$$

and the equilibrium constant is 0.69 at 400 K.

However, the reaction could also have been written the other way round:

$$N_2O_4(g) \rightleftharpoons 2NO_2(g)$$

The expression for the equilibrium constant in this case is:

$$K_c' = \frac{[NO_2(g)]^2}{[N_2O_4(g)]}$$

The value of the equilibrium constant at 400 K for this reaction is $\frac{1}{0.69}$ or 1.45.

It can be seen then that the value of the equilibrium constant depends on how the reaction is written, and therefore these constants are only useful when quoted in conjunction with their equilibrium reactions.

Consider the equilibrium:

$$H_2(g) + I_2(g) \rightleftharpoons 2HI(g)$$

The principal use of HI is to make hydriodic acid, which is used to make iodides and as a reducing agent. Hydriodic acid is a controlled chemical in the USA because it can be used in the production of methamphetamine (crystal meth).

which can also be written:

$$\tfrac{1}{2}H_2(g) + \tfrac{1}{2}I_2(g) \rightleftharpoons HI(g)$$

Both expressions are completely valid and accurate representations of the reaction.

The equilibrium constant expression for the first of these is:

$$K_c' = \frac{[HI(g)]^2}{[H_2(g)][I_2(g)]}$$

whereas for the second reaction it is:

$$K_c'' = \frac{[HI(g)]}{[H_2(g)]^{\frac{1}{2}}[I_2(g)]^{\frac{1}{2}}}$$

At 700 K, $K_c' = 54$, whereas for the second reaction at the same temperature $K_c'' = \sqrt{54}$ or 7.3. The relationship between the two equilibrium constants is $K_c'' = \sqrt{K_c'}$.

? Test yourself

4 Write expressions for the equilibrium constants for the following reactions:
  a $NO(g) + NO_2(g) \rightleftharpoons N_2O_3(g)$
  b $CH_4(g) + H_2O(g) \rightleftharpoons CO(g) + 3H_2(g)$
  c $2H_2O(g) \rightleftharpoons 2H_2(g) + O_2(g)$
  d $4NH_3(g) + 5O_2(g) \rightleftharpoons 4NO(g) + 6H_2O(g)$
  e $2NO(g) + O_2(g) \rightleftharpoons 2NO_2(g)$

5 Given the following equilibrium constants:
  $H_2(g) + Cl_2(g) \rightleftharpoons 2HCl(g) \qquad K_c = 10^{33}$
  $H_2(g) + Br(g) \rightleftharpoons 2HBr(g) \qquad K_c = 10^{19}$
  work out the values for the equilibrium constants for these reactions:
  a $\tfrac{1}{2}H_2(g) + \tfrac{1}{2}Cl_2(g) \rightleftharpoons 2HCl(g)$
  b $2HBr(g) \rightleftharpoons H_2(g) + Br_2(g)$
  c $HBr(g) \rightleftharpoons \tfrac{1}{2}H_2(g) + \tfrac{1}{2}Br_2(g)$

6 a Consider the equilibrium:
  $2SO_2(g) + O_2(g) \rightleftharpoons 2SO_3(g)$
  The value of the equilibrium constant at 500 K is 1.03×10^{12}. What is the value of the equilibrium constant for the reaction: $2SO_3(g) \rightleftharpoons 2SO_2(g) + O_2(g)$ at 500 K?
  b The value of the equilibrium constant for the reaction $N_2(g) + 3H_2(g) \rightleftharpoons 2NH_3(g)$ at 500 K is 59.8. What is the value of the equilibrium constant for the reaction:
  $\tfrac{1}{2}N_2(g) + \tfrac{3}{2}H_2(g) \rightleftharpoons NH_3(g)$ at the same temperature?

How changing the conditions affects the value of the equilibrium constant

The value of the equilibrium constant for a particular reaction is **only** affected by a change in **temperature**.

The effect of temperature

Exothermic reactions

$$CO(g) + 2H_2(g) \rightleftharpoons CH_3OH(g) \qquad \Delta H^\ominus = -90\,kJ\,mol^{-1}$$

The equilibrium constant expression for this reaction is:

$$K_c = \frac{[CH_3OH(g)]}{[CO(g)][H_2(g)]^2}$$

Temperature / K	K_c
298	1.7×10^{17}
500	1.1×10^{11}
1000	2.1×10^6

Table 7.2 K_c for an exothermic reaction.

$$\frac{[CH_3OH(g)]}{[CO(g)]\,[H_2(g)]^2}$$ ↓ concentration decreases ↑ concentration increases

We use Le Chatelier's principle to explain the effect of temperature changes on the equilibrium. The reaction is exothermic in the forward direction and so, according to Le Chatelier's principle, an increase in temperature causes the position of equilibrium to shift in the endothermic direction – that is, the position of equilibrium shifts to the left. The concentration on the top of the expression for K_c decreases and the concentrations on the bottom increase. Therefore, the overall value of K_c decreases (Table **7.2**).

For an exothermic reaction, the value of the equilibrium constant decreases as the temperature is increased.

Endothermic reactions

$$N_2(g) + O_2(g) \rightleftharpoons 2NO(g) \qquad \Delta H^\ominus = +181\,kJ\,mol^{-1}$$

$$K_c = \frac{[NO(g)]^2}{[N_2(g)][O_2(g)]}$$

Temperature / K	K_c
298	4.3×10^{-31}
500	2.7×10^{-18}
1000	7.5×10^{-9}
2000	4.0×10^{-4}
3000	0.015

Table 7.3 K_c for an endothermic reaction.

For an endothermic reaction, the value of the equilibrium constant increases as the temperature is raised.

As the temperature increases, the value of the equilibrium constant increases (Table **7.3**). This is an endothermic reaction and so, according to Le Chatelier's principle, as the temperature increases the position of equilibrium shifts in the endothermic direction in order to take in heat and reduce the effect of the change. The position of equilibrium is shifted to the right so that more NO and less N_2 and O_2 are present at equilibrium. This results in the value of the equilibrium constant increasing.

Catalysts and the equilibrium constant

Catalysts are substances that increase the rate of a chemical reaction without being permanently changed in the process. Because they are the same at the beginning as at the end of the reaction, catalysts do not appear in the chemical equation.

A potential energy profile for a reversible reaction is shown in Figure **7.9**. Catalysts work by providing an alternative pathway of lower activation energy (E_a) for the reaction (Figure **7.10**). In a reversible reaction, a catalyst not only reduces the activation energy for the forward reaction but also that for the reverse reaction. The lowering of the activation energy is the same for both forward and reverse reactions. This means that a catalyst speeds up forward and reverse reactions **equally** and reduces the time taken to reach equilibrium. This is the only change that results from the introduction of a catalyst and therefore a catalyst has no effect on the position of equilibrium or on the value of the equilibrium constant.

A catalyst increases the rate of forward and reverse reactions equally.

The presence of a catalyst does not affect the position of equilibrium or the value of the equilibrium constant; it only reduces the time taken to reach equilibrium.

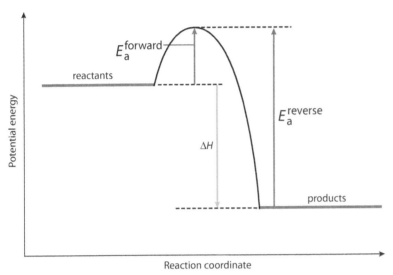

Figure 7.9 A potential energy profile for a reversible reaction.

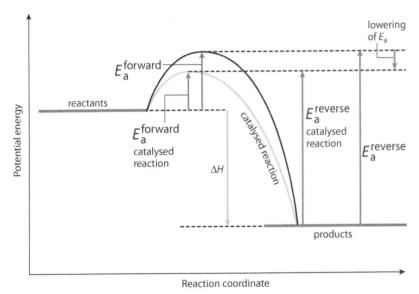

Figure 7.10 A potential energy profile for catalysed and uncatalysed reactions.

The equilibrium constant and rate

The equilibrium constant gives us information about how far a reaction goes towards completion (that is, about the extent of the reaction). It gives us absolutely no information about how quickly the reaction occurs. Kinetic data, such as the rate constant, indicate how quickly equilibrium is attained but provide no information whatsoever about the position of equilibrium and how far the reaction proceeds.

Summary

A summary of the effect of changes in conditions on the position of equilibrium and the value of the equilibrium constant is given in Table **7.4**.

Condition	Effect on position of equilibrium	Effect on value of K_c
Pressure	If a reaction involves a change in the number of gas molecules, an increase in pressure results in the position of equilibrium shifting in the direction that gives a decrease in the number of gas molecules.	no change
Concentration	The position of equilibrium will shift in order to use up any substance that has been added or replace any substance that has been removed from the equilibrium mixture.	no change
Catalyst	no effect	No effect because both forward and reverse reaction rates increase equally.
Temperature	If the temperature is increased, the position of equilibrium shifts in the endothermic direction. If the temperature is decreased, the position of equilibrium shifts in the exothermic direction.	For an exothermic reaction, K_c decreases as the temperature is increased. For an endothermic reaction, K_c increases as the temperature is increased.

Table 7.4 The effect of changes in conditions on the position of equilibrium and the value of the equilibrium constant.

? Test yourself

7 Explain the effect of the stated changes in conditions on the position of equilibrium and the value of the equilibrium constant. In each case state whether the value of the equilibrium constant increases, decreases or stays the same.

    a Increasing pressure in: $CH_4(g) + H_2O(g) \rightleftharpoons CO(g) + 3H_2(g)$ $\Delta H = +206\,kJ\,mol^{-1}$

    b Increasing temperature in: $CH_4(g) + H_2O(g) \rightleftharpoons CO(g) + 3H_2(g)$ $\Delta H = +206\,kJ\,mol^{-1}$

    c Decreasing temperature in: $H_2O(g) + CO(g) \rightleftharpoons H_2(g) + CO_2(g)$ $\Delta H = -40\,kJ\,mol^{-1}$

    d Increasing concentration of H_2 in: $N_2(g) + 3H_2(g) \rightleftharpoons 2NH_3(g)$ $\Delta H = -92\,kJ\,mol^{-1}$

    e Increasing pressure in: $N_2(g) + 3H_2(g) \rightleftharpoons 2NH_3(g)$ $\Delta H = -92\,kJ\,mol^{-1}$

    f Introducing a catalyst into the reaction: $2SO_2(g) + O_2(g) \rightleftharpoons 2SO_3(g)$ $\Delta H = -197\,kJ\,mol^{-1}$

7.2 The equilibrium law (HL)

7.2.1 Calculation of equilibrium constants

If the concentrations of all reaction components at equilibrium are given, calculating a value for the equilibrium constant simply involves putting these values into the equilibrium constant expression.

Learning objectives

- Solve problems involving equilibrium constants
- Understand the relationship between the equilibrium constant and the Gibbs free energy

Worked example

7.1 Use the data given to calculate the value of the equilibrium constant at 700 K for the reaction:

$$H_2(g) + I_2(g) \rightleftharpoons 2HI(g)$$

Substance	$H_2(g)$	$I_2(g)$	$HI(g)$
Equilibrium concentration / mol dm^{-3}	0.18	0.39	1.95

$$K_c = \frac{[HI(g)]^2}{[H_2(g)][I_2(g)]}$$

Substituting in the equilibrium concentrations we get:

$$K_c = \frac{1.95^2}{0.18 \times 0.39} = 54.$$

If we are given a value for the equilibrium constant, we can work out the concentration of any one of the species at equilibrium.

Equilibrium constants are normally worked out in terms of **activities** (effective concentrations). These are calculated relative to a standard concentration of 1 mol dm^{-3}. This results in equilibrium constants having no units. Equilibrium concentrations do have units (mol dm^{-3}), but in expressions for K_c what we are using is concentration / mol dm^{-3}, which does not have any units and therefore overall equilibrium constants have no units.

Worked examples

7.2 Given that the equilibrium constant for the reaction $H_2(g) + I_2(g) \rightleftharpoons 2HI(g)$ at 700 K is 54 and that the concentrations of H_2 and I_2 at equilibrium are 0.25 mol dm^{-3} and 0.50 mol dm^{-3}, respectively, what is the equilibrium concentration of HI?

$$K_c = \frac{[HI(g)]^2}{[H_2(g)][I_2(g)]}$$

Substituting the equilibrium concentrations we get $54 = \dfrac{[HI(g)]^2}{0.25 \times 0.50}$

Rearranging this $[HI(g)]^2 = 54 \times 0.25 \times 0.50$

$[HI(g)]^2 = 6.75$

Therefore, the equilibrium concentration of hydrogen iodide is 2.6 mol dm^{-3}.

A variation on this calculation is given in the next example.

7.3 Equal concentrations of hydrogen and iodine are mixed together in a closed container at 700 K and allowed to come to equilibrium. If the concentration of HI at equilibrium is 0.85 mol dm^{-3}, what are the equilibrium concentrations of H_2 and I_2? $K_c = 54$ at this temperature.

$$H_2(g) + I_2(g) \rightleftharpoons 2HI(g)$$

$$K_c = \frac{[HI(g)]^2}{[H_2(g)][I_2(g)]}$$

We can substitute the values we know: $54 = \dfrac{0.85^2}{[H_2(g)][I_2(g)]}$

From the chemical equation 1 mol H_2 reacts with 1 mol I_2, and since the initial concentrations of these two species were equal, their concentrations must stay equal as they react together. This means that the equilibrium concentration of H_2 is the same as that of I_2 and we can write:

$$54 = \frac{0.85^2}{[H_2(g)]^2}$$

Rearranging this $[H_2(g)]^2 = \dfrac{0.85^2}{54}$

$$[H_2(g)] = 0.12 \, mol \, dm^{-3}$$

Therefore the equilibrium concentrations of H_2 and I_2 are both 0.12 mol dm^{-3}.

> Note: if we had the equation $2SO_2(g) + O_2(g) \rightleftharpoons 2SO_3(g)$ and started off with equal concentrations of SO_2 and O_2, the concentrations would not be equal at equilibrium because 2 mol SO_2 react with 1 mol O_2 and therefore, as the reaction proceeds to equilibrium, the concentration of SO_2 drops by twice as much as the concentration of O_2.

Next we will look at questions in which we need to find the value of the equilibrium constant but have not been given all the equilibrium concentrations.

Worked examples

7.4 5.00 mol H_2 and 3.00 mol I_2 are mixed together in a vessel of volume 10.0 dm^3 and allowed to come to equilibrium at 1100 K. At equilibrium there were 0.43 mol I_2 present in the reaction mixture. Calculate the value of the equilibrium constant.

We need to consider the initial situation and see how the concentrations change as the system reaches equilibrium.

$$H_2(g) + I_2(g) \rightleftharpoons 2HI(g)$$

	$H_2(g)$	$I_2(g)$	$2HI(g)$
Initial number of moles / mol	5.00	3.00	0.00
Equilibrium number of moles / mol	?	0.43	?

The numbers of moles at equilibrium can be worked out as follows. 3.00 mol I_2 were originally present, and this became 0.43 mol I_2 at equilibrium. This means that 3.00 − 0.43, i.e. 2.57, mol I_2 reacted. From the chemical equation we can see that 1 mol I_2 reacts with 1 mol H_2, so 2.57 mol I_2 reacts with 2.57 mol H_2. If there were 5.00 mol H_2 originally present and 2.57 mol H_2 react, this leaves 5.00 − 2.57, i.e. 2.43 mol H_2 present at equilibrium. From the chemical equation, 1 mol I_2 reacts to form 2 mol HI. Therefore, 2.57 mol I_2 will react to form 2 × 2.57, i.e. 5.14 mol HI.

This can be summarised:

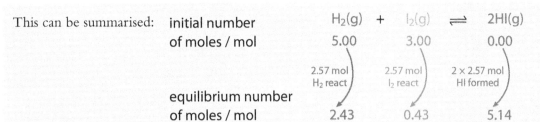

initial number of moles / mol

$H_2(g)$ + $I_2(g)$ $\rightleftharpoons$ $2HI(g)$

5.00   3.00   0.00

2.57 mol H_2 react   2.57 mol I_2 react   2 × 2.57 mol HI formed

equilibrium number of moles / mol   2.43   0.43   5.14

Now that we have the equilibrium number of moles, we must work out the equilibrium concentration of each species. To do this we use the equation:

$$\text{concentration} = \frac{\text{number of moles}}{\text{volume}}$$

> In this case, because there is the same number of molecules on both sides of the equation, all the concentrations will cancel, and so this step makes no difference to the overall answer.

Because the volume of the reaction vessel is $10.0\,dm^3$, we must divide each number of moles by 10.0, and therefore the equilibrium concentrations are:

$$H_2(g) + I_2(g) \rightleftharpoons 2HI(g)$$

equilibrium number of moles / mol	2.43	0.43	5.14
equilibrium concentration / mol dm^{-3}	0.243	0.043	0.514

The expression for the required constant is: $K_c = \dfrac{[HI(g)]^2}{[H_2(g)][I_2(g)]}$

The equilibrium concentrations are substituted into this equation:

$$K_c = \frac{0.514^2}{0.243 \times 0.043} = 25.3$$

The equilibrium constant for this reaction at $1100\,K$ is 25.3.

7.5 $3.00\,mol\ NO_2$ and $1.00\,mol\ N_2O_4$ are mixed together in a vessel of volume $1.00\,dm^3$ and allowed to come to equilibrium at $398\,K$. At equilibrium there were $1.74\,mol\ N_2O_4$ present in the reaction mixture. Calculate the value of the equilibrium constant. The equation for the reaction is: $2NO_2(g) \rightleftharpoons N_2O_4(g)$.

$$2NO_2(g) \rightleftharpoons N_2O_4(g)$$

Initial number of moles / mol	3.00	1.00
Equilibrium number of moles / mol	?	1.74

The equilibrium numbers of moles can be worked out as follows. $1.00\,mol\ N_2O_4$ was originally present, and this became $1.74\,mol\ N_2O_4$ at equilibrium. This means that $1.74-1.00$, i.e. 0.74, mol N_2O_4 were formed by NO_2 reacting. From the chemical equation we can see that $2\,mol\ NO_2$ reacts to form $1\,mol\ N_2O_4$. This means that 2×0.74, i.e. 1.48, mol NO_2 must have reacted to form $0.74\,mol\ N_2O_4$. There were $3.00\,mol\ NO_2$ originally present and $1.48\,mol\ NO_2$ reacted, so this leaves $3.00-1.48$, i.e. 1.52, mol NO_2 present at equilibrium.

$$2NO_2(g) \rightleftharpoons N_2O_4(g)$$

Initial number of moles / mol	3.00	1.00
Equilibrium number of moles / mol	1.52	1.74

Equilibrium concentrations are worked out using the equation:

$$\text{concentration} = \frac{\text{number of moles}}{\text{volume}}$$

The volume of the reaction vessel is $1.00\,dm^3$, so we must simply divide the numbers of moles by 1 and therefore the equilibrium concentrations are:

$$2NO_2(g) \rightleftharpoons N_2O_4(g)$$

equilibrium concentration / $mol\,dm^{-3}$ 1.52   1.74

The expression for K_c is:

$$K_c = \frac{[N_2O_4(g)]}{[NO_2(g)]^2}$$

The equilibrium concentrations are substituted into this equation:

$$K_c = \frac{1.74}{1.52^2} = 0.753$$

So the equilibrium constant for this reaction at $398\,K$ is 0.753.

Given the initial number of moles and the equilibrium constant it is also possible to work out the number of moles at equilibrium.

Worked example

7.6 Consider the reaction $H_2(g) + CO_2(g) \rightleftharpoons H_2O(g) + CO(g)$. In this reaction, $2.00\,mol\ H_2$ and $2.00\,mol\ CO_2$ are put into a container of volume $10.0\,dm^3$ together with $1.00\,mol\ H_2O$ and $1.00\,mol\ CO$. They are allowed to come to equilibrium at $1200\,K$. Given that the value of the equilibrium constant at $1200\,K$ is 2.10, work out the composition of the equilibrium mixture in terms of concentrations.

	$H_2(g)$	+	$CO_2(g)$	$\rightleftharpoons$	$H_2O(g)$	+	$CO(g)$
Initial number of moles / mol	2.00		2.00		1.00		1.00

We will assume that x mol H_2 react with x mol CO_2 to form x mol H_2O and x mol CO.

Equilibrium number of moles / mol $2.00 - x$ $2.00 - x$ $1.00 + x$ $1.00 + x$

We must divide each of these values by 10.0 to get equilibrium concentrations:

equilibrium concentration / $mol\,dm^{-3}$ $\dfrac{2.00-x}{10.0}$ $\dfrac{2.00-x}{10.0}$ $\dfrac{1.00+x}{10.0}$ $\dfrac{1.00+x}{10.0}$

The equilibrium concentrations and the value of K_c are substituted into the expression for K_c:

$$K_c = \frac{[H_2O(g)][CO(g)]}{[H_2(g)][CO_2(g)]}$$

$$2.10 = \frac{\dfrac{1.00+x}{10.0} \times \dfrac{1.00+x}{10.0}}{\dfrac{2.00-x}{10.0} \times \dfrac{2.00-x}{10.0}}$$

> The number of species on both sides of the equation is the same so the volumes cancel.

$$2.10 = \frac{(1.00+x)(1.00+x)}{(2.00-x)(2.00-x)}$$

$$2.10 = \frac{(1.00+x)^2}{(2.00-x)^2}$$

The right-hand side is a perfect square, and so the square root of each side can be taken:

$$\sqrt{2.10} = \frac{(1.00 + x)}{(2.00 - x)}$$

$$1.45 = \frac{(1.00 + x)}{(2.00 - x)}$$

Exam tip
It is stated specifically on the IB syllabus that you should not have to solve quadratic equations, but this way of finding the square root of each side seems to be acceptable – remember this method.

This can be rearranged to give:

$$1.45(2.00 - x) = 1.00 + x$$
$$2.90 - 1.45x = 1.00 + x$$
$$1.90 = 2.45x$$

Therefore, $x = 0.776\,\text{mol}$.

The equilibrium concentrations are worked out from:

	$H_2(g)$	+	$CO_2(g)$	⇌	$H_2O(g)$	+	$CO(g)$
equilibrium concentration / mol dm^{-3}	$\dfrac{2.00 - x}{10.0}$		$\dfrac{2.00 - x}{10.0}$		$\dfrac{1.00 + x}{10.0}$		$\dfrac{1.00 + x}{10.0}$

Substituting in the value of x we get:

equilibrium concentration / mol dm^{-3}	0.122		0.122		0.178		0.178

Explaining the effect of changes in concentration on the position of equilibrium

> The value of the equilibrium constant is not affected by a change in concentration.

Let us consider the equilibrium:

$$CH_3COOH(l) + C_2H_5OH(l) \rightleftharpoons CH_3COOCH_2CH_3(l) + H_2O(l)$$
ethanoic acid   ethanol   ethyl ethanoate   water

At 373 K, the equilibrium constant is 4.0. An equilibrium mixture at this temperature could be:

	$CH_3COOH(l)$ +	$C_2H_5OH(l)$ ⇌	$CH_3COOCH_2CH_3(l)$ +	$H_2O(l)$
Equilibrium concentrations / mol dm^{-3}	2.88	2.88	5.76	5.76

If some water is now added so that the concentration of water in the mixture changes to 10.40 mol dm^{-3}, the concentrations of the other species in the mixture will drop slightly because the total volume of the reaction mixture has increased.

The new values for the concentrations are:

	$CH_3COOH(l)$ +	$C_2H_5OH(l)$ ⇌	$CH_3COOCH_2CH_3(l)$ +	$H_2O(l)$
Concentrations / mol dm^{-3}	2.61	2.61	5.22	10.40

Working out the reaction quotient:

$$Q = \frac{[CH_3COOCH_2CH_3(l)][H_2O(l)]}{[CH_3COOH(l)][C_2H_5OH(l)]}$$

$$= \frac{5.22 \times 10.40}{2.61 \times 2.61}$$

$$= 8.0$$

This value of Q is higher than the equilibrium constant (4.0) at this temperature and so the system is not at equilibrium. The value of Q must decrease to bring the value back to 4.0 – so ethyl ethanoate must react with water (these are on the top of the reaction quotient expression). The reverse reaction is favoured over the forward reaction until a new equilibrium is established:

$$CH_3COOH(l) + C_2H_5OH(l) \rightleftharpoons CH_3COOCH_2CH_3(l) + H_2O(l)$$

Equilibrium concentrations / $mol\,dm^{-3}$ 3.31   3.31   4.52   9.70

The new reaction quotient is:

$$Q = \frac{4.52 \times 9.70}{3.31 \times 3.31}$$

$$= 4.0$$

Q is now equal to K_c at this temperature and so the system is at equilibrium. The concentrations of ethanoic acid and ethanol are higher than in the original equilibrium mixture. So the position of equilibrium has shifted to the left to use up the added water, which is what we would have predicted using Le Chatelier's principle.

Similar arguments can be used to explain the effect of changes in pressure on the position of equilibrium.

We can see from the above discussion that, if a system at equilibrium is subjected to a change in concentration (or pressure), the system will no longer be at equilibrium ($Q \neq K_c$). The reaction will then proceed in one direction or the other until the value of $Q = K_c$ and a new position of equilibrium is reached.

This means that we can re-state Le Chatelier's principle for changes in concentration (or pressure) as 'if a system at equilibrium is subjected to a change in concentration (or pressure), the position of equilibrium will shift so that the value of the reaction quotient, Q, is restored to the value of K_c again'.

Nature of science

Scientists often employ quantitative reasoning. For a simple reaction such as $A \rightleftharpoons B$ the rate of the forward and reverse reactions could be determined as $Rate_{forward} = k_f[A]$ and $Rate_{reverse} = k_r[B]$. At equilibrium we get $k_f[A] = k_r[B]$. If some B is added to the equilibrium mixture then the rate of the reverse reaction is increased and some of the B is used up, which is in accordance with Le Chatelier's principle.

8 Given the equilibrium concentrations, calculate the value of the equilibrium constant for each reaction:

a $2A(g) \rightleftharpoons Z(g)$

equilibrium concentrations are:

[A] $0.150 \, mol \, dm^{-3}$ [Z] $0.500 \, mol \, dm^{-3}$

b $2Q(g) + 3X(g) \rightleftharpoons 2Z(g) + 4E(g)$

equilibrium concentrations are:

[Q] $1.50 \times 10^{-3} \, mol \, dm^{-3}$

[X] $2.75 \times 10^{-3} \, mol \, dm^{-3}$

[Z] $7.86 \times 10^{-4} \, mol \, dm^{-3}$

[E] $9.37 \times 10^{-5} \, mol \, dm^{-3}$

9 In each of the following reactions use the given data to calculate the equilibrium number of moles of each substance.

a $2A(g) \rightleftharpoons Z(g)$

initial moles of A = 0.100 mol

initial moles of Z = 0.500 mol

moles of A at equilibrium = 0.0600 mol

b $2A(g) + X(g) \rightleftharpoons 2Z(g)$

initial moles of A = 1.00 mol

initial moles of X = 2.00 mol

initial moles of Z = 0.00 mol

moles of Z at equilibrium = 0.400 mol

10 Consider the reversible reaction:

$A(g) + X(g) \rightleftharpoons Q(g)$

a Write the expression for K_c for this reaction.

b In each of the following situations calculate a value for the equilibrium constant from the data given:

  i initial moles of A = 0.200 mol

    initial moles of X = 0.400 mol

    initial moles of Q = 0.000 mol

    moles of A at equilibrium = 0.100 mol

    volume of container = $1.00 \, dm^3$

    temperature = 300 K

  ii initial moles of A = 0.200 mol

    initial moles of X = 0.100 mol

    initial moles of Q = 0.100 mol

    moles of A at equilibrium = 0.150 mol

    volume of container = $10.0 \, dm^3$

    temperature = 400 K

c Use the values of K_c that you have calculated to work out whether the forward reaction is exothermic or endothermic.

11 Consider the reversible reaction:

$2A(g) + X(g) \rightleftharpoons 4Q(g) + Z(g)$

a Write the expression for K_c for this reaction.

b In each of the following situations calculate a value for the equilibrium constant from the data given:

  i initial moles of A = 0.800 mol

    initial moles of X = 0.400 mol

    initial moles of Q = 0.000 mol

    initial moles of Z = 0.000 mol

    moles of Z at equilibrium = 0.100 mol

    volume of container = $10.0 \, dm^3$

    temperature = 800 K

  ii initial moles of A = 0.000 mol

    initial moles of X = 0.000 mol

    initial moles of Q = 1.000 mol

    initial moles of Z = 2.000 mol

    moles of A at equilibrium = 0.200 mol

    volume of container = $20.0 \, dm^3$

    temperature = 500 K

c Use the values of K_c that you have calculated to work out whether the forward reaction is exothermic or endothermic.

12 Consider the equilibrium:

$A(g) + Z(g) \rightleftharpoons X(g) + Q(g)$

a At a certain temperature, the value of the equilibrium constant for this reaction is 9.00. At this temperature 0.100 mol A and 0.100 mol Z are placed in a container of volume $1.00 \, dm^3$ and allowed to come to equilibrium. Calculate the number of moles of X present at equilibrium.

b At a different temperature, the value of the equilibrium constant for this reaction is 16.0. At this temperature 0.200 mol A, 0.200 mol Z, 0.200 mol X and 0.200 mol Q are placed in a container of volume $2.00 \, dm^3$ and allowed to come to equilibrium. Calculate the concentration of A at equilibrium.

The position of equilibrium corresponds to the mixture of reactants and products that produces the minimum value of the Gibbs free energy and the maximum value of entropy.

When $K < 1$ the value of $\ln K$ is negative and so ΔG is positive. When $K > 1$ the value of $\ln K$ is positive and so ΔG is negative.

The relationship between ΔG and the position of equilibrium is discussed on pages **232–233**.

7.2.2 The relationship between equilibrium constants and Gibbs free energy

An equilibrium constant is related to the change in Gibbs free energy by the equation:

$$\Delta G = -RT \ln K$$

where R is the gas constant and T is the temperature in kelvin. The equilibrium constant in this equation has no subscript because there is more than one type of equilibrium constant and which one is obtained depends on the reaction involved. If this equation is used to determine the value for an equilibrium constant for a reaction in solution then the equilibrium constant obtained will be K_c, which we are familiar with from work above. However, for reactions involving gases, the equilibrium constant obtained is K_p, which is the equilibrium constant in terms of partial pressures.

Partial pressure refers to the pressure exerted by a particular gas in a mixture of gases. If the gases behave ideally the partial pressure is the same as the pressure that the same amount of a particular gas would exert if it were in the container by itself. If the pressure exerted by a mixture of 80% nitrogen and 20% hydrogen is $100\,kPa$, the partial pressure of nitrogen is $80\,kPa$ and that of hydrogen is $20\,kPa$.

The expression for K_p is the same as that for K_c except concentrations are replaced by partial pressures. For the reaction

$$N_2(g) + 3H_2(g) \rightleftharpoons 2NH_3(g)$$

the expression for K_p is

$$K_p = \frac{(P_{NH_3})^2}{(P_{N_2})(P_{H_2})^3}$$

where P_{N_2} represents the partial pressure of nitrogen at equilibrium.

The relationship between a partial pressure and a concentration can be worked out using $PV = nRT$. Concentration is number of moles per unit volume (n/V), therefore

$$P = \frac{n}{V} \times RT$$

or, partial pressure of a gas $=$ [concentration] $\times RT$

It can then be worked out that $K_p = K_c \times (RT)^{\Delta n}$ where Δn is the number of molecules of gas on the right hand side of an equation – the number of molecules of gas on the left hand side. For the reaction given above $\Delta n = -2$ and $K_p \neq K_c$ but for
$H_2(g) + I_2(g) \rightleftharpoons 2HI(g)$ $\Delta n = 0$ and $K_p = K_c$.

Worked example

7.7 Consider the reaction $H_2(g) + I_2(g) \rightleftharpoons 2HI(g)$

Given that the value of $\Delta G^{\ominus}$ at 298 K for this reaction is $+1.3\,kJ\,mol^{-1}$, calculate the value of the equilibrium constant.

$\Delta G = -RT \ln K$

Because R has units of $J\,K^{-1}\,mol^{-1}$ the value of ΔG must be converted into $J\,mol^{-1}$

$\Delta G = 1300\,J\,mol^{-1}$

Substituting in values we get:

$1300 = -8.31 \times 298 \times \ln K$

Rearranging:

$$\ln K = \frac{-1300}{8.31 \times 298}$$

$$= -0.525$$

The inverse function for $\ln x$ is e^x, therefore both sides need to be raised as powers of e

$K = e^{-0.525}$

$K = 0.59$

Exam tip
When doing this calculation the other way round, that is working out ΔG given a value for K, the units of ΔG will come out as $J\,mol^{-1}$.

Use the key combination shift/2nd function + ln on your calculator.

The equilibrium constant has a value of 0.59, which is less than 1, which implies that the position of equilibrium lies closer to reactants than products. This is also consistent with the value of ΔG being positive.

The relationship between ΔG and the value of K is summarised in Table **7.5**.

ΔG	K	Position of equilibrium
negative	>1	closer to products than reactants
positive	<1	closer to reactants than products

Table 7.5 The relationship between ΔG and K_c.

? Test yourself

The value of the gas constant is $R = 8.31\,J\,K^{-1}\,mol^{-1}$

13 Calculate the value of ΔG (in $kJ\,mol^{-1}$) given the following values:

    a $K = 2.78$ $T = 298\,K$

    b $K = 3.58 \times 10^5$ $T = 333\,K$

    c $K = 7.56 \times 10^{-4}$ $T = 455\,^{\circ}C$

14 Calculate the value of K given the following values:

    a $\Delta G = 50.0\,kJ\,mol^{-1}$ $T = 298\,K$

    b $\Delta G = -25.2\,kJ\,mol^{-1}$ $T = 350\,K$

    c $\Delta G = -154\,kJ\,mol^{-1}$ $T = 27\,^{\circ}C$

Nature of science

There are ethical and political sides to science and the story of Fritz Haber (Figure **7.11**) illustrates some aspects of this.

One of the most important industrial processes in the world is the **Haber (Haber–Bosch) process** for the production of ammonia. Fritz Haber was born in 1868 in Silesia, which was then part of Germany but is now part of Poland. His parents were Jewish, but Haber later renounced his religion in favour of Christianity because the opportunities available to Jews in Germany in the late 19th century were limited. He is most famous for developing a process for 'fixing' nitrogen – converting nitrogen in the air into a much more reactive chemical – ammonia.

Nitrogen is an important element for plant growth and, because most plants are unable to 'fix' nitrogen from the air, they rely on absorbing it in some soluble form from the soil. At the beginning of the 20th century, scientists were struggling to come up with a method of producing ammonia and Germany, along with other countries, were reliant on imported guano as a fertiliser.

It was Haber and his co-workers who came up with a solution. Their process was scaled up to industrial level by Carl Bosch of BASF, and the first industrial ammonia production plant opened in Germany in 1913. Ammonia can be further converted into fertilisers such as ammonium nitrate. Without artificial fertilisers it is likely that, in the last hundred years, millions of people around the world would have died of starvation. For his work on this process, which was regarded as benefiting the 'whole of humanity', Haber was awarded the Nobel Prize for Chemistry in 1918.

There is, however, another side to Fritz Haber and ammonia. As well as being the basis of artificial fertilisers, ammonia can be used to make explosives. Without Haber's process, Germany would have very quickly run out of explosives in World War I, the war would almost certainly have been shorter and millions of lives might been saved. To what extent is Haber personally responsible for lengthening the war? How far does the responsibility of a scientist stretch? Can he or she be expected to foresee all possible uses of their discoveries?

Haber also developed and supervised the use of chemical weapons, such as chlorine gas, in World War I and could perhaps be regarded as the founder of chemical warfare. He was a staunch patriot who fully supported the German war effort, but he did not make the ultimate decision to use these weapons. How far was Haber responsible for the deaths of people through the use of poison gas? Is he more responsible for this than for the deaths that arose due to the lengthening of the war through the use of ammonia to make explosives?

After World War I, Germany had to pay huge sums of money in war reparations and Haber yet again tried to use chemistry to save his country. He investigated processes for extracting gold from sea water, but eventually had to abandon the project when it became apparent that the concentration of gold in sea water was substantially lower than had been originally estimated.

Figure 7.11 Fritz Haber

When the Nazis took control of Germany in 1933, laws were passed to prevent Jews from holding positions in universities. Although Haber could have remained in his position as head of the Kaiser Wilhelm Institute, he resigned his post. Shortly afterwards he left Germany and died a few months later in Switzerland. It is a sad postscript to the life of a talented chemist that Zyklon B, which was developed as an insecticide at Haber's institute, was used in the gas chambers of the Nazi concentration camps in World War II. Could Haber be held in any way responsible for this?

Exam-style questions

1   Which of the following is a property of a system at equilibrium?

    A The concentrations of reactants and products are equal.
    B The rate of the forward reaction is zero.
    C The rate of the forward reaction is equal to the rate of the reverse reaction.
    D The rate of the reverse reaction is a maximum.

2   Consider the reaction:

$$CO_2(g) + 4H_2(g) \rightleftharpoons CH_4(g) + 2H_2O(g)$$

What is the expression for the equilibrium constant?

  A $K_c = \dfrac{[CO_2][H_2]}{[CH_4][H_2O]}$ **C** $K_c = \dfrac{[CH_4][2H_2O]}{[CO_2][4H_2]}$

  B $K_c = \dfrac{[CH_4][H_2O]^2}{[CO_2][H_2]^4}$ **D** $K_c = \dfrac{[CH_4] + [H_2O]^2}{[CO_2] + [H_2]^4}$

3   In which system will decreasing the pressure increase the equilibrium concentration of the species in bold?

    A $CH_4(g) + 2H_2O(g) \rightleftharpoons CO_2(g) + 4\mathbf{H_2(g)}$
    B $N_2O_5(g) + \mathbf{NO(g)} \rightleftharpoons 3NO_2(g)$
    C $H_2(g) + I_2(g) \rightleftharpoons 2\mathbf{HI(g)}$
    D $NO(g) + NO_2(g) \rightleftharpoons \mathbf{N_2O_3(g)}$

4   What is the effect of increasing the temperature on the position of equilibrium and the value of the equilibrium constant for the reaction:

$$CO(g) + 3H_2(g) \rightleftharpoons CH_4(g) + H_2O(g) \qquad\qquad \Delta H = -206 \, kJ \, mol^{-1}$$

	Position of equilibrium	Value of equilibrium constant
A	shifts to left	decreases
B	shifts to right	decreases
C	shifts to right	increases
D	shifts to left	increases

5 Consider the system: $A(g) \rightleftharpoons B(g)$

What is the effect of introducing a catalyst?

    A The rate of the forward reaction increases but the rate of the reverse reaction remains the same.
    B The position of equilibrium is shifted to the right.
    C The time taken to reach equilibrium is decreased.
    D The value of the equilibrium constant is reduced.

6 Which of the following changes will increase the equilibrium yield of ammonia according to the following equation?

$$N_2(g) + 3H_2(g) \rightleftharpoons 2NH_3(g) \qquad\qquad \Delta H = -92\,kJ\,mol^{-1}$$

    A decreasing the temperature from $200\,°C$ to $100\,°C$
    B introducing a catalyst
    C decreasing the pressure from $200\,atm$ to $100\,atm$
    D increasing the temperature from $100\,°C$ to $200\,°C$

HL 7 The equilibrium concentrations at a certain temperature for the reaction

$$2SO_2(g) + O_2(g) \rightleftharpoons 2SO_3(g)$$

are $0.20\,mol\,dm^{-3}$ for $[SO_2]$, $0.40\,mol\,dm^{-3}$ for $[O_2]$, $0.80\,mol\,dm^{-3}$ for $[SO_3]$.

The value of the equilibrium constant at this temperature is:

    A 0.025   **B** 2   **C** 10   **D** 40

HL 8 Consider the equilibrium:

$$A(g) + X(g) \rightleftharpoons 2Z(g)$$

$0.20\,mol$ Z are placed in a sealed $10\,dm^3$ container and allowed to come to equilibrium at $500\,K$. At equilibrium, $0.040\,mol$ X were present in the reaction vessel. The value of the equilibrium constant at this temperature is:

    A 5   **B** 9   **C** 16   **D** 25

HL 9 Consider the reversible reaction $N_2(g) + 3H_2(g) \rightleftharpoons 2NH_3(g)$.

The value of the equilibrium constant at certain temperature, T, is 2.0.
Given the following concentrations in the reaction mixture:

Reactant	Concentration / $mol\,dm^{-3}$
$N_2(g)$	1.0
$H_2(g)$	2.0
$NH_3(g)$	4.0

which of the following is correct?

    A $Q=0.50$ and the system is not at equilibrium
    B $Q=1.3$ and the system is not at equilibrium
    C $Q=2.0$ and the system is at equilibrium
    D $Q=2.0$ and the system is not at equilibrium

10 Which of the following describes the relationship between the change in Gibbs free energy, ΔG, and the equilibrium constant, K?

    A If $K>1$ then $\Delta G>0$.
    B If $K=0$ then $\Delta G=0$.
    C If $K<0$ then $\Delta G<0$.
    D If $K>1$ then $\Delta G<0$.

11 **a** Explain what is meant by 'dynamic equilibrium'. **[2]**

    b Consider the following system at equilibrium:

$$H_2(g) + CO_2(g) \rightleftharpoons H_2O(g) + CO(g) \qquad\qquad \Delta H = +41\,kJ\,mol^{-1}$$

    State and explain the effect of the following changes on the position of equilibrium and the value of the equilibrium constant.

      i Increasing the temperature at constant pressure. **[3]**
      ii Increasing the pressure at constant temperature. **[2]**
      iii Introducing a catalyst at constant temperature and pressure. **[2]**

12 Consider the reaction:

$$N_2(g) + O_2(g) \rightleftharpoons 2NO(g)$$

    a Write an expression for the equilibrium constant for this reaction. **[1]**

    b The value of the equilibrium constant for this reaction at $700\,K$ is 5×10^{-13}, but the value at 1100 K is 4×10^{-8}. Use these values to state and explain whether this reaction is exothermic or endothermic in the forward direction as written. **[3]**

13 Consider the reversible reaction between colourless N_2O_4 and brown NO_2:

$$N_2O_4(g) \rightleftharpoons 2NO_2(g) \qquad \Delta H^{\ominus} = +58\,kJ\,mol^{-1}$$

    a Write an expression for the equilibrium constant for this reaction. **[1]**

    b State and explain how the colour of the mixture will change as the temperature is increased. **[3]**

    c State and explain the effect of increasing the pressure on the colour of the equilibrium mixture and the value of the equilibrium constant. **[4]**

    HL   **d** $0.200\,mol\,N_2O_4$ is placed in a container of volume $4.00\,dm^3$ at a certain temperature and allowed to come to equilibrium. At equilibrium there were $0.140\,mol\,N_2O_4$ present. Calculate the equilibrium concentrations of N_2O_4 and NO_2 and hence the value of the equilibrium constant at this temperature. **[4]**

HL 14 The following reaction is an important stage in the Contact process for the production of sulfuric acid:

$$2SO_2(g) + O_2(g) \rightleftharpoons 2SO_3(g)$$

a 0.120 mol SO_2 and 0.120 mol O_2 are introduced into a reaction vessel of volume 2.00 dm^3 and allowed to come to equilibrium at 1100 K. At equilibrium there were 0.060 mol SO_3 present.
   i Write an expression for the equilibrium constant for this reaction.   **[1]**
   ii Calculate the concentration of O_2 at equilibrium.   **[2]**
   iii Calculate a value for the equilibrium constant for this reaction at 1100 K.   **[3]**

b The standard enthalpy changes of formation of $SO_2(g)$ and $SO_3(g)$ are -297 kJ mol^{-1} and -395 kJ mol^{-1}, respectively.
   i Calculate a value for the enthalpy change for the conversion of two moles of SO_2 to two moles of SO_3.   **[2]**
   ii State and explain whether the value of the equilibrium constant for this reaction at 1500 K will be higher or lower than the value you found in **a iii**.   **[2]**

15 Consider the reaction $H_2(g) + I_2(g) \rightleftharpoons 2HI(g)$

a Write an expression for the equilibrium constant for this reaction.   **[1]**

b The value of the equilibrium constant for this reaction at 700 K is 54.0. What is the value of the equilibrium constant for the following reaction?

$$2HI(g) \rightleftharpoons H_2(g) + I_2(g)$$ **[1]**

HL c 2.00 mol of HI are introduced into a closed reaction vessel of volume 4.00 dm^3 and allowed to come to equilibrium at 700 K. Calculate the equilibrium concentration of H_2.   **[3]**

Summary

EQUILIBRIUM —— In chemistry, equilibria are always **dynamic**.

—— *characteristics of chemical equilibria*

- rate of forward reaction = rate of reverse reaction
- concentrations of reactants and products are constant
- all species in equation present at equilibrium
- can be attained from either direction
- can only be attained in a closed system

in which there is no exchange of matter with the surroundings

A **phase equilibrium** involves a change of state rather than a chemical change. —— e.g. rate of vaporisation = rate of condensation

position of equilibrium – relative amount of reactants and products —— not affected by catalyst – this increases rate of forward and reverse reactions equally

Le Chatelier's principle – if a system at equilibrium is subjected to change, it will shift in order to minimise the effect of that change. —— *factors that can be changed*

- If **temperature** is increased, position of equilibrium will shift in endothermic direction.
- If **pressure** is increased, position of equilibrium will shift to side with fewer gas molecules.
- If **concentration** of a species is increased, position of equilibrium will shift in the direction away from that species.

equilibrium constant (K_c) for
$aA + bB \rightleftharpoons cC + dD$

$$K_c = \frac{[C]^c[D]^d}{[A]^a[B]^b}$$

(concentrations at equilibrium)

—— provides information about how far a reaction proceeds at a given temperature

If $K_c \gg 1$, the reaction proceeds almost completely towards the products.

If $K_c \ll 1$, the reaction proceeds hardly at all towards the products.

value only affected by change in temperature

For an exothermic reaction, K_c decreases as the temperature increases.

Reaction quotient, $Q = \dfrac{[C]^c[D]^d}{[A]^a[B]^b}$

(concentration not equilibrium concentrations)

If $Q = K_c$ the system is at equilibrium.
If $Q \neq K_c$ the system is not at equilibrium.

If $Q < K_c$ the reaction must proceed towards the products (to the right) to reach equilibrium.

HL

$\Delta G = -RT\ln K$
The position of equilibrium corresponds to the mixture of reactants and products that produces the minimum value of the Gibbs free energy and the maximum value of entropy.

8 Acids and bases

Figure 8.1 These all contain acids. Cola is a solution of phosphoric acid, among other things.

Acid	Formula
hydrochloric	HCl
sulfuric	H_2SO_4
nitric	HNO_3
carbonic	H_2CO_3
ethanoic	CH_3COOH
benzoic	C_6H_5COOH

Table 8.1 The formulas for some common acids, with the H that is lost as H^+ shown in red.

It was originally thought that all acids contain oxygen, and the names of this element in English, German (Sauerstoff) and other languages reflect this mistaken assumption.

Conjugate acid–base pairs always differ by one proton (H^+).

8.1 Theories of acids and bases

Acids and bases are substances that are familiar, both in the laboratory and in everyday life. A few everyday acids are shown in Figure **8.1**. Examples of acids include hydrochloric acid (HCl) and sulfuric acid (H_2SO_4), and bases include sodium hydroxide (NaOH) and ammonia (NH_3).

If we look at the formulas of some acids (Table **8.1**), we can see that all acids contain at least one hydrogen atom. It is this hydrogen atom that leads to the characteristic properties of an acid. In the simplest definition of acids and bases, an acid is defined as a substance that produces hydrogen ions in solution, and a base is a substance that produces hydroxide ions in solution.

Brønsted–Lowry acids and bases

The **Brønsted–Lowry** definition is:
- an **acid** is a proton (H^+) **donor**
- a **base** / alkali is a proton (H^+) **acceptor**

Consider the reaction of ethanoic acid with water (Figure **8.2**). In the forward direction (Figure **8.2a**), the CH_3COOH donates a proton (H^+) to the H_2O – so the CH_3COOH is an acid, because it donates a proton. The H_2O accepts a proton and therefore acts as a base. In the reverse direction (Figure **8.2b**), H_3O^+ donates a proton to CH_3COO^- – so H_3O^+ is an acid and CH_3COO^- is a base.

a $CH_3COOH(aq) + H_2O(l) \longrightarrow CH_3COO^-(aq) + H_3O^+(aq)$
H^+ donated

b $CH_3COOH(aq) + H_2O(l) \longleftarrow CH_3COO^-(aq) + H_3O^+(aq)$
H^+ donated

Figure 8.2 The reaction of ethanoic acid with water. **a** In the forward direction ethanoic acid acts as an acid because it donates a proton to water; **b** in the reverse direction, H_3O^+ acts as an acid because it donates a proton to CH_3COO^-.

When CH_3COOH acts as an acid and donates a proton, it forms a base, CH_3COO^-. CH_3COO^- is called the **conjugate base** of CH_3COOH. CH_3COOH and CH_3COO^- are called a **conjugate acid–base pair**. Similarly, when H_2O acts as a base and accepts a proton, it forms H_3O^+, which acts as an acid in the reverse direction. H_3O^+ is the **conjugate acid** of H_2O, and H_3O^+ and H_2O are also a conjugate acid–base pair.

We can label the conjugate acid–base pairs in the equation:

$$CH_3COOH(aq) + H_2O(l) \rightleftharpoons CH_3COO^-(aq) + H_3O^+(aq)$$
acid 1   base 2   base 1   acid 2

Species with the same number form a conjugate pair.

Let us look at another example:

$$NH_3(aq) + H_2O(l) \rightleftharpoons NH_4^+(aq) + OH^-(aq)$$
$$\text{base 1} \quad\quad \text{acid 2} \quad\quad\quad \text{acid 1} \quad\quad\quad \text{base 2}$$

In the forward direction: H_2O donates a proton to NH_3. H_2O therefore acts as an acid and as NH_3 accepts the proton it acts as a base.

When H_2O donates the proton, it forms OH^-. OH^- is the conjugate base of H_2O, and H_2O and OH^- are a conjugate acid–base pair. We could also say that H_2O is the conjugate acid of OH^-.

In the reverse direction: NH_4^+ donates a proton to OH^-. NH_4^+ therefore acts as an acid, and as OH^- accepts the proton, it acts as a base.

When NH_3 accepts a proton it forms NH_4^+. NH_4^+ is the conjugate acid of NH_3, and NH_3 and NH_4^+ are a conjugate acid–base pair. We could also say that NH_3 is the conjugate base of NH_4^+.

Amphiprotic and Amphoteric

These two terms are used to describe a substance can act as both an acid and a base. Amphiprotic refers to the Bronsted-Lowry definition of acids and bases and indicates a species that can donate (acting as an acid) or accept (acting as a base) a proton. Water is a substance that is amphiprotic and we have seen above that it acted as a proton donor (to form OH^-) in its reaction with NH_3 but as a proton acceptor (to form H_3O^+) in its reaction with CH_3COOH.

Amphoteric is a more general term and refers to a substance that can act as an acid and a base – all amphiprotic substances are also amphoteric but not all amphoteric substances are amphiprotic. The difference arises because there is another, more general, definition of acids and bases (the Lewis definition), which does not require the transfer of a proton. The most commonly encountered substances that are amphoteric but not amphiprotic are hydroxides of some metals such as aluminium or zinc.

$$Al(OH)_3(s) + 3H^+(aq) \rightarrow Al^{3+}(aq) + 3H_2O(l) \quad\quad \textbf{equation 1}$$

$$Al(OH)_3(s) + OH^-(aq) \rightarrow Al(OH)_4^-(aq) \quad\quad \textbf{equation 2}$$

Equation 1 shows aluminium hydroxide reacting with an acid – acting as a base and equation 2 shows it reacting with a base and therefore acting as an acid.

Nature of science

Many common acids (HNO_3, H_2SO_4 etc.) contain oxygen and in the late 18th and early 19th centuries it was thought that acids are substances that contain oxygen. However, the discovery, by Humphry Davy, that muriatic acid (hydrochloric acid) does not contain oxygen paved the way for new theories, based on hydrogen, about the nature of acids.

The hydrated proton may be written as $H^+(aq)$ or $H_3O^+(aq)$. H_3O^+ (called the hydronium ion, hydroxonium ion or oxonium ion) has the structure shown in Figure **8.3**. It is essentially just an aqueous hydrogen ion (proton) and is often just written as $H^+(aq)$.

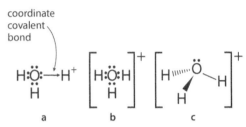

Figure 8.3 Three different representations of the structure of H_3O^+: **a** and **b** the electronic structure; **c** the trigonal pyramidal shape.

Another amphiprotic species, HCO_3^-, is discussed on page **321**.

1 Define acids and bases according to the Brønsted–Lowry definition.

2 Write an equation for the reaction between ammonia and ethanoic acid, classifying each species as either an acid or base and indicating the conjugate acid-base pairs.

3 Give the formula of the conjugate acid of each of the following:

  a NH_3 d CN^-

  b OH^- e HPO_4^{2-}

  c HSO_4^-

4 Give the formula of the conjugate base of each of the following:

  a HCO_3^- c $HCOOH$

  b H_2O d NH_3

5 In each of the following reactions state whether the species in **bold** is acting as an acid or a base according to the Brønsted–Lowry definition:

  a $\mathbf{HSO_4^-}(aq) + H_2O(l) \rightleftharpoons SO_4^{2-}(aq) + H_3O^+(aq)$

  b $\mathbf{HCO_3^-}(aq) + HSO_4^-(aq)$
                  $\rightleftharpoons H_2CO_3(aq) + SO_4^{2-}(aq)$

  c $\mathbf{CH_3COOH}(aq) + H_2SO_4(aq)$
                  $\rightleftharpoons CH_3COOH_2^+(aq) + HSO_4^-(aq)$

Learning objectives

- Understand the Lewis definition of acids and bases
- Recognise Lewis acids and bases

8.2 Lewis acids and bases (HL)

The Brønsted–Lowry definition of acids and bases is only one of several definitions of acids and bases – we will now consider the Lewis definition.

Brønsted–Lowry
- An **acid** is a proton (H^+) **donor**.
- A **base** / alkali is a proton (H^+) **acceptor**.

Lewis
- An **ACid** is an electron **pair ACceptor**.
- A **base** is an electron **pair donor**.

The Lewis definition is more general than the Brønsted–Lowry definition and can be applied to reactions that do not involve the transfer of protons. It is also useful when considering acid–base reactions in solvents other than water.

The Lewis definition of acids and bases covers all the Brønsted–Lowry reactions, because the acceptance of a proton by a base must involve the donation of an electron pair to the proton. Consider the protonation of ammonia (Figure **8.4**):

$$NH_3 + H^+ \rightleftharpoons NH_4^+$$

NH_3 is the Lewis base, because it donates an electron pair to H^+, which is the electron pair acceptor, i.e. Lewis acid.

The reaction between BF_3 and NH_3 (Figure **8.5**) is a Lewis acid–base reaction that does not involve the transfer of a proton:

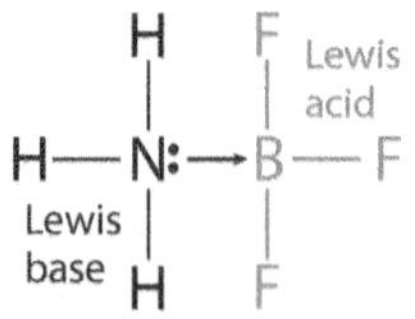

$$NH_3 + BF_3 \rightleftharpoons H_3N{:}BF_3$$

A coordinate (dative) covalent bond is always formed in a Lewis acid–base reaction.

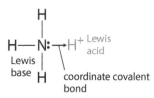

Figure 8.4 Protonation of ammonia.

Figure 8.5 An adduct is formed when NH_3 and BF_3 react together.

The formation of a complex ion by a transition metal ion (Figure **8.6**) is another example of a Lewis acid–base reaction. The transition metal ion is the Lewis acid and the ligand is the Lewis base. The ligands bond to the transition metal ion through the formation of coordinate covalent bonds. For example:

$$Fe^{2+} + 6H_2O \rightleftharpoons [Fe(H_2O)_6]^{2+}$$

The last two reactions would not be described as acid–base reactions according to the Brønsted–Lowry theory.

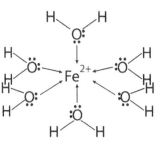

Figure 8.6 A transition metal complex ion.

> For a substance to act as a Lewis base, it must have a lone pair of electrons. For a substance to act as a Lewis acid, it must have space to accept a pair of electrons in its outer shell.

For example, NH_3 and H_2O can both act as Lewis bases, because they have a lone pair of electrons – H^+ and BX_3 (where X is a halogen) can act as Lewis acids because they have space in their outer shells to accept a pair of electrons.

The Lewis definition is used sometimes to describe reactions in organic chemistry – this will be discussed in Topic **10**.

In all further discussion of the behaviour and properties of acids, we will be using the Brønsted–Lowry definition.

The relationship between depth and simplicity

There are various theories of acids and bases. We have met the Brønsted–Lowry and Lewis theories. The Lewis theory is more sophisticated and extends the Brønsted–Lowry theory, which is limited to describing the acid–base behaviour of species in aqueous solution. The Lewis theory can be used to describe the acid–base reactions of substances, including those not containing hydrogen atoms, in a variety of different solvents. The Lewis theory can be applied to reactions in organic chemistry – a nucleophile is a Lewis base, and the reaction between an electrophile and a nucleophile is a Lewis acid–base reaction. The reaction of a transition metal with a ligand such as CO is a Lewis acid–base reaction. To what extent is it useful to describe these reactions as acid–base reactions? Before we knew about the Lewis theory, we probably had a fairly good picture in our minds as to what an acid–base reaction is. Have we got a more sophisticated theory at the expense of losing an understanding of what an acid–base reaction is?

? Test yourself

6 Define acids and bases according to the Lewis definition.

7 Classify each of the following as a substance that can act as a Lewis acid or as a Lewis base:
H_2O BF_3 HCO_3^- H^+ $AlCl_3$ NH_3 CO

8.3 Properties of acids and bases

Salts

The hydrogen ion (H^+) in an acid can be regarded as 'replaceable' and can be replaced by a metal ion (or ammonium ion) to form a **salt**. For example, if the hydrogen ions in sulfuric acid are replaced by sodium ions, the salt called sodium sulfate (Na_2SO_4) is formed. The names of the salts formed by some acids are given in Table **8.2**.

Acid	Formula	Name of salt	Example of salt
hydrochloric	**H**Cl	chloride	NaCl
sulfuric	**H**$_2$SO$_4$	sulfate	Na$_2$SO$_4$
nitric	**H**NO$_3$	nitrate	NH$_4$NO$_3$
carbonic	**H**$_2$CO$_3$	carbonate	K$_2$CO$_3$
ethanoic	CH$_3$COO**H**	ethanoate	Ca(CH$_3$COO)$_2$

Table 8.2 The formulas for some common acids and their salts – a hydrogen lost as H^+ is shown **bold**.

The reactions of acids and bases

The characteristic reactions of acids are described below. In each case we can write an ionic equation where the acid is represented by $H^+(aq)$.

Reactions of acids with metals

The general equation for these reactions is:

> metal + acid $\rightarrow$ salt + hydrogen

Acids react with reactive metals such as magnesium to produce a salt and hydrogen gas. For example:

$$Mg(s) + 2HCl(aq) \rightarrow MgCl_2(aq) + H_2(g)$$

Ionic equation: $Mg(s) + 2H^+(aq) \rightarrow Mg^{2+}(aq) + H_2(g)$

Unreactive metals such as copper do not react with dilute acids.

Reactions of acids with carbonates and hydrogencarbonates

> acid + carbonate / hydrogencarbonate $\rightarrow$ salt + carbon dioxide + water

For example:

$$Na_2CO_3(aq) + H_2SO_4(aq) \rightarrow Na_2SO_4(aq) + H_2O(l) + CO_2(g)$$

(ionic equation: $CO_3^{2-}(aq) + 2H^+(aq) \rightarrow H_2O(l) + CO_2(g)$)

and $NaHCO_3(aq) + HCl(aq) \rightarrow NaCl(aq) + H_2O(l) + CO_2(g)$
    sodium hydrogencarbonate

Reactions of acids with bases and alkalis

Bases are metal oxides or metal hydroxides, such as copper(II) oxide or calcium hydroxide. These react with acids to form a salt and water.

For example:

$$CuO(s) + H_2SO_4(aq) \rightarrow CuSO_4(aq) + H_2O(l)$$

Alkalis are solutions obtained when a metal hydroxide (such as sodium hydroxide) dissolves in water or when certain bases react with water. For instance, the base sodium oxide (Na_2O) reacts with water to produce the alkali sodium hydroxide:

$$Na_2O(s) + H_2O(l) \rightarrow 2NaOH(aq)$$

There are very few soluble metal hydroxides and so there are very few alkalis. The most common alkalis are the alkali metal hydroxides (e.g. sodium hydroxide solution), ammonia solution and barium hydroxide solution.

The terms 'alkali' and 'base' are often used interchangeably. 'Base' is the more general term and will be used in all of the discussions in subsequent sections.

The reaction between an acid and an alkali is similar to that between an acid and a base:

alkali + acid $\rightarrow$ salt + water

For example: $NaOH(aq) + HNO_3(aq) \rightarrow NaNO_3(aq) + H_2O(l)$

Ionic equation: $OH^-(aq) + H^+(aq) \rightarrow H_2O(l)$

The reactions between acids and bases or acids and alkalis are called **neutralisation** reactions and essentially just involve the H^+ ions from the acid reacting with the OH^- ions from the alkali to form the neutral substance water. Neutralisation reactions are exothermic and typically give out about 57 kJ per mole of water formed.

The reactions of acids with ammonia solution, an alkali, are often written slightly differently:

$$NH_3(aq) + HCl(aq) \rightarrow NH_4Cl(aq)$$

Ammonia solution is equivalent to ammonium hydroxide (NH_4OH) – in some laboratories bottles are labelled 'ammonia solution' and in others they are labelled 'ammonium hydroxide'. Ammonia is in equilibrium with the ammonium ion and the hydroxide ion:

$$NH_3(aq) + H_2O(l) \rightleftharpoons NH_4^+(aq) + OH^-(aq)$$

The reaction with hydrochloric acid could also have been written:

$$NH_4OH(aq) + HCl(aq) \rightarrow NH_4Cl(aq) + H_2O(l)$$

which corresponds to the general equation of alkali + acid $\rightarrow$ salt + water.

base + acid $\rightarrow$ salt + water

Neutralisation reactions are exothermic and produce a salt and water only.

Making salts from acids and bases

Generally the metal part of the salt comes from a metal oxide or hydroxide and the non-metal part from the acid. So, copper sulfate can be made from the reaction between the base copper oxide and sulfuric acid. Sodium nitrate can be made from the base sodium oxide (or the alkali sodium hydroxide) and nitric acid.

One common way of making a soluble salt is by titrating an acid against an alkali. A known amount of acid is measured out using a pipette and then an indicator is added. The alkali is then added from a burette until the indicator just changes colour. The experiment is then repeated using the same amounts of acid and alkali but without the indicator. The titration technique is covered on page **43**.

Indicators are essentially substances that have different colours in acidic and alkaline solutions – for example, bromothymol blue is yellow in acidic solution and blue in alkaline solution. Many different indicators can be used for titrations and they are chosen according to whether the acid or the alkali is strong or weak (see page **349**). Universal indicator contains a mixture of indicators and cannot be used to monitor titrations because it changes gradually from one colour to the next.

? Test yourself

8 Complete and balance the following equations:
   a $Zn + H_2SO_4 \rightarrow$
   b $CuO + HNO_3 \rightarrow$
   c $NH_3 + H_2SO_4 \rightarrow$
   d $Ca(HCO_3)_2 + HCl \rightarrow$
   e $Mg(OH)_2 + H_2SO_4 \rightarrow$
   f $Cu + H_2SO_4 \rightarrow$
   g $CaO + HCl \rightarrow$

9 State the name of an acid and a base that would react together to form each of these salts:
   a calcium nitrate
   b cobalt(II) sulfate
   c copper(II) chloride
   d magnesium ethanoate

Learning objectives

- Use pH values to distinguish between acidic, neutral and alkaline solutions
- Understand that pH provides a measure of the concentration of H^+ ions in aqueous solution
- Solve problems involving pH
- Write an equation for the dissociation of water and state the expression for K_w
- Use K_w values to work out the concentrations of $H^+(aq)$ and $OH^-(aq)$ and the pH of aqueous solutions

8.4 The pH scale

Measuring pH

The pH scale can be used to indicate whether a solution is acidic, alkaline or neutral. At 25 °C, a solution with pH lower than 7 is acidic, a solution with pH 7 is neutral and a solution with pH greater than 7 is alkaline (Figure **8.7**).

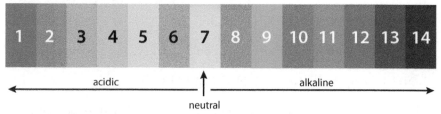

| 1 | 2 | 3 | 4 | 5 | 6 | 7 | 8 | 9 | 10 | 11 | 12 | 13 | 14 |

acidic ← | → alkaline

neutral

Figure 8.7 The pH scale showing the colours of universal indicator.

The pH of a solution can be determined by using a **pH meter** or by using universal indicator solution or paper.

pH is a measure of the concentration of $H^+(aq)$ ions in a solution. The concentration of H^+ ions can be stated in $mol\,dm^{-3}$, but the use of the $\log_{10}$ function simplifies the numbers involved.

> Definition: pH is the negative logarithm to base 10 of the hydrogen ion concentration in an aqueous solution.
>
> $$pH = -\log_{10}[H^+(aq)]$$

So, if the concentration of H^+ ions in a solution is $1.57 \times 10^{-3}\,mol\,dm^{-3}$, the pH is worked out as:

$$pH = -\log_{10}(1.57 \times 10^{-3}) = 2.80$$

pH has no units.

> Because pH is a log scale (to base 10), a 1 unit change in pH indicates a tenfold change in the H^+ ion concentration (see Table 8.3).

Note: the '10' is often omitted in '$\log_{10}$', giving $pH = -\log[H^+(aq)]$. To work out pH, use the 'log' button on a calculator.

$[H^+(aq)]/mol\,dm^{-3}$	pH
1	0
0.1	1
0.01	2
1×10^{-3}	3
1×10^{-4}	4
1×10^{-5}	5
1×10^{-6}	6
1×10^{-7}	7
1×10^{-8}	8

Table 8.3 pH values and the corresponding concentration of H^+ ions.

The pH for solutions in which the concentration of H^+ ions is a power of 10 can be worked out without using a calculator by understanding how the $\log_{10}$ function works.

$$\log_{10} 100 = 2 \qquad\qquad 10^2 = 100$$

The log of a number is the power that 10 must be raised to in order to equal that number.

0.01 can be written as a power of 10: $\quad 0.01 = 10^{-2}$

so $\log_{10} 0.01 = -2$

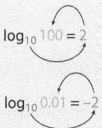

Therefore, if a solution has an $H^+(aq)$ concentration of $0.01\,mol\,dm^{-3}$, the pH of this solution is given by:

$$pH = -\log_{10} 0.01 = 2$$

Note: the power that 10 must be raised in order to equal 1 is 0; therefore $\log_{10} 1 = 0$.

Calculating $[H^+(aq)]$ from a pH

To calculate $[H^+(aq)]$ from a pH, the inverse function of $\log_{10}$ must be used. So 10 must be raised to the power of $-pH$.

> $$[H^+(aq)] = 10^{-pH}$$

This is usually done using the '2nd' (or 'shift') and 'log' key combination on a calculator.

pH is an artificial scale developed using a mathematical function that converts the concentration of H^+ ions into much simpler numbers. It is definitely easier to say 'the pH of the solution is 6' rather than 'the concentration of hydrogen ions in the solution is $1.0 \times 10^{-6} \, mol \, dm^{-3}$'.

Students of chemistry are introduced to the pH scale at an early age and are usually content to compare 'the acidity of solutions' in terms of pH and work out whether these solutions are acidic, alkaline or neutral. Most students do this without ever really understanding anything about what pH means. The idea of concentration of H^+ ions could not be introduced at such an early age; if it were, far fewer students would be able to work out whether a solution were acidic, alkaline or neutral and would almost certainly not have such a clear picture in their minds about the relative acidities/alkalinities of substances.

So is it better to have a scale that most people can use but do not understand, or to have a more accurate and in-depth description of the acidity of solutions that most people will not be able to understand? Do we lose or gain understanding by using the pH scale?

The dissociation of water

Water dissociates (or ionises) according to the equation:

$$H_2O(l) \rightleftharpoons H^+(aq) + OH^-(aq)$$

The degree of dissociation is very small and, at 25 °C, in pure water the concentrations of H^+ and OH^- ions are equal at $1.0 \times 10^{-7} \, mol \, dm^{-3}$.

An equilibrium constant, K_w, can be written for this reaction:

$K_w = [H^+(aq)][OH^-(aq)]$
K_w is called the **ionic product constant** (ionic product) for water. K_w has a value of 1.0×10^{-14} at 298 K.

K_w refers to the $H_2O(l) \rightleftharpoons H^+(aq) + OH^-(aq)$ equilibrium in all aqueous solutions – the product of the H^+ and OH^- concentrations in **any** aqueous solution at 298 K is 1.0×10^{-14}.

The concentration of water is not incorporated in the equilibrium expression because water is in excess and its concentration is essentially constant.
Also note that K_w has no units because the concentrations that we are using are compared to a standard concentration of $1 \, mol \, dm^{-3}$. When we write $[H^+(aq)]$ here, what we actually mean is 'the concentration of H^+ ions divided by $mol \, dm^{-3}$', which has no units – therefore K_w has no units. For advanced work, K_w should be worked out in terms of activities (effective concentrations), which are also compared to a standard concentration.

Consider a $0.10 \, mol \, dm^{-3}$ solution of hydrochloric acid. HCl is a strong acid and dissociates completely in solution – therefore the concentration of H^+ ions in solution will be $0.10 \, mol \, dm^{-3}$. Now, $[H^+(aq)][OH^-(aq)] = 1.0 \times 10^{-14}$ so the concentration of OH^- ions in this solution must be $1.0 \times 10^{-13} \, mol \, dm^{-3}$ because $0.10 \times (1 \times 10^{-13}) = 1.0 \times 10^{-14}$. So, even a solution of an acid contains some OH^- ions because of the ionisation of water.

Similarly, the concentration of H^+ ions in a $1.0 \times 10^{-3} \, mol \, dm^{-3}$ solution of sodium hydroxide (a strong base) is $1.0 \times 10^{-11} \, mol \, dm^{-3}$. The presence of H^+ ions from the ionisation of water in all aqueous solutions explains why we can always measure a pH.

We can understand whether solutions are acidic, alkaline or neutral in terms of the relative concentrations of H^+ and OH^-.

A solution is:
- neutral if $[H^+(aq)] = [OH^-(aq)]$
- acidic if $[H^+(aq)] > [OH^-(aq)]$
- alkaline if $[OH^-(aq)] > [H^+(aq)]$

Calculating pH values

For now, we will limit the calculation of pH to solutions of strong acids and bases – calculations involving weak acids and bases will be covered in Subtopic **8.7**.

Calculating the pH of a solution of a strong acid

Because strong acids can be assumed to dissociate fully in aqueous solution (see Subtopic **8.5**), the concentration of H^+ ions is the same as the concentration of the acid.

Example

We shall calculate the pH of a $0.001\,50\,mol\,dm^{-3}$ solution of hydrochloric acid. The acid is strong, so full dissociation produces an $H^+(aq)$ concentration of $0.001\,50\,mol\,dm^{-3}$ (the same as the original acid):

$$pH = -\log_{10}[H^+(aq)] = -\log_{10}[0.001\,50] = 2.82$$

Worked example

8.1 a What is the pH of $10\,cm^3$ of $0.10\,mol\,dm^{-3}$ hydrochloric acid?

    b If $90\,cm^3$ of water is added to the acid, what happens to the pH?

    c If the solution from part **b** is now diluted by a factor of a million (10^6), what is the approximate pH of the final solution?

a HCl is a strong acid and dissociates fully, so the concentration of $H^+(aq)$ is $0.10\,mol\,dm^{-3}$ and

    $$pH = -\log_{10}[H^+(aq)] = -\log_{10}[0.10] = 1.0$$

b If $90\,cm^3$ of water is added to $10\,cm^3$ of acid, the total volume of the solution becomes $100\,cm^3$. There are the same number of H^+ ions in 10 times the volume – the concentration of H^+ ions has been reduced by a factor of 10. Because pH is a log scale, a reduction in $[H^+(aq)]$ by a factor of 10 results in the pH of the solution increasing by 1. So the pH of the diluted solution is 2.0.

    We can also consider this in terms of numbers. The initial concentration of $H^+(aq)$ is $0.10\,mol\,dm^{-3}$. When this is diluted by a factor of 10, the concentration of $H^+(aq)$ decreases to $0.010\,mol\,dm^{-3}$.

    $$pH = -\log_{10}[H^+(aq)] = -\log_{10}[0.010] = 2.0$$

c Diluting a solution by a factor of 10 causes the pH to increase by 1 unit, so it might be expected that diluting the solution in part **b** by a factor of 10^6 would cause its pH to increase by 6 units to 8.0. However, it is not possible to dilute an acidic solution so that it becomes an alkaline solution; we have ignored the dissociation of water molecules into H^+ and OH^- ions. The actual pH of this solution will be slightly less than 7. If the solution is diluted further, the pH will get closer and closer to 7, but never reach it.

Calculating the pH of a solution of a strong base

The pH of an alkaline solution can be worked out using the ionic product constant, K_w, equal to 1.00×10^{-14} at 25 °C.

Worked example

8.2 Calculate the pH of a $0.250\,mol\,dm^{-3}$ solution of potassium hydroxide at 25 °C.

$$KOH(aq) \rightarrow K^+(aq) + OH^-(aq)$$

KOH is a strong base, so, the OH^- concentration is the same as that of the original base – i.e. $0.250\,mol\,dm^{-3}$.

> Strong bases are completely ionised in solution.

Now, $K_w = [H^+(aq)][OH^-(aq)]$ and, at 25 °C, $K_w = 1.00 \times 10^{-14}$.

Substituting the OH^- concentration and the value of K_w into this expression we obtain:

$$1.00 \times 10^{-14} = [H^+(aq)] \times 0.250$$

So $[H^+(aq)] = 4.00 \times 10^{-14}\,mol\,dm^{-3}$

and $pH = -\log_{10}[H^+(aq)]$

$$= -\log_{10}(4.00 \times 10^{-14})$$

$$= 13.4$$

Nature of science

It is important that scientists are able to communicate their findings and discuss issues with non-scientists. pH is an example of a concept that allows discussion of acidity without the requirement for a complete understanding of the theory – stating that the pH of acid rain is 4 is much simpler than saying that the concentration of hydrogen ions is $1 \times 10^{-4}\,mol\,dm^{-3}$.

? Test yourself

10 Arrange the following in order of increasing pH:
  $0.010\,mol\,dm^{-3}$ HCl
  $1.0\,mol\,dm^{-3}$ NaOH
  $0.10\,mol\,dm^{-3}$ H_2SO_4
  $0.10\,mol\,dm^{-3}$ HCl

11 State whether each of the following statements is **true** or **false**:
  a The pH of $0.10\,mol\,dm^{-3}$ HCl is 2.
  b The $[H^+]$ in a solution of pH = 3 is 100 times the $[H^+]$ in a solution of pH = 5.
  c The $[H^+]$ in a solution of pH = 13 is $1.0 \times 10^{13}\,mol\,dm^{-3}$.
  d The pH of $0.010\,mol\,dm^{-3}$ H_2SO_4 is 2.

12 A solution of $0.2\,mol\,dm^{-3}$ HCl has a pH of 0.7. Of this solution, $25.0\,cm^3$ is taken and made up to a total volume of $250\,cm^3$ with distilled water in a volumetric flask to form solution **X**. $10.0\,cm^3$ of solution **X** is taken and made up to a total volume of $1.00\,dm^3$ with distilled water in a volumetric flask; this is solution **Y**.

   a Work out the pH value for each solution.

   b Suggest an experimental method, other than measuring the pH or using universal indicator, that could be used to distinguish between solution **X** and solution **Y**.

13 Copy and complete the following table by working out pH values. The first three should be done without using a calculator.

a solution containing $0.001\,00\,mol\,dm^{-3}$ $H^+(aq)$	
a solution containing $1.00\times10^{-12}\,mol\,dm^{-3}$ $H^+(aq)$	
a solution of $1.00\,mol\,dm^{-3}$ $HCl(aq)$	
a solution of $2.00\times10^{-4}\,mol\,dm^{-3}$ $HNO_3(aq)$	
a solution of CH_3COOH of concentration $0.100\,mol\,dm^{-3}$ assuming 5% dissociation of the acid	

14 Calculate the concentration of H^+ ions in each of the following solutions:

   a pH = 3.5

   b pH = 7.9

   c pH = 12.8

15 Copy and complete the table for aqueous solutions at $25\,°C$ – try to do it without using a calculator. The first row has been completed for you.

$[H^+(aq)]/$ $mol\,dm^{-3}$	$[OH^-$ $(aq)]/$ $mol\,dm^{-3}$	pH	Acidic or alkaline?
1.0×10^{-3}	1.0×10^{-11}		acidic
1.0×10^{-5}			
	0.01		

16 Calculate the pH of these aqueous solutions assuming that both species ionise completely:

   a $0.0150\,mol\,dm^{-3}$ NaOH

   b $0.0500\,mol\,dm^{-3}$ $Ba(OH)_2$

17 Calculate the concentration of OH^- ions in:

   a $0.020\,mol\,dm^{-3}$ $HCl(aq)$

   b $1.4\times10^{-4}\,mol\,dm^{-3}$ $HNO_3(aq)$

8.5 Strong and weak acids and bases

Acids

When an acid (HA) reacts with water it **dissociates**, or **ionises**:

$$HA(aq) + H_2O(l) \rightleftharpoons H_3O^+(aq) + A^-(aq)$$

This can be understood using the Brønsted–Lowry theory – the acid donates a proton to water. Basically, though, what has happened is that the acid has dissociated into H^+ and A^- ions. This reaction is, therefore, often simplified to:

$$HA(aq) \rightleftharpoons H^+(aq) + A^-(aq)$$

This reaction is exactly the same as the one above, and the H^+ from the acid is still donated to H_2O, but the 'H_2O' has been omitted for simplicity ($H^+(aq)$ is equivalent to $H_3O^+(aq)$).

We can classify acids as strong or weak according to how much they dissociate in aqueous solution.

Learning objectives

- Understand the difference between strong and weak acids and bases
- Recognise examples of strong and weak acids and bases
- Describe and explain experiments to distinguish between strong and weak acids and bases

Strong acids

Extension

Actually, nothing ever dissociates **completely** because all these reactions involve an equilibrium. However, for strong acids this equilibrium lies a very long way to the right.

Strong acids such as hydrochloric (HCl), sulfuric (H_2SO_4) and nitric (HNO_3) acid dissociate completely in aqueous solution (see Figure **8.8**).

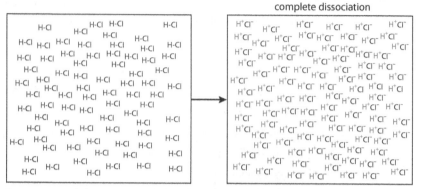

Figure 8.8 HCl dissociates completely in aqueous solution.

This can be represented as:

$$HA(aq) \rightarrow H^+(aq) + A^-(aq)$$

The non-reversible arrow ($\rightarrow$) is used to indicate that dissociation is essentially complete.

HCl is a **monoprotic acid** – it dissociates to form one proton per molecule. H_2SO_4 is a **diprotic acid** – it can dissociate to form two protons per molecule:

Sulfuric acid is a strong acid for the first dissociation only.

$$H_2SO_4(aq) + H_2O(l) \rightarrow HSO_4^-(aq) + H_3O^+(aq)$$

$$HSO_4^-(aq) + H_2O(l) \rightleftharpoons SO_4^{2-}(aq) + H_3O^+(aq)$$

Weak acids

Weak acids dissociate only partially in aqueous solution (see Figure **8.9**).

Examples of weak acids are carbonic acid (H_2CO_3) and carboxylic acids such as ethanoic acid (CH_3COOH).

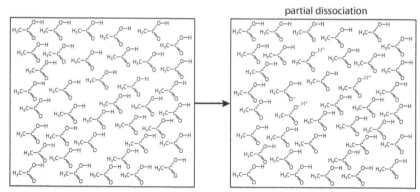

Figure 8.9 Only a few of the ethanoic acid molecules dissociate.

The dissociation of a weak acid is represented by:

$$HA(aq) \rightleftharpoons H^+(aq) + A^-(aq)$$

The equilibrium arrow is essential and indicates that the reaction is reversible and does not go to completion.

The dissociation of ethanoic acid is represented by:

$$CH_3COOH(aq) \rightleftharpoons CH_3COO^-(aq) + H^+(aq)$$

$$\text{or } CH_3COOH(aq) + H_2O(l) \rightleftharpoons CH_3COO^-(aq) + H_3O^+(aq)$$

Carbonic acid is formed when carbon dioxide dissolves in water (see Figure 8.10):

$$H_2O(l) + CO_2(g) \rightleftharpoons H_2CO_3(aq)$$

It is a diprotic acid and its dissociation can be shown as:

$$H_2CO_3(aq) \rightleftharpoons HCO_3^-(aq) + H^+(aq)$$

$$HCO_3^-(aq) \rightleftharpoons CO_3^{2-}(aq) + H^+(aq)$$

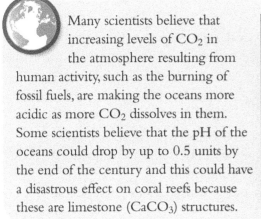

Many scientists believe that increasing levels of CO_2 in the atmosphere resulting from human activity, such as the burning of fossil fuels, are making the oceans more acidic as more CO_2 dissolves in them. Some scientists believe that the pH of the oceans could drop by up to 0.5 units by the end of the century and this could have a disastrous effect on coral reefs because these are limestone ($CaCO_3$) structures.

Figure 8.10 Carbonated water is acidic because of dissolved CO_2.

HCO_3^-, the hydrogencarbonate ion, is amphiprotic. It accepts a proton in the reverse reaction in the first equation (acts as a base) and donates a proton (acts as an acid) in the second.

Bases

When a base (B) reacts with water, it accepts a proton from the water and ionises according to:

$$B(aq) + H_2O(l) \rightleftharpoons BH^+(aq) + OH^-(aq)$$

Bases are defined as strong or weak depending on how much they ionise in aqueous solution.

Strong bases

Strong bases ionise completely in aqueous solution.

For example, sodium hydroxide ionises completely to produce OH^- ions:

$$NaOH(aq) \rightarrow Na^+(aq) + OH^-(aq)$$

Strong bases include the group 1 hydroxides (LiOH, NaOH etc.) and $Ba(OH)_2$.

Exam tip

$Ba(OH)_2$ ionises in aqueous solution to give $2OH^-$:

$$\begin{aligned}Ba(OH)_2(aq)\\ \rightarrow Ba^{2+}(aq) + 2OH^-(aq)\end{aligned}$$

So $1\,mol\,dm^{-3}\ Ba(OH)_2(aq)$ contains $2\,mol\,dm^{-3}\ OH^-$.

Week bases

> Weak bases ionise only partially in aqueous solution.

Ammonia is a typical weak base and ionises according to the equation:

$$NH_3(aq) + H_2O(l) \rightleftharpoons NH_4^+(aq) + OH^-(aq)$$

In a $0.10\,mol\,dm^{-3}$ solution of ammonia (at $25\,°C$) about 1.3% of the molecules are ionised.

Other weak bases are amines such as ethylamine (ethanamine), $CH_3CH_2NH_2$. Ethylamine ionises according to the equation:

$$CH_3CH_2NH_2(aq) + H_2O(l) \rightleftharpoons CH_3CH_2NH_3^+(aq) + OH^-(aq)$$

At $25\,°C$, a $0.10\,mol\,dm^{-3}$ solution of ethylamine has about 7.1% of the molecules ionised – so ethylamine is a stronger base than ammonia.

Extension

The higher base strength of ethylamine can be understood in terms of the electron-releasing effect (positive inductive effect) of the alkyl group attached to the nitrogen atom.

The relationship between the strength of an acid and the strength of its conjugate base

> The stronger an acid, the weaker its conjugate base.

A strong acid such as HCl dissociates completely in aqueous solution:

$$HCl(aq) \rightarrow H^+(aq) + Cl^-(aq)$$

The conjugate base of HCl is Cl^- – this is a very weak base because it has virtually no tendency to react with H_2O or H^+ to re-form HCl – the above reaction goes essentially to completion.

HCN, on the other hand, is a very weak acid and has very little tendency to dissociate:

$$HCN(aq) + H_2O(l) \rightleftharpoons H_3O^+(aq) + CN^-(aq)$$

The position of equilibrium lies a long way to the left – CN^- has a strong tendency to pick up H^+ from H_3O^+, that is, to act as a base.

The conjugate base of HCN is a much stronger base than that of HCl (Cl^-) and when it is added to water it reacts to re-form the parent acid:

$$CN^-(aq) + H_2O(l) \rightleftharpoons HCN(aq) + OH^-(aq)$$

> The stronger a base, the weaker its conjugate acid.

A weak base ionises partially in aqueous solution:

$$B(aq) + H_2O(l) \rightleftharpoons BH^+(aq) + OH^-(aq)$$

The stronger the base, the further the position of this equilibrium lies to the right and the less tendency the conjugate acid (BH^+) has to donate a proton to re-form B.

- A strong acid is a good proton donor that ionises completely in aqueous solution – it has a weak conjugate base.
- A strong base is a good proton acceptor that ionises completely in aqueous solution – it has a weak conjugate acid.

So, ethylamine (stronger base) has a weaker conjugate acid than ammonia (weaker base).

Distinguishing between strong and weak acids and between strong and weak bases

The methods for distinguishing between acids rely on the fact that strong acids dissociate more than weak acids.

In the following examples we will compare solutions of strong and weak acids of **equal concentrations**. When comparing strong and weak acids, it is essential that the solutions have the same concentration because the total concentration of H^+ ions in a concentrated solution of a weak acid could be higher than in a dilute solution of a strong acid.

Solutions of strong acids conduct electricity better than solutions of weak acids

A strong acid dissociates fully, so the concentration of ions is high and the solution is a good conductor of electricity. A weak acid dissociates only partially and so the concentration of ions is lower and the solution does not conduct electricity as well.

The conductivity of solutions can be measured using a conductivity meter or by looking at the brightness of a bulb in the experimental set-up shown in Figure **8.11**.

Strong acids may be described as **strong electrolytes**, whereas weak acids are **weak electrolytes**. This also applies to strong and weak bases – strong bases conduct electricity better than weak bases.

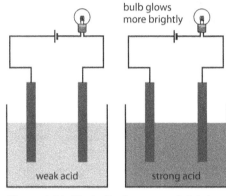

Figure 8.11 The bulb glows more brightly with a solution of a strong acid than it does with a solution of weak acid.

Strong acids have a lower pH than weak acids

pH is a measure of the concentration of H^+ ions in solution – the lower the pH, the higher the concentration of H^+ ions. Strong acids dissociate more, producing a higher concentration of H^+ ions in solution (a lower pH). This can be tested using universal indicator or a pH meter. Again, this method could also be used for distinguishing between strong and weak bases. Strong bases have a higher pH than weak bases of the same concentration.

Strong acids react more violently with metals or carbonates

Strong acids have a higher concentration of free H^+ ions and therefore react more rapidly with a metal such as magnesium to form hydrogen – this can be shown by more rapid bubbling (effervescence) when the metal is added (see Figure **8.12**). A similar effect is seen when a carbonate/hydrogencarbonate is added to an acid. Stronger acids would also have a faster rate of reaction with metal oxide/hydroxides, but this is harder to observe experimentally.

Figure 8.12 Magnesium reacting with a strong acid (hydrochloric acid, left) and a weak acid (ethanoic acid, right). The more violent reaction with the strong acid causes larger hydrogen bubbles, which keeps the magnesium at the surface of the fluid. The much smaller bubbles caused by the weaker acid reaction spread through the acid, making it appear milky.

Strength versus concentration

You should be aware of the difference between the concentration of an acid (or base) and its strength. The *concentration* of an acid refers to the number of moles of acid in a certain volume – e.g. in a solution of ethanoic acid of concentration $0.100 \, mol \, dm^{-3}$ the total number of moles of ethanoic acid before dissociation is $0.100 \, mol$ in every cubic decimetre (litre). The *strength* of an acid refers to how much it dissociates in aqueous solution.

A solution of ethanoic acid of concentration $1.00 \, mol \, dm^{-3}$ is a more concentrated solution than $0.100 \, mol \, dm^{-3}$ ethanoic acid. No matter how concentrated the solution of ethanoic acid is, it will never become a strong acid because it always dissociates partially. Similarly, diluting a strong acid such as hydrochloric acid does not make it anything other than a strong acid because it always dissociates fully.

pH is not a measure of acid or base strength

pH is simply a measure of the concentration of $H^+(aq)$ ions. It is possible for a dilute solution of a strong acid to have a higher pH than a concentrated solution of a weak acid. $1.00 \, mol \, dm^{-3}$ ethanoic acid has a pH of 2.38, but $1.00 \times 10^{-3} \, mol \, dm^{-3}$ hydrochloric acid has a pH of 3.00. Because the solution of ethanoic acid is much more concentrated, the partial dissociation of the weak acid produces a higher concentration of $H^+(aq)$ than the complete dissociation of the strong acid HCl.

Nature of science

Advances in technology and instrumentation have allowed for the generation of quantitative data about the relative strengths of acids and bases. This has allowed scientists to look for patterns and anomalies and allowed for development of new theories at the molecular level. For instance the relative basic strengths of NH_3, H_2NCH_3 and $HN(CH_3)_2$ can be explained in terms of the electron-releasing effect of alkyl groups but the fact that $N(CH_3)_3$ is a weaker base than $HN(CH_3)_2$ required modification of the theory and an alternative explanation.

> Titration using an indicator cannot be used to distinguish between a weak and a strong acid – if they have the same concentration they will have the same end point (react with the same number of moles of alkali).

> pH can be used to compare acid strength only if equal concentrations of acids are being compared.

? Test yourself

18 Classify the following as strong acid, weak acid, strong base, weak base or salt:

HCl	H_2SO_4	NH_4NO_3	NaOH
NH_3	HNO_3	Na_2SO_4	H_2CO_3
$Ba(OH)_2$	KNO_3	CH_3NH_2	HCOOH

19 A solution of $0.000\,100 \, mol \, dm^{-3}$ hydrochloric acid has a pH of 4.00, whereas a solution of $0.100 \, mol \, dm^{-3}$ ethanoic acid has a pH of 2.88. Explain what these values tell you about the usefulness of pH as a measure of acid strength.

20 For each pair of solutions, state which will have the higher electrical conductivity:
  a $0.10 \, mol \, dm^{-3}$ HCOOH and $0.10 \, mol \, dm^{-3}$ HCl
  b $0.10 \, mol \, dm^{-3}$ NH_3 and $0.10 \, mol \, dm^{-3}$ KOH
  c $0.10 \, mol \, dm^{-3}$ HCl and $0.10 \, mol \, dm^{-3}$ H_2SO_4
  d $0.10 \, mol \, dm^{-3}$ HNO_3 and $0.010 \, mol \, dm^{-3}$ HCl

8.6 Acid deposition

Oxides of non-metals are acidic and, if soluble in water, dissolve to produce acidic solutions.

Rain is naturally acidic because of dissolved carbon dioxide:

$$H_2O(l) + CO_2(g) \rightleftharpoons H_2CO_3(aq)$$
$$\text{carbonic acid}$$

H_2CO_3 is a weak acid and dissociates partially according to the equation:

$$H_2CO_3(aq) \rightleftharpoons H^+(aq) + HCO_3^-(aq)$$

Because of this reaction, the pH of rain water is about 5.6. This is a natural phenomenon – rain with a pH between 5.6 and 7 is not considered to be 'acid rain'. Acid rain is considered to be rain with a pH lower than 5.6. The average pH of rain in some areas can be as low as 4.

Acidic pollutants include the oxides of sulfur and nitrogen. Sulfur dioxide (sulfur(IV) oxide) can be formed by various natural and anthropogenic processes – such as the burning of sulfur-containing fuels:

$$S(s) + O_2(g) \rightarrow SO_2(g)$$

The process by which SO_2 is converted into SO_3 and H_2SO_4 in the atmosphere is complex and does not involve simple oxidation by atmospheric oxygen – interactions with hydroxyl radicals, ozone or hydrogen peroxide are involved. However, the reactions can be summarised:

$$2SO_2(g) + O_2(g) \rightarrow 2SO_3(g)$$
$$\text{sulfur(VI) oxide}$$

$$SO_3(g) + H_2O(l) \rightarrow H_2SO_4(aq)$$
$$\text{sulfuric(VI) acid}$$

SO_2 can also dissolve in water to produce sulfuric(IV) acid (sulfurous acid):

$$SO_2(g) + H_2O(l) \rightarrow H_2SO_3(aq)$$
$$\text{sulfuric(IV) acid}$$

The main anthropogenic sources of nitrogen oxides are the internal combustion engine, coal, gas, oil-fuelled power stations and heavy industry power generation. The combustion temperatures of fuels in these processes are very high and oxidation of atmospheric nitrogen occurs forming NO (nitrogen monoxide, nitric oxide or nitrogen(II) oxide):

$$N_2(g) + O_2(g) \rightarrow 2NO(g)$$

NO can be oxidised in the atmosphere to NO_2. Again, the exact nature of the process is complex but the reaction can be summarised as:

$$2NO(g) + O_2(g) \rightarrow 2NO_2(g)$$

Learning objectives

- Understand what is meant by acid deposition and what causes it
- Understand some of the problems associated with acid deposition
- Understand the difference between pre- and post-combustion methods of reducing sulfur dioxide emissions

Acid deposition is a more general term than **acid rain**. It refers to any process in which acidic substances (particles, gases and precipitation) leave the atmosphere to be deposited on the surface of the Earth. It can be divided into **wet deposition** (acid rain, fog and snow) and **dry deposition** (acidic gases and particles).

'Anthropogenic' means produced by human activities.

A **free radical** is a species with an unpaired electron – free radicals are very reactive. Many of the reactions that occur in the atmosphere involve free radicals.

The NO_2 can then react with a hydroxyl free radical (HO•) to form nitric(V) acid:

$$NO_2(g) + HO•(g) \rightarrow HNO_3(g)$$

Other reactions can occur and the formation of nitric acid can also be shown as:

$$4NO_2(g) + O_2(g) + 2H_2O(l) \rightarrow 4HNO_3(aq)$$

$$\text{or } 2NO_2(g) + H_2O(l) \rightarrow \underset{\text{nitric(III) acid}}{HNO_2(aq)} + \underset{\text{nitric(V) acid}}{HNO_3(aq)}$$

Problems associated with acid deposition

- **Effect on vegetation** – it is not necessarily the acid itself that causes problems. The acid (H^+ ions) can displace metal ions from the soil that are consequently washed away (particularly calcium, magnesium and potassium ions). Mg^{2+} ions are needed to produce chlorophyll, so plants could be prevented from photosynthesising properly. The acid rain also causes aluminium ions to dissolve from rocks, which damages plant roots and limits water uptake. This can cause stunted growth and thinning or yellowing of leaves on trees (Figure **8.13a**).
- **Lakes and rivers** – aquatic life is sensitive to the pH falling below 6. Insect larvae, fish and invertebrates, among others, cannot survive below pH 5.2. Below pH 4.0 virtually no life will survive. Acid rain can dissolve hazardous minerals from rocks, which can accumulate in lakes and damage aquatic life – in particular, Al^{3+} ions damage fish gills.
- **Buildings** – limestone and marble are eroded by acid rain and dissolve away exposing a fresh surface to react with more acid (Figure **8.13b**). A typical reaction is:

$$CaCO_3(s) + H_2SO_4(aq) \rightarrow CaSO_4(s) + H_2O(l) + CO_2(g)$$

- **Human health** – acids irritate mucous membranes and cause respiratory illnesses such as asthma and bronchitis. Acidic water can dissolve heavy metal compounds releasing poisonous ions such as Cu^{2+}, Pb^{2+} and Al^{3+} which may be linked to Alzheimer's disease.

Figure 8.13 Acid deposition can **a** kill trees and fish in lakes and **b** react with limestone buildings to cause erosion.

Methods of dealing with acid deposition

Methods include:
- improving the design of vehicle engines
- using catalytic converters
- removing sulfur before burning fuels
- using renewable power supplies
- making greater use of public transport
- designing more efficient power stations
- 'liming' of lakes – calcium oxide or hydroxide neutralises acidity:

$$CaO(s) + H_2SO_4(aq) \rightarrow CaSO_4(aq) + H_2O(l)$$

$$Ca(OH)_2(s) + H_2SO_4(aq) \rightarrow CaSO_4(aq) + 2H_2O(l)$$

Pre- and post-combustion methods of reducing sulfur dioxide emissions

Sulfur is present in fossil fuels such as coal and various fuels obtained from crude oil, such as gasoline (petrol). When such fuels are burnt sulfur dioxide is produced, which can contribute to acid rain. To produce a more environmentally-friendly fuel the sulfur can either be removed before the fuel is burnt (pre-combustion desulfurisation) or the SO_2 can be removed after the fuel has been burnt (post-combustion desulfurisation).

The regulations governing how much sulfur can be present in gasoline and diesel are quite strict and so sulfur is removed from these fuels before they are burnt. For instance, hydrodesulfurisation (hydrotreating) can be used – this involves heating crude oil fractions with hydrogen in the presence of a catalyst. This converts the sulfur to hydrogen sulfide (H_2S), which can be removed from the reaction mixture by bubbling it through an alkaline solution. The H_2S can be subsequently converted back into sulfur and sold to other companies to make sulfuric acid.

Post-combustion methods are commonly used in coal-fired power stations and involve passing the exhaust gases from the furnace through a vessel where the sulfur dioxide can react with alkalis/bases such as calcium oxide, calcium carbonate or calcium hydroxide. Calcium carbonate reacts with sulfur dioxide to form calcium sulfite (calcium sulfate(IV)):

$$CaCO_3(s) + SO_2(g) \rightarrow CaSO_3(s) + CO_2(g)$$

Nature of science

Acid rain is a problem that has been created by industrialisation. The work of chemists is vital in reducing the impact of acid rain by developing processes to limit the amounts of acidic gases released into the atmosphere. The problem of acid rain cannot however just be solved by scientists and it requires cooperation between scientists and politicians at a national and international level.

? Test yourself

21 Explain why rain is naturally acidic.

22 Give the formulas of two acids that arise from human activities and that are present in acid rain.

8.7 Calculations involving acids and bases (HL)

Learning objectives

- Understand what is meant by an acid dissociation constant and a base ionisation constant
- Arrange acids and bases in order of strength based on K_a, pK_a, K_b or pK_b values
- Carry out calculations involving K_a, K_b, pK_a, pK_b, pH and pOH
- Use the relationships $K_w = K_a \times K_b$ and $pK_a + pK_b = pK_w$

The acid dissociation constant – K_a

Consider the dissociation of ethanoic acid:

$$CH_3COOH(aq) + H_2O(l) \rightleftharpoons CH_3COO^-(aq) + H_3O^+(aq)$$

We can write an equilibrium expression for the dissociation of an acid:

$$K_c = \frac{[CH_3COO^-(aq)][H_3O^+(aq)]}{[CH_3COOH(aq)][H_2O(l)]}$$

However, the reaction occurs in aqueous solution, and so the concentration of water is very much higher than the other concentrations and is essentially constant ($55.5 \, mol \, dm^{-3}$). This can be incorporated in the equilibrium expression:

$$K_c \times [H_2O(l)] = \frac{[CH_3COO^-(aq)][H_3O^+(aq)]}{[CH_3COOH(aq)]}$$

$K_c \times [H_2O(l)]$ is a new constant called the **acid dissociation constant** (or acid ionisation constant) and is given the symbol K_a. So we can write:

$$K_a = \frac{[CH_3COO^-(aq)][H_3O^+(aq)]}{[CH_3COOH(aq)]}$$

The value of K_a for ethanoic acid at $25\,°C$ is 1.74×10^{-5}.

If we write the dissociation of ethanoic acid as:

$$CH_3COOH(aq) \rightleftharpoons CH_3COO^-(aq) + H^+(aq)$$

we can write an equilibrium expression for the dissociation of the acid as:

$$K_a = \frac{[CH_3COO^-(aq)][H^+(aq)]}{[CH_3COOH(aq)]}$$

This is entirely equivalent to the expression above.

Weak acids dissociate partially, and different weak acids dissociate to different extents. The acid dissociation constant is a measure of the extent to which an acid dissociates. The stronger the acid, the more it dissociates and the higher the value of K_a (the concentrations on the top of the expression are larger and that on the bottom is smaller).

> As with other equilibrium constants, the value of K_a depends on the temperature.

> In general, for the dissociation of acid, HA:
>
> $$HA(aq) \rightleftharpoons H^+(aq) + A^-(aq)$$
>
> The expression for the acid dissociation constant is:
>
> $$K_a = \frac{[A^-(aq)][H^+(aq)]}{[HA(aq)]}$$

> $$K_a = 10^{-pK_a}$$

pK_a

Acid dissociation constants can be expressed in a more convenient form by taking the negative logarithm to base 10 of K_a:

$$pK_a = -\log_{10} K_a$$

A low value of K_a corresponds to a high value of pK_a and vice versa, so that the lower a pK_a value is the stronger the acid (Table **8.4**).

You can think about this in the same way as for pH – the lower the pH value, the higher the $[H^+(aq)]$.

Acid	Formula	K_a	pK_a
hydrofluoric acid	HF	5.62×10^{-4}	3.25
benzoic acid	C_6H_5COOH	6.31×10^{-5}	4.20
propanoic acid	CH_3CH_2COOH	1.35×10^{-5}	4.87
chloric(I) acid	HOCl	3.72×10^{-8}	7.43
hydrocyanic acid	HCN	3.98×10^{-10}	9.40

Table 8.4 K_a and pK_a values of some acids – decreasing in strength downwards.

The acids in Table **8.4** are arranged in order of strength – HF is the strongest acid and dissociates to the greatest extent; HCN is the weakest acid and dissociates to the smallest extent in aqueous solution.

> The higher the value of K_a, the stronger the acid. The lower the value of pK_a the stronger the acid.

> K_a and pK_a are better measures of acid strength than pH because their values do not depend on the concentration of the acid – K_a and pK_a depend only on temperature.

Worked examples

8.3 Calculate the K_a value for methanoic acid at 25 °C if a $0.100 \, mol \, dm^{-3}$ solution dissociates to give a hydrogen ion concentration of $4.13 \times 10^{-3} \, mol \, dm^{-3}$.

Methanoic acid dissociates according to the equation: $HCOOH(aq) \rightleftharpoons HCOO^-(aq) + H^+(aq)$

If we consider the situation before dissociation occurs we have:

$$HCOOH(aq) \rightleftharpoons HCOO^-(aq) + H^+(aq)$$

initial concentration / $mol \, dm^{-3}$:   0.100   0   0

At equilibrium the concentration of H^+ ions is $4.13 \times 10^{-3} \, mol \, dm^{-3}$. To produce $4.13 \times 10^{-3} \, mol \, dm^{-3} \, H^+$, $4.13 \times 10^{-3} \, mol \, dm^{-3}$ HCOOH must dissociate. The concentration of HCOOH at equilibrium is thus $0.100 - 4.13 \times 10^{-3}$, i.e. $9.59 \times 10^{-2} \, mol \, dm^{-3}$.

Because one molecule of HCOOH dissociates to produce one H^+ ion and one $HCOO^-$ ion, the concentration of $HCOO^-$ at equilibrium is the same as that of H^I.

> we are assuming that we can ignore any contribution to $[H^+(aq)]$ from the dissociation of water.

So the equilibrium concentrations are:

$$HCOOH(aq) \rightleftharpoons HCOO^-(aq) + H^+(aq)$$

equilibrium concentration / $mol \, dm^{-3}$: 9.59×10^{-2} 4.13×10^{-3} 4.13×10^{-3}

These values can be put into the expression for K_a:

$$K_a = \frac{[HCOO^-(aq)][H^+(aq)]}{[HCOOH(aq)]} = \frac{(4.13 \times 10^{-3}) \times (4.13 \times 10^{-3})}{(9.59 \times 10^{-2})} = 1.78 \times 10^{-4}$$

If the degree of dissociation is very small compared with the concentration of the acid, the above calculation may be simplified, as shown in the next example.

8.4 Calculate the K_a value for HCN(aq) at 25 °C if a 0.500 mol dm^{-3} solution dissociates to give a hydrogen ion concentration of 1.41×10^{-5} mol dm^{-3}.

The dissociation of HCN is represented by the equation: $HCN(aq) \rightleftharpoons H^+(aq) + CN^-(aq)$

The concentration of CN^- at equilibrium will be the same as that of H^+. We will make the assumption that, because the dissociation of the acid is so small, the concentration of HCN at equilibrium is essentially the same as the initial concentration. We can now substitute these values into the expression for K_a:

$$K_a = \frac{[CN^-(aq)][H^+(aq)]}{[HCN(aq)]} = \frac{(1.41 \times 10^{-5}) \times (1.41 \times 10^{-5})}{(0.500)} = 3.98 \times 10^{-10}$$

This is the same as the literature value for K_a, and so it can be seen that the approximation works well when the degree of dissociation is so small. If we had made the same approximation in the calculation above on methanoic acid, the value obtained for K_a would have been 1.71×10^{-4}, which is reasonably close to the accepted value. This approximation will be discussed further below.

The base ionisation (dissociation) constant – K_b

Consider the ionisation of a weak base:

$$B(aq) + H_2O(l) \rightleftharpoons BH^+(aq) + OH^-(aq)$$

We can write an expression for the **base ionisation constant** (base dissociation constant), K_b:

$$K_b = \frac{[BH^+(aq)][OH^-(aq)]}{[B(aq)]}$$

As above, the reaction occurs in aqueous solution, so the concentration of water is very much higher than the other concentrations and is essentially constant. It is therefore omitted from the expression for K_b.

The value of K_b for ammonia is 1.78×10^{-5}. Similar to K_a for acids, K_b for bases provides a measure of the extent to which a base ionises, and hence the strength of the base.

pK_b

As for acids, the negative logarithm to base 10 of the base ionisation constant can be taken for a more convenient measure.

$K_b = 10^{-pK_b}$

$$pK_b = -\log_{10} K_b$$

> The higher the value of K_b, the more the base ionises and the stronger it is – the lower the value of pK_b, the stronger the base.

The K_b and pK_b values for some bases are shown in Table **8.5**. They are arranged in order of strength – dimethylamine is the strongest base and ionises to the greatest extent; phenylamine is the weakest base and ionises to the smallest extent in aqueous solution.

Base	Formula	K_b	pK_b
dimethylamine	$(CH_3)_2NH$	5.25×10^{-4}	3.28
methylamine	CH_3NH_2	4.37×10^{-4}	3.36
sodium carbonate	Na_2CO_3	2.09×10^{-4}	3.68
ammonia	NH_3	1.78×10^{-5}	4.75
phenylamine	$C_6H_5NH_2$	4.17×10^{-10}	9.38

Table 8.5 K_b and pK_b values for some bases – decreasing in strength downwards.

Worked example

8.5 Calculate the K_b value for ethylamine at $25\,°C$ if a $0.100\,mol\,dm^{-3}$ solution ionises to give a hydroxide ion concentration of $7.06 \times 10^{-3}\,mol\,dm^{-3}$.

The equation for the ionisation of ethylamine is:

$$CH_3CH_2NH_2(aq) + H_2O(l) \rightleftharpoons CH_3CH_2NH_3^+(aq) + OH^-(aq)$$

When one molecule of $CH_3CH_2NH_2$ reacts with water, it ionises to form one $CH_3CH_2NH_3^+$ ion and one OH^- ion. This means that, if $7.06 \times 10^{-3}\,mol\,dm^{-3}$ of OH^- are present at equilibrium, they must have come from the ionisation of $7.06 \times 10^{-3}\,mol\,dm^{-3}$ $CH_3CH_2NH_2$. The concentration of $CH_3CH_2NH_2$ at equilibrium is therefore $0.100 - 7.06 \times 10^{-3} = 9.29 \times 10^{-2}\,mol\,dm^{-3}$. The concentration of OH^- formed is the same as that of $CH_3CH_2NH_3^+$ (again, ignoring any contribution from the dissociation of water).

So at equilibrium, the concentrations are:

$$CH_3CH_2NH_2(aq) + H_2O(l) \rightleftharpoons CH_3CH_2NH_3^+(aq) + OH^-(aq)$$

equilibrium concentration / $mol\,dm^{-3}$: 9.29×10^{-2} 7.06×10^{-3} 7.06×10^{-3}

The expression for K_b is:

$$K_b = \frac{[CH_3CH_2NH_3^+(aq)][OH^-(aq)]}{[CH_3CH_2NH_2(aq)]}$$

The equilibrium concentrations can be substituted into this:

$$K_b = \frac{[7.06 \times 10^{-3}][7.06 \times 10^{-3}]}{[9.29 \times 10^{-2}]} = 5.37 \times 10^{-4}$$

As for the K_a calculation above, this can be simplified by making the approximation that the amount of ionisation of the base is small compared with its concentration, so the original concentration of the base could be used in the K_b expression.

Calculating the pH of a solution of a weak acid

A weak acid dissociates only partially, and therefore in order to calculate the pH we need to know how much the acid dissociates. This can be worked out using the K_a or pK_a value of the acid.

Worked example

8.6 Calculate the pH of a $0.100 \, mol \, dm^{-3}$ solution of ethanoic acid at $25\,°C$.

The pK_a of ethanoic acid at $25\,°C$ is 4.76.

$K_a = 10^{-pK_a}$

$ = 10^{-4.76}$

> '2nd/SHIFT log' (-4.76) on the calculator

$ = 1.74 \times 10^{-5}$

The equation for the dissociation of ethanoic acid is:

$$CH_3COOH(aq) \rightleftharpoons CH_3COO^-(aq) + H^+(aq)$$

and its equilibrium expression is:

$$K_a = \frac{[CH_3COO^-(aq)][H^+(aq)]}{[CH_3COOH(aq)]}$$

Because one molecule of CH_3COOH dissociates to form one CH_3COO^- ion and one H^+ ion, the concentration of CH_3COO^- and H^+ ions in the solution will be equal:

$$[CH_3COO^-(aq)] = [H^+(aq)]$$

We will make the assumption that the dissociation of the acid is negligible compared with the concentration of the acid – so the concentration of the acid at equilibrium is the same as its initial concentration, $0.100 \, mol \, dm^{-3}$ in this case.

These terms, together with the K_a value, are substituted into the K_a expression:

$$1.74 \times 10^{-5} = \frac{[H^+(aq)]^2}{0.100}$$

$$[H^+(aq)]^2 = 1.74 \times 10^{-5} \times 0.100 = 1.74 \times 10^{-6}$$

$$[H^+(aq)] = \sqrt{(1.74 \times 10^{-6})} = 1.32 \times 10^{-3} \, mol \, dm^{-3}$$

Now, $pH = -\log_{10}[H^+(aq)]$

So $pH = -\log_{10}(1.32 \times 10^{-3})$

$ = 2.88$

If c is the concentration of the acid, the hydrogen ion concentration can also be worked out using:

$$[H^+(aq)] = \sqrt{(c \times K_a)}$$

or the pH could be worked out using:

$$pH = \tfrac{1}{2}(pK_a - \log_{10} c) \text{ or } pH = -\tfrac{1}{2}\log_{10}(c \times K_a)$$

Calculating K_a for a weak acid from its pH

Worked example

8.7 Calculate the value of the acid dissociation constant for benzoic acid if a $0.250 \, \text{mol dm}^{-3}$ solution has a pH of 2.40.

The equation for the dissociation of benzoic acid is:

$$C_6H_5COOH(aq) \rightleftharpoons C_6H_5COO^-(aq) + H^+(aq)$$

$$K_a = \frac{[C_6H_5COO^-(aq)][H^+(aq)]}{[C_6H_5COOH(aq)]}$$

$[H^+(aq)]$ can be worked out from the pH:

$$[H^+(aq)] = 10^{-pH}$$

$$[H^+(aq)] = 10^{-2.40} = 3.98 \times 10^{-3} \, \text{mol dm}^{-3}$$

To produce $3.98 \times 10^{-3} \, \text{mol dm}^{-3}$ H^+, $3.98 \times 10^{-3} \, \text{mol dm}^{-3}$ C_6H_5COOH must dissociate. The concentration of C_6H_5COOH at equilibrium is thus $0.250 - 3.98 \times 10^{-3} \, \text{mol dm}^{-3}$, i.e. $0.246 \, \text{mol dm}^{-3}$.

Because one molecule of C_6H_5COOH dissociates to produce one H^+ ion and one $C_6H_5COO^-$ ion, the concentration of $C_6H_5COO^-$ at equilibrium is the same as that of H^+.

The equilibrium concentrations are thus:

$$C_6H_5COOH(aq) \rightleftharpoons C_6H_5COO^-(aq) + H^+(aq)$$

equilibrium concentration / mol dm^{-3}: 0.246 3.98×10^{-3} 3.98×10^{-3}

These values can be put into the expression for K_a:

$$K_a = \frac{[C_6H_5COO^-(aq)][H^+(aq)]}{[C_6H_5COOH(aq)]}$$

$$K_a = \frac{(3.98 \times 10^{-3}) \times (3.98 \times 10^{-3})}{0.246}$$

$$K_a = 6.44 \times 10^{-5}$$

If the approximation had been made that the effect of the dissociation of the acid on the concentration of the acid at equilibrium can be ignored, the answer 6.34×10^{-5} would have been obtained.

The answer 6.34×10^{-5} could also have been obtained using the equation:

$$K_a = \frac{10^{-2pH}}{c}$$

where c is the concentration of the acid.

Calculating the pOH or pH of a solution of a base

pOH

Calculating the pH of a solution of a base can be simplified using the idea of pOH.

$$pOH = -\log_{10}[OH^-(aq)]$$

Consider the expression for the ionic product constant for water:

$$K_w = [H^+(aq)][OH^-(aq)]$$

Working out $-\log_{10}$ for both sides gives:

$$-\log_{10}K_w = -\log_{10}([H^+(aq)][OH^-(aq)])$$

$$-\log_{10}K_w = -\log_{10}[H^+(aq)] + -\log_{10}[OH^-(aq)]$$

We can rewrite this:

$pK_w = -\log_{10}K_w$
'p' stands for '$-\log_{10}$'

$$pK_w = pH + pOH$$

Now, the value of pK_w at 25 °C is $-\log_{10}1.0\times10^{-14}$, which is 14; so we can rewrite the equation above:

$$pH + pOH = 14$$

Worked example

8.8 Calculate the pH of $0.0500\,mol\,dm^{-3}$ sodium hydroxide solution at 25 °C.

NaOH is a strong base and ionises completely to form $0.0500\,mol\,dm^{-3}\,OH^-(aq)$.

$$pOH = -\log_{10}(0.0500) = 1.30$$

At 25 °C, $pH + pOH = 14$

$$pH + 1.30 = 14$$

$$pH = 12.7$$

pH, pOH and neutrality

At 25 °C, neutral pH is 7. At that temperature, solutions that have a pH lower than 7 are acidic, and those that have a pH higher than 7 are alkaline.

Using pOH the situation is reversed – solutions with a pOH lower than 7 are alkaline, and those with a pOH higher than 7 are acidic.

Now $pH + pOH = pK_w$ so if $pH < pOH$ a solution is acidic, and if $pOH < pH$ the solution is alkaline.

In pure water (or a neutral solution), $pH = pOH$.

Calculating pH or pOH for a weak base

Because a weak base ionises only partially in aqueous solution, we must use K_b or pK_b to work out the concentration of OH^- ions present in a solution.

Worked example

8.9 Calculate pH and pOH of a $0.120 \, mol \, dm^{-3}$ solution of ammonia, given that its pK_b is 4.75 at 25 °C.

$$K_b = 10^{-pK_b}$$

$$K_b = 10^{-4.75} = 1.78 \times 10^{-5}$$

The ionisation of ammonia involves:

$$NH_3(aq) + H_2O(aq) \rightleftharpoons NH_4^+(aq) + OH^-(aq)$$

$$K_b = \frac{[NH_4^+(aq)][OH^-(aq)]}{[NH_3(aq)]}$$

One NH_3 molecule ionises to produce one NH_4^+ ion and one OH^- ion. This means that the concentration of NH_4^+ is equal to the concentration of OH^- and we can write:

$$K_b = \frac{[OH^-(aq)]^2}{[NH_3(aq)]}$$

We will make the approximation that the concentration of NH_3 at equilibrium is equal to the initial concentration – i.e. that the extent of ionisation of the base is negligible compared with its concentration. Therefore we take $[NH_3(aq)]$ as $0.120 \, mol \, dm^{-3}$. Substituting this value and the value for K_b into the expression for K_b:

$$1.78 \times 10^{-5} = \frac{[OH^-(aq)]^2}{0.120}$$

$$[OH^-(aq)]^2 = 1.78 \times 10^{-5} \times 0.120$$

$$= 2.14 \times 10^{-6}$$

Therefore, $[OH^-(aq)] = 1.46 \times 10^{-3} \, mol \, dm^{-3}$

and $pOH = -\log_{10}[OH^-(aq)]$

$$= -\log_{10}(1.46 \times 10^{-3})$$

$$= 2.84$$

But $pOH + pH = pK_w$

At 25 °C, $pOH + pH = 14$.

Therefore, $pH = 14 - 2.84$

$$= 11.16.$$

If c is the concentration of the base, the hydroxide ion concentration can also be worked out using:

$$[OH^-(aq)] = \sqrt{(c \times K_b)}$$

pOH can be worked out using:

$$pOH = -\tfrac{1}{2}\log_{10}(c \times K_b) \text{ or } pOH = \tfrac{1}{2}(pK_b - \log_{10} c)$$

The pH could be worked out using:

$$pH = 14 + \tfrac{1}{2}\log_{10}(c \times K_b) \text{ or } pH = 14 - \tfrac{1}{2}(pK_b - \log_{10} c)$$

Temperature/°C	K_w
0	1.14×10^{-15}
5	1.86×10^{-15}
10	2.93×10^{-15}
15	4.52×10^{-15}
20	6.81×10^{-15}
25	1.01×10^{-14}
30	1.47×10^{-14}
35	2.09×10^{-14}
40	2.92×10^{-14}
45	4.02×10^{-14}
50	5.48×10^{-14}

Table 8.6 The variation of K_w with temperature.

The variation of K_w with temperature

Like all other equilibrium constants, the value of K_w varies with temperature (Table **8.6** and Figure **8.14**). We can use the data in Table **8.6** to work out the pH of water at any temperature.

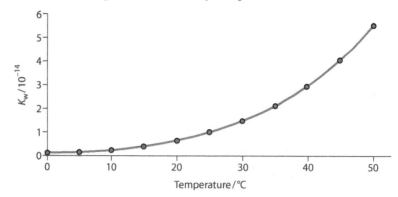

Figure 8.14 The variation of K_w with temperature.

Worked example

8.10 Using the values in Table **8.6**, calculate the pH of water at 40 °C.

From the table, the value of K_w at this temperature is 2.92×10^{-14}.

$$pK_w = -\log_{10}(2.92 \times 10^{-14}) = 13.53$$

$$pH + pOH = pK_w$$

$$pH + pOH = 13.53$$

Water dissociates according to the equation:

$$H_2O(l) \rightleftharpoons H^+(aq) + OH^-(aq)$$

Because one H_2O molecule dissociates to form one H^+ ion and one OH^- ion in pure water, the concentration of H^+ ions is equal to the concentration of OH^- ions, and therefore pH = pOH.

So $2 \times pH = 13.53$ and $pH = 6.77$

Therefore the pH of pure water at 40 °C is 6.77.

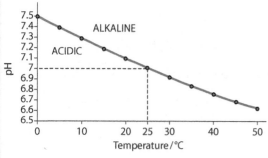

Figure 8.15 The variation of neutral pH with temperature.

This does not, however, mean that the water is acidic at 40 °C, and we must now modify our definition of 'neutral'. Pure water is neutral because $[H^+(aq)] = [OH^-(aq)]$. Because K_w equals 1.0×10^{-14} only at 25 °C, we can also see that pH 7 is neutral only at 25 °C!

The graph in Figure **8.15** shows the variation of neutral pH with temperature. Any pH value above the red line indicates an alkaline pH at that temperature; any pH below the line indicates an acidic pH at that temperature. So at 10 °C a pH of 7.1 is in the acidic region, below the line, and represents an acidic pH at this temperature!

A solution is neutral if $[H^+(aq)] = [OH^-(aq)]$; a solution is acidic if $[H^+(aq)] > [OH^-(aq)]$, and a solution is alkaline if $[OH^-(aq)] > [H^+(aq)]$.

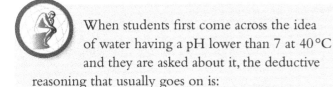

When students first come across the idea of water having a pH lower than 7 at 40°C and they are asked about it, the deductive reasoning that usually goes on is:

- all liquids with a pH lower than 7 are acidic
- at 40°C water is a liquid with a pH lower than 7
- therefore water is acidic at 40°C.

The conclusion of this reasoning is incorrect, but the argument is valid. The problem is, of course, with the first premise which, we have just learnt, is incorrect. Can you reword the first premise to make it true and construct a valid argument?

There is also another problem here – when I first thought about demonstrating this argument, I wanted to write the first premise as 'all solutions with pH less than 7 are acidic'. Why did I choose the word 'liquid' instead and does this introduce any extra difficulties in trying to come up with a completely true version of the first premise?

Another problem we encounter the first time we come across the idea of water having a pH other than 7 is, of course, that we 'know' that the neutral pH is 7 because we have been told it from an early age – it is even stated earlier in this chapter! The idea of water having a pH of anything other than 7 goes against everything we have learnt. If you obtained this result experimentally using a pH meter before you learnt about the theory of pH, would you dismiss the result as wrong (the pH meter had been poorly calibrated!) and ignore it?

The pH of strong acids/bases at different temperatures

It can be seen from the above discussion that pH varies with temperature and therefore we can work out the pH of any solution given the value of K_w or pK_w at that temperature.

Worked example

8.11 Calculate the pH of $0.0500 \, \text{mol dm}^{-3}$ sodium hydroxide solution at 50°C given that $pK_w = 13.3$ at this temperature.

$pOH = -\log_{10}(0.0500) = 1.30$

$pH + pOH = pK_w$

$pH + pOH = 13.3$

$pH + 1.30 = 13.3$

$pH = 12.0$ (it is 12.7 at 25°C)

23 HA to HF are all weak acids.

  a Copy and complete the following table:

Acid	Concentration / mol dm^{-3}	$[H^+]$/ mol dm^{-3}	pH	K_a	pK_a
HA	0.0100	2.00×10^{-6}			
HB	0.200		4.70		
HC	0.500	2.50×10^{-4}			
HD	2.20×10^{-2}		5.20		
HE	0.250			3.72×10^{-8}	
HF	0.0300			4.96×10^{-6}	

  b Use the values in part a to arrange the acids in order of acid strength (strongest first).

24 B_1, B_2 and B_3 are weak bases.

  a Copy and complete the table:

Base	Concentration of base/ mol dm^{-3}	$[OH^-]$/mol dm^{-3}	K_b	pK_b
B_1	0.100	1.33×10^{-3}		
B_2	0.250		5.75×10^{-5}	
B_3	0.0200	4.70×10^{-4}		

  b Arrange the bases in order of increasing strength (weakest first).

25 Arrange the following acids in order of increasing strength (weakest first):

Acid	pK_a
HOCl	7.4
$HClO_2$	2.0

Acid	K_a
HOI	3.0×10^{-11}
HNO_2	4.6×10^{-4}

26 Calculate pH values for the following solutions:

  a a 0.0200 mol dm^{-3} solution of propanoic acid, given that the pK_a value at 25 °C is 4.87.

  b a 0.500 mol dm^{-3} solution of HCN, given that the pK_a value at 25 °C is 9.40.

  c a 0.002 50 mol dm^{-3} solution of phenol (C_6H_5OH), given that the pK_a value at 25 °C is 10.00.

27 Calculate the pH of pure water at each of the following temperatures

  a at 5 °C, K_w is 1.86×10^{-15}

  b at 35 °C, K_w is 2.09×10^{-14}

  c at 100 °C, K_w is 5.13×10^{-13}

28 Calculate the pOH of the following solutions at 25 °C:

  a 0.0300 mol dm^{-3} KOH

  b 0.500 mol dm^{-3} NaOH

  c 0.0200 mol dm^{-3} HCl

29 Calculate the pH values for the following solutions:

  a a 0.0500 mol dm^{-3} solution of ethylamine ($CH_3CH_2NH_2$) at 25 °C, given that the K_b value is 5.37×10^{-4} at that temperature.

  b a 0.0100 mol dm^{-3} solution of phenylamine ($C_6H_5NH_2$) at 25 °C, given that the K_b value is 4.17×10^{-10} at that temperature.

30 Calculate the pH of each of the following solutions at the given temperature:

  a 0.1000 mol dm^{-3} NaOH at 40 °C ($pK_w = 13.53$)

  b 0.1000 mol dm^{-3} HCl at 30 °C ($pK_w = 13.83$)

  c 0.02000 mol dm^{-3} KOH at 30 °C ($pK_w = 13.83$)

The relationship between K_a and K_b

Consider the equilibria for an acid HA and its conjugate base A⁻:

$$HA(aq) \rightleftharpoons A^-(aq) + H^+(aq) \qquad \textbf{equation 1}$$

$$A^-(aq) + H_2O(l) \rightleftharpoons HA(aq) + OH^-(aq) \qquad \textbf{equation 2}$$

We can write the K_a expression for equation 1:

$$K_a = \frac{[A^-(aq)][H^+(aq)]}{[HA(aq)]}$$

and the K_b expression for equation 2:

$$K_b = \frac{[HA(aq)][OH^-(aq)]}{[A^-(aq)]}$$

If we multiply K_a by K_b we get:

$$K_a \times K_b = \frac{[\cancel{A^-}(aq)][H^+(aq)]}{[\cancel{HA}(aq)]} \times \frac{[\cancel{HA}(aq)][OH^-(aq)]}{[\cancel{A^-}(aq)]}$$

Certain terms cancel.

$$K_a \times K_b = [H^+(aq)] \times [OH^-(aq)]$$

But $[H^+(aq)] \times [OH^-(aq)] = K_w$, so we can write:

$$K_a \times K_b = K_w$$

This relationship works *only* for a conjugate acid–base pair, e.g. for NH_4^+ ($K_a = 5.62 \times 10^{-10}$) and for NH_3 ($K_b = 1.78 \times 10^{-5}$):

$$K_a \times K_b = 5.62 \times 10^{-10} \times 1.78 \times 10^{-5} = 1.00 \times 10^{-14}$$

which is the value for K_w at 25 °C.

If we take negative logarithms to base 10 of each side in the expression $K_a \times K_b = K_w$, we get the relationship:

$$pK_a + pK_b = pK_w$$

At 25 °C the value of pK_w is 14, so, at 25 °C, $pK_a + pK_b = 14$.

The pK_a for ethanoic acid (CH_3COOH) is 4.76 and the pK_b for its conjugate base, the ethanoate ion (CH_3COO^-), is 9.24 at 25 °C.

$4.76 + 9.24 = 14$

The relationship between the strength of an acid and the strength of its conjugate base

For a conjugate acid–base pair, $pK_a + pK_b = pK_w$ (or $K_a \times K_b = K_w$). This means that the lower the value of pK_a for an acid, the higher the value of pK_b for its conjugate base. A lower value of pK_a indicates a stronger acid and a higher value of pK_b indicates a weaker base. This leads to the conclusion that was discussed earlier:

The stronger the acid, the weaker its conjugate base.

Acid	pK_a		Conjugate base	pK_b	
HF	3.25		F$^-$	10.75	
C$_6$H$_5$COOH	4.20		C$_6$H$_5$COO$^-$	9.80	
CH$_3$CH$_2$COOH	4.87	increasing strength	CH$_3$CH$_2$COO$^-$	9.13	increasing strength
HOCl	7.43		OCl$^-$	6.57	
HCN	9.40		CN$^-$	4.60	

Table 8.7 The relationship between the strength of an acid and its conjugate base.

HCN is the weakest acid in Table **8.7** and has the strongest conjugate base. HCN has very little tendency to dissociate, according to the equation:

$$HCN(aq) \rightleftharpoons H^+(aq) + CN^-(aq)$$

and CN$^-$ has a strong tendency to react with water to re-form the parent acid:

$$CN^-(aq) + H_2O(l) \rightleftharpoons HCN(aq) + OH^-(aq)$$

Nature of science

We started the chapter with a qualitative discussion of acids and bases and have now considered a mathematical description of acidity. It is the goal of many areas of science to describe natural phenomena mathematically. Depending on the sophistication of the model, more or fewer assumptions/approximations are used. The development of more sophisticated equipment, such as the pH meter, can allow testing and refinement of the models.

? Test yourself

31 Copy and complete the following table, which contains values measured at 25 °C. The first row has been done for you.

Acid	K_a	pK_a	Conjugate base	K_b	pK_b
HCN	3.98×10^{-10}	9.40	CN$^-$	2.51×10^{-5}	4.60
HF	5.62×10^{-4}				
HIO$_3$		0.8			
NH$_4^+$					4.75
			CH$_3$COO$^-$	5.75×10^{-10}	
		10.64	CH$_3$NH$_2$		

8.8 pH curves (HL)

8.8.1 Acid–base titrations

We met the use of acid–base titrations in Topic **1** for determining the concentration of one solution by reacting it with another that we know the concentration of (a standard solution). The pH of the mixture does not vary linearly during titration and just how it changes during the process depends on whether the reactants are strong or weak acids/bases and what their concentrations are. Even when equivalent numbers of moles of acid and alkali have been added the pH is not necessarily 7. This is why different indicators, which change colour at different pH values, are required for different titrations – for instance, phenolphthalein changes colour around pH 9 and is usually used for weak acid–strong base titrations. This is explained in more detail on page **352**.

Here we will look at **titration curves**, which show how pH varies during titrations involving different combinations of strong and weak acids and bases. The pH can be monitored during a titration by using a pH meter or a pH probe attached to a data logger (Figure **8.16**).

Strong acid–strong base

The titration curve (pH curve) for adding $0.100\,mol\,dm^{-3}$ sodium hydroxide solution to $25.0\,cm^3$ of $0.100\,mol\,dm^{-3}$ hydrochloric acid is shown in Figure **8.17**.

The reaction that occurs is:

$$HCl(aq) + NaOH(aq) \rightarrow NaCl(aq) + H_2O(l)$$

The initial pH is 1.0 because the initial solution is $0.100\,mol\,dm^{-3}$ hydrochloric acid. Hydrochloric acid is a strong acid and dissociates completely:

$$HCl(aq) \rightarrow H^+(aq) + Cl^-(aq)$$

The concentration of H^+ ions is $0.100\,mol\,dm^{-3}$ and the $pH = -\log_{10} 0.10$, i.e. 1.00.

Learning objectives

- Sketch titration curves (pH curves) for titrations involving any combination of strong and weak acids and bases
- Explain the important features of titration curves
- Understand how to work out pK_a or pK_b values from a titration curve
- Explain how an acid–base indicator works
- Understand what is meant by the pH range of an indicator
- Select a suitable indicator for a titration

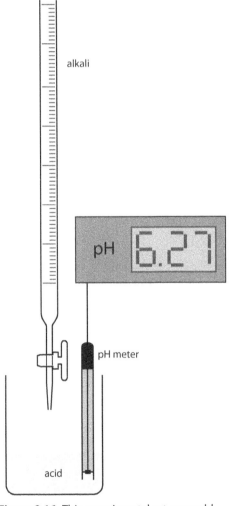

Figure 8.16 This experimental set-up could be used to track the pH of a solution in a titration.

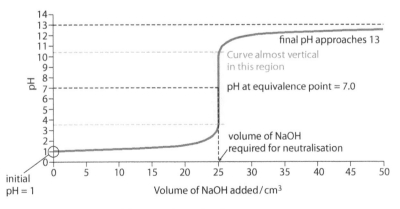

Figure 8.17 A strong acid–strong base titration curve.

From Figure **8.17** it can be seen that as NaOH is added, the pH initially changes very little. The pH after $20.0 \, cm^3$ of NaOH has been added is 1.95, so adding $20.0 \, cm^3$ of the sodium hydroxide has changed the pH of the solution by less than 1 unit. This is partly a consequence of the fact that the $H^+(aq)$ ion concentration is reasonably high in this region and therefore adding NaOH does not affect the overall pH very much. However, it also arises from the nature of the log scale – the $[H^+(aq)]$ in this region decreases by about $0.09 \, mol \, dm^{-3}$ from $0.100 \, mol \, dm^{-3}$ to $0.010 \, mol \, dm^{-3}$, but a ten fold change in the $[H^+(aq)]$ corresponds to a change in pH of only one unit.

The pH then increases much more rapidly, and between a pH of about 3.5 and 10.5 the curve is virtually vertical. In this region (shaded on the graph), the addition of about $0.3 \, cm^3$ of NaOH causes the pH to change by 7 units. In this region $[H^+(aq)]$ is low and adding NaOH has a larger effect on the pH. The fact that this is a log scale, however, magnifies the effect – in this region the $[H^+(aq)]$ actually decreases by only about $0.000 \, 31 \, mol \, dm^{-3}$, but because it decreases from $3.15 \times 10^{-4} \, mol \, dm^{-3}$ to $3.15 \times 10^{-11} \, mol \, dm^{-3}$, the pH changes by 7 units.

The **equivalence point** is the point at which equivalent numbers of moles of acid and base have been added. The pH at the equivalence point for a strong acid–strong base titration is 7.0.

In this case, because the concentrations of the acid and base are equal, the equivalence point occurs when equal volumes of the acid and base have been added.

After this region, the pH changes only gradually again, and as more and more NaOH is added the pH gets closer to 13, which is the pH of $0.100 \, mol \, dm^{-3}$ NaOH(aq). All the acid has been neutralised and the OH^- ions are in excess in this region. As more and more NaOH is added, the solution resembles more and more closely the original NaOH solution (when $200 \, cm^3$ have been added the pH is 12.89).

Variations on strong acid–strong base titrations

If, instead of adding $0.100 \, mol \, dm^{-3}$ NaOH to $25 \, cm^3$ of $0.100 \, mol \, dm^{-3}$ HCl, we add $0.200 \, mol \, dm^{-3}$ NaOH then we would get the titration curve shown in Figure **8.18a**. The equivalence point would now occur when $12.5 \, cm^3$ of the alkali has been added. The initial pH would be the same and the pH at the equivalence point still 7.0, but the final pH will tend towards 13.3, the pH of $0.200 \, mol \, dm^{-3}$ NaOH(aq).

If we used $1.00 \, mol \, dm^{-3}$ HCl and $1.00 \, mol \, dm^{-3}$ NaOH for the titration:

- the initial pH would be lower (0.0)
- the steep part would be longer
- the final pH would tend towards 14.0 – the pH of $1.00 \, mol \, dm^{-3}$ NaOH(aq)
- the pH at the equivalence point would still be 7.0.

If we had performed the original titration the other way round, adding the acid to the alkali, the curve would have been reversed, as shown in Figure **8.18b**. The initial pH would have been 13.0 – the pH of $0.100 \, mol \, dm^{-3}$ NaOH(aq) – and the curve would tend towards 1.0, the pH of $0.100 \, mol \, dm^{-3}$ HCl(aq).

Note: the equivalence point is **not** the same as the end point – the end point will be discussed in the section on indicators (page **349**).

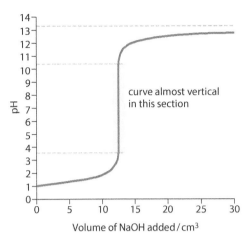
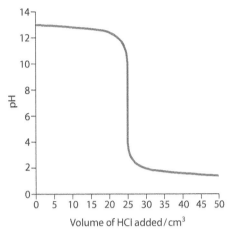

Figure 8.18 Titration curves: **a** adding $0.200\,mol\,dm^{-3}$ NaOH to $25\,cm^3$ of $0.100\,mol\,dm^{-3}$ HCl; **b** adding $0.100\,mol\,dm^{-3}$ HCl to $25\,cm^3$ of $0.100\,mol\,dm^{-3}$ NaOH.

Calculating the points on a strong acid–strong base titration curve

Consider again the original titration, i.e. the addition of $0.100\,mol\,dm^{-3}$ sodium hydroxide solution to $25.0\,cm^3$ of $0.100\,mol\,dm^{-3}$ hydrochloric acid (Figure **8.17**). We will look at how each point on the titration curve can be calculated.

Initial pH (no base added)

The initial pH is 1.00, because the initial solution is $0.100\,mol\,dm^{-3}$ hydrochloric acid. Hydrochloric acid is a strong acid and dissociates completely:

$$HCl(aq) \rightarrow H^+(aq) + Cl^-(aq)$$

The concentration of H^+ ions is $0.100\,mol\,dm^{-3}$ and the $pH = -\log_{10} 0.100$, i.e. 1.00.

In $25.0\,cm^3$ of $0.100\,mol\,dm^{-3}$ HCl, the number of moles of H^+ ions is given by:

no. moles = concentration × volume in dm^3

$$\text{no. moles of } H^+ = 0.100 \times \frac{25.0}{1000} = 2.50 \times 10^{-3}\,mol$$

Addition of $5.0\,cm^3$ of the base

The total volume of the solution is now:

$$25.0 + 5.0 = 30.0\,cm^3 \text{ or } 0.0300\,dm^3$$

The number of moles of OH^- ions in $5.0\,cm^3$ of solution is given by:

$$\text{no. moles of } OH^- = 0.100 \times \frac{5.0}{1000} = 5.0 \times 10^{-4}\,mol$$

These OH⁻ ions react with the H⁺ ions in the acid solution according to the equation:

$$H^+(aq) + OH^-(aq) \rightarrow H_2O(l)$$

H⁺ ions are in excess and 5.0×10^{-4} mol OH⁻ react with 5.0×10^{-4} mol H⁺. The original number of moles of H⁺ ions was 2.50×10^{-3}; therefore, if 5.0×10^{-4} mol react, the number of moles of H⁺ ions left over in the solution is $2.50 \times 10^{-3} - 5.0 \times 10^{-4}$ or 2.00×10^{-3} mol.

This number of moles of H⁺ is in $30.0 \, \text{cm}^3$ of solution; therefore the concentration of H⁺ ions in the solution is given by:

$$\text{concentration} = \frac{\text{no. moles}}{\text{volume in dm}^3}$$

$$[H^+(aq)] = \frac{2.00 \times 10^{-3}}{0.0300} = 6.67 \times 10^{-2} \, \text{mol dm}^{-3}$$

$$pH = -\log_{10}[H^+(aq)] = -\log_{10} 6.67 \times 10^{-2} = 1.18$$

This technique can be used to work out the pH of all the solutions up to the addition of just less than $25.0 \, \text{cm}^3$ NaOH.

If the calculation is attempted for $25.0 \, \text{cm}^3$ of NaOH added there is a problem because the number of moles of H⁺ ions in the original solution was $2.50 \times 10^{-3} \, \text{mol dm}^{-3}$ and the number of moles of OH⁻ that is added is $2.50 \times 10^{-3} \, \text{mol dm}^{-3}$. These will react together completely so that the number of moles of H⁺ left will be zero, but $-\log_{10} 0$ does not exist! The problem has arisen because, in all of the above calculations, we have ignored any H⁺ in the solution from the dissociation of water. In most cases we are totally justified in ignoring this. For instance, at $20.0 \, \text{cm}^3$ of NaOH added, the concentration of H⁺ ions in the solution from the hydrochloric acid is $0.0111 \, \text{mol dm}^{-3}$, whereas that from the dissociation of water is $9.00 \times 10^{-13} \, \text{mol dm}^{-3}$ – the dissociation of water accounts for $8.1 \times 10^{-9}\%$ of the H⁺ ion concentration of the solution. Even at $24.999 \, \text{cm}^3$ of NaOH added, this percentage has only risen to about 0.25%.

However, at $25.0 \, \text{cm}^3$ of NaOH added, the H⁺ ion concentration of the solution is entirely due to the dissociation of water and the H⁺ ion concentration is $1.00 \times 10^{-7} \, \text{mol dm}^{-3}$. So the pH of the solution is 7.0.

Beyond adding $25.0 \, \text{cm}^3$ of NaOH, the calculation changes slightly because the NaOH is in excess. Let us consider the situation when $30.0 \, \text{cm}^3$ of NaOH has been added. The total volume of the solution is:

$$25.0 + 30.0 = 55.0 \, \text{cm}^3 \text{ or } 0.0550 \, \text{dm}^3$$

The number of moles of OH⁻ ions in $30.0 \, \text{cm}^3$ of solution is given by:

$$\text{no. moles of OH}^- = 0.100 \times \frac{30.0}{1000} = 3.00 \times 10^{-3} \, \text{mol}$$

These OH⁻ ions reacts with the H⁺ ions in the solution according to the equation:

$$H^+(aq) + OH^-(aq) \rightarrow H_2O(l)$$

The original number of moles of H^+ ions was 2.50×10^{-3} mol.
   The number of moles of OH^- in excess is:

$$3.00 \times 10^{-3} - 2.50 \times 10^{-3} = 5.0 \times 10^{-4} \text{ mol}$$

$$\text{concentration} = \frac{\text{no. moles}}{\text{volume in dm}^3}$$

$$[OH^-(aq)] = \frac{5.0 \times 10^{-4}}{0.0550} = 9.09 \times 10^{-3} \text{ mol dm}^{-3}$$

$$pOH = -\log_{10}[OH^-(aq)]$$

$$pOH = -\log_{10} 9.09 \times 10^{-3} = 2.04$$

At 25 °C, pH + pOH = 14. Therefore, pH = 12.0 (to three significant figures). Table **8.8** lists the pH values at different stages of the strong acid–strong base titration.

Weak acid–strong base

Let us consider adding $0.100 \text{ mol dm}^{-3}$ NaOH to 25 cm^3 of $0.100 \text{ mol dm}^{-3}$ CH_3COOH (ethanoic acid) ($K_a = 1.74 \times 10^{-5}$). The titration curve is shown in Figure **8.19**.

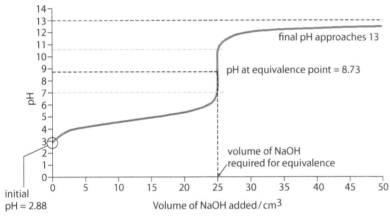

Figure 8.19 A titration curve for adding $0.100 \text{ mol dm}^{-3}$ NaOH to 25 cm^3 of $0.100 \text{ mol dm}^{-3}$ CH_3COOH.

The initial pH can be calculated using the K_a value for ethanoic acid and the method given on page **332**.
   The equivalence point is reached when equal numbers of moles of acid and alkali have been added. As the concentrations of the acid and bases are the same, this occurs when the same volumes have been added (that is, when 25.0 cm^3 of sodium hydroxide has been added). The pH at the equivalence point is not 7, however – it is greater than 7.
   The equation for the reaction is:

$$CH_3COOH(aq) + NaOH(aq) \rightarrow CH_3COONa(aq) + H_2O(l)$$

When equivalent amounts of ethanoic acid and sodium hydroxide have been added, a solution of sodium ethanoate ($CH_3COONa(aq)$) has been

pH values when 25.00 cm³ of 0.1000 mol dm⁻³ HCl is titrated with 0.1000 mol dm⁻³ NaOH	
Volume of NaOH added / cm³	**pH**
0.00	1.000
5.00	1.176
10.00	1.368
15.00	1.602
20.00	1.954
21.00	2.061
22.00	2.195
23.00	2.380
24.00	2.690
24.50	2.996
24.90	3.698
24.95	4.000
24.99	4.699
24.995	5.000
24.999	5.699
25.00	7.000
25.01	9.301
25.10	10.300
25.50	10.996
26.00	11.292
30.00	11.959
35.00	12.222
40.00	12.363
50.00	12.523
60.00	12.615
100.00	12.778
200.00	12.891

Table 8.8 Progressive addition of base and how the pH changes.

formed. The CH_3COO^- ion is the conjugate base of the weak acid CH_3COOH, and so, acting as a base, will react with some water molecules to accept a proton according to the equilibrium:

$$CH_3COO^-(aq) + H_2O(l) \rightleftharpoons CH_3COOH(aq) + OH^-(aq)$$

This reaction increases the concentration of OH^- ions in the solution and the pH at the equivalence point is therefore higher than 7.

The steep part of the curve, where it is virtually vertical, is more in the alkaline region than in the acidic region of the graph. As more and more NaOH is added, the pH approaches 13 – the pH of pure $0.100\,mol\,dm^{-3}$ NaOH.

In Figure **8.20**, the weak acid–strong base titration curve is compared with a strong acid–strong base titration curve.

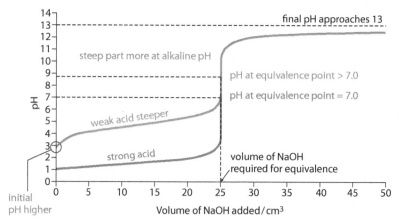

Figure 8.20 Comparison of a weak acid–strong base titration curve (blue) with a strong acid–strong base titration curve (red).

Note that the strength of the acid does not affect the volume of base required to reach the equivalence point for a titration. The volume of base required to reach the equivalence point depends only on the original volume of acid used, the concentration of the acid and the concentration of the base.

There are some important differences between strong acid–strong base titration curve and a weak acid–strong base titration curve. For a weak acid–strong base titration:

- The initial pH is higher. The weak acid is only partially dissociated.
- The initial part of the curve, up to a volume of about $24\,cm^3$ NaOH added, is steeper. Because the weak acid is only partially dissociated, the concentration of H^+ ions in the solution is lower, and adding a certain volume of NaOH has a greater relative effect on the pH than for a strong acid, in which the concentration of H^+ ions is higher.
- The steep part of the curve is more in the alkaline region.
- The pH at the equivalence point is higher than 7.

Determination of pKₐ from a titration curve

The pK_a value for a weak acid can be determined experimentally by constructing a titration curve.

In the above example of the titration of $0.100\,\text{mol}\,\text{dm}^{-3}$ ethanoic acid with $0.100\,\text{mol}\,\text{dm}^{-3}$ NaOH, $25.0\,\text{cm}^3$ of NaOH is required to reach the equivalence point. Here we will consider the point in the titration when half this amount of NaOH has been added, i.e. $12.5\,\text{cm}^3$.

The equation for the reaction is:

$$CH_3COOH(aq) + NaOH(aq) \rightarrow CH_3COONa(aq) + H_2O(l)$$

When $12.5\,\text{cm}^3$ of NaOH has been added, half of the CH_3COOH has been converted to CH_3COONa, and therefore the concentrations of CH_3COOH and CH_3COONa are equal. This can be seen if we work out the number of moles of each species:

$$\text{no. moles of } CH_3COOH \text{ originally in solution} = \frac{25.0}{1000} \times 0.1$$

$$= 2.50 \times 10^{-3}\,\text{mol}$$

$$\text{no. moles of NaOH added} = \frac{12.5}{1000} \times 0.1 = 1.25 \times 10^{-3}\,\text{mol}$$

$1.25 \times 10^{-3}\,\text{mol}$ NaOH will react with $1.25 \times 10^{-3}\,\text{mol}$ CH_3COOH to form $1.25 \times 10^{-3}\,\text{mol}$ CH_3COONa. There were originally $2.50 \times 10^{-3}\,\text{mol}$ CH_3COOH, and if $1.25 \times 10^{-3}\,\text{mol}$ reacted that leaves $1.25 \times 10^{-3}\,\text{mol}$ CH_3COOH in the reaction mixture – i.e. the same number of moles as CH_3COONa.

The expression for K_a is:

$$K_a = \frac{[CH_3COO^-(aq)][H^+(aq)]}{[CH_3COOH(aq)]}$$

Because the concentration of CH_3COO^- is equal to the concentration of CH_3COOH, we can write:

$$K_a = \frac{\cancel{[CH_3COO^-(aq)]}[H^+(aq)]}{\cancel{[CH_3COOH(aq)]}} = [H^+(aq)]$$

and if we take $-\log_{10}$ of both sides, we get $pK_a = pH$. So at the half-equivalence point we can read the value of pK_a from the graph shown in Figure **8.21**.

Strong acid–weak base titrations

Let us consider adding $0.100\,\text{mol}\,\text{dm}^{-3}$ $NH_3(aq)$ to $25\,\text{cm}^3$ $0.100\,\text{mol}\,\text{dm}^{-3}$ HCl(aq) (Figure **8.22**). The pH at the equivalence point for a strong acid–weak base titration is at a pH lower than 7. At the equivalence point the NH_3 has reacted with HCl to form NH_4Cl:

$$NH_3(aq) + HCl(aq) \rightarrow NH_4Cl(aq)$$

Exam tip
This seems to come up a lot in examinations!

Note: $[CH_3COONa(aq)]$ is equal to $[CH_3COO^-(aq)]$, as the ionic salt is fully dissociated in solution.

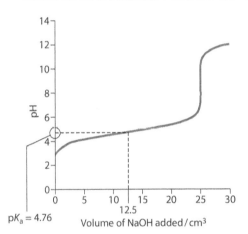

$pK_a = 4.76$

Figure 8.21 The pK_a value of the acid is equal to the pH at the half-equivalence point.

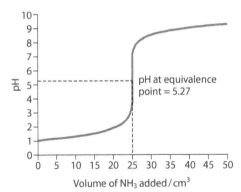

Figure 8.22 A strong acid–weak base titration curve.

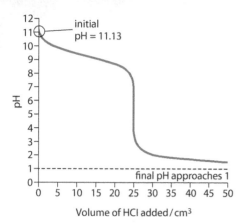

Figure 8.23 Titration curve for adding a weak base to a strong acid.

When a strong acid is added to a weak base, $pK_b = pOH$ at the half equivalence point.

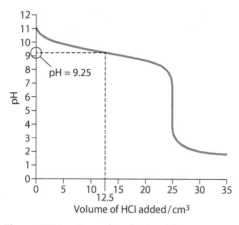

Figure 8.24 pK_b can be obtained from a strong acid–weak base titration curve.

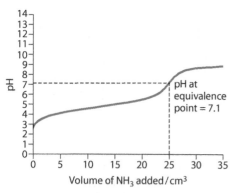

Figure 8.25 A weak acid–weak base titration curve.

The ammonium ion is, however, the conjugate acid of a weak base and as such will dissociate to a certain extent in aqueous solution according to the equation:

$$NH_4^+(aq) \rightleftharpoons NH_3(aq) + H^+(aq)$$

So the concentration of H^+ ions in the solution is increased, and the pH is lower than 7.

The titration could also have been performed the other way around – that is, by adding hydrochloric acid to the ammonia solution. The titration curve obtained is shown in Figure **8.23**. The initial pH can be worked out using the K_b value of ammonia, as already described.

At the half–equivalence point of this titration, 12.5 cm³ of HCl has been added. The equilibrium that exists in this solution is:

$$NH_3(aq) + H_2O(l) \rightleftharpoons NH_4^+(aq) + OH^-(aq)$$

The expression for K_b is:

$$K_b = \frac{[NH_4^+(aq)][OH^-(aq)]}{[NH_3(aq)]}$$

When 12.5 cm³ of HCl has been added, half of the NH_3 will have been converted into NH_4^+, and so the concentration of NH_3 will be equal to the concentration of NH_4^+.

$$K_b = \frac{\cancel{[NH_4^+(aq)]}[OH^-(aq)]}{\cancel{[NH_3(aq)]}}$$

Therefore, at the half equivalence point: $K_b = [OH^-]$.

Taking $-\log_{10}$ of both side we get: $pK_b = pOH$.

At 25 °C, $pH + pOH = 14$, so, at the half equivalence point, $14 - pH = pK_b$. So we can thus work out the pK_b for ammonia from the graph in Figure **8.24**.

The pH at the half–equivalence point is 9.25, so the pK_b of ammonia is given by:

$$14 - 9.25 = pK_b$$

pK_b for ammonia is 4.75.

Weak acid–weak base

An example of titration of a weak acid and a weak base is the addition of 0.100 mol dm⁻³ $NH_3(aq)$ to 25.0 cm³ of 0.100 mol dm⁻³ CH_3COOH (Figure **8.25**).

There is no very steep (almost vertical) part in this titration curve, and the change in pH throughout the titration is more gradual than in the other titrations we have considered.

The pH at the equivalence point may be lower than 7 or higher than 7, depending on the relative strength of the acid and the base. In this case, ethanoic acid ($K_a = 1.74 \times 10^{-5}$) is very similar in strength to ammonia

$(K_b = 1.78 \times 10^{-5})$, and the pH at the equivalence point is very close to 7.0 (it is just slightly higher than 7.0 because the base is very slightly stronger than the acid). If we had used a different acid, such as methanoic acid $(K_a = 1.78 \times 10^{-4})$, the pH at the equivalence point would have been lower than 7 because the acid is relatively stronger than the base.

Indicators

Acid–base titrations are carried out to establish the equivalent amounts of acid and base that react with each other, and hence the concentration of the acid or the alkali. We need some way of determining when equivalent amounts of acid and base have been mixed, and this can be done either using a pH meter and looking for the point of inflexion in the titration curve or, more usually and more conveniently, by using an acid–base indicator.

Indicators are either weak acids or weak bases. First, we consider an indicator that is a weak acid, represented by HIn. The indicator dissociates according to the equation:

$$HIn(aq) \rightleftharpoons H^+(aq) + In^-(aq)$$
colour I   colour II

The ionised (In⁻) and un-ionised (HIn) forms must have different colours for the substance to function as an indicator. In this case, we will take the colours as **red** and **blue**, as shown in the equation.

If we add some indicator to a solution and the colour of the indicator is blue, this indicates that the position of the above equilibrium lies mostly to the right – there is so much more of the blue form than the red form that the solution appears blue to our eyes.

If we now add some acid to the solution, the colour of the indicator changes to red (Figure **8.26**). This can be understood in terms of Le Chatelier's principle. If acid (H^+) is added to the system at equilibrium, the position of equilibrium will shift to the left to use up, as far as possible, the H^+ that has been added. As the position of equilibrium shifts to the left, there is now significantly more HIn (red) than In⁻ (blue) and the colour of the indicator appears red.

If we now add some base to this solution, the colour changes to blue (Figure **8.27**). When we add some base, the OH⁻ ions from the base react with the H^+ ions on the right-hand side of the equilibrium to produce water. The position of equilibrium shifts to the right to replace the H^+ as far as possible and restore the value of K_c.

If we now consider an indicator that is a weak base, this will be present in solution as, for example, the sodium salt, NaIn. The equilibrium in solution will be:

$$In^-(aq) + H_2O(l) \rightleftharpoons HIn(aq) + OH^-(aq)$$
colour I   colour II

The equivalence point is the **point of inflexion** on the curve. A point of inflexion is where the gradient (slope) of the curve stops increasing (or decreasing) and starts decreasing (or increasing).

The equivalence point of a titration is the point at which equivalent numbers of moles of acid and alkali have been added. The **end point of a titration** is the point at which the indicator changes colour – these are not necessarily the same.

Both colours are present, but we see the solution as if only one colour were present. Scientific reality is different to our everyday reality.

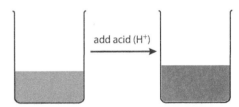

Figure 8.26 The indicator changes from colour II to colour I as acid is added.

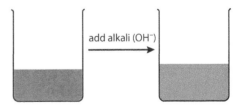

Figure 8.27 The indicator changes colour from colour I to colour II as alkali is added.

When some base (OH⁻) is added, the position of equilibrium will shift to the left and the indicator will appear as colour I. If some acid is added it will react with the OH⁻ ions causing the position of equilibrium to shift to the right, so that the indicator appears to have colour II.

The pH range of an indicator

If we imagine a different indicator, for which the colours of the un-ionised and ionised forms are yellow and blue, respectively:

$$HIn(aq) \rightleftharpoons H^+(aq) + In^-(aq)$$
colour I   colour II

Let us consider increasing the pH gradually from 1. We could get the results for the colour of the indicator shown in Figure **8.28**.

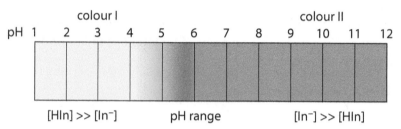

Figure 8.28 The variation of colour with pH for an indicator that is yellow in acidic solutions and blue in alkaline solution.

Up to pH 4 the concentration of HIn (yellow) is so much greater than the concentration of In⁻ (blue) that the indicator appears yellow. From pH 6 onwards, the concentration of In⁻ is so much greater than the concentration of HIn that the colour appears blue. Between pH 4 and 6 the concentrations of HIn and In⁻ are fairly evenly balanced, and the indicator is various shades of green in this region (green is a mixture of yellow and blue). In this region, if we gradually change the pH, we can see the indicator changing colour, and this is called the pH range of the indicator. In this case the pH range of the indicator would be quoted as 4–6.

The pH ranges of various indicators are given in Table **8.9**.

Indicator	pH range	Acid colour	Alkali colour
methyl orange	3.1–4.4	red	yellow
bromophenol blue	2.9–4.6	yellow	blue
bromocresol green	3.8–5.4	yellow	blue
methyl red	4.2–6.3	red	yellow
bromothymol blue	6.0–7.6	yellow	blue
phenol red	6.8–8.4	yellow	red
phenolphthalein	8.2–10.0	colourless	pink

Table 8.9 The pH ranges and colours of some indicators.

Indicators can be made from natural substances – for example red cabbage and some types of flowers. Litmus is extracted from lichen.

The **pH range of an indicator** is the pH values between which the indicator has intermediate colours because comparable amounts of the un-ionised and ionised forms are present.

Universal indicator solution and pH paper contain a mixture of indicators that change colour over different pH ranges.

If an indicator is to be suitable for a titration, we require that one drop of the solution being added from the burette (either the acid or the base) should change the indicator from colour I to colour II. An indicator is no good for a titration if it can be seen to change colour gradually. Another way of saying this is that the indicator must be chosen to give a sharp end point. In order to have a clear end point for a titration, the range of the indicator must occur completely within the very steep part of the titration curve.

Consider a strong acid–strong base titration, as shown in Figure **8.29**. In the vertical region of the curve, approximately one drop of alkali causes the pH to change by about 6–7 units. Therefore, if an indicator has its range in this part of the curve, adding one drop of alkali will cause the indicator to change from the acid colour to the alkali colour.

> Where possible, an indicator should be chosen so that the equivalence point of the titration occurs within the pH range of the indicator.

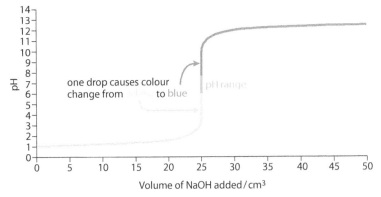

Figure 8.29 The colour changes of bromothymol blue during a strong acid–strong base titration.

Let us consider what happens if we use bromothymol blue for this titration. Before any base has been added, the colour of the indicator is yellow. As base is added, the colour remains yellow until the amount of base added is just less than $25.0\,\text{cm}^3$ and the pH of the mixture is about 4. At this point, adding one drop of base causes the pH to change from about 4 to 10 and the indicator changes colour from yellow to blue. So bromothymol blue is a suitable indicator for the titration because the pH range is entirely within the steep part of the titration curve and no intermediate colours are seen. We know that the equivalence point of the titration occurred somewhere within that one drop of base, and so we know the equivalence point to a good degree of precision.

If, we do the same titration with an indicator that changes colour from red to yellow and has a pH range of 1.2–2.8 we get the colour changes shown in Figure **8.30**. The indicator is red in the acid solution, but as we add alkali we see the colour changing gradually from red, through orange to yellow. It has changed colour to yellow before the equivalence point and so we are not able to detect the equivalence point using this indicator.

Depending on the concentrations of the solutions used, all the indicators in Table **8.9**, are suitable for a strong acid–strong base titration.

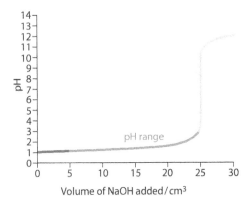

Figure 8.30 The colour changes during a strong acid–strong base titration using an indicator with a pH range of 1.2–2.8.

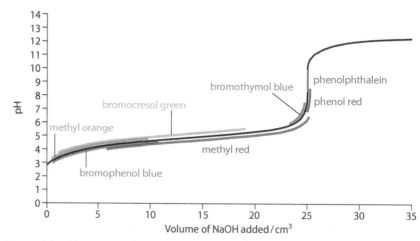

Figure 8.31 The ranges of various indicators superimposed on a weak acid–strong base titration curve.

Let us now consider a weak acid–strong base titration, as shown in Figure **8.31**. The pH ranges of various indicators are shown on the curve for such as titration, and it can be seen that phenolphthalein is the most suitable indicator for this type of titration, with its pH range occurring entirely within the steep part of the curve. Phenol red is also suitable. Although bromothymol blue is not a suitable indicator for the titration shown, it may be suitable for other weak acid–strong base titrations.

For titrating a strong acid with a weak base, methyl red will generally be the most suitable indicator, but bromocresol green, methyl orange or bromophenol may also be suitable depending on the weak base used and the concentration of the acid.

Because the change in pH is gradual for titrating a weak acid with a weak base (no very steep part in the curve), it is not generally possible to carry out this titration with an indicator and obtain a good estimate of the equivalence point – it is difficult to judge the end point of the indicator; the indicator is seen to change colour gradually during the titration.

The pK_a value of an indicator

Consider the dissociation of an indicator that is a weak acid:

$$HIn(aq) \rightleftharpoons H^+(aq) + In^-(aq)$$

The K_a expression for the indicator is:

$$K_a = \frac{[In^-(aq)][H^+(aq)]}{[HIn(aq)]}$$

At the midpoint of the pH range, half of the HIn will have dissociated and the concentrations of HIn and In$^-$ will be equal, in other words: $[HIn(aq)] = [In^-(aq)]$.

$$K_a = \frac{[\cancel{In^-(aq)}][H^+(aq)]}{[\cancel{HIn(aq)}]} = [H^+(aq)]$$

Because only strong acid–strong base titrations have pH = 7.0 at the end point, it is important to remember that the indicator is being used to determine the point at which equivalent numbers of moles of acid and alkali have been added, and not the point at which pH = 7.

Taking $-\log_{10}$ of each side we get $pK_a = pH$. Therefore, the pK_a of the indicator is equal to the pH at the midpoint of its pH range. As a rough rule of thumb, we can take the pH range of an indicator to be $pK_a \pm 1$.

This is a theoretical derivation of the pH range of an indicator. In practice, due to our ability to detect various colours and colour changes, the midpoint of the working pH range of an indicator may not correspond exactly to the pK_a of the indicator.

We can now decide on the best indicator to use for a particular titration using the pK_a value for the indicator. The indicator will be suitable if its pK_a value is close to the pH at the equivalence point of the titration.

For an indicator that is a weak base, the pH range will be $(14 - pK_b) \pm 1$ and so an indicator will be suitable for a titration if $14 - pK_b$ is close to the pH at the equivalence point of the titration.

Titration using an indicator cannot be used to distinguish between a weak and a strong acid – if they have the same concentration, they will require the same amount of alkali to reach the equivalence point.

> pK_a for an indicator is sometimes given the symbol pK_{in}.

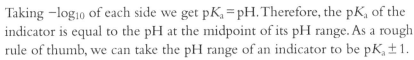 **Test yourself**

32 Select a suitable indicator, from the list given, for each of the following titrations in which the alkali is added to 25.0 cm³ of acid.

phenolphthalein	$pK_a = 9.3$
bromocresol green	$pK_a = 4.7$
bromothymol blue	$pK_a = 7.0$

Acid	Alkali	Indicator
0.100 mol dm⁻³ CH₃COOH	0.100 mol dm⁻³ NaOH	
0.010 mol dm⁻³ HNO₃	0.020 mol dm⁻³ KOH	
0.010 mol dm⁻³ HCl	0.010 mol dm⁻³ NH₃	

33 Consider adding $0.0100\,\text{mol dm}^{-3}$ sodium hydroxide solution to $25.0\,\text{cm}^3$ of $0.0200\,\text{mol dm}^{-3}$ ethanoic acid ($pK_a = 4.76$). Suggest values for the following important points on the titration curve.

initial pH	
volume of NaOH required to reach the equivalence point	
approximate pH at equivalence point	
approximate final pH (after approximately 200 cm³ of NaOH has been added)	

34 In the titration of $20.0\,\text{cm}^3$ of a $0.100\,\text{mol dm}^{-3}$ solution of a weak acid, HA, with $0.100\,\text{mol dm}^{-3}$ sodium hydroxide solution, the pH when $10.0\,\text{cm}^3$ of NaOH had been added was 4.80 and the pH when $20.0\,\text{cm}^3$ of NaOH had been added was 8.80. What is the value of pK_a for HA?

8.8.2 Salt hydrolysis

A salt will be formed when an acid reacts with a base:

$$acid + base \rightarrow salt + water$$

$$HNO_3 + NaOH \rightarrow NaNO_3 + H_2O$$

Sodium nitrate is the salt of a strong acid (HNO_3) and a strong base (NaOH).

When a solid salt is dissolved in water, the resulting solution may be acidic, basic or neutral, depending on the particular salt dissolving.

Salt of a weak acid and a strong base (pH > 7)

Consider the salt sodium ethanoate ($CH_3COO^-Na^+$), formed when NaOH reacts with ethanoic acid (CH_3COOH):

$$CH_3COOH(aq) + NaOH(aq) \rightarrow CH_3COONa(aq) + H_2O(l)$$

If some solid sodium ethanoate is dissolved in water, the pH of the solution will be greater than 7. A $0.500 \, mol \, dm^{-3}$ solution will have a pH of 9.23 (Figure **8.32**).

When the salt dissolves in water, the two ions separate from each other – the solution contains $CH_3COO^-(aq)$ and $Na^+(aq)$ ions.

The CH_3COO^- ion is the conjugate base of ethanoic acid (a weak acid) and so, acting as a base, will react with water molecules to accept a proton according to the equilibrium:

$$CH_3COO^-(aq) + H_2O(l) \rightleftharpoons CH_3COOH(aq) + OH^-(aq)$$

The concentration of OH^- ions in the solution has been increased and the solution is alkaline. This process is called **salt hydrolysis** – the salt has reacted with water.

The other ion from CH_3COONa present in this solution, the Na^+ ion, comes from a strong base (NaOH), which is fully ionised. There is therefore no tendency for Na^+ to react with the OH^- ions produced.

$$Na^+(aq) + OH^-(aq) \rightarrow\!\!\!\leftarrow NaOH(aq)$$

The pH of a $0.500 \, mol \, dm^{-3}$ solution of sodium ethanoate can be calculated as follows.

The pK_a of ethanoic acid is 4.76. pK_b for its conjugate base, the ethanoate ion, is given by:

$$pK_a + pK_b = pK_w$$

Therefore, at 25 °C, $pK_b = 14 - 4.76 = 9.24$. So K_b for the ethanoate ion is $10^{-9.24}$, or 5.75×10^{-10}.

The ethanoate ion reacts with water according to the equation:

$$CH_3COO^-(aq) + H_2O(l) \rightleftharpoons CH_3COOH(aq) + OH^-(aq)$$

Learning objectives

- Work out whether a solution of a salt will have a pH of lower than, equal to or higher than 7
- Understand that the pH of a solution of a salt depends on the charge (and size) of the cation

Exam tip
One way of **thinking** about this is that the base is stronger than the acid, so the pH will be basic.

Remember - the weaker the acid, the stronger its conjugate base.

The reason given here is the same reason why the pH at the equivalence point in a weak acid–strong base titration is greater than 7.

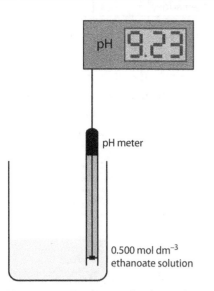

Figure 8.32 A solution of sodium ethanoate will have a pH > 7.

pH meter

0.500 mol dm^{-3} ethanoate solution

K_b for CH_3COO^- is:

$$K_b = \frac{[CH_3COOH(aq)][OH^-(aq)]}{[CH_3COO^-(aq)]}$$

One CH_3COO^- ion reacts with one H_2O molecule to form one molecule of CH_3COOH and one OH^- ion therefore:

$$[CH_3COOH(aq)] = [OH^-(aq)]$$

We will use the approximation that the extent to which the CH_3COO^- reacts is negligible compared with its concentration and take $[CH_3COO^-(aq)]$ to be $0.500\,mol\,dm^{-3}$.

Substituting known values into the K_b expression:

$$5.75 \times 10^{-10} = \frac{[OH^-(aq)]^2}{0.500}$$

$$[OH^-] = 1.70 \times 10^{-5}\,mol\,dm^{-3}$$

$$pOH = -log_{10}[OH^-(aq)] = -log_{10} 1.70 \times 10^{-5} = 4.77$$

$$pOH + pH = pK_w$$

At $25\,°C$, $pOH + pH = 14$, so $pH = 14 - 4.77 = 9.23$. Therefore the pH of a $0.500\,mol\,dm^{-3}$ solution of sodium ethanoate is 9.23.

Salt of a strong acid and a weak base (pH < 7)

Consider ammonium chloride (NH_4Cl). This is the salt of a strong acid (HCl) and a weak base (NH_3):

$$NH_3(aq) + HCl(aq) \rightarrow NH_4Cl(aq)$$

When this is dissolved in water, an acidic solution is formed (Figure **8.33**).

The NH_4^+ is the conjugate acid of the weak base NH_3 and will therefore dissociate according to the equation:

$$NH_4^+(aq) \rightleftharpoons NH_3(aq) + H^+(aq)$$

H^+ is generated, and so the solution of NH_4Cl is acidic. The pH of a $0.500\,mol\,dm^{-3}$ solution of NH_4Cl is 4.78.

Cl^- is the conjugate base of the strong acid, HCl. It is therefore an extremely weak base (the stronger the acid, the weaker the conjugate base). There is virtually no tendency for the following reaction to occur:

$$Cl^-(aq) + H_2O(l) \xrightarrow{\hspace{1cm}} HCl(aq) + OH^-(aq)$$

The pH of a solution of ammonium chloride can be worked out using a similar method to that used to work out the pH of a sodium ethanoate solution above.

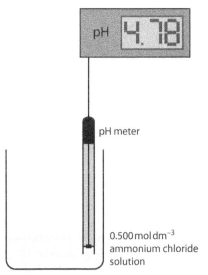

Figure 8.33 The pH of ammonium chloride solution is < 7.

Salt of a strong acid and a strong base (pH=7)

Consider sodium chloride, NaCl. This is made from the strong acid HCl and the strong base NaOH. These are both fully dissociated in aqueous solution, so a solution of NaCl is neutral (pH 7 at 25 °C). There is virtually no tendency for either of these reactions to occur:

$$Na^+(aq) + OH^-(aq) \rightleftharpoons NaOH(aq)$$

$$Cl^-(aq) + H_2O(l) \rightleftharpoons HCl(aq) + OH^-(aq)$$

Salt of a weak acid and a weak base (pH=?)

Consider ammonium ethanoate $CH_3COO^-NH_4^+$. This is the salt of the weak acid (CH_3COOH) and the weak base ($NH_3(aq)$). It is not possible to predict whether the solution is acidic or alkaline without looking at the pK_a values.

When ammonium ethanoate is dissolved in water, CH_3COO^- acts as a base and NH_4^+ acts as an acid:

$$CH_3COO^-(aq) + H_2O(l) \rightleftharpoons CH_3COOH(aq) + OH^-(aq) \qquad pK_b = 9.24$$

$$NH_4^+(aq) \rightleftharpoons NH_3(aq) + H^+(aq) \qquad pK_a = 9.25$$

Because the pK_b value is slightly lower (lower value = stronger) than the pK_a value, the solution will be slightly alkaline. However, the values are very close together, so the pH of a solution of ammonium ethanoate will be very close to 7.0.

A solution of ammonium methanoate would have a pH lower than 7:

$$HCOO^-(aq) + H_2O(l) \rightleftharpoons HCOOH(aq) + OH^-(aq) \qquad pK_b = 10.25$$

$$NH_4^+(aq) \rightleftharpoons NH_3(aq) + H^+(aq) \qquad pK_a = 9.25$$

because pK_a for the ammonium ion (stronger acid) is lower than pK_b for the methanoate ion.

A solution of ethylammonium ethanoate ($C_2H_5NH_3^+CH_3COO^-$) will have a pH higher than 7 because pK_b is lower (stronger base) than pK_a.

$$CH_3COO^-(aq) + H_2O(l) \rightleftharpoons CH_3COOH(aq) + OH^-(aq) \qquad pK_b = 9.24$$

$$C_2H_5NH_3^+(aq) \rightleftharpoons C_2H_5NH_2(aq) + H^+(aq) \qquad pK_a = 10.73$$

Acidity due to positive ions in solution

Let us consider a solution of iron(III) chloride. A $0.100 \, mol \, dm^{-3}$ solution of iron(III) will have a pH of less than 2 (Figure **8.34**).

Positive ions in solution are hydrated to form aqueous ions, e.g. $Fe^{3+}(aq)$. The formula of $Fe^{3+}(aq)$ is $[Fe(H_2O)_6]^{3+}$. This ion dissociates in solution according to the equation:

$$[Fe(H_2O)_6]^{3+}(aq) \rightleftharpoons [Fe(H_2O)_5(OH)]^{2+}(aq) + H^+(aq)$$

Hence the solution is acidic due to the H^+ ions produced.

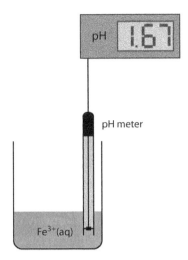

pH   1.67

pH meter

$Fe^{3+}(aq)$

Figure 8.34 A solution containing 3+ ions has a pH < 7.

The reason that this occurs with 3+ ions is because of the high charge density (or charge to radius ratio) of the ion, which causes the water molecule to be sufficiently **polarised** for H^+ to dissociate. The Fe^{3+} ion pulls electrons away from the H_2O molecules, which increases the $\delta+$ charge on the hydrogen so that it is more readily lost as H^+ (Figure **8.35**).

The larger an ion is and the lower its charge, then the smaller the charge density of the ion and the lower its tendency to polarise water molecules. So the Fe^{2+} ion will cause less polarisation of water molecules and, although a solution of $Fe^{2+}(aq)$ will still be acidic, it will have a higher pH than a solution of $Fe^{3+}(aq)$ of the same concentration.

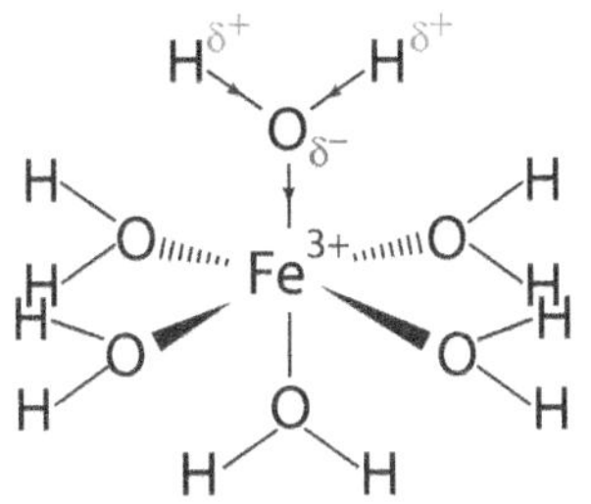

Exam tip
All 3+ ions in solution are acidic. This seems to come up often in questions and should be remembered.

Figure 8.35 The polarisation of one of the water ligands in $[Fe(H_2O)_6]^{3+}$.

? Test yourself

35 Predict whether the pH of each of the following solutions will be equal to 7, greater than 7 or less than 7.

	pH
$0.100\,mol\,dm^{-3}$ $CH_3CH_2CH_2CH_2COONa$	
$0.500\,mol\,dm^{-3}$ KNO_3	
$0.100\,mol\,dm^{-3}$ Na_2CO_3	
$0.100\,mol\,dm^{-3}$ $CH_3CH_2NH_3^+Cl^-$	
$0.200\,mol\,dm^{-3}$ $CrCl_3$	

36 Calculate pH values for the following salt solutions:
   a a $0.100\,mol\,dm^{-3}$ solution of potassium cyanide (KCN), given that pK_a for hydrocyanic acid (HCN) is 9.40.
   b a $0.200\,mol\,dm^{-3}$ solution of sodium propanoate (CH_3CH_2COONa), given that the pK_a for propanoic acid is 4.87.

Learning objectives

- Understand what is meant by a buffer solution and how both acidic and basic buffers can be made
- Describe how a buffer solution works when small amounts of acid/base are added

8.8.3 Buffer solutions

The blue line on the graph in Figure **8.36** shows the result of adding $10\,cm^3$ of $0.100\,mol\,dm^{-3}$ hydrochloric acid in stages to $100\,cm^3$ of water.

The orange line shows the effect of adding the hydrochloric acid to $100\,cm^3$ of a buffer solution formed by mixing $50\,cm^3$ of $1.00\,mol\,dm^{-3}$ ethanoic acid and $50\,cm^3$ of $1.00\,mol\,dm^{-3}$ sodium ethanoate. The pH of the water changes from 7.00 to 2.04 when $10\,cm^3$ of hydrochloric acid are added – the pH of the buffer solution changes from 4.76 to 4.74. When $10\,cm^3$ of $0.100\,mol\,dm^{-3}$ sodium hydroxide are added to $100\,cm^3$ of the ethanoic acid/sodium ethanoate buffer solution, the pH changes from 4.76 to 4.78. The pH of the buffer solution remains virtually constant when an acid or a base is added, which leads to the definition of a **buffer solution**:

> a solution that resists changes in pH when small amounts of acid or alkali are added.

The 'small amount' in this definition is important. If, for instance, $33.3\,cm^3$ of $2.00\,mol\,dm^{-3}$ hydrochloric acid is added to $100\;cm^3$ of the ethanoic acid/sodium ethanoate buffer solution we have just considered, the pH will change by about 4.5 units!

A buffer solution consists of two components and must always contain something to react with any acid added and something to react with any base added. In other words, a buffer solution always contains an acid and a base.

Buffers are important in many industrial processes and biological systems (for example, the blood – the pH of blood is about 7.4, and if it changes by about 0.5 in either direction you would die!). They are also used in products such as contact lens solutions, cosmetics and shampoos.

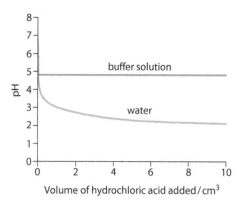

Figure 8.36 The pH of a buffer solution changes very little when hydrochloric acid is added. Water is not a good buffer.

Acid buffer solutions

An acid buffer solution consists of a weak acid (e.g. ethanoic acid) and a salt of that weak acid (e.g. sodium ethanoate). The equilibrium in this solution is:

$$CH_3COOH(aq) \rightleftharpoons CH_3COO^-(aq) + H^+(aq)$$

If some hydrochloric acid is added to this solution, the added H^+ reacts with the CH_3COO^- in the solution:

$$CH_3COO^-(aq) + H^+(aq) \rightarrow CH_3COOH(aq)$$

The H^+ added is 'mopped up' by reaction with the ethanoate ion (a base).

If some sodium hydroxide is added to the solution, the added OH^- reacts with the CH_3COOH in the solution:

$$CH_3COOH(aq) + OH^-(aq) \rightarrow CH_3COO^-(aq) + H_2O(l)$$

The OH^- added is 'mopped up' by reaction with the ethanoic acid.

If we consider the expression for K_a we can understand why the pH changes so little:

$$K_a = \frac{[CH_3COO^-(aq)][H^+(aq)]}{[CH_3COOH(aq)]}$$

This can be rearranged as:

$$[H^+(aq)] = \frac{[CH_3COOH(aq)]}{[CH_3COO^-(aq)]} \times K_a$$

Because K_a is a constant, this means that the H^+ concentration in the solution is proportional to the ratio of the ethanoic acid concentration to the ethanoate ion concentration.

For the solution to function as a buffer, both $[CH_3COO^-(aq)]$ and $[CH_3COOH(aq)]$ must be large so that any changes in their concentrations that occur when an acid or a base are added are small compared with these concentrations. This means that the value of the ratio $([CH_3COOH(aq)]:[CH_3COO^-(aq)])$ changes very little and the $[H^+(aq)]$ value (and pH) changes very little.

Let us consider a buffer solution in which the concentrations of ethanoic acid and sodium ethanoate are both $0.500\,mol\,dm^{-3}$. If we take $100.0\,cm^3$ of this solution, the number of moles of CH_3COOH and CH_3COO^- are given by:

no. moles = concentration × volume (in dm^3)

no. moles of CH_3COOH = no. moles of CH_3COO^-

$$= 0.500 \times \frac{100.0}{1000} = 0.0500\,mol$$

If $1.00\,cm^3$ of $0.100\,mol\,dm^{-3}$ HCl is added to this solution:

no. moles of $HCl = 0.100 \times \frac{1.00}{1000} = 1.00 \times 10^{-4}\,mol$

The acid will react with the ethanoate ions:

$$CH_3COO^-(aq) + H^+(aq) \rightleftharpoons CH_3COOH(aq)$$

	CH_3COO^-	CH_3COOH
initial no. moles / mol	0.0500	0.0500
no. moles after HCl added / mol	$0.0500 - 1.00 \times 10^{-4}$	$0.0500 + 1.00 \times 10^{-4}$
i.e.	0.0499	0.0501

The concentrations after the HCl has been added can be worked out by dividing the number of moles by the new volume ($100\,cm^3 + 1\,cm^3 = 101\,cm^3 = 0.101\,dm^3$)

$$CH_3COO^-(aq) + H^+(aq) \rightleftharpoons CH_3COOH(aq)$$

	CH_3COO^-	CH_3COOH
concentration after HCl added / mol dm^{-3}	0.494	0.496

We can now, using Table **8.10**, compare the ratio $[CH_3COOH(aq)]:[CH_3COO^-(aq)]$ before and after adding the HCl.

	$\dfrac{[CH_3COOH(aq)]}{[CH_3COO^-(aq)]}$	K_a	$\dfrac{[CH_3COOH(aq)]}{[CH_3COO^-(aq)]} \times K_a$	$[H^+(aq)]$ / mol dm^{-3}	pH
Before adding HCl	1.000	1.738×10^{-5}	1.738×10^{-5}	1.738×10^{-5}	4.760
After adding HCl	1.004	1.738×10^{-5}	1.745×10^{-5}	1.745×10^{-5}	4.758

Table 8.10 Comparing the [acid]:[base] ratio in a buffer solution.

It can be seen that because the concentrations of both CH_3COOH and CH_3COO^- are large compared with the amount of acid added, the addition of a small amount of HCl hardly changes the $[CH_3COOH(aq)]$: $[CH_3COO^-(aq)]$ ratio at all, and, to three significant figures, the pH does not change.

If we do the same calculation but add the hydrochloric acid to a solution of $0.500\,mol\,dm^{-3}$ ethanoic acid (i.e. no ethanoate is originally present, but a small amount of ethanoate is formed from the dissociation of the acid):

$$CH_3COOH(aq) \rightleftharpoons CH_3COO^-(aq) + H^+(aq)$$

initial equilibrium concentration / $mol\,dm^{-3}$	0.497	2.94×10^{-3}	2.94×10^{-3}
concentration after HCl added / $mol\,dm^{-3}$	0.493	1.92×10^{-3}	

	$\dfrac{[CH_3COOH(aq)]}{[CH_3COO^-(aq)]}$	K_a	$\dfrac{[CH_3COOH(aq)]}{[CH_3COO^-(aq)]} \times K_a$	$[H^+(aq)] / mol\,dm^{-3}$	pH
Before adding HCl	169	1.738×10^{-5}	2.93×10^{-3}	2.93×10^{-3}	2.53
After adding HCl	257	1.738×10^{-5}	4.47×10^{-3}	4.47×10^{-3}	2.35

Table 8.11 Comparing the [acid] : [base] ratio in a non-buffer solution.

In this case, for the same addition of HCl, the pH has changed by almost 100 times as much (Table **8.11**). This is because the initial concentration of the ethanoate ion was so small that any small change in concentration caused a significant percentage change in its concentration, and hence in the value of the ratio $[CH_3COOH(aq)]$: $[CH_3COO^-(aq)]$ and therefore in the pH.

Basic buffer solutions

A basic buffer solution consists of a weak base (e.g. ammonia) and a salt of that weak base (e.g. ammonium chloride). The equilibrium in this solution is:

$$NH_3(aq) + H_2O(l) \rightleftharpoons NH_4^+(aq) + OH^-(aq)$$

If some hydrochloric acid is added to this solution, the added H^+ reacts with the NH_3 in the solution:

$$NH_3(aq) + H^+(aq) \rightarrow NH_4^+(aq)$$

The H^+ added is 'mopped up' by reaction with the ammonia.

If some sodium hydroxide is added to the solution, the added OH^- reacts with the NH_4^+ in the solution:

$$NH_4^+(aq) + OH^-(aq) \rightarrow NH_3(aq) + H_2O(l)$$

The OH^- added is 'mopped up' by reaction with the ammonium ion.

As above, if the concentrations of NH_3 and NH_4^+ in the solution are both high, any small change in their concentrations has very little effect on the $[NH_4^+(aq)]$: $[NH_3(aq)]$ ratio in the K_b expression.

$$K_b = \frac{[NH_4^+(aq)][OH^-(aq)]}{[NH_3(aq)]}$$

The value of the OH^- concentration, and hence the pH, therefore changes very little.

Making buffer solutions by partial neutralisation of a weak acid/base

A buffer solution can be made by partial neutralisation of a weak acid with a strong base – for example by adding sodium hydroxide to ethanoic acid. The reaction that occurs is:

$$CH_3COOH(aq) + NaOH(aq) \rightarrow CH_3COONa(aq) + H_2O$$

When $10\,cm^3$ of $0.100\,mol\,dm^{-3}$ NaOH is added to $25\,cm^3$ of $0.100\,mol\,dm^{-3}$ CH_3COOH, some of the ethanoic acid will be converted to sodium ethanoate, but there will still be some ethanoic acid left. So the solution contains both ethanoic acid and sodium ethanoate, which are the components of a buffer system. As long as the number of moles of sodium hydroxide added is fewer than the number of moles of ethanoic acid present in the original solution, the solution will contain both ethanoic acid and sodium ethanoate and act as a buffer.

The graph in Figure **8.37** shows the titration curve of a weak acid (ethanoic acid) with a strong base (NaOH). The shaded region in Figure **8.37** represents the pH range over which this solution acts as buffer. In this region, significant amounts of both ethanoic acid and sodium ethanoate are present and adding sodium hydroxide does not cause the pH to change very much; addition of $20\,cm^3$ of NaOH in this region causes the pH to change by only about 1.5 units. Although this seems to be a much more significant increase than we have seen above, it must be remembered that here we are adding sodium hydroxide of the same concentration as the ethanoic acid to a similar volume of solution – we are not just adding small amounts.

A buffer solution can be also be made by partial neutralisation of a weak base with a strong acid – for example when hydrochloric acid is added to ammonia solution. The reaction that occurs is:

$$NH_3(aq) + HCl(aq) \rightarrow NH_4Cl(aq)$$

As long as the number of moles of hydrochloric acid added is lower than the number of moles of ammonia in the solution, the solution will contain some NH_3 and some NH_4Cl and will act as a buffer. The titration curve for adding $0.100\,mol\,dm^{-3}$ HCl to $25\,cm^3$ of $0.100\,mol\,dm^{-3}$ ammonia solution is shown in Figure **8.38** and the buffering region is marked.

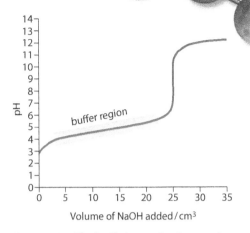

Figure 8.37 The buffering region in a weak acid–strong base titration curve.

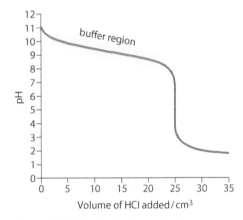

Figure 8.38 The buffering region in a strong acid–weak base titration curve.

? Test yourself

37 State whether or not each of the following mixtures would constitute a buffer solution:

   a $50\,cm^3$ $0.100\,mol\,dm^{-3}$ $CH_3CH_2COOH(aq)$ and $25\,cm^3$ $0.100\,mol\,dm^{-3}$ $CH_3CH_2COONa(aq)$

   b $50\,cm^3$ $0.100\,mol\,dm^{-3}$ $CH_3CH_2COOH(aq)$ and $25\,cm^3$ $0.100\,mol\,dm^{-3}$ $HCOONa(aq)$

   c $50\,cm^3$ $0.100\,mol\,dm^{-3}$ $HCOOH(aq)$ and $75\,cm^3$ $0.100\,mol\,dm^{-3}$ $HCOONa(aq)$

   d $25\,cm^3$ $0.010\,mol\,dm^{-3}$ $NaOH(aq)$ and $25\,cm^3$ $0.020\,mol\,dm^{-3}$ $CH_3COOH(aq)$

   e $50\,cm^3$ $0.200\,mol\,dm^{-3}$ $HCl(aq)$ and $25\,cm^3$ $0.400\,mol\,dm^{-3}$ $NaCl(aq)$

   f $50\,cm^3$ $0.100\,mol\,dm^{-3}$ $NH_3(aq)$ and $50\,cm^3$ $0.100\,mol\,dm^{-3}$ $NH_4NO_3(aq)$

   g $50\,cm^3$ $0.100\,mol\,dm^{-3}$ $NH_3(aq)$ and $25\,cm^3$ $0.100\,mol\,dm^{-3}$ $HCl(aq)$

Exam-style questions

1 Consider the dissociation of carbonic acid in aqueous solution:

$$H_2CO_3 + H_2O \rightleftharpoons HCO_3^- + H_3O^+ \qquad \textbf{reaction 1}$$

$$HCO_3^- + H_2O \rightleftharpoons CO_3^{2-} + H_3O^+ \qquad \textbf{reaction 2}$$

Which of the following is correct?

    A HCO_3^- acts as an acid in reaction 1 but as a base in reaction 2.
    B HCO_3^- is the conjugate base of H_2O.
    C HCO_3^- is the conjugate acid of CO_3^{2-}.
    D H_2CO_3 and CO_3^{2-} are a conjugate acid–base pair.

2 The pH of a solution of $0.0100\,mol\,dm^{-3}$ hydrochloric acid is 2. A $10\,cm^3$ sample of the acid is measured out and distilled water added to dilute it to a total volume of $100\,cm^3$. How do the hydrogen ion concentration and the pH change as this solution is diluted?

	hydrogen ion concentration	pH
A	decreases by a factor of 10	increases by 1
B	decreases by a factor of 100	decreases by 2
C	increases by a factor of 10	increases by 1
D	decreases by a factor of 10	decreases by 1

3 The electrical conductivity of five aqueous solutions is measured:

    I $0.100\,mol\,dm^{-3}$ $NaOH(aq)$
    II $0.100\,mol\,dm^{-3}$ $CH_3COOH(aq)$
    III $0.100\,mol\,dm^{-3}$ $NH_4Cl(aq)$
    IV $0.100\,mol\,dm^{-3}$ $NH_3(aq)$
    V $0.100\,mol\,dm^{-3}$ $HCl(aq)$

Which of the following is correct?

    A The conductivity of **I** is higher than that of **II** but lower than that of **IV**.
    B The conductivity of **III** is higher than that of **II** and higher than that of **IV**.
    C The conductivity of **V** is lower than that of **II** but higher than that of **III**.
    D The conductivity of **I** is lower than that of **III** and lower than that of **IV**.

4 HCN is a weak acid and dissociates in water according to the following equilibrium:

$$HCN(aq) + H_2O(l) \rightleftharpoons H_3O^+(aq) + CN^-(aq)$$

Which of the following statements is correct?

    A HCN dissociates completely in solution.
    B The concentration of CN^- ions is greater than that of HCN.
    C CN^- is a stronger base than H_2O.
    D H_3O^+ acts as a Brønsted–Lowry base.

5 What is the pH of a $1.00 \times 10^{-5}\,\text{mol dm}^{-3}$ solution of sodium hydroxide?

        A 5   **B** 9   **C** 13   **D** 14

HL 6 The acid HA has a K_a value of 1.00×10^{-7}. What is the pH of a $0.100\,\text{mol dm}^{-3}$ solution of HA?

        A 7   **B** 4   **C** 8   **D** 3

HL 7 The following salts are dissolved in water:

    I   NaCl   **III** Na_2CO_3
    II CH_3COONa **IV** $FeCl_3$

Which of these solutions will have a pH higher than 7?

    A **I** and **IV** only   **C** **IV** only
    B **II** and **III** only   **D** **II** only

HL 8 A series of solutions are made up:

    I $50\,\text{cm}^3$ $0.100\,\text{mol dm}^{-3}$ CH_3COOH and $25\,\text{cm}^3$ $0.100\,\text{mol dm}^{-3}$ NaOH
    II $50\,\text{cm}^3$ $0.050\,\text{mol dm}^{-3}$ CH_3COOH and $25\,\text{cm}^3$ $0.100\,\text{mol dm}^{-3}$ NaOH
    III $25\,\text{cm}^3$ $0.100\,\text{mol dm}^{-3}$ CH_3COOH and $50\,\text{cm}^3$ $0.100\,\text{mol dm}^{-3}$ NaOH
    IV $25\,\text{cm}^3$ $0.100\,\text{mol dm}^{-3}$ CH_3COOH and $50\,\text{cm}^3$ $0.100\,\text{mol dm}^{-3}$ CH_3COONa

Which solution(s) is/are buffer solutions?

    A **I** and **IV** only
    B **III** and **IV** only
    C **II, III** and **IV** only
    D **I** and **II** only

HL 9 The pK_a values for four indicators are shown here.

Indicator	pK_a
methyl violet	0.8
bromocresol green	4.7
phenol red	7.9
phenolphthalein	9.3

Which indicator would be most suitable for a titration in which $0.0100\,\text{mol dm}^{-3}$ ammonia is added gradually to $25.00\,\text{cm}^3$ of $0.0100\,\text{mol dm}^{-3}$ hydrochloric acid?

    A   methyl violet
    B   bromocresol green
    C   phenol red
    D   phenolphthalein

10 According to the Lewis definition, an acid is:

> **A** a proton donor
> **B** an electron pair donor
> **C** an electron acceptor
> **D** an electron pair acceptor

11 Ethanoic acid is a weak acid and hydrochloric acid is a strong acid.

  a Write an equation for the reaction of ethanoic acid with water, and identify the conjugate base of ethanoic acid. **[2]**

  b Explain the difference between a strong acid and a weak acid. **[2]**

  c **i** Calculate the pH of a $1.00 \times 10^{-3}\,mol\,dm^{-3}$ solution of hydrochloric acid.
     ii Suggest a value for the pH of a $1.00 \times 10^{-3}\,mol\,dm^{-3}$ solution of ethanoic acid. **[3]**

  d One method for distinguishing between a strong and a weak acid involves reacting them with a metal such as magnesium.
    i Write an equation for the reaction between magnesium and ethanoic acid and name the products of the reaction. **[2]**
    ii Explain how this method enables a strong acid to be distinguished from a weak acid. **[2]**

12 Water ionises slightly and the value of the ionic product constant at 298 K is 1.0×10^{-14}.

  a Write an equation for the ionisation of water. **[1]**

  b Write an expression for the ionic product constant for water. **[1]**

  c Calculate the pH of a $0.10\,mol\,dm^{-3}$ solution of sodium hydroxide at 298 K. **[3]**

  d Explain whether a $0.10\,mol\,dm^{-3}$ solution of barium hydroxide will have a higher or lower pH than a $0.10\,mol\,dm^{-3}$ solution of sodium hydroxide at the same temperature. **[2]**

13 **a** Define pH. **[1]**

  b The pK_a value of butanoic acid is 4.82. Determine the pH of a $0.150\,mol\,dm^{-3}$ solution of butanoic acid. **[4]**

  c Explain, using a chemical equation, whether the pH of a $0.100\,mol\,dm^{-3}$ solution of sodium butanoate will be lower than 7, higher than 7 or equal to 7. **[3]**

  d $25.00\,cm^3$ of $0.150\,mol\,dm^{-3}$ butanoic acid is titrated with sodium hydroxide solution. $27.60\,cm^3$ of sodium hydroxide is required to reach the equivalence point.
    i Calculate the concentration of the sodium hydroxide solution. **[3]**
    ii Calculate the pH of the sodium hydroxide solution. **[3]**
    iii Sketch a graph to show how the pH changes as the sodium hydroxide is added to the butanoic acid up to a total volume of $50\,cm^3$ of sodium hydroxide. **[3]**
    iv Suggest, with a reason, a suitable indicator for the titration. **[2]**

HL 14 Propanoic acid has a pK_a of 4.87 and ethanoic acid has a pK_a of 4.76.

    a Write an equation for the ionisation of propanoic acid in water. [2]

    b Calculate the values of K_a for propanoic acid and ethanoic acid and use them to explain which is the stronger acid. [4]

    c Calculate the concentration of H^+ ions and pH of a $0.250\,mol\,dm^{-3}$ solution of propanoic acid, stating any assumptions you make. [4]

    d Write an equation for the reaction of the propanoate ion ($CH_3CH_2COO^-$) with water. [1]

    e Calculate the value of K_b for the propanoate ion. [2]

    f Calculate the pOH and pH of a $0.200\,mol\,dm^{-3}$ solution of sodium propanoate. [4]

HL 15 The value of the ionic product constant for water, K_w, is 5.48×10^{-14} at 323 K.

    a Write an equation for the ionisation of water. [1]

    b Calculate the pH of water at 323 K. [3]

    c A solution has a pOH of 7.0 at 323 K. State and explain whether this solution is acidic, alkaline or neutral. [3]

    d pK_w for water is 14.34 at 288 K. Explain whether the ionisation of water is exothermic or endothermic. [3]

HL 16 A buffer solution can be made by mixing together ammonia solution and ammonium chloride solution. The pK_b of ammonia is 4.75.

    a State what is meant by a buffer solution. [2]

    b Explain, using equations, what happens when sodium hydroxide and hydrochloric acid are added separately to separate samples of this buffer solution. [2]

    c Calculate the pOH of $25.0\,cm^3$ of a $0.125\,mol\,dm^{-3}$ solution of ammonia. [3]

    d A buffer solution can also be made by adding hydrochloric acid to the solution in part **c**. State and explain the value of the pH of the buffer when $12.5\,cm^3$ of $0.125\,mol\,dm^{-3}$ hydrochloric acid have been added to the solution in part **c**. [3]

Summary

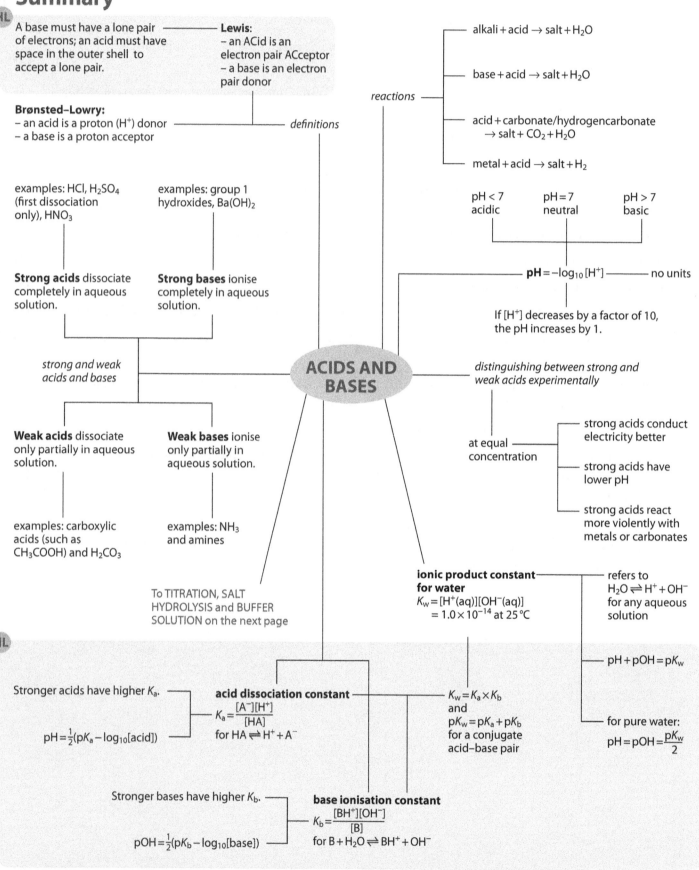

HL

A base must have a lone pair of electrons; an acid must have space in the outer shell to accept a lone pair.

Lewis:
– an ACid is an electron pair ACceptor
– a base is an electron pair donor

Brønsted–Lowry:
– an acid is a proton (H^+) donor
– a base is a proton acceptor

definitions

reactions

alkali + acid → salt + H_2O

base + acid → salt + H_2O

acid + carbonate/hydrogencarbonate → salt + CO_2 + H_2O

metal + acid → salt + H_2

examples: HCl, H_2SO_4 (first dissociation only), HNO_3

examples: group 1 hydroxides, $Ba(OH)_2$

| pH < 7 acidic | pH = 7 neutral | pH > 7 basic |

Strong acids dissociate completely in aqueous solution.

Strong bases ionise completely in aqueous solution.

$pH = -\log_{10}[H^+]$ —— no units

If $[H^+]$ decreases by a factor of 10, the pH increases by 1.

strong and weak acids and bases

ACIDS AND BASES

distinguishing between strong and weak acids experimentally

at equal concentration

strong acids conduct electricity better

strong acids have lower pH

strong acids react more violently with metals or carbonates

Weak acids dissociate only partially in aqueous solution.

Weak bases ionise only partially in aqueous solution.

examples: carboxylic acids (such as CH_3COOH) and H_2CO_3

examples: NH_3 and amines

To TITRATION, SALT HYDROLYSIS and BUFFER SOLUTION on the next page

ionic product constant for water
$K_w = [H^+(aq)][OH^-(aq)]$
$= 1.0 \times 10^{-14}$ at 25 °C

refers to $H_2O \rightleftharpoons H^+ + OH^-$ for any aqueous solution

HL

$pH + pOH = pK_w$

Stronger acids have higher K_a.

acid dissociation constant
$K_a = \dfrac{[A^-][H^+]}{[HA]}$
for $HA \rightleftharpoons H^+ + A^-$

$K_w = K_a \times K_b$
and
$pK_w = pK_a + pK_b$
for a conjugate acid–base pair

for pure water:
$pH = pOH = \dfrac{pK_w}{2}$

$pH = \frac{1}{2}(pK_a - \log_{10}[acid])$

Stronger bases have higher K_b.

base ionisation constant
$K_b = \dfrac{[BH^+][OH^-]}{[B]}$
for $B + H_2O \rightleftharpoons BH^+ + OH^-$

$pOH = \frac{1}{2}(pK_b - \log_{10}[base])$

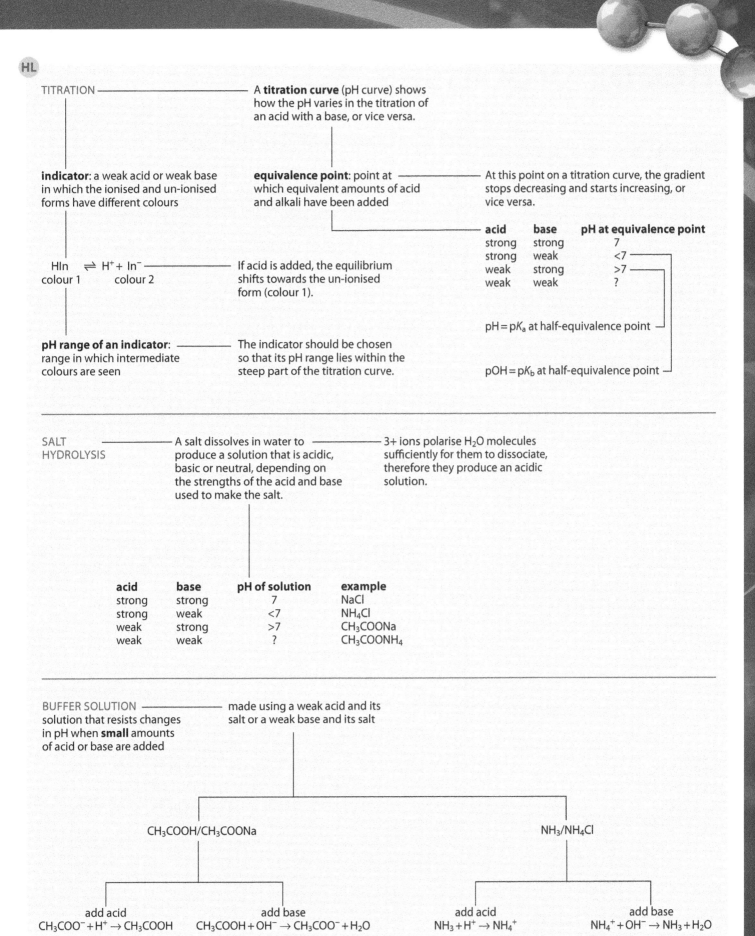

TITRATION — A **titration curve** (pH curve) shows how the pH varies in the titration of an acid with a base, or vice versa.

indicator: a weak acid or weak base in which the ionised and un-ionised forms have different colours

equivalence point: point at which equivalent amounts of acid and alkali have been added

At this point on a titration curve, the gradient stops decreasing and starts increasing, or vice versa.

acid	base	pH at equivalence point
strong	strong	7
strong	weak	<7
weak	strong	>7
weak	weak	?

$HIn \rightleftharpoons H^+ + In^-$
colour 1   colour 2

If acid is added, the equilibrium shifts towards the un-ionised form (colour 1).

$pH = pK_a$ at half-equivalence point

pH range of an indicator: range in which intermediate colours are seen

The indicator should be chosen so that its pH range lies within the steep part of the titration curve.

$pOH = pK_b$ at half-equivalence point

SALT HYDROLYSIS — A salt dissolves in water to produce a solution that is acidic, basic or neutral, depending on the strengths of the acid and base used to make the salt.

3+ ions polarise H_2O molecules sufficiently for them to dissociate, therefore they produce an acidic solution.

acid	base	pH of solution	example
strong	strong	7	NaCl
strong	weak	<7	NH_4Cl
weak	strong	>7	CH_3COONa
weak	weak	?	CH_3COONH_4

BUFFER SOLUTION — made using a weak acid and its salt or a weak base and its salt
solution that resists changes in pH when **small** amounts of acid or base are added

CH_3COOH/CH_3COONa

NH_3/NH_4Cl

add acid
$CH_3COO^- + H^+ \rightarrow CH_3COOH$

add base
$CH_3COOH + OH^- \rightarrow CH_3COO^- + H_2O$

add acid
$NH_3 + H^+ \rightarrow NH_4^+$

add base
$NH_4^+ + OH^- \rightarrow NH_3 + H_2O$

9 Redox processes

Learning objectives

- Understand the terms oxidation and reduction in terms of the loss/gain of hydrogen/oxygen and the loss/gain of electrons
- Work out oxidation numbers for elements in compounds and ions
- Understand that transition metals and some main group elements can have more than one oxidation number
- Understand how to name compounds and ions using oxidation numbers
- Understand what is meant by a redox reaction
- Understand oxidation and reduction in terms of changes in oxidation numbers
- Work out whether an element has been oxidised or reduced in a given reaction
- Understand what is meant by an oxidising agent and a reducing agent and identify them in reactions

O xidation
I s
L oss of electrons
R eduction
I s
G ain of electrons

Figure 9.1 A mnemonic to help you remember the definition of oxidation and reduction.

$$Zn(s) \longrightarrow Zn^{2+}(aq) + \boxed{2e^-} \quad \text{OXIDATION}$$

electrons lost

$$Cu^{2+}(aq) + \boxed{2e^-} \longrightarrow Cu(s) \quad \text{REDUCTION}$$

electrons gained

Figure 9.2 Breaking an oxidation–reduction reaction into two half-equations.

9.1 Oxidation and reduction

9.1.1 Definitions

Oxidation and **reduction** may be defined in many ways, such as:

> Oxidation is the loss of hydrogen or the gain of oxygen, and reduction is the gain of hydrogen or the loss of oxygen.

For instance, in the equation:

$$C_2H_5OH + [O] \rightarrow CH_3CHO + H_2O$$

we can see that the C_2H_5OH (ethanol) has been oxidised because it has lost H to form CH_3CHO (ethanal).

> [O] indicates oxygen added in an oxidation reaction.

In the equation:

$$C_6H_5NO_2 + 6[H] \rightarrow C_6H_5NH_2 + 2H_2O$$

$C_6H_5NO_2$ has been reduced because it has lost oxygen/gained hydrogen.

> [H] indicates hydrogen added in a reduction reaction.

These definitions are, however, only useful in a limited number of reactions and a more fundamental definition of oxidation and reduction is:

> Oxidation – loss of electrons
> Reduction – gain of electrons (Figure **9.1**)

For instance, in the reaction:

$$Zn(s) + Cu^{2+}(aq) \rightarrow Zn^{2+}(aq) + Cu(s)$$

the Zn has been oxidised to Zn^{2+}, because it has lost electrons, and the Cu^{2+} has been reduced to Cu, because it has gained electrons. This can be seen more clearly by splitting the overall reaction into two **half-equations**, as shown in Figure **9.2**.

If something loses electrons, something else must gain them, so oxidation and reduction always occur together – if something is oxidised, something else must be reduced. Reactions such as the one described above are called **redox reactions**, indicating that both reduction and oxidation occur.

Although it is fairly easy to understand which species has been oxidised and which reduced in ionic equations such as $Zn(s) + Cu^{2+}(aq) \rightarrow Zn^{2+}(aq) + Cu(s)$, it is more difficult to see just where electrons are being transferred in redox reactions such as $PCl_3 + Cl_2 \rightarrow PCl_5$. For discussing reactions such as these in terms of oxidation and reduction, we use the concept of **oxidation number** (oxidation state).

Oxidation number

Oxidation number is a purely formal concept, which regards all compounds as ionic and assigns charges to the components accordingly. It provides a guide to the distribution of electrons and relative charges on atoms in covalent compounds and allows us to understand redox processes more easily.

There are some general rules for working out oxidation numbers.

1. Treat the compound as totally ionic (if the compound is ionic then the charges on the ions are the oxidation numbers). So, CH_4 is a covalent compound but we assign oxidation numbers of -4 for C and $+1$ for H as if it were ionic. KCl is an ionic compound and the oxidation numbers are $+1$ for K and -1 for Cl, which correspond to the charges on the ions.

2. The most electronegative atom in a molecule is assigned a negative oxidation number according to how many electrons it needs to gain to have a full outer shell (see Table **9.1**). So in H_2O, oxygen is the more electronegative element and because an oxygen atom has six electrons in its outer shell it needs to gain two electrons to have a full outer shell, therefore the oxidation number of O in H_2O is -2.

3. Assign oxidation numbers accordingly to give the overall charge on the molecule/ion – the sum of the oxidation numbers, taking into account signs and coefficients, is equal to the overall charge on the molecule/ion.

4. The oxidation number of atoms in an element is zero – so the oxidation number of oxygen in O_2 is 0.

5. The elements in groups 1 and 2 virtually always have the group number as their oxidation number.

6. The maximum possible oxidation number for an element will be its group number for elements in groups 1 and 2 and the group number -10 for elements in groups 13–17 (see Table **9.2**). It is not possible to lose more electrons than there are in the outer shell – the maximum possible oxidation number for a group 16 element is $+6$ because there are six electrons in the outer shell.

Exam tip
Oxidation number is written with the sign first, e.g. -2, but a charge is written with the number first, e.g. $2-$. This may seem like a trivial point, but if you get it the wrong way round you could lose a mark in the exam!

Atom	Ionic charge	Oxidation number	Comment
F	F^-	-1	always, as most electronegative element
O	O^{2-}	-2	virtually always, but not in compounds with F or in peroxides and superoxides
Cl	Cl^-	-1	not in compounds with O and F; other oxidation numbers include $+1$, $+3$, $+5$, $+7$
H	H^+	$+1$	not in metal hydrides, e.g. NaH, where oxidation number is -1

Table 9.1 Assigning charge according to valency.

	Group 1	Group 2	Group 13	Group 14	Group 15	Group 16	Group 17
Maximum oxidation number	$+1$	$+2$	$+3$	$+4$	$+5$	$+6$	$+7$

Table 9.2 The connection between group number and the maximum possible oxidation number.

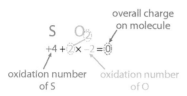

Figure 9.3 Assigning oxidation numbers in sulfur dioxide.

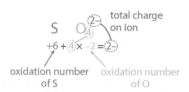

Figure 9.4 Assigning oxidation numbers in the sulfate(VI) ion.

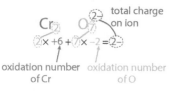

Figure 9.5 Assigning oxidation numbers in the dichromate(VI) ion.

Examples

SO_2

Sulfur dioxide is a covalent compound, but we will assign oxidation numbers as if it were an ionic compound. The more electronegative atom is oxygen, and therefore this is assigned its normal valency (the charge it would have in an ionic compound) – the oxidation number of oxygen is −2. The overall charge on the molecule is zero, and therefore the oxidation number of S must be +4 to cancel out the total oxidation number of two oxygen atoms (−4) (Figure **9.3**).

SO_4^{2-}

The more electronegative atom is oxygen and therefore this is assigned its normal valency – the oxidation number of oxygen is −2. The total oxidation number of four oxygen atoms is 4 × −2, or −8. Because the overall charge on the ion is 2−, the oxidation number of sulfur must be +6 to cancel out all but 2 of the total oxidation number of the four oxygen atoms (Figure **9.4**).

$Cr_2O_7^{2-}$

The more electronegative atom is oxygen and therefore the oxidation number of oxygen is −2. The total oxidation number of seven oxygen atoms is 7 × −2, or −14. Because the overall charge on the ion is 2−, the total oxidation number of two chromium atoms must be +12 to cancel out all but 2 of the total oxidation number of the seven oxygen atoms. Therefore the oxidation number of chromium is +12 / 2, or +6 (Figure **9.5**).

Hydrogen peroxide (H_2O_2)

The oxidation number of oxygen cannot be −2 here because the oxidation number of hydrogen would have to be +2. This is not possible because hydrogen has only one electron – its maximum oxidation number is +1. This means that the oxidation number of hydrogen in H_2O_2 is +1 and that of oxygen is −1. There are many ionic peroxides, such as Na_2O_2, which contain the O_2^{2-} ion.

Metal hydrides

Hydrogen forms hydrides with many metals – for example NaH and CaH_2. Hydrogen is more electronegative than metals and so its oxidation number will be −1. For many of these, the bonding is mostly ionic and they can be regarded as containing the H^- ion.

Some other oxidation numbers are shown in Table **9.3**.

Transition metals

The oxidation number of a transition metal in a complex ion is worked out using the charges on the ligands. Ligands may be either neutral or negatively charged – examples of each type are shown in Table **9.4**.

Let us consider $[Fe(H_2O)_6]^{2+}$. All the ligands are neutral, so the overall charge on the ion is just due to the iron and so the oxidation number of iron must be +2.

In $[Ni(CN)_4]^{2-}$, all the ligands have a 1− charge, so the total charge on all four ligands is 4−. The overall charge on the ion is 2−, so the oxidation number of Ni must be +2 to cancel out 2− from the 4− charge.

Extension

Superoxides, such as RbO_2, contain the O_2^- ion.

Exam tip

HL students may encounter questions on oxidation numbers of transition metals in complexes.

Neutral ligands	1− ligands
H_2O	Cl^-
NH_3	CN^-

Table 9.4 Examples of neutral and negatively charged ligands in transition metals.

	Oxidation numbers		
SO_3	S: +6	O: −2	
PCl_3	P: +3	Cl: −1	
Cl_2	Cl: 0	Cl: 0	
PO_4^{3-}	P: +5	O: −2	
OF_2	O: +2	F: −1	
CaO_2	Ca: +2	O: −1	
CO_2	C: +4	O: −2	
C_2H_6	C: −3	H: +1	
LiH	Li : +1	H: −1	
CaF_2	Ca: +2	F: −1	
HNO_3	H: +1	N: +5	O: −2
H_2SO_4	H: +1	S: +6	O: −2
$KMnO_4$	K: +1	Mn: +7	O: −2
$NaClO_3$	Na: +1	Cl: +5	O: −2

Table 9.3 The oxidation numbers of some elements in a variety of compounds.

Naming compounds using oxidation numbers

SO_2 and SO_3 are commonly called sulfur dioxide and sulfur trioxide, respectively. They are, however, more correctly named using the oxidation number of the sulfur atom. So SO_2 is sulfur(IV) oxide and SO_3 is sulfur(VI) oxide. Strangely enough, no-one ever calls carbon dioxide carbon(IV) oxide or carbon monoxide carbon(II) oxide!

> Roman numerals are used for the oxidation numbers in the names of compounds.

The transition metals can have variable oxidation numbers, and so the names of their compounds almost always contain the oxidation number – so $FeCl_2$ is called iron(II) chloride and contains the Fe^{2+} ion, whereas $FeCl_3$ is called iron(III) chloride and contains the Fe^{3+} ion.

Some other names of compounds and ions are shown in Table **9.5**.

Exam tip
In most cases, you will be able to work out the oxidation number of an atom in a molecule/ion by assuming that O has an oxidation number of −2 and that H has an oxidation number of +1.

CaO_2 is calcium peroxide.

$KMnO_4$ contains the ions K^+ and MnO_4^-.

What exactly are oxidation numbers? The sulfur atom in SO_2 definitely does not have a charge of 4+, so what does the oxidation number actually represent?

Why have scientists developed this systematic way of naming compounds – what has been gained by this? The old name for H_2SO_3 was sulfurous acid and that for $KMnO_4$ was potassium permanganate. Why do some people still prefer to use the old names? Will anything be lost when the old names become obsolete?

Compound/ion	Name
H_2SO_3	sulfuric(IV) acid
H_2SO_4	sulfuric(VI) acid
PCl_3	phosphorus(III) chloride
PCl_5	phosphorus(V) chloride
N_2O	nitrogen(I) oxide
NO_2	nitrogen(IV) oxide
$HClO_3$	chloric(V) acid
PO_4^{3-}	phosphate(V) ion
$CuSO_4$	copper(II) sulfate
SO_4^{2-}	sulfate(VI) ion
$Cr_2O_7^{2-}$	dichromate(VI) ion
CrO_4^{2-}	chromate(VI) ion
MnO_4^-	manganate(VII) ion

Table 9.5 These compounds/ions are named systematically using oxidation numbers.

Exam tip
Although you need to know how to name substances using oxidation numbers, it is still acceptable to use their more common names – like sulfur dioxide for SO_2 and sulfuric acid for H_2SO_4.

? Test yourself

1 Work out oxidation numbers for nitrogen in the following molecules/ions:
   a NF_3 **c** N_2O **e** N_2H_4
   b NO **d** N_2O_4 **f** NO_2^-

2 Work out the oxidation number of chlorine in the following species:
   a Cl_2O **c** ClO_4^- **e** $HClO_3$
   b HCl **d** ClF_3

3 Work out the oxidation number of the species in bold in each of the following compounds:
   a $\mathbf{Na_2O}$ **c** $Na_2\mathbf{S}O_3$ **e** $\mathbf{N}H_4NO_3$
   b $K\mathbf{Br}O_3$ **d** $K_2\mathbf{Cr}O_4$ **f** $K\mathbf{H}$

4 Name the following compounds using oxidation numbers:
   a NO **c** SeO_2 **e** Cr_2O_3
   b Cl_2O_7 **d** KIO_3

HL 5 Work out the oxidation number of the transition metal in each of the following complex ions or compounds:
   a $[CuCl_4]^{2-}$ **d** $[Fe(CN)_6]^{4-}$
   b $[Co(H_2O)_6]^{3+}$ **e** $[Ag(NH_3)_2]^+$
   c $[MnBr_4]^{2-}$ **f** $[Co(H_2O)_6]Cl_2$

Reduction – decrease in oxidation number
Oxidation – increase in oxidation number

Oxidation and reduction in terms of oxidation numbers

If an atom gains electrons, its oxidation number will become more negative. Therefore reduction involves a decrease in oxidation number. If an atom loses electrons, its oxidation number increases. Therefore oxidation involves an increase in oxidation number.

We can work out which species is oxidised or reduced in a redox reaction by analysing the oxidation numbers of the substances involved in the reaction.

Examples

$Br_2 + SO_2 + 2H_2O \rightarrow H_2SO_4 + 2HBr$

Br_2 is reduced, because its oxidation number decreases from 0 to −1. The sulfur in SO_2 is oxidised because its oxidation number increases from +4 to +6 (Figure **9.6**). We would usually say that the SO_2 is oxidised.

$6OH^- + 3Cl_2 \rightarrow 5Cl^- + ClO_3^- + 3H_2O$

In this reaction, five chlorine atoms have been reduced but one has been oxidised – therefore Cl_2 has been oxidised and reduced (Figure **9.7**). This type of reaction, in which the same species has been oxidised *and* reduced, is called a **disproportionation** reaction.

$Cr_2O_7^{2-} + H_2O \rightleftharpoons 2CrO_4^{2-} + 2H^+$

The oxidation number of Cr on both sides of the equation is +6, and no other atom undergoes a change in oxidation number. This reaction is, therefore, not a redox reaction.

> For a reaction to be a redox reaction, it must involve a change in oxidation number.

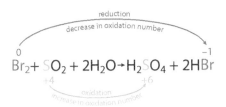

Figure 9.6 Reduction of Br_2 and oxidation of sulfur.

Figure 9.7 Oxidation and reduction of Cl_2, an example of a disproportionation reaction.

Oxidising and reducing agents

An oxidising agent is a substance that oxidises something else. A reducing agent is a substance that reduces something else.

Examples

$2Br^- + Cl_2 \rightarrow 2Cl^- + Br_2$

Cl_2 is the oxidising agent, because it oxidises the Br^- to Br_2. In terms of electrons, an oxidising agent oxidises a substance by removing electrons from it – here the Cl_2 removes electrons from the bromide ions, oxidising them. The Cl_2 has gained electrons and has been reduced.

The Br^- in this reaction causes the Cl_2 to be reduced by giving electrons to it to form Cl^- – the Br^- acts as a reducing agent in this reaction. The Br^- ions have lost electrons and been oxidised.

$Cr_2O_7^{2-}(aq) + 6Fe^{2+}(aq) + 14H^+(aq) \rightarrow 2Cr^{3+}(aq) + 6Fe^{3+}(aq) + 7H_2O(l)$

The Fe^{2+} is oxidised to Fe^{3+} by the $Cr_2O_7^{2-}$, so $Cr_2O_7^{2-}$ is the oxidising agent. We can tell that $Cr_2O_7^{2-}$ is the oxidising agent because it has been reduced in the reaction. The $Cr_2O_7^{2-}$ has been reduced by the Fe^{2+}, which means that Fe^{2+} is the reducing agent. We can tell that Fe^{2+} is a reducing agent because it has been oxidised.

> **Oxidising agents (oxidants)** oxidise other species and, in the process, are themselves reduced. An oxidising agent takes electrons away from something.

> **Reducing agents (reductants)** reduce other species and, in the process, are themselves oxidised. A reducing agent gives electrons to something.

Nature of science

Scientific knowledge is constantly changing and evolving and, wherever possible, scientists aim to make models/theories as fundamental and all-encompassing as possible. The original idea of defining oxidation and reduction in terms of oxygen and hydrogen has been broadened to include other reactions. The model of oxidation and reduction in terms of electrons is more fundamental than that in terms of oxygen/hydrogen loss/gain.

6 State whether each of the following half-equations involves oxidation or reduction:
   a $Cl_2 + 2e^- \rightarrow 2Cl^-$
   b $Mn^{3+} + e^- \rightarrow Mn^{2+}$
   c $Cu^+ \rightarrow Cu^{2+} + e^-$
   d $I_2 + 6H_2O \rightarrow 2IO_3^- + 12H^+ + 10e^-$
   e $Cr_2O_7^{2-} + 14H^+ + 6e^- \rightarrow 2Cr^{3+} + 7H_2O$

7 State which of the following reactions are redox reactions. For each redox reaction identify the element that has been oxidised and the element that has been reduced.
   a $Cu^{2+}(aq) + 2OH^-(aq) \rightarrow Cu(OH)_2(s)$
   b $2ZnS + 3O_2 \rightarrow 2ZnO + SO_2$
   c $2Na + 2H_2O \rightarrow 2NaOH + H_2$
   d $SO_3 + H_2O \rightarrow H_2SO_4$
   e $Na_2CO_3 + 2HCl \rightarrow 2NaCl + H_2O + CO_2$

f $2FeSO_4 + H_2SO_4 + H_2O_2 \rightarrow Fe_2(SO_4)_3 + 2H_2O$
g $3HgSO_4 \rightarrow Hg_2SO_4 + Hg + 2SO_2 + 2O_2$
h $2I^- + H^+ + HOCl \rightarrow I_2 + H_2O + Cl^-$

8 In each of the following redox reactions, identify the oxidising agent and the reducing agent:
   a $Zn + CuSO_4 \rightarrow ZnSO_4 + Cu$
   b $Cl_2 + 2Br^- \rightarrow 2Cl^- + Br_2$
   c $I_2O_5 + 5CO \rightarrow 5CO_2 + I_2$
   d $S + 6HNO_3 \rightarrow 2H_2O + H_2SO_4 + 6NO_2$
   e $2Na_2S_2O_3 + I_2 \rightarrow Na_2S_4O_6 + 2NaI$
   f $2KMnO_4 + 5Na_2C_2O_4 + 8H_2SO_4$
   $\rightarrow 2MnSO_4 + 10CO_2 + K_2SO_4$
   $+ 5Na_2SO_4 + 8H_2O$
   g $6FeSO_4 + K_2Cr_2O_7 + 7H_2SO_4$
   $\rightarrow 3Fe_2(SO_4)_3 + K_2SO_4 + Cr_2(SO_4)_3 + 7H_2O$

Learning objectives

- Understand what is meant by half-equations and be able to balance them
- Combine half-equations to produce an overall redox equation
- Balance redox equations in acidic or neutral solution using half-equations

9.1.2 Redox equations

Half-equations

As already mentioned, a redox equation may be broken down into two half-equations. These half-equations show the oxidation and reduction processes separately. For instance, from the equation:

$$2Br^-(aq) + Cl_2(aq) \rightarrow 2Cl^-(aq) + Br_2(aq)$$

we can separate two processes:

$2Br^-(aq) \rightarrow Br_2(aq) + 2e^-$ **oxidation**
$Cl_2(aq) + 2e^- \rightarrow 2Cl^-(aq)$ **reduction**

Examples

$$\mathbf{Cr_2O_7^{2-}(aq) + 6Fe^{2+}(aq) + 14H^+(aq) \rightarrow 2Cr^{3+}(aq) + 6Fe^{3+}(aq) + 7H_2O(l)}$$

This reaction can be separated as:

The number of electrons is the same in both half-equations.

$Cr_2O_7^{2-}(aq) + 14H^+(aq) + 6e^- \rightarrow 2Cr^{3+}(aq) + 7H_2O(l)$ **reduction**
$6Fe^{2+}(aq) \rightarrow 6Fe^{3+}(aq) + 6e^-$ **oxidation**

This second half-equation can be simplified to:

$Fe^{2+}(aq) \rightarrow Fe^{3+}(aq) + e^-$ **oxidation**

$Br_2 + SO_2 + 2H_2O \rightarrow H_2SO_4 + 2HBr$

This reaction can be separated as:

$$Br_2 + 2H^+ + 2e^- \rightarrow 2HBr \qquad \textbf{reduction}$$
$$SO_2 + 2H_2O \rightarrow H_2SO_4 + 2H^+ + 2e^- \qquad \textbf{oxidation}$$

It can be seen that all these half-equations balance, both in terms of number of atoms on both sides and the total charge on both sides. This must be true for all half-equations.

It should also be noted that the number of electrons lost in the oxidation reaction is equal to the number of electrons gained in the reduction reaction. This must be true because electrons cannot simply disappear or be created from nothing.

Balancing half-equations in neutral solution

Half-equations must balanced in terms of the number of atoms on both sides and in terms of the total charge on both sides. In some cases it is very straightforward to balance half-equations, and electrons must simply be added to one side or the other.

For example, consider the half-equation:

$$Ni^{2+} \rightarrow Ni$$

Although the number of nickel atoms on each side is the same, the total charge on the left-hand side is 2+ but the total charge on the right-hand side is 0. In order to balance the charges, we must add $2e^-$ to the left-hand side:

$$Ni^{2+} + 2e^- \rightarrow Ni$$

Now the number of atoms and the charges balance on each side.

Now, let's look at another half-equation:

$$Br_2 \rightarrow Br^-$$

In this example, neither the number of atoms nor the total charge balance. First of all we balance the atoms to get:

$$Br_2 \rightarrow 2Br^-$$

The total charge on the left-hand side is zero, but that on the right-hand side is 2−. To balance the charges we need to add $2e^-$ to the left-hand side so that the charge is equal on both sides:

$$Br_2 + 2e^- \rightarrow 2Br^-$$

This is now balanced.

These were fairly straightforward examples, but the process becomes a bit more difficult in some cases – we will now consider balancing more complex half-equations for reactions in acidic solution.

Although $2H^+$ and $2e^-$ are required to balance the individual half-equations, both cancel out when the half-equations are combined to give the overall redox equation.

Balancing half-equations in acidic solution

The following procedure should be followed for balancing these equations.
1. Balance all atoms except H and O.
2. Add H_2O to the side deficient in O to balance O.
3. Add H^+ to the side deficient in H to balance H.
4. Add e^- to the side deficient in negative charge to balance charge.

Examples

$Cr_2O_7^{2-} \rightarrow Cr^{3+}$

1. Balance all atoms except H and O – the Cr atoms must be balanced:

$$Cr_2O_7^{2-} \rightarrow 2Cr^{3+}$$

2. Add H_2O to the side deficient in O to balance O – there are 7O atoms on the left-hand side and none on the right-hand side, so $7H_2O$ must be added to the right-hand side:

$$Cr_2O_7^{2-} \rightarrow 2Cr^{3+} + 7H_2O$$

3. Add H^+ to the side deficient in H to balance H – there are 14H atoms on the right-hand side but none on the left-hand side, so $14H^+$ must be added to the left-hand side:

$$Cr_2O_7^{2-} + 14H^+ \rightarrow 2Cr^{3+} + 7H_2O$$

4. Add e^- to the side deficient in negative charge to balance the charge – the total charge on the left-hand side is **2− + 14+ = 12+**, and the total charge on the right-hand side is $2 \times 3+ = 6+$, so $6e^-$ must be added to the left-hand side to balance the charges:

$$Cr_2O_7^{2-} + 14H^+ + 6e^- \rightarrow 2Cr^{3+} + 7H_2O$$

The total charge on the left-hand side is now **2− + 14+ + 6− = 6+**, which is equal to the total charge on the right-hand side.

$S_4O_6^{2-} + H_2O \rightarrow H_2SO_3 + H^+ + e^-$

In this case the H_2O, H^+ and e^- have already been included on their correct sides. However, the overall process of balancing the equation is the same.

1. There are 4 S atoms on the left-hand side but only one on the right-hand side, so H_2SO_3 must be multiplied by 4:

$$S_4O_6^{2-} + H_2O \rightarrow 4H_2SO_3 + H^+ + e^-$$

2. There are 7 O atoms on the left-hand side and 12 on the right-hand side, so another 5 H_2O must be added to the left-hand side:

$$S_4O_6^{2-} + 6H_2O \rightarrow 4H_2SO_3 + H^+ + e^-$$

There are now 12 O atoms on each side.

3. There are 12 H atoms on the left-hand side and 9 on the right-hand side, so another 3 H^+ must be added to the right-hand side:

$$S_4O_6^{2-} + 6H_2O \rightarrow 4H_2SO_3 + 4H^+ + e^-$$

There are now 12 H atoms on each side.

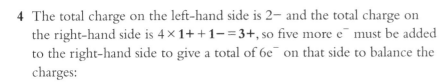

4 The total charge on the left-hand side is 2− and the total charge on the right-hand side is $4 \times \mathbf{1+} + \mathbf{1-} = \mathbf{3+}$, so five more e⁻ must be added to the right-hand side to give a total of 6e⁻ on that side to balance the charges:

$$S_4O_6{}^{2-} + 6H_2O \rightarrow 4H_2SO_3 + 4H^+ + 6e^-$$

The total charge on both sides is now 2−.

Combining half-equations to produce an overall redox equation

An oxidation half-equation may be combined with a reduction half-equation to produce an overall redox equation. When the half-equations are combined, the number of electrons lost in the oxidation reaction must be the same as the number gained in the reduction reaction.

Let us consider the two half-equations:

$$MnO_4{}^-(aq) + 8H^+(aq) + 5e^- \rightarrow Mn^{2+}(aq) + 4H_2O(l) \qquad \textbf{reduction}$$
$$Fe^{2+}(aq) \rightarrow Fe^{3+}(aq) + e^- \qquad \textbf{oxidation}$$

Five electrons are gained in the reduction half-equation, but only one is lost in the oxidation half-equation. The oxidation half-equation must therefore be multiplied by 5:

$$MnO_4{}^-(aq) + 8H^+(aq) + 5e^- \rightarrow Mn^{2+}(aq) + 4H_2O(l)$$
$$5Fe^{2+}(aq) \rightarrow 5Fe^{3+}(aq) + 5e^-$$

The number of electrons now balance and the two half-equations can be added together:

$$MnO_4{}^-(aq) + 8H^+(aq) + \mathbf{5e^-} \rightarrow Mn^{2+}(aq) + 4H_2O(l)$$
$$5Fe^{2+}(aq) \rightarrow 5Fe^{3+}(aq) + \mathbf{5e^-}$$
$$\overline{MnO_4{}^-(aq) + 8H^+(aq) + \cancel{\mathbf{5e^-}} + 5Fe^{2+}(aq) \rightarrow Mn^{2+}(aq) + 4H_2O(l) + 5Fe^{3+}(aq) + \cancel{\mathbf{5e^-}}}$$

The electrons cancel from each side to give the overall redox equation:

$$MnO_4{}^-(aq) + 8H^+(aq) + 5Fe^{2+}(aq) \rightarrow Mn^{2+}(aq) + 4H_2O(l) + 5Fe^{3+}(aq)$$

A final check can be done to see that this is indeed balanced by checking that the number of atoms of each type is the same on both sides and also that the total charge is the same on both sides.

Exam tip
Overall redox equations never contain electrons – only half-equations have electrons.

Worked example

9.1 What is the overall redox equation when the following two half-equations are combined?

$$MnO_4^-(aq) + 8H^+(aq) + 5e^- \rightarrow Mn^{2+}(aq) + 4H_2O(l) \qquad \textbf{reduction}$$
$$Re^{3+}(aq) + 4H_2O(l) \rightarrow ReO_4^{2-}(aq) + 8H^+(aq) + 3e^- \qquad \textbf{oxidation}$$

Five electrons are gained in the reduction half-equation, but only three are lost in the oxidation half-equation. Therefore, in order to balance the number of electrons lost with the number of electrons gained, the oxidation half-equation must be multiplied by 5 and the reduction half-equation by 3:

$$3MnO_4^-(aq) + 24H^+(aq) + \textbf{15e}^- \rightarrow 3Mn^{2+}(aq) + 12H_2O(l)$$
$$5Re^{3+}(aq) + 20H_2O(l) \rightarrow 5ReO_4^{2-}(aq) + 40H^+(aq) + \textbf{15e}^-$$

The numbers of electrons now balance and the two half-equations can be added together:

$$3MnO_4^-(aq) + 24H^+(aq) + \textbf{15e}^- \rightarrow 3Mn^{2+}(aq) + 12H_2O(l)$$
$$5Re^{3+}(aq) + 20H_2O(l) \rightarrow 5ReO_4^{2-}(aq) + 40H^+(aq) + \textbf{15e}^-$$

$$5Re^{3+}(aq) + 20H_2O(l) + 3MnO_4^-(aq) + 24H^+(aq) + \cancel{\textbf{15e}^-} \rightarrow 3Mn^{2+}(aq) + 12H_2O(l) + 5ReO_4^{2-}(aq) + 40H^+(aq) + \cancel{\textbf{15e}^-}$$

The electrons are cancelled from each side to give:

$$5Re^{3+}(aq) + 20H_2O(l) + 3MnO_4^-(aq) + 24H^+(aq) \rightarrow 3Mn^{2+}(aq) + 12H_2O(l) + 5ReO_4^{2-}(aq) + 40H^+(aq)$$

It can be seen that there are H_2O molecules and H^+ ions on both sides. These can also be cancelled to give:

$$5Re^{3+}(aq) + 8H_2O(l) + 3MnO_4^-(aq) \rightarrow 3Mn^{2+}(aq) + 5ReO_4^{2-}(aq) + 16H^+(aq)$$

This is the final redox equation.

Balancing redox equations for reactions in acidic or neutral solution

We can combine the above techniques to balance overall redox equations. The equation is first split up into two half-equations which are balanced separately and then combined to give the overall redox equation.

The procedure for doing this is as follows.

1 Separate the reaction equation into oxidation and reduction half-equations.
2 Balance each half-equation separately.
3 Combine the two half-equations after multiplying by the appropriate numbers to balance the electrons in each.
4 Subtract any duplications on left- and right-hand sides.

Example

$$MnO_4^-(aq) + C_2O_4^{2-}(aq) \rightarrow CO_2(g) + Mn^{2+}(aq)$$

Balance the above redox reaction equation, which happens in acidic solution.

The two half-equations are:

$$MnO_4^-(aq) \rightarrow Mn^{2+}(aq) \qquad\qquad C_2O_4^{2-}(aq) \rightarrow CO_2(g)$$
$$\text{reduction} \qquad\qquad\qquad\qquad \text{oxidation}$$

1 Balance all atoms except H and O:

$$MnO_4^- \rightarrow Mn^{2+} \qquad\qquad C_2O_4^{2-} \rightarrow 2CO_2$$

2 Add H_2O to side deficient in O to balance O:

$MnO_4^- \rightarrow Mn^{2+} + 4H_2O$ $\qquad\qquad\qquad\qquad$ $C_2O_4^{2-} \rightarrow 2CO_2$

3 Add H^+ to side deficient in H to balance H:

$MnO_4^- + 8H^+ \rightarrow Mn^{2+} + 4H_2O$ $\qquad\qquad\qquad$ $C_2O_4^{2-} \rightarrow 2CO_2$

4 Add e^- to the side deficient in negative charge to balance charge:

$MnO_4^- + 8H^+ + 5e^- \rightarrow Mn^{2+} + 4H_2O$ $\qquad$ $C_2O_4^{2-} \rightarrow 2CO_2 + 2e^-$

5 These must be combined so that the electrons balance. This is achieved by multiplying the reduction half-equation by 2 and the oxidation half-equation by 5:

$2MnO_4^- + 16H^+ + 10e^- \rightarrow 2Mn^{2+} + 8H_2O$ $\qquad$ $5C_2O_4^{2-} \rightarrow 10CO_2 + 10e^-$

The numbers of electrons now balance. The two half-equations are added together and the electrons cancelled:

$$2MnO_4^- + 16H^+ + 10e^- \rightarrow 2Mn^{2+} + 8H_2O(l)$$
$$\underline{5C_2O_4^{2-} \rightarrow 10CO_2 + 10e^-}$$
$$5C_2O_4^{2-}(aq) + 2MnO_4^-(aq) + 16H^+(aq) \rightarrow 2Mn^{2+}(aq) + 8H_2O(l) + 10CO_2(g)$$

There are no species the same on both sides, so this is the final equation.

Worked example

9.2 Balance the redox equation: $HAsO_2(aq) + BrO_3^-(aq) \rightarrow Br_2(aq) + H_3AsO_4(aq)$

The two half-equations are:

$HAsO_2(aq) \rightarrow H_3AsO_4(aq)$ $\qquad\qquad\qquad\qquad$ $BrO_3^-(aq) \rightarrow Br_2(aq)$

1 Balance all atoms except H and O:

$HAsO_2 \rightarrow H_3AsO_4$ $\qquad\qquad\qquad\qquad\qquad$ $2BrO_3^- \rightarrow Br_2$

2 Add H_2O to the side deficient in O to balance O:

$2H_2O + HAsO_2 \rightarrow H_3AsO_4$ $\qquad\qquad\qquad$ $2BrO_3^- \rightarrow Br_2 + 6H_2O$

3 Add H^+ to the side deficient in H to balance H:

$2H_2O + HAsO_2 \rightarrow H_3AsO_4 + 2H^+$ $\qquad$ $2BrO_3^- + 12H^+ \rightarrow Br_2 + 6H_2O$

4 Add e^- to the side deficient in negative charge to balance charge:

$2H_2O + HAsO_2 \rightarrow H_3AsO_4 + 2H^+ + 2e^-$ $\qquad$ $2BrO_3^- + 12H^+ + 10e^- \rightarrow Br_2 + 6H_2O$

5 Multiply by 5: $\qquad\qquad\qquad\qquad\qquad$ Multiply by 1:

$10H_2O + 5HAsO_2 \rightarrow 5H_3AsO_4 + 10H^+ + 10e^-$ $\quad$ $2BrO_3^- + 12H^+ + 10e^- \rightarrow Br_2 + 6H_2O$

The two half-equations can now be added together and the electrons cancelled:

$$10H_2O + 5HAsO_2 \rightarrow 5H_3AsO_4 + 10H^+ + 10e^-$$
$$\underline{2BrO_3^- + 12H^+ + 10e^- \rightarrow Br_2 + 6H_2O}$$
$$10H_2O + 5HAsO_2 + 2BrO_3^- + 12H^+ \rightarrow 5H_3AsO_4 + 10H^+ + Br_2 + 6H_2O$$

There are H_2O molecules and H^+ ions on both sides, and these can be cancelled to give the overall equation:

$4H_2O(l) + 5HAsO_2(aq) + 2BrO_3^-(aq) + 2H^+(aq) \rightarrow 5H_3AsO_4(aq) + Br_2(aq)$

A final check can be done to see that this is, indeed, balanced by checking that the number of atoms of each type is the same on both sides and also that the total charge is the same on both sides.

9 Balance the following half-equations for reactions that occur in neutral solution:
  a $Fe^{3+} \rightarrow Fe$
  b $Pb^{2+} \rightarrow Pb^{4+}$
  c $I_2 \rightarrow I^-$
  d $S_2O_3^{2-} \rightarrow S_4O_6^{2-}$
  e $C_2O_4^{2-} \rightarrow CO_2$

10 Balance the following half-equations for reactions that occur in acidic solution:
  a $I_2 + H_2O \rightarrow OI^- + H^+ + e^-$
  b $MnO_4^- + H^+ + e^- \rightarrow MnO_2 + H_2O$
  c $IO_3^- + H^+ + e^- \rightarrow I_2 + H_2O$
  d $N_2 + H_2O \rightarrow NO_3^- + H^+ + e^-$
  e $SO_4^{2-} + H^+ + e^- \rightarrow H_2SO_3 + H_2O$

11 Balance the following half-equations for reactions that occur in acidic solution:
  a $VO^{2+} \rightarrow V^{3+}$
  b $Xe \rightarrow XeO_3$
  c $NO_3^- \rightarrow NO$
  d $NO_3^- \rightarrow N_2O$
  e $VO_2^+ \rightarrow VO^{2+}$

12 Balance the following redox equations for reactions that occur in neutral solution:
  a $Cl_2 + Br^- \rightarrow Cl^- + Br_2$
  b $Zn + Ag^+ \rightarrow Zn^{2+} + Ag$
  c $Fe^{3+} + I^- \rightarrow Fe^{2+} + I_2$

13 Balance the following redox equations for reactions that occur in acidic solution:
  a $Fe^{2+} + Cr_2O_7^{2-} + H^+ \rightarrow Fe^{3+} + Cr^{3+} + H_2O$
  b $I^- + Cr_2O_7^{2-} + H^+ \rightarrow I_2 + Cr^{3+} + H_2O$
  c $Zn + VO^{2+} + H^+ \rightarrow V^{3+} + H_2O + Zn^{2+}$
  d $BrO_3^- + I^- \rightarrow Br_2 + I_2$
  e $NpO_2^{2+} + U^{4+} \rightarrow NpO_2^+ + UO_2^+$

9.1.3 The activity series

Metals can be arranged in an activity series (Figure **9.8**).

> The metals in the activity series are arranged in order of how easily they are oxidised to form positive ions.

Learning objectives

- Understand that metals can be arranged in an activity series according to how easily they can be oxidised
- Use the activity series to predict the feasibility of a redox reaction

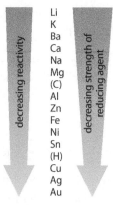

Li
K
Ba
Ca
Na
Mg
(C)
Al
Zn
Fe
Ni
Sn
(H)
Cu
Ag
Au

decreasing reactivity

decreasing strength of reducing agent

Figure 9.8 Activity series showing common metals. Carbon and hydrogen, although non-metals, are often included in the activity series.

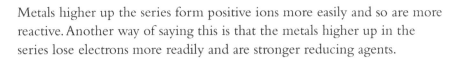

Metals higher up the series form positive ions more easily and so are more reactive. Another way of saying this is that the metals higher up in the series lose electrons more readily and are stronger reducing agents.

Extension

Lithium appears to be in the wrong place in this series – if you consider ionisation energies, it is easier to remove an electron from a potassium atom (it is bigger) than from a lithium atom. However, the activity series is arranged in terms of the ease of formation of aqueous ions and so the reaction $Li(s) \rightarrow Li^+(aq) + e^-$ is more favourable than $K(s) \rightarrow K^+(aq) + e^-$ because the Li^+ ion is much smaller than the K^+ ion and so has a much higher enthalpy change of hydration.

Predicting reactions using the activity series

The metals high in the activity series displace those lower in the series from solutions of their salts. For instance, when a piece of zinc is put in a solution containing copper(II) ions, the zinc becomes coated with copper; the blue colour of the solution fades and the solution gets warmer (Figure **9.9**).

Zinc is more readily oxidised than copper (more reactive than copper) and therefore displaces copper ions from solution. The reaction that occurs is:

$$Zn(s) + Cu^{2+}(aq) \rightarrow Zn^{2+}(aq) + Cu(s)$$

The zinc reduces the copper ions – Zn is a stronger reducing agent than Cu and is therefore able to reduce $Cu^{2+}(aq)$ ions.

Metals higher in the activity series are stronger reducing agents than metals lower in the series. So magnesium will reduce Zn^{2+} ions, zinc will reduce Fe^{2+} ions, iron will reduce Ni^{2+} ions and so on.

Metals above hydrogen in the activity series are stronger reducing agents than hydrogen and should reduce hydrogen ions to hydrogen gas. The reaction between magnesium and hydrogen ions is:

$$Mg(s) + 2H^+(aq) \rightarrow Mg^{2+}(aq) + H_2(g)$$

Acids release hydrogen ions in solution, so we can predict that metals above hydrogen in the series should liberate hydrogen gas when mixed with an acid – for example:

$$Mg(s) + 2HCl(aq) \rightarrow MgCl_2(aq) + H_2(g)$$

Metals lower than hydrogen in the activity series will not react with acids – copper and silver, for example, do not react with hydrochloric acid.

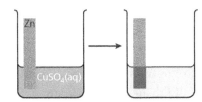

Figure 9.9 Zinc displaces copper from solution – this is called a displacement reaction or a single replacement reaction.

Learning objectives

- Solve problems involving redox titrations
- Understand what is meant by biochemical oxygen demand (BOD)
- Use the Winkler method to calculate biochemical oxygen demand

The amount of oxygen needed for fish to survive depends on the type of fish and the temperature. In summer, certain fish will not survive when the dissolved oxygen level falls below $6\,\text{mg}\,\text{dm}^{-3}$; if it drops below about $3\,\text{mg}\,\text{dm}^{-3}$ few fish are able to survive. The oxygen concentration needed for fish to survive is substantially higher at higher temperatures. During winter, certain fish may survive down to levels around $0.25\,\text{mg}\,\text{dm}^{-3}$.

9.1.4 Redox titrations

We met a redox titration earlier – in the worked example involving 'Linked reactions' on page **49**.

A common reaction in problems involving redox titrations is that between thiosulfate ions and iodine to determine the concentration of iodine in a sample:

$$I_2(aq) + 2S_2O_3^{2-}(aq) \rightarrow S_4O_6^{2-}(aq) + 2I^-(aq)$$

The thiosulfate ion $(S_2O_3^{2-})$ is oxidised and the 'average' oxidation number of sulfur changes from +2 in the thiosulfate ion to +2.5 in the tetrathionate ion $(S_4O_6^{2-})$. Sulfur is given a fractional oxidation number here because not all the sulfur atoms have the same oxidation number; +2.5 is an average.

Dissolved oxygen

The concentration of dissolved oxygen in water at $20\,^{\circ}\text{C}$ is about $9\,\text{mg}\,\text{dm}^{-3}$ (9 ppm) – this decreases with increasing temperature. Dissolved oxygen is important in the maintenance of life in aquatic systems. Dissolved oxygen comes from photosynthesis and also from the atmosphere.

Biochemical oxygen demand (also called biological oxygen demand), or BOD for short, is used as a measure of the quality of water. It is a measure of the amount of oxygen used by microorganisms to oxidise the organic matter in the water. Any organic pollutants in river water will be decomposed (oxidised) by **microorganisms** (aerobic bacteria) in the water and this process uses up dissolved oxygen. The higher the BOD, the more organic waste there is in water. If, for instance, sewage is released into a river or lake this will greatly increase the BOD – the water is more polluted. If the water is fast-flowing, new oxygen can be dissolved fairly quickly but this process is much slower in still water.

Organic matter in water might include leaves, animal manure, dead plants and animals. Effluent from water treatment plants will also contain organic matter.

> BOD is defined as the amount of oxygen used by the aerobic microorganisms in water to decompose the organic matter in the water over a fixed period of time (usually 5 days) at a fixed temperature (usually $20\,^{\circ}\text{C}$).

Good-quality river water will have a BOD of less than 1 ppm. Water is generally regarded as unpolluted if it has a BOD lower than 5 ppm. Untreated sewage can have a BOD of 500 ppm but treated sewage from water treatment plants should have a BOD lower than 20 ppm.

The basic principle in measuring BOD is to compare the initial amount of dissolved oxygen in a sample of water with the amount present when the sample has been incubated for 5 days at $20\,^{\circ}\text{C}$. If some water has

a dissolved oxygen concentration of 9 ppm, which after incubation for 5 days falls to 4 ppm, the BOD is $9 - 4$, or 5 ppm.

A typical method for determining the amount of dissolved oxygen is the Winkler titration method.

The basic chemistry behind the Winkler method is that manganese(II) sulfate is added to the water and the manganese(II) ions are oxidised under alkaline conditions to manganese(IV) by the oxygen in the water:

$$2Mn(OH)_2(s) + O_2(aq) \rightarrow 2MnO(OH)_2(s)$$
$$\text{manganese(II)} \qquad\qquad\qquad \text{manganese(IV)}$$

The sample is acidified with sulfuric acid to produce manganese(IV) sulfate:

$$MnO(OH)_2(s) + 4H^+(aq) \rightarrow Mn^{4+}(aq) + 3H_2O(l)$$

There is no change in oxidation number in this reaction.

Iodide ions are oxidised to I_2 by the manganese(IV) ions:

$$Mn^{4+}(aq) + 2I^-(aq) \rightarrow Mn^{2+}(aq) + I_2(aq)$$

This iodine can then be titrated against a standard sodium thiosulfate solution:

$$I_2(aq) + 2S_2O_3{}^{2-}(aq) \rightarrow S_4O_6{}^{2-}(aq) + 2I^-(aq)$$

The outcome from these equations is that the number of moles of dissolved oxygen is $\frac{1}{4}$ of the number of moles of sodium thiosulfate used in the titration – or the mass of oxygen is eight times the number of moles of sodium thiosulfate.

There is some disagreement about the reactions involved and an alternative set of equations is:

$$2Mn^{2+}(aq) + 4OH^-(aq) + O_2(aq) \rightarrow 2MnO_2(s) + 2H_2O(l)$$

$$MnO_2(s) + 2I^-(aq) + 4H^+(aq) \rightarrow Mn^{2+}(aq) + I_2(aq) + 2H_2O(l)$$

$$2S_2O_3{}^{2-}(aq) + I_2(aq) \rightarrow S_4O_6{}^{2-}(aq) + 2I^-(aq)$$

The stoichiometry of these reactions is the same as above.

Nature of science

Science is far from simple and trivial and it can sometimes be extremely difficult to work out exactly what is going on in a chemical reaction. In the reactions shown above it is not even certain that the Mn^{2+} is oxidised to manganese(IV) – it could be oxidised to manganese(III).

Water containing a high proportion of organic matter must be diluted before analysing in a BOD determination.

Manganese(II) sulfate is converted to manganese(II) hydroxide in the presence of hydroxide ions.

Only the manganese(IV) ion is shown here. The equation could also have been written as:
$$MnO(OH)_2(s) + 2H_2SO_4(aq) \rightarrow Mn(SO_4)_2(aq) + 3H_2O(l)$$

Worked example

9.3 The Winkler method was used to measure the concentration of dissolved oxygen in a sample of stream water. Manganese(II) sulfate, sulfuric acid and potassium iodide were added to $100.0\,cm^3$ of water. The iodine that was formed was titrated against a sodium thiosulfate solution with a concentration of $5.00 \times 10^{-3}\,mol\,dm^{-3}$. It was found that $16.00\,cm^3$ of sodium thiosulfate was required for the titration.

The equations for the reactions are:

$$2Mn(OH)_2(s) + O_2(aq) \rightarrow 2MnO(OH)_2(s) \qquad \textbf{equation 1}$$

$$MnO(OH)_2(s) + 4H^+(aq) \rightarrow Mn^{4+}(aq) + 3H_2O(l) \qquad \textbf{equation 2}$$

$$Mn^{4+}(aq) + 2I^-(aq) \rightarrow Mn^{2+}(aq) + I_2(aq) \qquad \textbf{equation 3}$$

$$I_2(aq) + 2S_2O_3{}^{2-}(aq) \rightarrow S_4O_6{}^{2-}(aq) + 2I^- (aq) \qquad \textbf{equation 4}$$

a Calculate the number of moles of sodium thiosulfate used.
b Calculate the number of moles of iodine in the solution.
c Calculate the number of moles of manganese(IV) that produced this amount of I_2.
d Calculate the concentration of dissolved oxygen in $mg\,dm^{-3}$ and in parts per million.
e Another sample of water from the same source was incubated for 5 days at $20\,°C$. At the end of the incubation the Winkler method was used to determine concentration of dissolved oxygen. It was found that the concentration of dissolved oxygen in the sample was $2.20\,mg\,dm^{-3}$. Calculate the biochemical oxygen demand.

a The number of moles of sodium thiosulfate can be calculated using:

no. of moles = volume (in dm^3) × concentration

$$= \left(\frac{16.00}{1000}\right) \times 5.00 \times 10^{-3}$$

$$= 8.00 \times 10^{-5}\,mol$$

b To calculate the number of moles of iodine, we must use equation **4**.

Two moles of thiosulfate ions react with one mole of iodine. The number of moles of iodine is therefore half the number of moles of thiosulfate.

$$\text{no. of moles of iodine} = \frac{8.00 \times 10^{-5}}{2} = 4.00 \times 10^{-5}\,mol$$

c The number of moles of manganese(IV) is obtained from equation **3**.

The number of moles of manganese(IV) is the same as the number of moles of I_2.

Number of moles of manganese(IV) is $4.00 \times 10^{-5}\,mol$.

d To determine the number of moles of oxygen we need to use the first three equations:

$$2Mn(OH)_2(s) + O_2(aq) \rightarrow 2MnO(OH)_2(s) \qquad\qquad \textbf{1}$$

$$MnO(OH)_2(s) + 4H^+(aq) \rightarrow Mn^{4+}(aq) + 3H_2O(l) \qquad\qquad \textbf{2}$$

$$Mn^{4+}(aq) + 2I^-(aq) \rightarrow Mn^{2+}(aq) + I_2(aq) \qquad\qquad \textbf{3}$$

The manganese(IV) that reacts in equation **3** is produced in equation **2** – so the number of moles of Mn^{4+}(aq) in equation **2** is 4.00×10^{-5} mol.

From equation **2**, the number of moles of $MnO(OH)_2$ is the same as the number of moles of Mn^{4+}(aq) – so there are 4.00×10^{-5} mol $MnO(OH)_2$.

This number of moles of $MnO(OH)_2$ are produced in equation **1** – therefore the number of moles of $MnO(OH)_2$ in equation **1** is 4.00×10^{-5} mol.

From equation **1** we can see that 1 mol O_2 produces 2 mol $MnO(OH)_2$, so the number of moles of O_2 is half the number of moles of $MnO(OH)_2$. Therefore the no. of moles of oxygen in the sample is $\dfrac{4.00\times10^{-5}}{2}$, that is 2.00×10^{-5} mol.

This is the number of moles of dissolved oxygen in $100.0\,cm^3$, so the number of moles of O_2 in $1.000\,dm^3$ of water is 10 times as much – that is 2.00×10^{-4} mol and so the concentration of dissolved oxygen in the water is $2.00\times10^{-4}\,mol\,dm^{-3}$.

The mass of oxygen present in $1.000\,dm^3$ of water is $2.00\times10^{-4}\times32.00 = 6.40\times10^{-3}\,g$

Therefore, we can say that the concentration of oxygen is $6.40\times10^{-3}\,g\,dm^{-3}$ or $6.40\,mg\,dm^{-3}$. This is equivalent to $6.40\,ppm$.

e The biochemical oxygen demand is determined by subtracting the final concentration of oxygen from its initial concentration.

$$BOD = 6.40 - 2.20 = 4.20\,mg\,dm^{-3} \text{ or } 4.20\,ppm.$$

? Test yourself

14 The Winkler method was used to measure the concentration of dissolved oxygen in a sample of river water. Manganese(II) sulfate, sulfuric acid and potassium iodide were added to $50.0\,cm^3$ of the water. The iodine that was liberated was titrated against sodium thiosulfate solution with a concentration of $2.00\times10^{-3}\,mol\,dm^{-3}$; $25.00\,cm^3$ of sodium thiosulfate was required for the titration.

  a Calculate the number of moles of sodium thiosulfate used.

  b Calculate the number of moles of iodine present in the solution.

  c Calculate the number of moles of manganese(IV) that produced this number of moles of I_2.

  d Calculate the concentration of dissolved oxygen in $mg\,dm^{-3}$.

  e Another sample of water from the same source was incubated for 5 days at $20\,°C$. At the end of this the Winkler method was used to determine the concentration of dissolved oxygen. It was found that the oxygen concentration in the sample was $4.60\,mg\,dm^{-3}$. Calculate the biochemical oxygen demand and comment on whether the water would be regarded as polluted.

9.2 Electrochemical cells

9.2.1 Voltaic cells

Voltaic cells (galvanic cells) provide us with a way of harnessing redox reactions to generate electricity. This is the basis of cells (batteries).

When a piece of zinc is put into a solution of copper(II) sulfate an exothermic reaction occurs and the zinc becomes coated with copper and the blue colour of the copper sulfate solution fades (Figure **9.9**, page **381**).

The overall reaction is:

$$Zn(s) + Cu^{2+}(aq) \rightarrow Zn^{2+}(aq) + Cu(s)$$

The half-equations involved are:

$$Zn(s) \rightarrow Zn^{2+}(aq) + 2e^{-} \qquad \textbf{oxidation}$$
$$Cu^{2+}(aq) + 2e^{-} \rightarrow Cu(s) \qquad \textbf{reduction}$$

When zinc metal is added to a solution containing Cu^{2+} ions, electrons are transferred from the zinc to the Cu^{2+} – the Cu^{2+} is reduced and the zinc is oxidised.

However, if the two reactions are separated, as in Figure **9.10**, exactly the same reaction occurs, except that instead of the electrons being transferred directly from the Zn to the Cu^{2+} they are transferred via the external circuit.

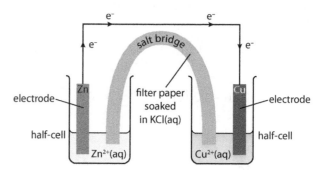

Figure 9.10 A voltaic cell.

In the left-hand beaker, zinc atoms are oxidised to Zn^{2+} ions and the electrons that are lost flow around the circuit to the other beaker, where they are gained by Cu^{2+} ions to form copper. If the reaction were allowed to keep going, we would see the zinc electrode getting smaller (as the zinc goes into solution as Zn^{2+}), the Cu electrode getting larger (as it is coated with copper) and the colour of the solution in the right-hand beaker becoming paler (as copper ions are converted to copper atoms).

The reaction between zinc and copper(II) ions is exothermic but when the two half-reactions are separated in a voltaic cell, instead of chemical energy (internal energy) being converted to heat it is converted to electrical energy which can do useful work – such as drive a motor to lift a mass.

Learning objectives

- Understand how electricity is produced in a voltaic cell

If we put a bulb in the circuit we would see the bulb light up because current flows.

The blue colour of the solution is due to the presence of $Cu^{2+}(aq)$ ions.

Why is a salt bridge necessary?

In the cell shown in Figure **9.10**, current will not flow unless the salt bridge is present. If the salt bridge were not present and the reaction were to proceed, there would be a build up of Zn^{2+} ions in the left-hand beaker, the solution would become positively charged overall and any further oxidation of zinc atoms to Zn^{2+} would be opposed. Similarly, there would be a decrease in the concentration of Cu^{2+} ions in the right-hand beaker which would mean that this solution would have a negative charge and any further reduction of Cu^{2+} ions would be opposed. The flow of electrons from the positively charged half-cell to the negatively charged half-cell would not occur. The salt bridge contains ions that can flow out of the salt bridge into the individual half-cells to prevent any build up of charge (Figure **9.11**). Similarly, any excess ions in the individual half-cells can flow into the salt bridge to prevent any build up of charge.

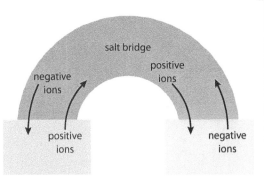

Figure 9.11 Ions flow into and out of the salt bridge to balance the charges in the half-cells.

| The **salt bridge** provides an electrical connection between the two half-cells to complete the circuit. It allows ions to flow into or out of the half-cells to balance out the charges in the half-cells.

| A salt bridge usually contains a concentrated solution of an ionic salt such as KCl.

Charge flow and the nature of the electrodes

Negative charge always flows in the same continuous direction around a complete circuit – in this case the electrons and the negative ions are all travelling clockwise around the circuit (Figure **9.12**).

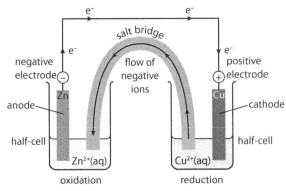

Figure 9.12 Direction of flow of charge. Positive ions flow in the opposite direction through the salt bridge.

In the zinc half-cell, an oxidation reaction occurs:

$$Zn(s) \rightarrow Zn^{2+}(aq) + 2e-$$

| An **anode** is defined as the electrode at which oxidation occurs.

So the zinc electrode is the anode in this cell. Because electrons are released at the zinc electrode, this is the negative electrode.
In the copper half-cell, a reduction reaction occurs:

$$Cu^{2+}(aq) + 2e^- \rightarrow Cu(s)$$

The more reactive metal will have the greater tendency to be oxidised and will be the negative electrode (anode) in a cell.

| In a voltaic cell, the anode is the negative electrode.

> The **cathode** is defined as the electrode at which reduction occurs.

So the copper electrode is the cathode in this cell. Because electrons flow through the external circuit towards this electrode, and are used up there, this is the positive electrode.

Cell notation for voltaic cells

Chemists often use a shorthand notation (cell–diagram convention) for describing the reactions that go on in a voltaic cell. The cell discussed above can be written as:

$$\underset{\text{oxidation}}{Zn(s)|Zn^{2+}(aq)}||\underset{\text{reduction}}{Cu^{2+}(aq)|Cu(s)}$$

A **single vertical line** represents a phase boundary (between solid and aqueous solution here) and the **double line** indicates the salt bridge. Reading from left to right we can see that Zn is oxidised to Zn^{2+} and that Cu^{2+} is reduced to Cu. The reaction at the anode is shown by convention on the left and that at the cathode on the right.

Any electrodes that the wires are actually joined to are put at the extremes on both sides. For example, some cells are set up using gases or metal ions but no solid metal. In this case we do not have anything that we can physically connect a wire to and so use platinum (a very inert metal) as the electrode (Figure **9.13**).

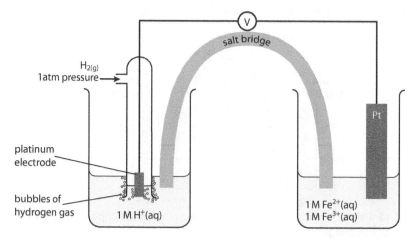

Figure 9.13 A voltaic cell with inert electrodes – the left half-cell involves a standard hydrogen electrode, which is discussed in more detail in the Higher Level section later. (Note: $1\,M = 1\,mol\,dm^{-3}$)

The overall reaction in this cell is:

$$H_2(g) + 2Fe^{3+}(aq) \rightarrow 2H^+(aq) + 2Fe^{2+}(aq)$$

The cell notation can be written as:

$$Pt|H_2(g), H^+(aq)||Fe^{3+}(aq), Fe^{2+}(aq)|Pt$$

Again, the oxidation reaction is written on the left, and the reduction on the right. Fe^{3+}(aq) and Fe^{2+}(aq) are separated by a comma because they are in the same phase (both aqueous solutions). The platinum electrodes are written at the extremes on both sides.

Worked example

9.4 Given the cell notation:

$$Zn(s)|Zn^{2+}(aq)||Fe^{3+}(aq),Fe^{2+}(aq)|Pt$$

write the half-equations for the reactions that occur in the cell, and hence the overall cell reaction.

The reaction at the anode is on the left-hand side: $Zn(s)|Zn^{2+}(aq)$

We now balance this half-equation by adding electrons:

$$Zn(s) \rightarrow Zn^{2+}(aq) + 2e^-$$ **oxidation**

Similarly, we can balance the half-equation for the reaction at the cathode:

$$Fe^{3+}(aq) + e^- \rightarrow Fe^{2+}(aq)$$ **reduction**

There are two electrons involved in the oxidation half-equation but only one in the reduction half-equation, therefore the reduction half-equation must be multiplied by 2 before combining with the oxidation half-equation for the overall cell reaction:

$$2Fe^{3+}(aq) + Zn(s) \rightarrow 2Fe^{2+}(aq) + Zn^{2+}(aq)$$

The size of the voltage

If a voltmeter is inserted across the $Zn|Zn^{2+}||Cu^{2+}|Cu$ cell described above, it will read a voltage of just over one volt (Figure **9.14a**). However, if the zinc electrode is replaced by a magnesium electrode, the voltage is much higher (Figure **9.14b**). This is because there is a bigger difference in reactivity between magnesium and copper than between zinc and copper – magnesium and copper are further apart in the activity series than magnesium and zinc. The magnesium has a greater tendency than zinc to donate electrons to copper ions.

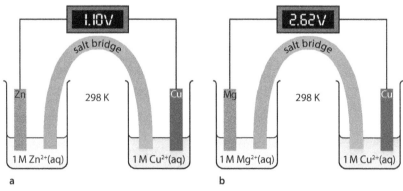

a   b

The bigger the difference in reactivity between the metal electrodes, the higher the voltage of the cell.

Figure 9.14 a The voltage in a $Zn|Zn^{2+}||Cu^{2+}|Cu$ cell is less than **b** in an $Mg|Mg^{2+}||Cu^{2+}|Cu$ cell. Note: $1\,M = 1\,mol\,dm^{-3}$.

Nature of science

Every year, more and more sophisticated electronic devices are being developed but they all rely on batteries and the development of better batteries is a huge area of research. This research involves collaboration between scientists with different specialisms and involves a great deal of creativity, imagination and investment. This type of scientific research is not, however, carried out for altruistic reasons and one of the main factors that drives this research is the desire to make money by the firms funding it.

? Test yourself

15 Consider four imaginary metals X, Z, A and Q.

a On the basis of the following data, arrange the metals X, Z and Q in order of reactivity (most reactive first):
- Z reduces Q^{2+} to Q
- X reduces Z^{2+} to Z

b The following reactions also occur:

$$A + Q^{2+} \rightarrow A^{2+} + Q$$
$$Z + A^{2+} \rightarrow Z^{2+} + A$$

 i Is A a stronger or weaker reducing agent than Q?

 ii Arrange all four metals in order of reducing ability (strongest reducing agent first).

 iii Which of the species, A, A^{2+}, Z, Z^{2+}, Q, Q^{2+}, X, X^{2+} is the strongest oxidising agent?

c An electrochemical cell is set up with a piece of A dipping into a solution of $A(NO_3)_2$ in one half-cell and a piece of X dipping into a solution of XSO_4 in the other half-cell.

 i Which way do electrons flow in the external circuit?

 ii Which is the anode in the cell?

 iii Write the cell notation for this cell.

 iv The X|XSO_4 half-cell is replaced by a Z|ZSO_4 half-cell. Will the voltage be higher or lower than that in the original cell?

Learning objectives

- Describe the features of an electrolytic cell
- Understand the differences between an electrolytic cell and a voltaic cell
- Understand how current is conducted during electrolysis
- Predict the products of electrolysis of a molten salt
- Write half-equations for the reactions occurring at the electrodes during electrolysis

9.2.2 Electrolytic cells

Electrolysis is the breaking down of a substance (in molten state or solution) by the passage of electricity through it.

Electrolysis of molten salts

The experimental set-up for the electrolysis of molten lead bromide is shown in Figure **9.15**. The overall reaction is:

$$PbBr_2(l) \xrightarrow[\text{energy}]{\text{electrical}} Pb(l) + Br_2(g)$$

The lead bromide is broken down into its elements, lead and bromine, by the passage of the electricity.

Positive electrode

At the positive electrode, bromide ions are oxidised and lose electrons to form bromine:

$$2Br^-(l) \rightarrow Br_2(g) + 2e^- \qquad \textbf{oxidation}$$

Because oxidation occurs at this electrode, the positive electrode is the anode in an electrolytic cell.

Negative electrode

At the negative electrode, lead ions are reduced as they gain electrons to form lead:

$$Pb^{2+}(l) + 2e^- \rightarrow Pb(l) \qquad \textbf{reduction}$$

Because reduction occurs at this electrode, the negative electrode is the cathode in an electrolytic cell.

Conduction of electricity in an electrolytic cell

In the external circuit, the current is carried by electrons (delocalised electrons in the metal wire) but in the molten salt (electrolyte), conduction involves the movement of **ions** (Figure **9.16**).

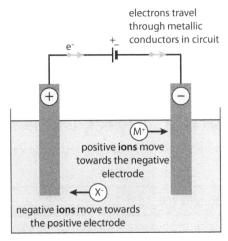

Figure 9.16 Conduction of electricity in an electrolytic cell.

Electrons travel from the negative pole of the battery to the negative electrode. The positive ions in the electrolyte move towards the negative electrode. At the negative electrode, electrons are transferred to the positive ions. The negative ions move towards the positive electrode. At the positive electrode the extra electron(s) from the negative ion is/are transferred to the electrode. The electrons released from the negative ions travel through the external circuit to the positive pole of the battery.

Electrons are taken from the external circuit at the negative electrode (by the positive ions) and given back to the external circuit at the positive electrode (by the negative ions). The circuit is completed, but the electrons that flow into the positive side of the battery are not the ones that flowed out of the negative side of the battery. No electrons travel through the electrolyte.

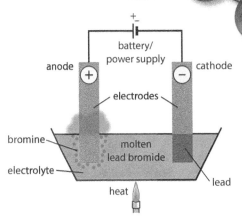

Figure 9.15 The electrolysis of molten lead bromide.

> The electrodes are usually made of graphite, a fairly inert non-metal which conducts electricity.

> The anode is the electrode at which oxidation occurs.

> The cathode is the electrode at which reduction occurs.

> Negative electrode:
> $M^+ + e^- \rightarrow M$ reduction
> Positive electrode:
> $X^- \rightarrow X + e^-$ oxidation

> An electrolyte is a solution or a molten compound that will conduct electricity, with decomposition at the electrodes as it does so. Electrolytes contain ions that are free to move towards the electrodes.

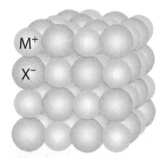
Figure 9.17 An ionic lattice.

Ionic salts will not conduct electricity when solid because the ions are held tightly in the lattice structure and are therefore not free to move (Figure **9.17**).

The products of electrolysis of a molten salt

When a molten salt is electrolysed, the products at the electrodes are the elements of which that salt is made up. A metal is formed at the negative electrode because metals form positive ions, and a non-metal is formed at the positive electrode because non-metals form negative ions.

Electrolysis of molten aluminium oxide (Al_2O_3)

A metal is formed at the negative electrode (cathode) – in this case, aluminium:

> Reduction occurs at the cathode.

$$Al^{3+} + 3e^- \rightarrow Al \qquad \textbf{reduction}$$

A non-metal is formed at the positive electrode – in this case, oxygen:

> Oxidation occurs at the anode.

$$2O^{2-} \rightarrow O_2 + 4e^- \qquad \textbf{oxidation}$$

Electrolysis of molten potassium chloride

Potassium is formed at the negative electrode (cathode) and chlorine is formed at the positive electrode (anode):

$$K^+ + e^- \rightarrow K \qquad \textbf{reduction}$$
$$2Cl^- \rightarrow Cl_2 + 2e^- \qquad \textbf{oxidation}$$

The differences between an electrolytic cell and a voltaic cell

The differences between a voltaic and an electrolytic cell are summarised in Table **9.6**.

Voltaic cell	Electrolytic cell
spontaneous redox reaction generates electricity	non-spontaneous redox reaction (one that would not happen by itself) is brought about by the passage of an electric current
conversion of chemical energy to electrical energy	conversion of electrical energy to chemical energy
anode is negative electrode and cathode is positive electrode	anode is positive electrode and cathode is negative electrode

Table 9.6 Comparison of voltaic and electrolytic cells.

? Test yourself

16 State the products at the anode and cathode when the following molten salts are electrolysed:
   a potassium bromide
   b copper(II) chloride
   c nickel(II) oxide
   d calcium chloride

17 Write equations for the reactions at the anode and cathode when the following molten salts are electrolysed:
   a sodium chloride
   b iron(III) oxide
   c magnesium bromide

Iron is the second most abundant metal in the Earth's crust and, because it is not very reactive, it can be extracted relatively easily from its ore by heating with carbon. It has been used since ancient times and, in the form of steel, is the most important construction metal. Mild steel has a very large number of uses, from building bridges to making car bodies. Although iron is not very reactive, it does undergo one very important redox reaction in the presence of air and water – rusting. Rusting of iron structures, machinery, vehicles etc. costs the world's economies billions of dollars each year.

Aluminium is the most abundant metal in the Earth's crust. It is much more reactive than iron and so is more difficult to extract from its compounds. It was not isolated until 1825. The advent of electricity allowed the commercial production of aluminium by electrolysis and has resulted in aluminium replacing iron for many uses. Aluminum has a low density (about one-third the density of iron) and so is useful in aeroplanes, for example. Aluminium is resistant to corrosion because, although it is significantly more reactive than iron, it has an impermeable oxide layer on the surface that prevents further reaction with air.

9.3 Electrochemical cells (HL)

9.3.1 Standard electrode potentials

Cell potentials

The voltmeter in the $Zn|Zn^{2+}||Cu^{2+}|Cu$ cell, shown on page 386, will read 1.10V if the concentrations of both solutions are $1\,mol\,dm^{-3}$ and the temperature is 298 K. This value is called the **standard cell potential** and gives us some idea of the favourability of the redox reaction:

$$Zn(s) + Cu^{2+}(aq) \rightleftharpoons Zn^{2+}(aq) + Cu(s)$$

A standard cell potential is the EMF (voltage) produced when two half-cells are connected under standard conditions. This drives the movement of electrons through the external circuit from the negative electrode to the positive electrode.

Learning objectives

- Describe the features of the standard hydrogen electrode
- Understand the terms standard electrode potential and standard cell potential
- Calculate the cell potential for a given redox reaction
- Use the cell potential to predict whether a redox reaction will be spontaneous
- Use standard cell potentials to calculate $\Delta G^{\ominus}$ for a reaction

EMF stands for 'electromotive force'.

The Zn/Cu^{2+} reaction has been shown with reversible equilibrium arrows above – indeed, $E^\ominus$ is related to the equilibrium constant, and this system will reach a state of equilibrium. However, in most cases the position of equilibrium lies a very long way to the right, and the reaction will go essentially to completion – K_c for the Zn/Cu^{2+} system at 298 K is about 1×10^{37}!

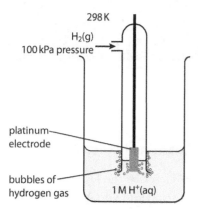

Figure 9.18 The standard hydrogen electrode. Note: 1 M = 1 mol dm^{-3}.

In the figure: 298 K / H$_2$(g) / 100 kPa pressure / platinum electrode / bubbles of hydrogen gas / 1 M H$^+$(aq)

Exam tip
You will need to recall the features of the standard hydrogen electrode.

Exam tip
1 mol dm^{-3} H$_2$SO$_4$(aq) is not suitable for use in the standard hydrogen electrode because H$_2$SO$_4$(aq) is a diprotic acid and the concentration of H+(aq) will be greater than 1 mol dm^{-3}.

A standard cell potential ($E^\ominus_{cell}$) is related to the value of $\Delta G^\ominus$ and K_c for the reaction. The higher the value (more positive) of the standard cell potential, the more favourable the reaction (ΔG is more negative and the value of K_c is larger).

When the zinc in this cell is replaced by magnesium, the value of the standard cell potential is higher, which indicates that the reaction:

$$Mg(s) + Cu^{2+}(aq) \rightarrow Mg^{2+}(aq) + Cu(s)$$

is more favourable (more spontaneous/ΔG is more negative) than the reaction involving zinc (see Figure **9.14**, page **389**). This suggests that Mg has a greater tendency to reduce Cu^{2+} ions than Zn does.

Using different half-cell combinations, we can measure the tendency for different redox reactions to occur and build up the activity series. However, what would be more useful is being able to predict the favourability of a particular redox reaction from a knowledge of the favourability of the individual oxidation and reduction half-reactions making up the redox reaction. This will provide us with a way of predicting in which direction a particular redox reaction will occur and how favourable it will be without setting up each cell physically.

However, it is not possible to measure the tendency for reactions such as:

$$Cu^{2+}(aq) + 2e^- \rightarrow Cu(s)$$

to occur in isolation, because if something is reduced then something else must be oxidised. The tendency of these reactions to occur can be measured only by connecting one half-cell to another half-cell and measuring the cell potential. However, if we always choose one particular half-cell as a reference half-cell in the system, then by measuring the cell potentials of lots of cells relative to this half-cell we will be able to build up a set of values that enables us to judge the relative oxidising and reducing abilities of various species. The reference half-cell that we choose is called the **standard hydrogen electrode**.

The standard hydrogen electrode

Individual half-cell electrode potentials cannot be measured in isolation, and so they are measured relative to a standard. The standard that is chosen is the standard hydrogen electrode (Figure **9.18**).

In the standard hydrogen electrode, hydrogen gas at 100 kPa (1 bar) pressure is bubbled around a platinum electrode of very high surface area in a solution of H$^+$ ions of concentration 1 mol dm^{-3}. Platinum is chosen because it is an inert metal, has very little tendency to be oxidised and does not react with acids. The reaction occurring in this half-cell is:

$$2H^+(aq) + 2e^- \rightleftharpoons H_2(g)$$

and this is arbitrarily assigned a standard electrode potential ($E^\ominus$) of 0.00 V.

Note: the reaction could also have been written as:

$$H^+(aq) + e^- \rightleftharpoons \tfrac{1}{2}H_2(g) \qquad\qquad E^\ominus = 0.00 V$$

Measuring standard electrode potentials

The standard electrode potential ($E^{\ominus}$) of copper could be measured by connecting a standard copper half-cell (1 mol dm^{-3} Cu^{2+}(aq)/Cu(s)) to the standard hydrogen electrode (Figure **9.19**).

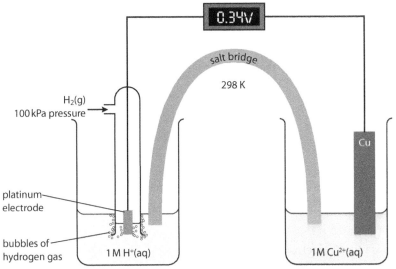

Figure 9.19 Measuring the standard electrode potential of the Cu^{2+}/Cu half-cell. Note: 1 M = 1 mol dm^{-3}.

A voltmeter with very high resistance is used so that the current is as low as possible. If current flows, the concentrations of the species in the half-cells will change and the system will no longer be under standard conditions.

The reactions that occur in the half-cells are:

$$H_2(g) \rightarrow 2H^+(aq) + 2e^- \qquad\qquad Cu^{2+}(aq) + 2e^- \rightarrow Cu(s)$$

The hydrogen is oxidised and the copper(II) ions are reduced.

The cell potential when copper is attached to a standard hydrogen electrode is 0.34 V and we can write:

$$Cu^{2+}(aq) + 2e^- \rightleftharpoons Cu(s) \qquad\qquad E^{\ominus} = +0.34\,V$$

So the standard electrode potential of copper is +0.34 V. The half-equations for standard electrode potentials are always written as reduction reactions and, because the reduction of Cu^{2+} occurs when it is attached to a standard hydrogen electrode, the standard electrode potential has a positive sign, which indicates that the reduction of copper(II) ions is favourable compared with the standard hydrogen electrode.

Figure **9.20** shows a cell in which a standard zinc half-cell is connected to a standard hydrogen electrode. The reactions are:

$$2H^+(aq) + 2e^- \rightarrow H_2(g) \qquad\qquad Zn(s) \rightarrow Zn^{2+}(aq) + 2e^-$$

The cell potential is 0.76 V. The zinc is oxidised when attached to the standard hydrogen electrode, but because a standard electrode potential is always written in terms of a reduction reaction, we write:

$$Zn^{2+}(aq) + 2e^- \rightleftharpoons Zn(s) \qquad\qquad E^{\ominus} = -0.76\,V$$

The negative sign indicates that it is the reverse reaction that is favourable when the zinc half-cell is connected to a standard hydrogen electrode.

The **standard electrode potential** is the EMF (voltage) of a half-cell connected to a standard hydrogen electrode, measured under standard conditions. All solutions must be of concentration $1 \, mol \, dm^{-3}$.

A standard electrode potential is always quoted for the reduction reaction.

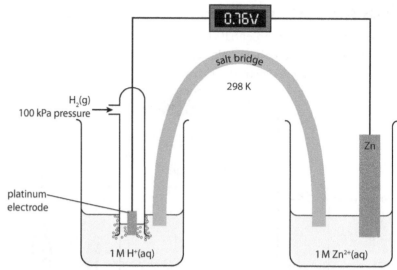

Figure 9.20 Measuring the standard electrode potential of the Zn^{2+}/Zn half-cell. Note: $1 \, M = 1 \, mol \, dm^{-3}$.

Some standard electrode potentials are given in Table **9.7**.

Half-equation	$E^{\ominus}/V$
$K^+(aq) + e^- \rightleftharpoons K(s)$	−2.92
$Na^+(aq) + e^- \rightleftharpoons Na(s)$	−2.71
$Mg^{2+}(aq) + 2e^- \rightleftharpoons Mg(s)$	−2.36
$Mn^{2+}(aq) + 2e^- \rightleftharpoons Mn(s)$	−1.18
$Zn^{2+}(aq) + 2e^- \rightleftharpoons Zn(s)$	−0.76
$Fe^{2+}(aq) + 2e^- \rightleftharpoons Fe(s)$	−0.44
$Ni^{2+}(aq) + 2e^- \rightleftharpoons Ni(s)$	−0.23
$Pb^{2+}(aq) + 2e^- \rightleftharpoons Pb(s)$	−0.13
$2H^+(aq) + 2e^- \rightleftharpoons H_2(g)$	0.00
$Cu^{2+}(aq) + 2e^- \rightleftharpoons Cu(s)$	+0.34
$I_2(aq) + 2e^- \rightleftharpoons 2I^-(aq)$	+0.54
$Fe^{3+}(aq) + e^- \rightleftharpoons Fe^{2+}(aq)$	+0.77
$Ag^+(aq) + e^- \rightleftharpoons Ag(s)$	+0.80
$Br_2(aq) + 2e^- \rightleftharpoons 2Br^-(aq)$	+1.09
$Cr_2O_7^{2-}(aq) + 14H^+(aq) + 6e^- \rightleftharpoons 2Cr^{3+}(aq) + 7H_2O(l)$	+1.33
$Cl_2(aq) + 2e^- \rightleftharpoons 2Cl^-(aq)$	+1.36
$MnO_4^-(aq) + 8H^+(aq) + 5e^- \rightleftharpoons Mn^{2+}(aq) + 4H_2O(l)$	+1.51
$2BrO_3^-(aq) + 12H^+(aq) + 10e^- \rightleftharpoons Br_2(aq) + 6H_2O(l)$	+1.52
$F_2(g) + 2e^- \rightleftharpoons 2F^-(aq)$	+2.87

Table 9.7 Some standard electrode potentials.

Note, in some tables of standard electrode potentials the half-equations for the F_2/F^- reaction, for example, are written as:

$$\tfrac{1}{2}F_2(g) + e^- \rightleftharpoons F^-(aq)$$

This makes absolutely no difference to the value of the standard electrode potential.

Working out cell potentials

We can use standard electrode potentials to work out the cell potential of the $Zn|Zn^{2+}||Ni^{2+}|Ni$ cell (Figure **9.21**).

The standard electrode potentials are:

$$Zn^{2+}(aq) + 2e^- \rightleftharpoons Zn(s) \qquad\qquad E^\ominus = -0.76\,V$$
$$Ni^{2+}(aq) + 2e^- \rightleftharpoons Ni(s) \qquad\qquad E^\ominus = -0.23\,V$$

However, both of these are written as reduction reactions and in any cell there must be a reduction reaction *and* an oxidation reaction. One of the reactions must occur in the reverse direction.

Because the value of the standard electrode potential is more negative for the Zn^{2+}/Zn reaction, this means that the oxidation reaction is more favourable for zinc than for nickel. We therefore reverse the zinc half-equation:

$$Zn(s) \rightarrow Zn^{2+}(aq) + 2e^- \qquad E^\ominus = +0.76\,V \qquad \textbf{oxidation}$$
$$Ni^{2+}(aq) + 2e^- \rightarrow Ni(s) \qquad E^\ominus = -0.23\,V \qquad \textbf{reduction}$$

The cell potential is just the sum of these electrode potentials:

$$E^\ominus_{cell} = 0.76 - 0.23 = +0.53\,V$$

An overall positive value indicates a spontaneous reaction.

The overall equation for the redox reaction is obtained by adding together the two half-equations (checking first that the numbers of electrons balance:

$$
\begin{array}{ll}
Zn(s) \rightarrow Zn^{2+}(aq) + 2e^- & \textbf{oxidation} \\
\underline{Ni^{2+}(aq) + 2e^- \rightarrow Ni(s)} & \textbf{reduction} \\
Zn(s) + Ni^{2+}(aq) \rightarrow Zn^{2+}(aq) + Ni(s) & \textbf{overall redox reaction}
\end{array}
$$

When a cell reaction is written using cell notation:

$$Zn|Zn^{2+}||Ni^{2+}|Ni$$

we can work out the cell potential using the equation:

$$E^\ominus_{cell} = E^\ominus_{rhs} - E^\ominus_{lhs}$$

where $E^\ominus_{rhs}$ is the standard electrode potential on the right-hand side of the cell notation and $E^\ominus_{lhs}$ is the standard electrode potential on the left-hand side of the cell notation.

In this case, $E^\ominus_{cell} = -0.23 - (-0.76) = +0.53\,V$

> We are always looking for a cell potential that is positive overall, as a positive value indicates a spontaneous reaction.

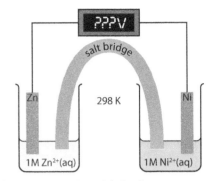

Figure 9.21 A zinc–nickel cell.

In general, the procedure for calculating a cell potential is:
1 Write down the half-equations and standard electrode potentials for the two reactions involved.
2 Change the sign of the more negative (less positive) standard electrode potential and add it to the other electrode potential.

Worked example

9.5 Work out the overall reaction and calculate the cell potential for a cell consisting of an Fe^{3+}/Fe^{2+} half-cell and a Zn^{2+}/Zn half-cell.

The two half-equations are:

$$Zn^{2+}(aq) + 2e^- \rightleftharpoons Zn(s) \qquad E^\ominus = -0.76\,V$$

$$Fe^{3+}(aq) + e^- \rightleftharpoons Fe^{2+}(aq) \qquad E^\ominus = +0.77\,V$$

The more negative value is $-0.76\,V$, and this half-equation is reversed:

$$Zn(s) \rightarrow Zn^{2+}(aq) + 2e^- \qquad E^\ominus = +0.76\,V \qquad \textbf{oxidation}$$
$$Fe^{3+}(aq) + e^- \rightarrow Fe^{2+}(aq) \qquad E^\ominus = +0.77\,V \qquad \textbf{reduction}$$

The electrode potential values are added together to give the cell potential:

$$E^\ominus_{cell} = 0.76 + 0.77 = +1.53\,V \qquad \text{positive cell potential} = \text{spontaneous}$$

In order to combine the two half-equations to produce the overall redox equation, the Fe^{3+}/Fe^{2+} half-equation must be multiplied by two so that the number of electrons is the same in the two half-equations.

$$Zn(s) \rightarrow Zn^{2+}(aq) + 2e^- \qquad E^\ominus = +0.76\,V \qquad \textbf{oxidation}$$
$$\underline{2Fe^{3+}(aq) + 2e^- \rightarrow 2Fe^{2+}(aq) \qquad E^\ominus = +0.77\,V \qquad \textbf{reduction}}$$
$$Zn(s) + 2Fe^{3+}(aq) \rightarrow Zn^{2+}(aq) + 2Fe^{2+}(aq) \qquad E^\ominus_{cell} = +1.53\,V \qquad \textbf{overall redox reaction}$$

Note: although the Fe^{3+}/Fe^{2+} half-equation is multiplied by two, the standard electrode potential is not. A standard electrode potential indicates the potential for a reaction to occur – it is never multiplied by anything when working out a standard cell potential.

The polarity of the electrodes and the direction of electron flow in the external circuit

Consider again the cell made from $Mg|Mg^{2+}$ and $Zn^{2+}|Zn$ half-cells (Figure **9.22**). In the right-hand half-cell, the magnesium is oxidised and therefore electrons are lost. This means that the magnesium electrode is the negative one because electrons are released there. The electrons move through the external circuit from the magnesium to the zinc electrode, where they combine with Zn^{2+} ions in a reduction reaction. The zinc electrode is the positive electrode because electrons are used up there.

The electrode at which oxidation occurs is the anode; therefore the magnesium electrode is the anode in Figure **9.22**. The electrode at which reduction occurs is the cathode; therefore the zinc electrode is the cathode.

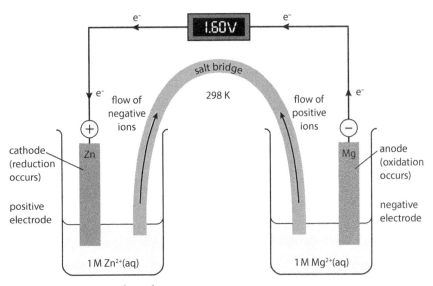

Figure 9.22 An Mg|Mg^{2+}||Zn^{2+}|Zn cell.

Another way to determine which are the negative and positive electrodes is to just look at the original electrode potentials:

$$Mg^{2+}(aq) + 2e^- \rightleftharpoons Mg(s) \qquad\qquad E^{\ominus} = -2.36\,V$$

$$Zn^{2+}(aq) + 2e^- \rightleftharpoons Zn(s) \qquad\qquad E^{\ominus} = -0.76\,V$$

The electrode potential is more negative for the Mg|Mg^{2+} half-cell – therefore the magnesium electrode is the negative electrode (and the zinc electrode is the positive electrode). The electrons flow from the negative electrode to the positive electrode – that is, from magnesium to zinc (Figure **9.22**). Negative charge always flows in the same continuous direction around the circuit, so negative ions will flow through the salt bridge from the zinc half-cell to the magnesium half-cell. Positive ions will flow in the opposite direction.

If the cell is written in terms of the cell notation, then the negative electrode is on the left and electrons flow through the external circuit from left to right (Figure **9.23**).

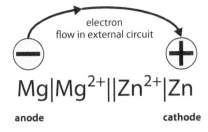

Figure 9.23 Cell notation is written so that the negative electrode is on the left.

> Note: these are the original standard electrode potentials (for the reduction reactions) – neither has been reversed.

> **Exam tip**
> Whichever half-cell has the more negative standard electrode potential will be the negative electrode in the cell, and the electrons will flow from this half-cell to the other one.

Worked example

9.6 Predict the cell potential for a cell made up of $Cu|Cu^{2+}$ and $Ni|Ni^{2+}$ half-cells. Indicate which is the positive and which is the negative electrode. Give the direction of electron flow in the external circuit and write an overall equation for the reaction that occurs.

The half-equations and standard electrode potentials are:

$$Cu^{2+}(aq) + 2e^- \rightleftharpoons Cu(s) \qquad\qquad E^\ominus = +0.34\,V$$

$$Ni^{2+}(aq) + 2e^- \rightleftharpoons Ni(s) \qquad\qquad E^\ominus = -0.23\,V$$

We can see immediately which is the negative and which the positive electrode. The Ni^{2+}/Ni standard electrode potential is more negative, and therefore this is the negative electrode. The electrons flow from the negative electrode (nickel) to the positive electrode (copper).

Exam tip
There is more than one half-equation in the data booklet involving copper – make sure that you select the correct one. You must always check that both the oxidised and reduced species are the same as in your half-equation.

The more negative electrode potential is that for the Ni^{2+}/Ni reaction and so this half-equation is reversed:

$$
\begin{array}{lll}
Cu^{2+}(aq) + 2e^- \rightarrow Cu(s) & E^\ominus = +0.34\,V & \textbf{reduction} \\
\underline{\qquad\qquad Ni(s) \rightarrow Ni^{2+}(aq) + 2e^-} & \underline{E^\ominus = +0.23\,V} & \textbf{oxidation} \\
Cu^{2+}(aq) + Ni(s) \rightarrow Ni^{2+}(aq) + Cu(s) & E^\ominus_{cell} = +0.57\,V & \textbf{overall redox reaction}
\end{array}
$$

Electrons are lost at the nickel electrode, so this is this negative electrode. Electrons are gained at the copper electrode and so this is the positive electrode. Electrons flow from the nickel electrode to the copper electrode.

Exam tip
In the IB Chemistry data booklet, the electrode potentials are arranged from negative to positive. The half-equation higher up the table is therefore the one that is reversed to give the overall spontaneous reaction.

? Test yourself

18 Predict cell potentials for the following cells (use Table **9.7**, page **396**). Indicate which is the positive and which is the negative electrode. Give the direction of electron flow in the external circuit and write an overall equation for the reaction that occurs.

   a $Ni^{2+}(aq) + 2e^- \rightleftharpoons Ni(s)$ $\qquad\qquad Fe^{2+}(aq) + 2e^- \rightleftharpoons Fe(s)$

   b $I_2(s) + 2e^- \rightleftharpoons 2I^-(aq)$ $\qquad\qquad Cl_2(g) + 2e^- \rightleftharpoons 2Cl^-(aq)$

   c $Ag^+(aq) + e^- \rightleftharpoons Ag(s)$ $\qquad\qquad Zn^{2+}(aq) + 2e^- \rightleftharpoons Zn(s)$

   d $Cr_2O_7^{2-}(aq) + 14H^+(aq) + 6e^- \rightleftharpoons 2Cr^{3+}(aq) + 7H_2O(l)$ $\qquad Fe^{3+}(aq) + e^- \rightleftharpoons Fe^{2+}(aq)$

   e $MnO_4^-(aq) + 8H^+(aq) + 5e^- \rightleftharpoons Mn^{2+}(aq) + 4H_2O(l)$ $\qquad Cl_2(g) + 2e^- \rightleftharpoons 2Cl^-(aq)$

   f $Fe(s)|Fe^{2+}(aq)||Br_2(l)|Br^-(aq)|Pt$

Using standard electrode potentials to predict the feasibility of a redox reaction

In chemistry, rather than being concerned with generating electricity and voltaic cells, we usually use standard electrode potentials to predict whether or not a particular redox reaction is likely to occur. For instance, will acidified potassium dichromate(VI) oxidise iron(II) ions? If we examine the relevant half-equations we get:

$$Cr_2O_7^{2-}(aq) + 14H^+(aq) + 6e^- \rightleftharpoons 2Cr^{3+}(aq) + 7H_2O(l) \qquad E^\ominus = +1.33\,V$$

$$Fe^{3+}(aq) + e^- \rightleftharpoons Fe^{2+}(aq) \qquad E^\ominus = +0.77\,V$$

We are looking at the possibility of Fe^{2+} being oxidised to Fe^{3+} and therefore we must reverse this half-equation:

$$Fe^{2+}(aq) \rightarrow Fe^{3+}(aq) + e^- \qquad E^\ominus = -0.77\,V \qquad \textbf{oxidation}$$

$$Cr_2O_7^{2-}(aq) + 14H^+(aq) + 6e^-$$
$$\rightarrow 2Cr^{3+}(aq) + 7H_2O(l) \qquad E^\ominus = +1.33\,V \qquad \textbf{reduction}$$

When we add up these values we get $E^\ominus_{cell} = 1.33 - 0.77 = +0.56\,V$.

The positive value indicates that the reaction will be spontaneous, so Fe^{2+} ions will be oxidised to Fe^{3+} ions by acidified dichromate(VI) ions.

The overall redox equation is obtained by multiplying the oxidation half-equation by 6 (so that the electrons balance) and then adding it to the reduction half-equation:

$$Cr_2O_7^{2-}(aq) + 14H^+(aq) + 6Fe^{2+}(aq) \rightarrow 2Cr^{3+}(aq) + 7H_2O(l) + 6Fe^{3+}(aq)$$

As with using ΔG to predict the spontaneity of a reaction, the fact that a reaction is spontaneous does not tell us anything about the speed of the reaction. Although this reaction occurs relatively rapidly at room temperature, this will not always be the case and some reaction mixtures will need heating to occur at a significant rate.

Predictions using standard electrode potentials are valid only under standard conditions – that is, at 298 K and 100 kPa with $1\,mol\,dm^{-3}$ solutions. However, so long as the values are not too close to zero, the predictions are likely to be useful even when a reaction is carried out under non-standard conditions.

Worked example

9.7 Predict, using standard electrode potentials, whether or not acidified potassium dichromate(VI) will oxidise Cl^- ions to Cl_2.

The half-equations are:

$$Cr_2O_7^{2-}(aq) + 14H^+(aq) + 6e^- \rightleftharpoons 2Cr^{3+}(aq) + 7H_2O(l) \qquad E^\ominus = +1.33\,V$$
$$Cl_2(aq) + 2e^- \rightleftharpoons 2Cl^-(aq) \qquad E^\ominus = +1.36\,V$$

We are looking at the possibility of Cl^- ions being oxidised and must therefore reverse this half-equation:

$$2Cl^-(aq) \rightarrow Cl_2(aq) + 2e^- \qquad E^\ominus = -1.36\,V \qquad \textbf{oxidation}$$
$$Cr_2O_7^{2-}(aq) + 14H^+(aq) + 6e^- \rightarrow 2Cr^{3+}(aq) + 7H_2O(l) \qquad E^\ominus = +1.33\,V \qquad \textbf{reduction}$$

When the electrode potentials are added together we get $E^\ominus_{cell} = 1.33 - 1.36 = -0.03\,V$. The value is negative, which indicates that the reaction will not be spontaneous under standard conditions, and acidified dichromate(VI) will not oxidise Cl^- to Cl_2.

Oxidising and reducing agents

A very positive value for $E^{\ominus}$, for example:

$$Cr_2O_7^{2-}(aq) + 14H^+(aq) + 6e^- \rightarrow 2Cr^{3+}(aq) + 7H_2O(l) \qquad E^{\ominus} = +1.33\,V$$

means that the reduction reaction is very favourable (relative to the standard hydrogen electrode), and the substance has a very strong tendency to pick up electrons from other species – i.e. it is a strong oxidising agent.

> The more positive the standard electrode potential, the stronger the oxidising agent.

> Oxidising agents have a strong tendency to remove electrons from other things substances – i.e. oxidise other species. Oxidising agents are substances that are readily reduced.

A very negative value of the standard electrode potential, for example:

$$Na^+(aq) + e^- \rightleftharpoons Na(s) \qquad\qquad\qquad E^{\ominus} = -2.71\,V$$

indicates that the reverse reaction is very favourable:

$$Na(s) \rightarrow Na^+(aq) + e^- \qquad\qquad\qquad E^{\ominus} = +2.71\,V$$

So, sodium has a very strong tendency to give electrons to other species – that is, to reduce other species. So, Na is a very good reducing agent.

> The more negative the standard electrode potential, the stronger the reducing agent.

In general we can say:

> a substance with a more positive standard electrode potential will oxidise a substance with a less positive electrode potential
> or
> a substance with a more negative standard electrode potential will reduce a substance with a less negative standard electrode potential.

Example 1

Consider the half-equations:

$$Cr_2O_7^{2-}(aq) + 14H^+(aq) + 6e^- \rightleftharpoons 2Cr^{3+}(aq) + 7H_2O(l) \qquad E^{\ominus} = +1.33\,V$$

$$I_2(aq) + 2e^- \rightleftharpoons 2I^-(aq) \qquad\qquad\qquad\qquad E^{\ominus} = +0.54\,V$$

$Cr_2O_7^{2-}$ has a more positive standard electrode potential than I_2, and therefore $Cr_2O_7^{2-}$ is a stronger oxidising agent than I_2 and will oxidise I^- to I_2.

The overall reaction is:

$$Cr_2O_7^{2-}(aq) + 14H^+(aq) + 6I^-(aq) \rightarrow 2Cr^{3+}(aq) + 7H_2O(l) + 3I_2(aq)$$

$$E^{\ominus}_{cell} = 1.33 - 0.54 = +0.79\,V$$

In this reaction, $Cr_2O_7^{2-}$ is the oxidising agent and I^- is the reducing agent.

Example 2

Consider the half-equations:

$$Mg^{2+}(aq) + 2e^- \rightleftharpoons Mg(s) \qquad\qquad E^{\ominus} = -2.36\,V$$

$$Pb^{2+}(aq) + 2e^- \rightleftharpoons Pb(s) \qquad\qquad E^{\ominus} = -0.13\,V$$

The standard electrode potential of the $Mg^{2+}|Mg$ half-cell is more negative than that of the $Pb^{2+}|Pb$ half-cell, which indicates that Mg is a stronger reducing agent than Pb and that Mg will therefore reduce Pb^{2+} to Pb.

The overall equation is:

$$Mg(s) + Pb^{2+}(aq) \rightarrow Mg^{2+}(aq) + Pb(s) \qquad E^{\ominus}_{cell} = +2.23\,V$$

In this reaction, Mg is the reducing agent and Pb^{2+} is the oxidising agent.

Let us consider the standard electrode potentials in Table **9.8**. Species on the right of the reversible arrow higher in the table reduce those on the left of the reversible arrow lower in the table. So, Na will reduce Mg^{2+} ions, and Zn will reduce Fe^{2+} ions. Species on the left of the reversible arrow lower in the table oxidise those on the right of the reversible arrow higher in the table – BrO_3^- will oxidise Cl^- and Br_2 will oxidise I^-.

Half-equation	$E^{\ominus}/V$
$K^+(aq) + e^- \rightleftharpoons K(s)$	−2.92
$Na^+(aq) + e^- \rightleftharpoons Na(s)$	−2.71
$Mg^{2+}(aq) + 2e^- \rightleftharpoons Mg(s)$	−2.36
$Zn^{2+}(aq) + 2e^- \rightleftharpoons Zn(s)$	−0.76
$Fe^{2+}(aq) + 2e^- \rightleftharpoons Fe(s)$	−0.44
$Ni^{2+}(aq) + 2e^- \rightleftharpoons Ni(s)$	−0.23
$Pb^{2+}(aq) + 2e^- \rightleftharpoons Pb(s)$	−0.13
$2H^+(aq) + 2e^- \rightleftharpoons H_2(g)$	0.00
$Cu^{2+}(aq) + 2e^- \rightleftharpoons Cu(s)$	+0.34
$I_2(s) + 2e^- \rightleftharpoons 2I^-(aq)$	+0.54
$Fe^{3+}(aq) + \rightleftharpoons Fe^{2+}(aq)$	+0.77
$Ag^+(aq) + e^- \rightleftharpoons Ag(s)$	+0.80
$Br_2(aq) + 2e^- \rightleftharpoons 2Br^-(aq)$	+1.09
$Cr_2O_7^{2-}(aq) + 14H^+(aq) + 6e^- \rightleftharpoons 2Cr^{3+}(aq) + 7H_2O(l)$	+1.33
$Cl_2(aq) + 2e^- \rightleftharpoons 2Cl^-(aq)$	+1.36
$MnO_4^-(aq) + 8H^+(aq) + 5e^- \rightleftharpoons Mn^{2+}(aq) + 4H_2O(l)$	+1.51
$2BrO_3^-(aq) + 12H^+(aq) + 10e^- \rightleftharpoons Br_2(aq) + 6H_2O(l)$	+1.52
$F_2(g) + 2e^- \rightleftharpoons 2F^-(aq)$	+2.87

Table 9.8 The connection between standard electrode potentials and oxidising/reducing ability.

strongest reducing agent (K)

decreasing strength of reducing agent

decreasing strength of oxidising agent

strongest oxidising agent (F₂)

The activity series and electrode potentials – the electrochemical series

We saw earlier in this topic that a more reactive metal will displace a less reactive metal from solution. For example, zinc is more reactive than copper and displaces copper(II) ions from solution:

$$Zn(s) + Cu^{2+}(aq) \rightarrow Zn^{2+}(aq) + Cu(s) \qquad E^{\ominus}_{cell} = +1.10\,V$$

In terms of standard electrode potentials, the more reactive a metal, the more negative its standard electrode potential. The more negative standard electrode potential indicates that the reduction reaction of the metal ion is very unfavourable and therefore that the oxidation of the metal is very favourable. More reactive metals have a greater tendency to be oxidised and thus give electrons to (reduce) other species.

In Table **9.9**, Mg is the most reactive metal and Ag the least reactive.

Metals above hydrogen in the activity series are stronger reducing agents than hydrogen and should displace hydrogen from a solution of its ions – that is, from acids:

$$Mg(s) + 2H^+(aq) \rightarrow Mg^{2+}(aq) + H_2(g) \qquad E^{\ominus}_{cell} = +2.36\,V$$

Metals lower than hydrogen in the activity series will not react with acids – so copper and silver do not react with hydrochloric acid.

In terms of standard electrode potentials, any metal with a negative standard electrode potential is a stronger reducing agent than hydrogen ($E^{\ominus} = 0.00\,V$) and reduces hydrogen ions to hydrogen gas. In other words, metals with negative standard electrode potentials should liberate hydrogen from acids.

We can consider the reactivity of the halogens in terms of their oxidising ability. Chlorine is a stronger oxidising agent than bromine and iodine and will oxidise bromide ions to bromine and iodide ions to iodine. Bromine is a stronger oxidising agent than iodine and will oxidise iodide ions to iodine. In terms of electrons, chlorine has the strongest affinity for electrons and will remove electrons from bromide ions and iodide ions.

Let us consider the standard electrode potentials in Table **9.10**. Chlorine has the most positive standard electrode potential and is therefore the strongest oxidising agent:

$$Cl_2(aq) + 2Br^-(aq) \rightarrow 2Cl^-(aq) + Br_2(aq) \qquad E^{\ominus}_{cell} = +0.27\,V$$

$$Cl_2(aq) + 2I^-(aq) \rightarrow 2Cl^-(aq) + I_2(aq) \qquad E^{\ominus}_{cell} = +0.82\,V$$

$$Br_2(aq) + 2I^-(aq) \rightarrow 2Br^-(aq) + I_2(aq) \qquad E^{\ominus}_{cell} = +0.55\,V$$

All these reactions have a positive cell potential and are, therefore, all spontaneous. All other possible reactions between the halogens and the halide ions would have a negative electrode potential and are therefore not spontaneous, for example:

$$Br_2(aq) + 2Cl^-(aq) \rightarrow 2Br^-(aq) + Cl_2(aq) \qquad E^{\ominus}_{cell} = -0.27\,V$$

	$E^{\ominus}$/V
$Mg^{2+}(aq) + 2e^- \rightleftharpoons Mg(s)$	−2.36
$Zn^{2+}(aq) + 2e^- \rightleftharpoons Zn(s)$	−0.76
$Fe^{2+}(aq) + 2e^- \rightleftharpoons Fe(s)$	−0.44
$Ni^{2+}(aq) + 2e^- \rightleftharpoons Ni(s)$	−0.23
$Cu^{2+}(aq) + 2e^- \rightleftharpoons Cu(aq)$	+0.34
$Ag^+(aq) + e^- \rightleftharpoons Ag(s)$	+0.80

Table 9.9 Activity series and standard electrode potentials.

	$E^{\ominus}$/V
$I_2(aq) + 2e^- \rightleftharpoons 2I^-(aq)$	+0.54
$Br_2(aq) + 2e^- \rightleftharpoons 2Br^-(aq)$	+1.09
$Cl_2(aq) + 2e^- \rightleftharpoons 2Cl^-(aq)$	+1.36

Table 9.10 Standard potentials for halogens.

19 Use standard electrode potentials to predict whether the following reactions will be spontaneous. If the reaction is spontaneous, state the oxidising agent and the reducing agent.

a $Cu(s) + Mg^{2+}(aq) \rightarrow Cu^{2+}(aq) + Mg(s)$

b $Cr_2O_7^{2-}(aq) + 14H^+(aq) + 6Br^-(aq)$
$\rightarrow 2Cr^{3+}(aq) + 7H_2O(l) + 3Br_2(l)$

c $Cr_2O_7^{2-}(aq) + 14H^+(aq) + 6F^-(aq)$
$\rightarrow 2Cr^{3+}(aq) + 7H_2O(l) + 3F_2(g)$

d $5Fe^{3+}(aq) + Mn^{2+}(aq) + 4H_2O(l)$
$\rightarrow 5Fe^{2+}(aq) + MnO_4^-(aq) + 8H^+(aq)$

20 Consider the following standard electrode potentials:

$U^{4+} + e^- \rightleftharpoons U^{3+}$	$E^\ominus = -0.61\,V$
$U^{3+} + 3e^- \rightleftharpoons U$	$E^\ominus = -1.79\,V$
$Eu^{3+} + e^- \rightleftharpoons Eu^{2+}$	$E^\ominus = -0.43\,V$
$Po^{2+} + 2e^- \rightleftharpoons Po$	$E^\ominus = +0.65\,V$
$In^{3+} + 3e^- \rightleftharpoons In$	$E^\ominus = -0.34\,V$
$Sm^{3+} + e^- \rightleftharpoons Sm^{2+}$	$E^\ominus = -1.15\,V$
$Np^{3+} + 3e^- \rightleftharpoons Np$	$E^\ominus = -1.86\,V$
$Np^{4+} + e^- \rightleftharpoons Np^{3+}$	$E^\ominus = +0.15\,V$

a Select from this list:
  i the strongest oxidising agent
  ii the strongest reducing agent

b Use the standard electrode potentials above to work out whether the following statements are **true** or **false**:
  i Eu^{2+} will reduce In^{3+} to In
  ii Sm^{3+} will oxidise Np to Np^{3+}
  iii Po will reduce Sm^{3+} to Sm^{2+}
  iv Np^{3+} is a stronger reducing agent than Po
  v U is a stronger reducing agent than Np
  vi Np^{3+} will reduce Po^{2+} to Po but will not oxidise Eu^{2+} to Eu^{3+}
  vii Sm^{2+} will not reduce U^{3+} to U but will reduce U^{4+} to U^{3+}

c Predict whether or not the following reactions will be spontaneous:
  i $3Eu^{3+} + In \rightarrow 3Eu^{2+} + In^{3+}$
  ii $2Eu^{2+} + Po^{2+} \rightarrow 2Eu^{3+} + Po$
  iii $3Np^{3+} + In \rightarrow 3Np^{4+} + In^{3+}$

Standard cell potentials and $\Delta G^\ominus$

The standard cell potential and the standard free energy change ($\Delta G^\ominus$) are related by the equation:

$$\Delta G^\ominus = -nFE^\ominus$$

where n is the number of electrons transferred in a particular redox reaction and F is the Faraday constant, which is equal to the charge on one mole of electrons and has a value of approximately $96\,500\,C\,mol^{-1}$.

Example
Calculate $\Delta G^\ominus$ for the reaction:

$$Zn(s) + 2Ag^+(aq) \rightarrow Zn^{2+}(aq) + 2Ag(s)$$

The cell potential can be worked out from the standard electrode potentials ($Zn^{2+}/Zn = -0.76\,V$ and $Ag^+/Ag = +0.80\,V$):

$$E^\ominus_{cell} = 0.76 + 0.80 = 1.56\,V$$

If the overall equation is split into its two half-equations:

$$Zn(s) \rightarrow Zn^{2+}(aq) + 2e^-$$

$$2Ag^+(aq) + 2e^- \rightarrow 2Ag(s)$$

It can be seen that two electrons are transferred from the zinc to the silver ions, therefore $n = 2$.

Using $\Delta G^{\ominus} = -nFE^{\ominus}$

$$\Delta G^{\ominus} = -2 \times 96\,500 \times 1.56 = -301\,000\,\text{J}\,\text{mol}^{-1}$$

We normally quote $\Delta G^{\ominus}$ values in $\text{kJ}\,\text{mol}^{-1}$, therefore $\Delta G^{\ominus} = -301\,\text{kJ}\,\text{mol}^{-1}$

So, both $E^{\ominus}_{cell}$ and $\Delta G^{\ominus}$ can be used to predict whether a reaction is spontaneous or not (Table **9.11**).

$E^{\ominus}_{cell}$	$\Delta G^{\ominus}$	Spontaneous?
positive	negative	spontaneous
negative	positive	non-spontaneous
0	0	at equilibrium

Table 9.11 The relationship between $E^{\ominus}_{cell}$, $\Delta G^{\ominus}$ and spontaneity.

? Test yourself

21 Work out the cell potential and $\Delta G^{\ominus}$ for each of the following and predict whether the reaction will be spontaneous or not:

a $Zn(s) + Ni^{2+}(aq) \rightarrow Zn^{2+}(aq) + Ni(s)$
b $Pb^{2+}(aq) + Cu(s) \rightarrow Pb(s) + Cu^{2+}(aq)$
c $Cr_2O_7{}^{2-}(aq) + 6I^-(aq) + 14H^+(aq) \rightarrow 2Cr^{3+}(aq) + 3I_2(aq) + 7H_2O(l)$

9.3.2 Electrolysis of aqueous solutions

Ionic salts conduct electricity when dissolved in water because the ions are free to move. These solutions can be electrolysed, but the products are not as straightforward as in electrolysis of molten salts. The general rule for the products formed at the electrodes when aqueous solutions are electrolysed is shown in Table **9.12**.

Learning objectives

- Understand the factors that affect the nature of the products of electrolysis of aqueous solutions
- Understand how electrolysis is used in electroplating
- Work out the relative amounts of products at the electrodes during electrolysis

Electrode	Product
positive (**anode**)	oxygen or a halogen
negative (**cathode**)	a metal or hydrogen

Table 9.12 Products at the electrodes when aqueous solutions are electrolysed.

The products obtained from the electrolysis of various $1\,\text{mol}\,\text{dm}^{-3}$ solutions using platinum electrodes are given in Table **9.13**.

Solution	Product at anode (+)	Product at cathode (−)
copper(II) chloride	chlorine gas	copper metal
copper(II) sulfate	oxygen gas	copper metal
sodium chloride	chlorine gas	hydrogen gas
hydrochloric acid	chlorine gas	hydrogen gas
water (acidified)	oxygen gas	hydrogen gas

Table 9.13 Electrolysis products of aqueous solutions using platinum electrodes.

There are three main factors that influence the products formed when an aqueous solution of a salt is electrolysed:
- the standard electrode potentials of the species in solution
- the concentration of the electrolyte
- the material from which the electrodes are made.

Using standard electrode potentials to predict the product of electrolysis

We will first of all look at the products at the cathode. When sodium chloride solution is electrolysed, a reduction reaction occurs at the cathode and hydrogen gas is formed. The two possible species that are present in the solution that could be reduced are Na^+ and H_2O. Let us consider the standard electrode potentials for the reduction of Na^+ and water:

$$Na^+(aq) + e^- \rightleftharpoons Na(s) \qquad\qquad E^\ominus = -2.71\,V$$

$$H_2O(l) + e^- \rightleftharpoons \tfrac{1}{2}H_2(g) + OH^-(aq) \qquad\qquad E^\ominus = -0.83\,V$$

It can be seen that the standard electrode potential for the reduction of water is much more positive than that for the reduction of Na^+. Therefore the reduction of water is more favourable than the reduction of Na^+, and hydrogen will be formed from the reduction of water rather than sodium metal from the reduction of Na^+.

If we then compare this with the electrolysis of copper(II) sulfate solution, the electrode potentials for the possible reduction reactions are:

$$H_2O(l) + e^- \rightleftharpoons \tfrac{1}{2}H_2(g) + OH^-(aq) \qquad\qquad E^\ominus = -0.83\,V$$

$$Cu^{2+}(aq) + 2e^- \rightleftharpoons Cu(s) \qquad\qquad E^\ominus = +0.34\,V$$

It can be seen that it is more favourable to reduce Cu^{2+} ions than to reduce water and therefore copper, from the reduction of Cu^{2+}, will be formed at the cathode rather than hydrogen.

In general, metals can be divided into three groups.
1 Very reactive metals (with very negative standard electrode potentials), such as sodium, potassium and magnesium – that is, metals above zinc in the activity series. These produce hydrogen when aqueous solutions of their ions are electrolysed.
2 Unreactive metals (with positive standard electrode potentials), such as copper and silver. These produce the metal when aqueous solutions of their ions are electrolysed.

'Acidified water' is water with a small amount of sulfuric acid added.

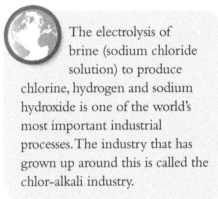
The electrolysis of brine (sodium chloride solution) to produce chlorine, hydrogen and sodium hydroxide is one of the world's most important industrial processes. The industry that has grown up around this is called the chlor-alkali industry.

Water is reduced here because the oxidation number of hydrogen changes from +1 in water to 0 in H_2. No other oxidation numbers have changed.

Predicting the products of electrolysis is more complex for metals lying between zinc and hydrogen in the reactivity series, i.e. for metals such as zinc ($E^\ominus = -0.76\,V$), nickel ($E^\ominus = -0.23\,V$) and lead ($E^\ominus = -0.13\,V$). Electrolysis of a solution of zinc sulfate produces zinc metal at the cathode, electrolysis of nickel chloride solution produces a mixture of nickel and hydrogen gas and electrolysis of lead(II) nitrate solution produces lead metal at the cathode.

You should not be asked about these metals in examination questions.

3 Metals of intermediate reactivity (with standard electrode potentials between $-0.83\,V$ and $0.00\,V$), such as zinc, nickel and lead. For these metals it is much more difficult to make predictions and we can get either the metal or a mixture of the metal and hydrogen, depending on the metal and the conditions used.

We can see that it is possible, to a certain extent, to predict the products of electrolysis of aqueous solutions based on the reactivity of the metals and their standard electrode potentials. Problems with using standard electrode potentials arise because they refer to standard conditions ($1\,mol\,dm^{-3}$ concentrations) and to solutions containing both the oxidised and the reduced species.

Formation of halogens during electrolysis of aqueous solutions

When a solution of sodium iodide is electrolysed, iodine is produced at the anode but when a solution of sodium fluoride is electrolysed, oxygen (from the oxidation of water) is produced at the anode. This can be explained in term of how easy it is to oxidise the halide ions to the element, compared to the oxidation of water – standard electrode potentials can be used as a guide to this.

Consider these standard electrode potentials:

$$I_2(s) + 2e^- \rightleftharpoons 2I^-(aq) \qquad\qquad E^{\ominus} = +0.54\,V$$

$$O_2(g) + 4H^+(aq) + 4e^- \rightleftharpoons 2H_2O(l) \qquad\qquad E^{\ominus} = +1.23\,V$$

$$F_2(s) + 2e^- \rightleftharpoons 2F^-(aq) \qquad\qquad E^{\ominus} = +2.87\,V$$

E_{ox} is the potential for oxidation.

At the anode, oxidation occurs and therefore we can reverse all these half-equations to get the oxidation reaction:

$$2I^-(aq) \rightarrow I_2(s) + 2e^- \qquad\qquad E_{ox} = -0.54\,V$$

$$2H_2O(l) \rightarrow O_2(g) + 4H^+(aq) + 4e^- \qquad\qquad E_{ox} = -1.23\,V$$

$$2F^-(aq) \rightarrow F_2(g) + 2e^- \qquad\qquad E_{ox} = -2.87\,V$$

Water is oxidised in this half-equation because the oxidation number of oxygen increases from -2 in water to 0 in O_2.

At the anode, two oxidation reactions are possible – either the halide ion can be oxidised to produce the halogen, or water can be oxidised to produce oxygen. From these values it can be seen that it is more favourable to oxidise I^- ions (more positive potential for oxidation) than to oxidise water, but it is less favourable to oxidise F^- ions (more negative potential for oxidation). So in $NaI(aq)$, iodide ions will be oxidised at the anode to produce iodine, but in $NaF(aq)$, water will be oxidised in preference to fluoride ions to produce oxygen gas at the anode.

The anode product in the electrolysis of copper(II) sulfate using graphite electrodes

Oxygen is produced at the anode from the oxidation of water:

$$2H_2O(l) \rightarrow O_2(g) + 4H^+(aq) + 4e^- \qquad\qquad \textbf{oxidation}$$

Electrolysis of sulfates and nitrates always produces oxygen gas at the anode.

Sulfates contain sulfur in its highest oxidation state and therefore it is not susceptible to oxidation.

It is actually possible to oxidise the SO_4^{2-} ion, but this is much more difficult than the oxidation of water:

$$2H_2O(l) \rightleftharpoons O_2(g) + 4H^+(aq) + 4e^- \qquad\qquad E_{ox} = -1.23\,V$$

$$SO_4^{2-}(aq) \rightleftharpoons \tfrac{1}{2}S_2O_8^{2-}(aq) + e^- \qquad\qquad E_{ox} = -2.01\,V$$

$S_2O_8^{2-}$ contains a peroxo group, so the oxidation state of S is still +6; the O is oxidised.

The effect of concentration of the electrolyte on the products at the electrodes

We will consider the products at the anode when a solution of sodium chloride is electrolysed. The two species in solution that can be oxidised at the anode are Cl^- and H_2O and the equations with the relevant standard electrode potentials are:

$$O_2(g) + 4H^+(aq) + 4e^- \rightleftharpoons 2H_2O(l) \qquad\qquad E^\ominus = +1.23\,V$$

$$Cl_2(g) + 2e^- \rightleftharpoons 2Cl^-(aq) \qquad\qquad E^\ominus = +1.36\,V$$

These are both reduction reactions.

We are, however, interested in the oxidation reactions and these equations must be reversed:

$$2H_2O(l) \rightarrow O_2(g) + 4H^+(aq) + 4e^- \qquad\qquad E_{ox} = -1.23\,V$$

$$2Cl^-(aq) \rightarrow Cl_2(g) + 2e^- \qquad\qquad E_{ox} = -1.36\,V$$

From these values it can be seen that it is slightly more favourable (E_{ox} more positive) to oxidise H_2O to form oxygen than to oxidise Cl^- to form chlorine. If a solution of sodium chloride of low concentration is electrolysed, the major product at the anode is oxygen. However, these values are very close together and at higher concentrations of sodium chloride, chlorine becomes the major product.

At very low chloride concentrations, the product of electrolysis is mainly oxygen, but with more concentrated solutions, chlorine is the major product.

The exact reasons for this are complex, but it can be explained in terms of the difficulty in transferring electrons from water across the electrode–solution interface – a higher voltage is required. The voltage in excess of the expected voltage is known as the overvoltage.

The nature of the electrodes affects the products formed

In order to illustrate this, we will consider the electrolysis of copper sulfate solution using two different sets of electrodes. Different products are obtained at the anode depending on the material of which the electrodes are made.

Electrolysis of copper(II) sulfate solution using inert electrodes (graphite or platinum)

The electrolysis of copper sulfate solution using platinum electrodes is shown in Figure **9.24**. The products and half-equations at each electrode are shown in Table **9.14**.

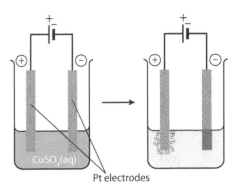

Figure 9.24 Electrolysis of copper sulfate solution using platinum electrodes.

Electrode	Product	Half-equation
anode (+)	oxygen	$2H_2O(l) \rightarrow O_2(g) + 4H^+(aq) + 4e^-$
cathode (−)	copper	$Cu^{2+}(aq) + 2e^- \rightarrow Cu(s)$

Table 9.14 Electrolysis of copper(II) sulfate solution with platinum electrodes.

During the experiment, the cathode (−) becomes coated in a brown metal (copper), bubbles of a colourless gas (oxygen) are given off at the anode (+), the blue colour of the solution fades and the solution becomes more acidic (test with a pH meter or pH paper).

The blue colour of the solution is due to the presence of $Cu^{2+}(aq)$ ions. The colour of the solution becomes paler as the Cu^{2+} ions are removed from the solution by reduction at the cathode.

The oxidation of water at the anode produces H^+ ions as well as O_2. This means that the solution becomes acidic.

The sulfate ions remain unchanged in the solution, and, because H^+ ions are also produced, the remaining solution after all the copper ions have been removed is sulfuric acid (H_2SO_4).

Electrolysis of copper(II) sulfate solution using copper electrodes

Electrolysis of copper sulfate solution using copper electrodes is shown in Figure **9.25**.

The positive electrode (anode) becomes smaller, the negative electrode (cathode) becomes coated with a brown metal (copper) and the solution remains the same colour.

The reaction at the cathode is exactly the same as when inert electrodes are used. However, at the anode the reaction is different and no oxygen is given off. The reaction at the anode is oxidation and two reactions are possible:

$$Cu(s) \rightarrow Cu^{2+}(aq) + 2e^- \qquad\qquad E_{ox} = -0.34\,V$$

$$2H_2O(l) \rightarrow O_2(g) + 4H^+(aq) + 4e^- \qquad\qquad E_{ox} = -1.23\,V$$

The potential for oxidation of copper is more positive than that for the oxidation of water, and therefore the oxidation of copper is more favourable than the oxidation of water. So copper ions pass into solution from the anode and oxygen is not produced. The oxidation of copper at the anode was not possible when inert electrodes were used.

The overall processes that occur are summarised in Table **9.15**.

Electrode	Product	Half-equation
anode (+)	copper ions pass into solution	$Cu(s) \rightarrow Cu^{2+}(aq) + 2e^-$
cathode (−)	copper	$Cu^{2+}(aq) + 2e^- \rightarrow Cu(s)$

Table 9.15 Electrolysis of copper(II) sulfate solution with copper electrodes.

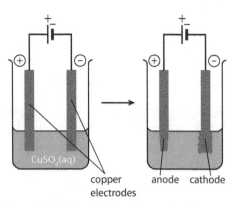

Figure 9.25 Electrolysis of copper sulfate solution using copper electrodes.

copper electrodes

anode   cathode

CuSO$_4$(aq)

This process is used in the purification of copper. The anode is made of impure copper and the cathode of pure copper. Copper dissolves from the anode and is deposited as pure copper at the cathode. Impurities are either left as a sludge below the anode or go into solution. The gain in mass of the cathode is less than the loss in mass of the anode as a result of the impurities.

The net process is therefore a transfer of copper from the anode to the cathode. If pure copper electrodes are used, the mass of copper lost from the anode is equal to the mass of copper deposited on the cathode.

Because one Cu^{2+} ion is removed from the solution at the cathode for every Cu^{2+} ion added to the solution at the anode, the overall concentration of Cu^{2+} ions in the electrolyte does not change and the blue colour of the solution remains constant.

Electroplating

Electroplating is the process of coating an object with a thin layer of a metal using electrolysis. The object to be coated should be used as the cathode, the anode should be made of the metal with which the object is to be plated and the electrolyte will normally be a solution containing the ions of the coating metal. The object to be plated must be the cathode (−) because metal ions are positively charged and will be attracted to the cathode. The object to be plated must be thoroughly cleaned before electroplating, otherwise the coating will not stick properly to the surface. Figure **9.26** shows the experimental set-up for coating a key with copper.

The reactions involved are:

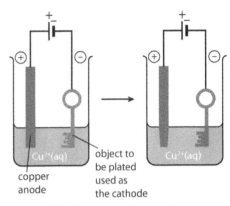

Electrode	Half-equation
anode (+)	$Cu(s) \rightarrow Cu^{2+}(aq) + 2e^-$
cathode (−)	$Cu^{2+}(aq) + 2e^- \rightarrow Cu(s)$

Figure 9.26 Experimental set-up for electroplating a key with copper.

The processes here are essentially the same as for the electrolysis of copper sulfate solution using copper electrodes. The copper anode gets smaller as copper ions go into solution and the colour of the electrolyte remains constant as the concentration of $Cu^{2+}(aq)$ ions remains constant.

Exam tip
The electrolyte must be a solution containing ions of the plating metal. To select a suitable solution, remember that all nitrates are soluble in water.

Steel objects can be electroplated with chromium to prevent rusting and to provide a decorative finish.

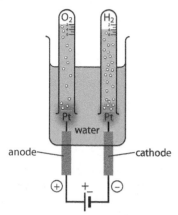

Figure 9.27 Electrolysis of acidified water to which some universal indicator has been added.

The half-equations have been written here so that the number of electrons is the same in each.

Electrolysis of water

Distilled water is a very poor conductor of electricity, and so the electrolysis of water is usually carried out on water to which small amount of sulfuric acid has been added ('acidified water'). Universal indicator has also been added to the water in Figure **9.27**.

The reaction that occurs at the anode (+) is oxidation of water:

$$2H_2O(l) \rightarrow O_2(g) + 4H^+(aq) + 4e^-$$

The product at the anode is oxygen gas. H^+ ions are also produced in this reaction and the universal indicator goes redder around the anode.

The reaction that occurs at the cathode (−) is reduction of water:

$$H_2O(l) + e^- \rightarrow \tfrac{1}{2}H_2(g) + OH^-(aq)$$

The product at the cathode is hydrogen gas. OH^- ions are also produced at the cathode, and the universal indicator goes blue around the cathode.

The H^+ ions produced at the anode combine with the OH^- ions formed at the cathode, so that the overall reaction when water is electrolysed is:

$$2H_2O(l) \rightarrow 2H_2(g) + O_2(g)$$

It can be seen from this equation that twice as much hydrogen as oxygen should be collected. Why this occurs can be understood by looking at the half-equations:

$$2H_2O(l) \rightarrow O_2(g) + 4H^+(aq) + \mathbf{4e^-} \qquad \mathbf{anode}$$
$$4H_2O(l) + \mathbf{4e^-} \rightarrow 2H_2(g) + 4OH^-(aq) \qquad \mathbf{cathode}$$

When four electrons are lost at the anode, one molecule of O_2 is formed, but when four electrons are gained at the cathode, two molecules of H_2 are formed. Electrons must be lost from the anode at the same rate at which they are gained at the cathode (continuous flow of electrons in the external circuit), so two molecules of H_2 are formed for every molecule of O_2 formed.

Electrolysis and $\Delta G^{\ominus}$

The overall equation for the electrolysis of aluminium oxide is:

$$2Al_2O_3 \rightarrow 4Al + 3O_2 \quad \Delta G \approx 2000 \text{ kJ mol}^{-1} \text{ at the melting point of } Al_2O_3$$

The value of ΔG at the melting point of Al_2O_3 is very positive, indicating that the reaction is definitely not spontaneous at this temperature. However, passing of a current continually through the molten salt causes the non-spontaneous reaction to occur.

The reaction that occurs in an electrolytic cell is non-spontaneous – electricity has to be supplied to cause it to occur (ΔG for the process is positive). This can be contrasted with a voltaic cell, in which a spontaneous reaction can be used to generate electricity (ΔG negative).

Quantitative electrolysis

In this section we will discuss the factors that affect the amounts of products formed at the electrodes.

The factors that affect the amount of product formed at an electrode are:

- the current
- the length of time of electrolysis
- the charge on the ion.

Current

The current that flows is directly proportional to the charge and hence to the number of electrons that flow around the circuit. This can be represented by the equation:

$$Q = It$$

where Q is the charge (in coulombs, C), I is the current (in amperes, A) and t is the time (in seconds, s).

If twice the current flows in a certain time, then twice as much charge will pass and twice as many electrons will be transferred to/from ions – therefore twice as many moles of substance will be produced.

Let us consider the half-equation:

$$Cu^{2+}(aq) + 2e^- \rightarrow Cu(s)$$

To produce one mole of copper, two moles of electrons must flow around the circuit. A current of about 2.23A must flow for 24 hours to produce 1 mol copper. If a current of 4.46A flows for 24 hours, twice as many electrons (4 mol) will flow around the circuit, and 2 mol copper will be formed.

> The number of moles of substance produced at the electrodes is proportional to the current and the time and inversely proportional to the charge on the ion.

Time of electrolysis

If current flows for twice the time, then twice as many electrons will be transferred and twice as much product will be formed.

Time, like current, is directly proportional to the charge, and hence the number of electrons, that flows around the circuit.

Charge on the ion

Consider the electrolysis of molten samples containing Na^+, Mg^{2+} and Al^{3+} ions. These ions will be reduced according to the equations:

$$Na^+ + e^- \rightarrow Na$$
$$Mg^{2+} + 2e^- \rightarrow Mg$$
$$Al^{3+} + 3e^- \rightarrow Al$$

Therefore if three moles of electrons are passed through each electrolyte, 3 mol Na, 1.5 mol Mg and 1 mol Al will be obtained. This can also be seen if we rewrite the above half-equations using $3e^-$ in each:

$$3Na^+ + 3e^- \rightarrow 3Na$$
$$\tfrac{3}{2}Mg^{2+} + 3e^- \rightarrow \tfrac{3}{2}Mg$$
$$Al^{3+} + 3e^- \rightarrow Al$$

Worked examples

9.8 A solution of copper(II) sulfate is electrolysed using platinum electrodes. If 0.636 g of copper is produced at the cathode, calculate the volume of oxygen (measured at standard temperature and pressure) produced at the anode.

0.636 g of copper is $\dfrac{0.636}{63.55} = 0.0100\,\text{mol}$

The half-equation for the production of copper at the cathode is:

$$Cu^{2+}(aq) + 2e^- \rightarrow Cu(s)$$

So two moles of electrons are required to produce one mole of copper. This means that the passage of 0.0200 mol electrons is required to produce 0.0100 mol copper.

The half-equation for the reaction at the anode is:

$$2H_2O(l) \rightarrow O_2(g) + 4H^+(aq) + 4e^-$$

So four moles of electrons are required to produce one mole of O_2.
The number of moles of oxygen produced when 0.0200 mol electrons are passed is $\dfrac{0.0200}{4}$, or $5.00 \times 10^{-3}\,\text{mol}$.

The volume of oxygen produced at standard temperature and pressure is given by:

number of moles × molar volume

$$\text{volume} = 5.00 \times 10^{-3} \times 22.7 = 0.114\,\text{dm}^3$$

> This is essentially the same approach as we used to work out moles problems in Topic **1**.

9.9 Two electrolytic cells are connected in series as shown in the diagram below. The first contains $1\,\text{mol}\,\text{dm}^{-3}$ sodium chloride solution and the second $1\,\text{mol}\,\text{dm}^{-3}$ copper(II) sulfate solution. Platinum electrodes are used throughout. Compare the volumes of each of the gases produced and the volumes of gas produced in each cell.

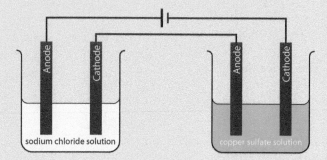

> Note that the anode in the copper sulfate cell is positive relative to the cathode in the same cell, but negatively charged relative to the cathode in the sodium chloride cell. Electrons flow from the anode of the copper sulfate cell to the cathode of the sodium chloride cell through the external circuit.

The products formed at the electrodes and the half-equations for their formation are as follows:

NaCl(aq)

$2Cl^-(aq) \rightarrow Cl_2(g) + 2e^-$	anode
$2H_2O(l) + 2e^- \rightarrow H_2(g) + 2OH^-(aq)$	cathode

CuSO$_4$(aq)

$2H_2O(l) \rightarrow O_2(g) + 4H^+(aq) + 4e^-$	anode
$Cu^{2+}(aq) + 2e^- \rightarrow Cu(s)$	cathode

Because the cells are connected in series the same current passes through each cell.

If we rewrite the equations involving gases so that the same number of electrons is transferred in each we get:

$$4Cl^-(aq) \rightarrow 2Cl_2(g) + 4e^-$$ **chlorine**
$$4H_2O(l) + 4e^- \rightarrow 2H_2(g) + 4OH^-(aq)$$ **hydrogen**
$$2H_2O(l) \rightarrow O_2(g) + 4H^+(aq) + 4e^-$$ **oxygen**

From this it can be seen that the volumes of chlorine and hydrogen produced would be the same, but the volume of oxygen produced would be half that of either of the other two gases. Therefore, for every mole of oxygen gas produced, two moles of hydrogen and two moles of chlorine are produced.

The total volume of gas produced in the first cell (chlorine and hydrogen) is thus four times the amount of gas produced in the second cell (oxygen).

> Actually, if the experiment is carried out in the laboratory, the results will not match this exactly, as the gases are soluble to different extents in water.

Nature of science

Scientists can use quantitative data to test theories and models but must be aware of possible sources of errors in experimental method. For example, when measuring the volumes of different gases collected during electrolysis they must take account of the gases having different solubilities in water. When looking at the decrease in mass of, say, a copper electrode they must be aware that the electrode may not have been pure copper to start with.

Hydrogen fuels cells are being increasingly used to power vehicles and are being promoted as an environmentally friendly alternative to petrol and diesel-powered vehicles. This technology is a major area of research but there are many factors to consider when deciding how environmentally friendly fuel cells are. For instance, the production of the materials used to construct the fuel cells and how the hydrogen is generated are major considerations.

> Hydrogen fuels cells and microbial fuels cells, which are another area of research with environmental implications, are discussed in Option C.

? Test yourself

22 Predict the products at the anode and cathode when the following aqueous solutions are electrolysed using platinum electrodes:
  a potassium iodide solution
  b calcium nitrate solution
  c concentrated potassium chloride solution
  d magnesium sulfate solution
  e silver nitrate solution

23 Write equations for the reactions at the anode and cathode when the following solutions are electrolysed using platinum electrodes:

  a sodium sulfate solution
  b silver fluoride solution
  c magnesium nitrate solution

24 0.20 mol electrons are passed through each of the following electrolytic cells. In each case state the number of moles of each product formed.
  a molten sodium chloride
  b aqueous (concentrated) copper chloride solution
  c aqueous potassium sulfate solution

Exam-style questions

1 Which compound contains chlorine with the lowest oxidation number?

    A NaCl   **B** HOCl   **C** Cl_2O_7 **D** ClF

2 Which of the following is **not** a redox reaction?

    A $Zn(NO_3)_2(aq) + Mg(s) \rightarrow Mg(NO_3)_2(aq) + Zn(s)$

    B $U(s) + 6ClF(l) \rightarrow UF_6(l) + 3Cl_2(g)$

    C $2NO_2(g) \rightarrow N_2O_4(g)$

    D $2SO_2(g) + O_2(g) \rightarrow 2SO_3(g)$

3 Which of the following half-equations represents a reduction reaction?

    A $CuCl \rightarrow Cu^{2+} + Cl^- + e^-$ **C** $VO^{2+} + 2H^+ + e^- \rightarrow V^{3+} + H_2O$

    B $N_2O_4 + 2H_2O \rightarrow 2NO_3^- + 4H^+ + 2e^-$ **D** $MnO_4^{2-} \rightarrow MnO_4^- + e^-$

4 Consider the following reactions of four metals – D, Z, Q and X.

    $Q(s) + X(NO_3)_2(aq) \rightarrow Q(NO_3)_2(aq) + X(s)$ $Q(s) + Z(NO_3)_2(aq) \rightarrow Q(NO_3)_2(aq) + Z(s)$

    $Z(s) + X(NO_3)_2(aq) \rightarrow Z(NO_3)_2(aq) + X(s)$ $D(s) + Z(NO_3)_2(aq) \rightarrow D(NO_3)_2(aq) + Z(s)$

From these data it can be deduced that:

    A X is more reactive than Q   **C** Z is a stronger reducing agent than X

    B Q is the most reactive metal   **D** D is a stronger reducing agent than Q

5 How many electrons are required when the following half-equation is balanced using the smallest possible integers?

    $Br_2 + H_2O \rightarrow BrO_3^- + H^+ + e^-$

    A 2   **B** 5   **C** 10   **D** 12

HL 6 Use the following standard electrode potentials to calculate the cell potential when an $Fe^{3+}|Fe^{2+}$ half-cell is connected to a $Cl_2|Cl^-$ half-cell:

    $Fe^{3+}(aq) + e^- \rightleftharpoons Fe^{2+}(aq)$ $E^\ominus = +0.77\,V$

    $Cl_2(aq) + 2e^- \rightleftharpoons 2Cl^-(aq)$ $E^\ominus = +1.36\,V$

    A $-0.18\,V$ **B** $+2.13\,V$ **C** $+0.59\,V$ **D** $+0.09\,V$

HL 7 Consider the following standard electrode potentials:

    $Mn^{2+} + 2e^- \rightleftharpoons Mn$ $E^\ominus = -1.18\,V$ $Pb^{2+} + 2e^- \rightleftharpoons Pb$ $E^\ominus = -0.13\,V$

    $Co^{2+} + 2e^- \rightleftharpoons Co$ $E^\ominus = -0.28\,V$ $Pd^{2+} + 2e^- \rightleftharpoons Pd$ $E^\ominus = +0.99\,V$

From these data it can be deduced that:

    A Co is a stronger reducing agent than Mn   **C** Mn is a stronger reducing agent than Pd

    B Pb^{2+} is a stronger oxidising agent than Pd^{2+} **D** Co^{2+} is a stronger reducing agent than Mn^{2+}

HL 8 Use the following standard electrode potentials to decide which reaction will be spontaneous:

$$Cr^{3+} + e^- \rightleftharpoons Cr^{2+} \qquad\qquad E^\ominus = -0.41\,V$$

$$Cr^{3+} + 3e^- \rightleftharpoons Cr \qquad\qquad E^\ominus = -0.74\,V$$

$$Po^{2+} + 2e^- \rightleftharpoons Po \qquad\qquad E^\ominus = +0.65\,V$$

$$UO_2^+ + 4H^+ + e^- \rightleftharpoons U^{4+} + 2H_2O \qquad E^\ominus = +0.62\,V$$

$$ReO_4^- + 8H^+ + 7e^- \rightleftharpoons Re + 4H_2O \qquad E^\ominus = +0.36\,V$$

    A $UO_2^+ + 4H^+ + 2Po^{2+} \rightarrow U^{4+} + 2H_2O + 2Po$

    B $2Cr + 3Po^{2+} \rightarrow 2Cr^{3+} + 3Po$

    C $Re + 4H_2O + 7Cr^{3+} \rightarrow ReO_4^- + 8H^+ + 7Cr^{2+}$

    D $3U^{4+} + 6H_2O + Cr^{3+} \rightarrow Cr + 3UO_2^+ + 12H^+$

HL 9 An experiment is carried out to electroplate a key with silver. Which of the following is correct about the way the experiment is set up?

    A The cathode is made of silver and the key is the anode.

    B The electrolyte is a solution of copper sulfate.

    C The key is the cathode and the electrolyte is a silver nitrate solution.

    D The cathode is made of silver and the electrolyte is a silver nitrate solution.

HL 10 A copper sulfate solution is electrolysed for 30 minutes using platinum electrodes. 0.010 mol copper is deposited on one of the electrodes. Which of the following is correct?

    A 0.020 mol oxygen is produced at the cathode.

    B 0.010 mol oxygen is produced at the anode.

    C 0.010 mol hydrogen is produced at the cathode.

    D 0.0050 mol oxygen is produced at the anode.

11 A student carried out a redox titration to determine the amount of iron present in some iron tablets. The manufacturer claimed that each tablet contained 65 mg of iron present as iron(II) sulfate. The student crushed five tablets and heated them with sulfuric acid to dissolve the iron. The mixture was filtered and made up to a total volume of 250.0 cm³ with distilled water. She then measured out 25.00 cm³ of this solution, added 25.00 cm³ of 1 mol dm⁻³ sulfuric acid to acidify the solution and titrated against $5.00 \times 10^{-3}\,mol\,dm^{-3}$ KMnO₄(aq) until the first permanent pink colour was seen. An average volume of 21.50 cm³ of KMnO₄(aq) was required for the titration.

    a Define oxidation in terms of electrons. [1]

    b The *unbalanced* ionic equation for the reaction between KMnO₄ and iron(II) is:

$$MnO_4^-(aq) + H^+(aq) + Fe^{2+}(aq) \rightarrow Mn^{2+}(aq) + Fe^{3+}(aq) + H_2O(l)$$

        i What is the oxidation number of Mn in MnO_4^-? [1]

        ii Identify the reducing agent in this reaction. [1]

        iii Balance the equation. [2]

    c **i** Calculate the number of moles of KMnO₄ that reacted. [1]

        ii Calculate the number of moles of iron(II) ions that reacted with the KMnO₄ [1]

        iii Determine the mass of iron in an iron tablet [3]

        iv Comment on the manufacturer's claim. [2]

12 a Define a **reducing agent** in terms of electrons. **[1]**

b A series of experiments was carried out in order to work out an activity series for some metals. Different metals were added to solutions of salts and the following experimental data were obtained:

	Metal	Salt solution	Observations
I	zinc	copper(II) sulfate	brown deposit formed and the blue colour of the solution fades
II	zinc	lead(II) nitrate	grey crystals formed on the piece of zinc
III	copper	lead(II) nitrate	no reaction
IV	zinc	magnesium nitrate	no reaction

   i Write an ionic equation for the reaction that occurs in experiment **I**. **[2]**
   ii What do the results of experiment **III** indicate about the relative reactivity of copper and lead? **[1]**
   iii Arrange the metals in order of reactivity, stating clearly which is the most reactive and which the least. **[1]**
   iv Explain which of the four metals is the strongest reducing agent. **[3]**
   v Write an ionic equation for the reaction between magnesium and lead nitrate. **[2]**

c A voltaic cell was set up with a piece of magnesium dipping into a solution of magnesium nitrate and a piece of zinc dipping into a solution of zinc nitrate.
   i Draw a labelled diagram of the voltaic cell. Label the anode in the cell and show the direction of electron flow in the external circuit. **[3]**
   ii Write an ionic equation for the reaction that occurs in the zinc half-cell and classify this reaction as oxidation or reduction. **[2]**

13 Sodium metal can be obtained by the electrolysis of molten sodium chloride.

a Explain why solid sodium chloride does not conduct electricity but molten sodium chloride does. **[2]**

b State the name of the product at the anode in this process and write half-equations for the reactions at each electrode, stating clearly which is which. **[3]**

HL 14 Standard electrode potentials are measured relative to the standard hydrogen electrode.

a Draw a labelled diagram showing the essential features of a standard hydrogen electrode. **[5]**

b The diagram shows a voltaic cell:

   i Use standard electrode potentials in Table **9.7** (page **396**) to calculate the standard cell potential. **[1]**
   ii Write an equation, including state symbols, for the overall cell reaction. **[2]**
   iii Explain which electrode is the anode. **[2]**
   iv Show on the diagram the direction of electron flow in the external circuit. **[1]**
   v Explain the role of the salt bridge in this cell. **[2]**
   vi Calculate $\Delta G^{\ominus}$ for the cell reaction and comment on whether the reaction is spontaneous or not at $25\,°C$. Faraday's constant $= 96\,500\,C\,mol^{-1}$ **[3]**

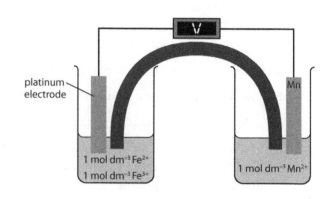

platinum electrode

Mn

1 mol dm⁻³ Fe²⁺
1 mol dm⁻³ Fe³⁺

1 mol dm⁻³ Mn²⁺

HL 15 Consider the following half-equations and standard electrode potentials:

$$MnO_4^-(aq) + 8H^+(aq) + 5e^- \rightleftharpoons Mn^{2+}(aq) + 4H_2O(l) \qquad E^\ominus = +1.51\,V$$

$$Cl_2(g) + 2e^- \rightleftharpoons 2Cl^-(aq) \qquad E^\ominus = +1.36\,V$$

$$Cr_2O_7^{2-}(aq) + 14H^+(aq) + 6e^- \rightleftharpoons 2Cr^{3+}(aq) + 7H_2O(l) \qquad E^\ominus = +1.33\,V$$

$$Br_2(l) + 2e^- \rightleftharpoons 2Br^-(aq) \qquad E^\ominus = +1.09\,V$$

  a Use these equations to identify, giving a reason, the strongest oxidising agent. **[2]**

  b A solution of acidified potassium dichromate(VI) is added to a solution containing equal concentrations of bromide ions and chloride ions. Deduce the balanced equation for the reaction that occurs. **[3]**

  c State and explain the sign of $\Delta G^\ominus$ for the following reaction:

$$2Cl^-(aq) + Br_2(l) \rightarrow Cl_2(g) + 2Br^-(aq) \qquad \textbf{[2]}$$

  d Consider the unbalanced half-equation:

$$\underset{\text{methanoic acid}}{HCOOH(aq)} + H^+(aq) + e^- \rightarrow \underset{\text{methanal}}{HCHO(aq)} + H_2O(l) \qquad E^\ominus = +0.06\,V$$

    i State the oxidation numbers of C in HCOOH and HCHO and use these to explain whether HCOOH is oxidised or reduced. **[3]**

    ii Balance the half-equation. **[1]**

    iii Write a balanced equation for the spontaneous reaction that occurs when a solution containing MnO_4^- and H^+ is added to a solution containing methanal. **[2]**

    iv Explain why, when the two solutions in **iii** are mixed, a reaction may not occur. **[1]**

HL 16 a A concentrated solution of sodium chloride is electrolysed using platinum electrodes. State the products formed at the anode and the cathode and write half-equations for their formation. **[4]**

  b When a solution of copper sulfate is electrolysed using platinum electrodes, a gas is formed at one of the electrodes.

    i Identify the electrode at which the gas is formed and state an equation for its formation. **[2]**

    ii State two ways in which the electrolyte changes during this experiment. **[2]**

  c When a solution of copper sulfate is electrolysed using copper electrodes, no gas is evolved.

    i Write equations for the half-equations occurring at the anode and cathode in this cell. **[2]**

    ii State and explain any changes in the appearance of the electrolyte during this experiment. **[2]**

Summary

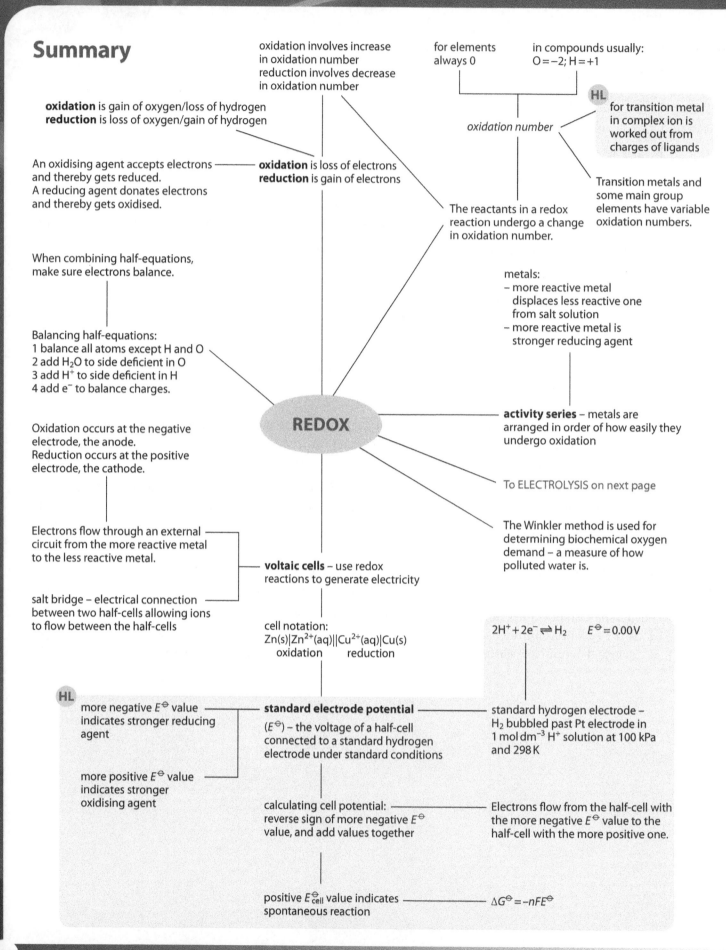

oxidation involves increase in oxidation number
reduction involves decrease in oxidation number

for elements always 0

in compounds usually: $O = -2$; $H = +1$

oxidation is gain of oxygen/loss of hydrogen
reduction is loss of oxygen/gain of hydrogen

An oxidising agent accepts electrons and thereby gets reduced.
A reducing agent donates electrons and thereby gets oxidised.

oxidation is loss of electrons
reduction is gain of electrons

oxidation number

HL for transition metal in complex ion is worked out from charges of ligands

Transition metals and some main group elements have variable oxidation numbers.

The reactants in a redox reaction undergo a change in oxidation number.

When combining half-equations, make sure electrons balance.

Balancing half-equations:
1 balance all atoms except H and O
2 add H_2O to side deficient in O
3 add H^+ to side deficient in H
4 add e^- to balance charges.

metals:
– more reactive metal displaces less reactive one from salt solution
– more reactive metal is stronger reducing agent

Oxidation occurs at the negative electrode, the anode.
Reduction occurs at the positive electrode, the cathode.

REDOX

activity series – metals are arranged in order of how easily they undergo oxidation

To ELECTROLYSIS on next page

Electrons flow through an external circuit from the more reactive metal to the less reactive metal.

The Winkler method is used for determining biochemical oxygen demand – a measure of how polluted water is.

voltaic cells – use redox reactions to generate electricity

salt bridge – electrical connection between two half-cells allowing ions to flow between the half-cells

cell notation:
$Zn(s)|Zn^{2+}(aq)||Cu^{2+}(aq)|Cu(s)$
oxidation   reduction

$2H^+ + 2e^- \rightleftharpoons H_2$ $E^{\ominus} = 0.00\,V$

HL more negative $E^{\ominus}$ value indicates stronger reducing agent

standard electrode potential

$(E^{\ominus})$ – the voltage of a half-cell connected to a standard hydrogen electrode under standard conditions

standard hydrogen electrode – H_2 bubbled past Pt electrode in $1\,mol\,dm^{-3}$ H^+ solution at 100 kPa and 298 K

more positive $E^{\ominus}$ value indicates stronger oxidising agent

calculating cell potential:
reverse sign of more negative $E^{\ominus}$ value, and add values together

Electrons flow from the half-cell with the more negative $E^{\ominus}$ value to the half-cell with the more positive one.

positive $E^{\ominus}_{cell}$ value indicates spontaneous reaction

$\Delta G^{\ominus} = -nFE^{\ominus}$

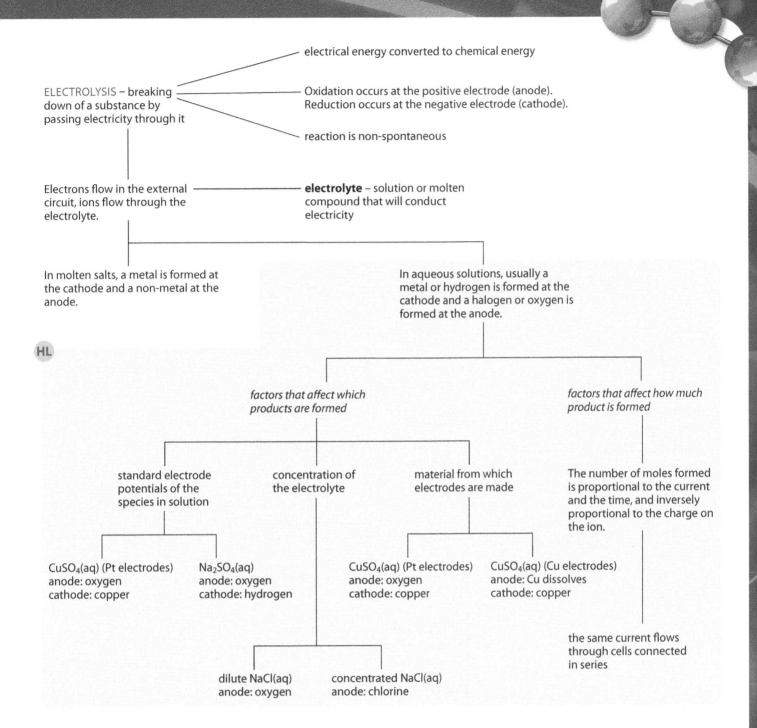

ELECTROLYSIS – breaking down of a substance by passing electricity through it

electrical energy converted to chemical energy

Oxidation occurs at the positive electrode (anode).
Reduction occurs at the negative electrode (cathode).

reaction is non-spontaneous

Electrons flow in the external circuit, ions flow through the electrolyte.

electrolyte – solution or molten compound that will conduct electricity

In molten salts, a metal is formed at the cathode and a non-metal at the anode.

In aqueous solutions, usually a metal or hydrogen is formed at the cathode and a halogen or oxygen is formed at the anode.

HL

factors that affect which products are formed

factors that affect how much product is formed

standard electrode potentials of the species in solution

concentration of the electrolyte

material from which electrodes are made

The number of moles formed is proportional to the current and the time, and inversely proportional to the charge on the ion.

$CuSO_4(aq)$ (Pt electrodes)
anode: oxygen
cathode: copper

$Na_2SO_4(aq)$
anode: oxygen
cathode: hydrogen

$CuSO_4(aq)$ (Pt electrodes)
anode: oxygen
cathode: copper

$CuSO_4(aq)$ (Cu electrodes)
anode: Cu dissolves
cathode: copper

the same current flows through cells connected in series

dilute $NaCl(aq)$
anode: oxygen

concentrated $NaCl(aq)$
anode: chlorine

10 Organic chemistry

Learning objectives

- Understand that the alkanes are a family of hydrocarbons
- Distinguish between the different types of formulas (empirical, molecular, structural)
- Understand the difference between a full and a condensed structural formula

Most of the world's oil reserves are controlled by just a few countries. Indeed it is estimated that more than 80% of the world's oil reserves are controlled by OPEC. The control of oil has already been the source of conflicts between countries – like the 1990 Gulf War – and is likely to be even more of a political issue in the future as oil reserves dwindle.

Empirical and molecular formulas are discussed more fully in Topic **1**.

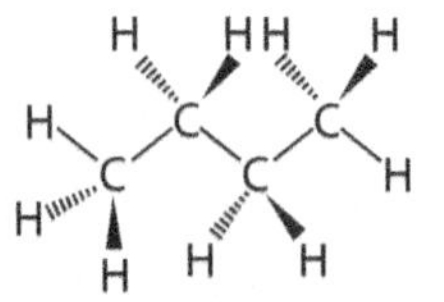

Figure 10.1 The full structural formula of butane showing the tetrahedral arrangement around the carbon atoms.

10.1 Fundamentals of organic chemistry

10.1.1 Introduction to organic chemistry

Organic chemistry is the study of carbon compounds, and specifically compounds containing the C–H bond. The chemistry of carbon is more extensive than that of any other element and there are more compounds of carbon than of all other elements put together. There are so many carbon compounds because of the fact that carbon atoms can join together to form chains and rings – a property called **catenation**.

The source of many carbon compounds is crude oil (petroleum) and many substances can be separated from this complex mixture using fractional distillation. These molecules form the basis of organic synthesis reactions, in which ever-more-complex molecules are made to be used as medicines, cosmetics, polymers etc.

Alkanes

The alkanes are a family of hydrocarbons. The first six members are shown in Table **10.1**.

These compounds are all straight-chain alkanes, and the molecular formula of all alkanes can be represented by the general formula C_nH_{2n+2}. For instance eicosane, the alkane with 20 carbon atoms, has $(20 \times 2) + 2$, i.e. 42 hydrogen atoms and its molecular formula is $C_{20}H_{42}$.

> **Hydrocarbon:** a compound containing only carbon and hydrogen.

> **Empirical formula:** the simplest whole number ratio of the elements present in a compound.

> **Molecular formula:** the total number of atoms of each element present in a molecule of the compound. A molecular formula is an integer multiple of the empirical formula.

A full structural formula, also called a displayed or graphic formula, shows all the atoms and bonds in a molecule. Although we draw the full structural formula with 90° bond angles, it must be remembered that with four electron pairs around each carbon atom, the shape is tetrahedral around each one and butane should be more correctly drawn as shown in Figure **10.1**. However, this form is significantly less convenient to draw.

A condensed structural formula is the simplest representation that shows how the atoms are joined together in a molecule.

Name	Molecular formula	Empirical formula	Structural formula	
			Condensed	**Full**
methane	CH_4	CH_4	CH_4	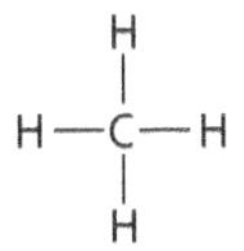 H | H—C—H | H
ethane	C_2H_6	CH_3	CH_3CH_3	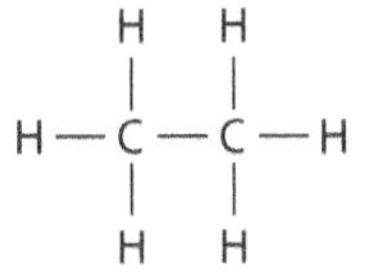 H H | | H—C—C—H | | H H
propane	C_3H_8	C_3H_8	$CH_3CH_2CH_3$	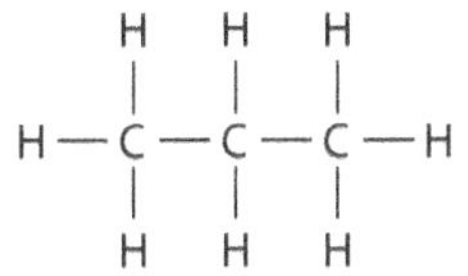 H H H | | | H—C—C—C—H | | | H H H
butane	C_4H_{10}	C_2H_5	$CH_3CH_2CH_2CH_3$	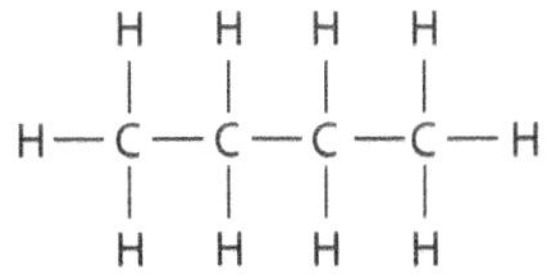 H H H H | | | | H—C—C—C—C—H | | | | H H H H
pentane	C_5H_{12}	C_5H_{12}	$CH_3CH_2CH_2CH_2CH_3$	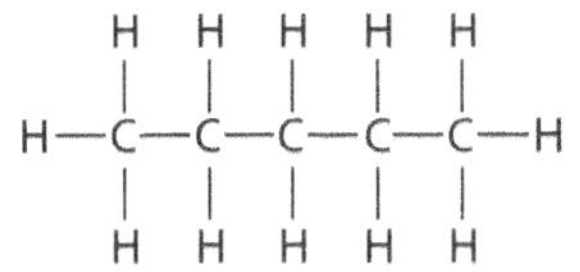 H H H H H | | | | | H—C—C—C—C—C—H | | | | | H H H H H
hexane	C_6H_{14}	C_3H_7	$CH_3CH_2CH_2CH_2CH_2CH_3$	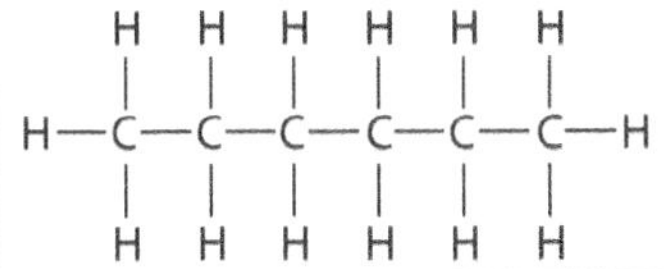 H H H H H H | | | | | | H—C—C—C—C—C—C—H | | | | | | H H H H H H

Table 10.1 The first six straight-chain alkanes.

$CH_3CH_2CH_2CH_3$ is a condensed structural formula for butane because it is an unambiguous representation of the structure of the molecule but C_4H_{10} is not a structural formula because it could represent either of the molecules shown in Figure **10.2**.

Molecules can also be represented using skeletal formulas like the one shown in Figure **10.3**. Skeletal formulas are very convenient and are widely used for drawing more complex molecules.

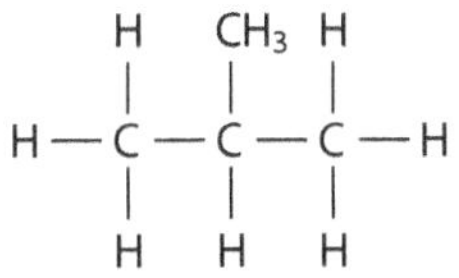

Exam tip
Skeletal formulas should not be used in examinations unless specifically requested.

H   CH₃ H
\| \| \|
H—C—C—C—H
\| \| \|
H   H   H
(CH₃)₃CH

or

H   H   H   H
\| \| \| \|
H—C—C—C—C—H
\| \| \| \|
H   H   H   H
CH₃CH₂CH₂CH₃ or CH₃(CH₂)₂CH₃

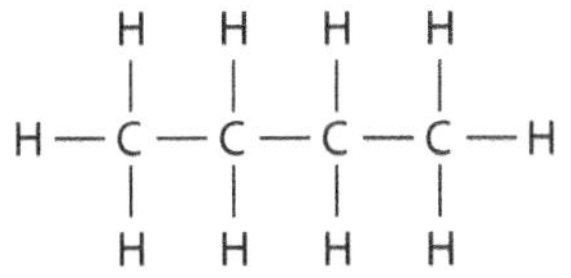

Figure 10.2 Same molecular formula (C_4H_{10}), different structural formulas.

Figure 10.3 The skeletal formula of butane.

Different ways of representing the structures of molecules have been shown here. Why is one way sometimes preferred to another? What is the relationship between ease of use and sophistication? Which is probably the most accurate way of representing a molecule? Which is the most useful if we need to draw a reaction scheme?

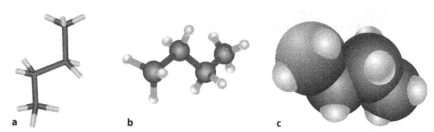

Figure 10.4 Computer-generated images: **a** a stick model, **b** a ball and stick model and **c** a space-filling model.

Some other ways of representing the structure of butane are shown in Figure **10.4**.

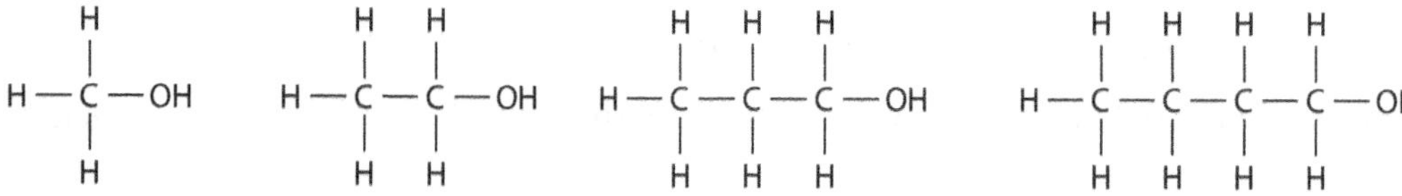

Learning objectives

- Understand what is meant by a homologous series
- Understand what is meant by a functional group
- Identify functional groups in molecules
- Understand that compounds containing a benzene ring are described as aromatic
- Explain trends in boiling points within homologous series

10.1.2 Homologous series

The alkanes represent a homologous series.

> A **homologous series** is a series of compounds that have the same functional group. Each member differs from the next by a common structural unit (usually $-CH_2-$).

> A **functional group** is the atom or group of atoms in a molecule that gives it its characteristic chemical properties – this is the reactive part of a molecule.

Chemical properties: how a substance reacts.

Although there is not really a functional group in alkanes, because only single C–C and C–H bonds are present, examples of functional groups in other homologous series are C=C in alkenes and –OH in alcohols.

The features of a homologous series are:

- they can usually be described by a general molecular formula – e.g. C_nH_{2n+2} for alkanes (non-cyclic) and C_nH_{2n} for alkenes
- members of the series have similar chemical properties
- members of the series show a gradation in physical properties such as boiling point.

Note that the diagrams in Figure **10.5** do not show the full structural formulas because the O–H bonds are not shown. However, it is a structural formula type that is often drawn to represent molecules.

Other homologous series

Table **10.2** shows the functional groups you are likely to meet during the IB Chemistry course.

Each of the functional groups in Table **10.2** could form the basis of a homologous series. For example, the first four members of the alcohol series, in which the –OH group is on the first carbon, are shown in Figure **10.5**.

methanol ethanol propan-1-ol butan-1-ol

Figure 10.5 The first four members of the alcohol series.

Homologous series / class name	Functional group	Functional group name	Example	General name	Name
alkane		alkyl	H₃C–CH₂–CH₂–CH₃ (butane skeletal)	alkane	butane
alkene	C=C	alkenyl	H₂C=CH–CH₂–CH₃	alk-*x*-**ene** (or *x*-alkene)	but-1-**ene** (or 1-butene)
alkynes	C≡C	alkynyl	HC≡C–CH₂–CH₃	alk-*x*-**yne** (or *x*-alkyne)	but-1-yne (or 1-butyne)
alcohol	–OH	hydroxyl	CH₃–CH₂–CH₂–OH	alkan-*x*-**ol** (or *x*-alkanol)	propan-1-**ol** (or 1-propanol)
ether	C–O–C	ether	CH₃–O–CH₂–CH₃	alk**oxy**alkane	meth**oxy**ethane
aldehyde	(C=O, –H)	carbonyl	CH₃–CH₂–CHO	alkan**al**	propan**al**
ketone	(C=O between C)	carbonyl	CH₃–CO–CH₂–CH₂–CH₃	alkan-*x*-**one** (or *x*-alkanone)	pentan-2-**one** (or 2-pentanone)
carboxylic acid	(C=O, O–H)	carboxyl	CH₃–CH₂–COOH	alkan**oic acid**	propan**oic acid**
halogenoalkane	–X X = Cl/Br/I	halo (chloro, bromo, iodo)	CH₃–CHBr–CH₂–CH₃	*x*-haloalkane	2-bromobutane
amine	–NH₂ –NHR –NR₂	amine	CH₃–CH₂–CH₂–NH₂	alkyl**amine** or *x*-aminoalkane or alkan-*x*-amine (or *x*-alkanamine)	propyl**amine** or 1-aminopropane or propan-1-amine (or 1-propanamine)
ester	(C–O–C=O)	ester	CH₃–O–CO–CH₂–CH₃	alkyl alkan**oate**	methyl propan**oate**
nitrile	–C≡N	nitrile	CH₃–CH₂–C≡N	alkane**nitrile** (C of C≡N included in chain)	propane**nitrile**
amide	(C=O, NH₂)	carboxamide	CH₃–CH₂–CO–NH₂	alkan**amide**	propan**amide**

Table 10.2 Functional groups that you are likely to meet. 'R' can be used to represent an alkyl group – so a general carboxylic acid may be represented as 'RCOOH' and an aldehyde as 'RCHO'. 'R' is occasionally also used to represent a phenyl group ($-C_6H_5$).

Name	Molecular formula	Empirical formula	Structural formula	
			Condensed	**Full**
ethene	C_2H_4	CH_2	CH_2CH_2	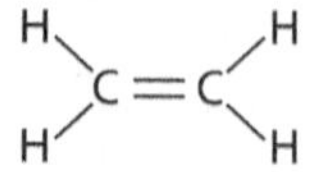 $\begin{array}{c}H\\ \end{array}C=C\begin{array}{c}H\\ \end{array}$ (H's top and bottom)
prop-1-ene	C_3H_6	CH_2	CH_2CHCH_3	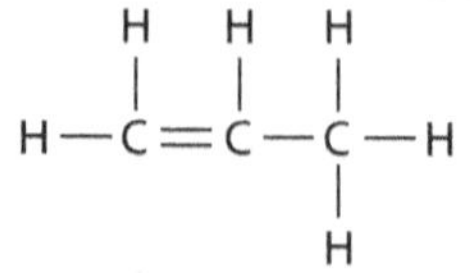 full structure shown
but-1-ene	C_4H_8	CH_2	$CH_2CHCH_2CH_3$	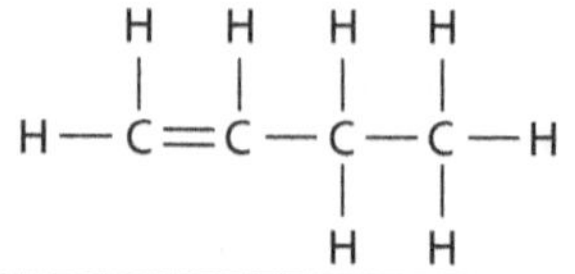 full structure shown
pent-1-ene	C_5H_{10}	CH_2	$CH_2CHCH_2CH_2CH_3$	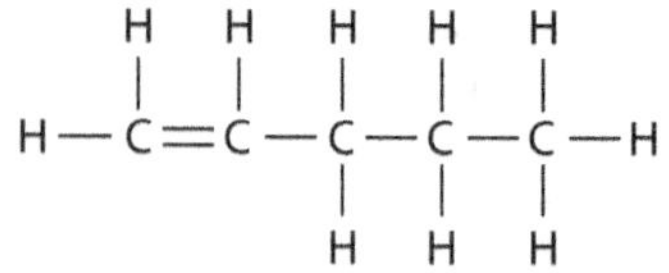 full structure shown
hex-1-ene	C_6H_{12}	CH_2	$CH_2CHCH_2CH_2CH_2CH_3$	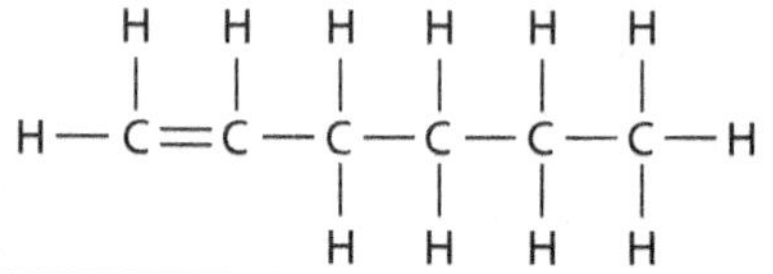 full structure shown

Table 10.3 The first five members of the alkene homologous series.

Amines are strong-smelling substances (said to smell like fish) and some of the trivial (non-systematic) names for them reflect this. Cadaverine (1,5–diaminopentane) and putrescine (1,4–diaminobutane) are produced in rotting flesh.

There are three different functional groups for amines in Table **10.2** because amines are classified as primary, secondary or tertiary – see Section **10.1.6**.

The first five members of the alkene homologous series, in which the C=C is between carbon 1 and carbon 2, are shown in Table **10.3**.

Compounds containing a benzene ring

Benzene has the molecular formula C_6H_6 and its molecules have a planar, hexagonal ring of six carbon atoms with one hydrogen atom joined to each. It is usually represented by the skeletal formula shown in Figure **10.6**.

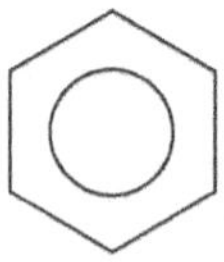

Compounds that contain a benzene ring are described as **aromatic** – compounds without benzene rings are called **aliphatic**. The functional group in aromatic compounds is the benzene ring because it performs characteristic chemical reactions.

When determining molecular formulas or condensed structural formulas for compounds that contain benzene rings, it must be remembered that there is a carbon atom and, if there is nothing else attached, a hydrogen atom at each vertex. So, for example, methylbenzene has the molecular formula C_7H_8 and can be represented in different ways (Figure **10.7**).

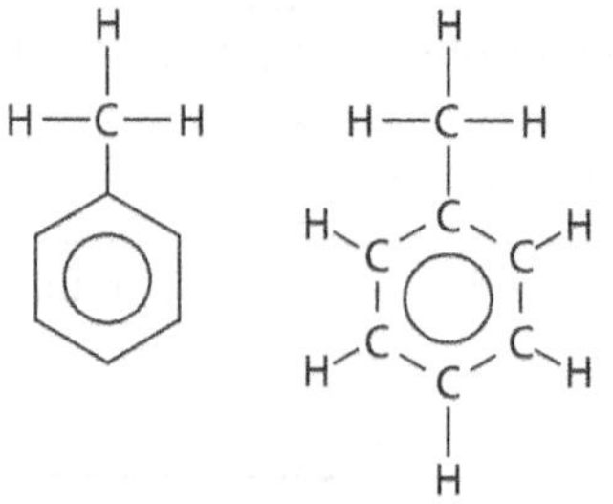

The condensed structural formula of methylbenzene can be written as $C_6H_5CH_3$. The C_6H_5 group in methylbenzene and other aromatic compounds is often called a phenyl group.

Figure 10.6 A common representation of a molecule of benzene.

Figure 10.7 Representations of methylbenzene.

Boiling point and homologous series

As the number of carbon atoms in a molecule in any homologous series increases, the boiling point increases.

The boiling point of straight-chain alkanes increases when a methylene ($-CH_2-$) group is added, because the strength of the London forces between molecules increases as the relative molecular mass (number of electrons in the molecule) of the alkanes increases. A similar trend is seen in the boiling points of other homologous series.

Figure **10.8** compares the boiling points of alcohols (with the $-OH$ group on the first carbon atom) with the boiling points of alkanes. It can be seen that both series show the same trend – the boiling point increases as the number of carbon atoms increases. However, the alcohol with one carbon atom (M_r 32.05) has a higher boiling point than the alkane with five carbon atoms (M_r 72.17). It is important to realise that comparisons based on the relative molecular mass (and hence strength of London forces) can be made *within* a particular homologous series but not *between* homologous series. The boiling points of the alcohols are higher because of hydrogen bonding between alcohol molecules.

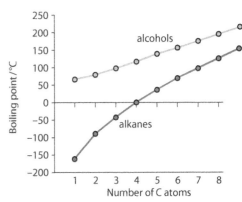

Figure 10.8 The boiling points of some alkanes and alcohols.

Boiling points of alcohols and ethers

Alcohols have higher boiling points than ethers and are more soluble in water because alcohol molecules can form hydrogen bonds to each other and also to water molecules. Ethers have no hydrogen atoms bonded to oxygen atoms and so cannot form hydrogen bonds - Figure **10.9**.

Ethanol is soluble in water in all proportions – this is because the $-OH$ group allows it to hydrogen bond with water. However, the solubility decreases as the length of the hydrocarbon chain increases (see page **157**) so that pentan-1-ol and hexan-1-ol are only sparingly soluble in water.

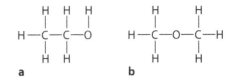

Figure 10.9 The structures of: **a** ethanol (an alcohol) and **b** methoxymethane (an ether).

Nature of science

Scientists share a common language in the form of mathematics and also in terms of organic structures. Organic structures drawn in skeletal form would be recognised and understood by any organic chemist anywhere in the world.

There are certain areas of scientific research which are more controversial and involve significant ethical implications. Organic chemists make a large range of substances and amongst these would be pesticides, drugs and food additives. These are all substances that are almost certainly being produced because a need exists and there is the potential to make money but they could also have an adverse effect on the environment and human health. Scientists, however, also have to earn a living and, if working for a large company (which is likely to be driven solely by the desire to make money) may not have the choice to follow their own interests and conscience when selecting projects. Regulatory authorities have to consider the risks and benefits of new substances before licensing them for use.

The role of serendipity in scientific discovery is discussed on page **483**.

1 What is the molecular formula of the alkane with 15 carbon atoms?

2 Give the molecular formula and empirical formula of each of the following:

3 Identify the functional group(s) in each of the following molecules.

Learning objectives

- Use IUPAC rules to name alkanes, alkenes and alkynes
- Distinguish between saturated and unsaturated compounds

Carbon atoms	Prefix
1	meth-
2	eth-
3	prop-
4	but-
5	pent-
6	hex-

Table 10.4 Prefixes indicating the number of carbon atoms in the longest continuous carbon chain.

10.1.3 Naming hydrocarbons

Naming alkanes

Organic molecules are named according to the International Union of Pure and Applied Chemistry (**IUPAC**) system. Although older, non-systematic names are still in use in some situations, molecules should, wherever possible, be named according to the following systematic set of rules.

1 Find the longest *continuous* carbon chain in the molecule.
2 Choose the prefix from Table **10.4** corresponding to the number of carbon atoms in the longest continuous carbon chain and add the ending 'ane' to indicate an alkane.
3 Look for substituent groups (alkyl groups) – the names for these are shown in Table **10.5**.

4 Number the positions of the substituent groups using the combination that has the lowest individual numbers (not the sum).

5 Choose a prefix from Table **10.6** to indicate the number of each substituent present.

6 Arrange the names of the substituent groups in alphabetical order (ignoring the prefix used in step **5**).

Examples

2-methylpentane

The longest continuous carbon chain has five carbon atoms, and this gives the basic name 'pentane'. A methyl group is present in the molecule, and this is on carbon number 2 if we start numbering from the right-hand carbon atom. If we had started numbering from the left-hand carbon atom, we would have a methyl group on carbon 4, which is a higher number than the 2 obtained if we number from the other side – therefore '4' is not used.

$-CH_3$	methyl
$-C_2H_5$	ethyl
$-C_3H_7$	propyl

Table 10.5 The names of substituent groups.

Number of identical substituents	Prefix
2	di-
3	tri-
4	tetra-

Table 10.6 Prefixes indicating the number of identical substituents.

methyl group

H H H (CH_3) H

H—C_5—C_4—C_3—C_2—C_1—H

H H | H H H

longest continuous carbon chain

2,2-dimethylbutane

The longest continuous carbon chain is of four carbon atoms, and this gives rise to the 'butane' part of the name. There are two methyl groups, so we use 'dimethyl'. These methyl groups are both on carbon 2 (this time we count from the left-hand side to generate the lowest numbers) and so we have '2,2-dimethyl' – both 2s are needed.

> When naming molecules we use commas between numbers, and dashes between numbers and letters.

longest continuous carbon chain

methyl group

H (CH_3) H H

H—C_1—C_2—C_3—C_4—H

H (CH_3) H H

methyl group

3-methylhexane

The longest continuous carbon chain may not always be shown as a horizontal, straight line of carbon atoms, as in this molecule:

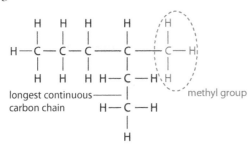

longest continuous carbon chain

methyl group

4-ethyl-2-methylhexane

This combination of numbers gives the lowest number possible – the alternative name would be 3-ethyl-5-methylhexane, but 2 is lower than 3. The substituent (alkyl) groups are arranged in alphabetical order.

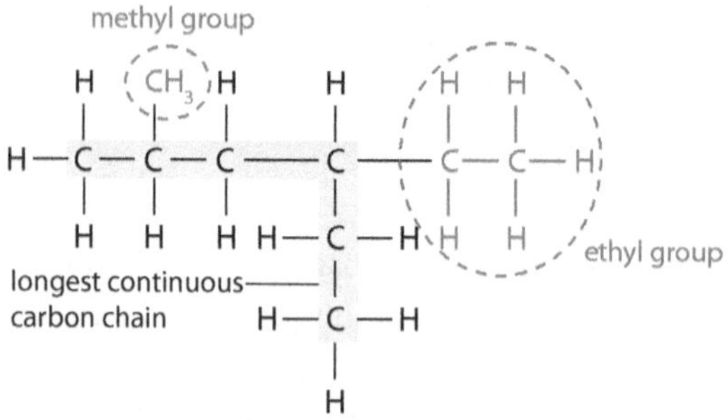

2,6,6-trimethyloctane

If we numbered the carbon atoms from the other direction, we would have obtained the name 3,3,7-trimethyloctane. Although the sum of the numbers in this name is lower, the lowest number here is higher than the lowest number in the accepted name, and so this name is not used.

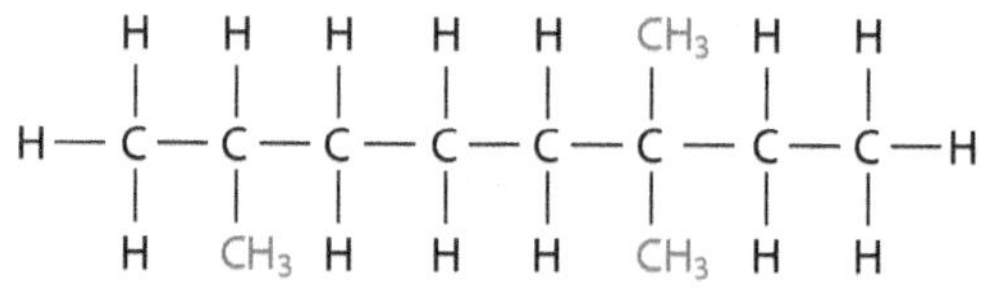

3,3,4-trimethylhexane

If the first number is the same after numbering from both directions then the second number is investigated and the name is chosen to give the lower number at the first position at which the names differ. The molecule shown below could be named 3,3,4-trimethylhexane or 3,4,4-trimethylhexane, depending on the side we start counting from. Both names start with 3 but they differ at the second digit, and so the name 3,3,4-trimethylhexane is chosen because it gives the lower number at the second position.

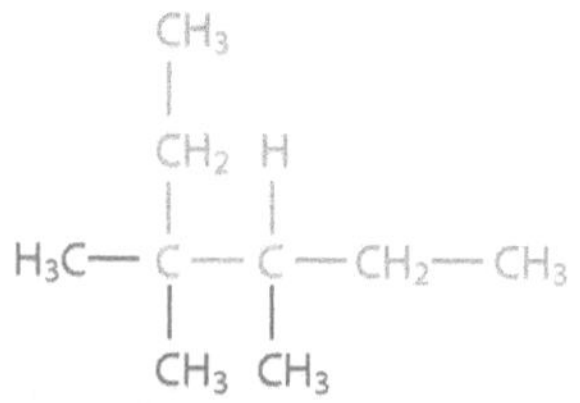

Exam tip

You should be able to name alkanes with up to a maximum of six C atoms in the longest carbon chain. Some of the examples here are for a bit of extra practice and to illustrate all the rules.

Naming alkenes and alkynes

The names of alkenes are of the form **alk-*x*-ene** (*x*-alkene is also acceptable). The number, *x*, indicates the position of the double bond. If there is a double bond between carbon 1 and carbon 2, the lower number is taken to generate a '-1-ene'. If there are alkyl groups present in the

molecule, the numbering is chosen to give the double bond the lowest possible number.

Alkynes are named in exactly the same way except the suffix **-yne** is used.

Examples

2-methylbut-1-ene

The molecule has four carbon atoms in the longest continuous carbon chain and produces the stem 'but-'. A double bond between carbon 1 and carbon 2 gives us '-1-ene' and the methyl group on carbon 2 produces '2-methyl'.

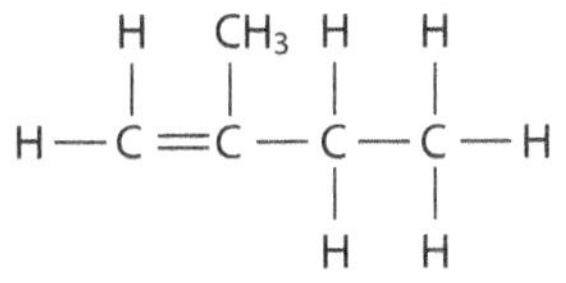

4-methylpent-2-yne

This alkyne could have been named 2-methylpent-3-yne, but the lowest possible number is given to the triple bond component – so it is more correctly named 4-methylpent-2-yne.

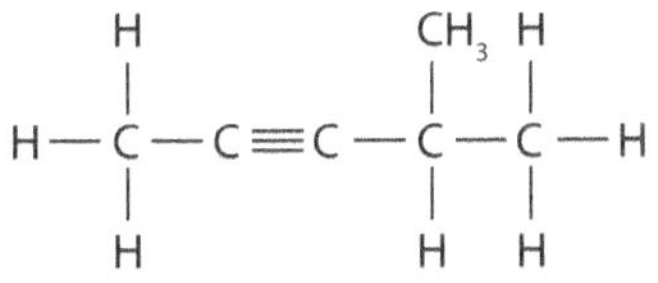

When more than one C=C bond is present in a molecule, the naming changes slightly – an extra 'a' is added to the stem. The alkene in Figure **10.10a** is called hex**a**-1,3-diene and that in **b** is called hex**a**-1,3,5-triene.

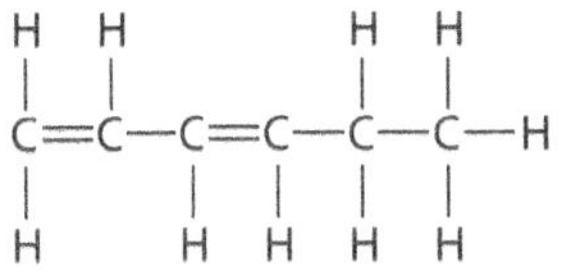

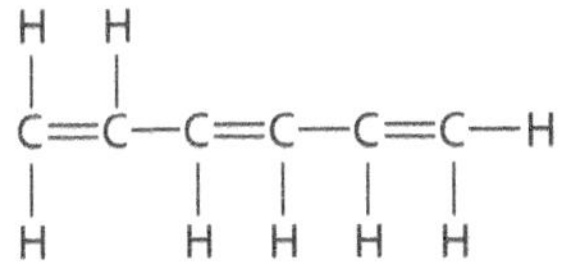

Figure 10.10 Compounds containing more than one C=C bond; **a** hexa-1,3-diene; **b** hexa-1,3,5-triene.

Saturated and unsaturated compounds

Alkenes and alkynes are called **unsaturated compounds** because they contain C=C bonds or C≡C bonds (Figure **10.11**). Alkanes and other compounds not containing multiple bonds are described as **saturated**.

Benzene is also described as an unsaturated hydrocarbon as the structure could be drawn with three C=C bonds.

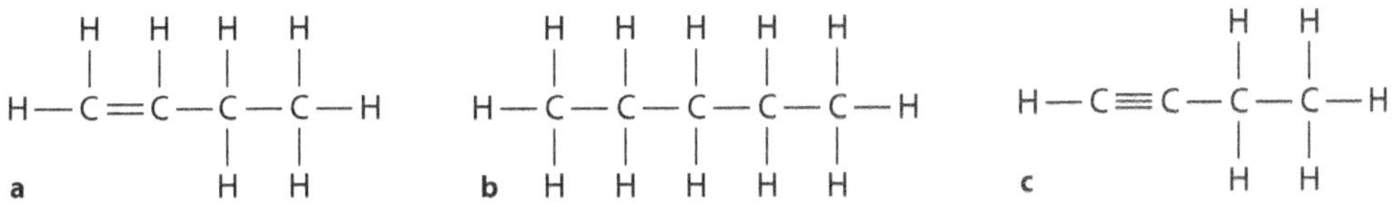

Figure 10.11 Hydrocarbons: **a** unsaturated; **b** saturated; **c** unsaturated.

4 Name the following molecules:

CH₃ H | | H—C—C—H | | CH₃ CH₂ | CH₃	CH₃ CH₃ | | H—C—C—CH₃ | | CH₃ CH₃	CH₃(CH₂)₂C(CH₃)₃	*(branched structure d)*
a	**b**	**c**	**d**

(structures e, f)

e	**f**

(structures g, h)

g	**h**

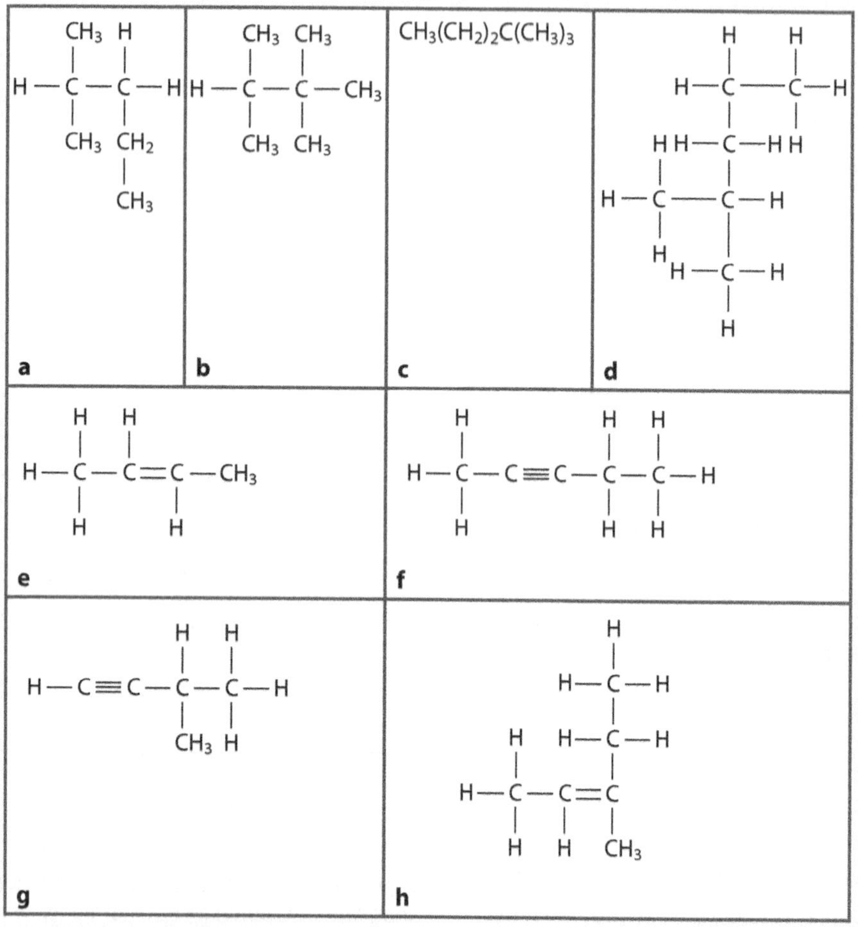

5 Draw the following molecules:
  a 3-methylhexane
  b 2,2,3-trimethylpentane
  c 2,4-dimethylhexane
  d 2,3-dimethylbut-1-ene

6 Give the correct name for each of the following:
  a 1,2-dimethylbutane
  b 1,2,3-trimethylpropane
  c 1-methyl-2,2-dimethylpropane

10.1.4 Naming halogenoalkanes

Halogenoalkanes have molecules that contain a halogen atom as the functional group. Halogenoalkanes are named using the format *x*-**halo**alkane, where *x* indicates the position of the halogen atom on the chain. Examples of some names of simple halogenoalkanes are given in Table **10.7**.

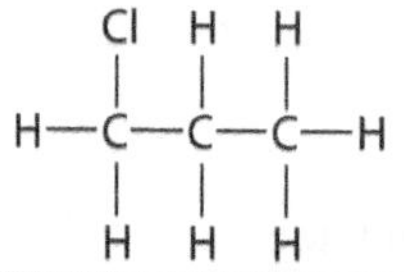 Cl H H | | | H—C—C—C—H | | | H H H	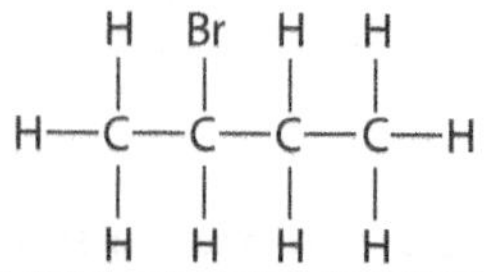 H Br H H | | | | H—C—C—C—C—H | | | | H H H H	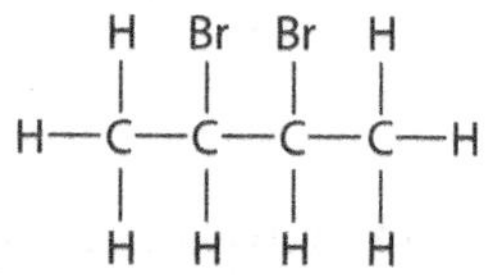 H Br Br H | | | | H—C—C—C—C—H | | | | H H H H
1-chloropropane	2-bromobutane	2,3-dibromobutane

Table 10.7 The names of some simple halogenalkanes.

Where more than one substituent is present, they are arranged in alphabetical order using the numbering system that gives the lowest possible number for any substituent. Where two or more substituents would have the same numbers when numbering from either side, the one that comes first in the alphabet (ignoring di-, tri- etc.) is allocated the lowest position number.

3-bromo-2-chloropentane

The two possible names for this are 3-bromo-2-chloropentane and 3-bromo-4-chloropentane. The first name gives the lowest individual number (2), and so the name is 3-bromo-2-chloropentane.

2-bromo-5-methylhexane

There are two substituents on the chain – a bromo group and a methyl group. The two possible names are 2-bromo-5-methylhexane and 5-bromo-2-methylhexane. Both names contain the same numbers, and so the name is chosen according to position in the alphabet. 'Bromo' comes before 'methyl' in the alphabet, and so the numbering is chosen to give the lower number to the 'bromo', which leads to the name 2-bromo-5-methylhexane.

Some more examples of halogenoalkane names are given in Table **10.8**.

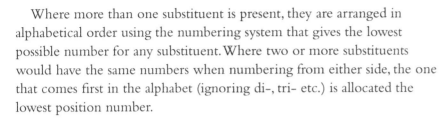

2-bromo-3-chlorobutane	2-chloro-3-methylpentane	3-bromo-2-methylpentane
2-chloro-4-methylpentane	1,1,1,2-tetrachloropropane	2-bromo-1,1-dichloropropane

Table 10.8 Examples of halogenoalkanes.

? Test yourself

7 Name the following compounds:

a	b	c	d

Learning objective

- Use IUPAC rules to name alcohols, ethers, aldehydes, ketones, carboxylic acids and esters

The World Health Organisation estimates that there are about 2.5 million alcohol-related deaths worldwide each year. This is more than the combined total of deaths through AIDS and malaria. The use and abuse of alcohol is a major social and economic problem in many countries and seems set to put an increasing strain on health systems in these countries in years to come.

a
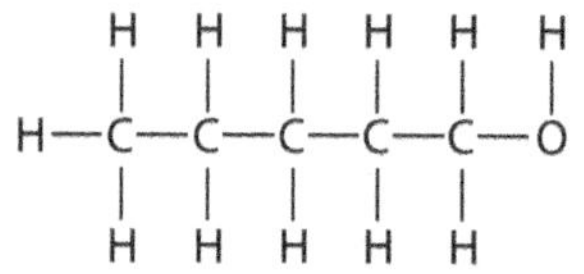

b
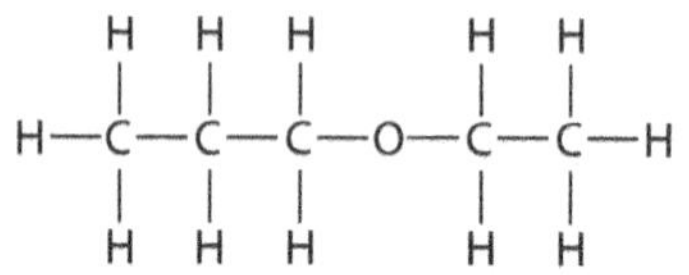

Figure 10.12 The alcohol **a** and ether **b** both have molecular formula $C_5H_{12}O$ and are isomers – functional group isomers.

The name changes slightly when more than one –OH group is present:

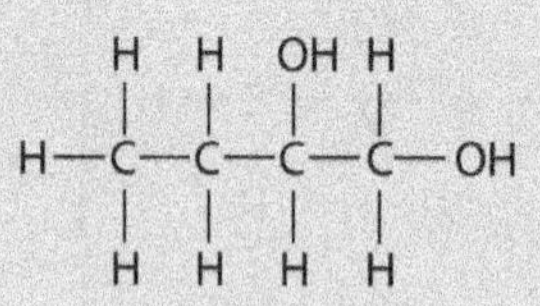

butane-1,2-diol

10.1.5 Naming organic compounds containing oxygen

Alcohols and ethers

Alcohols and ethers both contain oxygen. All alcohols contain the –OH functional group, whereas for ethers the functional group is C–O–C. Alcohols and ethers have the general molecular formula $C_nH_{2n+2}O$, and they can be isomers (see Subtopic **10.1.7**) as shown in Figure **10.12**.

Naming alcohols

Alcohols are named as **alkan-*x*-ol** (*x*-alkanol is also acceptable). Carbon atoms are always numbered to give the lowest possible value for *x*.

Table **10.9** lists the first six alcohols and Table **10.10** gives some examples where *x* is not equal to 1.

Name	Molecular formula	Structural formula	
		Condensed	Full
methanol	CH_3OH	CH_3OH	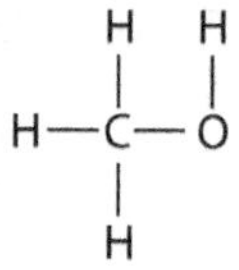
ethanol	C_2H_5OH	CH_3CH_2OH	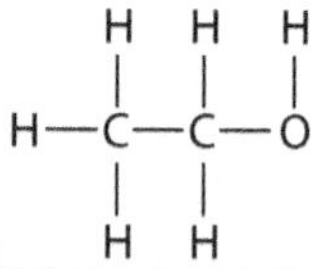
propan-1-ol	C_3H_7OH	$CH_3CH_2CH_2OH$	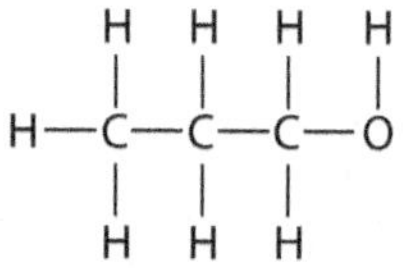
butan-1-ol	C_4H_9OH	$CH_3(CH_2)_2CH_2OH$	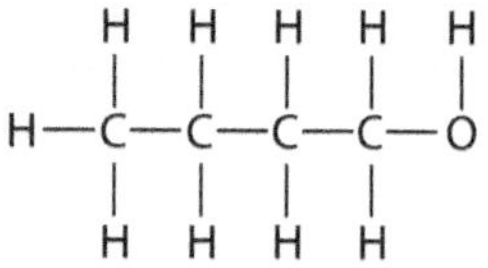
pentan-1-ol	$C_5H_{11}OH$	$CH_3(CH_2)_3CH_2OH$	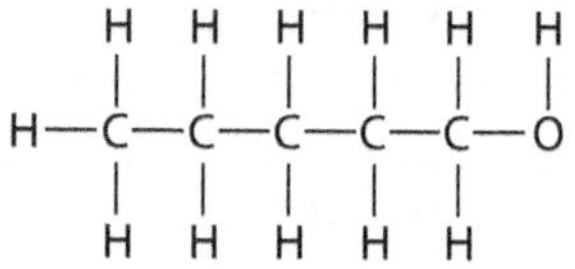
hexan-1-ol	$C_6H_{13}OH$	$CH_3(CH_2)_4CH_2OH$	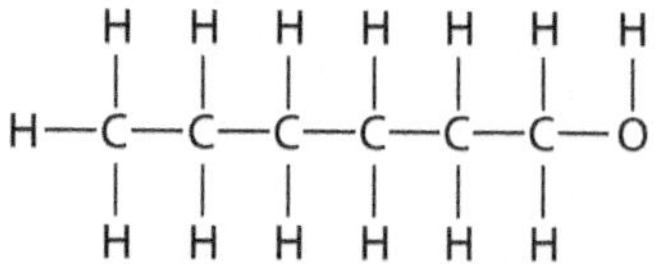

Table 10.9 The first six members of the alcohol homologous series.

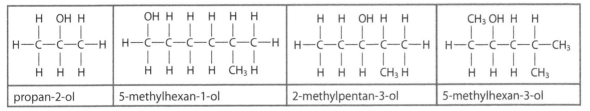

propan-2-ol	5-methylhexan-1-ol	2-methylpentan-3-ol	5-methylhexan-3-ol

Table 10.10 When x is not equal to 1 the numbering is chosen to give a lower number to the OH group than to an alkyl group.

Naming ethers

Ethers contain the C–O–C functional group and many ethers have two alkyl groups separated by an oxygen atom. The shorter group and the oxygen together is called an **alkoxy** group and is named as a substituent on the main chain.

In the molecule shown in Figure **10.13**, the longer carbon chain has three carbons and if the OCH_3 (methoxy) group were replaced by a hydrogen it would be called propane. The methoxy group is on carbon 1 and so the molecule is called 1–methoxypropane – this is often just called methoxypropane.

If the OC_2H_5 group in the molecule in Figure **10.14** were replaced by an H the molecule would be called 2-methylbutane. With the ethoxy group on carbon 2, it is called 2-ethoxy-2-methylbutane.

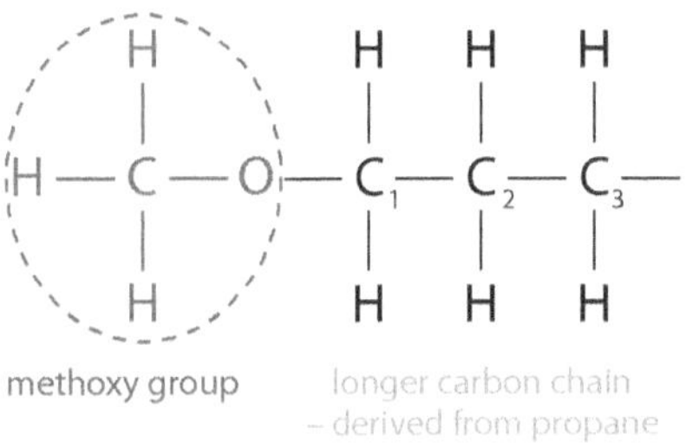

Figure 10.13 Naming an ether.

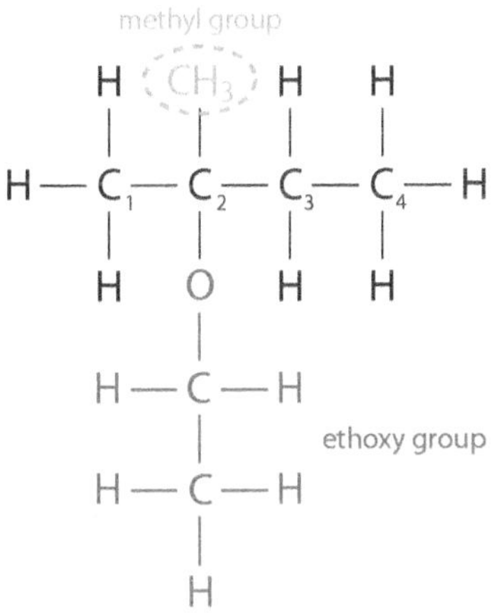

Figure 10.14 Naming an ether.

Test yourself

8 Draw and name all the isomers of $C_4H_{10}O$.

Naming aldehydes, ketones, carboxylic acids and esters

Aldehydes and ketones are known as **carbonyl** compounds and contain the carbonyl (C=O) functional group. Carboxylic acids contain the COOH functional group. Table **10.11** shows the functional group unit of each of these compounds.

- **Al**dehydes are named based upon alkan**al**.
- **Ket**ones are named as alkan-*x*-**one** (*x*-alkanone is also acceptable).
- Carboxylic acids have names based on alkan**oic acid**.

In aldehydes and carboxylic acids, the carbon of the CHO or COOH group is numbered carbon 1 and everything else is numbered relative to that. For ketones, the numbering scheme is chosen to give the C=O group the lowest possible number. Examples of some names of aldehydes and ketones are given in Table **10.12** and carboxylic acids in Table **10.13**.

Condensed structural formulas for aldehydes can be written using the CHO group – so ethanal can be written as CH_3CHO and butanal as $CH_3CH_2CH_2CHO$. The condensed structural formulas for ketones can be written using the CO group – so propanone can be written as CH_3COCH_3 or $(CH_3)_2CO$ and pentan-2-one as $CH_3COCH_2CH_2CH_3$.

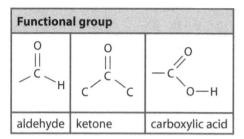

Functional group		
aldehyde	ketone	carboxylic acid

Table 10.11 The functional groups in aldehydes, ketones and carboxylic acids.

(structure)	methanal	
(structure)	propanal	
(structure)	4-methylpentanal	numbering starts from the aldehyde group
(structure)	propanone	
(structure)	pentan-2-one	
(structure)	4-methylpentan-2-one	the ketone group is given the lower number

aldehydes / ketones

Table 10.12 The names of some aldehydes and ketones.

H—C(=O)O—H	methanoic acid		carboxylic acids
CH₃ group (H—C(H)(H)—C(=O)O—H)	ethanoic acid		
propanoic acid structure	propanoic acid		
butanoic acid structure	butanoic acid		
4-methylpentanoic acid structure	4-methylpentanoic acid	numbering starts at the carboxylic acid group	
2,3-dimethylbutanoic acid structure	2,3-dimethylbutanoic acid		
butanedioic acid structure	butanedioic acid		

Table 10.13 The names of some carboxylic acids.

? Test yourself

9 Name the following compounds:

a, b, c, d (structural formulas shown)

10 Draw the structures of each of the following molecules:
   a butanal   **b** 2-methylpentanal   **c** pentan–3–one

Naming esters

Esters are formed by a reaction between a carboxylic acid and an alcohol and are named according to the carboxylic acid from which they are derived – two examples are given in Figure **10.15**.

When writing condensed structural formulas, the ester group can be represented as 'COOC' – so ethyl ethanoate can be written $CH_3COOCH_2CH_3$.

The names of some more complicated esters are shown in Table **10.13**.

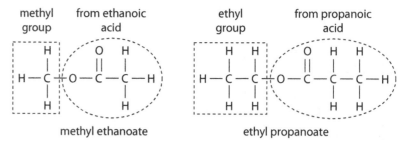

Figure 10.15 Naming esters.

Uses of esters

Many esters have a sweet, fruity smell and are used as artificial flavours and odours. Esters are often used to flavour sweets.
Probably the most controversial use of esters is as plasticisers (Option A), which are added to polymers such as PVC to make them more flexible. The use of phthalate esters such as DEHP is controlled in some countries due to fears that they could cause cancer.

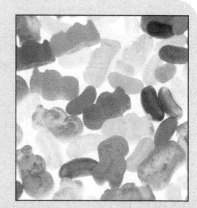

	ethyl 2-methylpropanoate	numbering in the acid derivative starts from the C=O group
	1-methylpropyl ethanoate	The longest carbon chain ending with the C attached to the O is taken. Thus, although there is a continuous carbon chain of 4 C atoms in the alkyl group, it does not have the C attached to the O at the end of the chain.

Table 10.14 Names of some esters.

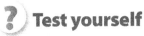

Test yourself

11 Name the following esters:

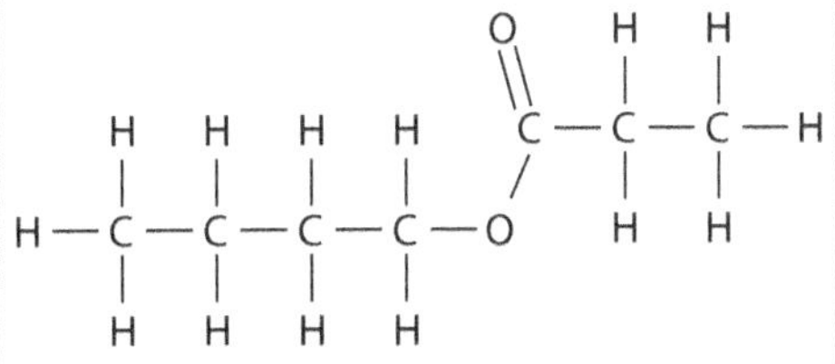 **a**	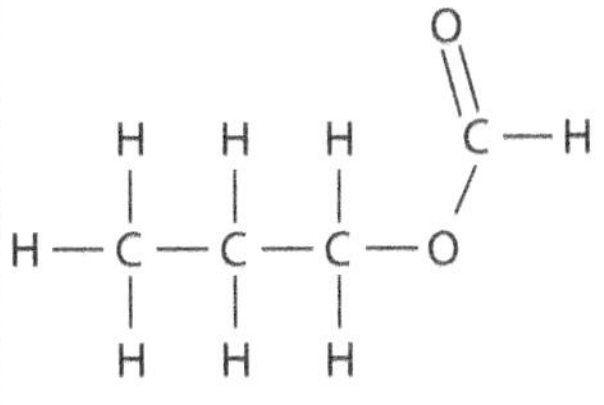 **b**	$CH_3CH_2CH_2CH_2COOCH_2CH_2CH_3$ **c**

10.1.6 Primary, secondary and tertiary organic compounds

Learning objectives

- Distinguish between primary, secondary and tertiary alcohols
- Distinguish between primary, secondary and tertiary halogenoalkanes
- Distinguish between primary, secondary and tertiary amines

Primary, secondary and tertiary alcohols

Alcohols are described as **primary**, **secondary** or **tertiary** depending on the number of carbon atoms attached to the carbon with the –OH group.

Ethanol is a primary alcohol because it has one carbon atom attached to the C with the –OH group (Figure **10.16**). Hence, a primary alcohol contains the $-CH_2OH$ group.

Propan-2-ol is a secondary alcohol because it has two carbon atoms attached to the C with the –OH group. Hence, secondary alcohols (Figure **10.17**) contain the –CHOH group.

2-methylpropan-2-ol is a tertiary alcohol (Figure **10.18**) because it has three C atoms attached to the C with the –OH group.

With methanol, even though there are no other carbon atoms attached to the C with the –OH group, it is worth noting that it is also a primary alcohol because it contains the $-CH_2OH$ group.

The symbol 1° is sometimes used for 'primary'. 2° can be used for 'secondary' and 3° for 'tertiary'.

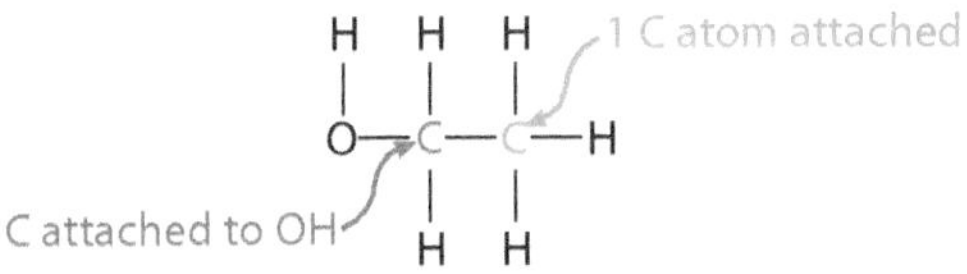

Figure 10.16 A primary alcohol – ethanol.

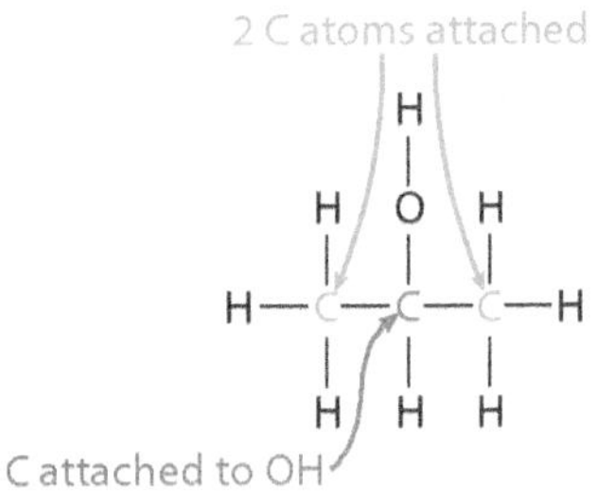

Figure 10.17 A secondary alcohol – propan-2-ol.

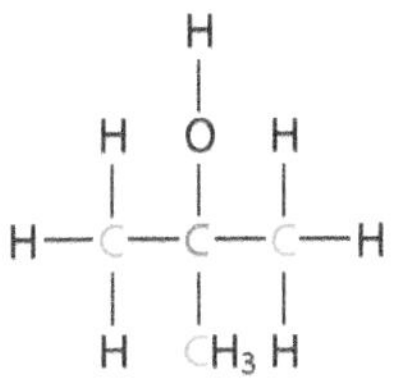

Figure 10.18 A tertiary alcohol – 2-methylpropan-2-ol.

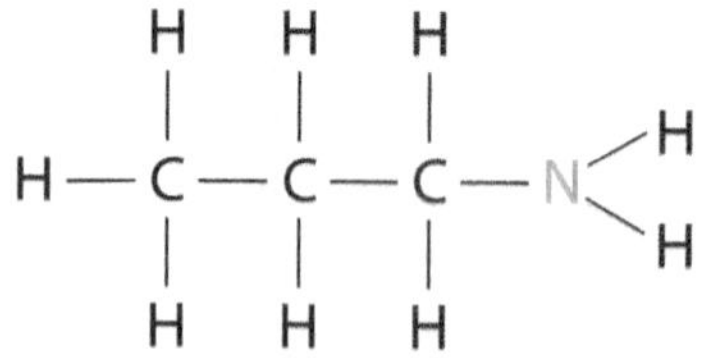

Primary, secondary and tertiary halogenoalkanes

Like alcohols, halogenoalkanes are also described as **primary**, **secondary** or **tertiary** depending on the number of carbon atoms (or alkyl groups) attached to the carbon atom of the C–X group. Table **10.15** summarises the three types.

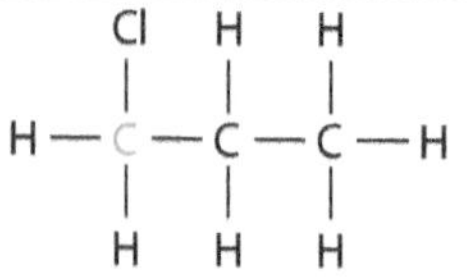 (structure: $H-C-C-C-H$ with Cl, H, H on top; H, H, H below)	primary	The C attached to the Cl has only one C atom attached – *one* C attached = 'primary'.
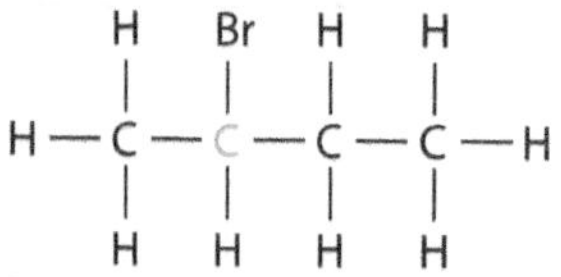 (structure: $H-C-C-C-C-H$ with H, Br, H, H on top; H, H, H, H below)	secondary	The C attached to the Br has two other C atoms attached – *two* Cs attached = 'secondary'.
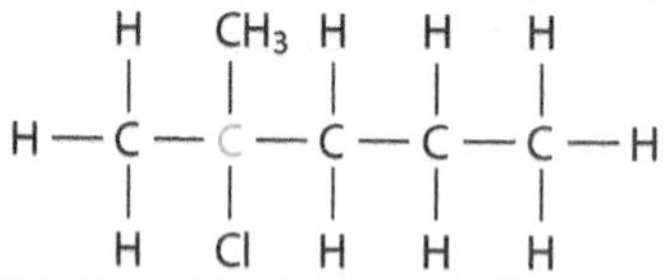 (structure: $H-C-C-C-C-C-H$ with H, CH$_3$, H, H, H on top; H, Cl, H, H, H below)	tertiary	The C attached to the Cl has three other C atoms attached – *three* Cs attached = 'tertiary'.

Table 10.15 Examples of primary, secondary and tertiary halogenalkanes.

Primary, secondary and tertiary amines

Amines are also classified as primary, secondary or tertiary – but using slightly different reasoning. The focus here is on the nitrogen atom in the amine functional group. If the nitrogen atom is attached to one carbon atom and two hydrogen atoms then it is part of a primary amine; if nitrogen is attached to two carbon atoms and one hydrogen atom it is a secondary amine, and if nitrogen is attached to three carbon atoms and has no hydrogen atoms it is a tertiary amine. This means that:

- primary amines contain the $-NH_2$ functional group
- secondary amines contain $-NHR$ (where 'R' represents an alkyl group)
- tertiary amines contain $-NR_2$ (see Figure **10.19**).

a primary amine

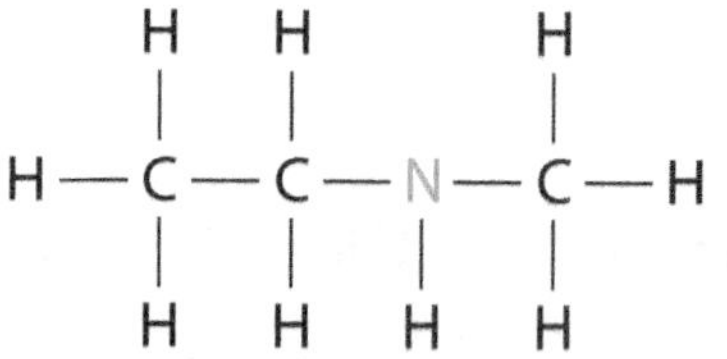

b secondary amine

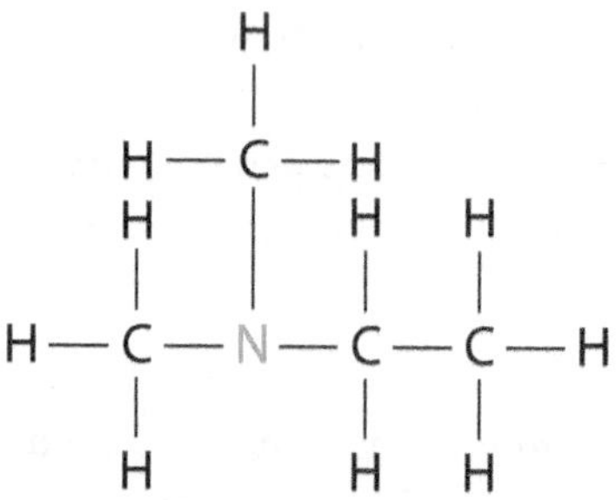

c tertiary amine

Figure 10.19 The structure of:
a propylamine; **b** ethylmethylamine;
c ethyldimethylamine.

? Test yourself

12 Classify each of the following molecules as primary, secondary or tertiary and name the halogenoalkanes and alcohols:

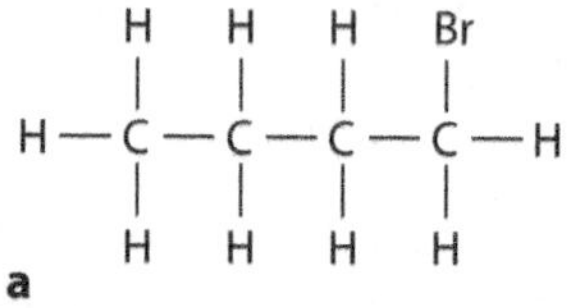 a	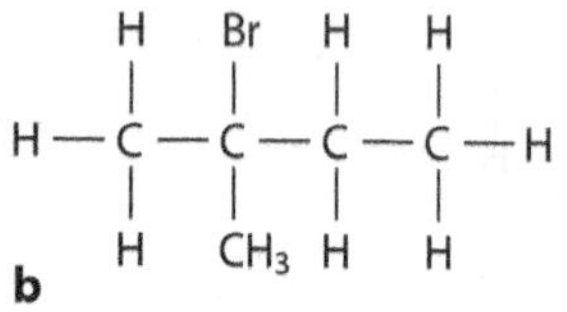 b	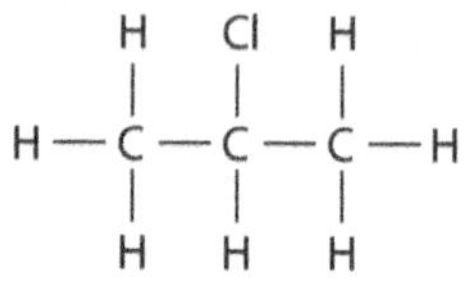 c	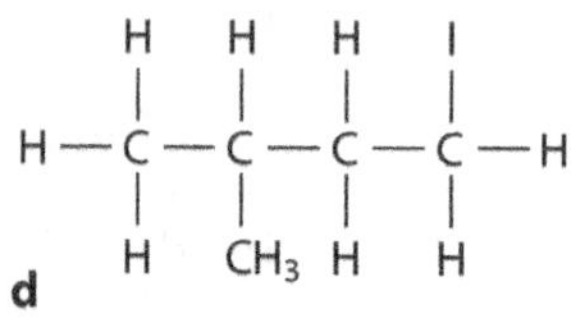 d
(structure with H, H, H, Br)	(structure with H, Br, H, H and CH$_3$)	(structure with H, Cl, H)	(structure with H, H, H, I and CH$_3$)
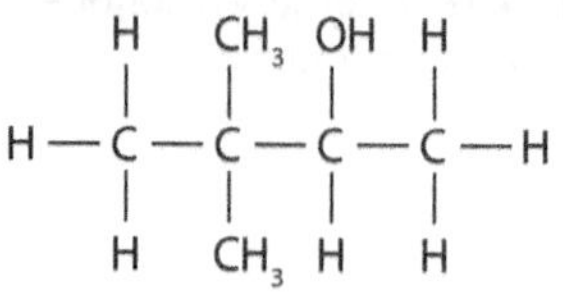 e	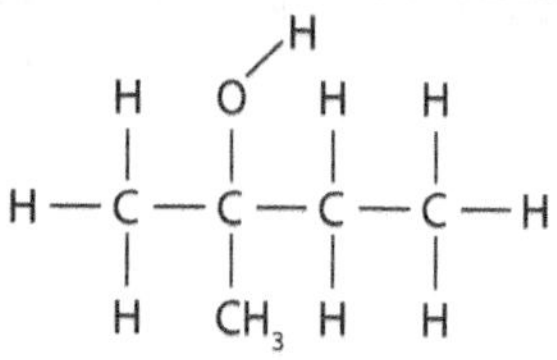 f	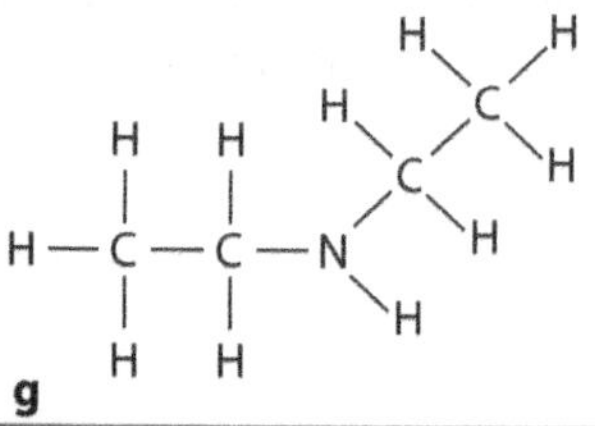 g	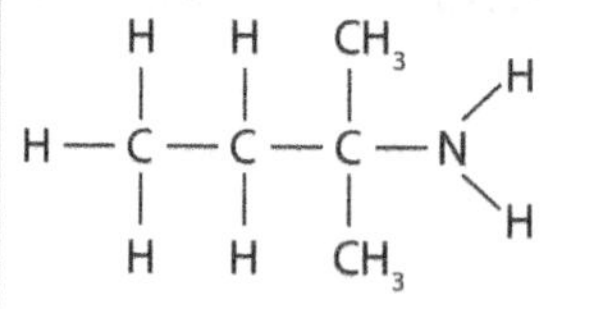 h
(structure with H, CH$_3$, OH, H and CH$_3$)	(structure with O, H, H, H and CH$_3$)	(amine structure)	(amine structure with CH$_3$ groups)

10.1.7 Isomers

Learning objectives

- Understand what is meant by structural isomerism
- Draw structural isomers for molecules with up to six carbon atoms

There are various forms of isomerism – the simplest is **structural isomerism**.

> **Structural isomers** are two or more compounds that have the same molecular formula but different structural formulas – the atoms are joined together in different ways.

The two structural isomers with the molecular formula C_4H_{10} are:

- butane

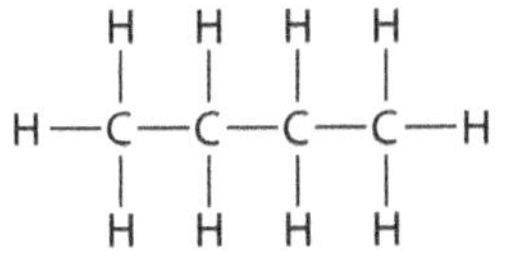

- 2-methylpropane

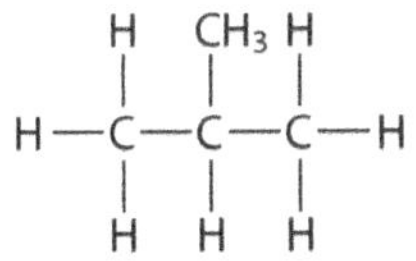

The '2' in this name is usually included, although the name would be unambiguous without it – there is only one possible structure for methylpropane. 1-methylpropane is butane!

The second molecule has a branched chain and a longest carbon chain of three.

The structural isomers of C_5H_{12} are:

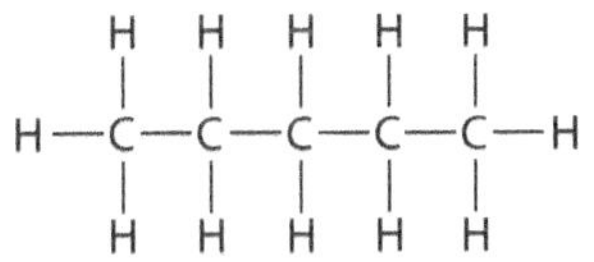

pentane

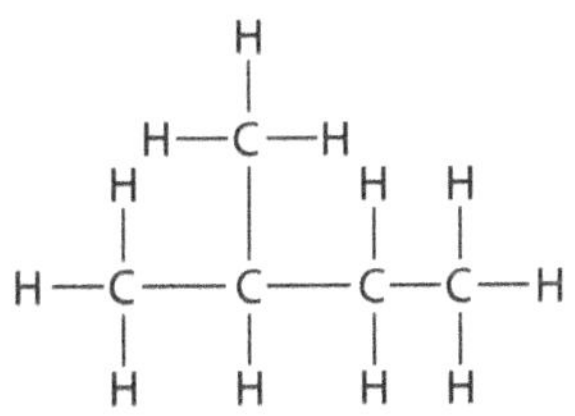

2-methylbutane

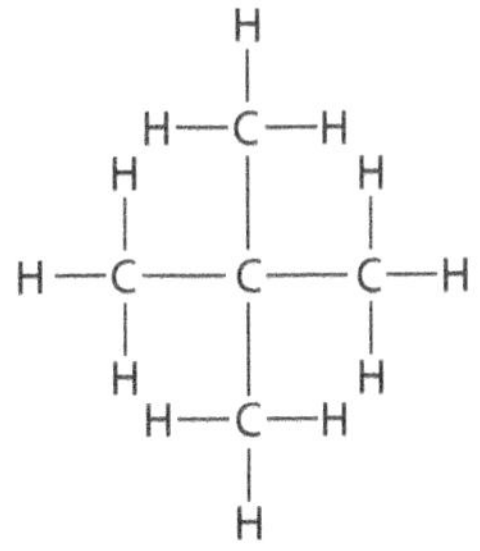

2,2-dimethylpropane

It is important to note that the structures shown in Figure **10.20** are *not* separate isomers of C_5H_{12} but are identical to each other. In a molecule in which only C–C single bonds are present, there is fairly free rotation about the single bond. Therefore, although we can show molecules bent into various conformations, if the atoms are joined together in the same way the molecules are *not* isomers.

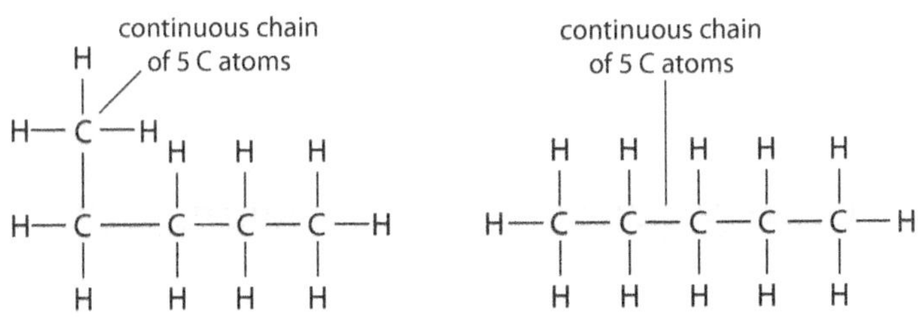

Figure 10.20 These two structures are drawn slightly differently. Careful examination reveals that all the atoms are joined together in exactly the same way and that there is a continuous chain of five carbon atoms in both structures. They are 'both' called pentane.

Longest carbon chain in the molecule	Isomers	
chain of six C atoms	CH₃CH₂CH₂CH₂CH₂CH₃ hexane	
chain of five C atoms	CH₃CH₂CH(CH₃)CH₂CH₃ 3-methylpentane	(CH₃)₂CHCH₂CH₂CH₃ 2-methylpentane
chain of four C atoms	(CH₃)₃CCH₂CH₃ 2,2-dimethylbutane	(CH₃)₂CHCH(CH₃)₂ 2,3-dimethylbutane

Table 10.16 Isomers of C_6H_{14}.

The isomers of C_6H_{14} are shown in Table **10.16**.

In general, branched–chain isomers have lower boiling points than straight–chain isomers. This is because the branches prevent the main chains from getting as close together, and so the London forces between the branched molecules are weaker.

Some isomers of C_6H_{12} are shown in Table **10.17**. Different isomers can be obtained simply by moving the double bond to a different position.

hex-2-ene	hex-3-ene	2-methylpent-1-ene
4-methylpent-1-ene	2-methylpent-2-ene	3-methylpent-2-ene

Table 10.17 Some isomers of C_6H_{12}.

Isomers may have different functional groups – Figure **10.21** shows another isomer of C_6H_{12} that is not an alkene but contains a ring of (four) carbon atoms.

Isomers with different functional groups are sometimes called **functional group isomers**. Some functional group isomers are shown in Table **10.18**.

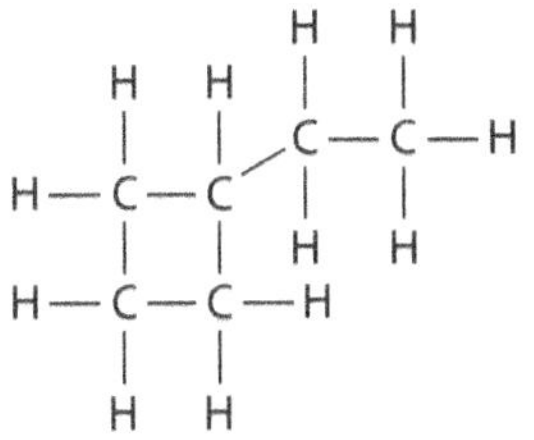

Figure 10.21 Ethylcyclobutane.

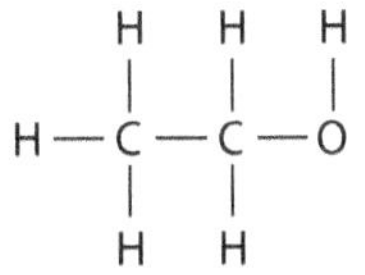	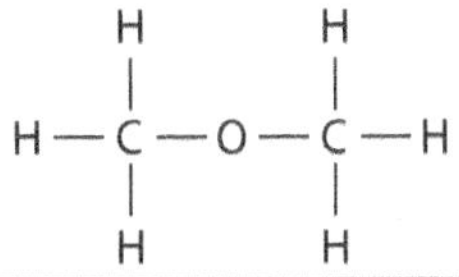
alcohol	ether
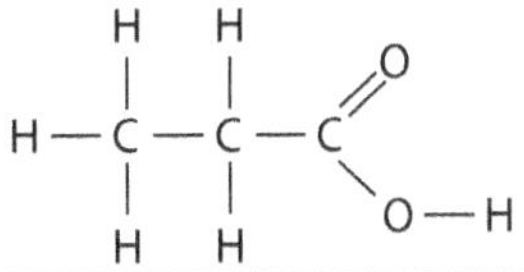	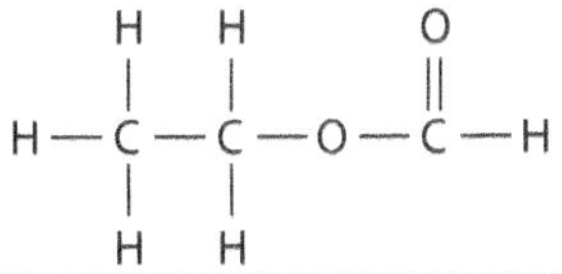
carboxylic acid	ester
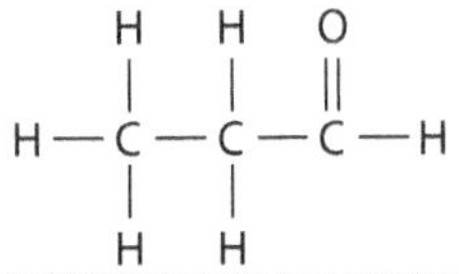	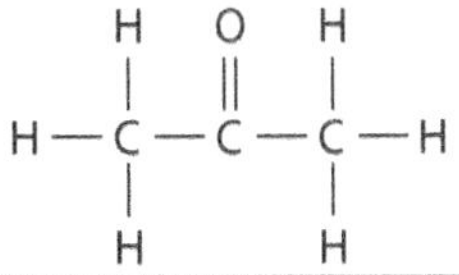
aldehyde	ketone

Table 10.18 Some pairs of functional group isomers.

Exam tip
Esters are isomeric with carboxylic acids. So ethanoic acid and methyl methanoate are isomers; propanoic acid, methyl ethanoate and ethyl methanoate are isomers.

? Test yourself

13 Draw and name all the isomers of C_4H_8 that are alkenes.

14 Draw and name all the isomers of C_5H_8 that are alkynes.

15 Draw all the esters that are isomeric with these carboxylic acids:

a butanoic acid

b pentanoic acid.

Chemists have agreed on a system for naming and classifying different molecules. Using various types of formula and this agreed system of names, chemists have their own language so that they can communicate with each other even when – in their everyday lives – they speak different languages.

How precise must the set of rules for this language be for chemists to make themselves understood? For instance, if you named pentane '1-methylbutane' or named 2-methylpentane '4-methylpentane' would everyone still understand what you meant? What if you named 2,2-dimethylbutane '2-dimethylbutane'?

Beyond school level, chemists often do not use systematic names when communicating with each other – why not?

Evidence for the structure of benzene

Benzene was first isolated in 1825 by Michael Faraday and its formula was known early on, but it was 40 years before a plausible structure was suggested. Kekulé proposed the original structure for benzene (Figure **10.22**) and so it is usually called 'Kekulé benzene'. The structure consists of a planar, hexagonal ring of carbon atoms, with a hydrogen atom joined to each carbon atom.

This structure has alternating single and double bonds between the carbon atoms. A systematic name for this molecule would be cyclohexa-1,3,5-triene or 1,3,5-cyclohexatriene.

The structure was accepted for many years but eventually the weight of evidence against it became too great and a modified structure was proposed.

The structure of benzene is nowadays better represented as in Figure **10.23**. Each carbon atom seems to form just three bonds – two to C atoms and one to an H atom.

The remaining electrons form a delocalised system of six electrons – this is represented by the circle in the centre of the structure. These six electrons are not localised between individual carbon atoms in double bonds but instead are spread over the whole ring.

The ring of electrons is formed when p orbitals overlap side-on to form a π delocalised system (Figure **10.24**).

Figure 10.22 Kekulé benzene.

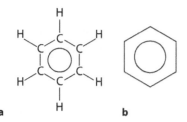

Figure 10.23 Benzene ring showing: **a** all the atoms; **b** the usual representation.

> Delocalisation: electrons shared between three or more atoms.

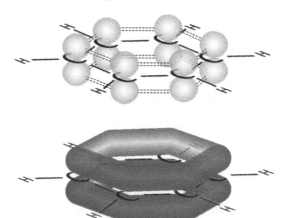

Figure 10.24 Formation of the benzene delocalised system.

Exam tip
The idea of delocalisation is not required at Standard Level but it helps to understand the structure of benzene. The 'resonance hybrid' approach to the structure of benzene is discussed in Topic 4.

Benzene is recognised as a human carcinogen which can cause leukemia. The main source of exposure to benzene in Europe is from car exhaust fumes, but benzene is used in some industrial applications and some people are exposed to it at work. Limits are usually set for occupational exposure to benzene but different countries set different limits on human exposure.

Evidence from carbon–carbon bond lengths

All C–C bond lengths are equal in benzene and also intermediate in length between a C–C single bond and a C=C double bond (Table **10.19**). If the structure of benzene were cyclohexa-1,3,5-triene (Kekulé benzene) then it would be expected that there would be three short C=C bonds (approximately 0.133 nm) and three longer C–C bonds (approximately 0.154 nm).

The delocalised structure of benzene suggests a C–C bond length between that of a C–C bond and a C=C double bond because there are, on average, three electrons (two from the single bond and one from the delocalised electrons) between each pair of carbon atoms compared with four electrons in a C=C bond and two in a C–C bond.

Bond	Compound	Bond length / nm
C=C	ethene	0.133
C–C	ethane	0.154
C≡C	benzene	0.140

Table 10.19 Bond lengths in ethene, ethene and benzene.

C–C bond lengths, and others, can be determined by X-ray crystallography.

Thermochemical evidence

When cyclohexene (C_6H_{10}) is heated with hydrogen in the presence of a nickel catalyst, cyclohexane (C_6H_{12}) is formed. This is an addition reaction in which hydrogen adds across the C=C bonds of the cyclohexene (Figure **10.25**) – the enthalpy change is approximately $-120\,\text{kJ}\,\text{mol}^{-1}$.

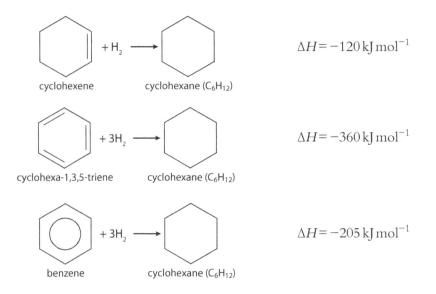

cyclohexene   cyclohexane (C_6H_{12}) $\Delta H = -120\,\text{kJ}\,\text{mol}^{-1}$

cyclohexa-1,3,5-triene   cyclohexane (C_6H_{12}) $\Delta H = -360\,\text{kJ}\,\text{mol}^{-1}$

benzene   cyclohexane (C_6H_{12}) $\Delta H = -205\,\text{kJ}\,\text{mol}^{-1}$

Figure 10.25 Comparing enthalpies of hydrogenation.

Cyclohexene molecules contain one C=C bond whereas cyclohexa-1,3,5-triene (Kekulé benzene) contains three. It would, therefore, be expected that the enthalpy change for the complete hydrogenation of cyclohexa-1,3,5-triene to cyclohexane would be 3×-120, i.e. $-360\,\text{kJ}\,\text{mol}^{-1}$. However, the enthalpy change when benzene undergoes complete hydrogenation to cyclohexane is only $-205\,\text{kJ}\,\text{mol}^{-1}$.

If an enthalpy level diagram (Figure **10.26**) is drawn for the hydrogenation reactions of benzene and cyclohexa-1,3,5-triene (Kekulé benzene), it is seen that benzene is $155\,\text{kJ}\,\text{mol}^{-1}$ more stable than 'expected' – i.e. $155\,\text{kJ}\,\text{mol}^{-1}$ more stable than would be predicted if it had the structure with alternating single and double carbon–carbon bonds.

The extra stability of benzene compared to the structure with alternating double and single bonds is due to the delocalisation of electrons.

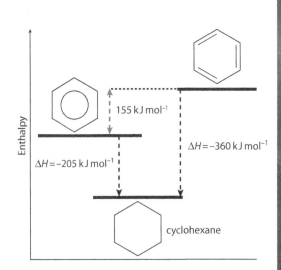

Figure 10.26 Comparative enthalpy changes for the hydrogenation reactions of benzene and Kekulé benzene.

Evidence from the number of isomers of $C_6H_4Cl_2$

Further evidence for a delocalised structure for benzene comes from examining the number of isomers that are possible for $C_6H_4X_2$ assuming a Kekulé structure and a delocalised structure (Figure **10.27**).

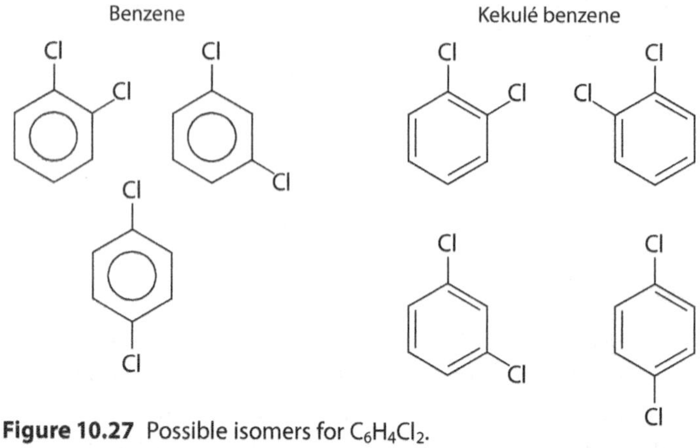

Figure 10.27 Possible isomers for $C_6H_4Cl_2$.

Only three isomers have ever been found for $C_6H_4Cl_2$ but the structure with alternating double and single bonds suggests that there should be four. The difference arises because two chlorine atoms that are on adjacent carbon atoms in the ring can either be separated by a C–C single bond or a C=C double bond, if benzene is assumed to have the the Kekulé structure.

Evidence from the reactions of benzene

Cyclohexa-1,3,5-triene (Kekulé benzene) would be expected to undergo addition reactions (like all other alkenes) and to decolorise bromine water. However, benzene does not react like alkenes – it does *not* undergo addition reactions under normal conditions and will *not* decolorise bromine water. An addition reaction would involve the destruction of the delocalised system and the loss of the extra stability associated with it.

Therefore benzene undergoes **substitution** reactions which involve replacement of one or more atoms on the ring with another atom or group.

For example, benzene reacts with chlorine in the presence of a catalyst such as aluminium chloride to form chlorobenzene, C_6H_5Cl:

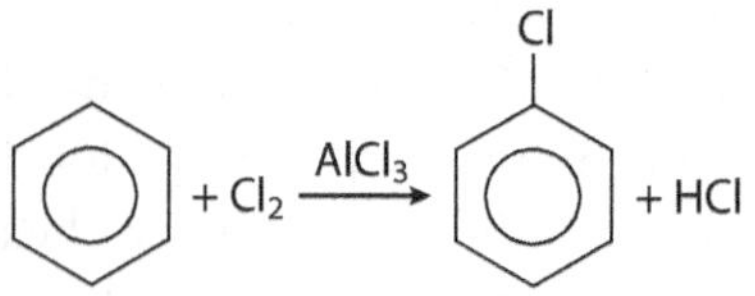

> Although the delocalised ring may be temporarily disrupted in reactions, it is restored so that the extra stability is not lost.

> **Exam tip**
> The evidence for the structure of benzene can be classified as either **chemical** (enthalpy changes of hydrogenation and products of substitution reactions) or **physical** (C–C bond lengths, number of isomers).

Nature of science

The structure of benzene will never be directly observed but can be inferred from experimental data. All the evidence taken together can be explained in terms of the delocalised structure of benzene. Science often works like this – scientists use reasoning to explain things we can see in terms of things we cannot.

446

10.2 Functional group chemistry

10.2.1 Alkanes

The alkanes are non-polar molecules with only London forces between molecules. This means that they are volatile (evaporate easily) – the first four members are gases at room temperature. Because of their non-polar nature, they are insoluble in water.

Reactions of alkanes

From a global perspective, the most important reaction of alkanes is **combustion** – these compounds form the basis of fuels such as petrol and natural gas (Figure **10.28**).

Complete combustion of hydrocarbons requires the presence of excess air/oxygen and produces carbon dioxide and water:

$$2C_2H_6(g) + 7O_2(g) \rightarrow 4CO_2(g) + 6H_2O(l)$$

When there is a limited supply of air/oxygen, **incomplete combustion** occurs – this produces carbon monoxide and soot (C) as well as water:

$$2C_2H_6(g) + 5O_2(g) \rightarrow 4CO(g) + 6H_2O(l)$$

$$2C_2H_6(g) + 3O_2(g) \rightarrow 4C(s) + 6H_2O(l)$$

The actual equation for the combustion of ethane is a combination of these three equations, and CO_2, CO and carbon are all produced if there is not sufficient oxygen present for complete combustion.

Incomplete combustion is dirtier than complete combustion, producing a smoky flame and the toxic gas carbon monoxide (which binds to hemoglobin more strongly than oxygen does, so oxygen is prevented from being transported by the blood).

Unreactivity of alkanes

Apart from combustion, alkanes are generally fairly unreactive. The reasons for this are:
- the high strengths of C–C and C–H bonds means that it is generally energetically unfavourable to break them in a reaction
- C–C and C–H bonds are essentially non-polar and so are unlikely to attract polar molecules or ions.

Reaction of alkanes with halogens

Alkanes react with halogens in the presence of sunlight or ultraviolet (UV) light. There is no reaction in the dark at room temperature. The equation for the reaction between methane and chlorine in the presence of UV light is:

$$CH_4 + Cl_2 \xrightarrow{\text{UV}} \underset{\text{chloromethane}}{CH_3Cl} + \underset{\text{hydrogen chloride}}{HCl}$$

Learning objectives

- Write equations for the complete and incomplete combustion of alkanes
- Explain why alkanes are not very reactive
- Write equations for the reactions of alkanes with halogens
- Explain the free radical substitution mechanism

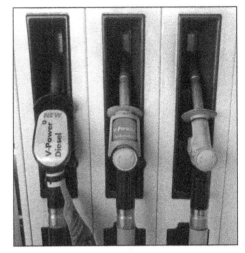

Figure 10.28 The carbon compounds in crude oil can be used to make medicines, plastics (polymers) and lots of other extremely useful chemicals and products. Is crude oil too valuable to burn?

Many deaths occur each year through accidental carbon monoxide poisoning, which occurs as a result of poorly serviced or poorly ventilated heating systems. Worldwide, carbon monoxide poisoning is probably the leading cause of death/injury through poisoning. The famous French author Émile Zola died of carbon monoxide poisoning.

Chloromethane is an important contributor to the destruction of the ozone layer (page 174). A significant source of chloromethane is from natural sources such as wood-rotting fungi.

That between ethane and bromine is:

$$C_2H_6 + Br_2 \xrightarrow{\text{UV}} \underset{\text{bromoethane}}{C_2H_5Br} + \underset{\text{hydrogen bromide}}{HBr}$$

If we look at this reaction in terms of full structural formulas (Figure **10.29**), we can see what is happening more clearly.

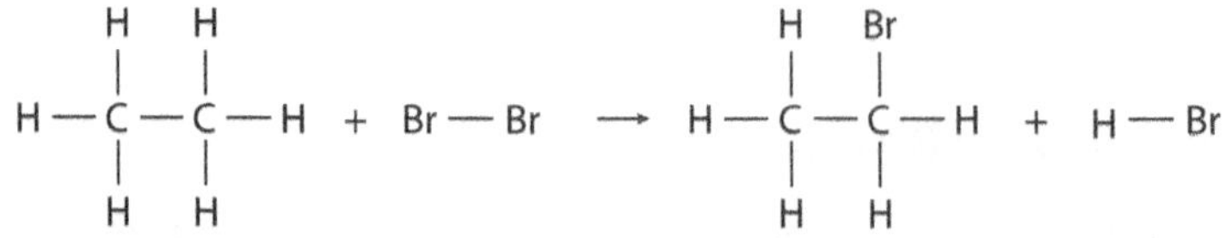

$$H-\overset{\displaystyle H}{\underset{\displaystyle H}{C}}-\overset{\displaystyle H}{\underset{\displaystyle H}{C}}-H + Br-Br \longrightarrow H-\overset{\displaystyle H}{\underset{\displaystyle H}{C}}-\overset{\displaystyle Br}{\underset{\displaystyle H}{C}}-H + H-Br$$

Figure 10.29 The reaction between ethane and bromine.

A **substitution** reaction is one in which an atom or a group is replaced by a different atom or group. In Figure **10.29**, a hydrogen atom in ethane is replaced by a bromine atom.

The reactions shown here all involve mono-substitution – the replacement of *one* hydrogen atom in the molecule by *one* halogen atom. However, the reaction is difficult to control and a mixture of products, some of which involve multiple substitutions, is formed:

$$CH_4 + Cl_2 \rightarrow \quad HCl \quad + \quad CH_3Cl$$

$$CH_3Cl + Cl_2 \rightarrow HCl \quad + \underset{\text{dichloromethane}}{CH_2Cl_2}$$

$$CH_2Cl_2 + Cl_2 \rightarrow HCl \quad + \underset{\text{trichloromethane}}{CHCl_3}$$

$$CHCl_3 + Cl_2 \rightarrow HCl \quad + \underset{\text{tetrachloromethane}}{CCl_4}$$

The mechanism for these reactions is called **free radical substitution**. It proceeds in a series of steps.

Free radical substitution

Consider the reaction between methane and chlorine:

$$CH_4 + Cl_2 \rightarrow CH_3Cl + HCl$$

Initiation

The first stage involves breaking apart chlorine molecules into separate chlorine atoms:

$$Cl_2 \xrightarrow{\text{UV}} 2Cl\bullet$$

The energy needed to do this is provided by the UV light. The Cl–Cl bond ($242\,\text{kJ}\,\text{mol}^{-1}$) is weaker than the C–H bond ($412\,\text{kJ}\,\text{mol}^{-1}$) and, therefore, it is the Cl–Cl bond that is broken.

The chlorine atoms produced have only seven electrons in their outer shell and so each has an unpaired electron – this is represented by the dot in the symbol. The chlorine atoms are described as **free radicals**.

> Free radicals are species (atoms or groups of atoms) with an unpaired electron. Free radicals are very reactive because of this unpaired electron.

This first step in the reaction mechanism involves an increase in the number of free radicals – it is called the **initiation step**.

This process is called **homolytic fission**. When the covalent bond (made up of two electrons) breaks, one electron goes back to each atom that made up the original molecule – in this case to form two species that are the same ('homo' means 'same').

This movement of one electron going back to each atom can be shown with single-headed curly arrows (fish-hooks):

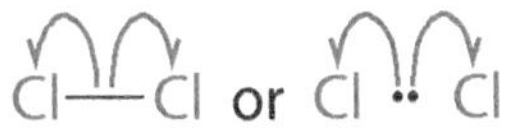

$$Cl\!-\!Cl \text{ or } Cl : Cl$$

Propagation

A chlorine free radical is a very reactive species – when it collides with a methane molecule in the reaction mixture, it will combine with a hydrogen atom to pair up its unpaired electron:

$$Cl\bullet + CH_4 \rightarrow HCl + \bullet CH_3$$
methyl free radical

A highly reactive methyl free radical is generated in this step, and this will react with a Cl_2 molecule to form a C–Cl bond:

$$\bullet CH_3 + Cl_2 \rightarrow CH_3Cl + Cl\bullet$$

The Cl• generated in this step can go on to react with another methane molecule, so that the **propagation** cycle starts again (Figure **10.30**).

This is an example of a **chain reaction** because one initial event causes a large number of subsequent reactions – with the reactive species being regenerated in each cycle of reactions.

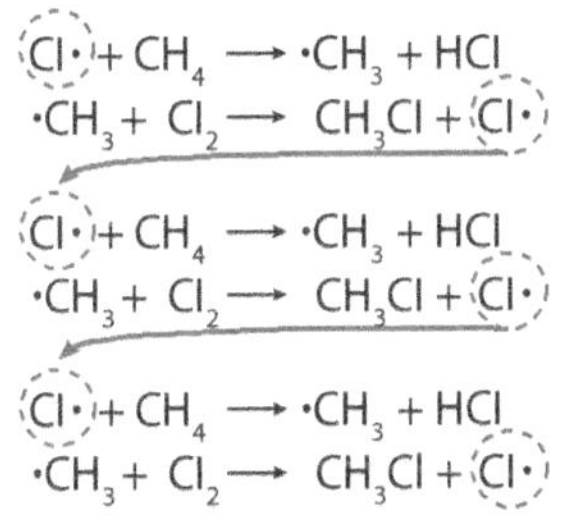

Termination

In the reaction mixture, the free radicals are always present in very low concentrations so the chance of two colliding is very low. However, they do collide sometimes and this brings the chain reaction to an end. There are several possible **termination** reactions:

$$Cl\bullet + Cl\bullet \rightarrow Cl_2$$
$$Cl\bullet + \bullet CH_3 \rightarrow CH_3Cl$$
$$\bullet CH_3 + \bullet CH_3 \rightarrow C_2H_6$$

Each termination reaction involves a decrease in the number of free radicals.

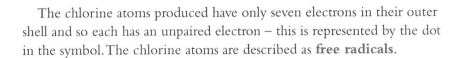

$$Cl\bullet + CH_4 \rightarrow \bullet CH_3 + HCl$$
$$\bullet CH_3 + Cl_2 \rightarrow CH_3Cl + Cl\bullet$$

$$Cl\bullet + CH_4 \rightarrow \bullet CH_3 + HCl$$
$$\bullet CH_3 + Cl_2 \rightarrow CH_3Cl + Cl\bullet$$

$$Cl\bullet + CH_4 \rightarrow \bullet CH_3 + HCl$$
$$\bullet CH_3 + Cl_2 \rightarrow CH_3Cl + Cl\bullet$$

Figure 10.30 The propagation step is repeated over and over again.

- Initiation: increase in the number of free radicals.
- Propagation: no change in the number of free radicals.
- Termination: decrease in the number of free radicals.

Formation of other products

Formation of multi-substituted products occurs when a chlorine free radical collides with a different molecule in the first propagation step:

$$Cl\bullet + CH_3Cl \rightarrow \bullet CH_2Cl + HCl \qquad \text{propagation}$$
$$\bullet CH_2Cl + Cl_2 \rightarrow CH_2Cl_2 + Cl\bullet \qquad \text{propagation}$$

$$Cl\bullet + CH_2Cl_2 \rightarrow \bullet CHCl_2 + HCl \qquad \text{propagation}$$
$$\bullet CHCl_2 + Cl_2 \rightarrow CHCl_3 + Cl\bullet \qquad \text{propagation}$$

$$Cl\bullet + CHCl_3 \rightarrow \bullet CCl_3 + HCl \qquad \text{propagation}$$
$$\bullet CCl_3 + Cl_2 \rightarrow CCl_4 + Cl\bullet \qquad \text{propagation}$$

Ethane and bromine

The mechanism for the reaction of ethane with bromine can be shown as follows:

$$Br_2 \xrightarrow{\text{UV}} 2Br\bullet \qquad \text{initiation}$$

$$C_2H_6 + Br\bullet \rightarrow \bullet CH_2CH_3 + HBr \qquad \text{propagation}$$
$$\bullet CH_2CH_3 + Br_2 \rightarrow CH_3CH_2Br + Br\bullet \qquad \text{propagation}$$

$$Br\bullet + Br\bullet \rightarrow Br_2 \qquad \text{termination}$$
$$Br\bullet + \bullet CH_2CH_3 \rightarrow CH_3CH_2Br \qquad \text{termination}$$
$$\bullet CH_2CH_3 + \bullet CH_2CH_3 \rightarrow CH_3CH_2CH_2CH_3 \qquad \text{termination}$$

Other alkanes

These reactions do not just occur with methane and ethane – all alkanes react in a similar way. Figure **10.31** shows what happens when butane reacts with chlorine in the presence of UV light; there are two possible mono-chlorinated products.

There are, of course, many multi-substituted products as well.

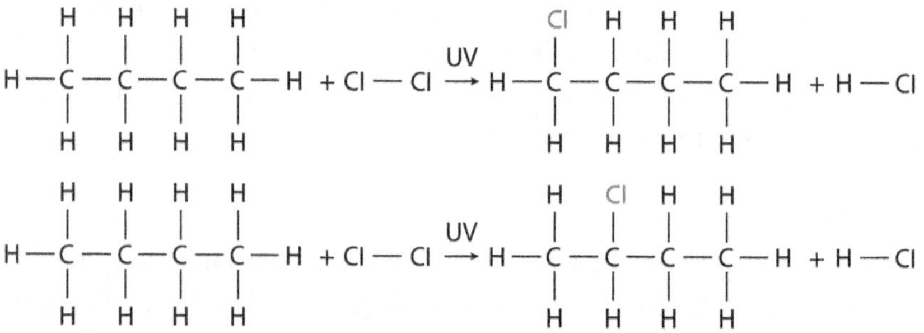

Figure 10.31 The formation of more than one mono-substituted product when butane reacts with chlorine.

Test yourself

16 Write an equation for the complete combustion of pentane.

17 Write an equation for the incomplete combustion of propane producing carbon monoxide.

18 Write an equation for the reaction of ethane with chlorine in the presence of UV light.

19 How many mono-chlorinated organic products are possible when chlorine reacts with each of the following molecules in the presence of UV light?

a hexane

b 2-methylpropane

c 2,2-dimethylpropane

20 Write the mechanism for the reaction between chlorine and ethane in the presence of UV light.

10.2.2 Reactions of alkenes and alkynes

Learning objectives

- Understand why alkenes are more reactive than alkanes
- Describe the addition reactions of alkenes
- Explain how to distinguish between alkanes and alkenes using bromine water
- Understand that benzene reacts differently to alkenes
- Understand the polymerisation reactions of alkenes

Alkenes have C=C double bonds – another way of saying this is that the functional group in alkenes is C=C.

The alkenes, like the alkanes, have non-polar molecules with only London forces between them. This means that they are volatile – the first four members are gases at room temperature. They are also insoluble in water because of their non-polar nature.

Saturated and unsaturated compounds

As mentioned above, alkenes and alkynes are called **unsaturated** compounds because they contain C=C bonds or C≡C bonds. Alkanes, and other compounds not containing multiple bonds, are described as **saturated**.

Reactions of alkenes

Alkenes are more reactive than alkanes. There are two reasons for this:

- the double bond (612 kJ mol^{-1}) is not twice as strong as a single bond. The reactions of alkenes usually involve the C=C bond breaking to form a C–C bond (348 kJ mol^{-1}). The second component (π bond) of the C=C bond is weaker than a normal C–C single bond (σ bond) and is, therefore, more easily broken.
- the double bond (four electrons) represents a region of high electron density, and therefore it attracts electrophiles.

An **electrophile** is a reagent (a positively charged ion or the positive end of a dipole) that is attracted to regions of high electron density and accepts a pair of electrons (Lewis acid) to form a covalent bond.

Addition reactions

Alkenes undergo **addition reactions.** The general reaction is shown in Figure **10.32** – 'R' represents either an alkyl group or hydrogen:

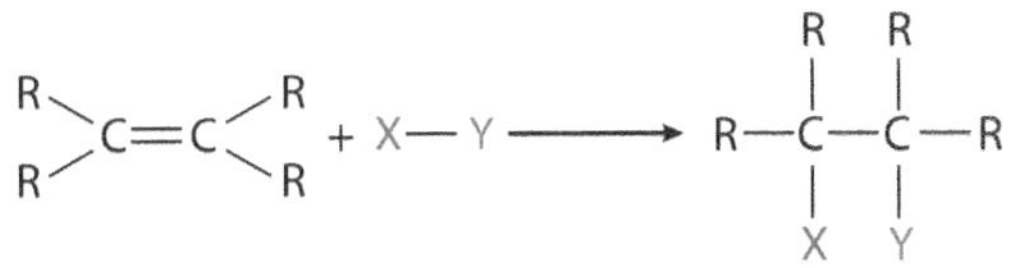

Figure 10.32 Addition reaction of alkenes.

The molecule X–Y is added across the double bond, which leaves a single bond between the carbon atoms.

Alkenes and halogens

Alkenes react with halogens at room temperature:

> The reaction of an alkene with iodine is very slow at room temperature.

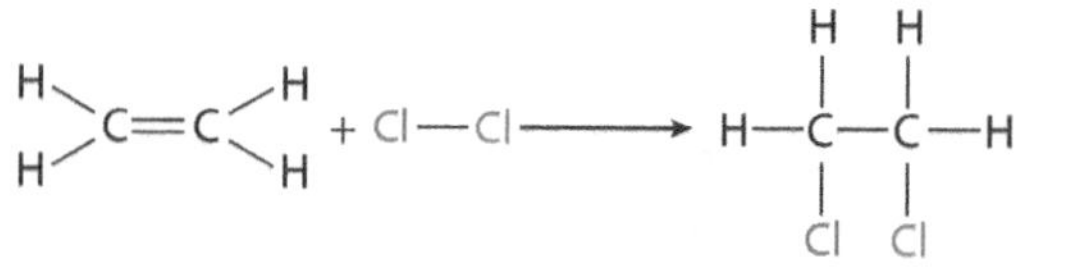

1,2-dichloroethane

Alkenes and hydrogen

Alkenes react with hydrogen when heated in the presence of a catalyst, such as finely divided nickel, to form alkanes. This reaction is called **hydrogenation**.

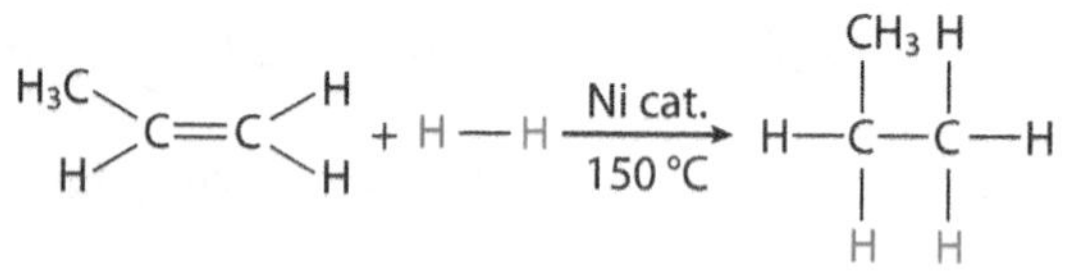

propane

Alkenes and hydrogen halides

Alkenes react with hydrogen halides, such as hydrogen bromide, by bubbling the alkene through a concentrated solution of the hydrogen halide at 100 °C.

> The reaction with HI is rapid, but a reaction with HCl is very slow and a catalyst (such as $AlCl_3$) is often used for that reaction.

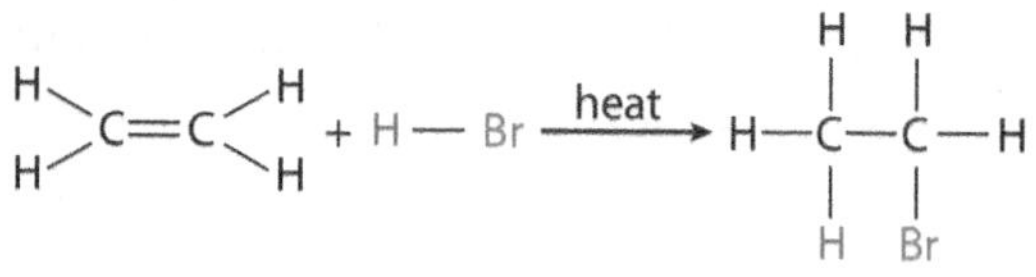

bromoethane

Alkenes and water

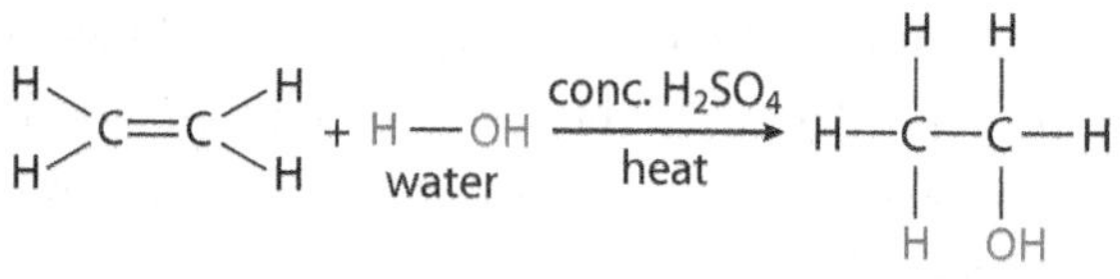

ethanol

In the laboratory, this reaction can be carried out by passing ethene through concentrated sulfuric acid at room temperature and then warming the product with water.

The reaction between an alkene and water is called **catalytic hydration** and is used in the industrial production of ethanol, an important solvent. The process is carried out by reacting ethene (from crude oil) with steam at 300 °C and 6000 kPa pressure with an H_3PO_4 catalyst. Ethanol can also be produced by fermentation, which uses a renewable raw material (sugar) and a much lower temperature (although a major energy input can come when ethanol is distilled). The proportion of ethanol (Figure **10.33**) produced by these processes varies from country to country and has also varied over time as the use of ethanol as a fuel additive has increased. The leading producers of ethanol for fuel use by fermentation are the US and Brazil. Brazil has pioneered the use of gasoline–ethanol blends as fuels for light vehicles and indeed pure gasoline is no longer available in Brazil as a fuel for cars.

Figure 10.33 Blends of ethanol and gasoline are becoming ever more important fuels for cars in some parts of the world.

Unsaturated fats (C=C bonds present) from vegetable sources tend to be liquids at room temperature and are called oils. Saturated fats (no C=C bonds) from animal sources are solids at room temperature.

Hydrogenation of alkenes is used in the manufacture of margarine. A polyunsaturated oil contains many C=C bonds (Figure **10.34**). The double bonds are all shown as *cis* (groups either side of the C=C bond are on the same side – see page **390**). It is the presence of *cis* double bonds that gives an oil its low boiling point.

Margarine is made from vegetable oils (liquids) and partial hydrogenation can be carried out to convert some of the C=C bonds into C–C single bonds. This increases the melting point of the margarine so that it is a solid at room temperature (Figure **10.35**). However, the process produces *trans*-fatty acids (groups either side of the C=C bond are on opposite sides) and these are believed to be bad for health. Partially hydrogenated vegetable oils have been banned in some countries.

Some countries have a very high consumption of saturated fat and this has been linked to increased rates of cardiovascular disease (but the data are not conclusive and the link not universally accepted). The so-called 'Mediterranean diet' is regarded by many people as a particularly healthy one and has been linked with low incidences of heart disease. This diet is low in saturated fats, with fat consumption mainly in the form of olive oil, which is rich in monounsaturated fats.

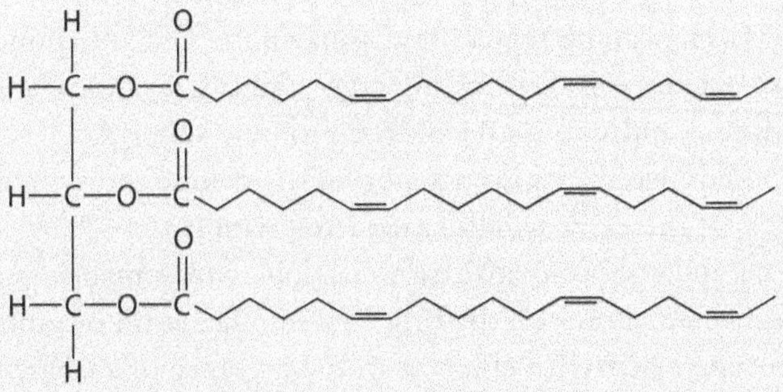

Figure 10.34 The basic structure of a polyunsaturated oil.

Figure 10.35 Different cooking oils and spreads contain different amounts of saturated/unsaturated fats.

The reactions described above can be regarded as being applicable to all alkenes and, although the conditions may vary slightly, the basic reactions are the same. So, but-2-ene would react with HBr according to the equation:

$$H_3C{\diagdown}{}{\diagup}CH_3 \atop H{\diagup}C{=}C{\diagdown}H \quad + \quad H{-}Br \xrightarrow{\text{HEAT}} \quad H_3C{-}\overset{\overset{\textstyle H}{|}}{\underset{\underset{\textstyle H}{|}}{C}}{-}\overset{\overset{\textstyle H}{|}}{\underset{\underset{\textstyle Br}{|}}{C}}{-}CH_3$$

and pent-2-ene would react with chlorine:

$$H_3C{\diagdown}{}{\diagup}CH_2CH_3 \atop H{\diagup}C{=}C{\diagdown}H \quad + \quad Cl{-}Cl \longrightarrow \quad H_3C{-}\overset{\overset{\textstyle H}{|}}{\underset{\underset{\textstyle Cl}{|}}{C}}{-}\overset{\overset{\textstyle H}{|}}{\underset{\underset{\textstyle Cl}{|}}{C}}{-}CH_2CH_3$$

Distinguishing between alkanes and alkenes

Alkenes can be distinguished from alkanes by shaking them with bromine water (Figure **10.36**). The orange bromine water is decolorised to colourless when shaken with an alkene, but there is no change in colour when it is shaken with an alkane.

> Alkene: bromine water – orange to colourless
> Alkane: bromine water – no colour change

The reaction that occurs with an alkene is:

$$H{\diagdown}{}{\diagup}H \atop H{\diagup}C{=}C{\diagdown}H \quad + \quad H_2O + Br_2 \longrightarrow \quad H{-}\overset{\overset{\textstyle H}{|}}{\underset{\underset{\textstyle Br}{|}}{C}}{-}\overset{\overset{\textstyle H}{|}}{\underset{\underset{\textstyle OH}{|}}{C}}{-}H + HBr$$

Comparison of the reactions of benzene and those of alkenes

Although benzene (C_6H_6) can be represented as having three C=C bonds, it does not react like alkenes – it does *not* undergo addition reactions under normal conditions and will *not* decolorise bromine water. An addition reaction would involve the destruction of the delocalised system and therefore the loss of the extra stability associated with it.

Therefore benzene undergoes **substitution** reactions, which involve replacement of one or more atoms on the ring with another atom or group.

For example, benzene reacts with chlorine in the presence of a catalyst such as aluminium chloride to form chlorobenzene, C_6H_5Cl.

The mechanism of the substitution reactions that benzene undergoes is electrophilic substitution and this is discussed on page **478**.

Exam tip
In this topic we are learning about the reactions of functional groups – not individual molecules. The reactions here are of the C=C functional group, and it makes little difference what the rest of the molecule is like.

Exam tip
The bromine water becomes colourless, not clear. 'Clear' and 'colourless' are not the same thing – the orange bromine water was originally clear because it was not cloudy and you could see through it!

To understand why the product of this reaction is not 1,2-dibromoethane, we have to consider the reaction mechanism. This is discussed in the Higher Level section on page **471**.

Figure 10.36 When shaken with an alkene, orange bromine water is decolorised.

Addition polymerisation

Alkenes undergo **addition polymerisation**, in which a large number of **monomer** molecules are joined together to form a polymer chain. The general equation for addition polymerisation is shown in Figure **10.37**.

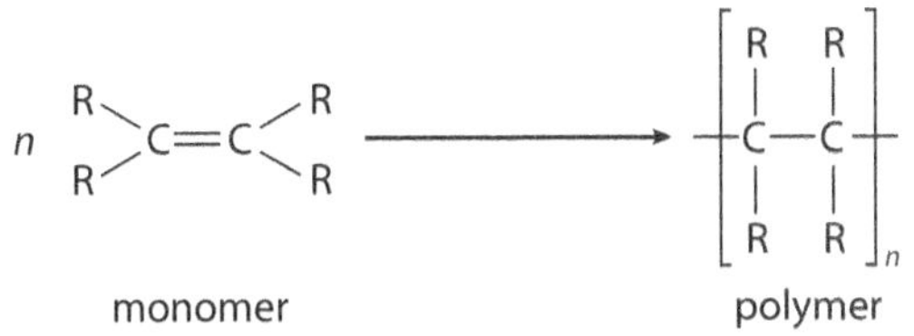

Figure 10.37 Addition polymerisation reaction.

It is important to realise that it is only the C=C group that reacts when polymerisation occurs – all the other groups attached to the C=C unit are unaffected.

Polyethene

In the production of low-density polyethene, the reaction takes place at high temperature and high pressure in the presence of a small amount of oxygen or an organic peroxide:

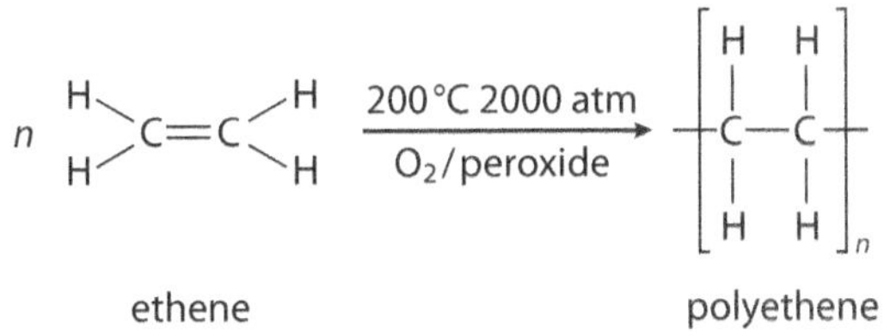

High-density polyethene is produced at much lower temperatures and lower pressures using a catalyst.

Addition polymers are named according to the monomer from which they were made. In this polymerisation reaction, ethene is the monomer and polyethene is the polymer. A section of the polymer chain is shown in Figure **10.38**.

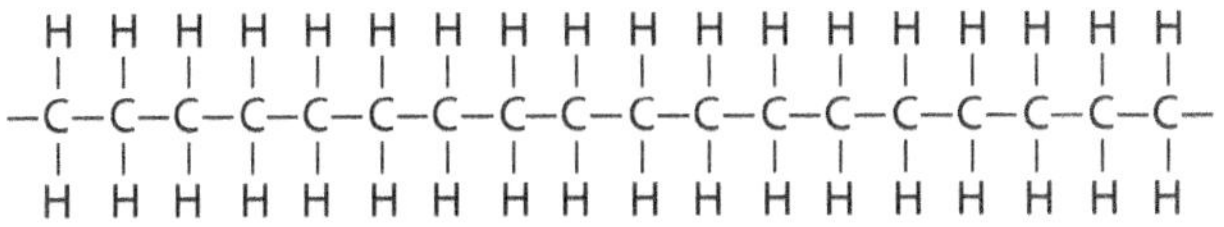

Figure 10.38 A section of a polyethene polymer chain.

This is an example of addition polymerisation, because monomer molecules are simply added to each other without anything else being produced.

n represents a large number.

Polyethene is also called polythene.

PVC – polyvinylchloride

This is more properly known as polychloroethene. It is made by the polymerisation of chloroethene monomers at a moderate temperature and pressure in the presence of an organic peroxide:

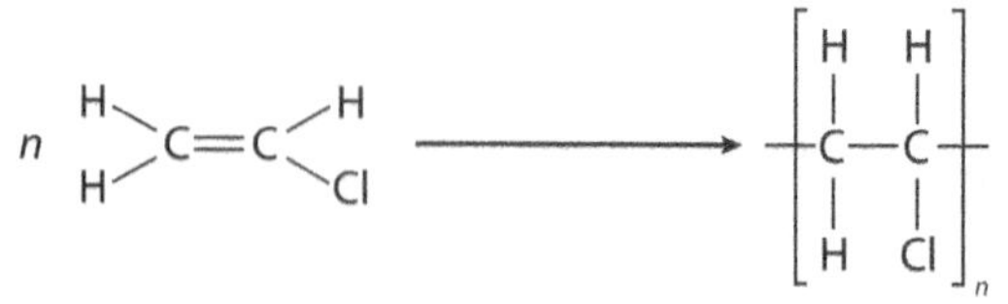

$$n \quad \underset{H}{\overset{H}{>}}C=C\underset{Cl}{\overset{H}{<}} \longrightarrow \left[\begin{array}{c} H \ \ H \\ | \ \ \ | \\ -C-C- \\ | \ \ \ | \\ H \ \ Cl \end{array} \right]_n$$

chloroethene   polychloroethene

The structure shown in square brackets as the product of this reaction is called the **repeating unit** of polychloroethene – many of these are joined to make a polymer chain:

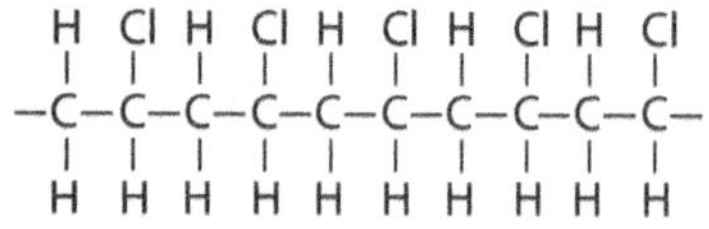

$$\begin{array}{c} H\ \ Cl\ \ H\ \ Cl\ \ H\ \ Cl\ \ H\ \ Cl\ \ H\ \ Cl \\ |\ \ \ |\ \ \ |\ \ \ |\ \ \ |\ \ \ |\ \ \ |\ \ \ |\ \ \ |\ \ \ | \\ -C-C-C-C-C-C-C-C-C-C- \\ |\ \ \ |\ \ \ |\ \ \ |\ \ \ |\ \ \ |\ \ \ |\ \ \ |\ \ \ |\ \ \ | \\ H\ \ H\ \ H\ \ H\ \ H\ \ H\ \ H\ \ H\ \ H\ \ H \end{array}$$

Polypropene

Remember that it is important to understand that it is only the C=C group that reacts when polymerisation occurs – all the other groups attached to the C=C unit are unaffected. This can be seen if we look at the polymerisation of propene (Figure **10.39**). The methyl group attached to the C=C unit does not become part of the main polymer chain – just a side group on the chain.

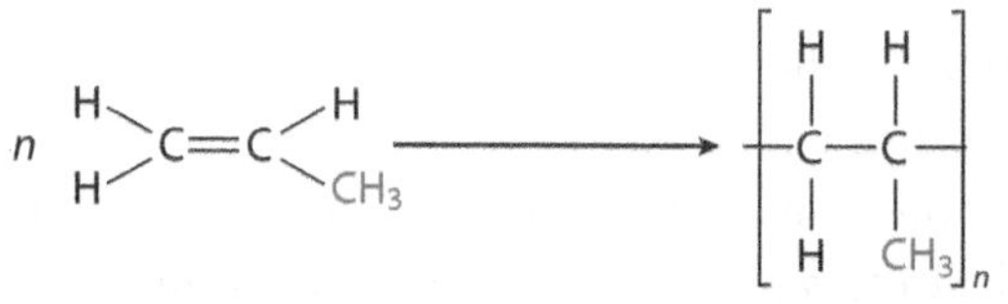

$$n \quad \underset{H}{\overset{H}{>}}C=C\underset{CH_3}{\overset{H}{<}} \longrightarrow \left[\begin{array}{c} H \ \ H \\ | \ \ \ | \\ -C-C- \\ | \ \ \ | \\ H \ \ CH_3 \end{array} \right]_n$$

propene   polypropene

$$\begin{array}{c} H\ \ CH_3\ H\ \ CH_3\ H\ \ CH_3\ H\ \ CH_3\ H\ \ CH_3\ H\ \ CH_3\ H\ \ CH_3\ H\ \ CH_3\ H\ \ CH_3 \\ |\ \ \ |\ \ \ |\ \ \ |\ \ \ |\ \ \ |\ \ \ |\ \ \ |\ \ \ |\ \ \ |\ \ \ |\ \ \ |\ \ \ |\ \ \ |\ \ \ |\ \ \ |\ \ \ |\ \ \ | \\ -C-C-C-C-C-C-C-C-C-C-C-C-C-C-C-C-C-C- \\ |\ \ \ |\ \ \ |\ \ \ |\ \ \ |\ \ \ |\ \ \ |\ \ \ |\ \ \ |\ \ \ |\ \ \ |\ \ \ |\ \ \ |\ \ \ |\ \ \ |\ \ \ |\ \ \ |\ \ \ | \\ H\ \ H\ \ H\ \ H\ \ H\ \ H\ \ H\ \ H\ \ H\ \ H\ \ H\ \ H\ \ H\ \ H\ \ H\ \ H\ \ H\ \ H \end{array}$$

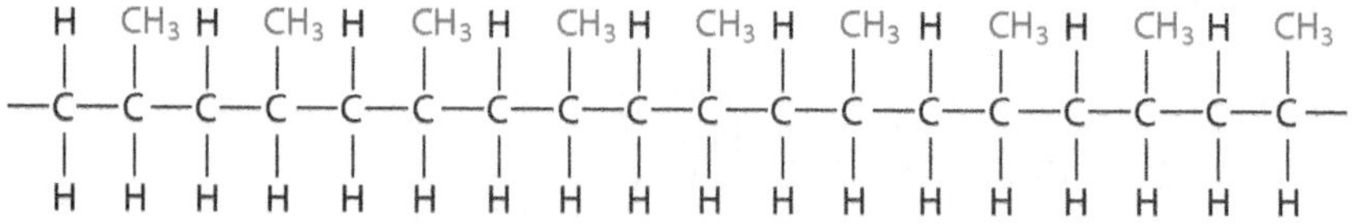

Figure 10.39 Polymerisation of propene.

The repeating unit of a polymer is the base from which the whole polymer chain can be made up. In each of the equations used so far, this has been shown in square brackets. The repeating unit for polypropene is shown in Figure **10.40**.

The repeating unit for any polymer (Figure **10.41**) can be identified simply by taking any two adjacent carbon atoms in the main polymer chain.

The monomer of a polymer can be derived (Figure **10.42**) by simply placing a double bond between the two carbon atoms of the repeating unit.

> The term 'repeat unit' is also used instead of 'repeating unit'.

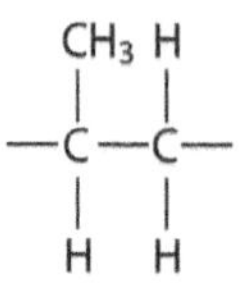

Figure 10.40 Repeating unit of polypropene.

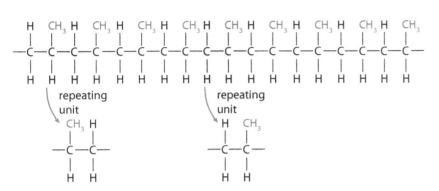

Figure 10.41 Identifying a repeating unit.

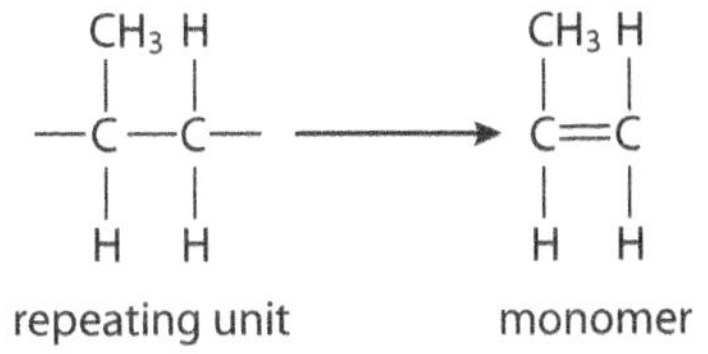

Figure 10.42 Identifying a monomer.

Economic importance of polymers

The polymers described here are more commonly known as 'plastics' and find extensive uses in everyday life. Among other things, polyethene is used for making plastic bags, washing-up bowls and bottles; polychloroethene is used for making insulation for electrical cables and window frames; and polypropene is used for making car bumpers and carpet fibres.

These substances are produced in vast quantities by the petrochemicals industry and present a huge environmental problem, both from the point of view of using crude oil (petroleum) – a limited natural resource – and in their disposal after use because they are non-biodegradable. Different countries tackle the problem of disposal of plastics in different ways – countries such as Denmark and Switzerland mostly incinerate plastics, whereas Finland and Ireland mainly use landfill sites. Recycling is growing in importance in some countries but attitudes towards environmental problems are far from uniform across the world.

? Test yourself

21 Draw the structures of the products of the following reactions (which occur under the appropriate conditions):
   a but-1-ene + hydrogen
   b hex-3-ene + hydrogen bromide
   c pent-2-ene + chlorine
   d but-2-ene + water

22 Write equations, using structural formulas, for the formation of the following compounds from appropriately chosen alkenes:
   a
   $$H-\overset{\displaystyle H}{\underset{\displaystyle H}{C}}-\overset{\displaystyle H}{\underset{\displaystyle H}{C}}-\overset{\displaystyle OH}{\underset{\displaystyle H}{C}}-\overset{\displaystyle H}{\underset{\displaystyle H}{C}}-\overset{\displaystyle H}{\underset{\displaystyle H}{C}}-\overset{\displaystyle H}{\underset{\displaystyle H}{C}}-H$$

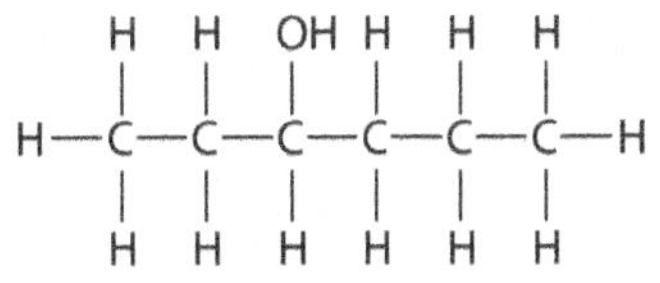

   b
   $$H-\overset{\displaystyle H}{\underset{\displaystyle H}{C}}-\overset{\displaystyle H}{\underset{\displaystyle CH_3}{C}}-\overset{\displaystyle Br}{\underset{\displaystyle CH_3}{C}}-\overset{\displaystyle Br}{\underset{\displaystyle H}{C}}-H$$

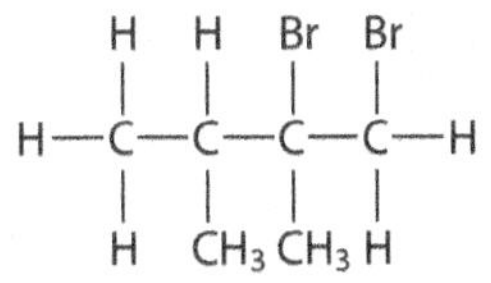

23 Draw the repeating units for the polymers formed from these alkenes:

a

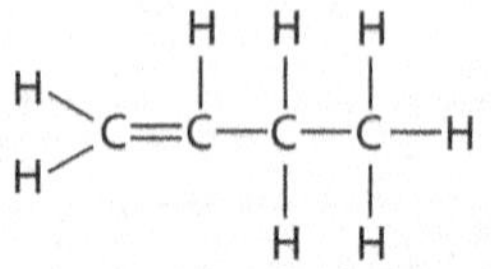

b

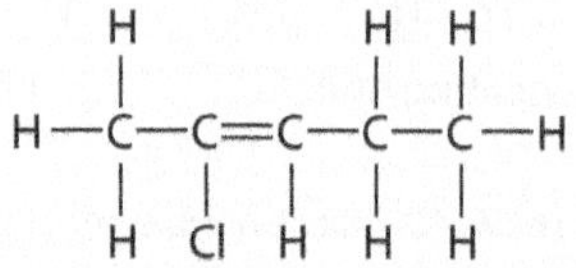

24 Draw the monomers used to make the following polymers:

a

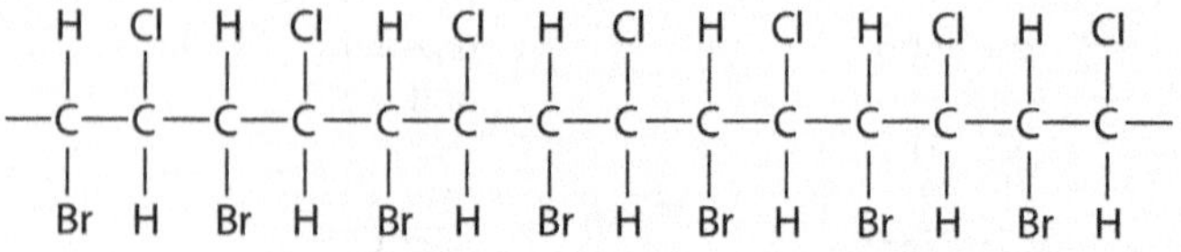

b

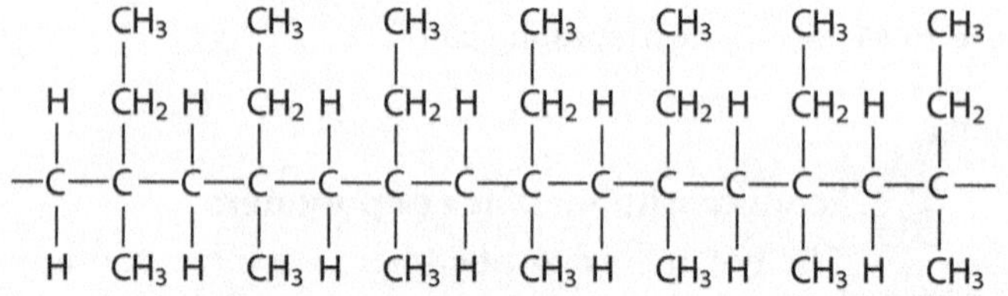

Learning objectives

- Write equations for the complete combustion of alcohols
- Describe the oxidation reactions of alcohols
- Understand that carboxylic acids can react with alcohols to form esters

Exam tip
Don't forget the word 'acidified' when describing the oxidising agent.

The potassium ions are not important in these reactions – the $Cr_2O_7^{2-}$ and MnO_4^- ions are the important parts – the sodium salts would work equally well.

10.2.3 Reactions of compounds containing oxygen

Reactions of alcohols

Combustion of alcohols

In a plentiful supply of oxygen, alcohols burn to produce carbon dioxide and water:

$$C_2H_5OH + 3O_2 \rightarrow 2CO_2 + 3H_2O$$

$$2C_3H_7OH + 9O_2 \rightarrow 6CO_2 + 8H_2O$$

Exam tip
When balancing these equations, don't forget the oxygen in the alcohol!

Oxidation

Alcohols can be classified as primary, secondary or tertiary (page **439**) according to the number of carbon atoms joined to the carbon attached to the OH group.

Primary and secondary alcohols can be oxidised using an oxidising agent such as acidified potassium dichromate(VI) ($K_2Cr_2O_7/H^+$) or acidified potassium manganate(VII) ($KMnO_4/H^+$).

Primary alcohols

Primary alcohols are oxidised, first to an aldehyde (partial oxidation). Then the aldehyde is oxidised further to a carboxylic acid (complete oxidation).

$$\text{primary alcohol} \xrightarrow[\text{heat}]{Cr_2O_7^{2-}/H^+} \text{aldehyde} \xrightarrow[\text{heat}]{Cr_2O_7^{2-}/H^+} \text{carboxylic acid}$$

For example:

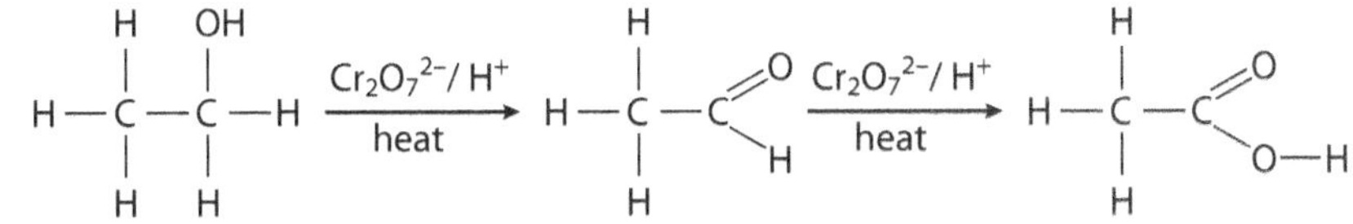

ethanol   ethanal   ethanoic acid

Each reaction can be shown as a balanced equation using [O] to represent oxygen from the oxidising agent:

$$CH_3CH_2OH + [O] \rightarrow CH_3CHO + H_2O$$

$$CH_3CHO + [O] \rightarrow CH_3COOH$$

However, if a fully balanced equation for the oxidation of an alcohol using acidified dichromate(VI), is required we must consider the half-equations involved. $Cr_2O_7{}^{2-}$ (orange) is the oxidising agent and is reduced during the reaction (to the green Cr^{3+}(aq) ion). The half-equation for the reduction is:

$$Cr_2O_7{}^{2-} + 14H^+ + 6e^- \rightarrow 2Cr^{3+} + 7H_2O$$
orange   green

Colour change: orange → green

In the first reaction, the oxidation half-equation is:

$$CH_3CH_2OH \rightarrow CH_3CHO + 2H^+ + 2e^-$$

The two half-equations can be combined by multiplying the oxidation half-equation by three to generate $6e^-$ and then adding them together. When H^+ ions are cancelled from both sides the overall equation obtained is:

$$Cr_2O_7{}^{2-} + 8H^+ + 3CH_3CH_2OH \rightarrow 3CH_3CHO + 2Cr^{3+} + 7H_2O$$

The half-equation for the reduction of manganate(VII) ions (purple) to colourless Mn^{2+}(aq) ions is:

$$MnO_4{}^- + 8H^+ + 5e^- \rightarrow Mn^{2+} + 4H_2O$$
purple   colourless

Colour change: **purple** → colourless

The half-equation for the oxidation of ethanal to ethanoic acid is:

$$CH_3CHO + H_2O \rightarrow CH_3COOH + 2H^+ + 2e^-$$

To get the overall redox equation for the oxidation of ethanal by acidified manganate(VII) we need to balance the electrons – so the oxidation half-equation is multiplied by five and the reduction half-equation by two:

$$2MnO_4{}^- + 5CH_3CHO + 6H^+ \rightarrow 5CH_3COOH + 2Mn^{2+} + 3H_2O$$

Aldehydes contain the

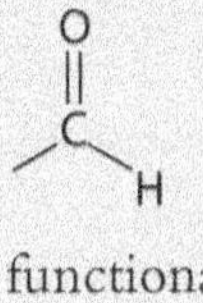

functional group.

Carboxylic acids have the

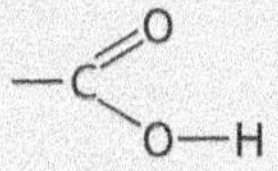

functional group.

Exam tip
The half-equations for the reduction of $Cr_2O_7{}^{2-}/H^+$ and $MnO_4{}^-/H^+$ are given in the IB Chemistry data booklet.

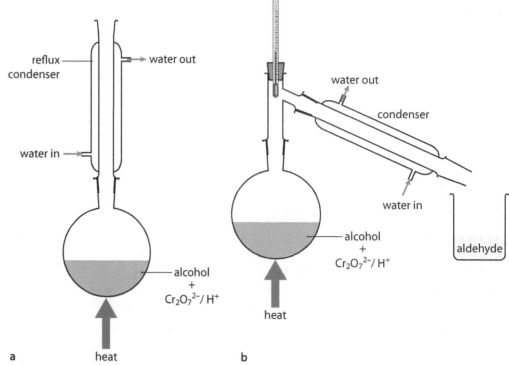

Figure 10.43 a Apparatus for reflux – this produces a carboxylic acid (complete oxidation); **b** apparatus for distillation – this produces an aldehyde (partial oxidation).

If a reaction mixture containing ethanol and acidified potassium dichromate(VI) is heated under **reflux**, ethanoic acid is obtained as the main product and the aldehyde is not usually isolated. However, it is possible to set up the apparatus so that the aldehyde is **distilled** off as soon as it is formed and before it can be oxidised further (Figure **10.43**). This technique works because aldehydes have lower boiling points than the equivalent alcohols (and all other components of the reaction mixture) because they do not have a hydrogen atom attached directly to an oxygen atom, and therefore there is no hydrogen bonding between molecules.

If we look at the oxidation of primary alcohols in terms of changes to the functional groups, it is easier to generalise the reaction to other compounds:

Consider the oxidation of 2-methylpropan-1-ol:

The only changes involve the groups highlighted in red – the rest of the molecule is unchanged.

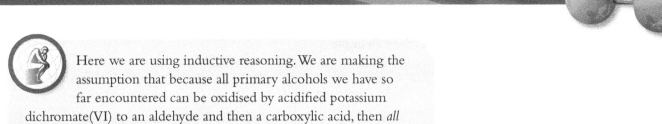

Here we are using inductive reasoning. We are making the assumption that because all primary alcohols we have so far encountered can be oxidised by acidified potassium dichromate(VI) to an aldehyde and then a carboxylic acid, then *all* primary alcohols will be oxidised in this way. Can we ever know this?

Secondary alcohols

Secondary alcohols are also oxidised by heating with acidified potassium dichromate(VI) (or acidified potassium manganate(VII)). They are oxidised to **ketones**, which cannot be oxidised any further.

$$\text{secondary alcohol} \xrightarrow[\text{heat}]{Cr_2O_7{}^{2-}/H^+} \text{ketone} \xcancel{\xrightarrow[\text{heat}]{Cr_2O_7{}^{2-}/H^+}}$$

For example:

$$
\underset{\text{propan-2-ol}}{
\begin{array}{ccc}
 & \text{OH} & \\
\text{H} & | & \text{H} \\
| & & | \\
\text{H}-\text{C}-\text{C}-\text{C}-\text{H} \\
| & | & | \\
\text{H} & \text{H} & \text{H}
\end{array}}
\xrightarrow[\text{heat}]{Cr_2O_7{}^{2-}/H^+}
\underset{\text{propanone}}{
\begin{array}{ccc}
 & \text{O} & \\
\text{H} & \| & \text{H} \\
| & & | \\
\text{H}-\text{C}-\text{C}-\text{C}-\text{H} \\
| & & | \\
\text{H} & & \text{H}
\end{array}}
$$

The balanced equation for the reaction is:

$$CH_3CH(OH)CH_3 + [O] \rightarrow (CH_3)_2CO + H_2O$$

or

$$Cr_2O_7{}^{2-} + 8H^+ + 3CH_3CH(OH)CH_3 \rightarrow 3(CH_3)_2CO + 2Cr^{3+} + 7H_2O$$

In terms of changes to the functional group, the reaction can be represented as:

$$
\begin{array}{c}
\text{OH} \\
| \\
\text{C}-\text{C}-\text{C} \\
| \\
\text{H}
\end{array}
\xrightarrow[\text{heat}]{Cr_2O_7{}^{2-}/H^+}
\begin{array}{c}
\text{O} \\
\| \\
\text{C}-\text{C}-\text{C}
\end{array}
$$

Looking at the reaction of another secondary alcohol:

$$
\underset{\text{3-methylbutan-2-ol}}{
\begin{array}{cccc}
\text{H} & \text{H} & \text{CH}_3 & \text{H} \\
| & | & | & | \\
\text{H}-\text{C}-\text{C}-\text{C}-\text{C}-\text{H} \\
| & | & | & | \\
\text{H} & \text{O} & \text{H} & \text{H} \\
 & | \\
 & \text{H}
\end{array}}
\xrightarrow[\text{heat}]{Cr_2O_7{}^{2-}/H^+}
\underset{\text{3-methylbutanone}}{
\begin{array}{cccc}
\text{H} & & \text{CH}_3 & \text{H} \\
| & & | & | \\
\text{H}-\text{C}-\text{C}-\text{C}-\text{C}-\text{H} \\
| & \| & | & | \\
\text{H} & \text{O} & \text{H} & \text{H}
\end{array}}
$$

The only change is to the group highlighted in red – the rest of the molecule is unchanged.

Ketones contain the

$$
\begin{array}{c}
\text{O} \\
\| \\
\text{C}-\text{C}-\text{C}
\end{array}
$$

functional group.

[O] represents oxygen from the oxidising agent.

Tertiary alcohols

Tertiary alcohols are resistant to oxidation.

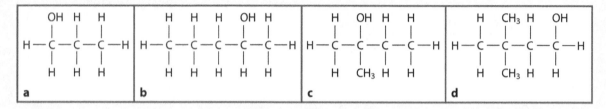

2-methylpropan-2-ol

25 Write a balanced equation for the complete combustion of butan-1-ol.

26 Name the following alcohols and classify each as primary, secondary or tertiary:

a	b	c	d

27 For each of the alcohols in question **26**, give the structural formula of the product (if any) of complete oxidation when they are heated with acidified potassium dichromate(VI).

28 Give the structure of the organic product formed when each of the following alcohols is heated with acidified potassium dichromate(VI) with the apparatus set up for distillation:

  a 3-methylbutan-1-ol

  b 2-methylpentan-3-ol

  c 2,3-dimethylpentan-1-ol

Esters

When an alcohol is heated with a carboxylic acid in the presence of a small amount of concentrated sulfuric acid as a catalyst, an ester is formed. The reaction is called **esterification**:

$$\text{alcohol} + \text{carboxylic acid} \underset{\text{heat}}{\overset{\text{conc. } H_2SO_4}{\rightleftharpoons}} \text{ester} + \text{water}$$

For example:

$$\text{ethanol} + \text{ethanoic acid} \underset{\text{heat}}{\overset{\text{conc. } H_2SO_4}{\rightleftharpoons}} \text{ethyl ethanoate} + \text{water}$$

Figure **10.44** shows the reaction more clearly using structural formulas.

Figure 10.44 An esterification reaction.

The alcohol and the carboxylic acid have been joined together and water has been eliminated – one H atom from the alcohol joins with the –OH group from the carboxylic acid.

The product formed when any alcohol and any carboxylic acid come together can be worked out simply by putting the alcohol and carboxylic acid together (Figure **10.45**) so that the two O–H groups are next to each other – the H from the OH of the alcohol along with the OH of the carboxylic acid are removed and the O of the alcohol is joined to the C=O group of the carboxylic acid.

Other examples of esterification reactions are shown in Figure **10.46**.

The reaction between an alcohol and a carboxylic acid can be described as a **nucleophilic substitution** reaction. The alcohol acts as a **nucleophile** and substitutes (replaces) the OH group of the carboxylic acid.

The esterification reaction can also be classified as a **condensation reaction** – a reaction in which two molecules join together with the elimination of the elements of water.

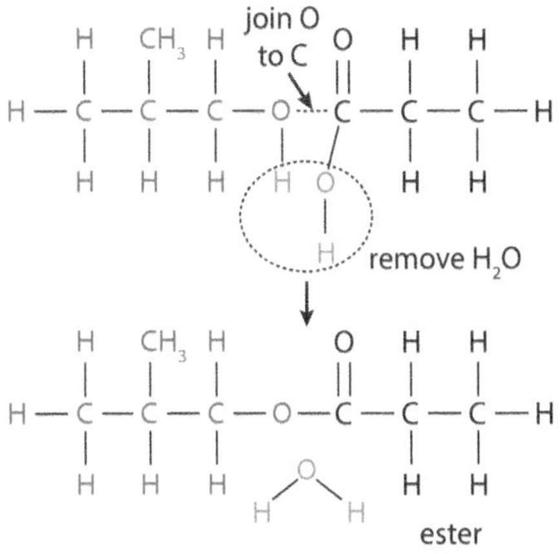

Figure 10.45 The esterification reaction dissected.

Nucleophilic substitution reactions are covered in more detail in the following pages.

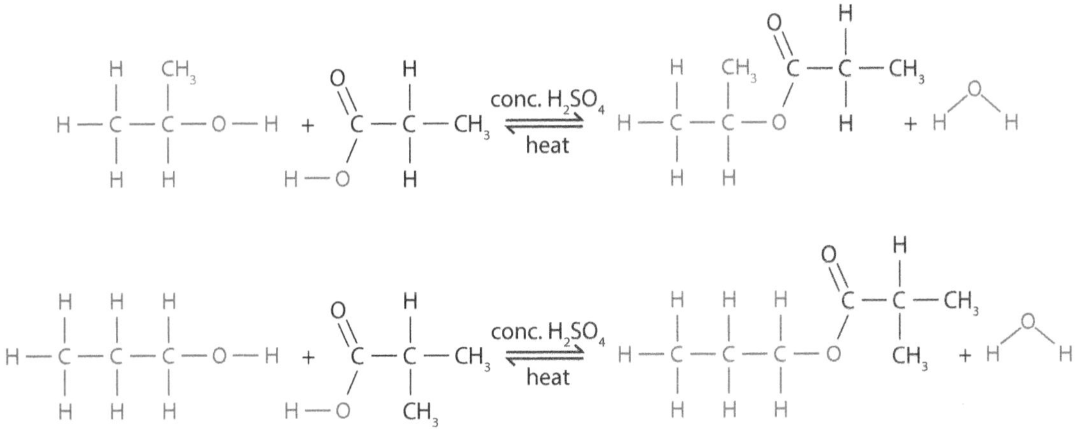

Figure 10.46 Typical esterification reactions.

29 Copy and complete the following table:

	Alcohol	Carboxylic acid	Ester
a	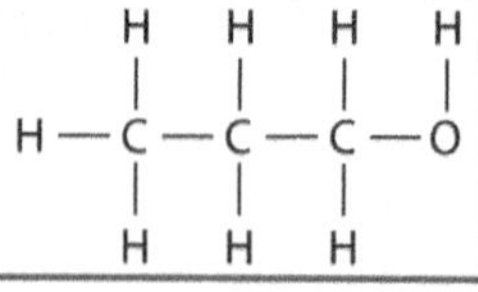 H₃C–CH₂–CH₂–O... (H—C—C—C—O, with H's)	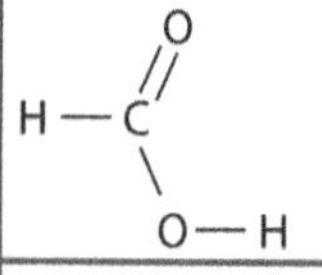 H—C(=O)—O—H	
b	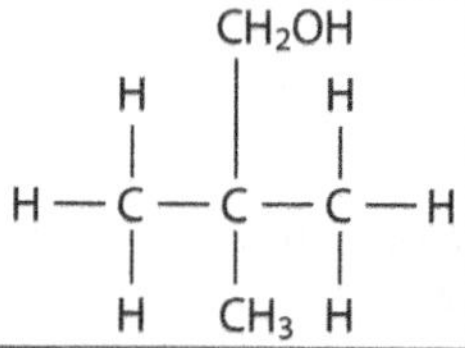 CH₂OH group; H—C—C—C—H with CH₃	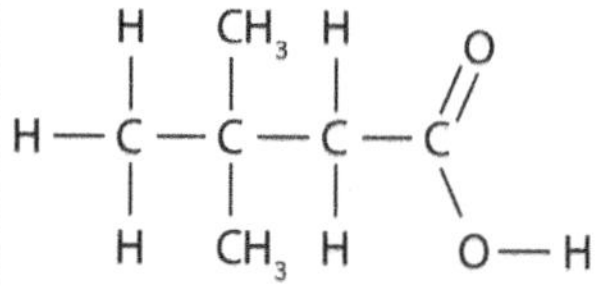 H—C—C—C—C(=O)—O—H with CH₃ groups	
c			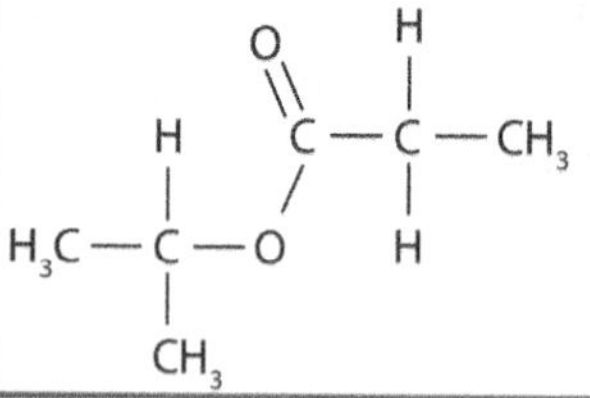 O=C–C(H)(CH₃); H₃C—C—O, CH₃
d			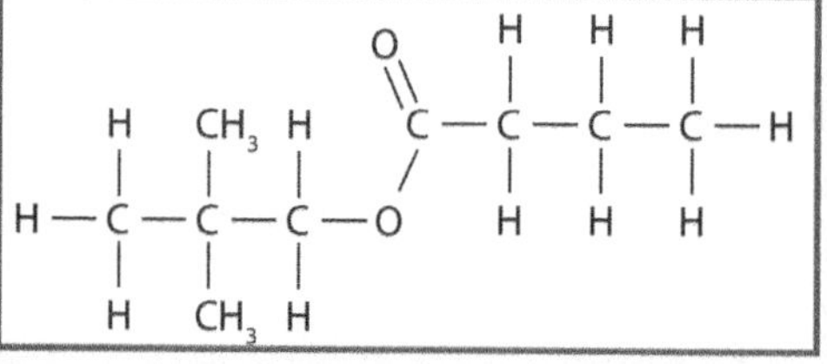 structure with CH₃ groups and C—C—C—C—H chain

30 Name the ester formed in question **29a**.

31 Write a balanced chemical equation for the reaction occurring in question **29a**.

Learning objectives

- Describe the reaction of halogenoalkanes with aqueous sodium hydroxide
- Understand that halogenoalkanes undergo nucleophilic substitution reactions

Substitution reaction: one atom or group is replaced by another atom or group.

10.2.4 Halogenoalkanes

Nucleophilic substitution reactions

Halogenoalkanes usually undergo **substitution reactions**. Figure **10.47** shows what happens when, for example, 1-bromopropane is heated with aqueous sodium hydroxide – the Br atom is replaced by an OH group to form propan-1-ol:

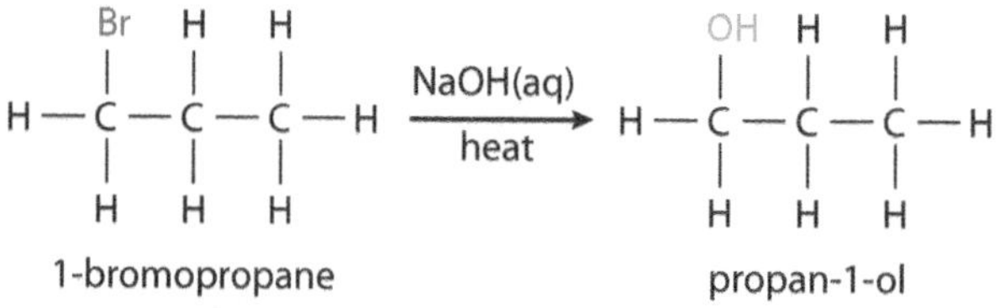

$$\text{1-bromopropane} \xrightarrow[\text{heat}]{\text{NaOH(aq)}} \text{propan-1-ol}$$

Figure 10.47 A nucleophilic substitution reaction.

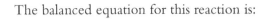

The balanced equation for this reaction is:

$$CH_3CH_2CH_2Br + NaOH \rightarrow CH_3CH_2CH_2OH + NaBr$$

or, as an ionic equation:

$$CH_3CH_2CH_2Br + OH^- \rightarrow CH_3CH_2CH_2OH + Br^-$$

This reaction is also known as **hydrolysis** because it essentially equivalent to breaking molecules apart using water.

The type of mechanism by which this reaction occurs is **nucleophilic substitution** – the halogenoalkane is attacked by a **nucleophile** and the nucleophile replaces the halogen atom. It is covered in more detail in the Higher Level section on the next page.

Halogenoalkanes can also react with ammonia to form amines:

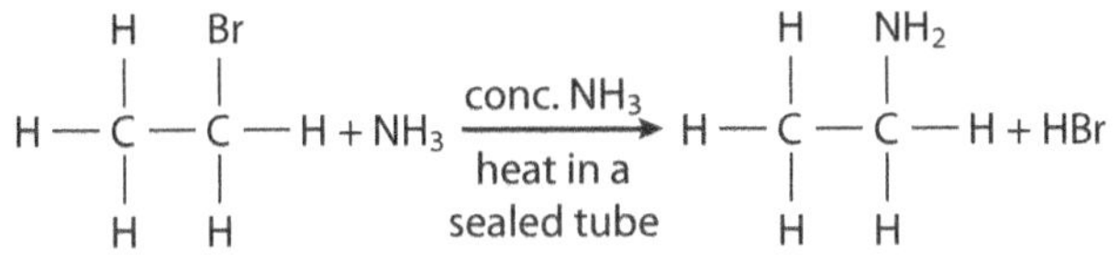

H — C — C — H + NH₃ (conc. NH₃ heat in a sealed tube) H — C — C — H + HBr

ethylamine

… and with potassium cyanide to form nitriles:

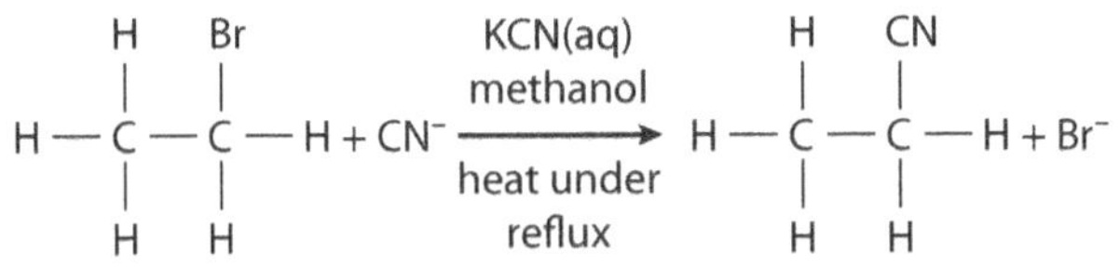

H — C — C — H + CN⁻ (KCN(aq) methanol heat under reflux) H — C — C — H + Br⁻

propanenitrile

Note that the nitrile contains the –CN functional group with a triple bond between the carbon atom and the nitrogen atom. The longest continuous carbon chain in the molecule is now three C atoms, and so the name is **propane**nitrile. This makes this type of reaction useful because it provides a way of increasing the length of the carbon chain.

We have seen in this section that halogenoalkanes are more reactive than alkanes – this is due to the presence of the halogen atom. For instance, if a chlorine atom (very electronegative) is present in a molecule, the C–Cl bond will be polar and a nucleophile will be attracted to the $\delta+$ carbon atom. Another factor is that Cl⁻ is classified as a 'good leaving group' – the C–Cl bond is weaker than a C–C bond or a C–H bond and Cl⁻ is stable in many solvents.

Nature of science

Data is extremely important in science. The collection of large amounts of data about functional group interconversions has been fundamental in developing synthetic routes to making vast numbers of key organic compounds.

A **nucleophile** is a molecule or negatively charged ion that has a lone pair of electrons – it is attracted to a relatively highly positively charged region in a molecule (a region with lower electron density) and donates a lone pair of electrons to form a covalent bond.

There are various ways of naming amines and the amine shown here could also be called ethanamine or aminoethane.

10.3 Types of organic reactions (HL)

10.3.1 Nucleophilic substitution mechanisms

The exact nature of a nucleophilic substitution reaction depends on which type of halogenoalkane is involved in the reaction – primary, secondary or tertiary.

Primary halogenoalkanes – the S_N2 mechanism

Consider the reaction of bromoethane with sodium hydroxide:

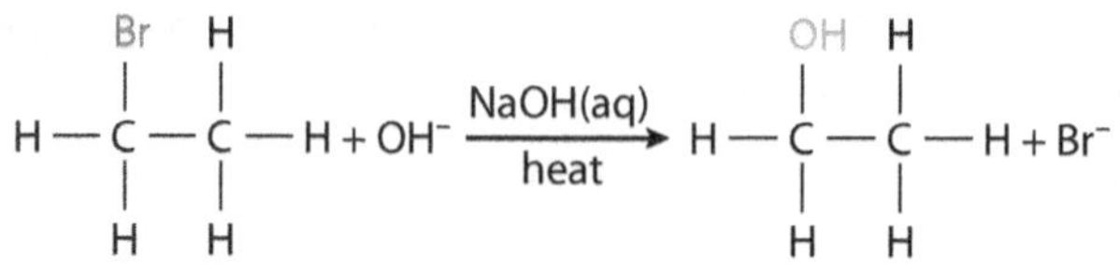

$$H-\underset{\underset{H}{|}}{\overset{\overset{Br}{|}}{C}}-\underset{\underset{H}{|}}{\overset{\overset{H}{|}}{C}}-H + OH^- \xrightarrow[\text{heat}]{\text{NaOH(aq)}} H-\underset{\underset{H}{|}}{\overset{\overset{OH}{|}}{C}}-\underset{\underset{H}{|}}{\overset{\overset{H}{|}}{C}}-H + Br^-$$

Br is more electronegative than C, so the carbon atom is slightly positive ($\delta+$) and the OH^- ion is attracted to it. The hydroxide ion is a **nucleophile** and donates a lone pair of electrons to the carbon atom to form a (coordinate) covalent bond (it acts as a Lewis base) – see Figure **10.48**. The curly arrow shows the lone pair on OH^- becoming a bonding pair of electrons between the oxygen and carbon. A carbon atom can have a maximum of eight electrons in its outer shell, and therefore as the C–O bond forms, the C–Br bond must break.

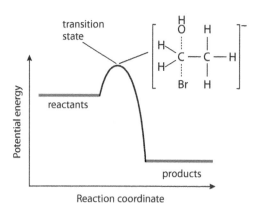

Figure 10.48 The nucleophile (OH^-) attacks.

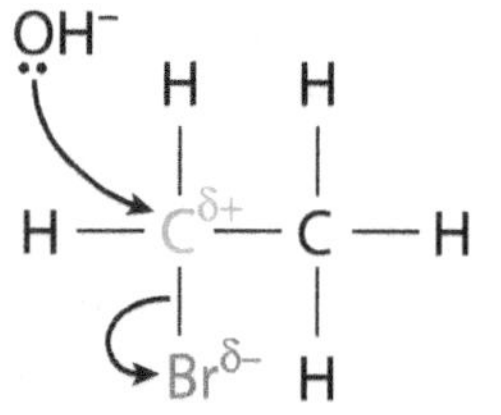

The arrows are called **curly arrows** and represent the movement of a pair of electrons.

The C–Br bond breaks such that both electrons from the C–Br bond go back to the bromine atom to form a bromide ion – as shown by the curly arrow from the C-Br bond to the Br. This is called **heterolytic fission** – the bond breaks so that both electrons go to the same atom. (This should be contrasted with the homolytic fission of the Cl–Cl bond during the free radical substitution mechanism described on page **448–449**.)

Part way through this nucleophilic substitution process, both the OH and the bromine are partially joined to the carbon (Figure **10.49**), and this is often shown in the mechanism. This is not an intermediate in the reaction but is the highest point on the potential energy profile – often called the **transition state** or **activated complex**.

Figure **10.50** shows the full mechanism for the reaction of bromoethane with aqueous sodium hydroxide.

Figure 10.49 Potential energy profile of nucleophilic substitution.

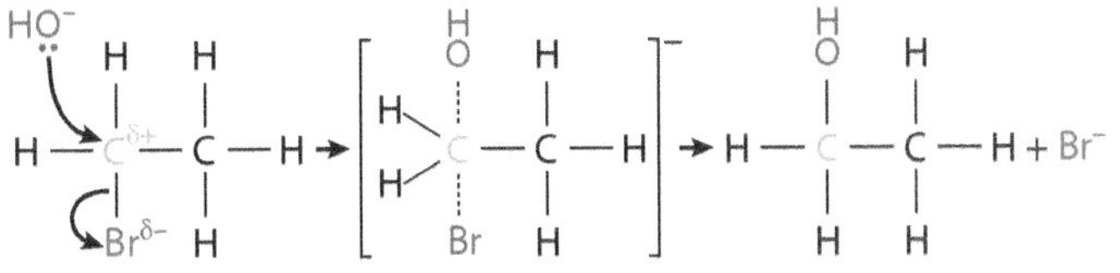

Figure 10.50 A typical S_N2 reaction mechanism.

Exam tip

In the transition state, you must show that the oxygen of the OH is joined to the carbon (OH----C is not correct).

This is a single-step mechanism and the formation of the transition state is just part of the way along this step. The mechanism could be shown just as well without this, as in Figure **10.51**.

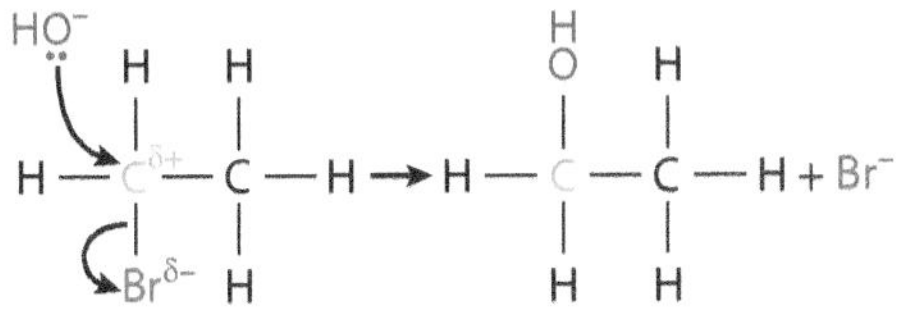

Figure 10.51 A simplified version of the S_N2 mechanism.

Exam tip

Note that the transition state must be shown in an answer to an exam question.

S_N2 is the main mechanism for substitution in reactions of primary halogenoalkanes. The 'shorthand' is explained in Figure **10.52**. The '2' indicates the molecularity of the reaction – this was introduced in Topic **6**. It refers to the number of molecules (or ions) that take part in a particular step (usually the rate-determining step) in a mechanism.

The rate-determining step is the slowest step and, therefore, the step that governs the overall rate of reaction. The S_N2 mechanism occurs in a single step, so this must be the rate-determining step. The rate equation for the above reaction is:

$$\text{rate} = k[CH_3CH_2Br][OH^-]$$

This means that if the concentration of either bromoethane or sodium hydroxide is doubled, the rate of reaction is doubled.

In the S_N2 mechanism the OH^- ion approaches the $C^{\delta+}$ atom from the opposite side to the Br, and this results in inversion of the tetrahedral configuration of the central carbon atom – like an umbrella turning inside out (Figure **10.53**).

This reaction is described as *stereospecific* because the stereochemistry (arrangement of groups around a chiral centre) of the reactant determines the stereochemistry of the product. If a nucleophile attacks a chiral centre, the reaction will occur with inversion of configuration. This will be discussed further in the section on optical isomers on page **495**.

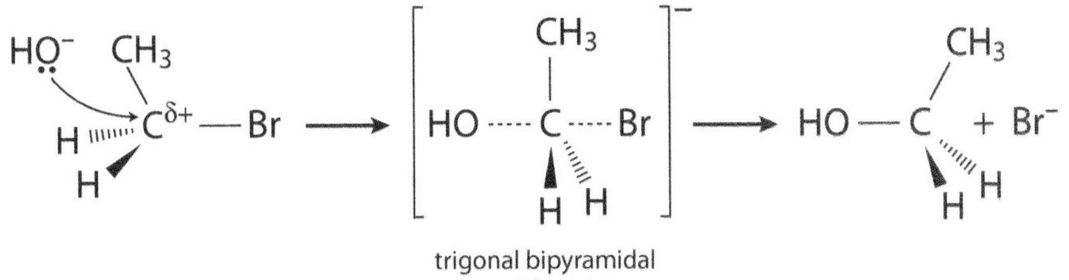

bimolecular – two molecules involved in a particular step, usually the rate-determining step.

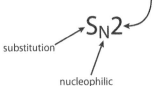

substitution
S_N2
nucleophilic

Figure 10.52 What S_N2 means.

Molecularity and order are not the same thing. 'Order' is an experimentally determined quantity that relates the concentrations of reactants to the rate. 'Molecularity' refers to the number of molecules (or ions) that take part in a particular step (usually the rate-determining step) in a mechanism.

trigonal bipyramidal transition state

Figure 10.53 A change in spatial configuration.

Tertiary halogenoalkanes

Consider the reaction of 2-bromo-2-methylpropane with aqueous sodium hydroxide. This occurs fairly rapidly at room temperature:

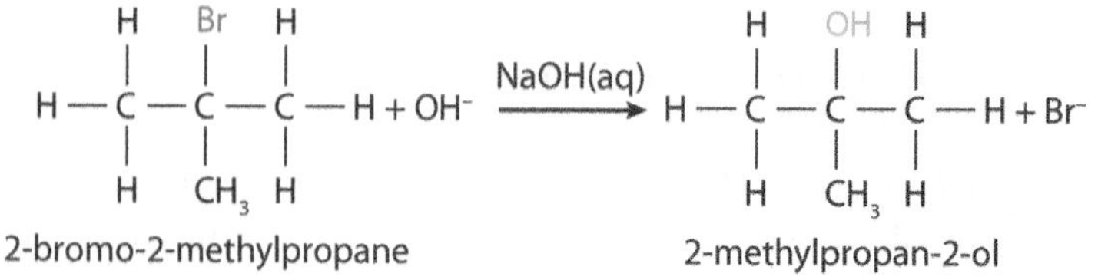

2-bromo-2-methylpropane → 2-methylpropan-2-ol

Studies of the rate of this reaction have determined that the rate does not depend on the concentration of the sodium hydroxide, unlike in the S_N2 mechanism discussed earlier. This reaction must, therefore, have a different mechanism.

In the first step, the C–Br bond breaks to form a positively charged carbocation (Figure 10.54). This is the rate-determining step in the mechanism.

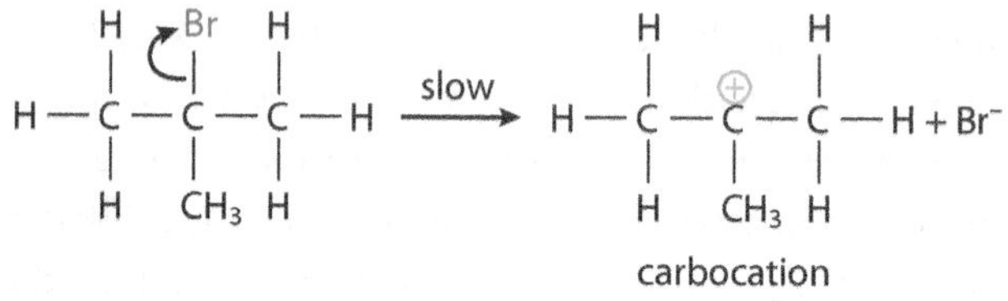

Figure 10.54 An intermediate species is formed.

Once formed, the carbocation is open to attack by nucleophiles, such as the OH⁻ ion (Figure 10.55).

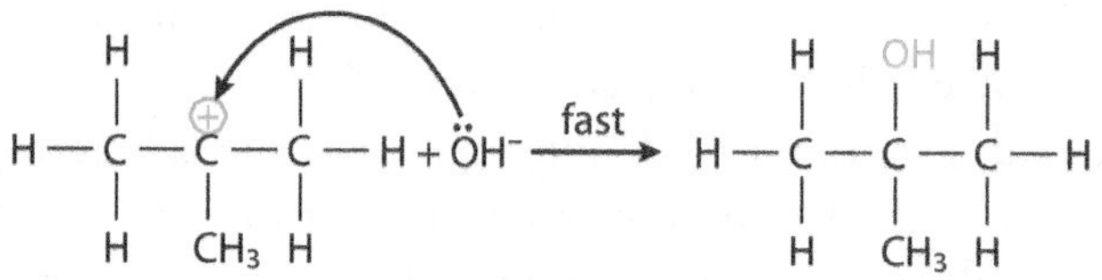

Figure 10.55 The intermediate is attacked by a nucleophile rapidly.

This mechanism is described as S_N1 – 'substitution nucleophilic unimolecular' because there is only one molecule (the halogenoalkane) involved in the rate-determining step.

The nucleophile is only involved in the mechanism in a fast step that happens after the rate-determining step and so does not appear in the rate equation. The rate equation for the above reaction is:

$$\text{rate} = k[(CH_3)_3CBr]$$

The effects of changing the concentration of a nucleophile on the rate of S_N1 and S_N2 mechanisms are summarised in Table 10.20.

Mechanism	Rate equation	Effect of doubling the concentration of the nucleophile
S_N1	rate = k[halogenoalkane]	No effect – nucleophile is only involved in a fast step after the rate-determining step.
S_N2	rate = k[halogenoalkane][nucleophile]	Rate doubles – nucleophile is involved in the rate-determining step.

Table 10.20 The effect of changing the concentration of the nucleophile on the rate of S_N1 and S_N2 mechanism reactions.

In the first step of the S_N1 mechanism, the carbocation formed is planar (Figure **10.56**) and so the nucleophile is equally likely to attack the central carbon from either side in the second step.

This reaction is not stereospecific – the stereochemistry of the final product is random. A racemic mixture would be formed if the original halogenoalkane were optically active (see page **499**).

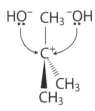

Figure 10.56 The nucleophile is equally likely to attack from either side.

The effect of the halogen on the rate of nucleophilic substitution

The rate of nucleophilic substitution of halogenoalkanes by hydroxide ions is fastest with iodo-compounds – the order of decreasing rate is R–I > R–Br > R–Cl > R–F.

Both S_N2 and S_N1 mechanisms involve the C–X bond breaking in the rate-determining step. The C–I bond is easiest to break (Table **10.21**), so the reaction will be fastest using iodoalkanes.

This factor is much more important than any effects due to electronegativity differences – the C–X bond is least polar in iodoalkanes, so it might be expected that the carbon atom should attract a nucleophile least strongly.

Bond	Bond energy / kJ mol^{-1}
C–F	484
C–Cl	338
C–Br	276
C–I	238

Table 10.21 C–halogen bond enthalpies.

The effect of the nucleophile on the rate of nucleophilic substitution

S_N2 reactions are generally faster when ions are involved rather than neutral molecules. OH$^-$ ions will react faster than H$_2$O molecules because the higher negative charge on the O in OH$^-$ means it will be more strongly attracted to the C$^{\delta+}$ atom in the halogenoalkane.

The rate of an S_N1 reaction is not affected by changing the nucleophile because the nucleophile attacks only *after* the rate-determining step.

Why two different mechanisms?

For primary halogenoalkanes, the dominant mechanism is S_N2, whereas for tertiary halogenoalkanes it is S_N1. Secondary halogenoalkanes undergo nucleophilic substitution via a mixture of the two mechanisms – the more dominant mechanism will depend on the specific conditions of the reaction.

Why is S_N2 more favourable for primary halogenoalkanes than for tertiary halogenoalkanes?

This is mainly because of **steric effects**. The alkyl groups surrounding the central carbon in a tertiary halogenoalkane make it much more difficult

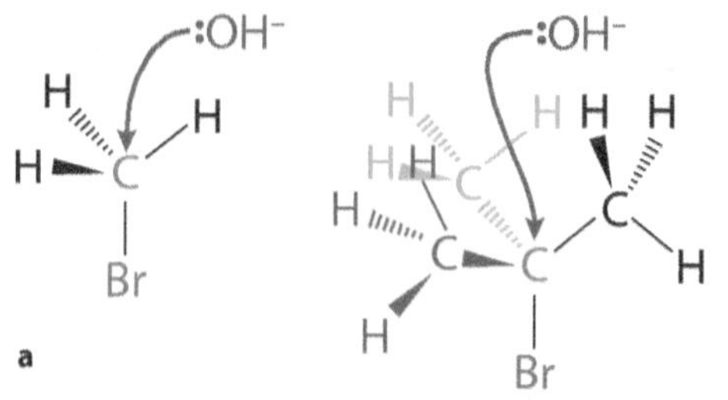

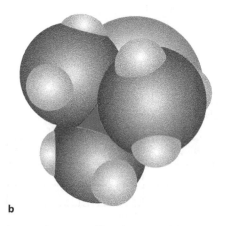

b

Figure 10.57 a Attack of OH⁻ on a primary and a tertiary halogenoalkane – steric effects are to do with the size/bulkiness of groups; **b** A space-filling model of 2-bromo-2-methylpropane – the nucleophile must attack the area highlighted pink.

- The more highly substituted the C, the slower the rate of S_N2.
- The more highly substituted the C, the faster the rate of S_N1.

Extension

The electron-releasing effect of alkyl groups can be understood in terms of donation of electron density from a σ bond of the carbon atom attached to the C⁺ into the vacant p orbital on the C⁺. This is called σ conjugation, or hyperconjugation.

for the nucleophile to get in to attack the central carbon atom as shown in Figure **10.57a**. This can also be seen if we look at the space-filling model of 2-bromo-2-methylpropane in Figure **10.57b**.

Another way of explaining this is that if the OH⁻ did get in to attack the carbon, there would be five large groups surrounding the central carbon atom in the transition state – this would lead to instability and the activation energy would be very high.

So the overall rate of reaction for S_N2 reactions of halogenoalkanes is: primary > secondary > tertiary.

Why is S_N1 more favourable for tertiary halogenoalkanes?

This is because of the stability of the intermediate carbocation. Methyl groups (and alkyl groups in general) have an **electron-releasing** effect (also called a positive inductive effect) so that they are able to stabilise a positively charged carbon atom to which they are bonded. The order of stability of carbocations is shown in Figure **10.58**.

This electron-releasing effect of alkyl groups means that the charge is spread out most in the tertiary carbocation – it has most alkyl groups around the positively charged carbon atom. This stabilises the ion, making it more likely to be formed. So the overall rate of reaction for S_N1 reactions of halogenoalkanes is: tertiary > secondary > primary.

$\overset{H\cdots}{\underset{H}{C^+}}-CH_3$ $\overset{H\cdots}{\underset{H_3C}{C^+}}-CH_3$ $\overset{H_3C\cdots}{\underset{H_3C}{C^+}}-CH_3$

primary carbocation   secondary carbocation   tertiary carbocation

increasing stability of carbocation →

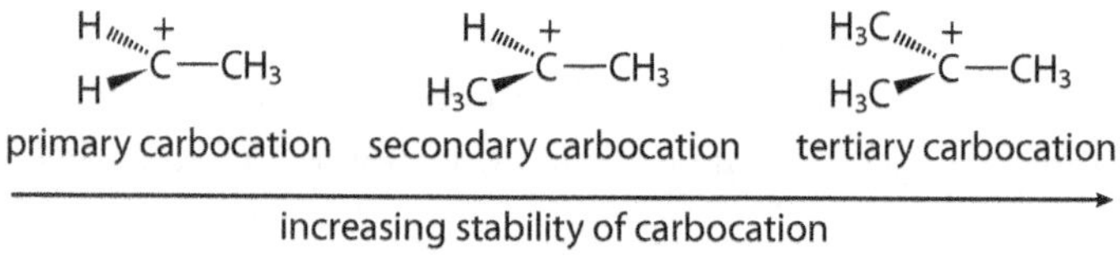

Figure 10.58 The order of stability of carbocations.

> In the reactions of aqueous sodium hydroxide with halogenoalkanes, the S_N1 reactions of tertiary halogenoalkanes are faster than S_N2 reactions of primary halogenoalkanes.

Extension

It is very difficult to draw an overall conclusion about the effect of the structure of a halogenoalkane (primary, secondary, tertiary) on the rate of nucleophilic substitution because it depends on so many factors. Measurement of the rate of reaction of a series of related compounds with dilute aqueous ethanolic sodium hydroxide by the pioneer in this field, Christopher Ingold, produced the order $(CH_3)_3CBr > CH_3Br > CH_3CH_2Br > (CH_3)_2CHBr$. However, it is also possible to carry out simple laboratory experiments, by adding silver nitrate solution to a series of isomeric primary, secondary and tertiary halogenoalkanes dissolved in ethanol, that produces the order tertiary > secondary > primary.

Worked example

10.1 Arrange the following in order of rate of nucleophilic substitution by aqueous sodium hydroxide:

$$CH_3CH_2Cl, \; CH_3CH_2Br, \; (CH_3)_3CBr, \; (CH_3)_3CCl.$$

The first two are primary halogenoalkanes and will react via an S_N2 mechanism; the final two are tertiary halogenoalkanes and will react via an S_N1 mechanism.

Tertiary halogenoalkanes react more quickly than primary halogenoalkanes.

A C–Br bond is weaker than a C–Cl bond and is therefore broken more easily meaning that, within each type of halogenoalkane, bromo-compounds react faster.

The overall order is $(CH_3)_3CBr > (CH_3)_3CCl > CH_3CH_2Br > CH_3CH_2Cl$.

The effect of the solvent on the rate of nucleophilic substitution

The nature of the solvent can have a big effect on the rate of nucleophilic substitution reactions. Solvents can be divided into two broad classes – non-polar and polar. Non-polar solvents include hydrocarbons such as hexane (C_6H_{14}) and tetrachloromethane (CCl_4). Polar solvents can be further subdivided into protic and aprotic solvents:

- **Protic** solvents have a hydrogen atom joined to N or O and can participate in hydrogen bonding – examples are water, ethanol (CH_3CH_2OH) and ammonia.
- **Aprotic** polar solvents do not have a hydrogen atom joined to N or O and cannot participate in hydrogen bonding – examples are propanone, [$(CH_3)_2CO$] and dimethyl sulfoxide (DMSO), $(CH_3)_2SO$. These solvents do not act as proton donors.

Protic polar solvents are able to solvate both negative ions and positive ions in solution. Negative ions are solvated by interaction (Figure **10.59**) with the H atoms that are attached to O or N using hydrogen bonding.

Aprotic polar solvents are good at solvating positive ions but the absence of very positive hydrogen atoms means that they are not very good at solvating negative ions.

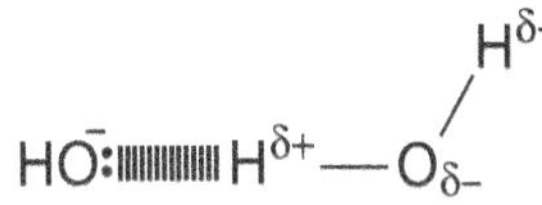

Figure 10.59 Negative ions (OH^-) are solvated by hydrogen bonding interactions with protic solvents.

- S_N1 reactions are favoured by protic, polar solvents.
- S_N2 reactions are favoured by aprotic, polar solvents.

Two ions are formed in the rate-determining step of an S_N1 reaction and these are stabilised by the presence of a protic solvent, which is able to solvate both ions effectively. The rate of an S_N1 reaction would be decreased by switching from a protic solvent to an aprotic polar solvent – the aprotic solvent is not as good at solvating the negative ion formed in the rate-determining step.

The rate-determining step of an S_N2 mechanism involves the attack of a negative ion on the halogenoalkane. In the presence of a protic solvent, this negative ion will be surrounded by solvent molecules and will not be a very effective nucleophile – it is being kept away from the $\delta+$ carbon atom by the solvent molecules that are surrounding it. In an aprotic polar solvent, the nucleophile is not surrounded by solvent molecules and is free to attack the $\delta+$ carbon atom. The rate of an S_N2 reaction would be decreased by switching from an aprotic polar solvent to a protic polar solvent – solvation by the protic solvent makes the nucleophile less effective.

When the nucleophile is a negative ion, the S_N2 mechanism is favoured by switching to a non-polar solvent. This does not solvate the nucleophile and also, because the negative charge is more spread out in the transition state than in the original nucleophile, the starting materials are destabilised more than the transition state by the presence of a non-polar solvent and the activation energy is lowered, speeding up the reaction. However, when choosing a solvent for a reaction, other factors must also be taken into account and the solvent must be able to dissolve the halogenoalkane *and* the nucleophile.

Nature of science

Experimental data have to be collected to allow the development of theories about reaction mechanisms. Determination of the rate equation and looking at the stereochemistry of reactants and products are both important in providing information to propose a mechanism for nucleophilic substitution.

The ideal situation for carrying out scientific experiments is to change just one variable while controlling all others. In order to determine the effect of the nucleophile, the solvent etc. on the rate of nucleophilic substitution, ideally only one factor should be changed each time. The limitations of the data must be considered when trying to draw general conclusions because all the measurements may not have been made under the same conditions. For instance, it is not easy to answer the question 'Is the S_N1 or S_N2 mechanism faster?' because several factors influence the reaction rate.

? Test yourself

32 State whether each of the halogenoalkanes below reacts with NaOH(aq) via an S_N1 mechanism or an S_N2 mechanism.

33 Give the structural formula of the products formed when each of the substances in question 32 reacts with NaOH(aq).

10.3.2 Electrophilic addition reactions of alkenes

The addition reactions of alkenes were covered earlier (page **451–452**). The mechanism for these reactions is **electrophilic addition**.

> An **electrophile** is an electron-deficient species (a positively charged ion or the positive end of a dipole) that is attracted to regions of relatively high electron density and accepts a pair of electrons to form a covalent bond – electrophiles are Lewis acids (page **310**).

Learning objectives

- Explain what is meant by an electrophile
- Explain the electrophilic addition reactions of alkenes
- Predict and explain the formation of the major product when a hydrogen halide reacts with an unsymmetrical alkene

The C=C bond has four electrons and is a region of relatively high electron density in a molecule. Electrophiles will be attracted to this.

The reaction of ethene with hydrogen bromide

The overall equation is shown in Figure **10.60**.

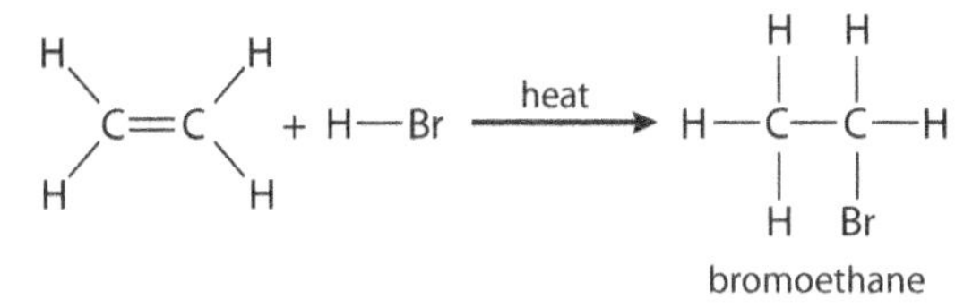

Figure 10.60 Addition to an alkene.

The mechanism is shown in Figure **10.61**.

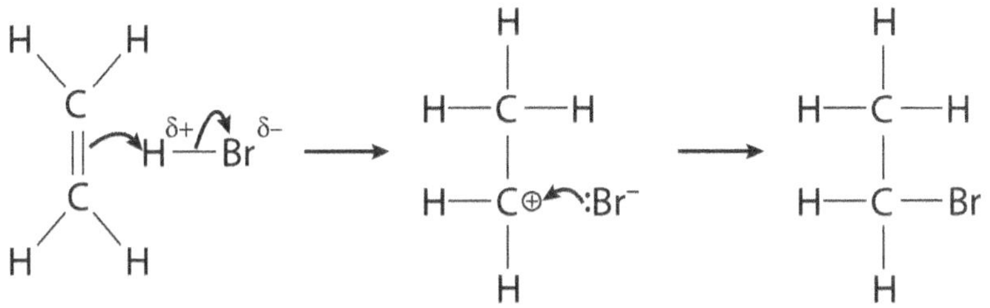

Figure 10.61 The mechanism of the reaction of ethene with hydrogen bromide.

H–Br is polar – as shown by $H^{\delta+}$ and $Br^{\delta-}$. The hydrogen atom is attracted to the high electron density region in the C=C bond.

The curly arrow from the π component of the C=C bond to $H^{\delta+}$ represents the movement of a pair of electrons.

In the first stage, a pair of electrons is donated from the π bond in C=C to form a bond between C1 (Figure **10.62**) and the hydrogen atom of the H–Br molecule. At the same time, the H–Br bond breaks and the pair of electrons from this bond goes to a bromine atom.

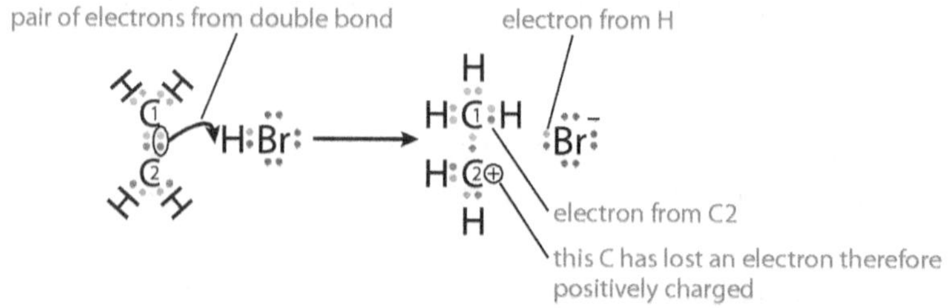

Figure 10.62 Movement of electrons in reaction of ethene with hydrogen bromide.

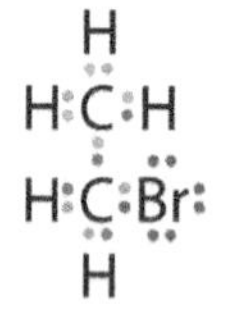

The alkene can be classified as a Lewis base (electron pair donor) and the hydrogen bromide as a Lewis acid (electron pair acceptor).

C2 has lost an electron (this electron is now in the C–H bond) and so C2 has a positive charge. The intermediate formed is called a **carbocation** (a positively charged organic ion with a positive charge on a carbon atom). The bromine atom gains an electron from the H–Br bond and therefore becomes a negatively charged bromide ion. In the second stage, the Br⁻ ion is attracted to the C⊕ atom and donates a pair of electrons to form a bond (Lewis acid–base reaction). The electron distribution in the final product is shown in Figure **10.63**.

The initial type of bond breaking, when the H–Br bond breaks, is called **heterolytic fission** because the bond breaks so that both electrons go to the same atom (Figure **10.64**).

This should be compared with homolytic fission of the Cl–Cl bond in the free radical substitution mechanism described on page **448**.

Figure 10.63 The location of the original electrons in the final product, bromoethane.

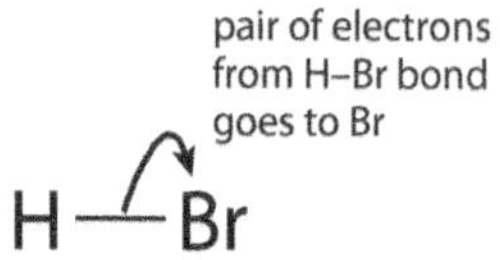

Figure 10.64 Heterolytic fission.

Reaction of propene with HBr

When HBr adds to propene, two products are possible – shown in Figure **10.65**.

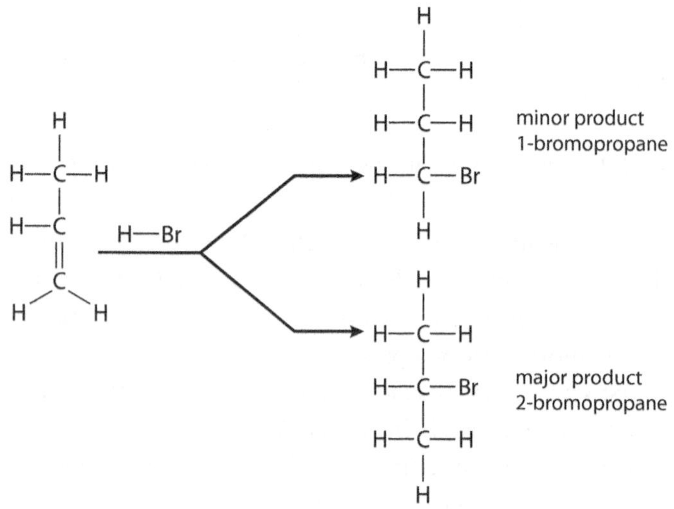

Figure 10.65 Addition of hydrogen bromide to propene.

The major product (the minor product is usually made in only very small amounts) can be predicted using **Markovnikov's rule**.

> **Markovnikov's rule:** when H–X adds across the double bond of an alkene, the H atom becomes attached to the C atom that has the larger number of H atoms already attached.

Markovnikov's rule can be used to predict the major product formed when H–X adds to any unsymmetrical alkene. When H–Cl adds to 2-methylpropene, the H atom of the HCl becomes attached to C1 (Figure **10.66**) because this has two hydrogen atoms already attached, rather than C2, which has no hydrogen atoms attached.

The explanation for Markovnikov's rule involves the stability of the intermediate carbocation.

The two possible carbocations that can be formed when propene reacts with HBr are shown in Figure **10.67**.

A secondary carbocation is more stable than a primary carbocation. This is because of the **electron-releasing effect** of alkyl groups (also called a **positive inductive effect**). There are two alkyl groups next to the positively charged carbon in the secondary carbocation in Figure **10.67** but only one in the primary carbocation. Two alkyl groups reduce the positive charge on the carbon more than one alkyl group and stabilise the ion more. Because the secondary carbocation is more stable, it is more likely to be formed.

The electron-releasing effect depends very little on the size of an alkyl group – a methyl group donates approximately the same amount of electron density as an ethyl group.

In general, the more alkyl groups there are attached to C⊕, the more stable the carbocation (Figure **10.68**).

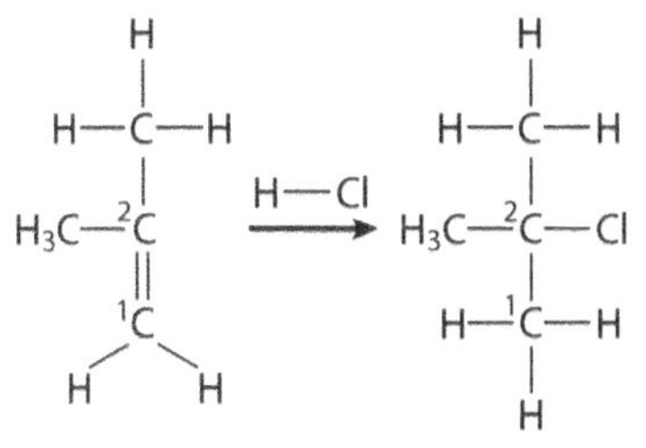

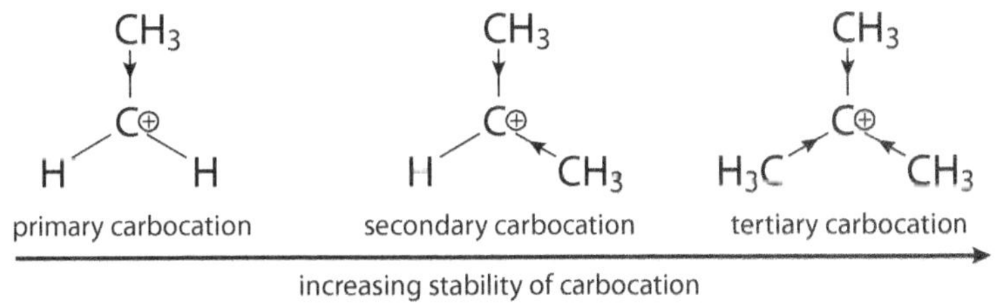

Figure 10.68 Relative stability of carbocations.

Exam tip
When asked to predict and explain the major product when a hydrogen halide adds to an alkene, always explain it in terms of the stability of the intermediate carbocation – it is never enough to just say 'Markovnikov's rule'. You must also make sure that you include a comparison – there are *more* electron-releasing alkyl groups around the positively charged carbon in a secondary carbocation than in a primary carbocation.

Markovnikov's rule could be paraphrased as 'Those who have shall get more'.

Figure 10.66 Predicting the major product of electrophilic addition.

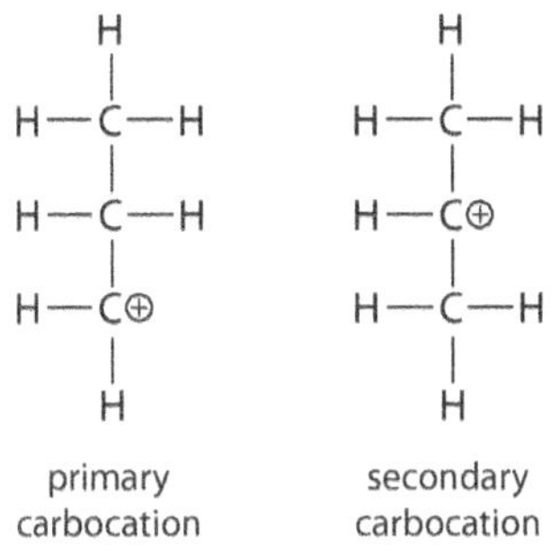

Figure 10.67 Stability of carbocation intermediates.

- A primary carbocation has one carbon atom attached to the C with the positive charge.
- A secondary carbocation has two carbon atoms attached to the C with the positive charge.

Markovnikov's rule can also be used to explain the products obtained when interhalogen compounds react with propene or other unsymmetrical alkenes. The less electronegative halogen atom behaves like hydrogen and, in the major product, becomes attached to the carbon atom that has more hydrogen atoms attached.

Figure **10.69** shows the products of the reaction of Br–Cl with propene.

Bromine is less electronegative than chlorine and becomes attached to C1, which has two hydrogen atoms attached, rather than C2, which only has one hydrogen atom attached.

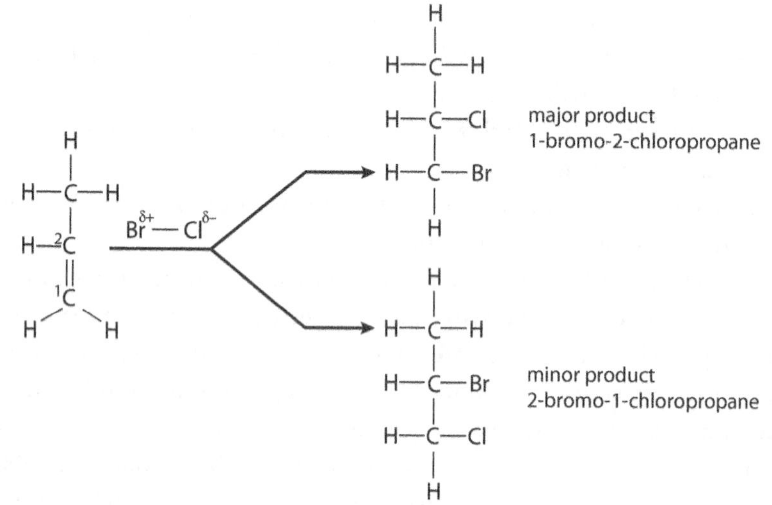

Figure 10.69 Addition of an interhalogen compound to propene.

Reaction of ethene with bromine

The reaction is summarised in Figure **10.70**.

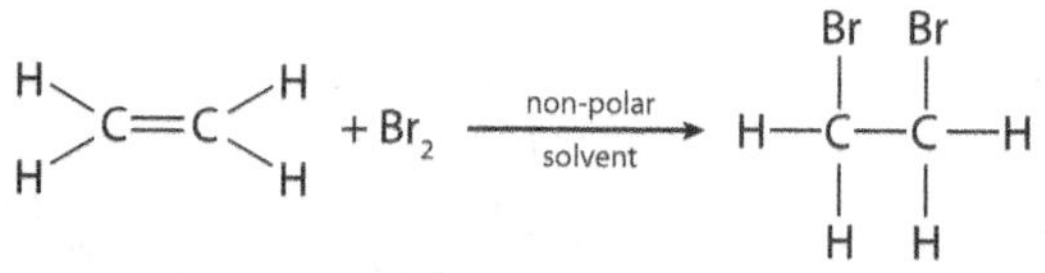

Figure 10.70 The addition reaction of ethene and bromine.

Although bromine molecules are not polar, as one approaches the high electron density in the C=C double bond it becomes polarised (there is an induced dipole). Electrons are repelled in the Br_2 molecule so that the bromine atom closest to the C=C bond has a slight positive charge and the bromine atom further away has a slight negative charge.

The mechanism can be shown in two different ways (Figures **10.71** and **10.72**). The first is exactly analogous to the reaction of HBr with ethene – via a carbocation intermediate.

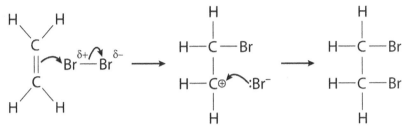

Figure 10.71 A mechanism for the reaction of ethene with bromine.

Exam tip
You need to learn only one of these mechanisms.

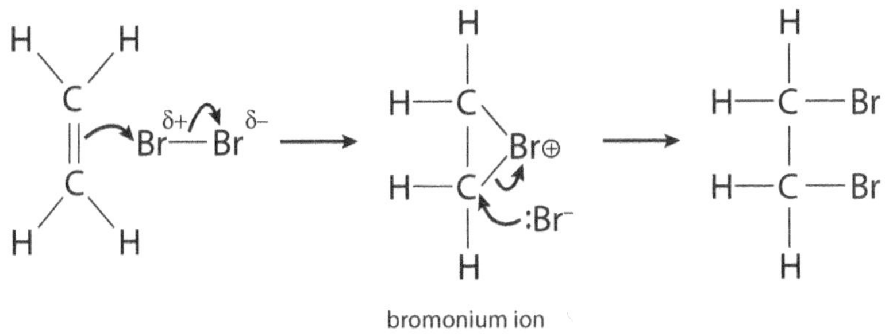

bromonium ion

Figure 10.72 Another mechanism for the reaction of ethene with bromine showing the formation of a bromonium ion.

However, the mechanism is probably better represented as shown in Figure **10.72**, where a **bromonium ion** rather than a carbocation is formed. In the bromonium ion, the bridging bromine has donated a lone pair to the C⊕, which stabilises the ion.

Nature of science

Nobel-prize-winning physicist Richard Feynman once said, 'I learned very early the difference between knowing the name of something and knowing something'. Science is about more than knowing the name of something – it is about understanding. There is a lot more to Markovnikov's rule than just knowing the name.

? Test yourself

34 State the names of the major products formed when the following alkenes react with HCl:
   a but–1–ene
   b 3–methylbut–1–ene
   c 2,4–dimethylpent–2–ene
35 State the name of the major product formed when but–1–ene reacts with ClBr.

The structure of benzene

The structure of benzene can be described as a resonance hybrid with equal contributions from the two structures (Figure **10.73**).

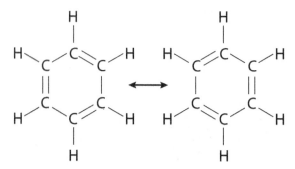

Figure 10.73 Benzene is a resonance hybrid of these two structures.

Because the π component of each double bond is shared between two C–C bonds atoms, the carbon–carbon bond order in benzene is 1.5.

The structure is much better explained by the idea of delocalisation (Figure **10.74**). Instead of the p orbitals just overlapping side-on between adjacent C atoms to give three π bonds, delocalisation can occur over the whole structure.

Benzene therefore has a π delocalised ring of electrons which extends all around the ring of carbon atoms. The structure of benzene is usually drawn as a hexagon, with a ring in the middle indicating the π delocalised system:

The π delocalised system, containing six electrons, represents a region of high electron density and, therefore, benzene is susceptible to attack by electrophiles.

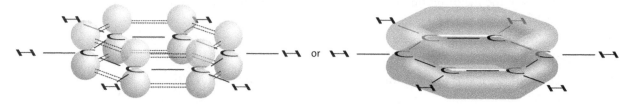

Figure 10.74 Delocalisation in benzene.

The nitration of benzene

When benzene is heated with a mixture of concentrated nitric and sulfuric acids, nitrobenzene is formed:

$$C_6H_6 + HNO_3 \rightarrow C_6H_5NO_2 + H_2O$$

Figure **10.75** shows the reaction in more detail.

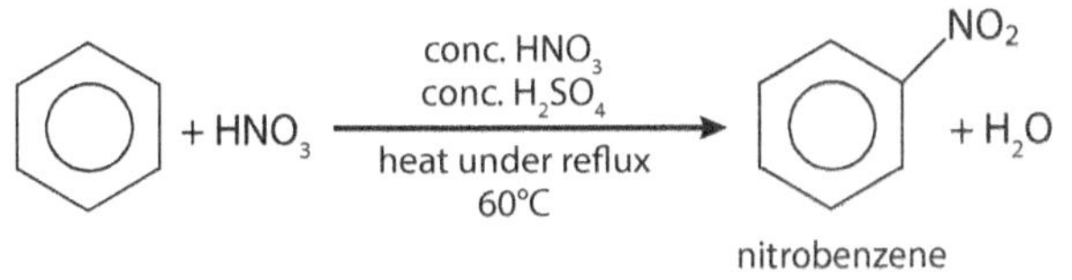

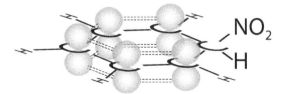

Figure 10.75 Nitration of benzene.

> This reaction occurs via **electrophilic substitution**.

The first stage in the reaction is the formation of the **electrophile** – this is the nitronium ion, NO_2^+, which is formed when concentrated sulfuric acid reacts with concentrated nitric acid:

$$HNO_3 + 2H_2SO_4 \rightarrow NO_2^+ + H_3O^+ + 2HSO_4^-$$

A simpler version of this equation is:
$$HNO_3 + H_2SO_4 \rightarrow NO_2^+ + H_2O + HSO_4^-$$

The sulfuric acid (the stronger acid) protonates the –OH group in nitric acid and the N–O bond breaks to form a water molecule.

The second stage starts with a collision between a benzene molecule and the electrophile (Figure **10.76**).

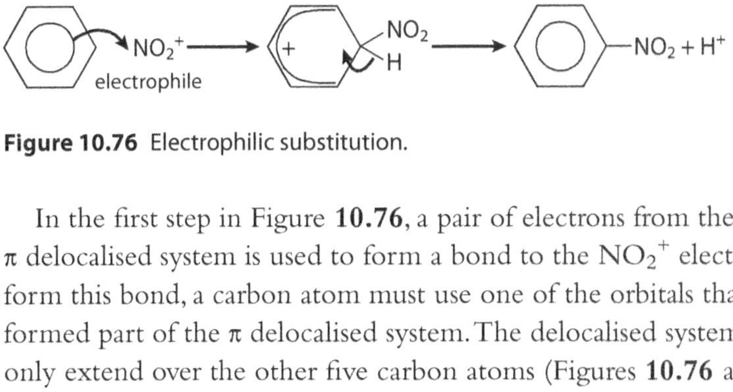

Figure 10.76 Electrophilic substitution.

Exam tip
The curly arrow in the first step in Figure **10.76** must come *from* the delocalised ring; and the partial ring in the second step must extend over five carbon atoms.

In the first step in Figure **10.76**, a pair of electrons from the benzene π delocalised system is used to form a bond to the NO_2^+ electrophile. To form this bond, a carbon atom must use one of the orbitals that previously formed part of the π delocalised system. The delocalised system can then only extend over the other five carbon atoms (Figures **10.76** and **10.77**) – this is shown by the partial ring.

Exam tip
The initial attack of the NO_2^+ on a benzene molecule can be classified as a Lewis acid–base reaction (page **310**). The NO_2^+ ion is an electron pair acceptor (Lewis acid) and the benzene molecule is an electron pair donor (Lewis base).

Figure 10.77 Partial delocalisation in the intermediate in electrophilic substitution.

Another way of thinking about this is that the initial reaction involved a neutral species reacting with a positively charged species – therefore the intermediate must also have a positive charge.

There were originally six electrons in the π delocalised system (one from each carbon atom) but two are now used to form the new C–N bond, which leaves only four electrons shared over five C atoms; so the ring has a positive charge.

In the second step in Figure **10.76** (and also in Figure **10.78**), the C–H bond breaks. This releases an orbital and a pair of electrons to restore the π delocalised system.

The second stage can also be shown slightly differently with the HSO_4^- ion (formed when the electrophile was formed) removing the H^+ ion (Figure **10.78**). The H_2SO_4 is re-formed and can be regarded as a catalyst in this reaction.

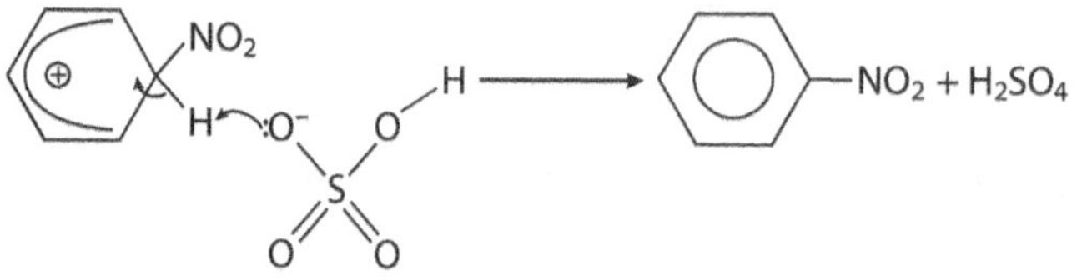

Figure 10.78 HSO_4^- removes H^+ from the intermediate to form nitrobenzene.

Nature of science

Science often involves the collection of large amounts of data and it can often prove useful to classify the data into different categories to make the data more manageable – for instance, classification of organic reactions according to type or mechanism or organic substances in terms of functional groups.

? Test yourself

36 Methylbenzene reacts in the same way as benzene with a mixture of concentrated nitric acid and concentrated sulfuric acid. Write a balanced equation, using condensed structural formulas for the reaction.

37 When methylbenzene reacts with a mixture of concentrated nitric acid and concentrated sulfuric acid, substitution occurs on C4 (with the methyl group on C1). Write the mechanism for this reaction.

10.3.4 Reduction reactions

Reduction of aldehydes, ketones and carboxylic acids

The oxidation reactions of alcohols, aldehydes and ketones that we met earlier in this topic (Subtopic **10.2.3**) can be made to work in the opposite direction by using reduction reactions. Common reducing agents are lithium aluminium hydride ($LiAlH_4$) and sodium borohydride ($NaBH_4$). $LiAlH_4$ is the stronger reducing agent and can be used to reduce carboxylic acids, aldehydes and ketones but $NaBH_4$ is only strong enough to reduce aldehydes and ketones. However, $NaBH_4$ is much easier to use than $LiAlH_4$ and so is usually preferred for the reduction of aldehydes and ketones.

Learning objectives

- Describe the reduction reactions of aldehydes, ketones and carboxylic acids to form alcohols
- Describe the reduction of nitrobenzene to phenylamine (aniline)

Aldehydes

The basic reaction is:

$$\text{aldehyde} \xrightarrow[\text{methanol}]{NaBH_4} \text{primary alcohol}$$

Figure **10.79** shows the reduction of ethanal.

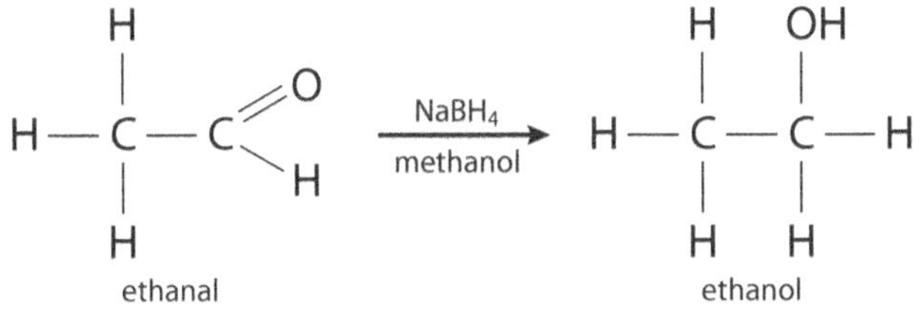

Figure 10.79 Reduction of ethanal to ethanol.

A balanced equation for the reaction can be written using [H] to represent hydrogen from the reducing agent:

$$CH_3CHO + 2[H] \rightarrow CH_3CH_2OH$$

Ketones

The basic reaction is:

$$\text{ketone} \xrightarrow[\text{methanol}]{NaBH_4} \text{secondary alcohol}$$

Figure **10.80** shows the reduction of propanone.

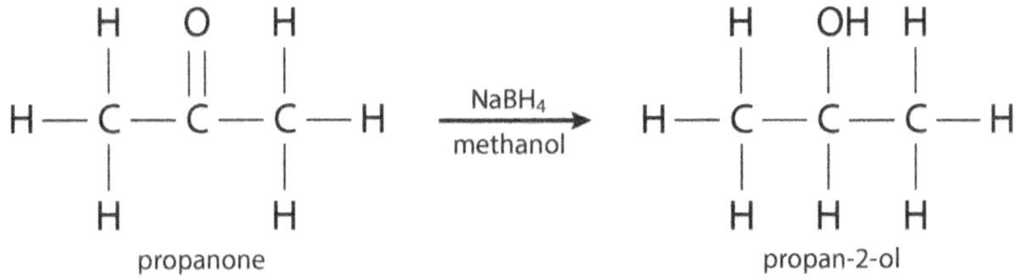

Figure 10.80 Reduction of propanone to propan-2-ol.

Carboxylic acids

A stronger reducing agent (LiAlH$_4$) must be used for the reduction of carboxylic acids. This means that the reduction cannot be stopped at the aldehyde stage and that carboxylic acids are reduced to primary alcohols. The basic reaction is:

$$\text{carboxylic acid} \xrightarrow[\textbf{(ii)}\ \text{H}^+/\text{H}_2\text{O}]{\textbf{(i)}\ \text{LiAlH}_4\ \text{in ethoxyethane}} \text{primary alcohol}$$

Figure **10.81** shows the reduction of ethanoic acid.

ethanoic acid — (i) LiAlH$_4$ in dry ethoxyethane / (ii) H$^+$/H$_2$O(aq) → ethanol

Figure 10.81 Reduction of ethanoic acid to ethanol.

This can also be written as a balanced equation using [H] to represent hydrogen from the reducing agent:

$$\text{CH}_3\text{COOH} + 4[\text{H}] \rightarrow \text{CH}_3\text{CH}_2\text{OH} + \text{H}_2\text{O}$$

If you need to make an aldehyde from a carboxylic acid, the carboxylic acid must be reduced to a primary alcohol using LiAlH$_4$ and then the primary alcohol must be oxidised back to an aldehyde as described on page **458** (partial oxidation with distillation).

Reduction of nitrobenzene

Nitrobenzene can be reduced to phenylamine (aniline) by heating with a reducing agent such as a mixture of tin and concentrated hydrochloric acid. This produces the protonated form of the amine C$_6$H$_5$NH$_3^+$ from which the free amine can be liberated (Figure **10.82**) by treatment with a stronger base such as sodium hydroxide.

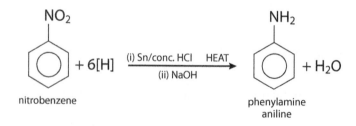

nitrobenzene + 6[H] — (i) Sn/conc. HCl HEAT / (ii) NaOH → phenylamine aniline + H$_2$O

Figure 10.82 Reduction of nitrobenzene to phenylamine (aniline).

> **Exam tip**
> The syllabus states that you should be able to write reactions for the reduction of carboxylic acids to aldehydes, so this is worth bearing in mind in the examination.

> 'Aniline' is the old (non-systematic) name for phenylamine.

> The A in 'BASF', the largest chemical company in the world, stands for 'anilin' (German for aniline).

The first step could be written as:

$$C_6H_5NO_2 + H^+ + 6[H] \rightarrow C_6H_5NH_3^+ + 2H_2O$$

and the second as:

$$C_6H_5NH_3^+ + OH^- \rightarrow C_6H_5NH_2 + H_2O$$

Other reducing agents, such as hydrogen in the presence of a palladium or platinum catalyst, can also be used.

Extension

The product at the end of the first step is actually $(C_6H_5NH_3^+)_2[SnCl_6]^{2-}$, which reacts with sodium hydroxide according to the equation:

$$(C_6H_5NH_3^+)_2[SnCl_6]^{2-} + 8NaOH \rightarrow$$
$$2C_6H_5NH_2 + Na_2SnO_3 + 6NaCl + 5H_2O$$

Nature of science

There is not really a 'scientific method' and many great discoveries have been accidental in the sense that scientists did not start out to make the thing they eventually discovered. There is, however, nothing accidental about the observational skills of the scientists involved and their skill in recognising the fact that something unusual and interesting had happened.

Phenylamine (aniline) was the chemical that sparked off the artificial dye industry. At the age of 18, William Henry Perkin was already involved in chemical research. He was given the task of making quinine (an antimalarial drug) from aniline and attempted to oxidise it using acidified potassium dichromate(VI). He ended up with a black sludge at the bottom of the flask (like many organic preparations do!). However, when he washed out the flask with ethanol he noticed a purple colour (later named mauveine or aniline purple) and had discovered, by accident, the first artificial organic dye, which was to become the basis of a huge industry. Other substances discovered by accident include PTFE (polytetrafluoroethene – used as non-stick coating) and superglue.

? Test yourself

38 Give the name of the organic product formed when each of the following is reduced using lithium aluminium hydride:
 a propanoic acid
 b 2-methylpentanal
 c pentan-3-one
 d 3-methylbutanoic acid

Learning objectives

- Work out reaction pathways for the formation of organic compounds
- Understand that chemists often adopt a retrosynthetic approach to the synthesis of organic compounds

10.4 Synthetic routes (HL)

Reaction pathways

We can use the reactions shown in Figure **10.83** to design syntheses for organic compounds.

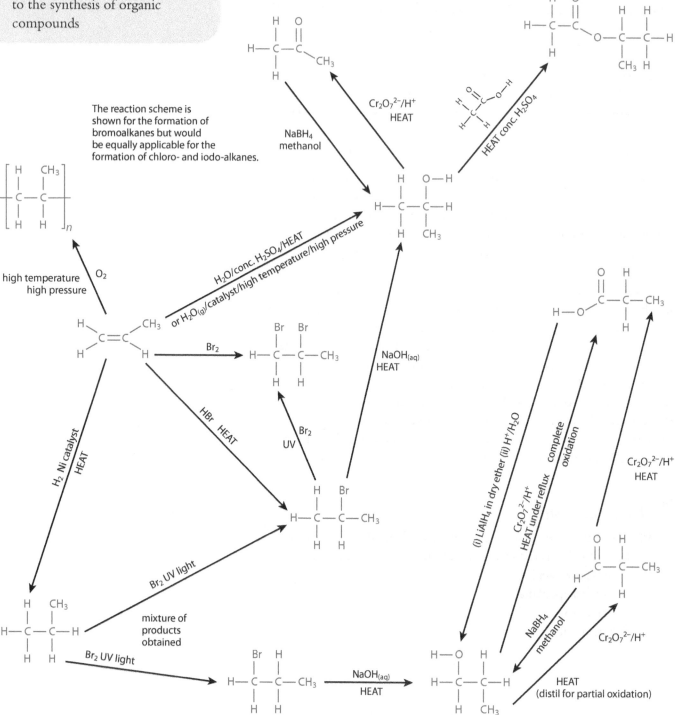

Figure 10.83 Reaction pathways.

The important thing when working out a particular reaction scheme is to concentrate on only the **functional groups** in molecules – the carbon skeleton should make very little difference to the products of a reaction. This means that 1-bromo-4-methylpentane and bromoethane react in basically the same way with aqueous sodium hydroxide to form alcohols:

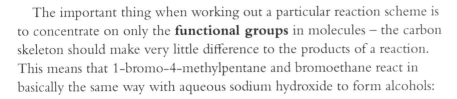

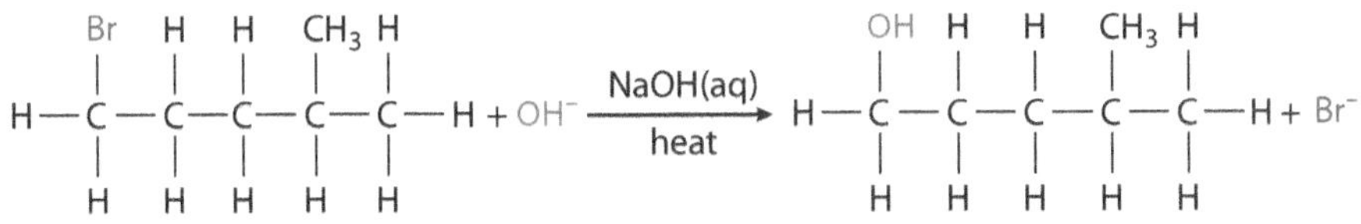

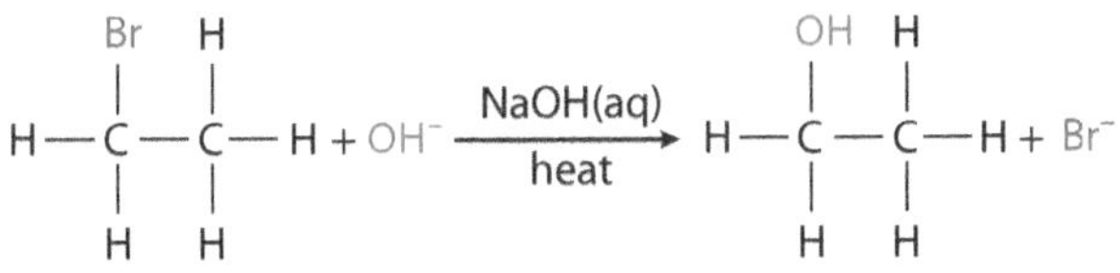

When designing a synthesis for a particular organic compound, chemists often adopt a **retrosynthetic** approach. They start with the target molecule they are trying to make and work backwards, using known reactions, to get to an appropriate and readily available starting material. For example, if it is required to make butanone from an alkane, the logic could go something like this:

- butanone can be made from butan-2-ol in an oxidation reaction
- butan-2-ol can be made from 2-chlorobutane in a (nucleophilic) substitution reaction
- 2-chlorobutane can be made from butane in a (free radical) substitution reaction.

A reaction sequence like that in Figure **10.84** could, therefore, be used to make butanone.

> Note that free radical substitution produces a mixture of products and some 1-chlorobutane will also be formed.

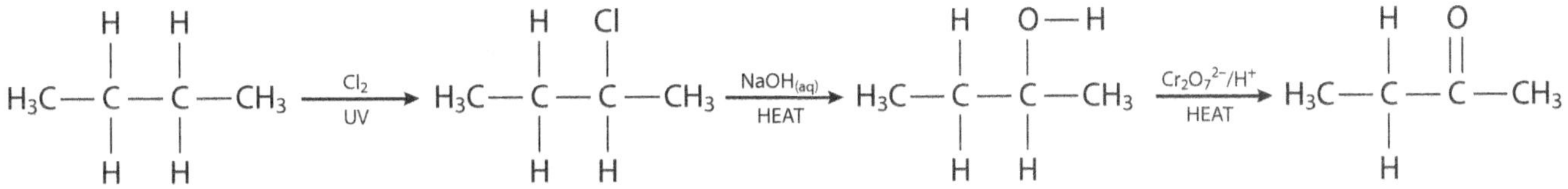

$$\underset{\substack{H\ H\\ \big|\ \big|\\ H_3C-C-C-CH_3\\ \big|\ \big|\\ H\ H}}{} \xrightarrow[\text{UV}]{Cl_2} \underset{\substack{H\ Cl\\ \big|\ \big|\\ H_3C-C-C-CH_3\\ \big|\ \big|\\ H\ H}}{} \xrightarrow[\text{HEAT}]{NaOH_{(aq)}} \underset{\substack{H\ O-H\\ \big|\ \big|\\ H_3C-C-C-CH_3\\ \big|\ \big|\\ H\ H}}{} \xrightarrow[\text{HEAT}]{Cr_2O_7^{2-}/H^+} \underset{\substack{H\ O\\ \big|\ \big\|\\ H_3C-C-C-CH_3\\ \big|\\ H}}{}$$

Figure 10.84 Synthesis of butanone.

This can also be written as a series of balanced equations:

$$CH_3CH_2CH_2CH_3 + Cl_2 \xrightarrow[\text{UV}]{} CH_3CH_2CHClCH_3 + HCl \qquad \text{free radical substitution}$$

$$CH_3CH_2CHClCH_3 + OH^- \xrightarrow[\text{HEAT}]{NaOH(aq)} CH_3CH_2CH(OH)CH_3 + Cl^- \qquad \text{nucleophilic substitution}$$

$$CH_3CH_2CH(OH)CH_3 + [O] \xrightarrow[\text{HEAT}]{Cr_2O_7^{2-}/H^+} CH_3CH_2COCH_3 + H_2O \qquad \text{oxidation}$$

Worked examples

10.2 Design a reaction pathway for the formation of the compound shown below. Use alkenes as the only organic starting materials.

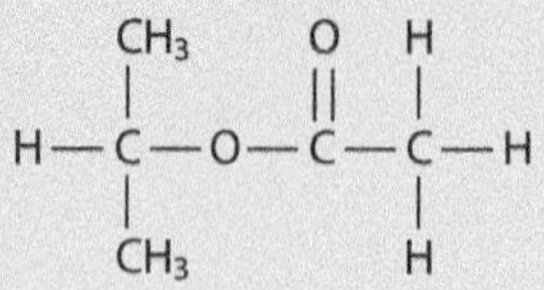

We can recognise this molecule as an ester – and therefore know that it can be made from an alcohol and a carboxylic acid. In the figure below, the right-hand part of the ester (with the C=O group) has two carbon atoms and suggests that it comes from ethanoic acid. So the left-hand part must be from an alcohol – the only alcohol that could give this arrangement of carbon atoms is propan-2-ol.

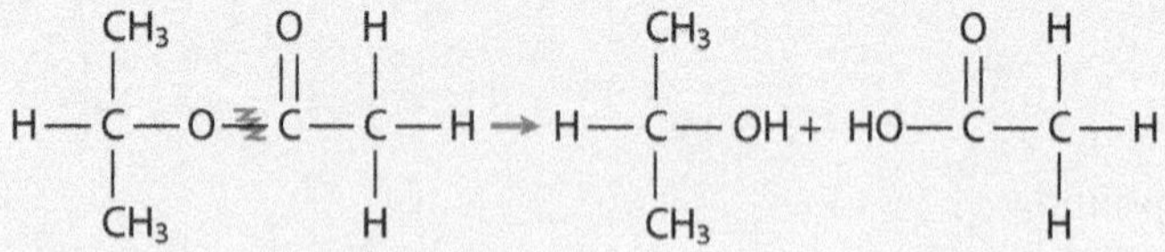

Neither of these two compounds are alkenes. Propan-2-ol can be made by the hydration of propene – so this is one of the starting compounds.

Ethanoic acid can be made using complete oxidation of ethanol, which can be made by hydration of ethene, which is the other starting compound.

The reaction sequence is shown below.

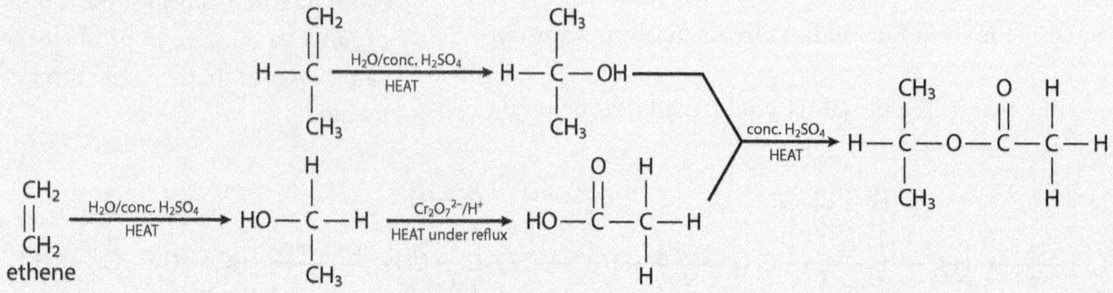

Presented as a series of balanced equations, this looks like:

$$C_2H_4 + H_2O \xrightarrow[\text{heat}]{\text{conc. } H_2SO_4} CH_3CH_2OH$$

$$CH_3CH_2OH + 2[O] \xrightarrow[\text{heat / reflux}]{Cr_2O_7^{2-}/H^+} CH_3COOH + H_2O$$

$$CH_2CHCH_3 + H_2O \xrightarrow[\text{heat}]{\text{conc. } H_2SO_4} CH_3CH(OH)CH_3$$

$$CH_3CH(OH)CH_3 + CH_3COOH \underset{\text{heat}}{\overset{\text{conc. } H_2SO_4}{\rightleftharpoons}} CH_3COOCH(CH_3)_2 + H_2O$$

10.3 Design a reaction sequence, using balanced chemical equations, for the conversion of propene to propanoic acid.

A problem here is that if water is added to propene, it will produce propan-2-ol (Markovnikov's rule), which cannot be oxidised to propanoic acid. Therefore, the alkene must first be converted to an alkane:

$$CH_3CHCH_2 + H_2 \xrightarrow[\text{heat}]{\text{Ni}} CH_3CH_2CH_3$$

When this is reacted with bromine, a significant amount of 1-bromopropane will be formed.

$$CH_3CH_2CH_3 + Br_2 \xrightarrow{\text{UV}} CH_3CH_2CH_2Br + HBr$$

This is followed by this reaction with aqueous sodium hydroxide:

$$CH_3CH_2CH_2Br + OH^- \xrightarrow[\text{heat}]{\text{NaOH(aq)}} CH_3CH_2CH_2OH + Br^-$$

Finally the oxidation of propan-1-ol:

$$CH_3CH_2CH_2OH + 2[O] \xrightarrow[\text{heat / reflux}]{Cr_2O_7^{2-} / H^+} CH_3CH_2COOH + H_2O$$

Benzene and other aromatic compounds

So long as a particular functional group is not attached *directly* to a benzene ring, the reactions of aromatic compounds will be basically the same as the reactions met in other sections. For example, we get the reaction scheme shown in Figure **10.85** for some conversions, starting with ethylbenzene.

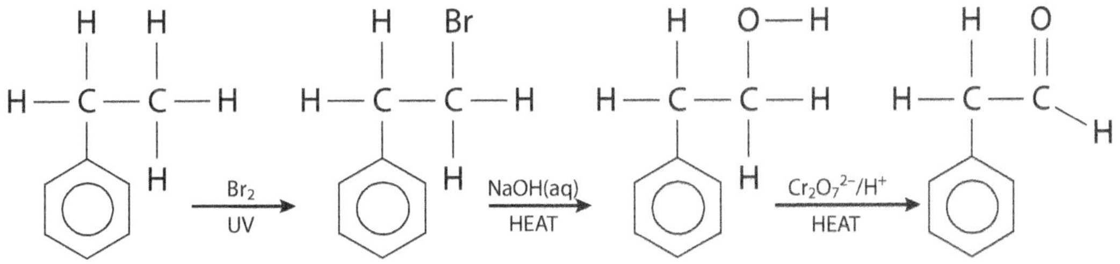

Figure 10.85 A reaction scheme starting with ethylbenzene.

Nature of science

Designing organic synthetic pathways involves a great deal of creativity and imagination – as well as detailed knowledge about organic reactions and suitable starting materials.

All science must be funded and commercial companies finance much research on organic synthesis to make new drugs to target important diseases. The amount of funding available is likely to depend on how many people are affected by a disease – there will be a lot more money available to do research on a disease such as malaria but less on so-called rare diseases because a commercial company is not likely to get back the money it spends on research by selling the drug if the disease only affects a small number of people. In 1983, the US government passed the Orphan Drug Act, which provides financial support to companies carrying out research to develop drugs to target rare diseases.

The principles of green chemistry (see Option D) are being used increasingly in designing organic syntheses as scientists become more aware of the environmental impact of their work.

? Test yourself

39 Draw out reaction pathways showing structural formulas and essential conditions for the following conversions:

a ethane to ethanol

b propene to propanone

c 1-chloropropane to propanoic acid

d

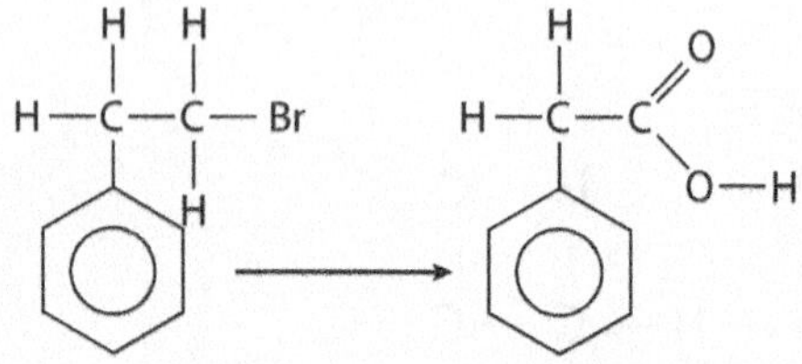

e ethene to ethyl ethanoate using ethene as the only organic starting material

40 Design a reaction sequence for the conversion of molecule **A** into molecule **B**:

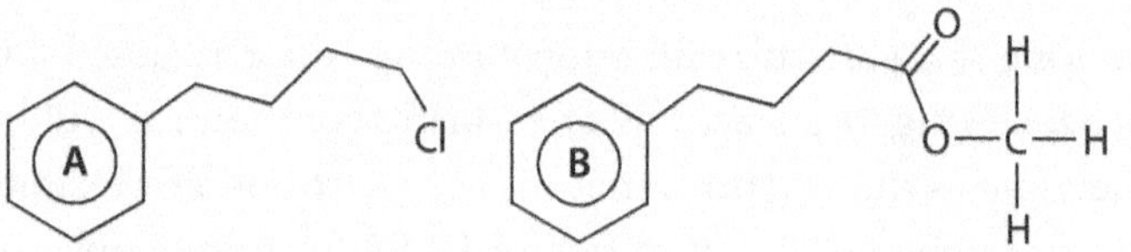

41 Design a reaction sequence, using balanced chemical equations, for the synthesis of 3-methylbutan-2-one from a suitable alkene.

10.5 Stereoisomerism (HL)

Types of stereoisomerism

We have already met structural isomerism (page **441**), in which molecules have the same molecular formula but the atoms are joined together differently. We will now consider **stereoisomerism**.

> **Stereoisomers** have the same structural formula (the atoms are joined together in the same way – same connectivity) but the atoms are arranged differently in space.

The different forms of stereoisomerism are summarised in Figure **10.86**.

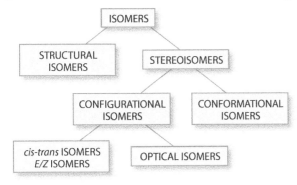

Figure 10.86 Different types of isomerism and how they are linked.

cis–trans isomerism

This type of isomerism occurs in alkenes and in cyclic (ring) compounds.

> **cis–trans isomerism**: two compounds have the same structural formula, but the groups are arranged differently in space around a double bond or a ring.

Before we can understand *cis–trans* isomerism in alkenes, we must remind ourselves about the nature of the C=C double bond. A double bond consists of two components (Figure **10.87**).

$$H_{\text{\tiny{IIII}}} \diagdown C \overset{\pi}{\underset{\sigma}{=\!=}} C \diagdown^{\text{\tiny{IIIII}}H}_{H}$$

Figure 10.87 A double bond is composed of σ and π components.

Fairly free rotation is possible about a C–C single bond because the nature of the sigma bond does not restrict rotation. However, with C=C, the π component of the bond prevents the groups either side of it from rotating relative to each other (Figure **10.88**). The π bond would have to be broken to allow rotation to occur, and this takes a lot of energy.

The term 'geometric isomerism', which is often used for *cis–trans* isomerism, is now strongly discouraged by IUPAC.

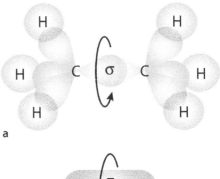

a

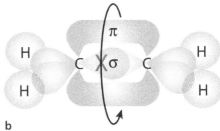

b

Figure 10.88 a Free rotation of groups around a C–C bond. **b** Restricted rotation of groups about a C=C bond.

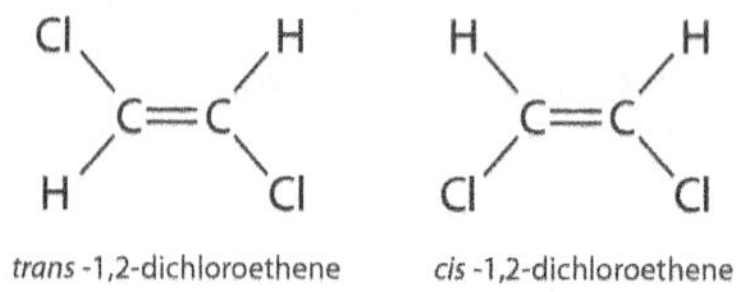

Figure 10.89 Isomers of 1,2-dichloroethene.

cis-but-2-ene   trans-but-2-ene

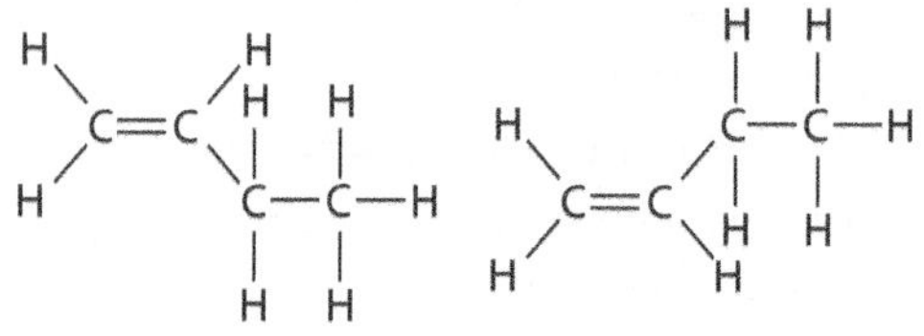

but-1-ene   but-1-ene

Figure 10.90 Isomers of butene.

cis: same side
trans: opposite sides

The fact that rotation of groups around a double bond is restricted gives rise to *cis–trans* isomerism. For example, two structures can be drawn for 1,2-dichloroethene (Figure **10.89**).

The chlorine atoms are arranged either on the **same** side of the double bond – forming the *cis* isomer – or on **opposite** sides of the double bond – forming the *trans* isomer. These cannot be interconverted easily because the π component of the C=C bond **restricts rotation** of groups around the bond.

> For a molecule to exhibit *cis–trans* isomerism, there must be two different groups on **both** sides of the double bond.

Figure **10.90** shows that *cis–trans* isomers are possible for but-2-ene, but not for but-1-ene.

The two molecules of but-1-ene shown in Figure **10.90** are identical – the second is simply the first turned upside down.

Other examples of molecules that do and do not exhibit *cis–trans* isomerism are shown in Table **10.22**.

Compound	*cis-trans* isomers?	*cis*-isomer	*trans*-isomer
1-bromobut-1-ene	yes	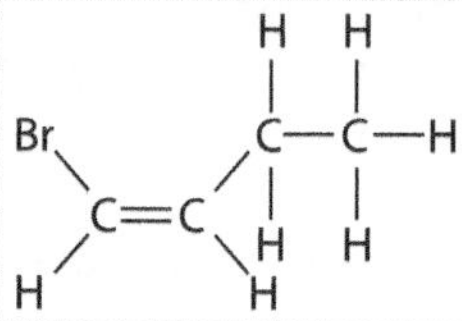	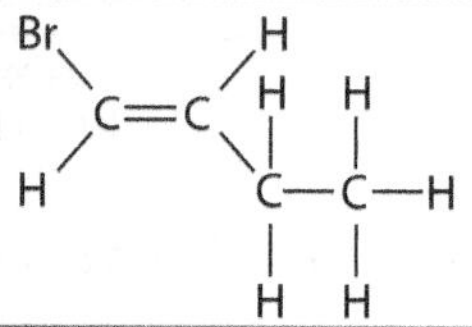
2-bromobut-1-ene	no	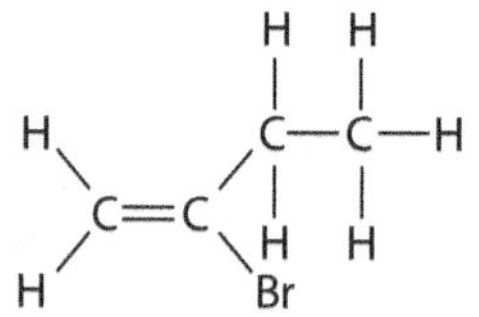	
3-methylpent-2-ene	yes	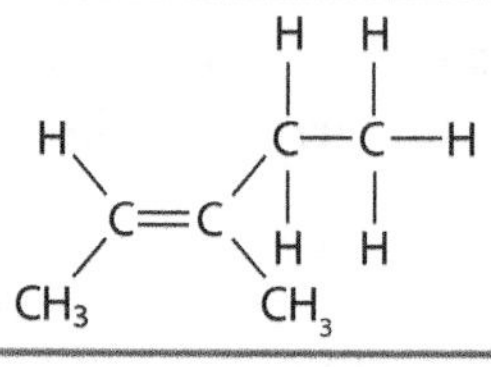	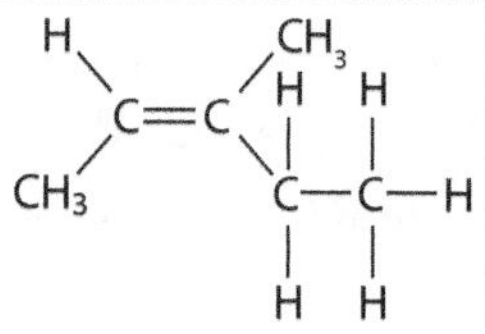
2-methylpent-2-ene	no	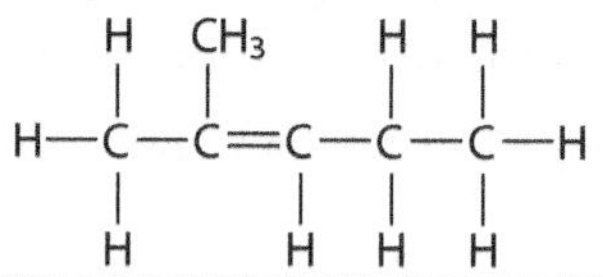	

Table 10.22 Examples of molecules that do and do not exhibit *cis–trans* isomerism.

The *cis* and *trans* nomenclature has limited value – when there are four different groups around the C=C bond it gives no insight into the structure at all. A more general way of naming the isomers that can arise because of the lack of rotation about a C=C bond is using the *E/Z* naming system.

To do this, we give each group attached to the C=C bond a priority according to a set of rules called the Cahn–Ingold–Prelog priority rules (CIP). These give higher priority to atoms attached to the C=C bond that have a higher atomic number.

We look at each side of the C=C bond separately and assign a priority (1 or 2) to each of the atoms or groups attached. If the two groups with highest priority (labelled '1') are on the same side of a plane perpendicular to and passing through the double bond, the isomer is labelled '*Z*' (German: *zusammen* – together); if they are on opposite sides of the plane, the isomer is labelled '*E*' (German: *entgegen* – against or *entgegengesetzt* – opposite).

Consider the molecule shown in Figure **10.91**. The priorities of the atoms and groups joined directly to the carbon atoms of the C=C bond are assigned – fluorine has a higher atomic number than H, therefore F has higher priority (1) than H (2). On the other side of the C=C bond, Cl has a higher priority than C. The two groups with the higher priority (labelled '1') are on the *same* side of the plane perpendicular to and through the C=C bond, so this is the *Z* isomer and it can be named (*Z*)-1-fluoro-2-chloroprop-1-ene.

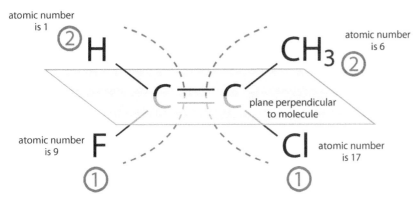

Figure 10.91 (*Z*)-1-fluoro-2-chloroprop-1-ene.

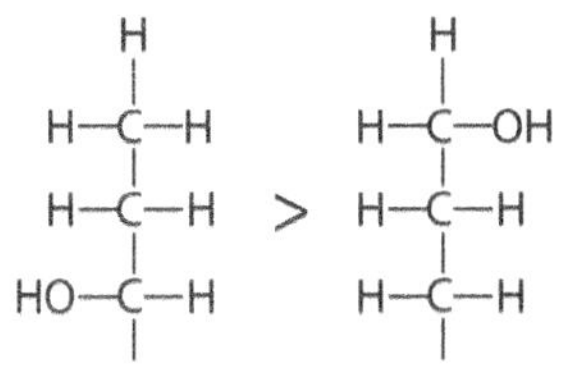

Priority of other groups

With larger groups, which have the same atom attached to the central C, we keep going along the chain until a difference between the groups is spotted. So, −CH$_2$CH$_3$ has a higher priority than −CH$_3$ because the first C has a another C (atomic number 6) atom attached.

In the groups shown in Figure **10.92**, the group on the left has higher priority because the C with O (higher atomic number) is closer to the C=C bond.

Double-bonded atoms (Figure **10.93**) count as two groups attached to an atom – so CHO has higher priority than CH$_2$OH because it is counted as if there are two O atoms attached to the C.

Table **10.23** shows some other molecules classified as *E* or *Z* isomers.

$$
\begin{array}{ccc}
& H & \quad\quad H \\
& | & \quad\quad | \\
H-\!\!&C\!\!-H & \quad H-C-OH \\
& | & \quad\quad | \\
H-\!\!&C\!\!-H \quad > & \quad H-C-H \\
& | & \quad\quad | \\
HO-\!\!&C\!\!-H & \quad H-C-H \\
& | & \quad\quad |
\end{array}
$$

Figure 10.92 The group on the left has higher priority.

$$
\underset{\text{decreasing priority} \longrightarrow}{\overset{O}{\underset{|}{C}}\!\!\diagdown^{OH} \; > \; \overset{O}{\underset{|}{C}}\!\!\diagdown^{H} \; > \; HO\!-\!\overset{H}{\underset{|}{C}}\!-\!H}
$$

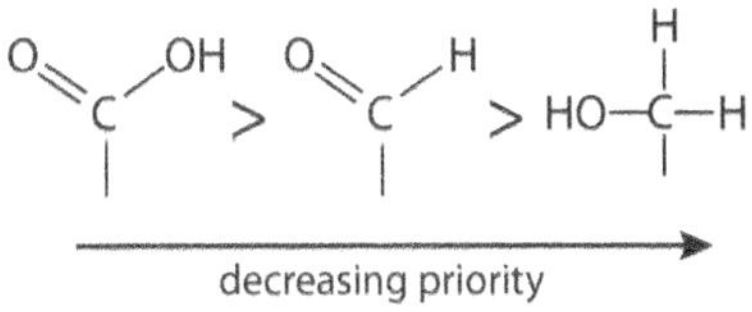

Figure 10.93 Double-bonded atoms count as two groups attached.

HO　　　　CH$_2$OH 　＞C＝C＜ H　　　　CH$_2$CH$_3$	HO　　　　CH$_3$ 　＞C＝C＜ H　　　　CH$_2$CH$_3$
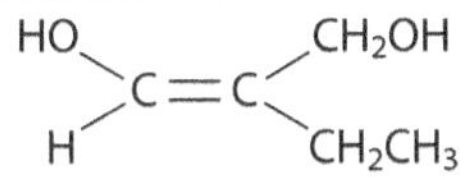 *Z*	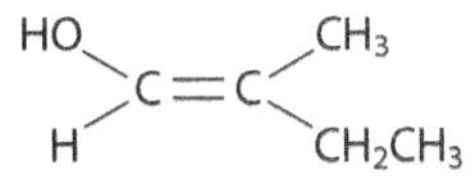 *E*
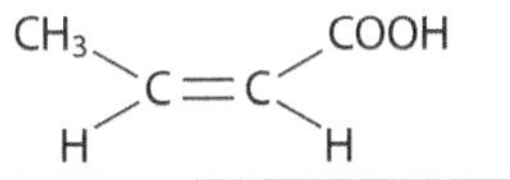 CH$_3$　　　　COOH 　＞C＝C＜ H　　　　H	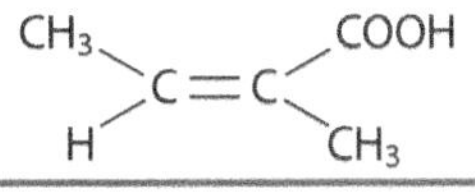 CH$_3$　　　　COOH 　＞C＝C＜ H　　　　CH$_3$
Z	*Z*

Table 10.23 Classification of some isomers as *E* or *Z*.

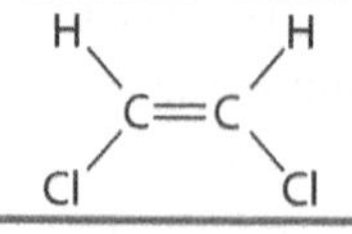

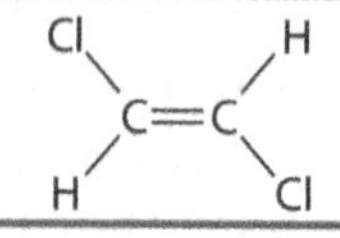

cis-1,2-dichloroethene	**trans**-1,2-dichloroethene
M_r	96.94	96.94
Boiling point/°C	60.5	48.7

Table 10.24 Boiling points of a pair of isomers.

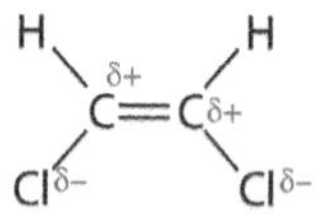

cis-1,2-dichloroethene (polar)

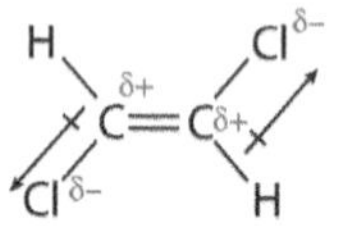

trans-1,2-dichloroethene (non-polar)

Figure 10.94 Polarity differences between similar molecules.

Properties of *cis–trans* isomers

Table **10.24** shows that *cis–trans* (*E/Z*) isomers have different physical properties.

The structures of *cis*-1,2-dichloroethene and *trans*-1,2–dichloroethene differ only in the orientation of their chlorine atoms around the C=C bond – Figure **10.94**. This leads to the *cis* form being polar and the *trans* form being non-polar. The *trans* form is non-polar because the orientation of the chlorine atoms around the double bond means that the dipoles cancel. So there are permanent dipole–dipole interactions for the *cis* form but not for the *trans* form. Both molecules have the same relative molecular mass (and hence very similar London forces) so the difference between the boiling points is due to the permanent dipole–dipole interactions between the *cis* molecules.

cis–trans (*E/Z*) isomers may also have different chemical properties.

cis–trans isomerism in substituted cycloalkanes

A cycloalkane is a ring (cyclic) compound containing only single C–C bonds (and hydrogen). The simplest members of the homologous series are cyclopropane (C_3H_6) and cyclobutane (C_4H_8). Cycloalkanes are structural isomers of the corresponding alkene.

Two different ways of drawing these molecules are shown in Table **10.25**.

A ring containing three atoms, such as in cyclopropane, is called a three-membered ring; that in cyclobutane is a four-membered ring.

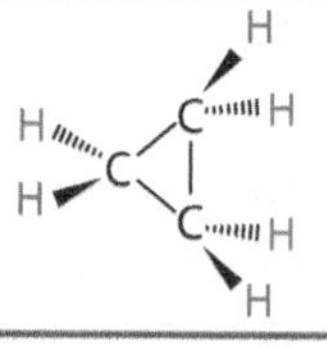

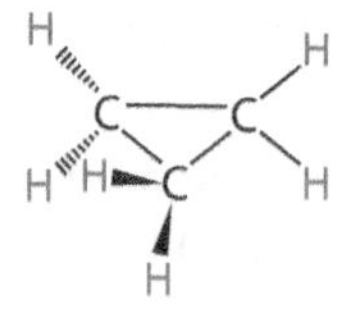

Extension

These ring compounds belong to the general group of **cyclic** compounds – those that contain a ring of carbon or other atoms. If there is another atom, such as O or N, in the ring as well as carbon atoms, the compound is described as **heterocyclic**.

cyclopropane	
cyclobutane	

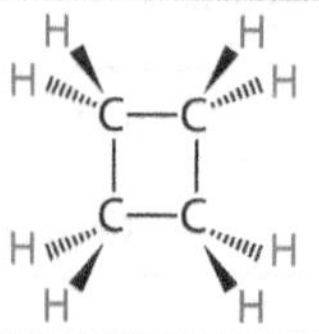

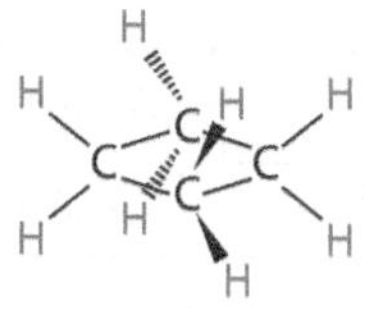

Table 10.25 Two ways of drawing cyclopropane and cyclobutane. Hydrogen atoms above the plane of the ring are red and those below are blue.

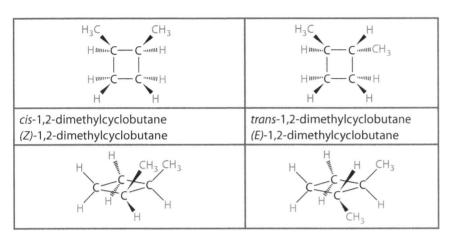

cis-1,2-dimethylcyclobutane (Z)-1,2-dimethylcyclobutane	trans-1,2-dimethylcyclobutane (E)-1,2-dimethylcyclobutane

Table 10.26 Different ways of showing the isomers of 1,2-dimethylcyclobutane.

The ring structure prevents rotation of a group from above the plane of the ring to below the ring. So *cis–trans* isomers are possible in substituted cycloalkanes. Groups can be rotated from above to below the ring only if the ring is broken – this requires a lot of energy.

> The condition for a cycloalkane to exhibit *cis–trans* isomerism is that at least two carbon atoms must have two different groups attached.

Take 1,2-dimethylcyclobutane as an example. Table **10.26** shows the structures in two different ways – the colours indicate the same convention as used in Table **10.25**.

As before, in the *cis* form the two methyl groups are on the **same** side of the ring and in the *trans* form they are on **opposite** sides of the ring. The *E/Z* nomenclature is used in the same way as for alkenes – if the two groups with the higher priority are on the same side of the ring then it is a *Z* isomer.

Note:

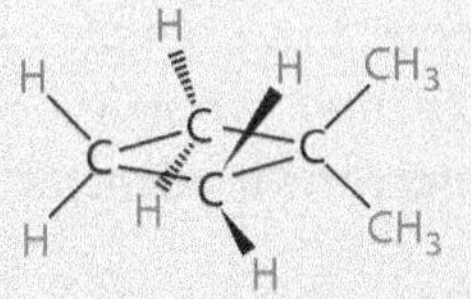

is not a stereoisomer of 1,2-dimethylcyclobutane but a structural isomer. The atoms are joined together differently, with two methyl groups on the same carbon atom – so it has a different name, 1,1-dimethylcyclobutane. This molecule does not show *cis–trans* isomerism.

1,3-dimethylcylobutane is a structural isomer of 1,1-dimethylcyclobutane and 1,2-dimethylcyclobutane and does exhibit *cis–trans* isomerism.

42 Which of the following will exhibit *cis–trans* isomerism? If the molecule exhibits *cis–trans* isomerism, draw the *cis* and *trans* forms.

a
b

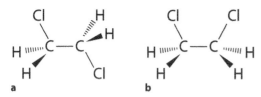

c 2,3-dimethylpent-2-ene
d 3,4-dimethylpent-2-ene
e 1,2,3-trimethylcyclopropane
f 1,3-dimethylcyclobutane

43 Classify each of the following isomers as *E* isomers or *Z* isomers.

a   **c**

b   **d**

Conformational isomers

Conformational isomers are forms of the same molecule that have different conformations because of rotation about a σ bond. For example, 1,2-dichloroethane can exist in different conformations according to how the chlorine atoms are arranged relative to each other. Two different conformations of 1,2-dichloroethane are shown in Figure **10.95**.

Molecule **10.95a** has a lower potential energy than molecule **10.95b** because the chlorine atoms are further apart (and there is therefore less repulsion between them). There is a (relatively low) barrier to rotation from one form to the other due to changes in repulsion between the Cl atoms as they get further apart/closer together. Another conformation is shown in two different ways in Figure **10.96**.

It is important to realise that all the conformations in Figures **10.95** and **10.96** represent the same molecule and that individual conformational forms cannot be isolated because the barrier to rotation is so low. Different forms will exist simultaneously in a sample of the substance – and these forms will be constantly interconverting. The rate of interconversion between the different conformations depends on the barrier to rotation and the temperature.

A common example of conformational isomers is the chair and boat forms (Figure **10.97**) of cyclohexane (C_6H_{12}).

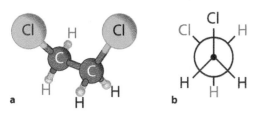

Figure 10.95 Different conformations of 1,2-dichloroethane.

Figure 10.96 a Ball and stick model of a conformation of 1,2-dichloroethane; **b** Newman projection of 1,2-dichloroethane – the atoms at the back are shown in red.

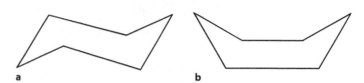

Figure 10.97 Cyclohexane: **a** chair form; **b** boat form.

Conformational isomers can be interconverted without breaking chemical bonds. Interconversion of configurational isomers requires the breaking and subsequent re-forming of chemical bonds.

Optical isomerism

Optical isomerism is another type of stereoisomerism. Butan-2-ol exhibits optical isomerism. There are two forms of this compound – they are mirror images (Figure **10.98**).

Although they look identical, if we try to put one molecule 'on top' of the other (Figure **10.98b**), it can be seen that only two of the groups can be made to correspond – the other two groups are distributed differently around the central carbon atom. Therefore the molecules are not identical but isomers of each other – we say that the two isomers are **non-superimposable**.

In order to exhibit this property of optical isomerism – that the mirror images are not superimposable – there must be **four different** groups attached to a carbon atom. If you look at butane (Figure **10.99**), you can see that it has a maximum of only three different groups attached to any one carbon atom, and also that the mirror images *are* superimposable.

A carbon atom with four different groups attached to it is called a **chiral centre** (chirality centre or stereogenic unit) and molecules that exhibit optical isomerism are often described as **chiral**. We talk about the *configuration* of the chiral centre – that is the spatial arrangement of the groups at the chiral centre.

Some molecules that exhibit optical isomerism are shown in Table **10.27**.

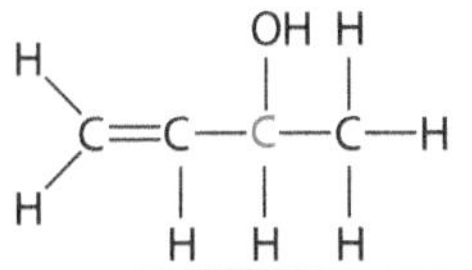

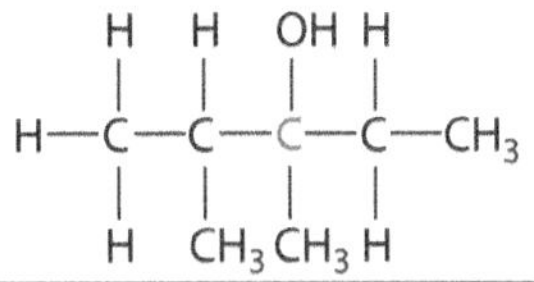

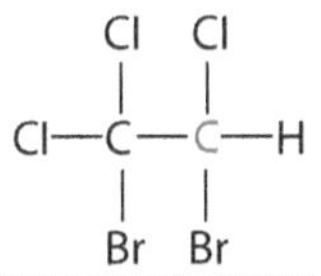

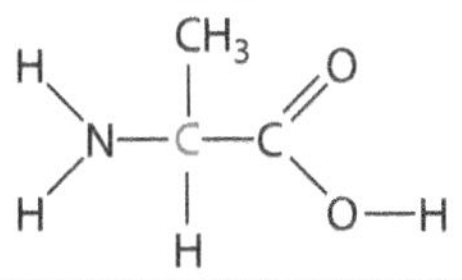

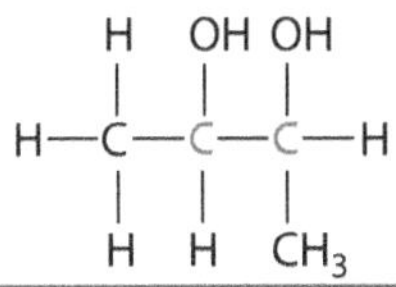

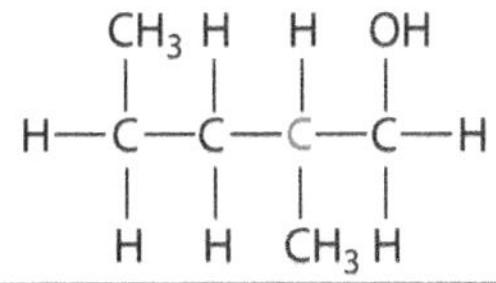

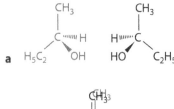

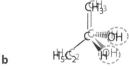

Figure 10.98 a Mirror image molecules; **b** non-superimposable.

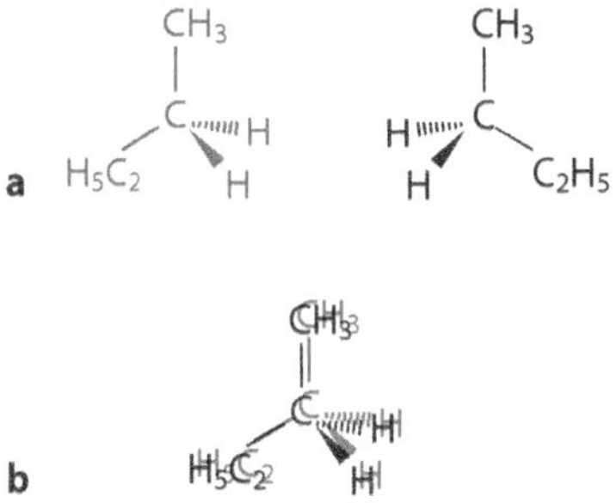

Figure 10.99 There is only one form of butane: **a** these two molecules are identical; **b** butane does not have optical isomers.

> The word 'chiral' is derived from the Greek word for hand – your hands are non-superimposable mirror images of each other.

> A carbon atom with four different atoms or groups attached is sometimes called an **asymmetric** carbon atom.

Table 10.27 Molecules that exhibit optical isomerism – the chiral centres are shown in red.

Some molecules that do *not* exhibit optical isomerism are shown in Table **10.28**.

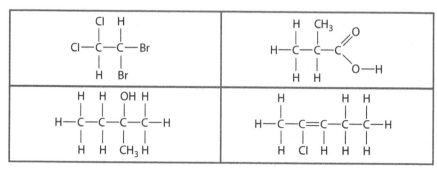

Table 10.28 Molecules that do not exhibit optical isomerism.

Table 10.29 The three-dimensional representation of optical isomers showing two pairs of mirror images.

The fact that these carbon compounds exhibit optical isomerism allows us to reason that these molecules are tetrahedral. If they were square planar, optical isomers would not be possible (if you draw 2-bromobutane with a square planar structure you should be able to see that the mirror images are superimposable). This is an example of how reasoning can provide us with information about the microscopic world. Do we know or believe that these molecules are tetrahedral?

To show the optical isomerism exhibited by some compounds more clearly, optical isomers are usually drawn in three dimensions, as shown in the diagrams in Table **10.28**. A solid wedge indicates a bond coming out of the plane of the paper; a dashed wedge goes into the paper.

The individual optical isomers of a compound are called **enantiomers** – so, for example, the bottom row of Table **10.29** shows the two enantiomers of 2-bromobutane.

Optical isomerism and ring compounds

Optical isomerism can also occur with ring (cyclic) compounds. Figure **10.100** shows that there are two chiral centres in *trans*–1,2–dichlorocyclopropane, each shown in red.

To determine if a ring compound has optical isomers, it is necessary to look at whether or not the molecule has a plane of symmetry. If a molecule *does* have a plane of symmetry then it will *not* have optical isomers. For example, *cis*–1,2–dichlorocyclopropane has a plane of symmetry and does not exhibit optical isomerism – Figure **10.101**.

> If a molecule has a plane of symmetry it will *not* exhibit optical isomerism.

Figure 10.100 *trans*-1,2-dichlorocyclopropane

cis-1,2-dichlorocyclopropane

Figure 10.101 A molecule with a plane of symmetry.

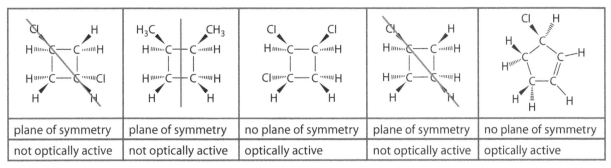

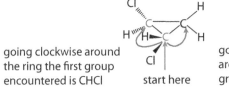				
plane of symmetry	plane of symmetry	no plane of symmetry	plane of symmetry	no plane of symmetry
not optically active	not optically active	optically active	not optically active	optically active

Table 10.30 Some ring compounds with optical isomers and some without.

going clockwise around the ring the first group encountered is CHCl

start here

going anti-clockwise around the ring the first group encountered is CH₂

Figure 10.102 In addition to the H and Cl directly attached to the C indicated with the blue arrow, if we go clockwise around the ring, the group attached to this C is CHCl, but in the anticlockwise direction it is a CH_2 group.

The fact that the chiral centres in the molecule in Figure **10.101** are indeed carbon atoms with four different groups attached can be seen by starting at the point shown and going around the ring in two different directions (Figure **10.102**).

Some ring compounds that have optical isomers and others that do not are shown in Table **10.30**.

Optical isomers and plane-polarised light

The two enantiomers of an optically active compound have the property that they rotate **plane-polarised light** in opposite directions. Or, more precisely, they rotate the plane of polarisation of plane-polarised light in opposite directions.

Normal, non-polarised, light vibrates in all planes (Figure **10.103**). If non-polarised light is passed through a polarising filter, plane-polarised light is produced (Figure **10.104**).

If plane-polarised light is passed through samples of the two isomers of butan-2-ol, shown below, we find that one of the isomers rotates the plane of the plane-polarised light to the right (clockwise), and the other isomer rotates the plane of the plane-polarised light to the left (anticlockwise). The two enantiomers rotate the plane of the plane-polarised light by equal amounts.

Figure 10.103 If we imagine being able to look at the vibrations in a beam of non-polarised light coming towards us, then it could be represented as shown here.

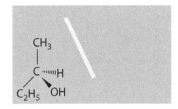

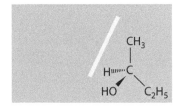

> Optical isomers rotate the plane of plane-polarised light in opposite directions.

There are various ways of labelling the two enantiomers. They may be labelled according to the direction in which they rotate plane-polarised light using +/− or d/l, or they may be labelled according to the absolute configuration (the arrangement of the groups around the chiral centre) using D/L or R/S.

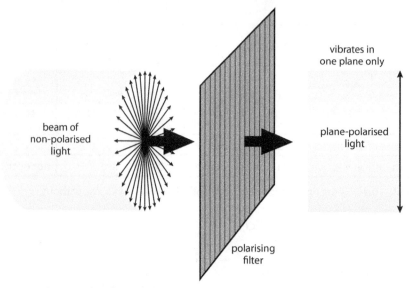

Figure 10.104 Plane-polarised light is light that vibrates in one plane only.

Which configuration corresponds to the direction in which light is rotated can be worked out only by determining the absolute configuration using X-ray crystallography and the rotation of the light using a polarimeter. We cannot just look at a particular enantiomer's three-dimensional structure and say that it rotates plane-polarised light to the right or to the left.

Using a polarimeter to determine the direction in which light is rotated

A simple polarimeter consists of a source of light (usually a sodium lamp producing one specific wavelength), two polarising filters, a sample tube and a scale to measure the degree of rotation of the plane-polarised light (Figure 10.105).

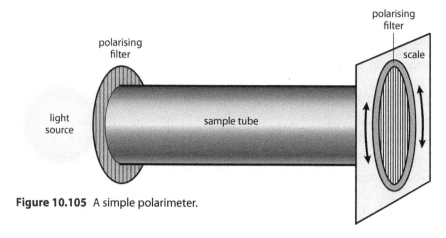

Figure 10.105 A simple polarimeter.

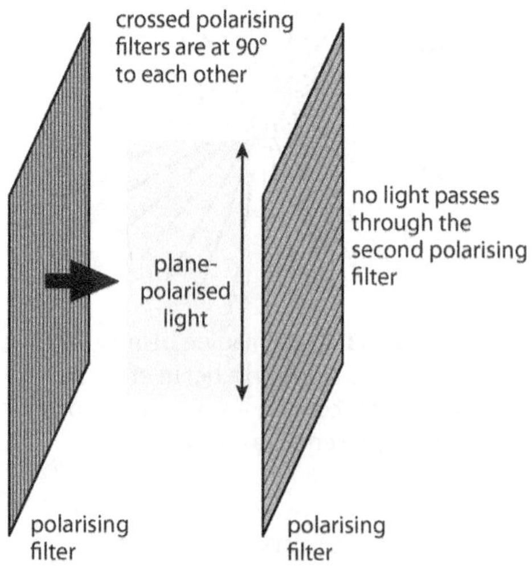

The solvent in which the test substance is to be dissolved is placed in the sample tube, and the second polarising filter is rotated until no light can be seen. At this point the polarising filters are exactly crossed (Figure 10.106). The solvent is then replaced by the sample dissolved in the solvent and the polarising filter rotated again until no light can be seen. The angle through which the light is rotated by the sample is the difference between the readings with and without the sample.

It is possible to distinguish between optical isomers of a particular substance using a polarimeter – one enantiomer rotates the plane-polarised light clockwise and the other rotates it in the opposite direction.

Figure 10.106 When polarising filters are crossed no light can pass through unless it has been rotated by an optically active sample between the first and second filter.

Racemic mixtures

An equimolar mixture of the two enantiomers of a chiral compound is called a **racemic mixture** (or racemate). A racemic mixture has no effect on plane-polarised light (is optically inactive) because the rotation effects of the two enantiomers cancel each other out.

Because the enantiomers are equally stable, reactions that produce molecules containing a chiral carbon atom will usually produce a racemic mixture – the resulting reaction mixture will have no effect on plane-polarised light.

When attempting to make one specific enantiomer of an optically active compound, either the synthetic route has to be carefully designed to produce this isomer (stereospecific synthesis) or the racemic mixture may be resolved into its different enantiomers.

An example of a stereospecific reaction is the S_N2 nucleophilic substitution reaction of 2-bromobutane with aqueous sodium hydroxide (Figure **10.107**). If the reaction is carried out with just one enantiomer of 2-bromobutane, we will be able to predict which enantiomer of the product will be made.

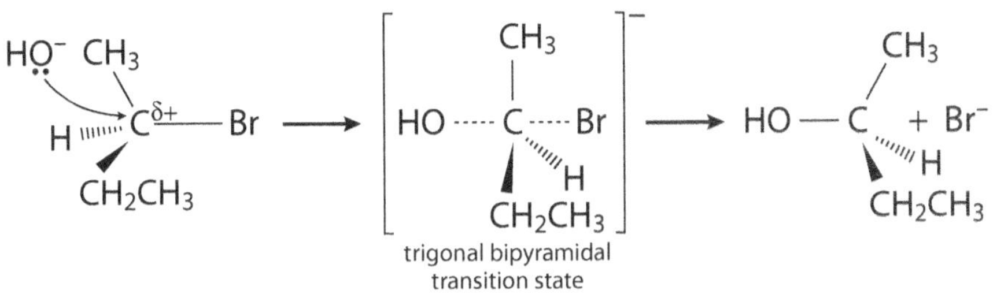

Figure 10.107 Inversion of configuration during an S_N2 reaction.

The product formed is basically the mirror image of the starting material (ignoring the fact that one molecule contains Br and the other OH). Therefore the reaction occurs with inversion of the configuration at the chiral centre.

The S_N1 mechanism is not stereospecific (page **469**) and produces a racemic mixture if the product has a chiral centre.

Exam tip
If the product of a reaction has a chiral centre and you are asked why it is optically inactive - the answer is 'because a racemic mixture is formed'.

Diastereomers

Diastereomers are stereoisomers that are not mirror images of each other – so *cis–trans* isomers are diastereomers. Diastereomers can also arise when more than one chiral centre is present in a molecule.

There are two chiral centres in the molecule shown in Figure **10.108** The compounds are mirror images of each other and are enantiomers.

However, in Figure **10.109** the two compounds are stereoisomers (same structural formula) but are not mirror images of each other, and are therefore diastereomers.

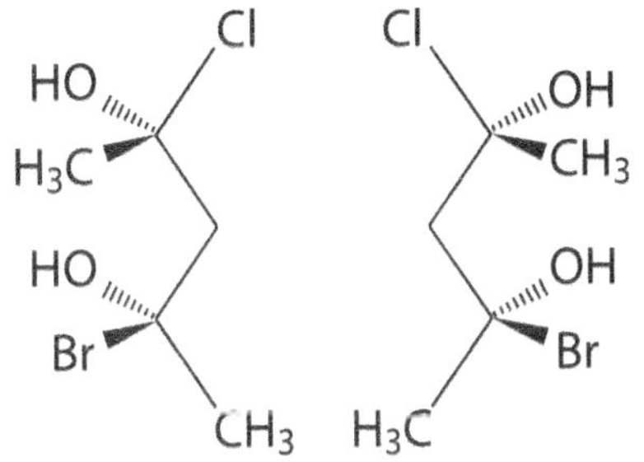

Figure 10.108 Skeletal structures of two enantiomers.

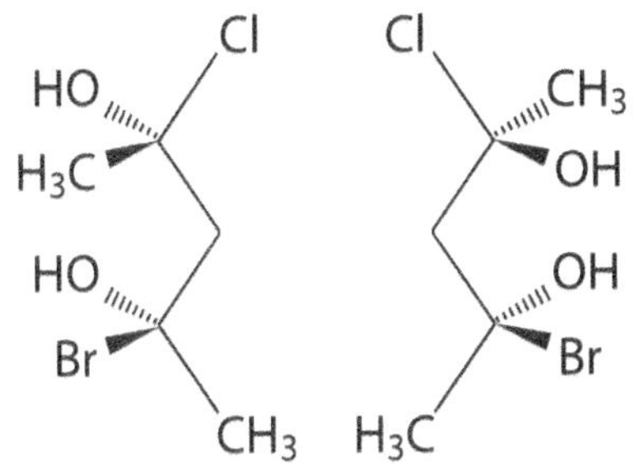

Figure 10.109 Skeletal structures of two diastereomers.

Physical properties (except the direction of rotation of the plane of plane-polarised light) such as melting point, boiling point and solubility are identical in enantiomers, but not for diastereomers. The chemical properties of enantiomers are identical for reactions with compounds that are not optically active. Diastereomers may have different chemical properties.

The sedative called thalidomide was introduced in the late 1950s and was used in various countries throughout the world – it was thought to be safe with very few side-effects. As well as being a sedative, it was also effective at relieving sickness and so was widely given to pregnant women in the first three months of pregnancy to alleviate morning sickness. The result was devastating, with thousands of children being born with missing or malformed limbs, and a number dying in infancy. This tragedy could have been avoided if the drug had been evaluated for teratogenicity (toxicity to the foetus) before it was marketed.

Thalidomide was given as a mixture of enantiomers (racemic mixture) but one of the enantiomers (the S-enantiomer) was responsible for producing the teratogenic effect and caused limb deformities in the foetus (Figure **10.110**).

Giving the single enantiomer would not have helped in the case of thalidomide because the enantiomers interconvert in the body producing the racemic mixture.

Nowadays, if a new drug is going to be marketed as the racemic mixture, testing must be carried out on each enantiomer separately and also on the racemic mixture. Pharmaceutical companies now tend to either synthesise or separate out the active single enantiomer of a drug and develop this instead of the racemic mixture. Different countries have different standards for approval of drugs – thalidomide was not licensed for sale in the US.

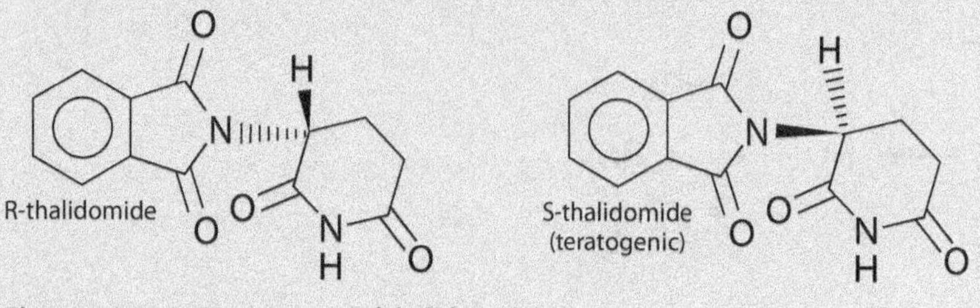

Figure 10.110 Enantiomers of thalidomide.

Nature of science

A great deal of Chemistry can be applied in other areas of science. The three dimensional shape of a molecule and stereochemistry are extremely important to reactions going on in the body and a knowledge of this is critical in some areas of biochemistry and other biological sciences.

Test yourself

44 Which of the following molecules exhibit optical isomerism? For those that do, identify the chiral centre and draw three-dimensional diagrams showing the optical isomers.

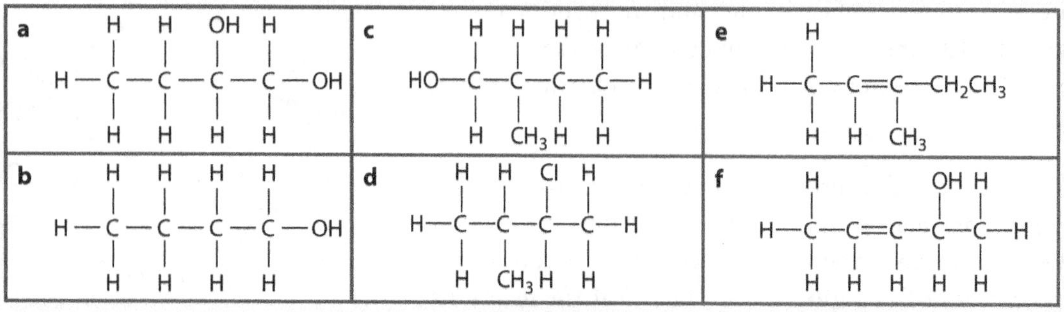

45 Identify which of the following molecules is/are optically active:

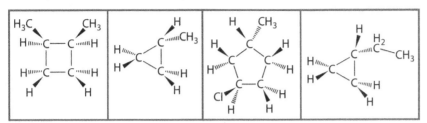

46 Identify which of the following are pairs of enantiomers or pairs of diastereomers:

a

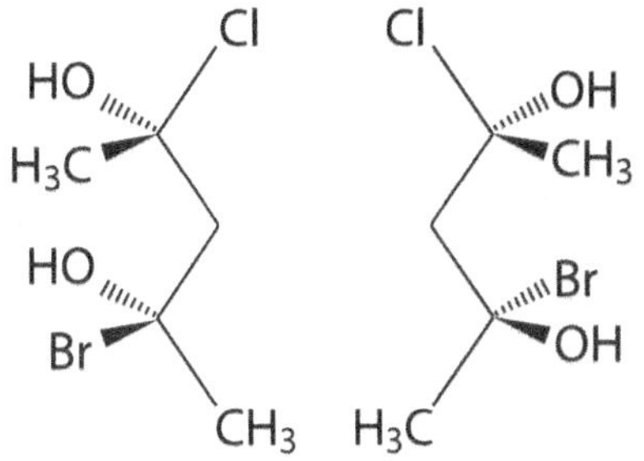

HO,,,, Cl   Cl OH
H₃C   CH₃
HO,,,,   Br
Br   OH
CH₃   H₃C

b

HO,,,, Cl   Cl OH
H₃C   H   H CH₃
      F   F
H₃C   CH₃

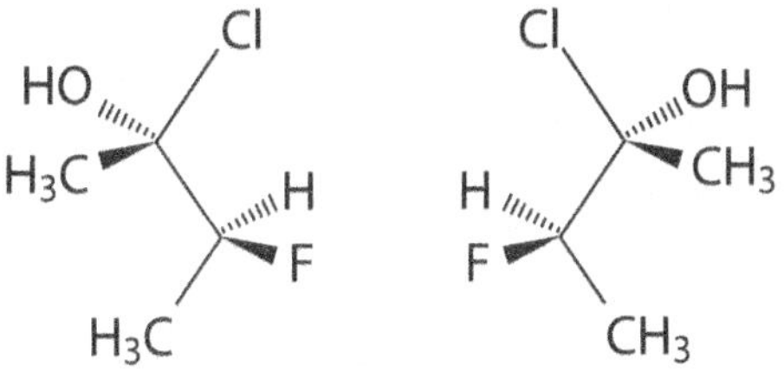

Exam-style questions

1 The functional groups present in this molecule are:

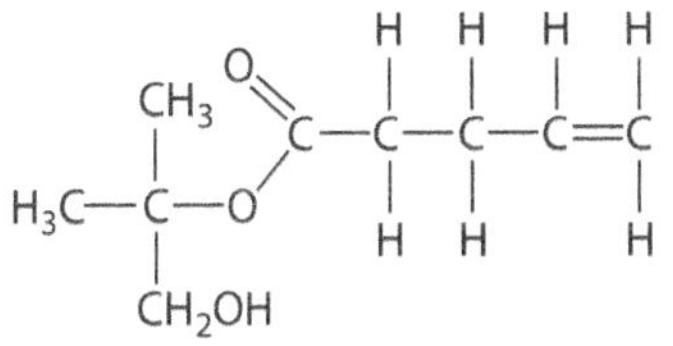

    A hydroxyl, ester, alkenyl
    B ketone, carboxy, alkenyl
    C ester, aldehyde, alkenyl
    D alkenyl, ketone, hydroxyl

2 Which of the following is not a possible product when ethane reacts with chlorine in the presence of UV light?

    A HCl
    B CH_3CH_2Cl
    C H_2
    D $CH_3CH_2CH_2CH_3$

3 The product made when but-2-ene reacts with bromine is:

    A $CH_3CHBrCH_2CH_3$
    B $CH_3CHBrCHBrCH_3$
    C $CH_3CBr_2CH_2CH_3$
    D $CH_3CHBrCH_2CH_2Br$

4 The condensed structural formulas of three alcohols are shown below:

I $CH_3CH_2CH_2CH_2OH$

II $CH_3CH_2CH(OH)CH_3$

III $(CH_3)_3COH$

Which would be oxidised by heating with acidified potassium dichromate(VI)?

    A   I, II and III

    B   I only

    C   I and II only

    D   III only

5 The name of this molecule is:

    A   5-methylpentanal

    B   hexan-1-one

    C   hexanal

    D   1-methylpentan-5-one

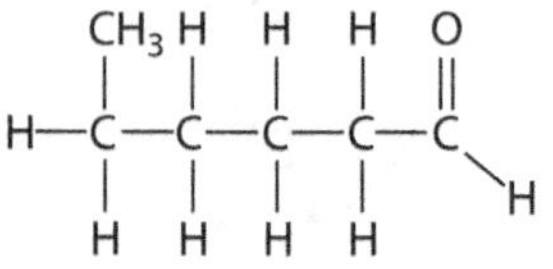

6 $CH_3CH_2CH_2CH_2Br$ reacts with aqueous sodium hydroxide. The name of the product formed and the type of reaction are:

	product	type of reaction
A	butanal	oxidation
B	butan-1-ol	electrophilic substitution
C	butan-2-ol	electrophilic addition
D	butan-1-ol	nucleophilic substitution

7 The ester shown could be formed from the reaction between:

    A   butanoic acid and propan-1-ol

    B   propanoic acid and butan-2-ol

    C   butanoic acid and propan-2-ol

    D   butanoic acid and ethanol

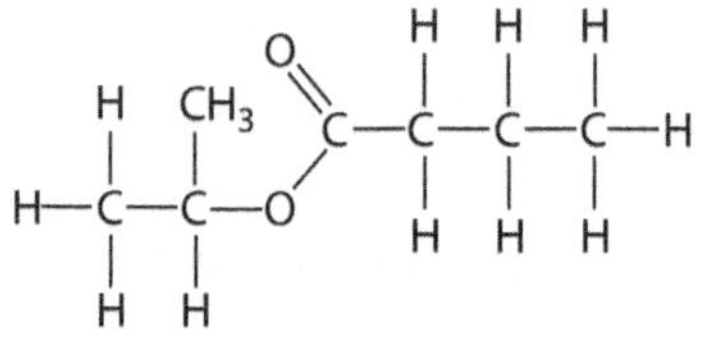

HL 8 Which of the following substitution reactions is likely to occur most rapidly?

    A $(CH_3)_3CBr + OH^- \rightarrow (CH_3)_3COH + Br^-$

    B $(CH_3)_3CCl + OH^- \rightarrow (CH_3)_3COH + Cl^-$

    C $CH_3(CH_2)_2CH_2Br + OH^- \rightarrow CH_3(CH_2)_2CH_2OH + Br^-$

    D $CH_3(CH_2)_2CH_2Cl + OH^- \rightarrow CH_3(CH_2)_2CH_2OH + Cl^-$

HL 9 Which of the following will exist as optical isomers?

    A   1-bromobutane

    B   but-2-ene

    C   pentan-2-ol

    D   pentan-3-ol

HL 10 Butan-2-ol can be formed in each of the following reactions:

      I   reduction of compound **X** with $NaBH_4$

      II   heating compound **Y** with aqueous sodium hydroxide

      III heating **Z** with concentrated sulfuric acid

X, **Y** and **Z** could be:

	X	**Y**	**Z**
A	butanoic acid	2-chlorobutane	butane
B	butanone	2-chlorobutane	but–1–ene
C	butanone	but–2–ene	butan–1–ol
D	butanal	butane	but–2–ene

11 Four of the structural isomers of $C_4H_{10}O$ are alcohols.

  a Draw the structures and give the names of these alcohols.   **[4]**

  b Two of the alcohols can be oxidised to carboxylic acids. Give the name and formula of a suitable oxidising agent, the structures of the carboxylic acids formed and describe any colour change that occurs.   **[6]**

  c Draw and name an isomer of $C_4H_{10}O$ that is not an alcohol.   **[2]**

12 Ethane can react with chlorine in the presence of UV light to form chloroethane.

  a Write an equation for this reaction.   **[1]**

  b **i** State the name of the mechanism by which this reaction occurs.   **[1]**

      ii Use the mechanism of this reaction to explain the terms 'homolytic fission', 'free radical' and 'termination step'.   **[4]**

  c Chloroethane reacts with aqueous sodium hydroxide.

      i Write an equation for the reaction that occurs.   **[1]**

    HL **ii** Draw the mechanism for this reaction.   **[3]**

13 **a** Describe a chemical test that can be used to distinguish between butane and but–2–ene.   **[3]**

  b Draw the structure of the compound formed when but-2-ene reacts with hydrogen bromide under appropriate conditions.   **[1]**

  c But-2-ene can undergo polymerisation under suitable conditions. Draw three repeating units of the polymer formed.   **[2]**

HL d But-2-ene can be converted to butanone in a two-step reaction sequence. Draw out the reaction sequence showing all structures and giving essential conditions.   **[5]**

14 An organic compound has the percentage composition 48.6% carbon, 8.2% hydrogen, 43.2% oxygen. The relative molecular mass of the compound is approximately 74.

    a   Determine the empirical formula and the molecular formula of the compound.   **[3]**

    b   **i**   Two of the isomers of this compound are esters. Draw out full structural formulas and name these compounds.   **[4]**

        ii Write an equation for the formation of **one** of the esters from a carboxylic acid and an alcohol.   **[2]**

    c   A third isomer of this compound reacts with magnesium to form hydrogen gas. Draw the structure of this isomer and write an equation for the reaction with magnesium.   **[3]**

HL 15 This question is about 2-bromobutane.

    a   2-bromobutane exhibits optical isomerism. Explain what structural feature of 2-bromobutane allows it to exhibit optical isomerism and draw clear diagrams showing the optical isomers.   **[3]**

    b   **i**   2-bromobutane can react with sodium hydroxide via an S_N2 mechanism. Explain what is meant by the 'N' in S_N2.   **[1]**

        ii Explain why the S_N2 reaction of 2-bromobutane with sodium hydroxide is described as 'stereospecific'.   **[2]**

        iii Explain whether you would expect 1-bromobutane or 2-bromobutane to react more quickly with sodium hydroxide via an S_N2 mechanism.   **[2]**

HL 16 **a**   Draw the full structural formula of 2-methylpent-2-ene.   **[1]**

    b   **i**   Draw the full structural formula of the major product formed when 2-methylpent-2-ene reacts with hydrogen bromide in the dark.   **[2]**

        ii Write the mechanism for this reaction and explain the formation of the major product.   **[5]**

    c   The major product of the reaction in part **b i** reacts with sodium hydroxide via an S_N1 mechanism to form the alcohol, **X**. Write the mechanism for the formation of **X**.   **[3]**

    d   Explain whether **X** can be oxidised or not when it is heated with acidified potassium dichromate(VI).   **[1]**

Summary

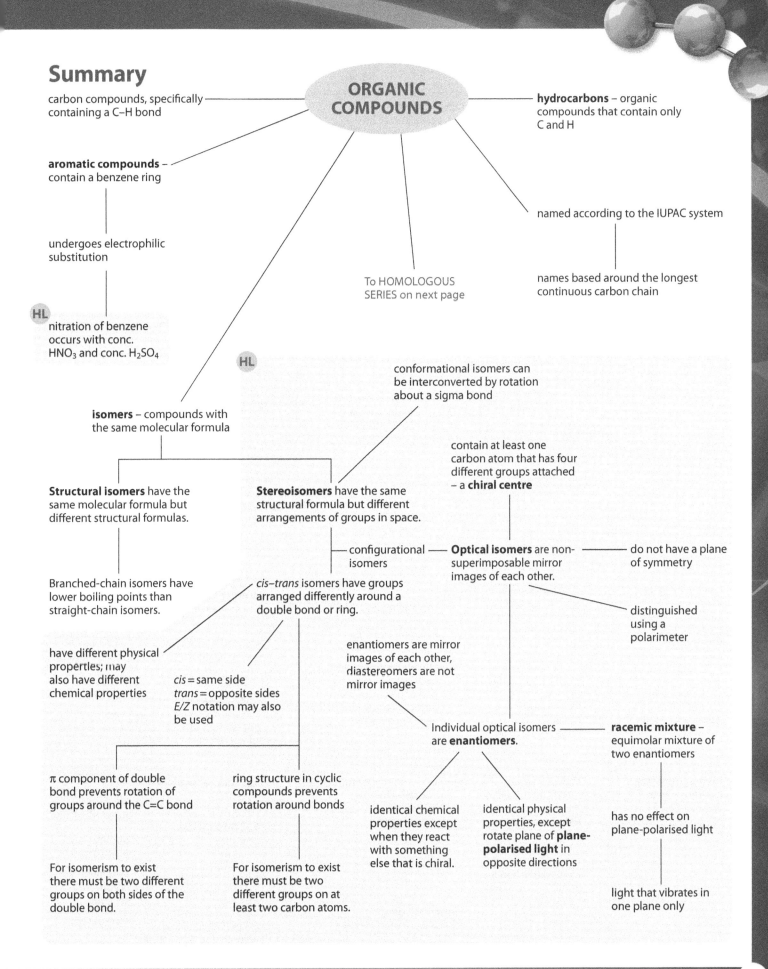

carbon compounds, specifically containing a C–H bond

ORGANIC COMPOUNDS

hydrocarbons – organic compounds that contain only C and H

aromatic compounds – contain a benzene ring

named according to the IUPAC system

undergoes electrophilic substitution

To HOMOLOGOUS SERIES on next page

names based around the longest continuous carbon chain

HL nitration of benzene occurs with conc. HNO_3 and conc. H_2SO_4

HL

conformational isomers can be interconverted by rotation about a sigma bond

isomers – compounds with the same molecular formula

contain at least one carbon atom that has four different groups attached – a **chiral centre**

Structural isomers have the same molecular formula but different structural formulas.

Stereoisomers have the same structural formula but different arrangements of groups in space.

configurational isomers

Optical isomers are non-superimposable mirror images of each other.

do not have a plane of symmetry

Branched-chain isomers have lower boiling points than straight-chain isomers.

cis–trans isomers have groups arranged differently around a double bond or ring.

distinguished using a polarimeter

have different physical properties; may also have different chemical properties

cis = same side
trans = opposite sides
E/Z notation may also be used

enantiomers are mirror images of each other, diastereomers are not mirror images

Individual optical isomers are **enantiomers**.

racemic mixture – equimolar mixture of two enantiomers

π component of double bond prevents rotation of groups around the C=C bond

ring structure in cyclic compounds prevents rotation around bonds

identical chemical properties except when they react with something else that is chiral.

identical physical properties, except rotate plane of **plane-polarised light** in opposite directions

has no effect on plane-polarised light

For isomerism to exist there must be two different groups on both sides of the double bond.

For isomerism to exist there must be two different groups on at least two carbon atoms.

light that vibrates in one plane only

Summary – continued

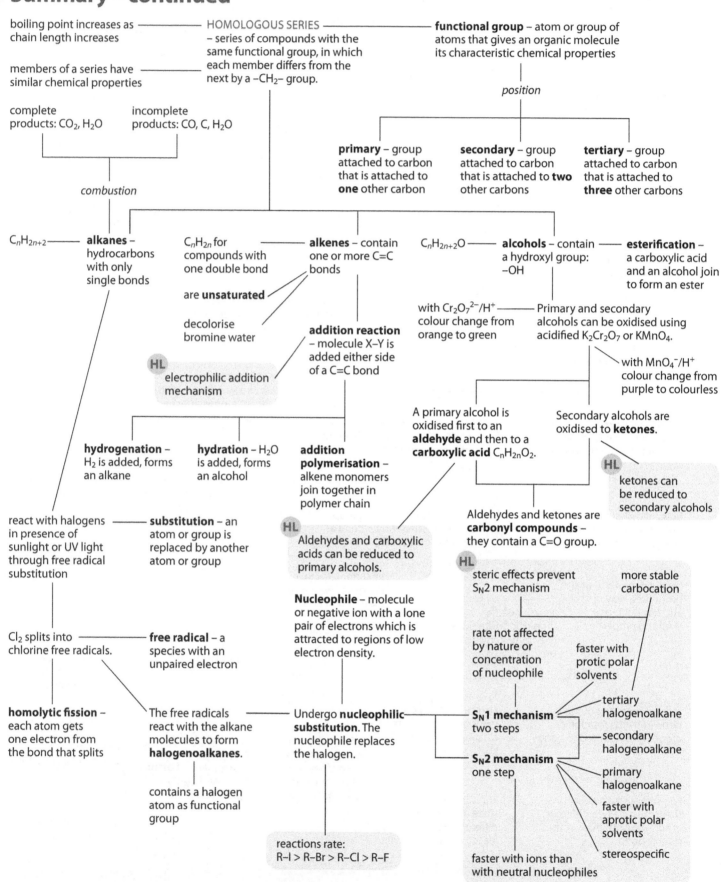

boiling point increases as chain length increases

members of a series have similar chemical properties

HOMOLOGOUS SERIES – series of compounds with the same functional group, in which each member differs from the next by a $-CH_2-$ group.

functional group – atom or group of atoms that gives an organic molecule its characteristic chemical properties

position

primary – group attached to carbon that is attached to **one** other carbon

secondary – group attached to carbon that is attached to **two** other carbons

tertiary – group attached to carbon that is attached to **three** other carbons

complete products: CO_2, H_2O

incomplete products: CO, C, H_2O

combustion

C_nH_{2n+2} — alkanes – hydrocarbons with only single bonds

C_nH_{2n} for compounds with one double bond

alkenes – contain one or more C=C bonds

are **unsaturated**

decolorise bromine water

HL electrophilic addition mechanism

addition reaction – molecule X–Y is added either side of a C=C bond

$C_nH_{2n+2}O$ — alcohols – contain a hydroxyl group: –OH

esterification – a carboxylic acid and an alcohol join to form an ester

with $Cr_2O_7^{2-}/H^+$ colour change from orange to green

Primary and secondary alcohols can be oxidised using acidified $K_2Cr_2O_7$ or $KMnO_4$.

with MnO_4^-/H^+ colour change from purple to colourless

A primary alcohol is oxidised first to an **aldehyde** and then to a **carboxylic acid** $C_nH_{2n}O_2$.

Secondary alcohols are oxidised to **ketones**.

HL ketones can be reduced to secondary alcohols

hydrogenation – H_2 is added, forms an alkane

hydration – H_2O is added, forms an alcohol

addition polymerisation – alkene monomers join together in polymer chain

Aldehydes and ketones are **carbonyl compounds** – they contain a C=O group.

HL steric effects prevent S_N2 mechanism

more stable carbocation

react with halogens in presence of sunlight or UV light through free radical substitution

substitution – an atom or group is replaced by another atom or group

HL Aldehydes and carboxylic acids can be reduced to primary alcohols.

rate not affected by nature or concentration of nucleophile

faster with protic polar solvents

Nucleophile – molecule or negative ion with a lone pair of electrons which is attracted to regions of low electron density.

tertiary halogenoalkane

Cl_2 splits into chlorine free radicals.

free radical – a species with an unpaired electron

HL

S_N1 mechanism two steps

secondary halogenoalkane

homolytic fission – each atom gets one electron from the bond that splits

The free radicals react with the alkane molecules to form **halogenoalkanes**.

Undergo **nucleophilic substitution**. The nucleophile replaces the halogen.

S_N2 mechanism one step

primary halogenoalkane

faster with aprotic polar solvents

contains a halogen atom as functional group

reactions rate: R–I > R–Br > R–Cl > R–F

faster with ions than with neutral nucleophiles

stereospecific

Measurement and data processing 11

11.1 Uncertainties and errors in measurements and results

11.1.1 Uncertainties in measurements

Qualitative and quantitative data

Qualitative data are non-numerical data. These would normally be observations made during an experiment. For instance, when determining the empirical formula of magnesium oxide by burning magnesium in air you might make the following observations:

- the piece of magnesium is grey and not very shiny
- the electronic balance fluctuated as the magnesium was weighed
- the crucible was dirty
- the magnesium burned with a bright, white flame
- white smoke escaped as the lid of the crucible was lifted.

These are all pieces of qualitative data.

Quantitative data are numerical data. So, for example, all the measurements of mass made during the experiment above constitute quantitative data.

Measurements in science

Although in mathematics it is possible to have pure numbers and to write a number to any number of decimal places, in chemistry we are concerned with real quantities obtained by making measurements in a laboratory.

It is not possible to measure the actual, or true, value of a particular quantity. The true mass of a piece of magnesium ribbon could be 0.257 846 368 246 89 g, but we have no way of actually measuring that. The best we can ever get is an estimate. If we put the piece of magnesium on an electronic balance that gives values to one decimal place, we might measure 0.3 g. If we had a balance that could measure to three decimal places, then we might get 0.258 g. Neither of these values is the true mass of the piece of magnesium ribbon.

Consider the following argument.
Experimental data can never provide a true value for a quantity. The knowledge we get from science is ultimately obtained from experimental data. The knowledge we obtain from science is never true.

Figure 11.1 This electronic balance reads to two decimal places.

Reading	Mass / g
1	0.27
2	0.28
3	0.28
4	0.27
5	0.27
6	0.27
7	0.26
8	0.28
9	0.28
10	0.27

Table 11.1 The results obtained from weighing ten strips of magnesium ribbon.

The effect of random uncertainties can be reduced by repeating the measurements more often.

The random uncertainties can never be completely eliminated.

The effects of random uncertainties should mean that the measurements taken will be distributed either side of the mean – i.e. fluctuations will be in both directions.

Random uncertainties

Suppose we need to find the mass of a piece of magnesium ribbon 20 cm long. To do this we can measure a piece of magnesium 20 cm long using a ruler on which the smallest division is 1 mm. The strip of magnesium ribbon is cut from a roll and weighed on an electronic balance that reads to two decimal places. Suppose that mass of the first strip of magnesium ribbon is 0.27 g (Figure **11.1**). We then cut and measure nine more strips in the same way from the same roll, and the results are shown in Table **11.1**.

The first thing we should notice is that not all the readings are the same. This is because of **random uncertainties**.

> Random uncertainties are caused by the limitations of the measuring apparatus and other uncontrollable variables that are inevitable in any experiment.

There are several sources of random uncertainty, even in this simple procedure. Firstly, there will be slight variations in the length of the strips of magnesium ribbon that we cut because of the limitations of the ruler that we use to measure it, the thickness of the scissors etc.; secondly, when we measure the mass using the electronic balance, there could be variations due to air currents in the room, the heating effects of the current in the circuits, friction between various mechanical parts etc.

We can give an indication of the size of the random uncertainty by quoting the measured value along with an uncertainty. For instance, the mass of a piece of magnesium ribbon 20 cm long could be quoted as 0.27 ± 0.01 g, where '± 0.01 g' is the uncertainty in the measurement. This means that we are reasonably confident that the mass of a piece of magnesium ribbon 20 cm long is somewhere between 0.26 g and 0.28 g. If we kept cutting strips from this roll we would expect each mass to lie between 0.26 g and 0.28 g.

Taking more repeats gives a more reliable mean value. If we had just taken one measurement in the above example, we could have obtained 0.26 g, 0.27 g or 0.28 g as the value of the mass of the strip of magnesium. If, however, we take the reading 100 times and work out the mean value, we should get a value that more closely represents the true mean.

The effect of random uncertainties can also be minimised by careful design of an experiment. For instance, if you plan to carry out an experiment using 0.05 g magnesium and to measure the mass with a two decimal place balance then the uncertainty of the mass will be ± 0.01 g, which is 20% of the mass. If, however, you carry out the experiment with 0.20 g magnesium then the percentage uncertainty due to random error is reduced to 5%.

No matter to how many decimal places a piece of measuring apparatus is able to measure a quantity, there will always be an uncertainty in the value.

Estimating the random uncertainties associated with measuring apparatus

Analogue instruments

As a rule of thumb, the uncertainty of a measurement is half the smallest division to which you take a reading. This may be the division on the piece of apparatus used or it may be an estimate if the divisions are sufficiently far apart that you are able to estimate between them.

For instance, the smallest division on the $100\,\text{cm}^3$ measuring cylinder shown in Figure **11.2** is $1\,\text{cm}^3$; however, the divisions are sufficiently far apart that we can probably estimate between these divisions to the nearest $0.5\,\text{cm}^3$. We can thus estimate that our actual value is greater than 68.75 and less than 69.25. The uncertainty is half the smallest division, i.e. $0.25\,\text{cm}^3$, or $0.3\,\text{cm}^3$ to one significant figure (uncertainties are usually quoted to only one significant figure). We can therefore quote the reading as $69.0 \pm 0.3\,\text{cm}^3$, i.e. the volume of water is somewhere between 68.7 and $69.3\,\text{cm}^3$.

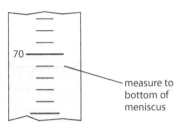

Figure 11.2 Measuring a liquid volume.

Digital instruments

On an electronic balance reading to two decimal places, the smallest division is 0.01 g, so the uncertainty associated with this is ± 0.005 g. So if a particular reading is taken as 2.46 g, this means that the value is somewhere between 2.455 g and 2.465 g.

However, the balance will have been zeroed beforehand and this zero value also has an uncertainty associated with it. That is, when the reading on the balance is 0.00 g it indicates a value between -0.005 and $+0.005$ g. The largest possible value of the measured mass would be from -0.005 to 2.465 g, i.e. 2.47 g, and the smallest possible value is from 0.005 to 2.455 g, i.e. 2.45 g. So, the uncertainty of this measurement is ± 0.01 g, and the measurement should be quoted as 2.46 ± 0.01 g.

> In general, the uncertainty of a measurement made on a digital instrument should be quoted as $\pm$ the smallest division.

Note: this is only an estimate, so if you quoted this value as 69.0 ± 0.5 cm, this would not be incorrect.

Human limitations

There will also be random errors associated with the person doing the measurement. Imagine an experiment in which a student is measuring the time taken for a reaction mixture to change colour in order to determine the rate of reaction. If the student is using a digital stopwatch that can measure to 1/100 second then the above discussion would lead to an uncertainty of ± 0.01 s. However, if you actually try doing this experiment there will always be a delay between noticing the colour change and stopping the stopwatch – this is called **reaction time**. In this case, you should estimate the uncertainty of the measurement of time and not just use that from the stopwatch. A suitable uncertainty for this experiment might be ± 0.5 s or ± 1 s.

Precision

> **Precision** relates to the reproducibility of results. If a series of readings is taken with high precision, it indicates that the repeated values are all very close together and close to the mean (average) value.

The uncertainties here are worked out from the range of the data, i.e. looking at how far the maximum and minimum values are from the mean.

Consider the values in Table **11.2**. The results in the first column could be reported as $21.2 \pm 0.1\,cm^3$, whereas the second column would yield $21.4 \pm 0.7\,cm^3$. The larger uncertainty in the second set of readings indicates the greater spread of the values, i.e. lower precision.

If a single reading is taken, the uncertainty gives us an indication of the precision of the reading – for instance a temperature recorded as $21.33 \pm 0.01\,°C$ is more precise than $21.3 \pm 0.1\,°C$. In the first case, repeat values of the quantity would be expected to mostly lie between 21.32 and 21.34 °C, whereas in the second (less precise) case, the values would be expected to lie between 21.2 and 21.4 °C.

It is not really correct to talk about the precision of a single reading, but the term is sometimes used. For a single reading, a more precise value is a value to more significant figures.

Extension

Standard deviation is often used to give an idea of the precision of a set of measurements. The larger the standard deviation, the less precise the measurements.

Reading	Temperature/°C	Temperature/°C
1	21.2	21.0
2	21.3	21.9
3	21.1	22.1
4	21.3	21.2
5	21.1	20.7
6	21.2	21.5
mean value	21.2	21.4

Table 11.2 Measurements of the same quantity using two different pieces of apparatus: the first set of readings is more precise than the second set, because the values in the first set are much closer to each other and to the mean value.

Accuracy

> **Accuracy** refers to how close a measurement is to the actual value of a particular quantity.

It is possible for a measurement to have great precision but to not be very accurate.

In the example discussed above the true value of the temperature was 19.57 °C – although a value of $21.33 \pm 0.01\,°C$ is quite precise, it is not very accurate because it is not very close to the true value.

In this case, there appears to be some sort of **systematic error** with the procedure (for instance, perhaps the thermometer was not originally calibrated properly). In this case, repeating the readings will not improve the accuracy of the measurements because all the values would be expected to be around $21.33 \pm 0.01\,°C$.

Systematic errors

A systematic error is an error introduced into an experiment by the apparatus or the procedure. Systematic errors result in a loss of accuracy, i.e. the measured value is further away from the true value.

Systematic errors are always in the same direction. For instance, you might measure the mass of some sodium chloride using a balance that has not been calibrated recently and the balance might always record a mass that is 1.00 g too high. The actual value of the mass is 23.25 g, but the balance reads 24.25 g. No matter how many times this reading is repeated, the mass will always be 1.00 g too high.

The above experiment could probably have been improved by using several different balances and taking an average of the values from all of them. (However, this is still no guarantee of greater accuracy because each might have a systematic error associated with it!)

Consider an experiment carried out to measure an enthalpy change of neutralisation by reacting $50 \, cm^3$ of $0.10 \, mol \, dm^{-3}$ sodium hydroxide and $50 \, cm^3$ of $0.10 \, mol \, dm^{-3}$ hydrochloric acid in a beaker.

Systematic errors can be identified by comparison with accepted literature values. The above experiment might give a calculated heat of neutralisation of $-55.8 \pm 0.1 \, kJ \, mol^{-1}$. The ± 0.1 indicates the uncertainties due to random errors. The accepted literature value for this quantity is $-57.3 \, kJ \, mol^{-1}$. We can use the **percentage error** to compare the experimental value with the accepted literature value.

> The effect of a systematic error cannot be reduced by repeating the readings.

$$\text{percentage error} = \frac{|\text{experimental value} - \text{accepted value}|}{|\text{accepted value}|} \times 100$$

> $|value|$ indicates the modulus of the value; i.e. ignore the overall sign

In this case, the percentage error $= \dfrac{|55.8 - 57.3|}{57.3} \times 100 = 2.6\%$

The percentage uncertainty due to random uncertainties is

$\dfrac{0.1}{55.8} \times 100 = 0.2\%$

The percentage error in this experiment is greater than the percentage random uncertainty. This suggests that the experiment involves some systematic errors.

Some of the systematic errors in this experiment could be:
- the beaker is not that well insulated so heat will escape – the measured temperature rise will be less than the actual value
- the reaction does not occur instantaneously and the thermometer does not respond instantaneously and so the measured temperature rise will be less than the actual value
- the concentration of the sodium hydroxide is less than $0.10 \, mol \, dm^{-3}$ and so the measured temperature rise will be less than the actual value.

These systematic errors can be reduced by changing the way the experiment is carried out, for instance by using an insulated container

with a lid in which to do the reaction or by checking the concentration of the sodium hydroxide by titration against a standard solution of an acid.

When compared to the literature value, if the percentage error is smaller than that due to random errors then the experiment has worked well as a way of obtaining a value for this particular quantity. If the random errors are, however, quite large then the experiment can be further refined by using more precise measuring apparatus and/or taking more repeat measurements.

If the percentage error is larger than the percentage uncertainty due to random errors then this suggests that there are systematic errors in the experimental procedure – these must be eliminated as far as possible to obtain a more accurate value for the measured quantity. In the case of systematic errors redesign of the experiment must be considered.

If your experiment involves finding a value for a quantity for which no literature value exists, it can be very difficult to spot systematic errors.

The difference between accuracy and precision

As we have seen, precision refers to the reproducibility of results (how close repeat readings are to each other and to the mean value), whereas accuracy refers to how close a value is to the true value of the measurement. The diference between accuracy and precision can be seen in Figure **11.3**.

Similarily, think about a tennis player serving the ball (Figure **11.4**). If he can hit nearly the same spot each time then he is serving with a great deal of precision. If this spot is not in the service box, however, then the serve is not very accurate. In this case (and in the case of scientific experiments) precision without accuracy is useless – it doesn't win any points!

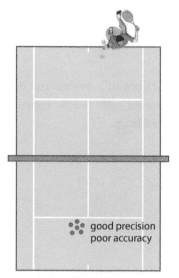

Figure 11.4 Like a game of tennis, the goal of a scientific experiment is high precision and high accuracy.

The Hubble Space Telescope, a collaborative project of NASA and the European Space Agency (ESA), was launched in 1990 but it was very soon noticed that there was a problem with the mirror because the quality of the images was not as high as expected. Although the mirror was shaped to a high degree of precision (very small uncertainty), the device used to control the shape was not constructed correctly, which resulted in a systematic error in the shape of the mirror. This problem was corrected in a subsequent space mission. Time for using the Hubble telescope and data it collects has been made available to scientists throughout the world and has contributed enormously to our understanding of astronomy.

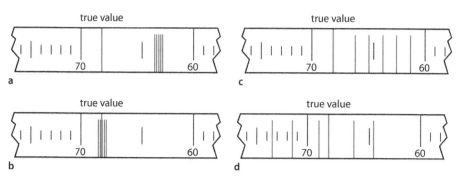

Figure 11.3 a A set of readings (shown in red) with high precision but poor accuracy; **b** a set of readings with high precision and high accuracy; **c** a set of readings with low precision and low accuracy; **d** a set of readings with low precision and high accuracy. In this last set, the accuracy is high because the mean value is close to the true value. However, it is a poor set of data and the high accuracy is probably more a case of luck than good experimental design!

? Test yourself

1 The table shows data from five trials of two separate experiments. Which set of values is more precise?

Trial	Experiment	
	A	B
1	1.34	1.37
2	1.48	1.22
3	1.40	1.58
4	1.27	1.18
5	1.38	1.44

2 A series of experiments was carried out to determine the value for a particular quantity. The results are shown in the table. The literature value of this quantity is 47.0 J. Which experiment produced the most accurate value?

Experiment	Value / J
1	45.276
2	48
3	44.2
4	49.57

Quoting values with uncertainties

Consider a value of 1.735 ± 0.1 obtained from an experiment.

The uncertainty is in this decimal place, so no figures should be quoted beyond it. This quantity should then be quoted as 1.7 ± 0.1. Table **11.3** shows some more examples.

With 363.2 ± 8, the random uncertainties mean that the value is known only to within 8 either way – the 0.2 is meaningless. With 0.0650 ± 0.0001, the final zero is required because the uncertainty is in the fourth decimal place. If the value were quoted as 0.065, this would imply that the value lies between 0.0645 and 0.0655 rather than between 0.0651 and 0.0649, and therefore there would be a loss of precision.

The uncertainty in a mean value

There are various more or less complicated ways of quoting the uncertainty in a mean (average) value. Consider the data given in Table **11.4**. One way of quoting the mean is as $21.4 \pm 0.7 \, \text{cm}^3$. Here we are taking the uncertainty as the deviation of the maximum and minimum values from the mean. The largest value is $22.1 \, \text{cm}^3$, which is $0.7 \, \text{cm}^3$ greater than the mean value, and the smallest value is $20.7 \, \text{cm}^3$, which is $0.7 \, \text{cm}^3$ smaller than the mean value. However, if we have a large number of readings, quoting the uncertainty based on the range is probably pessimistic, because most of the values would be expected to lie closer to the mean than that.

A rough rule of thumb is to take the uncertainty of the mean to be two-thirds of the deviation from the mean. For example, with the data in Table **11.4**, the deviation from the mean is $\pm 0.7 \, \text{cm}^3$ and two-thirds of this is approximately $0.5 \, \text{cm}^3$ – so we could quote our average value as $21.4 \pm 0.5 \, \text{cm}^3$ and be reasonably confident that, if we make any further measurements, most of them will lie between 20.9 and $21.9 \, \text{cm}^3$.

The uncertainty is usually quoted to one significant figure, and your measurement should be stated so that the uncertainty is in the last significant figure – no figures should be quoted after the uncertainty.

Measured value ± uncertainty	Value you should quote
151.3 ± 0.2	151.3 ± 0.2
157.47 ± 0.1	157.5 ± 0.1
0.06500 ± 0.0001	0.0650 ± 0.0001
363.2 ± 8	363 ± 8
363.2 ± 20	360 ± 20
363.2 ± 100	400 ± 100

Table 11.3 Measured values with uncertainties and how they should be quoted.

Reading	Volume / cm³
1	21.0
2	21.9
3	22.1
4	21.2
5	20.7
6	21.5
mean value	21.4

Table 11.4 The same quantity has been measured six times and the average taken.

Value	Number of significant figures	Number of decimal places
23.14	4	2
0.012	2	3
1.012	4	3
100.35	5	2
0.001 005 0	5	7
50.0	3	1

Table 11.5 The difference between significant figures and decimal places.

Value	Number of significant figures	Rounded value
27.346	3	27.3
27.346	4	27.35
0.03674	2	0.037
0.03674	3	0.0367
0.399 967 2	3	0.400
0.399 967 2	4	0.4000
0.399 967 2	5	0.399 97

Table 11.6 Rounding to the appropriate number of significant figures.

Significant figures and decimal places

When counting significant figures, we start counting from the left with the first non-zero digit. For example, 0.012 has two significant figures because we do not count the first two zeros. 0.001 005 0 has five significant figures because we do not count the first three zeros but we must count any zeros after the first non-zero digit and after a decimal point (Table **11.5**).

Problems come when numbers are quoted as 500, for example, because this could represent one, two or three significant figures. This ambiguity is avoided by quoting the number in standard form. If it is quoted as 5.0×10^2 there are two significant figures, and if it is quoted as 5.00×10^2 there are three significant figures.

> The number could also be quoted with an uncertainty, e.g. 500 ± 1, so that it is clear that the value is known to three significant figures.

It is important to realise that 5 g, 5.0 g and 5.00 g are not the same. 5 g implies that the mass is between 4.5 and 5.5 g; 5.0 g implies that the mass is between 4.95 and 5.05 g; 5.00 g implies that the mass is between 4.995 and 5.005 g. A value of 5 g measured using a two decimal place electronic balance must be quoted as 5.00 g and not 5 g, otherwise there will be a loss of precision in your results.

Rounding to the appropriate number of significant figures

If a number is to be quoted to a certain number of significant figures, then we must look at the next figure after the last one that we wish to quote. If the next figure to the right is five or greater, the last significant figure should be rounded up; if the next figure to the right is less than five then the last significant figure stays the same. Some examples of rounding to the appropriate number of significant figures are shown in Table **11.6**.

Recurring decimals such as $0.\dot{3}$ should never be used in chemistry, as they imply infinite precision. A value should always be rounded to the appropriate number of significant figures, e.g. as 0.33 or 0.333.

? Test yourself

3 Copy and complete the table:

Measured value ± uncertainty	Value you should quote
71.7 ± 0.2	
3.475 ± 0.01	
0.065 06 ± 0.001	
63.27 ± 5	
593.2 ± 30	
783.28 ± 100	

4 State the number of significant figures for each of the following numbers:

  a 2.78   **d** 3.95×10^4

  b 0.057   **e** 80.00

  c 0.003 480

5 Round each of the following numbers to three significant figures:

  a 6.7863   **d** $8.245 7 \times 10^5$

  b 0.000 079 835   **e** $1.783 39 \times 10^{-3}$

  c 0.004 999 31

11.1.2 Uncertainties in calculations

Decimal places and calculations

> If a calculation involves just adding or subtracting numbers, the final answer should be quoted to the same number of **decimal places** as the piece of original data that has the fewest **decimal places**.

For example, $23.57 - 8.4 = 15.17$, but the answer should be quoted as 15.2 because 8.4 has only one decimal place.

Other examples of calculations are shown in Table **11.7**.

Calculation			Actual result	Value to quote
23.5	−	14.8	8.7	8.7
0.786	+	0.0367	0.8227	0.823
5.234×10^3	−	1.2×10^3	4.034×10^3	4.0×10^3

Table 11.7 Some examples of calculations involving addition and subtraction.

Significant figures and calculations

> When carrying out calculations involving multiplication and/or division, the general rule is that the final answer should be quoted to the number of **significant figures** of the piece of data with the fewest significant figures.

Example

Sulfuric acid is titrated against $25.00 \, cm^3$ of $0.200 \, mol \, dm^{-3}$ sodium hydroxide solution. $23.20 \, cm^3$ of sulfuric acid are required for neutralisation. Calculate the concentration of the sulfuric acid.

$$2NaOH(aq) + H_2SO_4(aq) \rightarrow Na_2SO_4(aq) + 2H_2O(l)$$

The volumes are quoted to four significant figures but the concentration is only to three significant figures – therefore the final answer should be quoted to three significant figures.

Carrying all the numbers through in the calculation gives an answer of $0.1077586 \, mol \, dm^{-3}$ so the concentration of the sulfuric acid should be quoted as $0.108 \, mol \, dm^{-3}$

Other examples of calculations are shown in Table **11.8**.

Calculation			Actual result	Value to quote
23.5	×	14.87	349.445	349
0.79	÷	0.0367	21.52588556	22
5.234×10^3	×	1.2×10^3	6.2808×10^6	6.3×10^6

Table 11.8 Some examples of calculations involving multiplication or division.

Learning objectives

- Quote the result of a calculation involving multiplication/ division or involving addition/ subtraction to the appropriate number of decimal places
- Quote the result of a calculation to the appropriate number of significant figures
- Understand what is meant by absolute uncertainties and percentage uncertainties
- Understand how to combine uncertainties in calculations

'25.00' implies that the volume is somewhere between 24.995 and $25.005 \, cm^3$.

$$\frac{c_1 V_1}{n_1} = \frac{c_2 V_2}{n_2}$$

Rounding values in calculations

When carrying out multi-stage calculations, it is important to avoid rounding errors that could, after several stages, introduce large inaccuracies into the calculation. As a general rule, all numbers should be carried through in a calculation and rounding should only happen when an answer to a particular part of a question is required.

Absolute and percentage uncertainties

An uncertainty can be reported either as an absolute value – e.g. $1.23 \pm 0.02\,\text{g}$ – or as a percentage value – e.g. $1.23\,\text{g} \pm 2\%$.

The **percentage uncertainty** is worked out using the equation:

> A percentage uncertainty has no units.

$$\text{percentage uncertainty} = \frac{\text{absolute uncertainty}}{\text{value}} \times 100$$

For example, for $0.257 \pm 0.005\,\text{cm}$:

$$\text{percentage uncertainty} = \frac{0.005}{0.257} \times 100 = 2\%$$

The absolute uncertainty can be worked out from the percentage uncertainty using the equation:

$$\text{absolute uncertainty} = \frac{\text{percentage uncertainty}}{100} \times \text{value}$$

For example, if the final value of a calculation is $0.518 \pm 1\%$:

$$\text{absolute uncertainty} = \frac{1}{100} \times 0.518 = 0.005 \text{ (to one significant figure).}$$

Therefore the final answer is 0.518 ± 0.005.

Propagating uncertainties in calculations

Adding or subtracting

When quantities with uncertainties are added or subtracted, the **absolute** uncertainties are **added**.

Worked example

11.1 Calculate the change in temperature from the following data:

	Value	Uncertainty
Maximum temperature/°C	57.58	±0.02
Initial temperature/°C	23.42	±0.02

change of temperature = maximum temperature − initial temperature

change of temperature = $57.58 - 23.42 = 34.16\,°\text{C}$

The uncertainty in the change of temperature is obtained by adding the uncertainty in the initial temperature to the uncertainty in the maximum temperature:

uncertainty in change of temperature $= 0.02 + 0.02 = 0.04\,°C$

Therefore the change of temperature is quoted as $34.16 \pm 0.04\,°C$.

That this is appropriate can be seen by subtracting the minimum possible value for the initial temperature $(23.42 - 0.02 = 23.40)$ from the maximum possible value for the final temperature $(57.58 + 0.02 = 57.60)$ and vice versa.

Multiplying or dividing

When multiplying or dividing quantities with uncertainties, the **percentage** uncertainties should be added.

Worked example

11.2 What is the absolute uncertainty when 2.57 ± 0.01 is multiplied by 3.456 ± 0.007 and to how many significant figures should the answer be quoted?

$2.57 \times 3.456 = 8.881\,92$

Percentage uncertainties:

$$\frac{0.01}{2.57} \times 100 = 0.39\% \qquad\qquad \frac{0.007}{3.456} \times 100 = 0.20\%$$

total percentage uncertainty $= 0.39 + 0.20 = 0.59\%$

To work out the absolute uncertainty of the final value, it is multiplied by its percentage uncertainty:

absolute uncertainty $= \dfrac{0.59}{100} \times 8.88192 = 0.05$ (to one significant figure).

The absolute uncertainty is in the second decimal place, and therefore no figures should be quoted beyond that. The final answer should be quoted as 8.88 ± 0.05.

> When multiplying or dividing a quantity with an uncertainty by a pure number, the absolute uncertainty is multiplied/divided by that number so that the percentage uncertainty stays the same.

So, if 12.12 ± 0.01 (percentage uncertainty $= 0.083\%$) is multiplied by 3, the answer is 36.36 ± 0.03 (percentage uncertainty $= 0.083\%$). If 2.00 ± 0.03 (percentage uncertainty $= 1.5\%$) is divided by 3, the answer is 0.67 ± 0.01 (percentage uncertainty $= 1.5\%$).

> Sometimes the uncertainty of one quantity is so large relative to the uncertainties of other quantities that the uncertainty of the final value can be considered as arising just from this measurement.

This will be considered in the next worked example.

Worked example

11.3 Weigh out accurately approximately $100\,g$ of water in a polystyrene cup. Measure the initial temperature of the water. Weigh out accurately approximately $6\,g$ of potassium bromide. Add the potassium bromide to the water, stir rapidly until it has all dissolved and record the minimum temperature reached. Use the data below to work out the enthalpy change of solution to the appropriate number of significant figures.

mass of polystyrene cup/g	5.00 ± 0.01
mass of polystyrene cup + water/g	105.23 ± 0.01
initial temperature of water/°C	21.1 ± 0.1
minimum temperature of water/°C	19.0 ± 0.1
mass of weighing boat/g	0.50 ± 0.01
mass of weighing boat + potassium bromide/g	6.61 ± 0.01

The specific heat capacity of water
$= 4.18\,J\,g^{-1}\,°C^{-1}$.

mass of water $= (105.23 \pm 0.01) - (5.00 \pm 0.01) = 100.23 \pm 0.02\,g$

percentage uncertainty of mass of water $= \dfrac{0.02}{100.23} \times 100 = 0.02\%$

mass of KBr $= (6.61 \pm 0.01) - (0.50 \pm 0.01) = 6.11 \pm 0.02\,g$

percentage uncertainty of mass of KBr $= \dfrac{0.02}{6.11} \times 100 = 0.3\%$

change in temperature of water $= (21.1 \pm 0.1) - (19.0 \pm 0.1) = 2.1 \pm 0.2\,°C$

percentage uncertainty of temperature change $= \dfrac{0.2}{2.1} \times 100 = 9.5\%$

The percentage uncertainty of the change in temperature is much larger than the other two uncertainties and therefore we can assume that the uncertainty of the final value is also going to be about 9.5%.

heat taken in $= mc\Delta T = 100.23 \times 4.18 \times 2.1 = 879.82\,J$

number of moles of KBr $= \dfrac{6.11}{119.00} = 0.0513\,mol$

enthalpy change $= \dfrac{879.82}{0.0513} = 17150\,J\,mol^{-1} = 17.150\,kJ\,mol^{-1}$

We must now consider to how many significant figures the answer can be quoted. We will take the percentage uncertainty of the final answer as 9.5% and must work out 9.5% of 17.150.

absolute uncertainty $= \dfrac{9.5}{100} \times 17.150 = 2$ (to one significant figure).

Therefore, the enthalpy change of solution should be quoted as $17 \pm 2\,kJ\,mol^{-1}$.

Working out the total percentage error as $(0.02 + 0.3 + 9.5)\%$ would have made no difference to the final value or uncertainty.

Nature of science

Scientific work often involves the collection of quantitative data, but scientists must always be aware of the limitations of their data. When publishing, for instance, a value for the relative atomic mass of sulfur, to how many decimal places can the value be quoted?

Scientists often repeat experiments to reduce random uncertainties but the effects of systematic errors can be more difficult to spot, especially when reporting the value of a quantity for the first time, because there is nothing with which to compare it.

? Test yourself

6 Copy and complete the table:

0.345 ± 0.001	+	0.216 ± 0.002	=	
23.45 ± 0.03	−	15.23 ± 0.03	=	
0.0034 ± 0.0003	+	0.0127 ± 0.0003	=	
1.103 ± 0.004	−	0.823 ± 0.001	=	
1.10 ± 0.05	+	17.20 ± 0.05	=	

7 Copy and complete the table:

Value	Percentage uncertainty %
27.2 ± 0.2	
0.576 ± 0.007	
4.46 ± 0.01	
$7.63 \times 10^{-5} \pm 4 \times 10^{-7}$	

8 a What is the absolute uncertainty when 2.13 ± 0.01 is multiplied by 4.328 ± 0.005? Give the final answer to the appropriate number of significant figures.

b What is the absolute uncertainty when 48.93 ± 0.02 is divided by 0.567 ± 0.003. Give the final answer to the appropriate number of significant figures.

9 Use the equation $E = mc\Delta T$ and the values in the table to calculate the energy released, to the appropriate number of significant figures, when a sample of a solution cools:

Mass of solution (m)	43.27 ± 0.01 g
Temperature change (ΔT)	22.8 ± 0.2
Specific heat capacity (c)	$4.2 \, J \, g^{-1} \, {}^{\circ}C^{-1}$

11.2 Graphical techniques

A graph is a very useful way of presenting the relationship between two quantities. Consider a graph of rate of reaction against concentration.

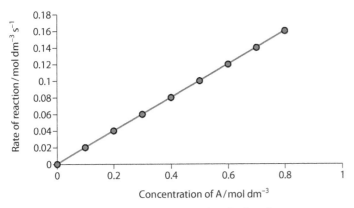

Figure 11.5 Rate of reaction against concentration for A → B.

Learning objectives

- Understand that graphs are an effective way of communicating the relationship between two variables
- Plot graphs of experimental results and interpret the graphs
- Calculate the gradient and intercept in graphs

$$y = mx + c$$

The relationship between volume and pressure, shown in Figure **11.7**, is an inversely proportional one:

$$\text{pressure} \propto \frac{1}{\text{volume}}$$

This means that doubling the volume causes the pressure to be halved, and vice versa.

That this is indeed an inversely proportional relationship can be confirmed by plotting pressure against $\frac{1}{\text{volume}}$. Because pressure is proportional to $\frac{1}{\text{volume}}$, this graph should produce a straight line through the origin (Figure **11.8**).

The graph in Figure **11.9** shows that the rate of reaction increases as the concentration of A increases. The relationship between the two quantities is, however, not immediately obvious, other than that it is not a

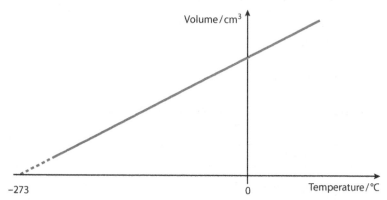

Figure 11.6 Volume versus temperature at constant pressure.

proportional relationship and that the rate of reaction more than doubles when the concentration of A doubles.

We can further analyse these data by trying out some relationships between the rate of reaction and concentration. If we try plotting the rate of reaction against (concentration of A)^2 (Figure **11.10**), we get a straight line through the origin.

The proportional relationship indicated by the straight line through the origin tells us that rate is proportional to (concentration of A)^2.

If this graph had not resulted in a straight line, we could have tried plotting rate against (concentration of A)^3 and so on until a proportional relationship was obtained.

Drawing graphs

There are several general rules for drawing graphs.
1 Make the graph as large as possible. Choose your scales and axes to retain the precision of your data as far as you can and to make the graph

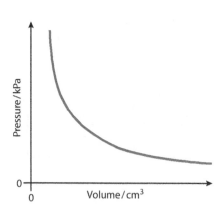

Figure 11.7 Pressure versus volume at constant temperature.

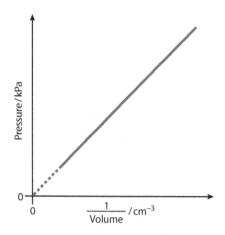

Figure 11.8 Pressure versus $\frac{1}{\text{volume}}$.

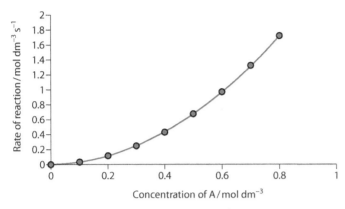

Figure 11.9 Rate of reaction against concentration of A.

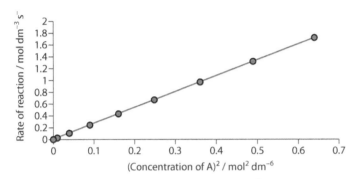

Figure 11.10 Rate of reaction against (concentration of A)^2.

as easy as possible to interpret. A graph would usually be expected to fill most of a piece of graph paper. Suppose the data in Table **11.9** were collected in an experiment.

Choosing a scale for the temperature in which the smallest square is one unit would make it very difficult to plot 23.12 precisely. Ideally, if possible, one small square should represent one unit of the last significant figure of your data, i.e. 0.01 °C in this case. This may not be possible with the size of the piece of graph paper available, however!

Extension

This trial and error process can be laborious, and a shortcut is to plot log(rate of reaction) against log(concentration of A). In this case, a straight-line graph with gradient 2 will be obtained. The gradient indicates that the rate is proportional to the concentration of A to the power 2. This works for relationships of the form $y = x^n$. Either $\log_{10}$ or ln may be used.

Time/s	Temperature/°C
0	23.12
30	25.56
60	28.78
90	29.67
120	30.23

Table 11.9 Sample data.

The independent variable is what is changed in an experiment to investigate its effect.

The dependent variable is what is measured in an experiment.

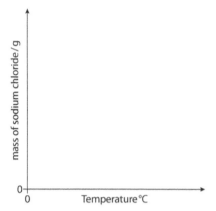

Figure 11.11 The dependent variable goes on the y-axis and the independent variable on the x-axis.

If the volume to be plotted is 23.0 cm^3, dividing by cm^3 gives the number 23.0, which will be plotted on the graph.

The graphs shown here are for continuous data – that is, the quantities plotted can take any value. When discrete data are plotted a best-fit line should not be drawn, and points may be joined or the data could be plotted as a bar chart.

If the pressure to be plotted is 5.34×10^5 Pa, then dividing that by 10^5 Pa means that the number can be plotted on the graph as 5.34.

$$\text{gradient} = \frac{\text{change in } y}{\text{change in } x}$$

2   The independent variable (what you change) should be plotted along the horizontal (x) axis and the dependent variable (what you measure) should be plotted along the vertical (y) axis.

For instance, you are given the task of investigating how the solubility of common salt (sodium chloride) is affected by temperature. You could carry out a series of experiments varying the temperature (independent variable) to determine what mass of salt (dependent variable) dissolves in a fixed volume of water at each temperature. A graph could now be drawn with mass of sodium chloride on the y-axis and temperature on the x-axis – Figure **11.11**.

Another graph could then also be drawn of the processed data – in this case solubility could be plotted on the y-axis and temperature on the x-axis.

3   Label the axes with the quantity and units. There are various conventions for laying out the units. One way that is commonly used is to have the quantity divided by the units, e.g. volume / cm^3 and pressure / 10^5 Pa. You could also put the units in brackets, e.g. volume (cm^3) and pressure (Pa).

4   Plot the points, which may be marked by crosses or by dots in circles.

5   Draw a best-fit line (do not join the points!). This may be a straight line or a curve (Figure **11.12**) and should represent, as well as possible, the trend in the data. The points should be evenly distributed about the line. A line of best fit is very much a matter of judgement and no two lines of best fit drawn by different people will be identical.

6   Give the graph a title describing what has been plotted.

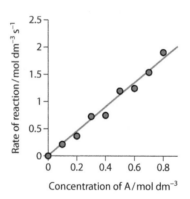 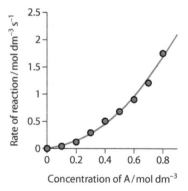

Figure 11.12 Best-fit lines.

Deriving quantities from graphs

The gradient (slope) is one of the most important quantities that can be derived from a graph. A gradient gives us an idea how much one quantity (the dependent variable) is affected by another quantity (the independent variable). If the gradient is large, a small change in the independent variable has a large effect on the dependent variable. Sometimes the gradient can also have an important physical meaning – e.g. if a graph of the amount of a reactant or a product of a chemical reaction against time is plotted, the gradient represents the rate of the reaction.

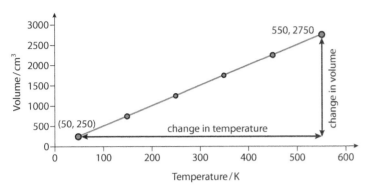

Figure 11.13 Volume versus temperature for a fixed mass of an ideal gas.

Consider the graph of the volume of a fixed mass of gas against temperature shown in Figure **11.13**.

When determining the gradient from a straight-line graph, a triangle is drawn as shown in Figure **11.13**, and the change in the quantity on the vertical axis is divided by the change in the quantity on the horizontal axis. The triangle should be made as large as possible to minimise any uncertainties caused by reading from the graph.

$$\text{gradient} = \frac{\text{change in volume}}{\text{change in temperature}} = \frac{2750 - 250}{550 - 50} = 5$$

The units of the gradient are obtained by dividing the units of volume (*y*-axis) by the units of temperature (*x*-axis), i.e. cm^3/K or $cm^3\,K^{-1}$ in this case. Therefore the gradient (slope) of the line is $5\,cm^3\,K^{-1}$.

Another important quantity that can be obtained from a graph is the intercept on the *y*-axis. An example of a calculation involving determination of the gradient and intercept is given on page **268** (in Topic **6**).

Finding the gradient of a curve

A curved line has a constantly changing gradient, and therefore we can work out the gradient only at specific points. This is done by drawing a tangent (straight line) to the curve at the desired point and working out the gradient of the tangent as usual.

In this graph the tangent is drawn at the initial point (Figure **11.14**) and the initial rate of reaction is the gradient of this tangent.

$$\text{gradient} = \frac{70 - 0}{32 - 0} = 2.2\,cm^3\,s^{-1}$$

The initial rate of reaction is therefore $2.2\,cm^3\,s^{-1}$.

Nature of science

The idea of the correlation between variables can be tested experimentally and the results displayed on a graph.

It is important that the public have an understanding of certain scientific issues and graphs can often provide a convenient way of presenting data and showing a correlation between factors. Scientists, however, have a responsibility to ensure that their data are reported fairly and accurately – incomplete graphs or graphs without scales can be misleading for example.

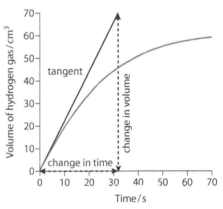

Figure 11.14 Calculating a gradient.

A tangent should be drawn as large as possible to minimise uncertainties caused by reading values off the graph.

? Test yourself

10 Describe the relationship between the two variables shown in each of the following graphs.

a

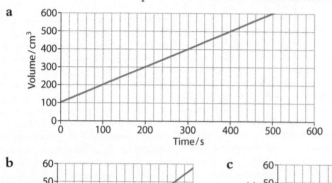

b

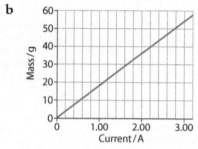

c

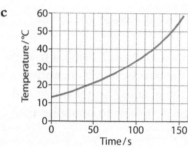

11 Calculate the gradients, with units, for the graphs in question **10a** and **b**.

Learning objectives

- Understand how the number of double bonds/rings in a compound can be worked out from a molecular formula
- Understand how to use mass spectrometry to gain information about the structure of an organic molecule
- Understand how to use infrared spectroscopy to gain information about the structure of an organic molecule
- Understand how to use proton NMR spectroscopy to gain information about the structure of an organic molecule

11.3 Spectroscopic identification of organic compounds

Determination of the number of multiple bonds/rings in a molecule

If we compare the molecular formulas of some organic compounds, we can see the relationship between the formula and the number of multiple bonds/rings present – Table **11.10**.

It can be seen that:

> for every two hydrogen atoms fewer than in the alkane with the same number of carbon atoms, there is one double bond or ring present (*double bond equivalent*)

So, a molecule with the formula C_4H_8 has two hydrogen atoms fewer than the parent alkane (C_4H_{10}) and can contain one double bond or one ring. C_4H_6 has four hydrogens fewer than in the alkane and therefore has two double bond equivalents – which could mean that there are two double bonds, one triple bond, two rings (not possible with four carbon atoms) or one ring and one double bond present.

The number of double bond equivalents is sometimes called the *degree of unsaturation* or *the index of hydrogen deficiency* (IHD).

Molecule				
Name	butane	but-1-ene	but-1-yne	cyclobutane
Molecular formula	C_4H_{10}	C_4H_8	C_4H_6	C_4H_8
Double bond equivalents	0	1	2	1
IHD	0	1	2	1

Table 11.10 The formulas of some compounds containing four carbon atoms.

We can extend the idea to compounds containing atoms other than carbon and hydrogen.

The rule works in the same way with compounds containing oxygen atoms – so C_2H_6O has the same number of hydrogens as ethane (C_2H_6) and contains no double bonds (IHD = 0), whereas C_4H_8O has two hydrogens fewer than butane and contains either one double bond or one ring (IHD = 1).

When a compound contains a halogen atom, we simply count this as a hydrogen atom and work out the number of double bond equivalents – so in $C_4H_8Cl_2$ we replace the Cl atoms with hydrogen atoms to make C_4H_{10}, which has the same number of hydrogen atoms as butane and therefore there are no double bond equivalents (IHD = 0).

The most complicated situation you are likely to meet is when a compound contains nitrogen atoms. To work out the number of double bond equivalents, we must subtract one hydrogen for every nitrogen atom and then calculate as above.

Worked example

11.4 Work out the index of hydrogen deficiency for C_3H_5N and suggest a possible structure for the molecule.

Subtracting 1 (because there is one nitrogen atom present) from the number of hydrogen atoms gives us C_3H_4. This has four hydrogens fewer than the parent alkane (C_3H_8) and so there are two double bond equivalents – therefore IHD = 2.

Possible structural formulas are $CH_3CH_2C≡N$, $HC≡CCH_2NH_2$ etc.

An IHD can also be worked out using an equation. In a molecule with the formula $C_cH_hN_nO_oX_x$, where X is a halogen atom, the IHD is given by:

$$IHD = \tfrac{1}{2} \times [2c + 2 - h - x + n]$$

Exam tip
Benzene (C_6H_6) has an IHD of 4 because it contains one ring and the equivalent of three double bonds.

12 Work out the index of hydrogen deficiency for each of the following:

    a C_3H_6 c C_6H_9Cl e C_6H_6

    b $C_4H_8O_2$ d $C_5H_{12}N_2$

Structure determination using infrared spectroscopy

> **Wavenumbers** (cm^{-1}) are used in infrared spectroscopy. This represents the number of wavelengths per cm. Wavenumbers are proportional to frequency and to energy.

Infrared spectroscopy is a very useful tool when determining the structure of an organic compound. In an infrared spectrometer, electromagnetic radiation in the range 400–4000 cm^{-1} is passed through a sample. The printout of the spectrum then shows which frequencies (wavenumbers) are absorbed.

Infrared spectra are always looked at with the baseline (representing 100% transmittance/zero absorbance of infrared radiation) at the top. So the troughs (usually called 'bands'; sometimes 'peaks') represent wavenumbers at which radiation is absorbed (Figure **11.15**).

> Even at absolute zero, the atoms in a molecule are vibrating relative to each other. Just as the electrons in an atom or molecule can exist only in certain energy levels (the energy of an electron is quantised), the vibrational energy of a molecule is quantised. This means that the vibrational energy of a molecule can take only certain allowed values. So a molecule can exist, for instance, in either the level with vibrational energy V_1 or that with V_2. The molecule can absorb a certain frequency of infrared radiation to move it from the lower vibrational energy level to the higher one. In the higher energy level, the molecule vibrates more violently – with a bigger amplitude.

An infrared spectrum can be used to determine which bonds are present in a molecule. In the infrared spectrum of propanone there are two bands in the region above 1500 cm^{-1}, corresponding to absorptions by the C–H bond and the C=O bond.

The segment of the spectrum below 1500 cm^{-1} is called the 'fingerprint region' and is characteristic of the molecule as a whole. Comparison of the fingerprint region of a spectrum with infrared spectra in databases can be used to identify molecules. For example, butanone and propanone both

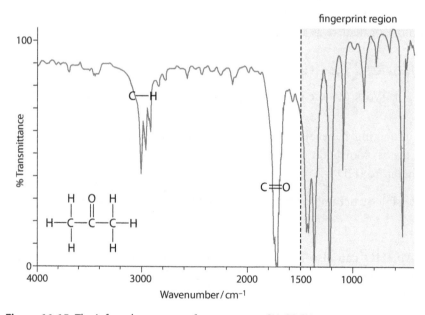

Figure 11.15 The infrared spectrum of propanone – CH_3COCH_3.

Bond	Functional group/class of compound	Characteristic range of wavenumber/cm^{-1}
C–Cl	chloroalkane	600–800
C–O	alcohol, ether, ester, carboxylic acid	1000–1300
C=C	alkene	1610–1680
C=O	aldehyde, ketone, carboxylic acids, ester	1700–1750
C≡C	alkyne	2100–2260
O–H	hydrogen bonded in carboxylic acids	2400–3400
C–H	alkane, alkene, arene	2840–3100
O–H	hydrogen bonded in alcohols, phenols	3200–3600
N–H	primary amine	3300–3500

Table 11.11 The characteristic ranges of wavenumbers at which some bonds vibrate.

Exam tip
A table of infrared absorption frequencies is given in the IB Chemistry data booklet. The values in that table differ slightly from those given here – you will use the values in the data booklet for the examination.

show very similar bands in the region above 1500 cm^{-1}, because they have the same functional group (C=O), but they can be distinguished using their fingerprint regions, which are very different.

To a good approximation, the various bonds in a molecule can be considered to vibrate independently of each other. The wavenumbers at which some bonds absorb are shown in Table **11.11**.

The precise wavenumber at which infrared radiation is absorbed by a particular functional group depends on the adjacent atoms. So a C=O bond in a ketone absorbs at a slightly different frequency to a C=O bond in an ester.

We can use infrared spectra to identify the bonds present in molecules but we cannot always distinguish between functional groups. For example, using Table **11.11** we could identify the presence of C=O in a molecule but would not be able to distinguish between an aldehyde and a ketone.

Consider the infrared spectrum of butanoic acid, shown in Figure **11.16**. To identify the bonds present in the molecule, we first of all look at the region above 1500 cm^{-1}.

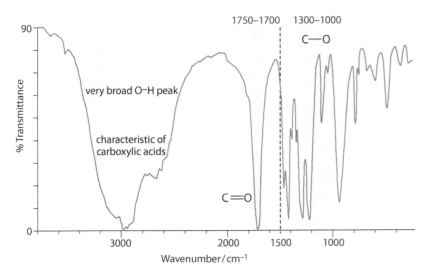

The broadness of the O–H band is due to hydrogen bonding between molecules. If the infrared spectrum of a carboxylic acid in the gas phase is examined, the O–H absorption band is much narrower.

Figure 11.16 The infrared spectrum of butanoic acid – CH$_3$CH$_2$CH$_2$COOH.

The C–H stretch also occurs in the region around $3000\,\text{cm}^{-1}$ but this is usually significantly obscured by the broad O–H absorption.

We now need to match up the wavenumbers of bands in the spectrum with the wavenumbers given in Table **11.11**. The presence of an absorption band in the $1700\text{–}1750\,\text{cm}^{-1}$ region suggests that there is a C=O bond in the molecule. The very broad absorption band between about 2400 and $3400\,\text{cm}^{-1}$ is due to the O–H bond – a very broad band in this region is characteristic of carboxylic acids.

The functional group of a carboxylic acid has a C–O bond, and so we should now look in the fingerprint region to confirm the presence of an absorption in the region $1000\text{–}1300\,\text{cm}^{-1}$, which is indeed the case. If there were no band in this region, we would have to review our hypothesis that the molecule is a carboxylic acid.

The region below $1500\,\text{cm}^{-1}$ contains many absorptions due to C–C bonds and C–H bonds and is difficult to interpret. We usually look only at the fingerprint region to confirm the presence of a particular vibration once we have a good idea of the structure of the molecule. For example, a band in the $1000\text{–}1300\,\text{cm}^{-1}$ region does not confirm the presence of a C–O bond in a molecule – but the absence of a band in this region means that a C–O bond is not present. For example, in the infrared spectrum of propanone (Figure **11.15**) there are peaks in the $1000\text{–}1300\,\text{cm}^{-1}$ region but no C–O bond.

The infrared spectrum of propan-1-ol is shown in Figure **11.17**.

Again, by comparison of the bands with the values in Table **11.11**, we can identify an O–H absorption (about $3350\,\text{cm}^{-1}$) and a C–H absorption (about $2900\text{–}3000\,\text{cm}^{-1}$). Evidence for a C–O bond should also be present and we can see that there is a band in the region $1000\text{–}1300\,\text{cm}^{-1}$.

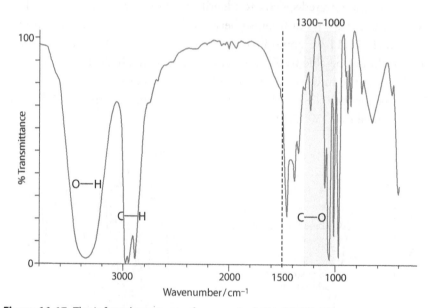

Figure 11.17 The infrared spectrum of propan-1-ol, $CH_3CH_2CH_2OH$.

Worked example

11.5 The infrared spectrum shown is for one of these compounds below.

Identify the compound and explain how you arrive at your choice.

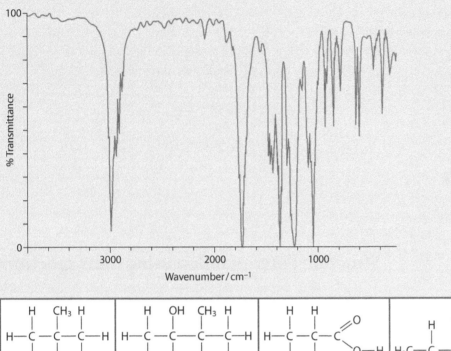

Looking in the region above 1500 cm^{-1}, we can identify an absorption band around 2840–3100 cm^{-1} that is due to a C–H bond, and an absorption in the region 1700–1750 cm^{-1} that is due to a C=O bond. This eliminates molecules **A** and **B** because neither contains a C=O bond.

C is a carboxylic acid, which would be expected to have a very broad O–H absorption band in the region 2400–3400 cm^{-1}. This band is not present so the spectrum cannot be for molecule **C**.

This means that we are left with molecule **D**. Molecule **D** contains C–H bonds and C=O bonds and should give rise to a band in the region 1000–1300 cm^{-1} because it contains a C–O bond – the spectrum shows that there is an absorption in this region.

Nature of science

Sciences use models routinely to provide a framework in which we can understand process on the atomic/molecular level. Our interpretation of infrared spectra is based on a simple model in which the bonds in different functional groups are assumed to vibrate completely independently of each other and absorption of infrared radiation causes the promotion of the bond to a higher vibrational energy level.

Understanding and knowledge in many areas of Chemistry has developed enormously as more and more sophisticated analytical

instruments became available throughout the twentieth century and into the twenty-first century. This has allowed chemists to work out the structures of simple molecules fairly routinely and to also derive the structures of some very complex systems such as proteins.

? Test yourself

13 Which of the following compounds will have an infrared absorption band in the 1700–1750 cm^{-1} region?

A but-2-ene

B propanal

C CH_3COCH_3

D $CH_3CH_2CH_2CH_2Cl$

E CH_3COOCH_3

F $CH_3CH_2CH(OH)CH_3$

G CH_3CH_2COOH

14 Predict the infrared absorption bands and the bonds responsible for them in the region above 1500 cm^{-1} for the following:

a propane

b propan-2-ol

c propene

d propanoic acid

Structure determination using mass spectrometry

A sample of the organic compound to be investigated is injected into a **mass spectrometer**. First, it is bombarded with high-energy electrons to produce positive ions. These pass on through the mass spectrometer where they are separated according to mass (actually the mass : charge ratio) and detected.

When a sample of propane (C_3H_8) is introduced into a mass spectrometer, $C_3H_8^+$ positive ions are produced:

$$C_3H_8(g) + \underset{\text{high-energy electron}}{e^-} \rightarrow \underset{\text{molecular ion}}{C_3H_8^+(g)} + 2e^-$$

The ion produced when just one electron is removed from a molecule is called the **molecular ion**, M^+.

> The peak in the spectrum at the highest mass (m/z value) corresponds to the molecular ion and indicates the relative molecular mass of the molecule (Figure **11.18**).

> The mass of an electron is negligible.

> The mass spectrometer sorts molecules according to their mass : charge (m/z or m/e) ratio. We assume here that only 1+ ions are formed so that the mass : charge ratio is equal to the mass.

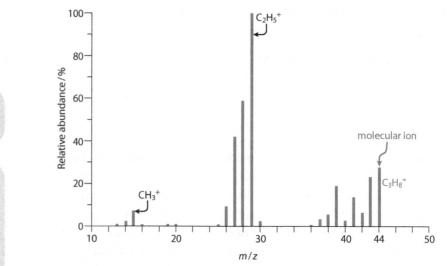

Figure 11.18 The mass spectrum of propane – $CH_3CH_2CH_3$.

We can deduce from Figure **11.18** that the relative molecular mass of the compound being investigated is 44. It is important to note that the molecular ion peak is not necessarily the biggest peak in terms of abundance, but it has the highest m/z value.

It can be seen from the mass spectrum of propane in Figure **11.18** that there are lots of peaks other than the molecular ion peak. These are called the **fragmentation pattern** and arise because a molecule can break apart into smaller fragments when it is bombarded by high-energy electrons – for example:

$$C_3H_8^+ \rightarrow C_2H_5^+ + CH_3$$

All the positive ions resulting from fragmentation will produce a peak in the mass spectrum. The peak at m/z 29 in the mass spectrum of propane is due to the $C_2H_5^+$ ion. Only positive ions can pass through a mass spectrometer, so the CH_3 radical produced in this process does not give rise to a peak. However, there is a peak due to CH_3^+ at m/z 15, which is produced in a different fragmentation process.

A mass spectrum may also show a peak with a mass one unit higher than the molecular ion – an $(M+1)^+$ peak. This is caused by the presence of an atom of ^{13}C in some molecules. ^{13}C is an isotope of carbon – its natural abundance is low at 1.1%.

Fragmentation patterns

We can think of fragmentation in two ways – we can look at:
- the ion formed when a molecular ion breaks apart
- the group lost from a molecular ion.

Look at the mass spectrum of propanoic acid in Figure **11.19**. The molecular ion peak occurs at m/z 74, so the relative molecular mass is 74.

There is a peak in the spectrum at m/z 57, which corresponds to the loss of OH (mass 17) from the molecular ion. So the fragment responsible for the peak at m/z 57 is $(C_2H_5COOH^+ - OH)$, or $C_2H_5CO^+$.

Only positive ions give peaks in a mass spectrum.

Exam tip
The fragmentation pattern in a mass spectrum can be very complicated and you should not try to identify every single peak.

Molecules can undergo fragmentation in many different ways – they can also undergo rearrangement as they fragment.

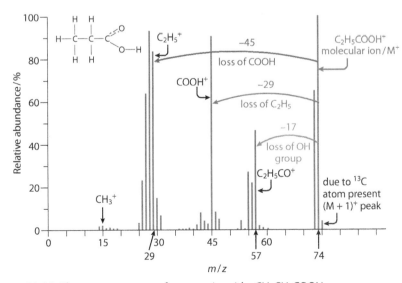

Figure 11.19 The mass spectrum of propanoic acid – CH_3CH_2COOH.

Fragment	Mass of fragment ion	Possible inference
CH_3^+	15	
OH^+	17	alcohol / carboxylic acid
$C_2H_5^+$ / CHO^+	29	
CH_3O^+ / CH_2OH^+	31	methyl ester / ether / primary alcohol
CH_3CO^+	43	ketone with C=O on second C
$COOH^+$	45	carboxylic acid
$C_6H_5^+$	77	arene

Table 11.12 Formulas of some common fragment ions. Peaks are also be formed when fragments are lost from a molecule – in this case the *m/z* value of the peak will be the relative molecular mass minus the mass of the fragment.

The peak at m/z 45 corresponds to the loss of 29 mass units from the molecular ion. C_2H_5 is a group with mass 29 and so we can deduce that C_2H_5 is lost from $C_2H_5COOH^+$ to form the $COOH^+$ ion.

The peak at m/z 29 is due to the $C_2H_5^+$ ion, which is formed by loss of COOH from $C_2H_5COOH^+$.

The formulas of some common fragment ions are shown in Table **11.12**.

The mass spectrum of chloroethane is shown in Figure **11.20**.

The relative molecular mass of chloroethane can be calculated to be 64.5, but it can be seen that there is no peak at 64.5. The relative atomic mass of chlorine, 35.45, arises because chlorine has two isotopes – ^{35}Cl and ^{37}Cl, with an abundance ratio of 3:1. So some chloroethane molecules contain a ^{35}Cl atom and others have a ^{37}Cl atom – therefore we get molecular ion peaks at m/z 64 ($C_2H_5{}^{35}Cl^+$) and m/z 66 ($C_2H_5{}^{37}Cl^+$).

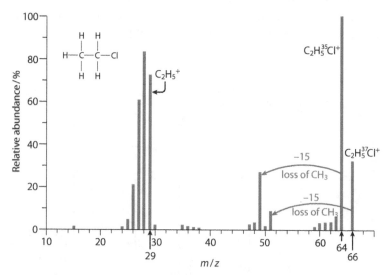

Figure 11.20 The mass spectrum of chloroethane – CH_3CH_2Cl.

The misuse of drugs is a major issue in sport and international sporting organisations have developed policies and testing regimes to make this as difficult as possible. Gas chromatography–mass spectrometry (GC–MS) is an important tool in analysing urine and blood samples for the presence of banned substances and their metabolites. The work of scientists is essential in ensuring that the extent of this cheating is minimised and that nations and individuals can compete fairly against each other in events such as the Olympic Games and the Tour de France.

15 The molecular ion peaks for a series of compounds, containing carbon, hydrogen and oxygen only, occur at the values shown below. Suggest possible molecular formulas for the compounds.

    a 44   **b** 60   **c** 72   **d** 88

16 The following peaks occur in the mass spectra of compounds containing carbon, hydrogen and oxygen only. Suggest possible identities for the fragments responsible.

a 15	**c** 29	**e** 43	**g** 59
b 28	**d** 31	**f** 45	**h** 77

Structure determination using nuclear magnetic resonance spectroscopy

A hydrogen nucleus has a property called **spin**. A spinning nucleus acts like a tiny bar magnet. This bar magnet can either align itself with (lower energy) or against (higher energy) an externally applied magnetic field. Energy in the **radio frequency** range of the electromagnetic spectrum can be used to cause a hydrogen nucleus to change its orientation relative to the applied magnetic field. It is these changes in energy state that occur in nuclear magnetic resonance (NMR) spectra. In analysing NMR spectra, we are looking at absorptions due to ^1H nuclei (usually just called **protons**) – this technique provides information about the environments of protons in molecules.

The low-resolution NMR spectrum of propanal is shown in Figure **11.21**.

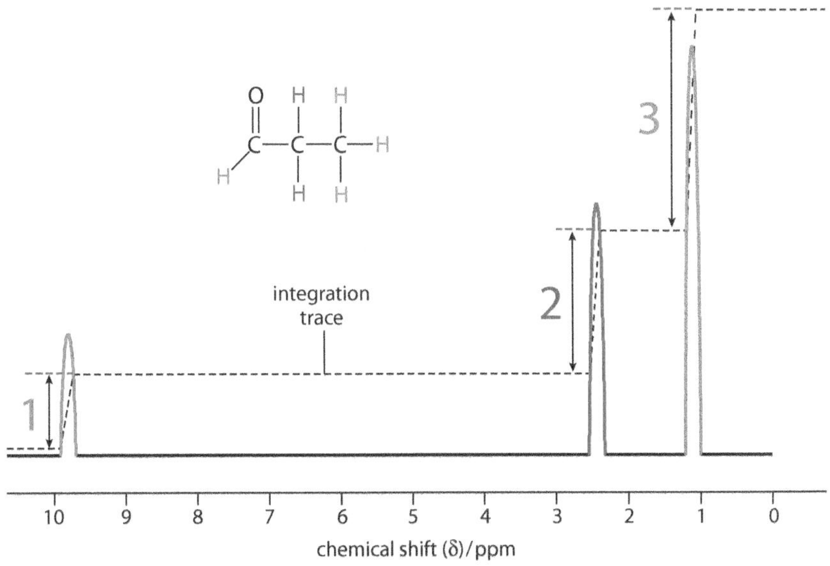

In all the discussion in this section, 'proton' refers to the nucleus of a hydrogen (^1H) atom.

Only nuclei with an odd value for the mass number and/or atomic number have the property of spin.

Figure 11.21 The low-resolution NMR spectrum of propanal – CH_3CH_2CHO. The integration trace allows us to work out the ratio of the numbers of protons in each environment.

Peaks in an NMR spectrum correspond to groups of protons (hydrogen atoms) in different chemical environments.

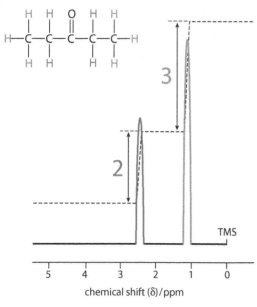

Figure 11.22 – $CH_3CH_2COCH_2CH_3$.

Figure 11.23 Butanone showing the different environments for protons.

It can be seen that there are three peaks in the spectrum – these correspond to three different chemical environments for the protons (hydrogen atoms) in one molecule of the aldehyde. The protons in the different environments are coloured differently in Figure **11.21**. Hydrogen atoms joined to the same carbon atom are said to be chemically equivalent (or just 'equivalent').

It can also be seen from Figure **11.21** that the peaks have different sizes. The area underneath a peak is proportional to the number of hydrogen atoms in that environment. So, the area under the red peak is twice (2 H in this environment) the area under the blue peak (1 H in this environment); and the area under the green peak (3 H in this environment) is three times that under the blue peak.

An NMR spectrometer can also work out the area under each peak to produce an **integration trace**. The vertical heights of the steps in the integration trace are proportional to the number of hydrogen atoms in each environment.

The horizontal scale on an NMR spectrum is the **chemical shift**, which is given the symbol δ and has units of parts per million (ppm). This quantity gives information about the environments that the protons (hydrogen atoms) are in – protons in different chemical environments have different chemical shifts.

The number of different hydrogen (proton) environments and the relative numbers of hydrogen atoms in each

To work out the number of hydrogen environments, you must first decide whether or not the molecule is symmetrical. If the molecule is not symmetrical then the hydrogen atoms on the different atoms in the carbon chain will be in different chemical environments.

The NMR spectrum for pentan-3-one (Figure **11.22**) shows two peaks (not counting any due to TMS – see page **536**).

There are only two different chemical environments for the hydrogen atoms in pentan-3-one because the molecule is symmetrical. The six hydrogens shown in green are equivalent – all in the same chemical environment; and the four hydrogens shown in red are also equivalent to each other.

The heights of the steps in the integration trace are in the ratio 2:3 because there are four hydrogen atoms in one environment and six hydrogen atoms in the other.

Butanone has three different environments for its protons (Figure **11.23**) and the ratio of the numbers of hydrogens in each environment is 3:2:3.

There are only three peaks in the NMR spectrum of propan-2-ol (Figure **11.24a**) because the molecule is symmetrical. The six hydrogens in the two CH_3 groups are equivalent. The ratio of the numbers of protons in each environment is 6:1:1. In general, the hydrogen atoms in CH_3 groups that are attached to the same carbon atom will be equivalent, so that there are only two environments in 2-methylpropan-2-ol (Figure **11.24b**). Table **11.13** shows some further examples.

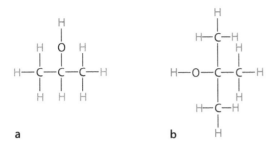

a   b

Figure 11.24 Proton environments: **a** propan-2-ol; **b** 2-methylpropan-2-ol.

	No. of different chemical environments for H	Ratio of no. of H atoms in each environment
H Cl Cl | | | H—C—C—C—H | | | H H H	3	3:1:2
H | H O H H | | | | H—C—C—C—C—H | | | | H H H H	5	3:1:1:2:3
H | H—C—H H Cl | H | | | | H—C—C—C—C—H | | | | H H H H	4	3:1:1:6

Table 11.13 Proton environments.

? Test yourself

17 Suggest the numbers of peaks and the ratio between the areas under the peaks in the NMR spectrum of each of the following:

Cl H H | | | Cl—C—C—C—H | | | H H H **a**	H H H H | | | | H—C—C—C—O | | | H H H **b**	CH₂OH H | H | | | H—C—C—C—H | | | H CH₃ H **c**
O CH₃ H H || | | | H—C—O—C—C—C—H | | | H H H **d**	H | H O H—C—H | || | H—C—C—O—C—H | | H H—C—H | H **e**	H H H H O | | | | || H—C—C—C—C—C | | | | \\ H CH₃ H H H **f**

- Understand how chemical shifts can provide information about the structure of organic molecules
- Use high-resolution ^{1}H NMR to determine the structure of molecules
- Understand how to use information from a combination of spectroscopic techniques to determine the structure of molecules.

11.4 Spectroscopic identification of organic compounds (HL)

Chemical shift

The horizontal scale on a nuclear magnetic resonance spectrum is the **chemical shift**, which is given the symbol δ and has units of parts per million (ppm). Chemical shift data provide information about the environments that protons (hydrogen nuclei) are in – protons in different chemical environments have different chemical shifts.

Chemical shift values are measured relative to tetramethylsilane (TMS) (Figure **11.25**).

The protons in TMS are assigned an arbitrary chemical shift of 0.00 ppm – all chemical shifts are measured relative to this.

TMS was chosen as the standard because it has 12 protons all in the same environment and so gives a strong signal when only a small amount is added. Also the chemical shift of the protons in TMS is at a lower value than the protons in virtually all organic molecules – therefore TMS fixes the lower end of the chemical shift scale. The position of the chemical shift is also such that it is well away from the chemical shifts of protons in nearly all organic molecules – so the TMS signal does not overlap with the protons signals that we are interested in. Other reasons for using TMS are that it is non-toxic and inert.

Some typical values for the chemical shifts of protons in different environments are given in Table **11.14**.

The values in Table **11.14** are approximate and can vary depending on any groups attached. For example, it makes a difference how many other hydrogen atoms are attached to the carbon atom to which the hydrogen of interest is attached. However, in most cases, we can assume that the ranges given in the table include all possibilities, and we just look at the environment of the proton rather than the number of hydrogen atoms in that environment. So we can assume that the chemical shift of a hydrogen on a carbon next to the C=O group of an aldehyde or ketone comes in the range 2.1–2.7 no matter how many other hydrogens are attached – Figure **11.26** shows this clearly.

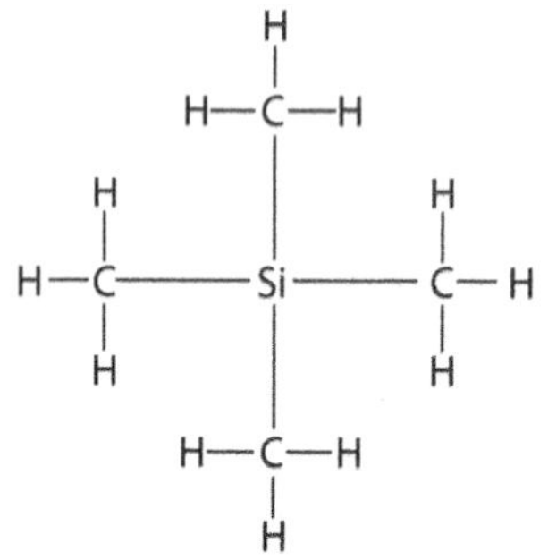

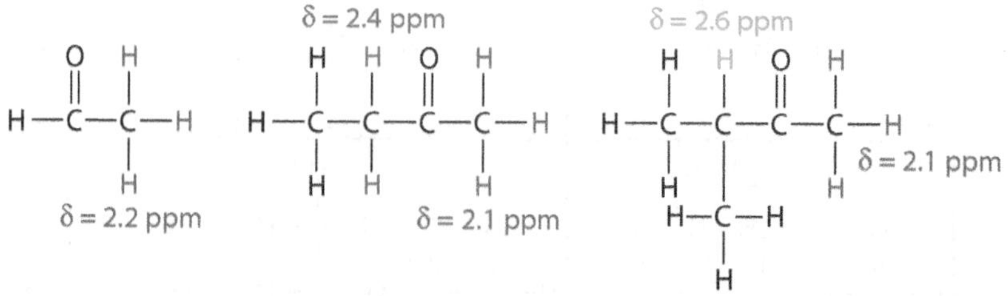

Figure 11.25 Tetramethylsilane.

TMS is a reference standard used to fix the chemical shift scale. A small amount of TMS is added to the sample before the NMR spectrum is recorded.

Another way of saying this is that the chemical shift for a particular hydrogen depends on the number of alkyl groups that are attached to the carbon to which it is bonded.

Figure 11.26 Chemical shifts are influenced by the number of hydrogen atoms on a particular carbon atom but the range in the Table **11.14** includes all possibilities.

Type of proton	Chemical shift/ppm	Comments
H | —C— |	0.9–1.7	H on a carbon chain but not next to any other functional groups
O || R—O—C—C—H |	2.0–2.5	H on a C next to C=O of an ester
O || R—C—C—H |	2.1–2.7	H on a C next to C=O of an aldehyde or ketone
H | —C— (benzene ring)	2.3–3.0	H on a C attached to a benzene ring
H | —C— | X X is a halogen	3.2–4.4	H attached to a C that also has a halogen atom attached
H | R—O—C— |	3.3–3.7	H attached to a C that has an O attached
O || R—C—O—C—H |	3.7–4.8	H on a C next to C–O of an ester
R—O—H	0.5–5.0	H attached to O in an alcohol
H (benzene ring)	6.7–8.2	H attached to a benzene ring
O || R—C—H	9.4–10.0	H attached to C=O of an aldehyde
O || R—C—O—H	9.0–13.0	H on an O in a carboxylic acid

Table 11.14 Chemical shifts for protons in different environments.

Exam tip

The representation of groups and chemical shifts here is different from those in the IB Chemistry data booklet. You should practise using those as well.

When using a table of chemical shift values, you must try to find the best match to the proton environments in the molecule you are analysing. So in the spectrum of propanal (Figure **11.27**), the H (blue) attached directly to the C=O group would be expected to have a chemical shift in the range 9.4–10.0 ppm; in fact the peak for this proton occurs at a chemical shift of 9.8 ppm.

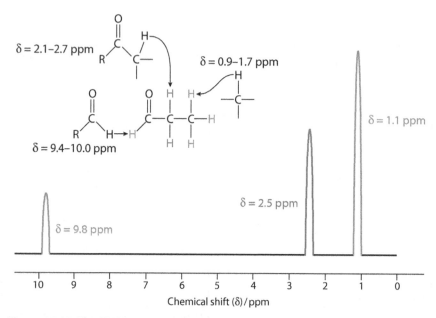

Figure 11.27 The NMR spectrum of propanal – CH_3CH_2CHO.

The ranges in which the chemical shifts of the other protons would be expected to occur are also shown in Figure **11.27** and it can be seen that all the chemical shifts occur within the expected ranges.

High-resolution NMR spectra

The low- and high-resolution NMR spectra of 1,1,2-trichloroethane are shown in Figure **11.28**.

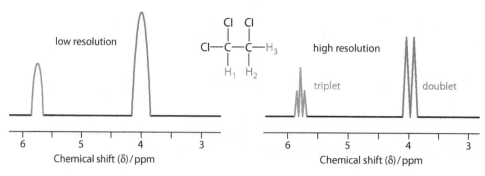

Figure 11.28 The low- and high-resolution NMR spectra of 1,1,2-trichloroethane.

There are two peaks in the low-resolution spectrum because there are two different chemical environments for protons (hydrogen atoms). However, in the high-resolution spectrum each of these peaks is split. The splitting is due to the hydrogen atoms on adjacent carbon atom(s). This results in the signal due to H_1 being split into three (a triplet), because there are two Hs on the adjacent carbon atom. The signal due to H_2 and H_3 is split into two (a doublet), because there is one H on the adjacent carbon atom. This **splitting** is called spin–spin splitting or spin–spin coupling.

We sometimes talk about the **multiplicity** of a peak – the number of smaller peaks it is split into. The multiplicity of a triplet is 3 and that of a doublet is 2.

In general, if there are n protons (hydrogen atoms) on an adjacent atom, the signal for a particular proton will be split into $(n+1)$ peaks. Another way of saying this is that if the multiplicity of a peak is x then the number of hydrogen atoms on the adjacent atom is $x - 1$.

The NMR spectrum of chloroethane is shown in Figure **11.29**.

There are two sets of peaks in the NMR spectrum of chloroethane because there are two different chemical environments for protons (hydrogen atoms). The total area under the peaks at δ 1.5 ppm is larger than that under the peaks at δ 3.5 ppm, because there are more protons in this environment.

The signal at δ 3.5 ppm is split into a quartet – a quartet consists of four peaks and, therefore, has a multiplicity of 4. Now, $4 - 1 = 3$ and so we can deduce from the presence of the quartet that there are three hydrogen atoms on the adjacent carbon atom.

The signal at δ 1.5 ppm is split into a triplet – this has a multiplicity of 3. Now, $3 - 1 = 2$ so there must be two hydrogen atoms on the adjacent carbon atom.

Splitting of the signal of a particular group of protons is due to the protons on *adjacent* (carbon) atoms.

Exam tip
When a signal due to two hydrogen atoms is split into a quartet and the signal due to three hydrogen atoms is split into a triplet, this indicates the presence of an ethyl group (CH_3CH_2) in a molecule – well worth remembering!

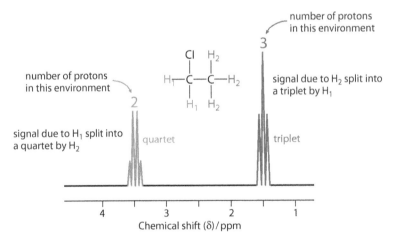

Figure 11.29 The NMR spectrum of chloroethane.

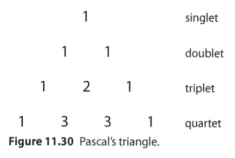

1				singlet
	1	1		doublet
1		2	1	triplet
1	3	3	1	quartet

Figure 11.30 Pascal's triangle.

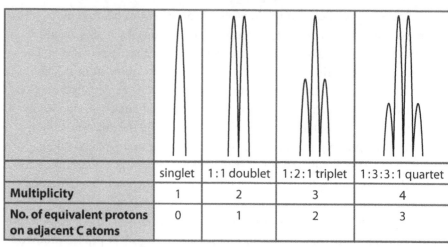

	singlet	1:1 doublet	1:2:1 triplet	1:3:3:1 quartet
Multiplicity	1	2	3	4
No. of equivalent protons on adjacent C atoms	0	1	2	3

Table 11.15 Different types of splitting patterns.

The intensities of the lines in a splitting pattern are given by Pascal's triangle (Figure **11.30**). So we talk about a 1:1 doublet (the areas under the peaks are equal) or a 1:2:1 triplet etc.

A summary of the different types of splitting patterns is given in Table **11.15**.

'Rules' for spin–spin coupling

- Protons on the same atom (e.g. CH_3, CH_2) do not split each other – they are **chemically equivalent** and behave as a group.
- Splitting generally only occurs with protons on adjacent atoms – see Figure **11.31**.
- Protons attached to oxygen atoms do not usually show or cause splitting – this is because the protons exchange with each other and with the solvent and experience an 'average' environment.

Figure 11.31 H_a protons couple only with H_b protons – H_c and H_d are too far away.

Working out structures from NMR spectra

Worked examples

11.6 The NMR spectrum for a compound with molecular formula C_4H_8O is shown below. Deduce the structure of the compound.

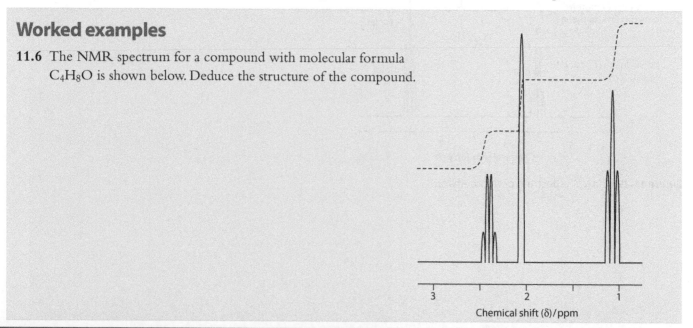

Chemical shift (δ)/ppm

First, we must measure the vertical heights of the steps in the integration trace – these are 1 cm, 1.5 cm and 1.5 cm. Multiplying by 2 to get whole numbers we get 2, 3, 3 – this gives us the ratio of the number of protons (hydrogen atoms) in the different environments. Because there is a total of eight hydrogen atoms in the molecule, we also know the actual number of hydrogen atoms in each environment.

Chemical shift / ppm	2.4	2.1	1.1
No. of protons in environment	2	3	3

The signal at δ 2.4 ppm is split into a quartet (multiplicity 4). Now $4 - 1 = 3$, so we know that there must be three hydrogen atoms on the adjacent C atom. Analysing all the peaks in the spectrum gives the following data:

Chemical shift / ppm	2.4	2.1	1.1
No. of protons in environment	2	3	3
Splitting	quartet	singlet	triplet
Multiplicity	4	1	3
No. of protons on adjacent C atom	3	0	2

Remember: multiplicity − 1 = number of hydrogen atoms on adjacent carbon atom.

The peaks at δ 2.4 ppm and δ 1.1 ppm are caused by an ethyl group – the signal due to 2Hs is split by 3Hs on the adjacent C and the signal due to 3Hs is split by 2Hs on the adjacent C.

We know that there are four C atoms in the molecule but only three sets of peaks. There are also three Hs that have no Hs on the adjacent C. These two pieces of information together suggest that there is a carbon atom with no hydrogen atoms attached.

Also the molecular formula C_4H_8O has two fewer hydrogen atoms than the alkane with four carbon atoms, so there must be a double bond (or a ring) in the molecule (IHD = 1).

The only structure that fits with all this information is:

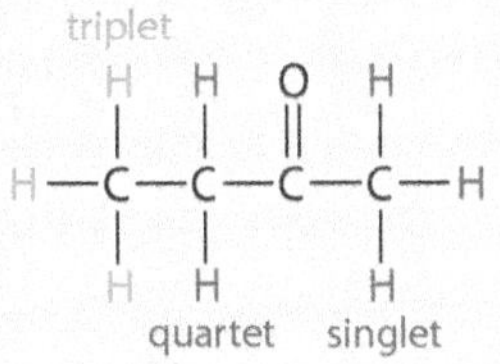

As a final check, we should try to match up the chemical shifts of the protons with the values given in Table **11.14**.

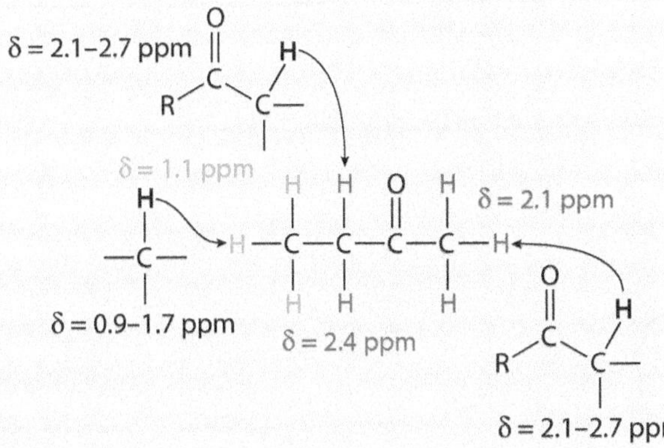

All protons have chemical shifts in the expected ranges.

11.7 The NMR spectrum for an ester with molecular formula $C_4H_8O_2$ is shown. Deduce the structure of the compound.

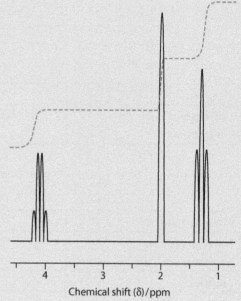

Chemical shift (δ)/ppm

Following the procedure from the previous example we get:

Chemical shift / ppm	4.1	2.0	1.3
No. of protons in environment	2	3	3
Splitting	quartet	singlet	triplet
Multiplicity	4	1	3
No. of protons on adjacent C atom	3	0	2

As in the previous example, there is an ethyl group present and a group of three hydrogen atoms with no hydrogens on the adjacent carbon. The only two possible esters that would have this splitting pattern are:

We can distinguish between these two possibilities using the chemical shift values. If we consider the singlets, we can use the values in Table **11.14** to predict the chemical shifts for these protons:

$\delta = 2.0-2.5$ ppm

$\delta = 3.7-4.8$ ppm

The singlet in the spectrum occurs at δ 2.0 ppm, so this molecule is the one on the left in the diagram above – ethyl ethanoate:

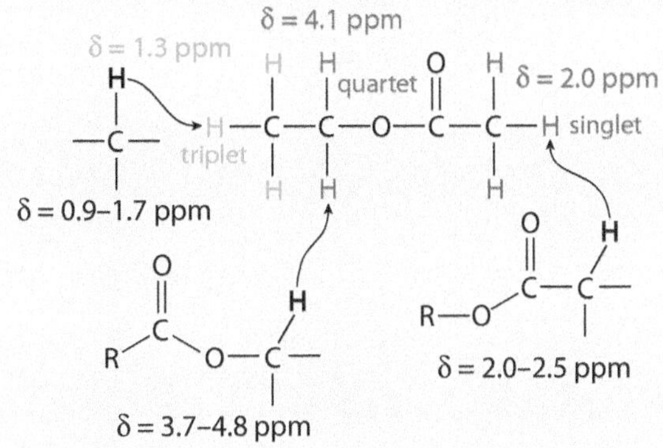

Notes on NMR spectra

When a molecule is non-symmetrical and there are non-equivalent protons on both sides of a particular group, the spectrum becomes complex. Figure **11.32** shows the structural formula of propan-1-ol. In the NMR spectrum of this compound the H_b signal is split by H_a and H_c. Depending on the strength of the coupling to each set of protons, the signal for H_b could either be described as a quartet of triplets or a triplet of quartets. This signal is very complicated, and it is difficult to see the exact nature of the splitting. One way of describing this peak is as a **complex multiplet**.

For a molecule such as methylbenzene (Figure **11.33**), the protons on the ring are not all equivalent, but because they are in very similar environments they could show up as just one peak in the NMR spectrum – unless a very high-resolution spectrum is generated.

The presence of highly electronegative atoms in a molecule (Figure **11.34**) can cause chemical shifts to move to higher values. The closer the protons are to the very electronegative atom, the greater the effect.

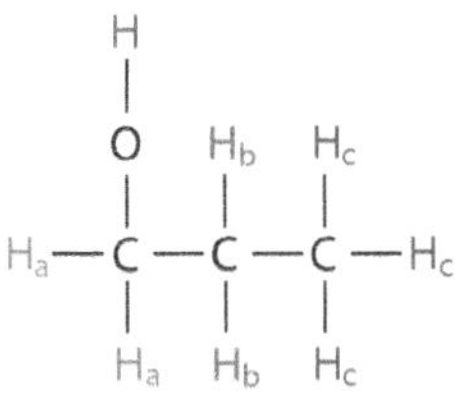

Figure 11.32 Propan-1-ol.

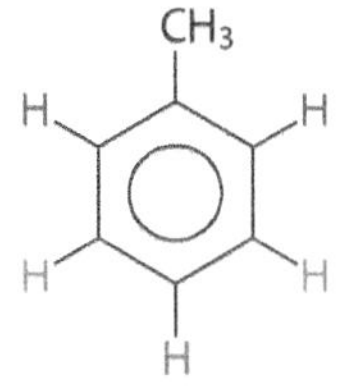

Figure 11.33 Methylbenzene.

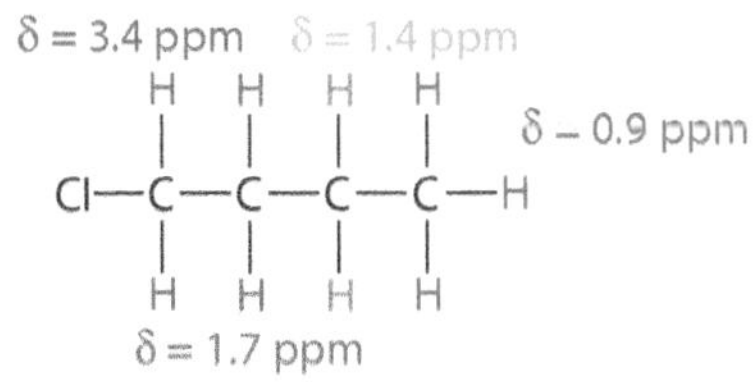

Figure 11.34 Chlorine is more electronegative than hydrogen and carbon.

18 Suggest the splitting pattern for each of the following:

a 1,1-dibromo-2,2-dichloroethane

b 1,1,3,3-tetrachloropropane

c propanoic acid

d
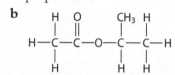

19 The NMR spectrum of each of the following contains a singlet. Suggest the chemical shift range for the singlet.

a propan-2-ol

b
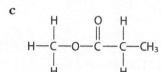

c

20 The NMR spectrum shown below is of an alcohol.

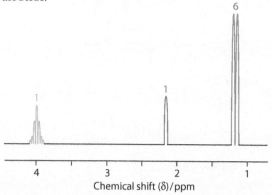

Work out the structure of the alcohol and state the multiplicity of the peak at $\delta = 4.0$ ppm

Single crystal X-ray crystallography

Further evidence for the structures of molecules can be obtained from **single crystal X-ray crystallography**. This involves irradiating a crystal with X-rays and looking at the positions and intensities of the diffracted beams. This is an extremely powerful technique and gives a three-dimensional picture of the molecule with bond lengths and bond angles (Figure **11.35**). If a single crystal can be grown, X-ray crystallography usually provides the final word on the structure.

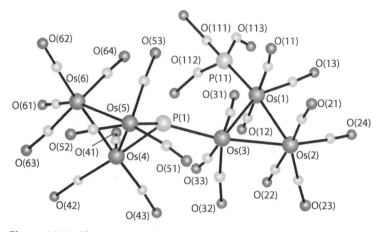

Figure 11.35 The structure of a molecule obtained by X-ray crystallography.

Using combined spectroscopic techniques to determine the structure of a molecule

In practice, information from more than one spectroscopic technique is often used to determine the structure of a molecule. This is best illustrated using some worked examples.

Worked examples

11.8 The mass spectrum below is for a compound, **X**, that contains carbon, hydrogen and oxygen. The infrared spectrum of **X** has an absorption band at $1725 \, \text{cm}^{-1}$.

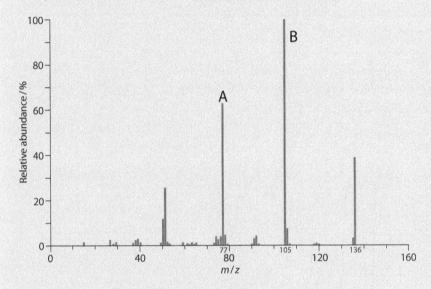

a What is the relative molecular mass of **X**?
b Given that **X** contains two oxygen atoms, suggest the molecular formula for **X**.
c Identify which group is lost to form the peak marked **B**.
d Suggest the identity of the species responsible for the peak marked **A**.
e Calculate the IHD for **X**.
f Deduce the structural formula of **X**.

a Using the mass spectrum, the peak at highest m/z value represents the molecular ion – so the relative molecular mass is 136.

b The only possible molecular formula containing two oxygen atoms that adds up to 136 is $C_8H_8O_2$.
  Two oxygen atoms have a combined mass of 32 and if we subtract this from 136 we get 104. The only other elements in the molecule are carbon and hydrogen, so we must make 104 from a combination of the masses of these. Seven carbon atoms gives a combined mass of $7 \times 12 = 84$, which would require there to be $104 - 84 = 20$ hydrogen atoms in the molecule. However, the alkane with seven carbon atoms has only 16 hydrogen atoms, therefore $C_7H_{20}O_2$ is not possible

c Peak **B** occurs at m/z 105. Now, $136 - 105 = 31$. So we must think of combinations of C, H and O that add up to 31. The only possible combination is CH_3O, so this group is lost from the molecular ion to produce peak **B**. (Note that C_2H_7 also adds up to 31, but the maximum number of hydrogen atoms for two carbon atoms is six.)

d Peak **A** occurs at m/z 77. The fragment responsible for this is $C_6H_5^+$.

e $IHD = \frac{1}{2} \times [2c + 2 - h - x + n]$ for a molecule with formula $C_cH_hN_nO_oX_x$, where X is a halogen atom. The molecular formula is $C_8H_8O_2$, so $c = 8$ and $h = 8$.

$$IHD = \frac{1}{2} \times [(2 \times 8) + 2 - 8] = 5$$

f A benzene ring has an IHD of 4 because it has a ring and three double bonds – so we can conclude that, apart from the benzene ring, there is one other double bond or ring present.

The peak at m/z 77 indicates that the compound contains a benzene ring substituted in one position. The loss of 31 mass units from the group suggests that it contains the $-O-CH_3$ group. The presence of an absorption band at $1725 \, cm^{-1}$ in the infrared spectrum suggests (using Table **11.11**) that the molecule contains a C=O group. The only possible molecule that contains all these elements is:

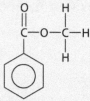

11.9 An organic compound, **Y**, is shown by elemental analysis to contain only carbon, hydrogen and oxygen. The following information about the structure of **Y** is available:

- **mass spectrum:** the molecular ion peak occurs at m/z 86
- **IHD** = 1
- **infrared spectrum:**

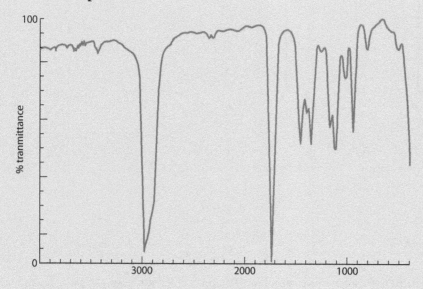

- **nuclear magnetic resonance spectrum:** shows two peaks – one quartet (δ 2.4 ppm) and one triplet (δ 1.1 ppm)

Determine the structure of the molecule from these data.

Adding up possible combinations of C(12), H(1) and O(16) to make the molecular ion peak at m/z 86 produces $C_5H_{10}O$, $C_4H_6O_2$, $C_3H_2O_3$ as possible formulas.

The IHD can be worked out for each of these molecular formulas using $IHD = \frac{1}{2} \times [2c + 2 - h - x + n]$ for a molecule with formula $C_cH_hN_nO_oX_x$, where X is a halogen atom. The IHD corresponding to the possible formulas are:

$C_5H_{10}O$: $\frac{1}{2} \times [(2 \times 5) + 2 - 10] = 1$
$C_4H_6O_2$: $\frac{1}{2} \times [(2 \times 4) + 2 - 6] = 2$
$C_3H_2O_3$: $\frac{1}{2} \times [(2 \times 3) + 2 - 2] = 3$

Because the IHD is given as 1, the molecular formula must be $C_5H_{10}O$.

The infrared spectrum has an absorption band in the range 1700–1750 cm^{-1} indicating a C=O group in an aldehyde, ketone, carboxylic acid or ester. Because there is only one oxygen atom in the molecule, we can rule out esters and carboxylic acids because they both have two oxygen atoms in the functional group. Therefore the compound must be an aldehyde or a ketone.

The NMR spectrum indicates that there are only two different chemical environments for the hydrogen atoms in the molecule. There are several different isomers that can be drawn for a compound with molecular formula $C_5H_{10}O$ containing a C=O group – but only the two below have just two different chemical environments for protons:

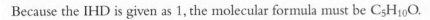

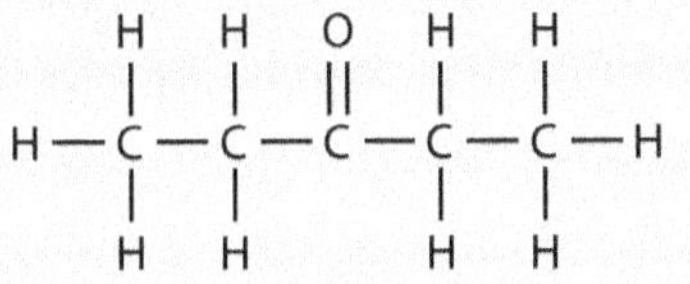

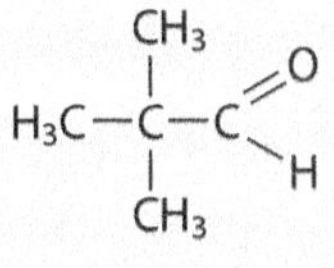

The splitting pattern from the NMR data can be used to distinguish between these two structures:
- a quartet signal indicates the presence of hydrogen atoms adjacent to a C atom with three H atoms attached
- a triplet signal indicates the presence of hydrogen atoms adjacent to a C atom with two H atoms attached.

These two signals together indicate the presence of an ethyl group. This group is present in only the left-hand structure – so we can conclude that **Y** is pentan-3-one:

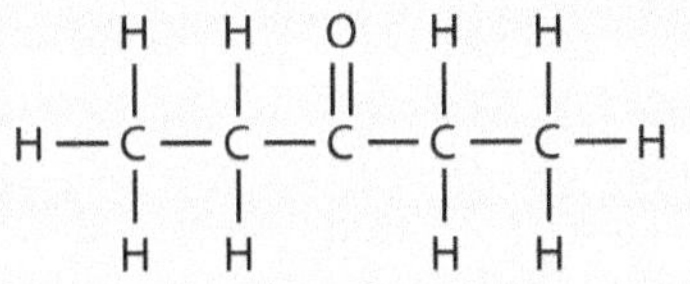

The NMR spectrum of the right-hand molecule would consist of two singlets because all the hydrogen atoms have no hydrogen atoms on the adjacent carbon.

As a final check you should use Table **11.14** to compare the chemical shift values given with what you would expect for pentan-3-one.

Nature of science

Advances in technology have allowed scientists to derive ever more detailed information about the structure of compounds. Theories of structure and bonding have developed further as more information has become available.

Exam-style questions

1 Rosie carried out an experiment in which she measured a temperature change. Her data are shown in the table.

Initial temperature / °C	18.7	±0.5
Maximum temperature / °C	37.6	±0.5

The temperature change should be quoted as:

    A 18.9 ± 0.5 **C** $18.9 \pm 1.0\,°C$

    B $19 \pm 1\,°C$ **D** $19.0 \pm 1.0\,°C$

2 Aazaish obtained the value $0.002\,560\,m^3$ from an experiment. The number of significant figures and decimal places is:

	Significant figures	Decimal places
A	4	6
B	6	4
C	6	6
D	3	6

3 Molly carried out an experiment to measure the enthalpy change of solution of a salt. In order to calculate a final value, the following calculation was carried out:

$$\frac{[(50 \pm 1) \times 4.2 \times (20 \pm 1)]}{1000 \times (0.10 \pm 0.01)}$$

Quantities without uncertainties can be assumed to be exact. How should the final value be quoted?

    A $-42 \pm 7\,kJ\,mol^{-1}$ **C** $-42.00 \pm 2.01\,kJ\,mol^{-1}$

    B $-42 \pm 2\,kJ\,mol^{-1}$ **D** $-42.1 \pm 7.1\,kJ\,mol^{-1}$

4 Which of the following would be a good method for reducing the random uncertainty in an experiment to measure the enthalpy change of neutralisation when $50\,cm^3$ of $0.50\,mol\,dm^{-3}$ sodium hydroxide reacts with $50\,cm^3$ of $0.50\,mol\,dm^{-3}$ hydrochloric acid?

    A Insulate the reaction vessel with cotton wool.

    B Stir the mixture more rapidly.

    C Repeat the experiment.

    D Measure out the liquids using a $50\,cm^3$ measuring cylinder instead of a burette.

5 The graph shows the results of a series of experiments to investigate how the rate of the reaction X → Y varies with the concentration of X.

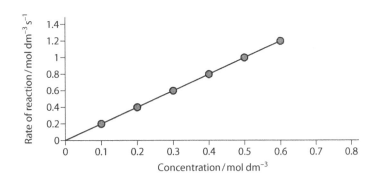

The gradient (slope) of this graph is:

A $2.0\,mol\,dm^{-3}\,s^{-1}$ C $2.0\,s^{-1}$

B $0.5\,s$ D $0.5\,mol^{-1}\,dm^{3}$

6 Which of the following would produce only a single peak in the low-resolution ^1H NMR spectrum?

A CH_3CH_2OH C CH_3CHO

B $(CH_3)_3CCl$ D $CH_2ClCHClCH_2Cl$

7 Which of the following has an index of hydrogen deficiency of 2?

A $C_3H_6O_2$ C C_4H_7N

B C_6H_6 D $C_5H_8Cl_2$

8 Which of the following is unlikely be a peak in the mass spectrum of propanoic acid?

A $m/z\ 45$ C $m/z\ 74$

B $m/z\ 15$ D $m/z\ 50$

HL 9 The following information is provided about a compound **M**:

- There is only one peak in the NMR spectrum.
- IHD = 1.
- The molecular ion peak occurs at $m/z\ 56$.
- There is only one major band above $1500\,cm^{-1}$ in the IR spectrum.

Which of the following could be the structure of compound M?

A   H₂C—CH₂
    | |
    H₂C—CH₂

C   H H H H
    | | | |
    H—C=C—C—C—H
          | |
          H H

B   H O H
    | || |
    H—C—C—C—H
    | |
    H   H

D   NH₂
    |
    H—C—C≡N
    |
    H

10 Which of the molecules below will **not** have a triplet in its NMR spectrum?

A

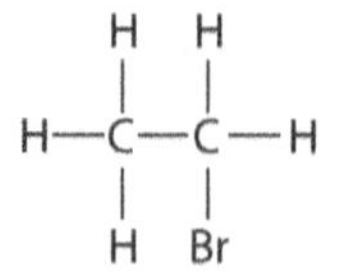

$$H-\overset{\overset{\displaystyle H}{|}}{\underset{\underset{\displaystyle H}{|}}{C}}-\overset{\overset{\displaystyle H}{|}}{\underset{\underset{\displaystyle Br}{|}}{C}}-H$$

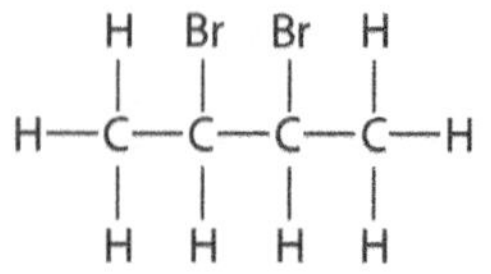

C

B

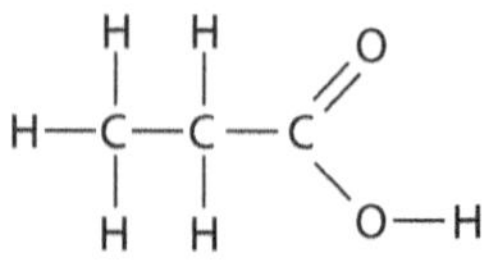

D

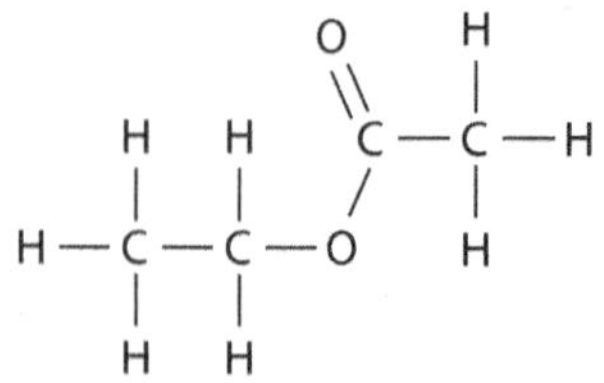

11 Two separate experimental methods were used to determine the value of a particular experimental quantity. Each experiment was repeated five times. The values obtained from these experiments are shown in the table.

Experiment	A	B
Trial	Value	Value
1	49.7	50.6
2	53.2	51.2
3	51.5	51.1
4	52.3	50.8
5	49.2	51.0

The literature value for this quantity is 50.9.

a Explain which set of experimental values is more precise. [2]

b Work out a mean value for each experiment and use this to explain which set of data is more accurate. [3]

12 Yi Jia carried out an experiment to measure a certain quantity. The value she obtained was $56.1 \pm 0.5\,kJ$. The literature value for this quantity is $55.2\,kJ$.

a Calculate the percentage error for this experiment. [1]

b The student maintained that any errors could be explained solely by random uncertainties.
Is she correct? Explain your answer. [2]

13 Jerry carried out an experiment to measure the rate of the reaction between magnesium and hydrochloric acid. He did this by recording the volume of hydrogen gas collected every 15 s. His data are shown in the table.

a Plot a graph of these data. [3]

b Use your graph to determine the initial rate of reaction – include units. [3]

Time / s	Volume / cm^3
0	0
15	19
30	33
45	44
60	50
75	54
90	56
105	57
120	57

14 Certain molecules absorb infrared (IR) radiation.

a IR spectroscopy is one of the techniques that is used to determine the structure of organic molecules. A student recorded the IR spectrum of a compound, **X**:

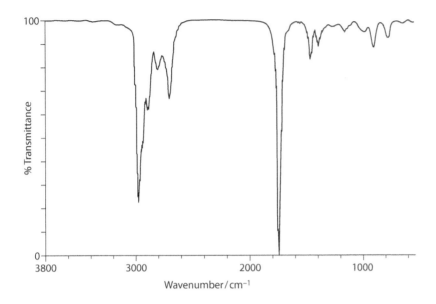

The student knew that **X** was one of the following compounds:

H   H   H   H
  | | | |
HO—C—C—C—C—H
   | | | |
   H CH₃ H H
	I

H H H | | | H—C—C—C—C=O | | | | H H CH₃ H	II
H H H | | | H—C—C—C—C=O | | | O—H H H CH₃	III
H H H H O H H H | | | | || | | | H—C—C—C—C—O—C—C—C—C—H | | | | | | | H H CH₃ H CH₃ H H	IV

Deduce which of the molecules is **X** and explain your choice by reference to the spectrum. **[4]**

b Explain why NMR is more useful than IR spectroscopy in distinguishing between propanal and propanone. **[4]**

15 The NMR spectrum of propan-2-ol is shown below.

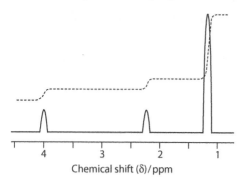

Chemical shift (δ)/ppm

a Draw the full structural formula of propan-2-ol. [1]

b Explain why there are three peaks in the NMR spectrum of propan-2-ol. [2]

c The integration trace is shown on the spectrum. Explain what information can be obtained from the integration trace. [2]

d **i** Draw the full structural formula of butan-2-ol. [1]

ii State the number of peaks and the relative areas under the peaks in the NMR spectrum of butan-2-ol. [2]

HL 16 Compounds may be identified using a combination of spectroscopic techniques. The mass spectrum for a compound, **Z**, is shown below. Elemental analysis showed that **Z** contains C, H and O.

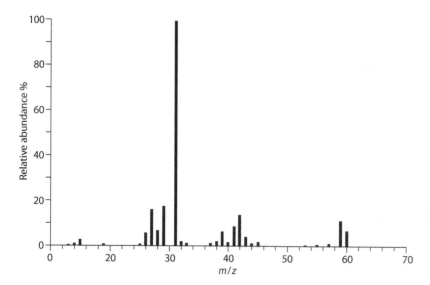

a Determine the relative molecular mass of **Z**. [1]

b Work out two possible molecular formulas for **Z**. [2]

c Deduce the formula of the fragments responsible for the peaks at m/z 31 and m/z 29. [2]

d The IR spectrum of Z shows a strong absorption band at about $3350\,cm^{-1}$ but no absorptions in the range $1600–1800\,cm^{-1}$. Explain what conclusions about the structure of **Z** can be drawn from these data. [2]

e On the basis of the information above, suggest two possible structural formulas for **Z**. [2]

f The low-resolution NMR spectrum of **Z** is shown below.

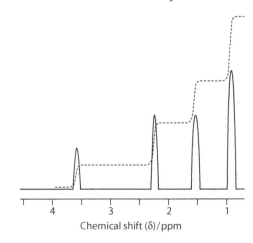

Chemical shift (δ)/ppm

Explain how information from this spectrum can be used to deduce the structural formula of **Z** and draw the structural formula of **Z**. [4]

HL **17** NMR spectroscopy is a very powerful tool in the identification of organic compounds.

a Predict and explain the splitting pattern of the hydrogen atom marked **bold** in the molecule below. [2]

H   H   H
| | |
H—C —— C —— C—O
| | \
H | H
    H—C—H
        |
        H

b A compound with the formula $C_4H_{10}O$ has the NMR spectrum shown below. Draw the structural formula of the compound and explain your reasoning. [4]

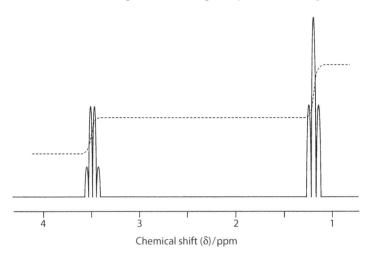

Chemical shift (δ)/ppm

c There are two esters that have the molecular formula $C_3H_6O_2$. Draw the structural formulas of these esters and explain the differences between their NMR spectra. **[5]**

d There are four esters with the molecular formula $C_4H_8O_2$. Draw the structural formulas of these esters and explain which will give rise to the NMR spectrum shown below. **[6]**

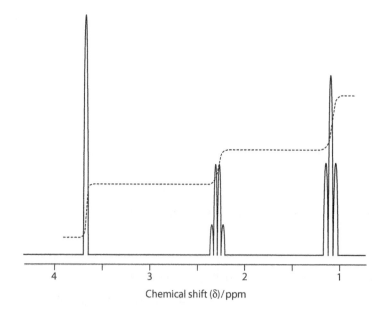

Chemical shift (δ)/ppm

Summary

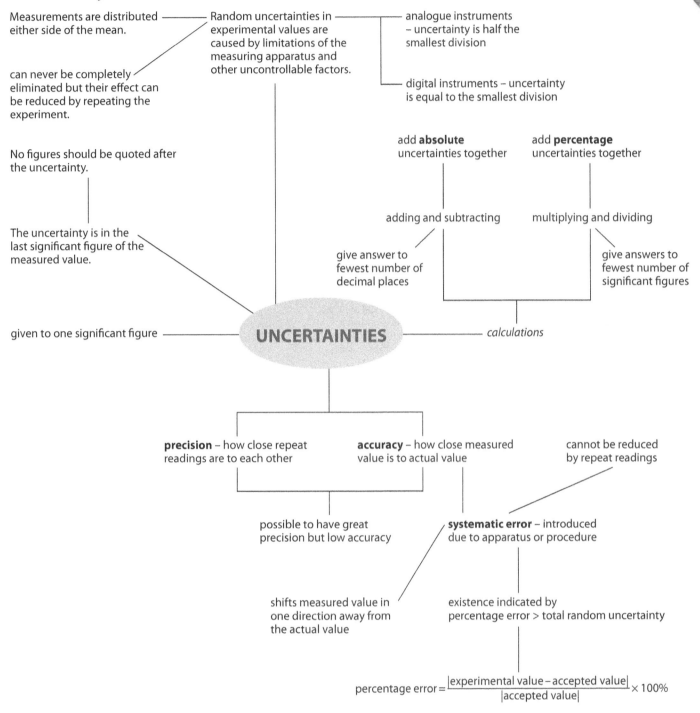

Measurements are distributed either side of the mean.

can never be completely eliminated but their effect can be reduced by repeating the experiment.

No figures should be quoted after the uncertainty.

The uncertainty is in the last significant figure of the measured value.

given to one significant figure

Random uncertainties in experimental values are caused by limitations of the measuring apparatus and other uncontrollable factors.

analogue instruments – uncertainty is half the smallest division

digital instruments – uncertainty is equal to the smallest division

add **absolute** uncertainties together

add **percentage** uncertainties together

adding and subtracting

multiplying and dividing

give answer to fewest number of decimal places

give answers to fewest number of significant figures

UNCERTAINTIES

calculations

precision – how close repeat readings are to each other

accuracy – how close measured value is to actual value

cannot be reduced by repeat readings

possible to have great precision but low accuracy

systematic error – introduced due to apparatus or procedure

shifts measured value in one direction away from the actual value

existence indicated by percentage error > total random uncertainty

$$\text{percentage error} = \frac{|\text{experimental value} - \text{accepted value}|}{|\text{accepted value}|} \times 100\%$$

Summary – continued

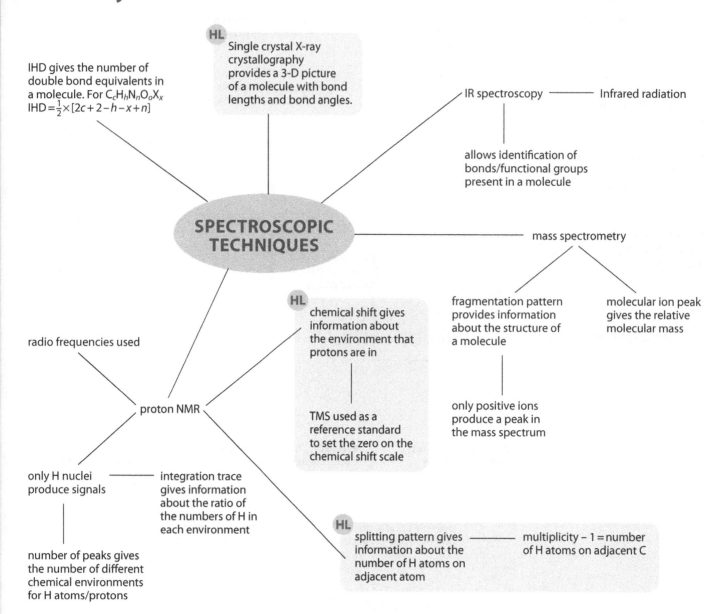

IHD gives the number of double bond equivalents in a molecule. For $C_cH_hN_nO_oX_x$
$IHD = \frac{1}{2} \times [2c + 2 - h - x + n]$

HL Single crystal X-ray crystallography provides a 3-D picture of a molecule with bond lengths and bond angles.

IR spectroscopy ——— Infrared radiation

allows identification of bonds/functional groups present in a molecule

SPECTROSCOPIC TECHNIQUES

mass spectrometry

fragmentation pattern provides information about the structure of a molecule

molecular ion peak gives the relative molecular mass

only positive ions produce a peak in the mass spectrum

HL chemical shift gives information about the environment that protons are in

TMS used as a reference standard to set the zero on the chemical shift scale

radio frequencies used

proton NMR

only H nuclei produce signals ——— integration trace gives information about the ratio of the numbers of H in each environment

number of peaks gives the number of different chemical environments for H atoms/protons

HL splitting pattern gives ——— multiplicity – 1 = number of H atoms on adjacent C information about the number of H atoms on adjacent atom

Appendix: the periodic table

Key

relative atomic mass
atomic symbol
name
atomic number

Group

1	2	3	4	5	6	7	8	9	10	11	12	13	14	15	16	17	18
1.01 **H** hydrogen 1																	4.00 **He** helium 2
6.94 **Li** lithium 3	9.01 **Be** beryllium 4											10.81 **B** boron 5	12.01 **C** carbon 6	14.01 **N** nitrogen 7	16.00 **O** oxygen 8	19.00 **F** fluorine 9	20.18 **Ne** neon 10
22.99 **Na** sodium 11	24.31 **Mg** magnesium 12											26.98 **Al** aluminium 13	28.09 **Si** silicon 14	30.97 **P** phosphorus 15	32.07 **S** sulfur 16	35.45 **Cl** chlorine 17	39.95 **Ar** argon 18
39.10 **K** potassium 19	40.08 **Ca** calcium 20	44.96 **Sc** scandium 21	47.87 **Ti** titanium 22	50.94 **V** vanadium 23	52.00 **Cr** chromium 24	54.94 **Mn** manganese 25	55.85 **Fe** iron 26	58.93 **Co** cobalt 27	58.69 **Ni** nickel 28	63.55 **Cu** copper 29	65.38 **Zn** zinc 30	69.72 **Ga** gallium 31	72.63 **Ge** germanium 32	74.92 **As** arsenic 33	78.96 **Se** selenium 34	79.90 **Br** bromine 35	83.80 **Kr** krypton 36
85.47 **Rb** rubidium 37	87.62 **Sr** strontium 38	88.91 **Y** yttrium 39	91.22 **Zr** zirconium 40	92.91 **Nb** niobium 41	95.96 **Mo** molybdenum 42	(98) **Tc** technetium 43	101.07 **Ru** ruthenium 44	102.91 **Rh** rhodium 45	106.42 **Pd** palladium 46	107.87 **Ag** silver 47	112.41 **Cd** cadmium 48	114.82 **In** indium 49	118.71 **Sn** tin 50	121.76 **Sb** antimony 51	127.60 **Te** tellurium 52	126.90 **I** iodine 53	131.29 **Xe** xenon 54
132.91 **Cs** caesium 55	137.33 **Ba** barium 56	138.91 **La** * lanthanum 57	178.49 **Hf** hafnium 72	180.95 **Ta** tantalum 73	183.84 **W** tungsten 74	186.21 **Re** rhenium 75	190.23 **Os** osmium 76	192.22 **Ir** iridium 77	195.08 **Pt** platinum 78	196.97 **Au** gold 79	200.59 **Hg** mercury 80	204.38 **Tl** thallium 81	207.2 **Pb** lead 82	208.98 **Bi** bismuth 83	(209) **Po** polonium 84	(210) **At** astatine 85	(222) **Rn** radon 86
(223) **Fr** francium 87	(226) **Ra** radium 88	(227) **Ac** ** actinium 89	(267) **Rf** rutherfordium 104	(268) **Db** dubnium 105	(269) **Sg** seaborgium 106	(270) **Bh** bohrium 107	(269) **Hs** hassium 108	(278) **Mt** meitnerium 109	(281) **Ds** darmstadtium 110	(281) **Rg** roentgenium 111	(285) **Cn** copernicium 112	(286) **Uut** ununtrium 113	(289) **Fl** flerovium 114	(288) **Uup** ununpentium 115	(293) **Lv** livermorium 116	(294) **Uus** ununseptium 117	(294) **Uuo** ununoctium 118

lanthanoids *

140.12 **Ce** cerium 58	140.91 **Pr** praseodymium 59	144.24 **Nd** neodymium 60	145 **Pm** promethium 61	150.36 **Sm** samarium 62	151.96 **Eu** europium 63	157.25 **Gd** gadolinium 64	158.93 **Tb** terbium 65	162.50 **Dy** dysprosium 66	164.93 **Ho** holmium 67	167.26 **Er** erbium 68	168.93 **Tm** thulium 69	173.05 **Yb** ytterbium 70	174.97 **Lu** lutetium 71

actinoids **

232.04 **Th** thorium 90	231.04 **Pa** protactinium 91	238.03 **U** uranium 92	(237) **Np** neptunium 93	(244) **Pu** plutonium 94	(243) **Am** americium 95	(247) **Cm** curium 96	(247) **Bk** berkelium 97	(251) **Cf** californium 98	(252) **Es** einsteinium 99	(257) **Fm** fermium 100	(258) **Md** mendelevium 101	(259) **No** nobelium 102	(262) **Lr** lawrencium 103

Answers to test yourself questions

Topic 1

1 a $2NO + O_2 \rightarrow 2NO_2$

  b $C_3H_8 + 5O_2 \rightarrow 3CO_2 + 4H_2O$

  c $CaCO_3 + 2HCl \rightarrow CaCl_2 + CO_2 + H_2O$

  d $C_2H_5OH + 3O_2 \rightarrow 2CO_2 + 3H_2O$

  e $WO_3 + 3H_2 \rightarrow W + 3H_2O$

  f $2H_2O_2 \rightarrow O_2 + 2H_2O$

  g $4CrO_3 \rightarrow 2Cr_2O_3 + 3O_2$

  h $Al_4C_3 + 6H_2O \rightarrow 3CH_4 + 2Al_2O_3$

  i $8HI + H_2SO_4 \rightarrow H_2S + 4H_2O + 4I_2$

  j $4PH_3 + 8O_2 \rightarrow P_4O_{10} + 6H_2O$

2 a compound   f element

  b element   g compound

  c compound   h mixture

  d mixture   i compound

  e mixture   j compound

3 a compound; gas   c mixture; compound;

  b element; solid   element; gas

                      d mixture; elements; gas

4

Compound	Relative molecular mass
SO_2	64.07
NH_3	17.04
C_2H_5OH	46.08
$MgCl_2$	95.21
$Ca(NO_3)_2$	164.10
$CH_3(CH_2)_5CH_3$	100.23
PCl_5	208.22
$Mg_3(PO_4)_2$	262.87
$Na_2S_2O_3$	158.12
$CH_3CH_2CH_2COOCH_2CH_3$	116.18

5

Compound	Molar mass / $g\,mol^{-1}$	Mass / g	Number of moles / mol
H_2O	18.02	9.01	0.500
CO_2	44.01	5.00	0.114
H_2S	34.09	3.41	0.100
NH_3	17.04	59.6	3.50
Q	28.6	1.00	0.0350
Z	51.61	0.0578	1.12×10^{-3}
$Mg(NO_3)_2$	148.33	1.75	0.0118
C_3H_7OH	60.11	2500	41.59
Fe_2O_3	159.70	9.07×10^{-3}	5.68×10^{-5}

6 a 2.99×10^{-23} g   c 7.31×10^{-23} g

  b 2.83×10^{-23} g

7 a 1.20×10^{24} c 9.03×10^{22}

  b 4.82×10^{23}

8 a 2.41×10^{22} c 2.17×10^{23}

  b 9.63×10^{23}

9 a 0.8 mol   c 0.12 mol

  b 0.7 mol

10 a 34.72%   c 61.24%

  b 43.19%

11 a 1.60 g   c 5.64 g

  b 2.50 g

12 a 2.00 g   c 2.29 g

  b 1.67 g

13 CO_2 CH   HO C_3H_8 H_2O PCl_5 $C_6H_5CH_3$

14

Empirical formula	Relative molecular mass	Molecular formula
HO	34.02	H_2O_2
ClO_3	166.90	Cl_2O_6
CH_2	84.18	C_6H_{12}
BNH_2	80.52	$B_3N_3H_6$

15 a CH_3 b C_2H_6

16 Cl_2O

17 I_2O_5

18 C_3H_8O

19 Fe_2O_3

20 a 0.2 mol **d** 0.9 mol
   b 0.05 mol **e** 0.6 mol
   c 0.01 mol **f** 3.6×10^{-3} mol

21 a 0.363 g **c** 4.62 g
   b 2.67 g **d** 2.97 g

22 a 80.0% **c** 65.1%
   b 94.7%

23 a H_2SO_4 **c** HNO_3
   b H_2O **d** NaCl

24 0.15 g

25 3.10 tonnes

26 a 200 cm^3 **b** 1.6 dm^3

27 a 0.0106 mol **d** 0.0176 mol
   b 0.0881 mol **e** 0.0110 mol
   c 0.00441 mol

28 a 2.27 dm^3 **d** 19.3 dm^3
   b 2270 dm^3 **e** 13.6 dm^3
   c 6.13 dm^3

29 114 cm^3

30 1910 cm^3

31 0.4669 g

32 a 290 cm^3 **b** 300 cm^3

33 21.7 cm^3

34 196 °C

35 0.0655 mol

36 43.2 g mol^{-1}

37 28.2 dm^3

38 a 0.115 dm^3 **b** 573 cm^3
39 0.782 g

40 a 3.55 g **b** 0.200 mol dm^{-3}

41 a 4.40×10^{-3} mol **c** 0.0108 mol
   b 2.34×10^{-3} mol

42 0.475 mol dm^{-3}

43 a 1.25 g **b** 0.284 dm^3

44 42.8 cm^3

45 56.8 cm^3

Topic 2

1

$^{238}_{92}U$	92 protons	146 neutrons	92 electrons
$^{75}_{33}As$	33 protons	42 neutrons	33 electrons
$^{81}_{35}Br$	35 protons	46 neutrons	35 electrons

2

$^{40}_{20}Ca^{2+}$	20 protons	20 neutrons	18 electrons
$^{127}_{53}I^{-}$	53 protons	74 neutrons	54 electrons
$^{140}_{58}Ce^{3+}$	58 protons	82 neutrons	55 electrons

3 The only element is ^1_1H, which has one proton and no neutrons.

4 a D and L are isotopes; Q and M are isotopes
   b D, X, L

5 52.06

6 28.11

7 a 91% indium-115; 9% indium-113
   b 63.85% gallium-69; 36.15% gallium-71

8 a microwaves infrared radiation orange light green light ultraviolet radiation
   b microwaves infrared radiation orange light green light ultraviolet radiation

9 An electron in a hydrogen atom that has been promoted to a higher energy level falls down to energy level 1. As it does so, it gives out energy in the form of a photon of light in the ultraviolet region of the electromagnetic spectrum.

10

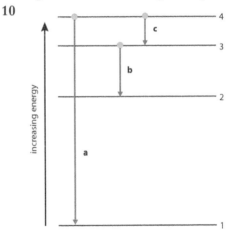

   a any transition to level 1
   b any transition to level 2
   c any transition to level 3

11 a $1s^2 2s^2 2p^3$
   b $1s^2 2s^2 2p^6 3s^2 3p^2$
   c $1s^2 2s^2 2p^6 3s^2 3p^6$
   d $1s^2 2s^2 2p^6 3s^2 3p^6 4s^2 3d^{10} 4p^3$
   e $1s^2 2s^2 2p^6 3s^2 3p^6 4s^2 3d^3$

Left column

12 a

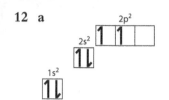

b

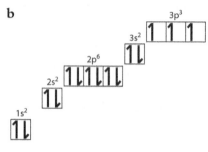

c

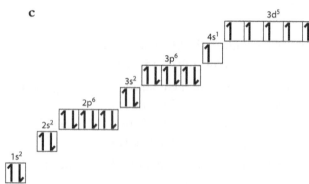

13 a 4.85×10^{-19} J; **b** 2.31×10^{15} Hz

14 X is in group 14; Z is in group 2; Q is in group 15.

15 a $1s^2 2s^2 2p^6 3s^2 3p^6$

b $1s^2 2s^2 2p^6 3s^2 3p^6 3d^3$

c $1s^2 2s^2 2p^6 3s^2 3p^6 3d^7$

d $1s^2 2s^2 2p^6 3s^2 3p^6 4s^2 3d^{10} 4p^6$

Topic 3

1 a silicon   **d** iodine

  b tellurium   **e** arsenic

  c polonium

2 a increase   **b** decrease

3 a Mg < Ca < Sr < Ba

  b $Na^+ < F^- < O^{2-}$

  c $Al^{3+} < Na^+ < Na < K$

  d $Cl < S < Cl^- < S^{2-} < I^-$

4 a false   **d** false

  b false   **e** true

  c true

5 Element X

Right column

6 a $2Rb(s) + 2H_2O(l) \rightarrow 2RbOH(aq) + H_2(g)$

  b $2K(s) + Br_2(l) \rightarrow 2KBr(s)$

  c $Cl_2(aq) + 2KBr(aq) \rightarrow 2KCl(aq) + Br_2(aq)$

  d $Na_2O(s) + H_2O(l) \rightarrow 2NaOH(aq)$

  e $SO_3(g) + H_2O(l) \rightarrow H_2SO_4(aq)$

7 a same   **c** different

  b different   **d** same

8 a acidic   **c** alkaline

  b alkaline

9 a $1s^2 2s^2 2p^6 3s^2 3p^6 3d^8$ **c** $1s^2 2s^2 2p^6 3s^2 3p^6 3d^2$

  b $1s^2 2s^2 2p^6 3s^2 3p^6 3d^6$ **d** $1s^2 2s^2 2p^6 3s^2 3p^6 3d^3$

10 a +2   **b** +2   **c** +2   **d** +3

  e +3   **f** +3   **g** +6   **h** 0

11 Cu^{2+}, Fe^{2+} and Co^{3+}

12 MnF_3, CoF_2

13 a $[Co(H_2O)_6]^{2+}$ **b** $[Fe(H_2O)_6]^{2+}$

Topic 4

1 a ionic   **c** ionic   **e** covalent

  b covalent   **d** covalent   **f** ionic

2 a MgO   **g** Li_3N **m** $AgNO_3$

  b $BaSO_4$ **h** $Mg_3(PO_4)_2$ **n** NH_4Cl

  c $Ca(OH)_2$ **i** MgF_2 **o** $Cu(NO_3)_2$

  d Na_2O **j** K_2SO_4 **p** Rb_2CO_3

  e SrS   **k** $(NH_4)_2CO_3$

  f Al_2O_3 **l** Ag_2S

3 Na < H < Br < Cl < O

4 a (Lewis structure of H_2S)

  g (Lewis structure F—N=O)

  b (Lewis structure of PCl_3)

  h (Lewis structure H—N—N—H with H)

  c (Lewis structure of CCl_4)

  i (Lewis structure H—O—O—H)

  d (Lewis structure of COF_2)

  j $[PCl_4]^+$ (Lewis structure)

  e H—C≡N

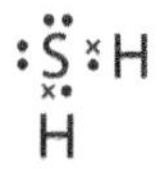

  k $[N≡O]^+$

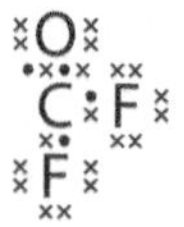

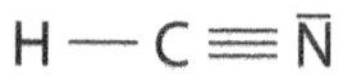

  f S=C=S

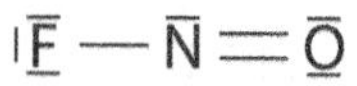

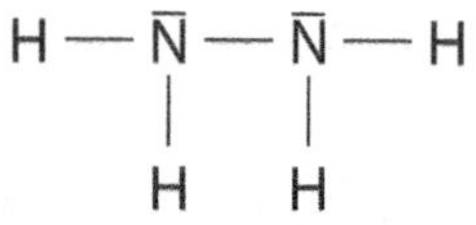

  l $[O=C=N]^-$

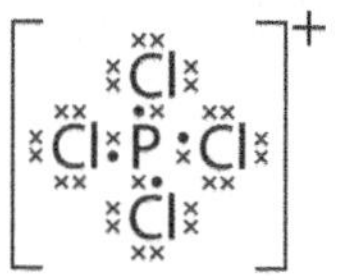

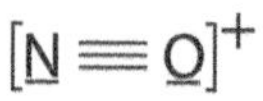

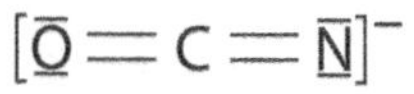

5 a

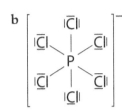

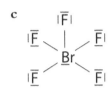

b

c

d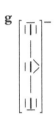

e N=N=Ō

or

N≡N—Ōl

f

g

h [N̲=N=N̲]⁻

or

[N̲≡N—N̲]⁻

6 a bent/V-shaped/angular; predict 100–108° (actual value 92°)

b trigonal pyramidal; predict 100–108° (actual value 100°)

c tetrahedral; 109.5°

d linear; 180°

e trigonal planar; predict F–C–F bond angle of about 118–120° (greater repulsion due to double bond) and O–C–F bond angle of about 120–122° (actual values: F–C–F bond angle = 108°, O–C–F bond angle = 126°)

f linear; 180°

g bent/V-shaped/angular; predict about 118° (based on trigonal planar) (actual value 110°)

h tetrahedral; 109.5°

i linear; 180°

j bent/V-shaped/angular; predict about 118° (based on trigonal planar) (actual value 117°)

k trigonal planar about each C; predict F–C–F bond angle of about 118° and F–C–C bond angle of about 122° (actual value for F–C–C bond angle about 124°)

l linear; 180°

7

HBr	polar	$\overset{\delta+}{H}\!-\!\overset{\delta-}{Br}$
HCN	polar	$\overset{\delta+}{H}\!-\!C\!\equiv\!\overset{\delta-}{N}$
PH_3	polar	
SCl_2	polar	
CF_4	non-polar	
N_2	non-polar	
OCl_2	polar	
BCl_3	non-polar	
C_2Cl_2	non-polar	
H_2S	polar	
CH_2Cl_2	polar	

8 a CH_4 CF_4 CCl_4

b PH_3 AsH_3 NH_3

c CH_4 NH_3 N_2H_4

d C_2H_4 CH_3F CH_3OH

e H_2S H_2O H_2O_2

f $CH_3CH_2CH_2CH_2CH_3$ $CH_3CH_2OCH_2CH_3$
$CH_3CH_2CH_2CH_2OH$

g Ne N_2 F_2 HF

9 a HCl CCl_4 $SiCl_4$ NaCl

b HBr Br_2 PBr_3 $CaBr_2$

c C_3H_8 C_4H_{10} C_3H_7OH C_4H_9OH
CH_3CH_2COOH

d CO_2 SO_2 SiO_2

10 a C_6H_{12} $C_5H_{11}OH$ NaCl

b CH_4 CH_3Cl $CaCl_2$

11 a K   Na   Li

b Na   Mg   Al

12 a Lewis structures: Formal charges:

preferred

b Lewis structures: Formal charges:

$$\left[\overset{\overset{\displaystyle |\bar{O}|}{\|}}{|\bar{O}=\overset{}{Cl}-\bar{O}|} \right]^{-} \quad \left[\overset{\overset{\displaystyle |\bar{O}|}{|}}{|\bar{O}-\overset{}{Cl}-\bar{O}|} \right]^{-} \quad |\bar{O}=\overset{\overset{\displaystyle |\bar{O}}{\|}}{Cl}-\bar{O}|^{\oplus} \quad |\bar{O}-\overset{\overset{\displaystyle |\bar{O}|^{\oplus}}{|}}{Cl}-\bar{O}|^{\oplus}$$

preferred

c Lewis structures: Formal charges:

$$|\bar{O}=\overset{\overset{\displaystyle |\bar{O}|}{\|}}{Xe}=\bar{O}| \quad |\bar{O}-\overset{\overset{\displaystyle |\bar{O}|}{|}}{Xe}-\bar{O}| \quad |\bar{O}=\overset{\overset{\displaystyle |\bar{O}}{\|}}{Xe}=\bar{O}| \quad |\bar{O}^{\oplus}-\overset{\overset{\displaystyle |\bar{O}|^{\oplus}}{|}}{Xe}-\bar{O}|^{\oplus}$$

preferred

13 **a** T-shaped (arrow-shaped); predict bond angles of about 88° (actual value 86°)

b square-based pyramid; predict bond angle of about 88°

c (distorted) tetrahedral; predict bond angles O–S–O of about 112° and Cl–S–Cl of about 107° (angles of 109° are fine!) (actual values: O–S–O = 120° and Cl–S–Cl = 111°)

d see-saw shaped; predict bond angles of about 88° and 118° (actual values 85° and 101°)

e linear; 180°

f octahedral; 90°

g square-based pyramid; predict bond angle of about 88° (actual value 79°)

h see-saw shaped; predict bond angles of about 88° and 118°

i bent/V-shaped/angular; predict bond angles of about 105° (actual value 102°)

j linear; 180°

14 XeF$_6$ non-polar
    XeF$_4$ non-polar
    SF$_4$ polar

    PCl$_5$ non-polar
    SF$_2$ polar

    SF$_6$ non-polar
    ClF$_5$ polar

BrF$_3$ polar

SOCl$_2$ polar

15 **a** 1σ 1π   **e** 3σ 1π   **i** 8σ 1π
    b 1σ 2π   **f** 2σ 2π   **j** 3σ 1π
    c 3σ   **g** 3σ 1π   **k** 6σ 2π
    d 2σ 2π   **h** 1σ 2π   **l** 6σ 2π

16 **a** 1.5
    b 1.25
    c 1.75

17 **a** sp^2 **e** sp   **i** sp^2
    b sp^3 **f** sp^3 **j** sp^2
    c sp^3 **g** sp^3
    d sp^3 **h** sp

18 **a** sp^3 **b** sp^2

19 **a** sp^2 **c** sp^3
    b sp   **d** sp^2

Topic 5

1 **a** 0.128 J g^{-1} °C^{-1} **b** 0.236 J g^{-1} °C^{-1}

2 **a** −4200 kJ mol^{-1} **b** −5490 kJ mol^{-1}

3 −1080 kJ mol^{-1}

4 Heat energy loss to the surroundings; incomplete combustion (other, more minor, factors include evaporation of water and/or propan-1-ol)

5 **a** −56.8 kJ mol^{-1} **b** **i** 1.36 °C
                            ii 2.72 °C
                            iii 2.72 °C

6 **a** −151 kJ mol^{-1} **c** 15.2 °C
    b 7.6 °C

7 **a** +44 kJ mol^{-1} **b** +31 kJ mol^{-1}

8 **a** −400 kJ mol^{-1} **b** −91 kJ mol^{-1}

9 −312 kJ mol^{-1}

10 −126 kJ mol^{-1}

11 **a** $\frac{1}{2}H_2(g) + \frac{1}{2}F_2(g) \rightarrow HF(g)$
    b $\frac{3}{2}H_2(g) + C(s) + \frac{1}{2}Cl_2(g) \rightarrow CH_3Cl(g)$
    c $H_2(g) + \frac{1}{2}O_2(g) \rightarrow H_2O(l)$
    d $6H_2(g) + 5C(s) + \frac{1}{2}O_2(g) \rightarrow C_5H_{11}OH(l)$

12 $+20\,\text{kJ}\,\text{mol}^{-1}$

13 $-574.1\,\text{kJ}\,\text{mol}^{-1}$

14 $+33\,\text{kJ}\,\text{mol}^{-1}$

15 **a** $-115\,\text{kJ}\,\text{mol}^{-1}$ **c** $-107\,\text{kJ}\,\text{mol}^{-1}$
    b $-287\,\text{kJ}\,\text{mol}^{-1}$

16 $147\,\text{kJ}\,\text{mol}^{-1}$

17 $124.5\,\text{kJ}\,\text{mol}^{-1}$

18 $-273\,\text{kJ}\,\text{mol}^{-1}$

19 $-344\,\text{kJ}\,\text{mol}^{-1}$

20 $2307\,\text{kJ}\,\text{mol}^{-1}$

21 $KCl < LiF < CaCl_2 < CaS < CaO$

22 **a** always endothermic
    b always endothermic
    c always endothermic
    d sometimes exothermic and sometimes endothermic
    e always exothermic
    f always endothermic

23 $14\,\text{kJ}\,\text{mol}^{-1}$

24 $-155\,\text{kJ}\,\text{mol}^{-1}$

25 **a** decrease   **c** increase
    b decrease   **d** increase

26 **a** $-38\,\text{J}\,\text{K}^{-1}\,\text{mol}^{-1}$ **c** $411\,\text{J}\,\text{K}^{-1}\,\text{mol}^{-1}$
    b $763\,\text{J}\,\text{K}^{-1}\,\text{mol}^{-1}$

27 $\Delta G^{\ominus} = -121\,\text{kJ}\,\text{mol}^{-1}$
    The reaction is spontaneous; $\Delta G^{\ominus}$ is negative.

28 $\Delta G^{\ominus} = -2108\,\text{kJ}\,\text{mol}^{-1}$
    The reaction is spontaneous; $\Delta G^{\ominus}$ is negative.

29 **a** $889\,\text{J}\,\text{K}^{-1}\,\text{mol}^{-1}$ **b** $399\,°\text{C}$

30 **a** more spontaneous
    b less spontaneous
    c more spontaneous
    d more spontaneous

31 Because $\Delta G^{\ominus}$ is positive, the position of equilibrium lies closer to the reactant (N_2O_4)

Topic 6

1 **a**

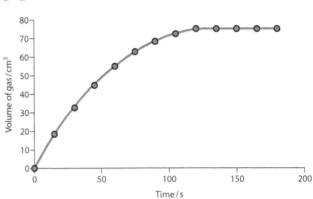

The gradient is steepest at the beginning – the reaction is fastest; the concentration of reactants is highest at the beginning, so the collision frequency is highest.
The gradient decreases as the reaction goes on – the reaction rate decreases; reactants are being used up – there is a lower concentration of reactants, so lower collision frequency.
The graph becomes horizontal – the reaction has finished; the acid is in excess, so the reaction finishes when all the magnesium has been used up.

b The initial rate is approximately $1.3\,\text{cm}^3\,\text{s}^{-1}$.

c The average rate is $0.62\,\text{cm}^3\,\text{s}^{-1}$.

d, e, f

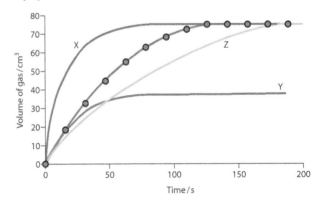

g

At the higher temperature, there are more particles with energy greater than the activation energy. Therefore there is a greater chance that a collision will result in a reaction and more successful collisions per unit time.

2 **a** First order with respect to A; second order with respect to B

   b 3

   c $10\,mol^{-2}\,dm^6\,s^{-1}$

   d $2.50 \times 10^{-3}\,mol\,dm^{-3}\,s^{-1}$

3 **a** rate $= k[X][Y]$

   b $k = 3.96$

   c $mol^{-1}\,dm^3\,s^{-1}$

4 The rate of reaction would be quadrupled (increased by a factor of 4).

5 **a** First order

   b Zero order

   c rate $= k[D]$

   d $k = 0.171\,s^{-1}$

   e $0.0103\,mol\,dm^{-3}\,s^{-1}$

6 • The mechanism does not agree with the stoichiometric equation (the overall equation from this mechanism is $2A + 3B \rightarrow 3C + 2D$).

   • The rate-determining step contains A, but A does not appear in the rate equation.

   • Step 2 involves three molecules colliding, which is extremely unlikely.

7 **a** rate $= k[Q]^2$

   b Because it does not agree with the stoichiometric equation – the overall equation from this mechanism would be: $2Q + P + Z \rightarrow Y + R + S$

   c Mechanism 3

8 $NO + NO \overset{fast}{\rightleftharpoons} N_2O_2$

   $N_2O_2 + Br_2 \overset{RDS}{\longrightarrow} 2NOBr$

   or

   $NO + Br_2 \overset{fast}{\rightleftharpoons} NOBr_2$

   $NOBr_2 + NO \overset{RDS}{\longrightarrow} 2NOBr$

   or

   $NO + Br_2 \overset{fast}{\rightleftharpoons} NOBr + Br$

   $NO + Br \overset{RDS}{\longrightarrow} NOBr$

9 $X + Y \overset{RDS}{\longrightarrow} XY$

   $XY + X \overset{fast}{\rightleftharpoons} 2Z$

10 $56\,kJ\,mol^{-1}$; $A = 1.97 \times 10^8\,mol^{-1}\,dm^3\,s^{-1}$

11 $90\,kJ\,mol^{-1}$; $A = 9.74 \times 10^9\,mol^{-1}\,dm^3\,s^{-1}$

Topic 7

1 **a** shifts to left   **c** shifts to right

   b shifts to left

2 **a** shifts to left   **c** shifts to left

   b shifts to right

3 **a** shifts to right

   b shifts to left

   c shifts to right

   d shifts to right (the products formed are acids and will react with NaOH – their concentration is lowered)

4 **a** $K_c = \dfrac{[N_2O_3(g)]}{[NO(g)][NO_2(g)]}$

   b $K_c = \dfrac{[CO(g)][H_2(g)]^3}{[CH_4(g)][H_2O(g)]}$

   c $K_c = \dfrac{[H_2(g)]^2[O_2(g)]}{[H_2O(g)]^2}$

   d $K_c = \dfrac{[NO(g)]^4[H_2O(g)]^6}{[NH_3(g)]^4[O_2(g)]^5}$

   e $K_c = \dfrac{[NO_2(g)]^2}{[NO(g)]^2[O_2(g)]}$

5 **a** $\sqrt{10^{33}}$ or 3.2×10^{16}

   b 10^{-19}

   c $\sqrt{10^{-19}}$ or 3.2×10^{-10}

6 **a** 9.71×10^{-13} **b** 7.73

7 **a** shifts to left; K_c stays the same

   b shifts to right; K_c increases

   c shifts to right; K_c increases

   d shifts to right; K_c stays the same

   e shifts to right; K_c stays the same

   f no effect; K_c stays the same

8 **a** 22.2   **b** 1.02×10^{-9}

9 **a** $0.520\,mol\,Z$

   b $0.600\,mol\,A$; $1.800\,mol\,X$

10 a $K_c = \dfrac{[Q(g)]}{[A(g)][X(g)]}$

   b i 3.33
   ii 200

   c endothermic

11 a $K_c = \dfrac{[Q(g)]^4[Z(g)]}{[A(g)]^2[X(g)]}$

   b i 2.37×10^{-4}
   ii 0.154

   c exothermic

12 a 0.075 mol X

   b $0.080\,\text{mol dm}^{-3}$ of A

13 a $-2.53\,\text{kJ mol}^{-1}$
   b $-35.4\,\text{kJ mol}^{-1}$
   c $43.5\,\text{kJ mol}^{-1}$

14 a 1.70×10^{-9}
   b 5.79×10^3
   c 6.72×10^{26}

Topic 8

1 An acid is a proton (H^+) donor; a base is a proton (H^+) acceptor.

2 $\underset{\text{acid 1}}{CH_3COOH(aq)} + \underset{\text{base 2}}{NH_3(aq)}$

$\rightleftharpoons \underset{\text{base 1}}{CH_3COO^-(aq)} + \underset{\text{acid 2}}{NH_4^+(aq)}$

3 a NH_4^+ **c** H_2SO_4 **e** $H_2PO_4^-$
   b H_2O **d** HCN

4 a CO_3^{2-} **c** $HCOO^-$
   b OH^- **d** NH_2^-

5 a acid
   b base
   c base

6 An acid is an electron pair acceptor; a base is an electron pair donor.

7 H_2O = base; BF_3 = acid; HCO_3^- = base; H^+ = acid; $AlCl_3$ = acid; NH_3 = base; CO = base

8 a $Zn + H_2SO_4 \rightarrow ZnSO_4 + H_2$
   b $CuO + 2HNO_3 \rightarrow Cu(NO_3)_2 + H_2O$
   c $2NH_3 + H_2SO_4 \rightarrow (NH_4)_2SO_4$
   d $Ca(HCO_3)_2 + 2HCl \rightarrow CaCl_2 + 2CO_2 + 2H_2O$
   e $Mg(OH)_2 + H_2SO_4 \rightarrow MgSO_4 + 2H_2O$
   f $Cu + H_2SO_4 \rightarrow$ no reaction
   g $CaO + 2HCl \rightarrow CaCl_2 + H_2O$

9 a calcium oxide/hydroxide and nitric acid
   b cobalt(II) oxide/hydroxide and sulfuric acid
   c copper(II) oxide/hydroxide and hydrochloric acid
   d magnesium oxide/hydroxide and ethanoic acid

10 $0.10\,\text{mol dm}^{-3}\,H_2SO_4 < 0.10\,\text{mol dm}^{-3}\,HCl$
   $< 0.010\,\text{mol dm}^{-3}\,HCl < 1.0\,\text{mol dm}^{-3}\,NaOH$

11 a false **b** true **c** false **d** false

12 a X = 1.7; Y = 3.7

   b measure electrical conductivity: X higher; add magnesium or calcium carbonate: X will react more vigorously

13

a solution containing $0.001\,00\,\text{mol dm}^{-3}\,H^+(aq)$	3
a solution containing $1.00 \times 10^{-12}\,\text{mol dm}^{-3}\,H^+(aq)$	12
a solution of $1.00\,\text{mol dm}^{-3}\,HCl(aq)$	0
a solution of $2.00 \times 10^{-4}\,\text{mol dm}^{-3}\,HNO_3(aq)$	3.70
a solution of CH_3COOH of concentration $0.100\,\text{mol dm}^{-3}$ assuming 5% dissociation of the acid	2.30

14 a $3.16 \times 10^{-4}\,\text{mol dm}^{-3}$
   b $1.26 \times 10^{-8}\,\text{mol dm}^{-3}$
   c $1.58 \times 10^{-13}\,\text{mol dm}^{-3}$

15

$[H^+(aq)]/$ mol dm^{-3}	$[OH^-(aq)]/\text{mol dm}^{-3}$	pH	Acidic or alkaline?
1×10^{-3}	1×10^{-11}	3	acidic
1×10^{-5}	1×10^{-9}	5	acidic
1×10^{-12}	0.01	12	basic

16 a 12.2 **b** 13.0

17 a $5.0 \times 10^{-13}\,\text{mol dm}^{-3}$
   b $7.1 \times 10^{-11}\,\text{mol dm}^{-3}$

18

Strong acid	Weak acid	Strong base	Weak base	Salt
HCl	H_2CO_3	NaOH	NH_3	NH_4NO_3
H_2SO_4	HCOOH	$Ba(OH)_2$	CH_3NH_2	Na_2SO_4
HNO_3				KNO_3

19 Hydrochloric acid is a strong acid and ethanoic acid is a weak acid, but HCl has the higher pH. The pH is just a measure of the concentration of H^+ ions in solution and therefore depends on concentration. pH is only a useful measure of acid strength when solutions of equal concentration are being compared.

20 a $0.10\,\text{mol dm}^{-3}$ HCl
  b $0.10\,\text{mol dm}^{-3}$ KOH
  c $0.10\,\text{mol dm}^{-3}$ H_2SO_4
  d $0.10\,\text{mol dm}^{-3}$ HNO_3

21 Because carbon dioxide in the atmosphere dissolves in rain to form carbonic acid, a weak acid.

22 H_2SO_3, H_2SO_4, HNO_2, HNO_3, (H_2CO_3)

23 a

Acid	Concentration of acid / mol dm^{-3}	$[H^+]$/ mol dm^{-3}	pH	K_a	pK_a
HA	0.0100	2.00×10^{-6}	5.70	4.00×10^{-10}	9.40
HB	0.200	2.00×10^{-5}	4.70	2.00×10^{-9}	8.70
HC	0.500	2.50×10^{-4}	3.60	1.25×10^{-7}	6.90
HD	2.20×10^{-2}	6.31×10^{-6}	5.20	1.81×10^{-9}	8.74
HE	0.250	9.64×10^{-5}	4.02	3.72×10^{-8}	7.43
HF	0.0300	3.86×10^{-4}	3.41	4.96×10^{-6}	5.30

  b HF > HC > HE > HB > HD > HA

24 a

Base	Concentration of base / mol dm^{-3}	$[OH^-]$/ mol dm^{-3}	K_b	pK_b
B_1	0.100	1.33×10^{-3}	1.77×10^{-5}	4.75
B_2	0.250	3.79×10^{-3}	5.75×10^{-5}	4.24
B_3	0.0200	4.70×10^{-4}	1.10×10^{-5}	4.96

  b $B_3 < B_1 < B_2$

25 HOI < HOCl < HNO_2 < $HClO_2$

26 a 3.28   **b** 4.85   **c** 6.30

27 a 7.37   **b** 6.84   **c** 6.14

28 a 1.52   **b** 0.301   **c** 12.3

29 a 11.7   **b** 8.31

30 a 12.53
  b 1.000 (pK_w makes no difference to the calculation because we ignore any dissociation of water)
  c 12.13

31

Acid	K_a	pK_a	Conjugate base	K_b	pK_b
HCN	3.98×10^{-10}	9.40	CN^-	2.51×10^{-5}	4.60
HF	5.62×10^{-4}	3.25	F^-	1.78×10^{-11}	10.75
HIO_3	0.158	0.8	IO_3^-	6.31×10^{-14}	13.2
NH_4^+	5.62×10^{-10}	9.25	NH_3	1.78×10^{-5}	4.75
CH_3COOH	1.74×10^{-5}	4.76	CH_3COO^-	5.75×10^{-10}	9.24
$CH_3NH_3^+$	2.29×10^{-11}	10.64	CH_3NH_2	4.37×10^{-4}	3.36

32

Acid	Alkali	Indicator
$0.100\,\text{mol dm}^{-3}$ CH_3COOH	$0.100\,\text{mol dm}^{-3}$ NaOH	phenolphthalein
$0.010\,\text{mol dm}^{-3}$ HNO_3	$0.020\,\text{mol dm}^{-3}$ KOH	bromothymol blue
$0.010\,\text{mol dm}^{-3}$ HCl	$0.010\,\text{mol dm}^{-3}$ NH_3	bromocresol green

33

initial pH	3.23
volume of NaOH required to reach the equivalence point	$50.0\,\text{cm}^3$
approximate pH at equivalence point	8–9 (8.3)
approximate final pH (after approximately $200\,\text{cm}^3$ of NaOH has been added)	approaches 12

34 4.80

35

	pH
$0.100\,\text{mol dm}^{-3}$ $CH_3CH_2CH_2CH_2COONa$	greater than 7
$0.500\,\text{mol dm}^{-3}$ KNO_3	7
$0.100\,\text{mol dm}^{-3}$ Na_2CO_3	greater than 7
$0.100\,\text{mol dm}^{-3}$ $CH_3CH_2NH_3^+Cl^-$	less than 7
$0.200\,\text{mol dm}^{-3}$ $CrCl_3$	less than 7

36 a 11.2   **b** 9.09

37 a yes   **c** yes   **e** no   **g** yes
  b no   **d** yes   **f** yes

Topic 9

1 a +3   **c** +1   **e** −2
  b +2   **d** +4   **f** +3

2 a +1   **c** +7   **e** +5
  b −1   **d** +3

3 a +1   **c** +4   **e** −3 ($NH_4^+ NO_3^-$)
  b +5   **d** +6   **f** −1

4 a nitrogen(II) oxide   **d** potassium iodate(V)
  b chlorine(VII) oxide   **e** chromium(III) oxide
  c selenium(IV) oxide

5 a +2   **c** +2   **e** +1
  b +3   **d** +2   **f** +2

6 a reduction   **d** oxidation
  b reduction   **e** reduction
  c oxidation

7 a not redox
  b redox: S oxidised; O reduced
  c redox: Na oxidised; H reduced
  d not redox
  e not redox
  f redox: Fe oxidised; O reduced (oxidation number of O in H_2O_2 is −1)
  g redox: O oxidised; Hg reduced; S reduced
  h redox: I oxidised; Cl reduced

8 a $CuSO_4$ is the oxidising agent; Zn is the reducing agent
  b Cl_2 is the oxidising agent; Br^- is the reducing agent
  c I_2O_5 is the oxidising agent; CO is the reducing agent
  d HNO_3 is the oxidising agent; S is the reducing agent
  e I_2 is the oxidising agent; $Na_2S_2O_3$ is the reducing agent
  f $KMnO_4$ is the oxidising agent; $Na_2C_2O_4$ is the reducing agent
  g $K_2Cr_2O_7$ is the oxidising agent; $FeSO_4$ is the reducing agent

9 a $Fe^{3+} + 3e^- \rightarrow Fe$
  b $Pb^{2+} \rightarrow Pb^{4+} + 2e^-$
  c $I_2 + 2e^- \rightarrow 2I^-$
  d $2S_2O_3^{2-} \rightarrow S_4O_6^{2-} + 2e^-$
  e $C_2O_4^{2-} \rightarrow 2CO_2 + 2e^-$

10 a $I_2 + 2H_2O \rightarrow 2OI^- + 4H^+ + 2e^-$
  b $MnO_4^- + 4H^+ + 3e^- \rightarrow MnO_2 + 2H_2O$
  c $2IO_3^- + 12H^+ + 10e^- \rightarrow I_2 + 6H_2O$

  d $N_2 + 6H_2O \rightarrow 2NO_3^- + 12H^+ + 10e^-$
  e $SO_4^{2-} + 4H^+ + 2e^- \rightarrow H_2SO_3 + H_2O$

11 a $VO^{2+} + 2H^+ + e^- \rightarrow V^{3+} + H_2O$
  b $Xe + 3H_2O \rightarrow XeO_3 + 6H^+ + 6e^-$
  c $NO_3^- + 4H^+ + 3e^- \rightarrow NO + 2H_2O$
  d $2NO_3^- + 10H^+ + 8e^- \rightarrow N_2O + 5H_2O$
  e $VO_2^+ + 2H^+ + e^- \rightarrow VO^{2+} + H_2O$

12 a $Cl_2 + 2Br^- \rightarrow 2Cl^- + Br_2$
  b $Zn + 2Ag^+ \rightarrow Zn^{2+} + 2Ag$
  c $2Fe^{3+} + 2I^- \rightarrow 2Fe^{2+} + I_2$

13 a $6Fe^{2+} + Cr_2O_7^{2-} + 14H^+ \rightarrow 6Fe^{3+} + 2Cr^{3+} + 7H_2O$
  b $6I^- + Cr_2O_7^{2-} + 14H^+ \rightarrow 3I_2 + 2Cr^{3+} + 7H_2O$
  c $Zn + 2VO^{2+} + 4H^+ \rightarrow Zn^{2+} + 2V^{3+} + 2H_2O$
  d $2BrO_3^- + 10I^- + 12H^+ \rightarrow Br_2 + 5I_2 + 6H_2O$
  e $2H_2O + NpO_2^{2+} + U^{4+} \rightarrow NpO_2^+ + UO_2^+ + 4H^+$

14 a 5.00×10^{-5} mol
  b 2.50×10^{-5} mol
  c 2.50×10^{-5} mol
  d 8.00 mg dm^{-3}
  e 3.40 mg dm^{-3}; <5 ppm therefore not polluted

15 a X > Z > Q
  b **i** stronger
     ii X > Z > A > Q
    iii Q^{2+}
  c **i** from X to A
     ii X
    iii $X(s)|X^{2+}(aq)||A^{2+}(aq)|A(s)$
    iv lower

16 a anode: bromine   cathode: potassium
  b anode: chlorine   cathode: copper
  c anode: oxygen   cathode: nickel
  d anode: chlorine   cathode: calcium

17 a anode: $2Cl^- \rightarrow Cl_2 + 2e^-$
    cathode: $Na^+ + e^- \rightarrow Na$
  b anode: $2O^{2-} \rightarrow O_2 + 4e^-$
    cathode: $Fe^{3+} + 3e^- \rightarrow Fe$
  c anode: $2Br^- \rightarrow Br_2 + 2e^-$
    cathode: $Mg^{2+} + 2e^- \rightarrow Mg$

18 a +0.21V; direction of electron flow from iron half-cell to nickel half-cell
    $Ni^{2+}(aq) + Fe(s) \rightarrow Ni(s) + Fe^{2+}(aq)$
  b +0.82V; from iodine half-cell to chlorine half-cell
    $Cl_2(g) + 2I^-(aq) \rightarrow 2Cl^-(aq) + I_2(aq)$
  c +1.56 V; from zinc half-cell to silver half-cell
    $Zn(s) + 2Ag^+(aq) \rightarrow Zn^{2+}(aq) + 2Ag(s)$

d +0.56 V; from iron half-cell to chromium half-cell
$$6Fe^{2+}(aq) + Cr_2O_7^{2-}(aq) + 14H^+(aq)$$
$$\rightarrow 6Fe^{3+}(aq) + 2Cr^{3+}(aq) + 7H_2O(l)$$

e +0.15 V; from chlorine half-cell to manganese half-cell
$$MnO_4^-(aq) + 16H^+(aq) + 10Cl^-(aq)$$
$$\rightarrow 2Mn^{2+}(aq) + 8H_2O(l) + 5Cl_2(g)$$

f +1.53 V; from iron half-cell to bromine half-cell
$$Fe(s) + Br_2(l) \rightarrow Fe^{2+}(aq) + 2Br^-(aq)$$

19 a cell potential negative, therefore not spontaneous
b cell potential +0.24 V, therefore spontaneous; $Cr_2O_7^{2-}$ is the oxidising agent and Br^- is the reducing agent
c cell potential negative, therefore not spontaneous
d cell potential negative, therefore not spontaneous

20 a **i** Po^{2+} **ii** Np
  b **i** true   **v** false
      ii true   **vi** true
      iii false   **vii** true
      iv true
  c **i** not spontaneous
      ii spontaneous
      iii not spontaneous (both oxidation reactions)

21 a $E^\ominus_{cell} = +0.53$ V;
     $\Delta G^\ominus = -100$ kJ mol^{-1};
     spontaneous
  b $E^\ominus_{cell} = -0.47$ V;
     $\Delta G^\ominus = +91$ kJ mol^{-1};
     non–spontaneous
  c $E^\ominus_{cell} = +0.79$ V;
     $\Delta G^\ominus = -460$ kJ mol^{-1};
     spontaneous

22 a anode: iodine   cathode: hydrogen
  b anode: oxygen   cathode: hydrogen
  c anode: chlorine   cathode: hydrogen
  d anode: oxygen   cathode: hydrogen
  e anode: oxygen   cathode: silver

23 a anode: $2H_2O(l) \rightarrow O_2(g) + 4H^+(aq) + 4e^-$
     cathode: $2H_2O(l) + 2e^- \rightarrow H_2(g) + 2OH^-(aq)$
  b anode: $2H_2O(l) \rightarrow O_2(g) + 4H^+(aq) + 4e^-$
     cathode: $Ag^+(aq) + e^- \rightarrow Ag(s)$
  c anode: $2H_2O(l) \rightarrow O_2(g) + 4H^+(aq) + 4e^-$
     cathode: $2H_2O(l) + 2e^- \rightarrow H_2(g) + 2OH^-(aq)$

24 a 0.20 mol Na and 0.10 mol Cl_2
  b 0.10 mol Cu and 0.10 mol Cl_2
  c 0.10 mol H_2 and 0.050 mol O_2

Topic 10

1 $C_{15}H_{32}$

2

	Molecular formula	Empirical formula
a	C_6H_{14}	C_3H_7
b	C_5H_{10}	CH_2
c	C_8H_8O	C_8H_8O
d	$C_{10}H_{12}O_2$	C_5H_6O

3

a	alcohol / hydroxyl / –OH group
b	alkene / C=C / alkenyl; carboxylic acid / carboxyl / COOH group
c	ketone / carbonyl
d	ester
e	ether; nitrile
f	benzene ring / phenyl; carboxylic acid / carboxyl
g	aldehyde / carbonyl; alkyne / alkynyl

4

a	2-methylpentane
b	2,2,3-trimethylbutane
c	2,2-dimethylpentane
d	2-methylpentane
e	but-2-ene
f	pent-2-yne
g	3-methylbut-1-yne
h	3-methylpent-2-ene

5 a

b

c

d

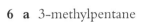

6 a 3-methylpentane
 b 3-methylpentane
 c 2,2-dimethylbutane

7

a	1-bromobutane
b	2-bromo-2-methylbutane
c	2-chloropropane
d	1-iodo-3-methylbutane

8

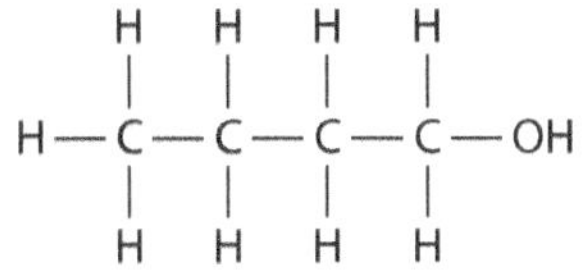 H—C—C—C—C—OH butan-1-ol	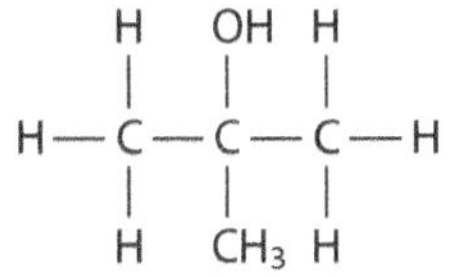 H—C—C—C—H (OH, CH₃) 2-methylpropan-2-ol
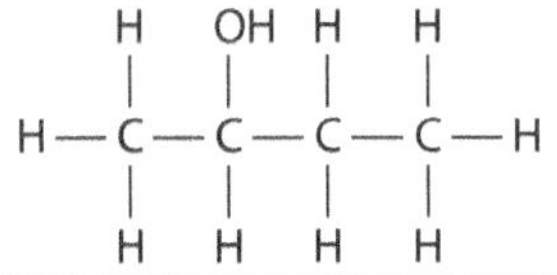 H—C—C—C—C—H (OH) butan-2-ol*	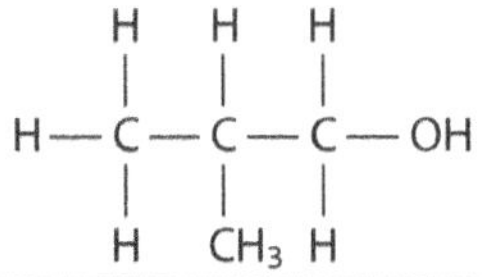 H—C—C—C—OH (CH₃) 2-methylpropan-1-ol
H—C—O—C—C—C—H 1-methoxypropane	H—C—C—C—H (CH₃, O) 2-methoxypropane
H—C—C—O—C—C—H ethoxyethane	

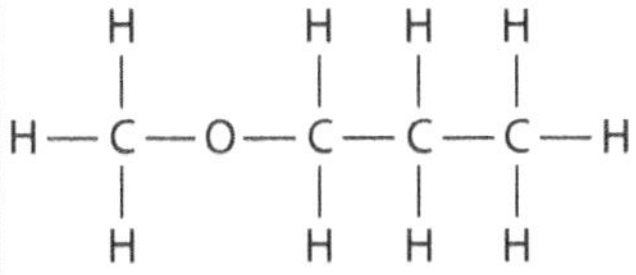

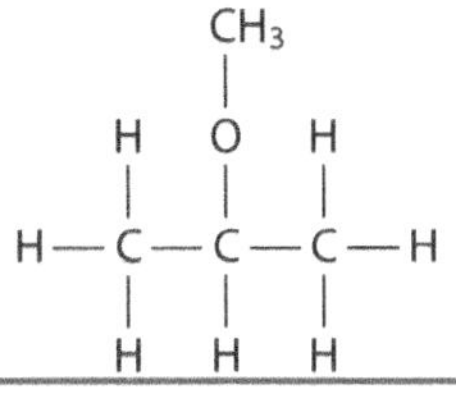

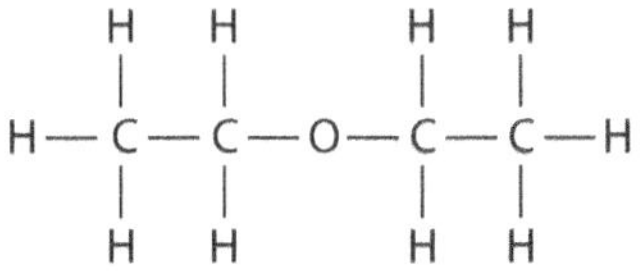

*Note: butan–2-ol also has optical isomers – see later in the Higher Level section, page **495**.

9

a	butanone
b	3-methylbutanal
c	2-methylpentanoic acid
d	3,3-dimethylbutanoic acid

10

	Structure
a	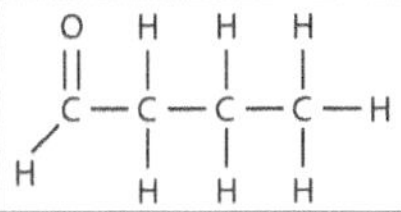 O H H H ‖ ‖ ‖ ‖ C—C—C—C—H / ‖ ‖ ‖ H H H H
b	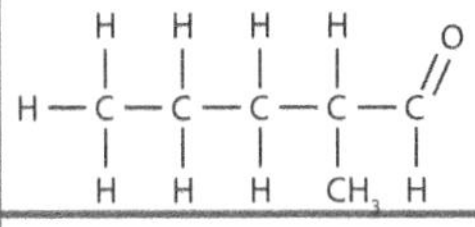 H H H H O ‖ ‖ ‖ ‖ // H—C—C—C—C—C ‖ ‖ ‖ ‖ ‖ H H H CH₃ H
c	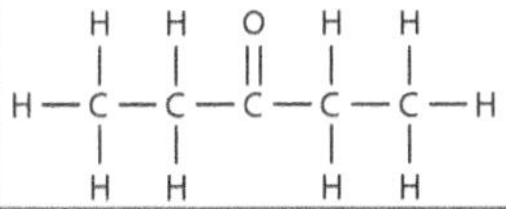 H H O H H ‖ ‖ ‖ ‖ ‖ H—C—C—C—C—C—H ‖ ‖ ‖ ‖ H H H H

11 a butyl propanoate

b propyl methanoate

c propyl pentanoate

12

a	1-bromobutane	primary halogenoalkane
b	2-bromo-2-methylbutane	tertiary halogenoalkane
c	2-chloropropane	secondary halogenoalkane
d	1-iodo-3-methylbutane	primary halogenoalkane
e	3,3-dimethylbutan-2-ol	secondary alcohol
f	2-methylbutan-2-ol	tertiary alcohol
g		secondary amine
h		primary amine

16 $C_5H_{12} + 8O_2 \rightarrow 5CO_2 + 6H_2O$

17 $2C_3H_8 + 7O_2 \rightarrow 6CO + 8H_2O$

18 $C_2H_6 + Cl_2 \rightarrow C_2H_5Cl + HCl$

19 a 3   **b** 2   **c** 1

20

$Cl_2 \rightarrow 2Cl\bullet$ initiation

$Cl\bullet + C_2H_6 \rightarrow \bullet CH_2CH_3 + HCl$ propagation

$\bullet CH_2CH_3 + Cl_2 \rightarrow CH_3CH_2Cl + Cl\bullet$ propagation

$Cl\bullet + Cl\bullet \rightarrow Cl_2$ termination

$Cl\bullet + \bullet CH_2CH_3 \rightarrow CH_3CH_2Cl$ termination

$\bullet CH_2CH_3 + \bullet CH_2CH_3 \rightarrow C_4H_{10}$ termination

13

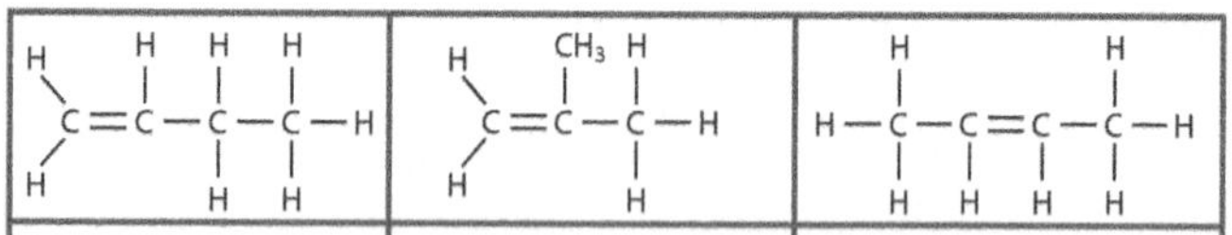

but-1-ene	2-methylpropene	but-2-ene

Note: *cis–trans* isomers are possible for but-2-ene – see page **490**.

14

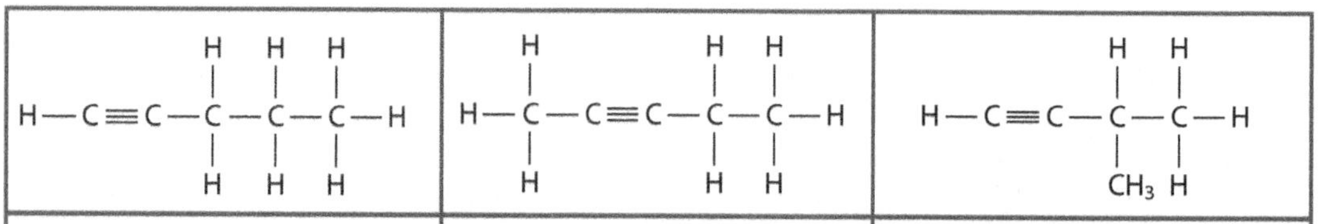

pent-1-yne	pent-2-yne	3-methylbut-1-yne

15 a

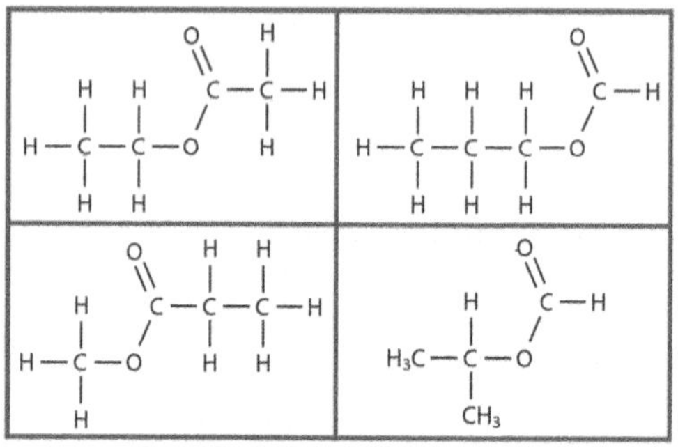

b

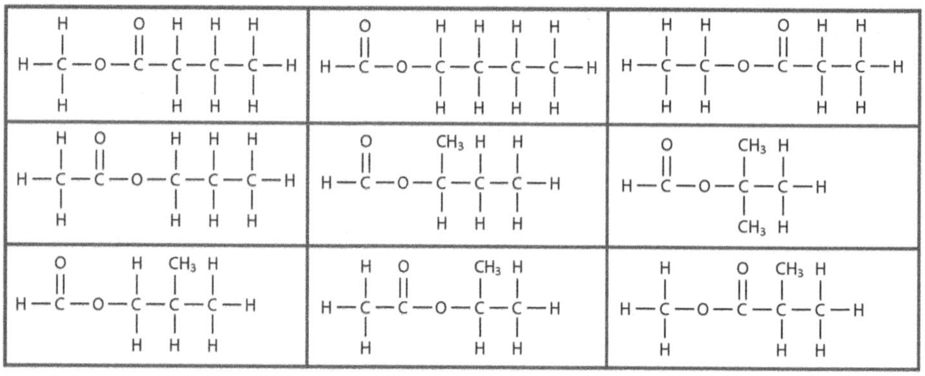

Note: one of these esters also exhibits optical isomerism.

21 a

```
    H   H   H   H
    |   |   |   |
H — C — C — C — C — H
    |   |   |   |
    H   H   H   H
```

b

```
    H   H   H   Br  H   H
    |   |   |   |   |   |
H — C — C — C — C — C — C — H
    |   |   |   |   |   |
    H   H   H   H   H   H
```

c

```
    H   H   Cl  Cl  H
    |   |   |   |   |
H — C — C — C — C — C — H
    |   |   |   |   |
    H   H   H   H   H
```

d

```
    H   OH  H   H
    |   |   |   |
H — C — C — C — C — H
    |   |   |   |
    H   H   H   H
```

22 a

```
    H   H       H   H              H   H   OH  H   H   H
    |   |       |   |              |   |   |   |   |   |
H — C — C — C = C — C — H + H₂O → H — C — C — C — C — C — C — H
    |   |       |   |              |   |   |   |   |   |
    H   H       H   H              H   H   H   H   H   H
```

b

```
    H   H                    H   H   Br  Br
    |   |                    |   |   |   |
H — C — C — C = C — H + Br₂ → H — C — C — C — C — H
    |   |                    |   |   |   |
    H  CH₃ CH₃ H             H  CH₃ CH₃ H
```

23 a

```
         CH₃
         |
     H   CH₂
     |   |
   — C — C —
     |   |
     H   H
```

b

```
        CH₃
        |
   CH₃  CH₂
    |   |
  — C — C —
    |   |
    Cl  H
```

24 a

```
  H    Cl
  |    |
  C  = C
  |    |
  Br   H
```

b

```
       CH₃
       |
   H   CH₂
   |   |
   C = C
   |   |
   H  CH₃
```

25 $CH_3CH_2CH_2CH_2OH + 6O_2 \rightarrow 4CO_2 + 5H_2O$

26

a	propan-1-ol	primary
b	pentan-2-ol	secondary
c	2-methylbutan-2-ol	tertiary
d	3,3-dimethylbutan-1-ol	primary

27

a	`H — C — C — C(=O) — O — H` (with H, H on carbons)
b	`H — C — C — C — C(=O) — C — H` (with H, H, H, H)
c	Not oxidised
d	`H — C — C(CH₃) — C — C(=O) — O — H` (with CH₃, CH₃)

28 a

```
   O   H   H   H
   ‖   |   |   |
   C — C — C — C — H
   |   |   |   |
   H   H  CH₃  H
```

c

```
   H   H   CH₃  H    O
   |   |   |    |    ‖
H — C — C — C — C — C
   |   |   |    |
   H   H   H   CH₃  H
```

b

```
    H   H   O   H   H
    |   |   ‖   |   |
H — C — C — C — C — C — H
    |   |       |   |
    H  CH₃      H   H
```

29

	Alcohol	Carboxylic acid	Ester
a			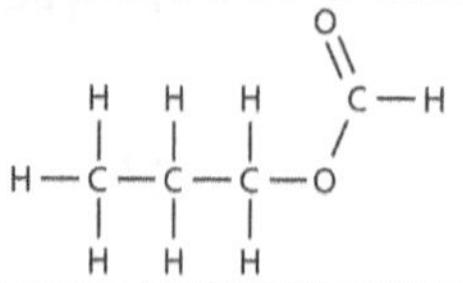 (ester structure: propyl group $H-C-C-C-O$ bonded to $C(=O)-H$)
b			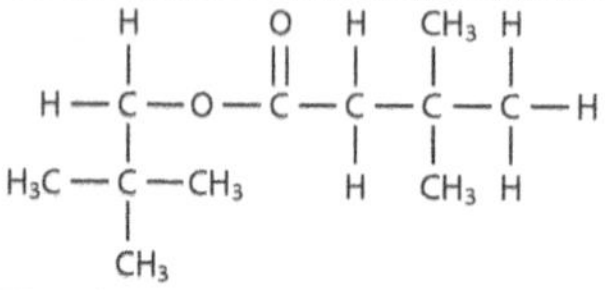 (ester structure with $H-C-O-C(=O)-C-C-C-H$ chain, branched with CH_3 groups)
c	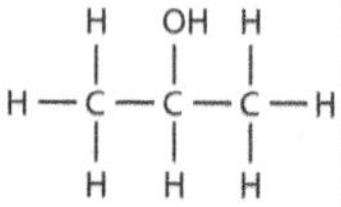 (alcohol: $H-C-C(OH)-C-H$)	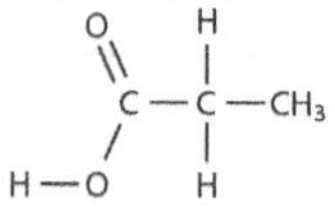 (carboxylic acid: $C(=O)(OH)-C-CH_3$)	
d	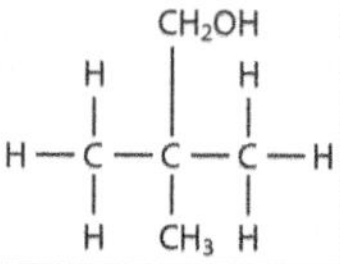 (alcohol: CH_2OH on $H-C-C-C-H$ with CH_3 branch)	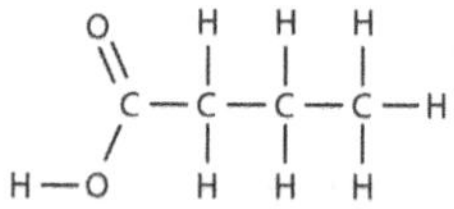 (carboxylic acid: $C(=O)(OH)-C-C-C-H$)	

30 propyl methanoate

31

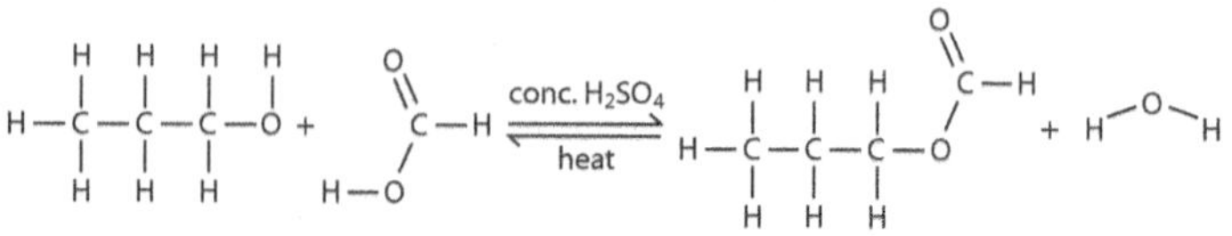

$$H-C-C-C-O + \overset{O}{C}-H \underset{heat}{\overset{conc.\ H_2SO_4}{\rightleftharpoons}} H-C-C-C-O-\overset{O}{C}-H + H-O-H$$

32

a	S_N2
b	S_N1
c	S_N1 and S_N2
d	S_N2

33

a	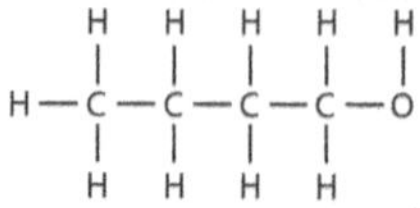 (structure: $H-C-C-C-C-C-O$ straight chain)
b	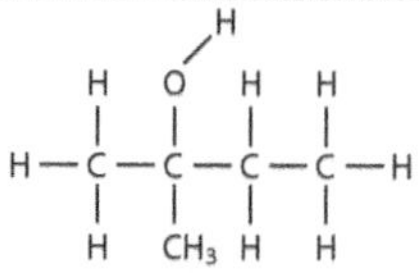 (structure: $H-C-C-C-C-H$ with $O-H$ and CH_3 branches)
c	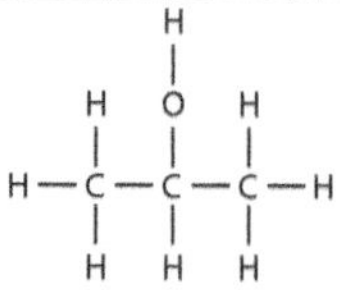 (structure: $H-C-C-C-H$ with $O-H$ branch)
d	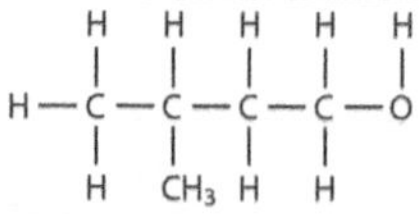 (structure: $H-C-C-C-C-O$ with CH_3 branch)

34
a 2-chlorobutane
b 2-chloro-3-methylbutane
c 2-chloro-2,4-dimethylpentane

35 1-bromo-2-chlorobutane

36 $C_6H_5CH_3 + HNO_3 \rightarrow CH_3C_6H_4NO_2 + H_2O$

37 Formation of electrophile:
$$HNO_3 + 2H_2SO_4 \rightarrow NO_2^+ + H_3O^+ + 2HSO_4^-$$

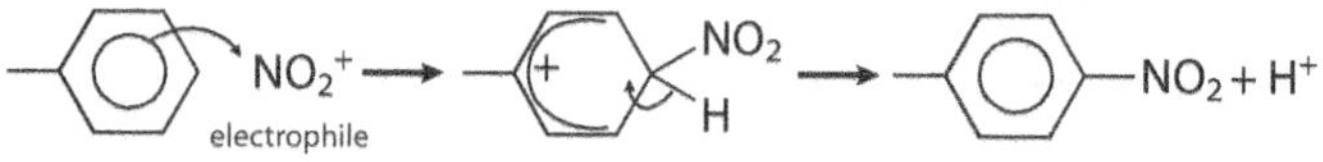

(reaction scheme: benzene ring attacks NO_2^+ electrophile → intermediate with $+$ charge and NO_2, H → product ring with $NO_2 + H^+$)

38
a propan-1-ol
b 2-methylpentan-1-ol
c pentan-3-ol
d 3-methylbutan-1-ol

39 a

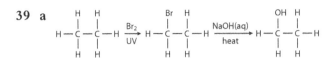

b

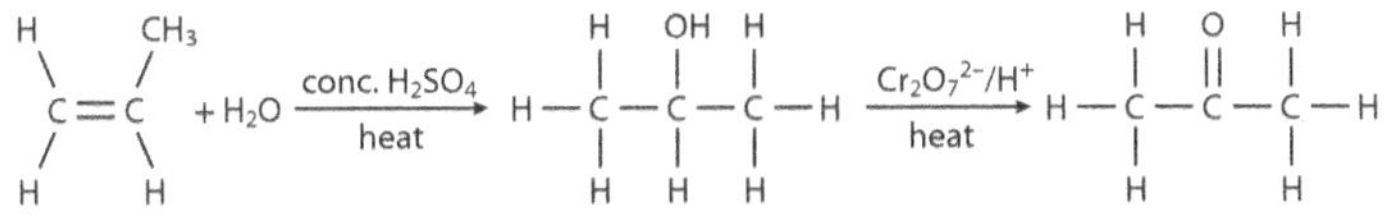

c

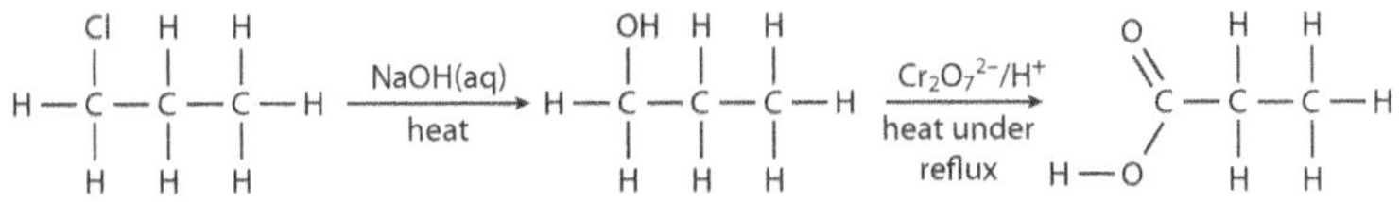

d

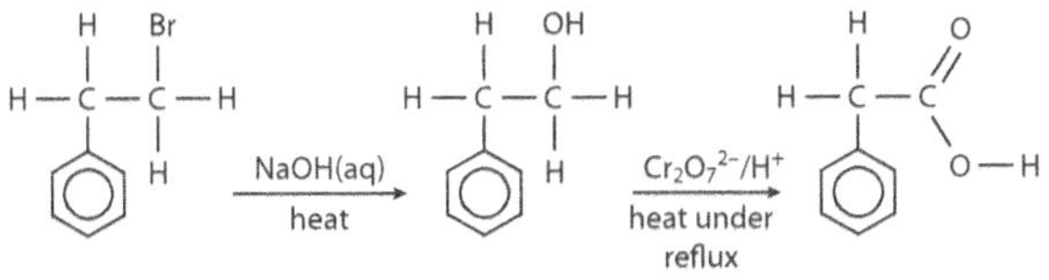

e

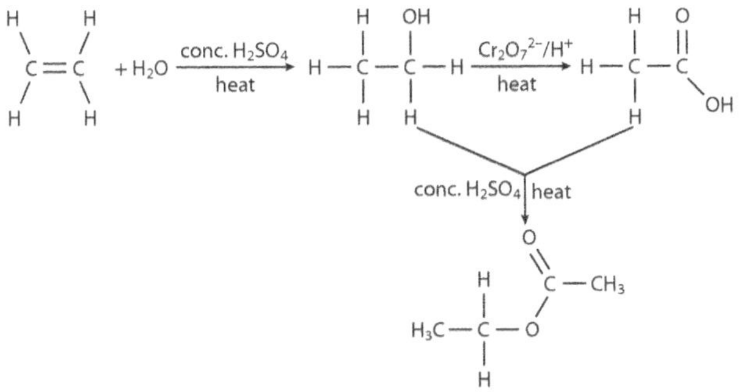

40

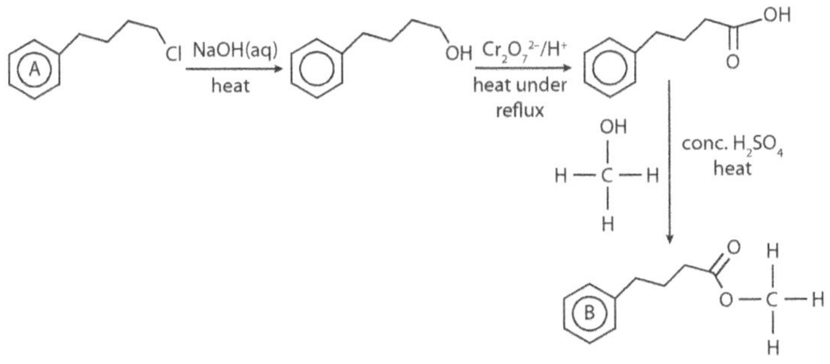

41 The alkene is 3-methylbut-1-ene: CH_2=$CHCH(CH_3)_2$

$$CH_2CHCH(CH_3)_2 + H_2O \xrightarrow[\text{heat}]{\text{c. } H_2SO_4} CH_3CH(OH)CH(CH_3)_2$$

$$CH_3CH(OH)CH(CH_3)_2 + [O] \xrightarrow[\text{heat / reflux}]{Cr_2O_7{}^{2-} / H^+} CH_3COCH(CH_3)_2 + H_2O$$

42

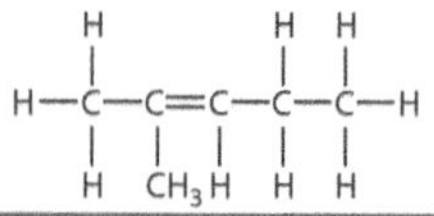 H—C—C=C—C—C—H (with H, CH₃, H, H, H substituents)	no		
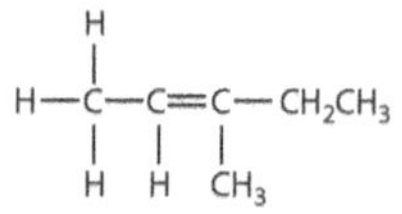 H—C—C=C—CH₂CH₃ (with H, H, CH₃)	yes	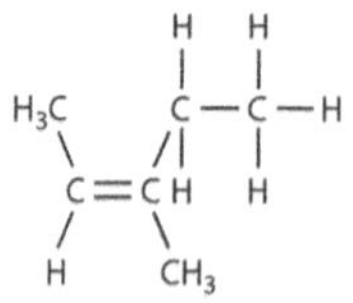 H₃C / C=CH structure with C—C—H, CH₃	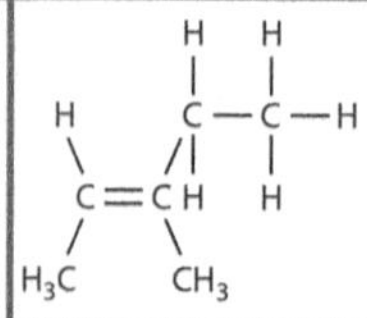 H / C=CH structure with C—C—H, H₃C CH₃
2,3-dimethylpent-2-ene	no		
3,4-dimethylpent-2-ene	yes	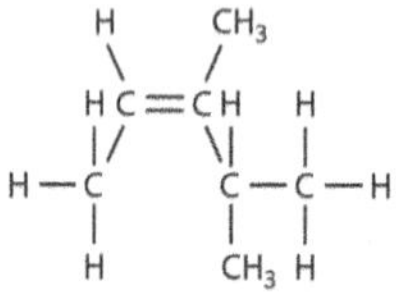 HC=CH structure (H, CH₃) with H—C—C—C—H, CH₃	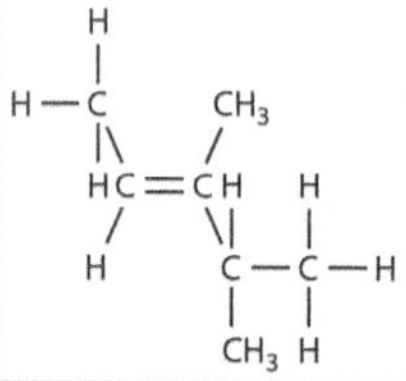 H—C structure HC=CH (CH₃) with C—C—H, CH₃ H
1,2,3-trimethylcyclopropane	yes	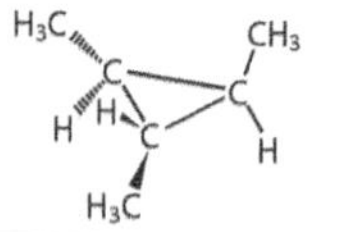 cyclopropane ring with H₃C, CH₃, H, H₃C, H substituents	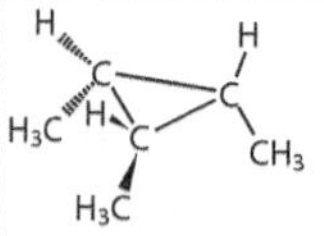 cyclopropane ring with H, H, H₃C, CH₃, H₃C substituents
1,3-dimethylcyclobutane	yes	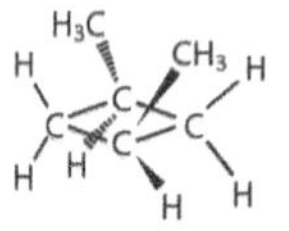 cyclobutane ring with H₃C, CH₃, H substituents	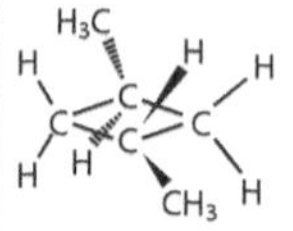 cyclobutane ring with H₃C, CH₃, H substituents

43 **a** Z; **b** E; **c** Z; **d** E

44

a	yes	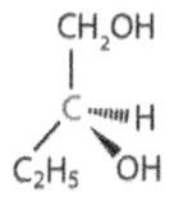 CH₂OH, C with H, C₂H₅, OH	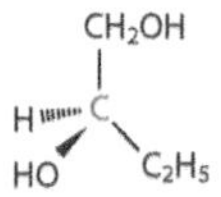 CH₂OH, C with H, HO, C₂H₅
b	no		
c	yes	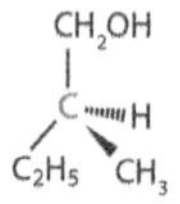 CH₂OH, C with H, C₂H₅, CH₃	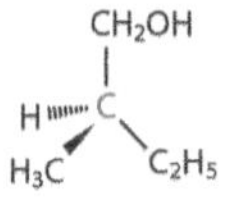 CH₂OH, C with H, H₃C, C₂H₅
d	yes	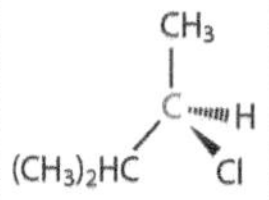 CH₃, C with H, (CH₃)₂HC, Cl	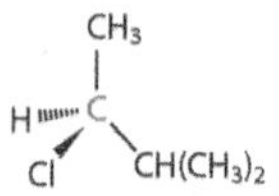 CH₃, C with H, Cl, CH(CH₃)₂
e	no		
f	yes	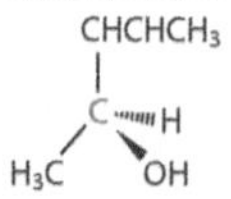 CHCHCH₃, C with H, H₃C, OH	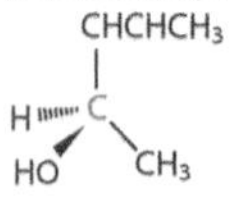 CHCHCH₃, C with H, HO, CH₃

Note: chiral centres are shown in red.

45 Only

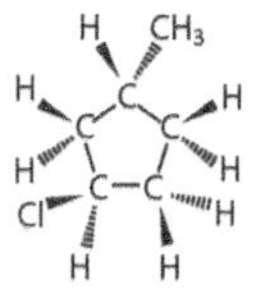

cyclopentane ring with H, CH₃, H, H, H, H, Cl, H, H, H substituents

The others all have planes of symmetry.

46 **a** diastereomers; **b** enantiomers

Topic 11

1 Set 1 is more precise, because there is less variation from the mean.

2 Experiment 2, because this value is closest to the literature value.

3

Measured value ± uncertainty	Value you should quote
71.7 ± 0.2	71.7 ± 0.2
3.475 ± 0.01	3.48 ± 0.01
0.065 06 ± 0.001	0.065 ± 0.001
63.27 ± 5	63 ± 5
593.2 ± 30	590 ± 30
783.28 ± 100	800 ± 100

4 a 3   c 4   e 4
  b 2   d 3

5 a 6.79
  b 0.000 079 8
  c 0.005 00
  d 8.25×10^5
  e 1.78×10^{-3}

6

0.345 ± 0.001	+	0.216 ± 0.002	=	0.561 ± 0.003
23.45 ± 0.03	−	15.23 ± 0.03	=	8.22 ± 0.06
0.0034 ± 0.0003	+	0.0127 ± 0.0003	=	0.0161 ± 0.0006
1.103 ± 0.004	−	0.823 ± 0.001	=	0.280 ± 0.005
1.10 ± 0.05	+	17.20 ± 0.05	=	18.3 ± 0.1

7

Value	Percentage uncertainty (%)
27.2 ± 0.2	0.74
0.576 ± 0.007	1.2
4.46 ± 0.01	0.22
$7.63 \times 10^{-5} \pm 4 \times 10^{-7}$	0.52

8 a The absolute uncertainty is 0.05; the final value should be quoted as 9.22 ± 0.05.

  b The absolute uncertainty is 0.5; the final value should be quoted as 86.3 ± 0.5.

9 4140 ± 40 J (assuming that the value for the specific heat capacity is exact)

10 a Linear relationship – as time increases the volume increases linearly; not a proportional relationship

  b Directly proportional – mass is directly proportional to the current

  c Non-linear relationship – as time increases the temperature increases; as time increases the increase in temperature per second gets larger

11 a $1.0 \, \text{cm}^3 \, \text{s}^{-1}$
  b 17.5–$18.4 \, \text{g} \, \text{A}^{-1}$

12 a 1   c 2   e 4
  b 1   d 1

13 B, C, E and G

14 a 2840–$3100 \, \text{cm}^{-1}$ by C–H
  b 2840–$3100 \, \text{cm}^{-1}$ by C–H; 3200–$3600 \, \text{cm}^{-1}$ by O–H
  c 2840–$3100 \, \text{cm}^{-1}$ by C–H; 1610–$1680 \, \text{cm}^{-1}$ by C=C
  d 2840–$3100 \, \text{cm}^{-1}$ by C–H; 2400–$3400 \, \text{cm}^{-1}$ by O–H in COOH; 1700–$1750 \, \text{cm}^{-1}$ by C=O; (1000–$1300 \, \text{cm}^{-1}$ by C–O)

15 a C_2H_4O
  b C_3H_8O; $C_2H_4O_2$
  c C_4H_8O; $C_3H_4O_2$
  d $C_5H_{12}O$; $C_4H_8O_2$; $C_3H_4O_3$

16 a CH_3^+
  b CO^+; $C_2H_4^+$
  c $C_2H_5^+$; HCO^+
  d OCH_3^+; CH_2OH^+
  e $C_3H_7^+$; $C_2H_3O^+$
  f $COOH^+$; $C_2H_5O^+$
  g $C_3H_7O^+$; CH_2COOH^+; $HCOOCH_2^+$
  h $C_6H_5^+$

17 a 3 peaks; 1:2:3
  b 4 peaks; 3:2:2:1
  c 3 peaks; 9:2:1
  d 5 peaks; 1:3:1:2:3
  e 3 peaks; 3:1:6
  f 5 peaks; 6:1:2:2:1

18 a two doublets
  b one doublet; one triplet
  c one triplet; one quartet; one singlet
  d one triplet; one quartet; one singlet

19 Using the values in Table 11.14:
  a 0.5–5.0 ppm
  b 2.0–2.5 ppm
  c 3.7–4.8 ppm

20 Multiplicity of peak at $\delta = 4.0$ ppm is 7

Glossary

Terms in **bold italic** refer to keywords from online material

absolute scale of temperature Kelvin scale of temperature, which starts at absolute zero; 1 °C is the same as 1 K and so 0 °C is equivalent to 273 K

absolute zero the temperature at which everything would be in its lowest energy state; 0 K or −273 °C

accuracy how close a measurement is to the actual value of a particular quantity

acid deposition a more general term than acid rain; refers to any process in which acidic substances (particles, gases and precipitation) leave the atmosphere to be deposited on the surface of the Earth; can be divided into wet deposition (acid rain, fog and snow) and dry deposition (acidic gases and particles)

acid dissociation constant (K_a) in general, for the dissociation of an acid HA: $HA(aq) \rightleftharpoons H^+(aq) + A^-(aq)$ the expression for the acid dissociation constant is:

$$K_a = \frac{[A^-(aq)][H^+(aq)]}{[HA(aq)]}$$

the higher the value of K_a, the stronger the acid

activation energy (E_a) the minimum energy that colliding species must have before collision results in a chemical reaction

actual yield the actual amount of product formed in the reaction

addition polymerisation alkenes undergo addition polymerisation, in which a large number of monomers are joined together into a polymer chain; no other groups/molecules are lost in the process

addition reaction in organic chemistry, a reaction in which a molecule is added to a compound containing a multiple bond without the loss of any other groups

alkali a base that is dissolved in water

alkali metals the elements in group 1 of the periodic table

allotropes different forms of the same element; e.g. diamond, graphite and fullerene are allotropes of carbon

alloy homogeneous mixture of two or more metals, or of a metal with a non-metal

amphiprotic a substance that can donate a proton (acting as a Bronsted-Lowry acid) and accept a proton (acting as a Bronsted-Lowry base), e.g. HCO_3^-

amphoteric a substance that can act as an acid and a base

anabolism process of synthesising molecules needed by cells; requires energy

analgesics drugs that reduce pain

anode the electrode at which oxidation occurs

Arrhenius equation an equation showing the variation of the rate constant with temperature:

$$k = Ae^{-E_a/RT}$$

'A' is the frequency factor or pre-exponential factor and takes account of the frequency of collisions and the orientation of the collisions

atactic polymer a polymer that has side groups orientated randomly on both sides of the main chain

atom the smallest part of an element that can still be recognised as the element; in the simplest picture of the atom, the electrons orbit around the central nucleus; the nucleus is made up of protons and neutrons (except for a hydrogen atom, which has no neutrons)

atom economy $= \dfrac{\text{molar mass of desired products}}{\text{total molar mass of all reactants}} \times 100\%$

atomic number (Z) the number of protons in the nucleus of an atom

atomic radius half the internuclear distance between two atoms of the same element covalently bonded; atomic radius is usually called 'covalent radius' in more advanced work; it is also possible for an element to have a 'van der Waals' radius'

Aufbau principle the process of putting electrons into atoms to generate the electronic configuration

average bond enthalpy the average amount of energy required to break one mole of covalent bonds, in a gaseous molecule under standard conditions; 'average' refers to the fact that the bond enthalpy is different in different molecules, and therefore the value quoted is the average amount of energy to break a particular bond in a range of molecules; bond breaking requires energy (endothermic) ΔH +ve; bond making releases energy (exothermic) ΔH −ve

Avogadro's constant (L) $6.02 \times 10^{23}\,\text{mol}^{-1}$

Avogadro's law equal volumes of ideal gases measured at the same temperature and pressure contain the same number of molecules

base ionisation constant (K_b) consider the ionisation of a weak base, $B(aq) + H_2O(l) \rightleftharpoons BH^+(aq) + OH^-(aq)$:

$$K_b = \frac{[BH^+(aq)][OH^-(aq)]}{[B(aq)]}$$

the higher the value of K_b, the stronger the base

biochemical oxygen demand the amount of oxygen used by the aerobic microorganisms in water to decompose the organic matter in the water over a fixed period of time (usually 5 days) at a fixed temperature (usually 20 °C)

biofuel a fuel produced from organic matter obtained from plants, waste material etc.

biomagnification the increase in concentration of a substance as it passes up a food chain

bond enthalpy the enthalpy change when one mole of covalent bonds, in a gaseous molecule, are broken under standard conditions

Born–Haber cycle   an enthalpy level diagram breaking down the formation of an ionic compound into a series of simpler steps

breeder reactor   a nuclear reactor that produces more fissionable material than it consumes

Brønsted–Lowry acids and bases   an acid is a proton (H^+) donor; a base/alkali is a proton (H^+) acceptor

buffer solution   one that resists changes in pH when small amounts of acid or alkali are added

carbon footprint   a measure of the total amount of greenhouse gases (primarily carbon dioxide and methane) emitted as a result of human activities

catabolism   breakdown of larger molecules into smaller ones with the release of energy

catalyst   a substance that increases the rate of a chemical reaction without itself being used up in the reaction; a catalyst acts by allowing the reaction to proceed by an alternative pathway of lower activation energy

cathode   the electrode at which reduction occurs

ceramics   inorganic solid engineering materials that are neither metals nor polymers; they are usually described as inorganic substances that contain at least one metallic and one non-metallic element although substances such as silicon carbide, diamond and graphite are also usually classified as ceramics

chain reaction   one initial event causes a large number of subsequent reactions – the reactive species is regenerated in each cycle of reactions

chelate complex (chelate)   a complex formed between a metal ion and a polydentate ligand that results in the formation of a ring that includes the transition metal ion

chemical properties   how a substance behaves in chemical reactions

chiral centre   a carbon atom with four different atoms or groups attached to it; sometimes called an asymmetric carbon atom

cis-trans **isomerism**   two compounds have the same structural formula, but the groups are arranged differently in space around a double bond or a ring

closed system   no exchange of matter with the surroundings

collision theory   a reaction can occur only when two particles collide in the correct orientation and with $E \geq E_a$

composite material   mixture containing two or more different materials present as distinct, separate phases; synthetic composite materials consist of a reinforcing phase embedded in a matrix

concentration   amount of solute dissolved in a unit volume of solution; the volume that is usually taken is $1\,dm^3$ (one litre); the amount of solute may be expressed in g or mol, so the units of concentration are $g\,dm^{-3}$ or $mol\,dm^{-3}$

condensation   joining together of two molecules with the formation of a covalent bond and the elimination of the elements of water

condensation polymers   polymers formed when monomers, each containing two functional groups, join together with the elimination of small molecules such as water or hydrogen chloride

conjugate acid–base pair   these differ by one proton (H^+); when an acid donates a proton it forms the conjugate base (CH_3COO^- is the conjugate base of CH_3COOH); when a base gains a proton it forms the conjugate acid (H_3O^+ is the conjugate acid of H_2O)

conjugated system   a sequence of alternating single and double bonds in a molecule - equivalent to a π delocalised system

continuous spectrum   a spectrum consisting of all frequencies/wavelengths of light

convergence limit   the point in a line emission spectrum where the lines merge to form a continuum; may be used to determine the ionisation energy

covalent bond   the electrostatic attraction between a shared pair of electrons and the nuclei of the atoms making up the bond

coordinate covalent bond   a type of covalent bond in which both electrons come from the same atom – also called a dative covalent bond

coordination number   the number of nearest neighbours for an atom or ion in a crystal

cytoplasm   the thick solution inside a cell but outside the nucleus of the cell

degenerate   describes orbitals with the same energy

delocalisation   the sharing of a pair of electrons between three or more atoms

diamagnetism   caused by paired electrons – diamagnetic substances are repelled slightly by a magnetic field

diastereomers   stereoisomers that are not mirror images of each other

dipole moment   the product of one of the charges making up a dipole and the distance between the charges; non-polar molecules have a zero dipole moment

diprotic acid H_2SO_4 is a diprotic acid, as it can dissociate to generate two protons per molecule:
$$H_2SO_4(aq) + H_2O(l) \rightarrow HSO_4^-(aq) + H_3O^+(aq)$$
$$HSO_4^-(aq) + H_2O(l) \rightleftharpoons SO_4^{2-}(aq) + H_3O^+(aq)$$

drug   any substance that, when applied to or introduced into a living organism, brings about a change in biological function through its chemical action

dynamic equilibrium   macroscopic properties are constant; rate of the forward reaction is equal to the rate of the reverse reaction

ED_{50}   the dose of a drug required to produce a therapeutic effect in 50% of the test population ('ED' stands for effective dose)

efficiency $\% \text{ efficiency} = \dfrac{\text{useful energy out}}{\text{total energy in}} \times 100$

effusion   the process by which a gas escapes through a very small hole in a container

elastomers polymers that display rubber-like elasticity, they are flexible and can be stretched to many times their original dimensions by the application of a force; they will then return to (nearly) their original size and shape once the force is removed

electrolysis the breaking down of a substance (in molten state or solution) by the passage of electricity through it

electrolyte a solution, or a molten compound, that will conduct electricity with decomposition at the electrodes as it does so; electrolytes contain ions that are free to move towards the electrodes

electronegativity a measure of the attraction of an atom in a molecule for the electron pair in the covalent bond of which it is a part

electrophile a reagent (a positively charged ion or the positive end of a dipole) that is attracted to regions of high electron density and accepts a pair of electrons to form a covalent bond; an electrophile is a Lewis acid

electroplating the process of coating an object with a thin layer of a metal using electrolysis

electrostatic attraction attraction between positive and negative charges

element a substance containing just one type of atom (although see isotopes)

emission spectrum electromagnetic radiation given out when an electron falls from a higher energy level to a lower one; only certain frequencies of electromagnetic radiation are emitted; each atom has a different emission spectrum

empirical formula the simplest whole number ratio of the elements present in a compound

enantiomers optical isomers

endothermic a chemical reaction in which heat is taken in from the surroundings – the reaction vessel gets colder; ΔH is positive for an endothermic reaction

end point of a titration the point at which an indicator changes colour

energy density $= \dfrac{\text{energy released from fuel}}{\text{volume of fuel consumed}}$

enthalpy change (ΔH) the heat energy exchanged with the surroundings at constant pressure

entropy (S) a measure of how the available energy is distributed among the particles; standard entropy ($S^{\ominus}$) is the entropy of a substance at $100\,\text{kPa}$ and $298\,\text{K}$; units are $\text{J}\,\text{K}^{-1}\,\text{mol}^{-1}$; $\Delta S^{\ominus}$ is the entropy change under standard conditions – a positive value indicates an increase in entropy, i.e. the energy is more spread out (less concentrated)

equivalence point the point at which equivalent numbers of moles of acid and alkali have been added in a titration

exothermic a chemical reaction that results in the release of heat to the surroundings – the reaction vessel gets hotter; ΔH for an exothermic reaction is negative

first electron affinity enthalpy change when one electron is added to each atom in one mole of gaseous atoms under standard conditions:
$$X(g) + e^- \rightarrow X^-(g)$$

the first electron affinity is exothermic for virtually all elements

first ionisation energy the minimum amount of energy required to remove an electron from a gaseous atom/the energy required to remove one electron from each atom in one mole of gaseous atoms under standard conditions

fossil fuels fuels formed from things that were once alive and have been buried underground for millions of years, e.g. coal, oil and gas

free energy change (ΔG) or Gibbs free energy change; ΔG is related to the entropy change of the Universe and can be defined using the equation:
$$\Delta G = \Delta H - T\Delta S$$
for a reaction to be spontaneous, ΔG for the reaction must be negative; $\Delta G^{\ominus}$ is the standard free energy change

free radical a species (atom or groups of atoms) with an unpaired electron; free radicals are very reactive because of this unpaired electron

fuel cell a type of electrochemical cell that uses the reaction between a fuel (such as hydrogen or methanol) and an oxidising agent (e.g. oxygen) to produce electrical energy directly; it uses a continuous supply of reactants from an external source

functional group an atom or group of atoms that gives an organic molecule its characteristic chemical properties

genetic code how the four-base code in DNA determines the sequence of 20 amino acids in proteins; each three base codon codes for only one amino acid and this code is the same in all organisms – it is universal

genetically modified organisms (GMO) have genetic material that has been changed in some way by genetic engineering

giant structure bonding extends fairly uniformly throughout the whole structure; there are no individual molecules

green chemistry (also called 'sustainable chemistry') an approach to chemical research and chemical industrial processes that seeks to minimise the production of hazardous substances and their release to the environment

group vertical column in the periodic table

Haber (Haber–Bosch) process an industrial process for the manufacture of ammonia

half-life the time it takes for the number of radioactive nuclei present in a sample at any given time to fall to half its value

halogens the elements in group 17 of the periodic table

heat a form of energy that flows from something at a higher temperature to something at a lower temperature.

Hess's law the enthalpy change accompanying a chemical reaction is independent of the pathway between the initial and final states

heterogeneous catalyst a catalyst that is in a different phase (state) from the reactants

heterogeneous mixture a mixture that does not have uniform composition and consists of separate phases; can be separated by mechanical means

heterolytic fission a covalent bond breaks so that both electrons go to the same atom

homogeneous catalyst   a catalyst that is in the same phase (state) as the reactants

homogeneous mixture   a mixture that has the same (uniform) composition throughout the mixture and consists of only one phase

homologous series   a series of compounds with the same functional group, in which each member differs from the next by $-CH_2-$

homolytic fission   a covalent bond breaks such that one electron goes back to each atom making up the original covalent bond

Hund's rule   electrons fill orbitals of the same energy (degenerate orbitals) so as to give the maximum number of electrons with the same spin

hybridisation   the mixing of atomic orbitals when a compound forms to produce a new set of orbitals (the same number as originally), which are better arranged in space for covalent bonding

hydrocarbon   compound containing carbon and hydrogen only

hydrogenation   addition of hydrogen (H_2) to a compound containing multiple bonds

hydrogen bonding   an intermolecular force resulting from the interaction of a lone pair on a very electronegative atom (N/O/F) in one molecule with an H atom attached to N/O/F in another molecule; these forces may also occur intramolecularly

hydrolysis   a reaction in which a covalent bond in a molecule is broken by reaction with water; most commonly hydrolysis reactions occur when a molecule is reacted with aqueous acid or aqueous alkali

ideal gas   a theoretical model that approximates the behaviour of real gases; it can be defined in terms of macroscopic properties (a gas that obeys the equation $PV = nRT$) or in terms of microscopic properties (the main assumptions that define an ideal gas on a microscopic scale are that the molecules are point masses – their volume is negligible compared with the volume of the container – and that there are no intermolecular forces except during a collision)

indicator   an acid–base indicator has different colours according to the pH of the solution; indicators are usually weak acids (HIn); they dissociate according to the equation:

$$HIn(aq) \rightleftharpoons H^+(aq) + In^-(aq)$$
$$\quad Colour\ I \qquad\qquad\qquad Colour\ II$$

the ionised (In^-) and un-ionised (HIn) forms must have different colours. Indicators may also be weak bases

initiation step   a step that starts off a chain reaction; it involves an increase in the number of free radicals

intermolecular forces   forces between molecules

internal energy   the total amount of energy (kinetic and potential) in a sample of a substance

intramolecular forces   forces within a molecule – usually covalent bonding

iodine number   a measure of the degree of unsaturation in a fat or oil; it is the number of grams of iodine that reacts with 100 g of fat or oil

ion   a charged particle that is formed when an atom loses or gains electron(s); a positive ion is formed when an atom loses (an) electron(s) and a negative ion is formed when an atom gains (an) electron(s)

ionic bonding   the electrostatic attraction between oppositely charged ions

ionic product constant (K_w)   a modified equilibrium constant for the dissociation of water:
$$K_w = [H^+(aq)][OH^-(aq)]$$
K_w has a value of 1.0×10^{-14} at $25\,°C$

isoelectronic   describes species with the same number of electrons

isotactic polymer   a polymer in which all the side groups are on the *same* side of the polymer chain

isotopes   different atoms of the same element with different mass numbers, i.e. different numbers of neutrons in the nucleus

kinetic energy   the energy a body has because of its motion; $K.E. = \frac{1}{2}mv^2$

lattice structure   regular 3D arrangement of particles

Le Chatelier's principle   if a system at equilibrium is subjected to some change, the position of equilibrium will shift in order to minimise the effect of the change

Lewis acids and bases   an acid is an electron pair acceptor; a base is an electron pair donor

Lewis (electron dot) structure   diagram showing all the valence (outer shell) electrons in a molecule (or ion)

ligands   negative ions or neutral molecules that use lone pairs of electrons to bond to a transition metal ion to form a complex ion; coordinate covalent bonds (dative bonds) are formed between the ligand and the transition metal ion

limiting reactant   the reactant that is used up first in a chemical reaction; when the number of moles of each species is divided by their coefficient in the stoichiometric equation, the limiting reagent is the one with the lowest number; all other reactants are in excess

line spectrum   the emission spectrum of an atom consists of a series of lines that get closer together at higher frequency; only certain frequencies/wavelengths of light are present

lipid   a broad class of biological molecules including steroids, triglycerides and phospholipids; mainly non-polar and insoluble in water

liquid crystal   a phase of matter in which the properties of a compound may exhibit the characteristics of both a liquid and a solid

lone pair   a pair of electrons in the outer shell of an atom that is not involved in covalent bonding

London (dispersion) force   intermolecular forces resulting from temporary (instantaneous) dipole–induced dipole interactions

mass defect (Δm)   is the difference between the mass of a nucleus and the sum of the masses of the individual nucleons

mass number (A)   the number of protons + neutrons in the nucleus of an atom

Markovnikov's rule   when H–X adds across the double bond of an alkene, the H atom becomes attached to the C atom that has the larger number of H atoms already attached

matter   something that has mass and occupies space

Maxwell–Boltzmann distribution   a graph showing the distribution of molecular kinetic energies in a sample of gas at a particular temperature

medicine   something that treats, prevents or alleviates the symptoms of disease

metabolism   chemical reactions that go on in cells – it involves the breakdown of molecules with the release of energy and the synthesis of molecules that are required by cells

metallic bonding   the electrostatic attraction between the positive ions in a metallic lattice and the delocalised electrons

Michaelis constant (K_m)   in enzyme kinetics is the concentration of substrate when the rate is equal to one half of V_{max} (maximum rate)

molar volume   the volume occupied by one mole of a gas; the molar volume of an ideal gas at STP is $22.7\,dm^3\,mol^{-1}$

mole   the amount of substance that contains the same number of particles (atoms, ions, molecules, etc.) as there are carbon atoms in 12 g of carbon-12 (6.02×10^{23})

molecular formula   the total number of atoms of each element present in a molecule of the compound; the molecular formula is a multiple of the empirical formula

molecular self-assembly   a bottom–up approach to producing nanoparticles, where molecules come together reversibly and spontaneously to create a larger structure

molecularity   the number of 'molecules' that react in a particular step (usually the rate-determining step) in a chemical reaction

molecule   an electrically neutral particle consisting of two or more atoms chemically bonded together

monomer   a molecule from which a polymer chain may be built up, e.g. ethene is the monomer for polyethene

monoprotic acid   HCl is a monoprotic acid as it dissociates to form one proton per molecule

nanotechnology   the production and application of structures, devices and systems at the nanometre scale; generally, involves man-made particles or structures that have at least one dimension smaller than 100 nm

nematic liquid crystal phase   molecules point, on average, in the same direction but are positioned randomly relative to each other (no positional order)

noble gases   the elements in group 18 of the periodic table; also sometimes called the 'inert gases'

non-renewable energy sources   sources of energy that are finite – they will eventually run out, e.g. coal

nuclear binding energy (ΔE)   is the energy required to break apart a nucleus into individual protons and neutrons

nuclear fission   the breakdown of a larger nucleus into two smaller fragments of comparable masses

nuclear fusion   the joining together of smaller nuclei to make a larger one

nucleophile   a molecule/negatively charged ion, possessing a lone pair of electrons, which is attracted to a more positively charged region in a molecule (region with lower electron density) and donates a lone pair of electrons to form a covalent bond; a nucleophile is a Lewis base

nucleophilic substitution   a nucleophile replaces an atom/group in a molecule

octane number   a measure of the tendency of a fuel to not undergo auto-ignition (not cause knocking) in an engine; the higher the octane number, the lower the tendency to undergo autoignition/cause knocking

opiates   natural narcotic (sleep-inducing) analgesics derived from the opium poppy

optical isomersism   when molecules have the same molecular and structural formula, but groups are arranged differently in space and the individual optical isomers are non-superimposable mirror images of each other; optical isomers rotate the plane of plane-polarised light in opposite directions

orbital   a region of space in which there is a high probability of finding an electron; it represents a discrete energy level; there are s, p, d and f orbitals; any orbital can contain a maximum of two electrons

order of a reaction   the power of the concentration of a particular reactant in the experimentally determined rate equation

osmosis   movement of water (or other solvents) across a semipermeable membrane from a less concentrated solution to a more concentrated one

overall order of reaction   the sum of the powers of the concentration terms in the experimentally determined rate equation

oxidation number   a purely formal concept that regards all compounds as ionic and assigns charges to the components accordingly; it provides a guide to the distribution of electrons in covalent compounds

oxidation   loss of electrons or increase in oxidation number

oxidising agent (oxidant)   oxidises other species and, in the process, is itself reduced; an oxidising agent takes electrons away from something

paramagnetism   caused by unpaired electrons – paramagnetic substances are attracted by a magnetic field

Pauli exclusion principle   two electrons in the same orbital must have opposite spins

percentage error $= \dfrac{|\,experimental\ value - accepted\ value\,|}{|\,accepted\ value\,|} \times 100$

percentage uncertainty $= \dfrac{uncertainty}{measured\ value} \times 100$

when multiplying or dividing quantities with uncertainties, the percentage uncertainties should be added to give the percentage uncertainty of the final value

period   horizontal row in the periodic table

pH   a measure of the concentration of H^+ ions in an aqueous solution; it can be defined as the negative logarithm to base 10 of the hydrogen ion concentration in aqueous solution:

$$pH = -\log_{10}[H^+(aq)]$$

pH meter an electronic device for measuring the pH of a solution

pH range of an indicator the pH range over which intermediate colours for an indicator can be seen

physical properties properties such as melting point, solubility and electrical conductivity, relating to the physical state of a substance and the physical changes it can undergo

pi (π) bond bond formed by the sideways overlap of parallel p orbitals; the electron density in the pi bond lies above and below the internuclear axis

plane-polarised light light that vibrates in one plane only; optical isomers rotate the plane of plane-polarised light in opposite directions

plasma a fully or partially ionised gas consisting of positive ions and electrons

plasticisers small molecules that are added to a polymer to increase its flexibility

polar molecule molecule in which one end is slightly positive relative to the other; whether a molecule is polar or not depends on the differences in electronegativity of the atoms and the shape of the molecule

polydentate ligand a ligand that binds to a transition metal ion through more than one donor atom

polymers long-chain molecules, usually based on carbon, which are formed when smaller molecules (monomers) join together

precision relates to the reproducibility of results; if a series of readings is taken with high precision, it indicates that the repeated values are all very close together and close to the mean (average) value

primary cell a cell (battery) that cannot usually be recharged using mains electricity – the reaction in the cell is non-reversible

primary structure of a protein the linear sequence of amino acids in a polypeptide chain

propagation step a step in a free radical substitution reaction that involves production of products and no change in the number of free radicals

prosthetic group a non-peptide part of a protein that is bound tightly to the protein and is required for correct function; for example, the heme group in haemoglobin

racemic mixture an equimolar mixture of the two enantiomers of a chiral compound; it has no effect on plane-polarised light

radioisotopes radioactive isotopes

random uncertainty uncertainty in a measurement due to the limitations of the measuring apparatus and other uncontrollable variables that are inevitable in any experiment; the effects of random uncertainties should mean that the measurements taken will be distributed either side of the mean, i.e. fluctuations will be in both directions; the effect of random uncertainties can be reduced by repeating the measurements more often, but random uncertainties can never be completely eliminated

rate-determining step the slowest step in a reaction mechanism

rate constant (k) a constant of proportionality relating the concentrations in the experimentally determined rate expression to the rate of a chemical reaction; the rate constant is only a constant for a particular reaction at a particular temperature

rate expression or rate equation an experimentally determined equation that relates the rate of reaction to the concentrations of substances in the reaction mixture, e.g.: rate = $k[A]^m[B]^n$

rate of reaction the speed at which reactants are used up or products are formed; or, more precisely, the change in concentration of reactants or products per unit time

redox reaction a reaction that involves both oxidation and reduction; if something is oxidised, something else must be reduced

reducing agent (reductant) reduces other species and, in the process, is itself oxidised; a reducing agent gives electrons to something

reduction gain of electrons or decrease in oxidation number

relative atomic mass (A_r) the average of the masses of the isotopes in a naturally occurring sample of the element relative to the mass of $\frac{1}{12}$ of an atom of carbon-12

relative formula mass if a compound contains ions, the relative formula mass is the mass of the formula unit relative to the mass of $\frac{1}{12}$ of an atom of carbon-12

relative molecular mass (M_r) the mass of a molecule of a compound relative to the mass of $\frac{1}{12}$ of an atom of carbon-12; the M_r is the sum of the relative atomic masses for the individual atoms making up a molecule

renewable energy sources sources of energy that are naturally replenished – they will not run out, e.g. solar energy, wind power

repeating unit (repeat unit) of a polymer the basic unit from which the whole polymer chain can be made up

resonance hybrid the actual structure of a molecule/ion for which resonance structures can be drawn can be described as resonance hybrid made up of contributions (not necessarily equal) from all possible resonance structures

resonance structure one of several Lewis structures that can be drawn for some molecules/ion

salt bridge completes the circuit in a voltaic cell by providing an electrical connection between two half cells, allowing ions to flow into or out of the half cells to balance out the charges in the half cells; the salt bridge contains a concentrated solution of an ionic salt such as KCl

saturated compounds organic compounds containing only single bonds

second electron affinity the enthalpy change for the process:
$$X^-(g) + e^- \rightarrow X^{2-}(g)$$
the second electron affinity is always endothermic

second ionisation energy the enthalpy change for the process:
$$M^+(g) \rightarrow M^{2+}(g) + e^-$$

secondary cell a cell (battery) that can be recharged using mains electricity and is often called a rechargeable battery; the chemical reactions in a rechargeable battery are reversible and can be reversed by connecting them to an electricity supply

side effect an unintended secondary effect of a drug on the body; it is usually an undesirable effect

sigma (σ) bond bond formed by the axial (head-on) overlap of atomic orbitals; the electron distribution in a sigma bond lies mostly along the axis joining the two nuclei

S_N1 a unimolecular nucleophilic substitution reaction – only one species involved in the rate-determining step

S_N2 a bimolecular nucleophilic substitution reaction – two species involved in the rate-determining step

solute a substance that is dissolved in another (the solvent)

solution that which is formed when a solute dissolves in a solvent

solvent a substance that dissolves another substance (the solute); the solvent should be present in excess of the solute

specific energy $= \dfrac{\text{energy released from fuel}}{\text{mass of fuel consumed}}$

specific heat capacity the energy required to raise the temperature of 1 g of substance by 1 K (1 °C) – units $J\,g^{-1}\,°C^{-1}$; it can also be defined in terms of 'unit mass' with different units

spectrochemical series a series of ligands arranged in order of the extent to which they cause splitting of d orbitals in a transition metal complex ion

spontaneous reaction one that occurs without any outside influence, i.e. no input of energy; ΔG is negative

stability usually refers to the relative energies of reactants and products – if the products are at lower enthalpy (energy) than the reactants, then they are more stable; it is also possible to define kinetic stability

standard cell potential a standard cell potential is the EMF (voltage) produced when two half-cells are connected under standard conditions. This drives the movement of electrons through the external circuit from the negative electrode to the positive electrode

standard conditions a common set of conditions used to compare enthalpy changes; the pressure is 100 kPa and for reactions involving solutions, all solutions should have a concentration of 1 mol dm^{-3}; a temperature is not stated in the definition and should be specified – when it is not specifed we will assume in this course that it is 298 K (25 °C)

standard electrode potential the emf (voltage) of a half-cell connected to a standard hydrogen electrode, measured under standard conditions; all solutions must be of concentration 1 mol dm^{-3}

standard enthalpy change of atomisation ($\Delta H_{at}^{\ominus}$) this is the enthalpy change when one mole of gaseous atoms is formed from an element under standard conditions

standard enthalpy change of combustion ($\Delta H_c^{\ominus}$) the enthalpy change (heat given out) when one mole of a substance is completely burnt in oxygen under standard conditions

standard enthalpy change of formation ($\Delta H_f^{\ominus}$) the enthalpy change when one mole of the substance is formed from its elements in their standard states under standard conditions; $\Delta H_f^{\ominus}$ for any element in its standard state is zero

standard enthalpy change of hydration (ΔH_{hyd}) the enthalpy change when one mole of gaseous ions are surrounded by water molecules to form an 'infinitely dilute solution' under standard conditions:

$$Na^+(g) \xrightarrow[\text{excess } H_2O]{\Delta H_{hyd}} Na^+(aq)$$

an exothermic process

standard enthalpy change of neutralisation (ΔH_n) the enthalpy change when one mole of H_2O molecules are formed when an acid (H^+) reacts with an alkali (OH^-) under standard conditions, i.e.:

$$H^+(aq) + OH^-(aq) \rightarrow H_2O(l)$$

standard enthalpy change of reaction ($\Delta H_r^{\ominus}$) the enthalpy change (heat given out or taken in) when molar amounts of reactants as shown in the stoichiometric equation react together under standard conditions to give products the enthalpy change of neutralisation is always exothermic

standard enthalpy change of solution (ΔH_{sol}) the enthalpy change when one mole of solute is dissolved in excess solvent to form a solution of 'infinite dilution' under standard conditions, e.g.:

$$NH_4NO_3(s) \xrightarrow{\text{excess } H_2O} NH_4^+(aq) + NO_3^-(aq)$$

'infinite dilution' means that any further dilution of the solution produces no further enthalpy change, i.e. the solute particles are assumed not to interact with each other in the solution; the enthalpy change of solution may be exothermic or endothermic

standard enthalpy change of vaporisation ($\Delta H_{vap}^{\ominus}$) the energy needed to convert one mole of a liquid to vapour under standard conditions

standard hydrogen electrode the standard half-cell relative to which standard electrode potentials are measured

standard lattice enthalpy ($\Delta H_{latt}^{\ominus}$) the enthalpy change when one mole of ionic compound is broken apart into its constituent gaseous ions under standard conditions, e.g. for NaCl:

$$NaCl(s) \rightarrow Na^+(g) + Cl^-(g) \qquad \Delta H_{latt} = +771\,kJ\,mol^{-1}$$

lattice enthalpy can be defined in either direction, i.e. as the making or breaking of the lattice

standard state the pure substance at 100 kPa and a specified temperature (assume 298 K if one is not given). It is often used to refer the state in which a substance exists under standard conditions, e.g. for iodine it is $I_2(s)$ but for nitrogen it is $N_2(g)$

standard temperature and pressure (STP) 273 K, 100 kPa

state symbols used to indicate the physical state of an element or compound; these may be either written as subscripts after the chemical formula or in normal type (aq) = aqueous (dissolved in water); (g) = gas; (l) = liquid; (s) = solid

stereoisomers molecules with the same molecular formula and structural formula but the atoms are arranged differently in space; *cis-trans* isomers, conformational isomers and optical isomers are stereoisomers

strong acid an acid such as HCl, H_2SO_4, HNO_3 that dissociates completely in aqueous solution:
$$HCl(aq) \rightarrow H^+(aq) + Cl^-(aq)$$
strong acids are also strong electrolytes

strong base a base that ionises completely in aqueous solution; strong bases are the group 1 hydroxides (LiOH, NaOH, etc.) and $Ba(OH)_2$; strong bases are also strong electrolytes

strong electrolyte a substance that dissolves in water with complete ionisation; e.g. NaCl separates completely into its ions when it dissolves in water; strong acids are strong electrolytes because they dissociate completely into ions

structural isomers two or more compounds that have the same molecular formula but different structural formulas, i.e. the atoms are joined together in a different way

substitution reaction a reaction in which one atom or group is replaced by another atom or group

superconductors materials that have zero electrical resistance below a critical temperature

sub-energy level a group of degenerate orbitals in an atom; a p subshell is made up of the p_x, p_y and p_z orbitals

syngas (synthesis gas) a mixture of carbon monoxide and hydrogen

systematic error an error introduced into an experiment due to the apparatus used or the procedure; systematic errors result in a loss of accuracy, i.e. the measured value being further away from the true value; systematic errors are always in the same direction; the effect of a systematic error cannot be reduced by repeating the readings

TD_{50} the dose of a drug required to produce a toxic effect in 50% of the test population ('TD' stands for toxic dose)

temperature a measure of the average kinetic energy of particles

termination step the step that ends a chain reaction – involves a decrease in the number of free radicals

theoretical yield the maximum possible amount of product formed

therapeutic effect a desirable and beneficial effect of a drug; one that alleviates symptoms or treats a particular disease

therapeutic index (TI) the ratio of the toxic dose to the therapeutic dose of a drug
$$TI = \frac{LD_{50}}{ED_{50}} \quad \text{or} \quad TI = \frac{TD_{50}}{ED_{50}}$$

therapeutic window the range of dosage of a drug between the minimum required to cause a therapeutic effect and the level which produces unacceptable toxic effects

thermoplastic a type of polymer that softens when it is heated and hardens when it is cooled; it can be repeatedly heated and cooled and remoulded into different shapes

thermosetting polymer a prepolymer in a soft solid or viscous state that changes irreversibly into a polymer network (thermoset) by curing

titration a technique that involves adding measured amounts of a solution (from a burette) to another solution to determine the amounts that react exactly with each other

tolerance when the body becomes less responsive to the effects of a drug – larger and larger doses are needed to produce the same effect which means that the patient may be at higher risk of toxic side effects

transition elements the elements in the central part of the periodic table; there are various ways of defining a transition element; the definition used in the IB course is 'an element which forms at least one stable ion with a partially filled d subshell'

transition state (activated complex) the highest energy species on the reaction pathway between reactants and products; the highest point on potential energy profile

unit cell the simplest repeating unit from which a whole crystal can be built up

unsaturated compounds organic compounds containing multiple bonds; this term is often just applied to compounds containing C=C and C≡C, but it is more widely applicable and compounds containing e.g. C=O are also unsaturated

valence shell electron pair repulsion (VSEPR) theory a technique for working out the shapes of molecules/ions; pairs of electrons in the valence (outer) shell of an atom repel each other and therefore take up positions in space to minimise these repulsions, i.e. to be as far apart in space as possible

van der Waals' forces the collective name given to the forces between molecules and includes London (dispersion) forces, dipole–dipole interactions and dipole–induced dipole interactions but not hydrogen bonding and ion-dipole interactions

vapour pressure the pressure exerted by a vapour in equilibrium with a liquid (or a solid)

viscosity a measure of the resistance of a liquid to flow

volatility how readily a substance evaporates

water of crystallisation water that is present in definite proportions in the crystals of hydrated salts, e.g. $CuSO_4 \cdot 5H_2O$. The water may or may not be directly bonded to the metal (in hydrated copper sulfate four water molecules are bonded to the copper ion and one is not)

weak acid an acid such as a carboxylic acid (ethanoic acid, propanoic acid, etc.) or carbonic acid (H_2CO_3) that dissociates partially in aqueous solution:
$$CH_3COOH(aq) \rightleftharpoons H^+(aq) + CH_3COO^-(aq)$$
weak acids are also weak electrolytes

weak base a base that ionises partially in aqueous solution, e.g. ammonia and amines:
$$NH_3(aq) + H_2O(l) \rightleftharpoons NH_4^+(aq) + OH^-(aq)$$
weak bases are also weak electrolytes

weak electrolyte a substance that is partially ionised when dissolved in water

xenobiotics compounds that are present in living organisms but should not normally be found there

yield the amount of product obtained from a chemical reaction; see also actual yield and theoretical yield

zeolites usually refers to an aluminosilicate, porous crystalline mineral (either natural or synthetic)

Index

Acknowledgements

The authors and publishers acknowledge the following sources of copyright material and are grateful for the permissions granted. While every effort has been made, it has not always been possible to identify the sources of all the material used, or to trace all copyright holders. If any omissions are brought to our notice, we will be happy to include the appropriate acknowledgements on reprinting.

Artwork illustrations throughout © Cambridge University Press

The chapter on Nature of Science was prepared by Dr. Peter Hoeben.

The publisher would like to thank Leigh Byrne of Cambridge House Grammar School, Ballymena, Northern Ireland for reviewing the content of this second edition.

The publisher would like to thank Caroline Ahmed, Chris Martin and Roger Woodward to their contribution to this material.

Cover, Leigh Prather/Shutterstock; pp. 3, 454 Martyn F. Chillmaid/SPL; p. 60 Andrew Brookes, National Physical Laboratory/SPL; p. 103 Tony Craddock/SPL; p. 161 Photoshot Holdings Ltd/Alamy; p. 241*t* Doug Steley – C/Alamy; p. 241*b* Tom Wood/Alamy; p. 302 Emilio Segre Visual Archives/American Institute of Physics/SPL; p. 308 Geoff Jones; p. 321*t* Maximilian Stock Ltd/SPL; p. 321*b* Michael McCoy/SPL; p. 393*l* Simon Fraser/SPL; p. 321*b* Mark Leach/Alamy; p. 393*l* catnap/Alamy; p. 393*r* David Nunuk/SPL; p. 411 Jorgen Udvang/Alamy; p. 438 Stephen Giardin/Alamy; pp. 447, 453*b* Cordelia Molloy/SPL; 453*t* Roger Job/Reporters/SPL

Key
l = left, *r* = right, *t* = top, *b* = bottom, c = centre
SPL = Science Photo Library